Advances in Intelligent Systems and Computing

Volume 902

The series "Advances in Intelligent Systems and Computing" contains publications on theory, applications, and design methods of Intelligent Systems and Intelligent Computing. Virtually all disciplines such as engineering, natural sciences, computer and information science, ICT, economics, business, e-commerce, environment, healthcare, life science are covered. The list of topics spans all the areas of modern intelligent systems and computing such as: computational intelligence, soft computing including neural networks, fuzzy systems, evolutionary computing and the fusion of these paradigms, social intelligence, ambient intelligence, computational neuroscience, artificial life, virtual worlds and society, cognitive science and systems, Perception and Vision, DNA and immune based systems, self-organizing and adaptive systems, e-Learning and teaching, human-centered and human-centric computing, recommender systems, intelligent control, robotics and mechatronics including human-machine teaming, knowledge-based paradigms, learning paradigms, machine ethics, intelligent data analysis, knowledge management, intelligent agents, intelligent decision making and support, intelligent network security, trust management, interactive entertainment, Web intelligence and multimedia.

The publications within "Advances in Intelligent Systems and Computing" are primarily proceedings of important conferences, symposia and congresses. They cover significant recent developments in the field, both of a foundational and applicable character. An important characteristic feature of the series is the short publication time and world-wide distribution. This permits a rapid and broad dissemination of research results.

** Indexing: The books of this series are submitted to ISI Proceedings, EI-Compendex, DBLP, SCOPUS, Google Scholar and Springerlink **

More information about this series at http://www.springer.com/series/11156

Zhengbing Hu · Sergey V. Petoukhov ·
Matthew He
Editors

Advances in Artificial Systems for Medicine and Education II

 Springer

Editors
Zhengbing Hu
School of Educational Information
Technology
Central China Normal University
Wuhan, Hubei, China

Sergey V. Petoukhov
Mechanical Engineering Research Institute
Russian Academy of Sciences
Moscow, Russia

Matthew He
Halmos College of Natural Sciences and
Oceanography
Nova Southeastern University
Davie, FL, USA

ISSN 2194-5357   ISSN 2194-5365   (electronic)
Advances in Intelligent Systems and Computing
ISBN 978-3-030-12081-8   ISBN 978-3-030-12082-5   (eBook)
https://doi.org/10.1007/978-3-030-12082-5

Library of Congress Control Number: 2018968110

This Springer imprint is published by the registered company Springer Nature Switzerland AG
The registered company address is: Gewerbestrasse 11, 6330 Cham, Switzerland

Organization

Honorary Chairs

Prof. R. F. Ganiev, Mechanical Engineering Research Institute of the Russian Academy of Sciences, Moscow, Russia
Prof. F. Chin, The University of Hong Kong, Hong Kong

General Co-Chairs

Prof. V. A. Glazunov, Mechanical Engineering Research Institute of the Russian Academy of Sciences, Moscow, Russia
Prof. C. Z. Wang, Hubei University of Technology, China

International Program Committee

Prof. Janusz Kacprzyk, Systems Research Institute, Polish Academy of Sciences
Prof. A. I. Ageev, Institute for Economic Strategies of RAS, Russia
Dr. O. K. Ban, IBM, USA
G. Darvas, Ph.D., Institute Symmetrion, Hungary
Prof. A. S. Dmitriev, Institute of Radioengineering and Electronics of RAS, Moscow, Russia
Prof. B. Dragovich, Institute of Physics, Belgrade, Serbia
Dr. K. Du, National University of Defense Technology, China
Prof. N. A. Balonin, Institute of Computer Systems and Programming, St. Petersburg, Russia
Prof. E. Fimmel, Institute of Mathematical Biology of the Mannheim University of Applied Sciences, Germany

Prof. S. S. Ge, National University of Singapore, Singapore
Dr. I. V. Gorbulina, Russian Academy of Business, Moscow, Russia
Prof. Yu. V. Gulyaev, Institute of Radioengineering and Electronics of RAS, Moscow, Russia
Prof. M. He, Nova Southeastern University, Fort Lauderdale, USA
Prof. A. S. Holevo, Steklov Mathematical Institute of the RAS, Moscow, Russia
Dr. J. W. Hu, Wuhan Ordnance Petty Officer School, China
Prof. A. U. Igamberdiev, Memorial University of Newfoundland, Canada
Prof. V. V. Kozlov, Steklov Mathematical Institute of the RAS, Moscow, Russia
Prof. N. N. Kudryavtsev, Moscow Institute of Physics and Technology, Moscow, Russia
Prof. O. P. Kuznetsov, Institute of Control Sciences of RAS, Moscow, Russia
Prof. O. P. Kuzovlev, Kolomensky Medical Center, Moscow, Russia
Dr. X. J. Ma, Huazhong University of Science and Technology, China
Prof. Yu. I. Manin, Max Planck Institute of Mathematics, Bonn, Germany
Prof. V. E. Mukhin, National Technical University of Ukraine, Ukraine
Prof. D. A. Novikov, Institute of Control Sciences of RAS, Moscow, Russia
Prof. G. S. Osipov, Program Systems Institute of RAS, Moscow, Russia
Prof. Yu. S. Popkov, Institute for Systems Analysis of RAS, Russia
Prof. A. A. Potapov, Institute of Radioengineering and Electronics of RAS, Moscow, Russia
Prof. S. C. Qu, Central China Normal University, China
Prof. Ig. A. Sokolov, Institute of Informatics Problems of RAS, Russia
Prof. Y. Shi, Bloomsburg University of Pennsylvania, USA
Dr. Eng. P. L. Simeonov, Charité University Medicine, Digital Pathology and IT, Institute of Pathology, Berlin, Germany
Dr. J. Su, Hubei University of Technology, China
Prof. V. V. Voevodin, Lomonosov Moscow State University
Prof. J. Q. Wu, Central China Normal University, China
Prof. Q. Wu, Harbin Institute of Technology, China
Prof. Z. W. Ye, Hubei University of Technology, China
Prof. K. V. Zenkin, Tchaikovsky Moscow State Conservatory, Moscow, Russia
Prof. C. C. Zhang, Feng Chia University, Taiwan

Steering Chairs

Prof. Z. B. Hu, Central China Normal University, China
Prof. S. V. Petoukhov, Mechanical Engineering Research Institute (MERI) of RAS, Moscow, Russia

Local Organizing Committee

N. S. Azikov, MERI of RAS, Moscow, Russia
A. V. Borisov, Udmurt State University, Russia
G. S. Filippov, MERI of RAS, Moscow, Russia
L. V. Hazina, MERI of RAS, Moscow, Russia
N. A. Ismailova, MERI of RAS, Moscow, Russia
I. S. Javelov, MERI of RAS, Moscow, Russia
I. D. Kireev, MERI of RAS, Moscow, Russia
A. V. Kirichenko, MERI of RAS, Moscow, Russia
I. S. Mamaev, MERI of RAS, Moscow, Russia
E. N. Petyukov, MERI of RAS, Moscow, Russia
T. A. Rakcheeva, MERI of RAS, Moscow, Russia
H. F. Saberov, MERI of RAS, Moscow, Russia
G. Zh. Sakhvadze, MERI of RAS, Moscow, Russia
A. M. Sergeev, MERI of RAS, Moscow, Russia
T. V. Silova, MERI of RAS, Moscow, Russia
V. O. Soloviev, MERI of RAS, Moscow, Russia
I. V. Stepanyan, MERI of RAS, Moscow, Russia

Publication Chair

Prof. Sergey V. Petoukhov, Mechanical Engineering Research Institute of the Russian Academy of Sciences, Moscow, Russia

Preface

The development of artificial intelligence systems and their applications in various fields belongs to the most urgent tasks of modern science and technology. One of these areas is medical engineering, in which their application is aimed at increasing the effectiveness of diagnosing various diseases and selecting appropriate ways of treatment. The rapid development of artificial intelligence systems requires the intensification of training of a growing number of relevant specialists. At the same time, artificial intelligence systems have significant perspectives of their application inside education technologies themselves for improving the quality of training of specialists taking into account personal characteristics of such specialists and also the emergence of new computer devices.

In digital systems of artificial intelligence, scientists try to reproduce inherited intellectual abilities of human and other biological organisms. The profound study of genetic systems and inherited biological processes can reveal bio-information patents of living nature and give new approaches to create more and more effective methods of artificial intelligence. For this reason, intensive development of bio-mathematical studies of genetic and sensory-motor systems is required on the basis of contemporary achievements of mathematics, computer and quantum informatics, biology, physics, etc. In other words, study of genetic systems and creation of methods of artificial intelligence should go in a parallel manner to enrich each other.

The Second International Conference of Artificial Intelligence, Medical Engineering, Education (October 6–8, 2018, Moscow, Russia) has its purpose to present new thematic approaches, methods, and achievements of mathematicians, biologists, physicians, and technologists and also to attract additional interest of different specialists to this perspective theme. Its proceedings additionally includes articles on specific tasks in various fields, where artificial intelligence systems can be applied in the future with great benefit.

The Conference is organized jointly by Mechanical Engineering Research Institute of the Russian Academy of Sciences (IMASH RAN, http://eng.imash.ru/), and the "International Research Association of Modern Education and Computer Science" (RAMECS, http://www.ramecs.org/part.html).

The organization of such conferences is one of examples of growing Russian–Chinese cooperation in different fields of science and education.

The 70 contributions to the conference were selected by the program committee for inclusion in this book out of all submissions.

Our sincere thanks and appreciation to the board members as listed below:

Rivner F. Ganiev, Russia
F. Chin, Hong Kong
Viktor A. Glazunov, Russia
C. Z. Wang, China
Janusz Kacprzyk, Poland
O. K. Ban, USA
Nikolay A. Balonin, Russia
Alexey V. Borisov, Russia
Gyorgy Darvas, Hungary
Branko Dragovich, Serbia
Alexandr S. Dmitriev, Russia
K. Du, China
Elena Fimmel, Germany
S. S. Ge, Singapore
Irina V. Gorbulina, Russia
Yu. V. Gulyaev, Russia
M. He, USA
Alexander S. Holevo, Russia
J. W. Hu, China
Abir U. Igamberdiev, Canada
Valery V. Kozlov, Russia
Nikolay N. Kudryavtsev, Russia
Oleg P. Kuznetsov, Russia
Oleg P. Kuzovlev, Russia
X. J. Ma, China
Yuri I. Manin, Germany
V. E. Muchin, Ukraine
Dmitriy A. Novikov, Russia
Gennadiy S. Osipov, Russia
Yuriy S. Popkov, Russia
Alexander A. Potapov, Russia
S. C. Qu, China
Igor A. Sokolov, Russia
Y. Shi, USA
Plamen L. Simeonov, Germany
J. Su, China
Vladimir V. Voevodin, Russia
Tatiana A. Rakcheeva, Russia
J. Q. Wu, China

Q. Wu, China
Z. W. Ye, China
Konstantin V. Zenkin, Russia
C. C. Zhang, Taiwan

Finally, we are grateful to Springer-Verlag and Janusz Kacprzyk as the editor responsible for the series "Advances in Intelligent Systems and Computing" for their great support in publishing this conference proceedings.

Wuhan, China   Zhengbing Hu
Moscow, Russia   Sergey V. Petoukhov
Davie, FL, USA   Matthew He
October 2018

Contents

Part I   Advances in Mathematics and Bio-mathematics

**The Technology for Development of Decision-Making Support
Services with Components Reuse**   3
Valeriya Gribova, Alexander Kleschev, Philip Moskalenko,
Vadim Timchenko and Elena Shalfeeva

A Life Ecosystem Management with DNA Base Complementarity   15
Moon Ho Lee, Sung Kook Lee and K. M. Cho

Standard Genetic Code and Golden Ratio Cubes   25
Matthew He, Z. B. Hu and Sergey V. Petoukhov

The Technique for Data Parallelism in Neural Processing Units   37
Vitaliy A. Romanchuk and Ruslan I. Bazhenov

**Algebra for Transforming Genetic Codes Based on Matrices Applied
in Quantum Field Theories**   47
György Darvas and Sergey V. Petoukhov

**Development of Methods for Solving Ill-Posed Inverse Problems
in Processing Multidimensional Signals in Problems of Artificial
Intelligence, Radiolocation and Medical Diagnostics**   57
Alexander A. Potapov, Andrey A. Pakhomov and Vladimir I. Grachev

**Numerical Approach to Solving the Problem of Choosing the Rational
Values of Parameters of Dynamic Systems**   69
I. N. Statnikov and G. I. Firsov

Algorithms for Topological Analysis of Spatial Data   81
Sergey Eremeev and Ekaterina Seltsova

Code Biology and Kolmogorov Complexity   93
Sergei V. Kozyrev

On Genetic Unitary Matrices and Quantum-Algorithmic Genetics 103
Sergey V. Petoukhov and Elena S. Petukhova

**Genetic Code Modelling from the Perspective
of Quantum Informatics**...................................... 117
Elena Fimmel and Sergey V. Petoukhov

Focal Model in the Pattern Recognition Problem................. 127
T. Rakcheeva

**Experimental Detection of the Parallel Organization of Mental
Calculations of a Person on the Basis of Two Algebras Having
Different Associativity** 139
A. V. Koganov and T. A. Rakcheeva

Determinant Optimization Method............................ 151
Nikolay A. Balonin and Mikhail B. Sergeev

Analysis of Oscillator Ensemble with Dynamic Couplings 161
M. M. Gourary and S. G. Rusakov

**The Genetic Language: Natural Algorithms, Developmental Patterns,
and Instinctive Behavior** 173
Nikita E. Shklovskiy-Kordi, Victor K. Finn, Lev I. Ehrlich
and Abir U. Igamberdiev

Part II   Advances in Medical Approaches

**Planck–Shannon Classifier: A Novel Method to Discriminate Between
Sonified Raman Signals from Cancer and Healthy Cells** 185
Sungchul Ji, Beum Jun Park and John Stuart Reid

**ECG Signal Spectral Analysis Approaches for High-Resolution
Electrocardiography**... 197
A. V. Baldin, S. I. Dosko, K. V. Kucherov, Liu Bin, A. Yu. Spasenov,
V. M. Utenkov and D. M. Zhuk

Intelligent System for Health Saving 211
V. N. Krut'ko, V. I. Dontsov and A. M. Markova

**Novel Approach to a Creation of Probe to Stop Bleeding from
Esophageal Varicose Veins Based on Spiral-Compression Method** 221
Sergey N. Sayapin, Pavel M. Shkapov, Firuz G. Nazyrov
and Andrey V. Devyatov

Parallel Mechanisms in Layout of Human Musculoskeletal System 231
A. V. Kirichenko and P. A. Laryushkin

**Methods of Mutual Analysis of Random Oscillations in the Problems
of Research of the Vertical Human Posture** 241
G. I. Firsov and I. N. Statnikov

**Verification of Mathematical Model for Bioimpedance Diagnostics
of the Blood Flow in Cerebral Vessels** 251
Anna A. Kiseleva, Petr V. Luzhnov and Dmitry M. Shamaev

**Evaluation of Adhesive Bond Strength of Dental Fiber Posts
by "Torque-Out" Test** 261
Anna S. Bobrovskaia, Sergey S. Gavriushin and Alexander V. Mitronin

**Image Segmentation Method Based on Statistical Parameters
of Homogeneous Data Set** 271
Oksana Shkurat, Yevgeniya Sulema, Viktoriya Suschuk-Sliusarenko
and Andrii Dychka

Steganographic Protection Method Based on Huffman Tree 283
Yevgen Radchenko, Ivan Dychka, Yevgeniya Sulema,
Viktoriya Suschuk-Sliusarenko and Oksana Shkurat

**The Role of Hybrid Classifiers in Problems of Chest Roentgenogram
Classification** 293
Rimma Tomakova, Sergey Filist, Roman Veynberg, Alexey Brezhnev
and Alexandra Brezhneva

**Patient-Specific Biomechanical Analysis in Computer Planning
of Dentition Restoration with the Use of Dental Implants** 305
I. N. Dashevskiy and D. A. Gribov

**A Viscoelastic Model of the Long-Term Orthodontic
Tooth Movement** 315
Eduard B. Demishkevich and Sergey S. Gavriushin

Metric Properties of the Visual Illusion of Intersection 323
T. Rakcheeva

**Method of the Data Adequacy Determination of Personal
Medical Profiles** 333
Yuriy Syerov, Natalia Shakhovska and Solomiia Fedushko

**Experimental Method for Biologically Active Frequencies
Determination** 345
Victor A. Panchelyuga, Victor L. Eventov and Maria S. Panchelyuga

**From the Golem to the Robot and Beyond
to the Smart Prostheses** 355
Anatoly K. Skvorchevsky, Alexander M. Sergeev and Nikita S. Kovalev

Part III   Advances in Technological and Educational Approaches

**Lifelong Education of Sports Media Professionals Based
on System Theory**   369
Ziye Wang, Qingying Zhang and Mengya Zhang

**Fuzzy Classification on the Base of Convolutional
Neural Networks**   379
A. Puchkov, M. Dli and M. Kireyenkova

**Modeling of Intellect with the Use of Complex Conditional Reflexes
and Selective Neural Network Technologies**   393
M. Mazurov

**Phase Equilibria in Fractal *Core-Shell* Nanoparticles
of the $Pb_5(VO_4)_3Cl$–$Pb_5(PO_4)_3Cl$ System: The Influence
of Size and Shape**   405
Alexander V. Shishulin, Alexander A. Potapov and Victor B. Fedoseev

**Application of Neural Networks for Controlling the Vibrational
System Based on Electric Dynamic Drive**   415
R. F. Ganiev, S. S. Panin, M. S. Dovbnenko and E. A. Bryzgalov

**A Dynamic Power Flow Model Considering the Uncertainty
of Primary Frequency Regulation of System**   425
Daojun Chen, Nianguang Zhou, Chenkun Li, Hu Guo and Ting Cui

A Data Model of the Internet Social Environment   439
Andriy Peleshchyshyn and Oleg Mastykash

**Anomaly Detection of Distribution Network Synchronous
Measurement Data Based on Large Dimensional Random Matrix**   449
Zhongming Chen, Yaoyu Zhang, Chuan Qing, Jierong Liu, Jiaqi Tang
and Jingzhi Pang

**The Hurst Exponent Application in the Fractal Analysis
of the Russian Stock Market**   459
Alexander Laktyunkin and Alexander A. Potapov

**Development of Models and Methods of Virtual Community Life
Cycle Organization**   473
Olha Trach and Andriy Peleshchyshyn

**A Formal Approach to Modeling the Characteristics of Users of Social
Networks Regarding Information Security Issues**   485
Andriy Peleshchyshyn, Volodymyr Vus, Solomiia Albota
and Oleksandr Markovets

**Advanced Morphological Approach for Knowledge-Based
Engineering (KBE) in Aerospace** 495
A. Bardenhagen, M. Pecheykina and D. Rakov

**Problems of Intelligent Automation of Unmanned Underground
Coal Mining** 507
Andrey M. Valuev and Ludmila P. Volkova

**Structural and Parametric Control of a Signalized Intersection
with Real-Time "Education" of Drivers** 517
Anatoliy A. Solovyev and Andrey M. Valuev

Some Aspects of the Method for Tourist Route Creation 527
Nataliya Shakhovska, Khrystyna Shakhovska and Solomia Fedushko

Experimental Substantiation of Soft Cutting Modes Method 539
P. A. Eremeykin, A. D. Zhargalova and S. S. Gavriushin

**Design of Oxidative Pyrolysis Control Algorithm Based on Fuzzy
Safety Area and Center Definition** 549
G. N. Sanayeva, I. E. Kirillov, A. E. Prorokov, V. N. Bogatikov
and D. P. Vent

**Secure Hash Function Constructing for Future Communication
Systems and Networks** 561
Sergiy Gnatyuk, Vasyl Kinzeryavyy, Karina Kyrychenko,
Khalicha Yubuzova, Marek Aleksander and Roman Odarchenko

**Code Obfuscation Technique for Enhancing Software Protection
Against Reverse Engineering** 571
Sergiy Gnatyuk, Vasyl Kinzeryavyy, Iryna Stepanenko, Yana Gorbatyuk,
Andrii Gizun and Vitalii Kotelianets

**Modern Method and Software Tool for Guaranteed Data Deletion
in Advanced Big Data Systems** 581
Sergiy Gnatyuk, Vasyl Kinzeryavyy, Tetyana Sapozhnik, Iryna Sopilko,
Nurgul Seilova and Anatoliy Hrytsak

**Computer Implementation of the Fuzzy Model for Evaluating
the Educational Activities of the University** 591
N. Yu. Mutovkina

**Temperature Field Simulation of Gyro Unit-Platform Assembly
Accounting for Thermal Expansion and Roughness of Contact
Surfaces** 601
Mikhail V. Murashov

**The Primary Geo-electromagnetic Data Preprocessing Received
from a Modified Geophysical Automatic Station** 617
Roman Kaminskyj, Nataliya Shakhovska and Lidia Savkiv

Statistical Analysis of Probability Characteristics of Precipitation
in Different Geographical Regions 629
Maria Vasilieva, Andrey Gorshenin and Victor Korolev

The Knowledge Management Approach to Digitalization
of Smart Education 641
Natalia V. Dneprovskaya, Nina V. Komleva and Arkadiy I. Urintsov

Algorithms for Agreement and Harmonization the Creative Solutions
of Agents in an Intelligent Active System 651
N. Yu. Mutovkina and V. N. Kuznetsov

Designing an Information Retrieval System in the Archival
Subdivision of Higher Educational Institutions 661
Tetiana Bilushchak, Andriy Peleshchyshyn, Ulyana Yarka
and Zhanna Myna

The Manner of Spacecraft Protection from Potential Impact of Space
Debris as the Problem of Selection with Fuzzy Logic 673
B. V. Paliukh, V. K. Kemaykin, Yu. G. Kozlova and I. V. Kozhukhin

Structural Synthesis of Spatial *l*-Coordinate Mechanisms with
Additional Links for Technological Robots 683
V. A. Glazunov, G. V. Rashoyan, A. K. Aleshin, K. A. Shalyukhin
and S. A. Skvortsov

Characteristics Analysis for Corporate Wi-Fi Network Using the
Buzen's Algorithm 693
Elena V. Kokoreva and Ksenia I. Shurygina

The Application of Elements of Information Theory to the Problem
of Rational Choice of Measuring Instruments 705
I. A. Meshchikhin and S. S. Gavriushin

Power System Transient Voltage Stability Assessment Based on
Kernel Principal Component Analysis and DBN 713
Zhang Guoli, Gong Qingwu, Qian Wenxiao, Lu Jianqiang, Zheng Bowen,
Gao He, Zheng Tingting, Wu Liuchuang, Chen Wenhui, Liu Xu, Wang Bo
and Qiao Hui

Simulation and Analysis of Operating Overvoltage of AC System
at ±800 kV Converter Station Based on EMTS/EMTPE 729
Zheng Ren, Jiaqi Fan, Xiaolu Chen, He Gao, Jianqiang Lu
and Bowen Zheng

Simulation Study and Experimental Analysis of Current Closure
Overvoltage Caused by High Reactance on the AC Bus at ±800 kV
Converter Station 739
Pengwe Yang, Rui Wang, Cai Xu, Weimng Liu and Yingkun Han

Study of Dynamic Reactive Power Support of Synchronous Condenser on UHV AC/DC Hybrid Power System 749
Bowen Zheng, Jianqiang Lu and He Gao

Wide-Area Feedback Signal and Control Location Selection for Low-Frequency Oscillation 759
Xiaolu Chen, Xiangxin Li, Zheng Ren and Tingting Zheng

Voltage Quality Evaluation of Distribution Network and Countermeasure 771
Wei Deng, Yuan Ji Huang, Liang Zhu, Miaomiao Yu and Rongtao Liao

Author Index 785

Part I
Advances in Mathematics and Bio-mathematics

The Technology for Development of Decision-Making Support Services with Components Reuse

Valeriya Gribova, Alexander Kleschev, Philip Moskalenko, Vadim Timchenko and Elena Shalfeeva

Abstract The article presents the methodology and technology of development and assembling of cloud decision-making support systems which use knowledge bases in network representation. The system analysis of problem area and identification of known subtasks allows to reuse tasks ontologies and domain ontologies, to find similar problem-solving method for these subtasks. The IACPaaS technology allows to reuse software components which implement certain solving methods in development process of solvers of intelligent tasks. It also allows to reuse knowledge bases and other information components, integrating them with the solvers at assembling of cloud decision-making support services.

Keywords Knowledge base · Disease · Intelligent task · Decision-making support system · Ontology · Cloud technology · Software service · Component

1 Introduction

An automation of daily intelligent activity and its quality management at the present stage demands updating of technologies of intelligent systems' development, focused on assembly by reused semantic compatible components, using of the known problem solver methods (task solution methods) and its software realizations [1–3]. The modern platforms provide developers with sets of tools and services for different problem and task solution, including the scientific problems and focusing on big data processing (the Clavire) [4, 5]. The majority of the known cloud decisions for application development support relational submission of databases [5, 6]. There are cloud editors of knowledge bases and ontologies, supporting object-oriented submission of information (Protege) [7, 8]. The basis of many known tools for ontology-based development (Ontowiki, WebProtégé) is the paradigm of object-oriented ontologies

V. Gribova · A. Kleschev · P. Moskalenko · V. Timchenko · E. Shalfeeva (✉)
Institute for Automation and Control Processes, Februs, Russia
e-mail: shalf@dvo.ru

© Springer Nature Switzerland AG 2020
Z. Hu et al. (eds.), *Advances in Artificial Systems for Medicine and Education II*,
Advances in Intelligent Systems and Computing 902,
https://doi.org/10.1007/978-3-030-12082-5_1

[9, 10]. This paradigm is unfamiliar to most of real and potential experts, which causes considerable difficulties in formation of knowledge bases.

In some domains, not only knowledge, but also ontologies of this knowledge can be changed. Based on that, new classes of expert problems can be formulated.

However, the technologies and tools of development of specialized shells that would allow to modify an ontology and related problem for decision-making support are not known from the literature.

The IACPaaS platform (Intelligent Applications, Control and Platform as a Service) which developed in IACP FEB RUS focused on working with knowledge bases and/or data in network representation and intellectual services creation [11]. Rather independent development of knowledge bases and software parts of such services is supported. The platform has a convenient editor for formation of knowledge bases, databases and another stored information in terms of their meta-information [11]. Nowadays there are supporting tools for three technologies of cloud services development. Each technology promotes a reuse of program (software) and information components created earlier [12].

This paper aims to describe of decision-making support cloud software system development technology allowing to reuse information and program components by the IACPaaS technology.

2   Decision-Making Problems and Used Knowledge

A system analysis is the important stage for decisions of intelligent activity support and automatization. All intelligent tasks solved by specialists' correctness of knowledge using are discussed. We will call intelligent task the task which statement (formal description of task) includes knowledge base (KB) and situation model (its mathematical abstraction—algebraic system AS) as input or output. KB and situation model must be consensual with domain ontology.

At the system analysis stage, analysts seek to reveal in intelligent activity those intellectual tasks which statements are known (or to decompose to known subtasks). The different classes of intellectual tasks (called briefly further by tasks) are known, and the formal statements are proposed for many of them [13, 14]. Allocation of famous (known) subtasks allows to reuse problem ontologies, structure of processed information, methods of this information processing and reuse ready components implementing such processing and some methods these subtasks decisions (problem-solving methods).

It is shown in the offered multilevel tasks classification [10] that "genetic linkage" takes place between some tasks. The "genetic linkage" is based on the principle of complication of domain properties (Fig. 1). The complication of domain properties is meant there are new kinds of data which should be considered during tasks solution in "more difficult" domain (e.g., time, space, classes, cause-and-effect relations, events, internal processes in system, not inherent to it) and additional restrictions.

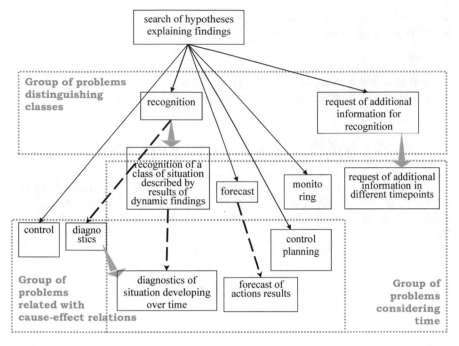

Fig. 1 Fragment of multilevel task classification—detailing of problem of search hypotheses, explaining findings [14]

In many domains, there are problems considering classifications of situations, and KB includes assertions about properties of situations of each class.

Some tasks are related to ontologies which contain time-dependent functional compliances (for representation of time-dependent attributes of dynamic system or a situation, e.g., a state).

In many domains, there are problems considering the actions (ordered at least partially), leading\conducting to some purpose.

So in multilevel classification of tasks, each complication of a domain is considered as "a layer of domains with similar features" where new tasks or expanded statements of basic tasks are expected. We can see the example of "genetic relation" between task of search of the hypotheses explaining findings, recognition task and task of diagnostics (Fig. 1).

In the task of search of the hypotheses explaining findings, it is required to find all hypotheses corresponding to results of observations (findings [16]) of a system or a situation and KB, if those observations (findings) and KB are available.

The recognition task is considered as specification of a task of search of the hypotheses explaining findings where it is required to find all hypotheses of a class of the situation described by the findings.

The major components of recognition task can be summarized as follows:

Givens:

- *R* (results of observations and characteristics);
- KB (knowledge base).

  Goals:

- To fit case into possible classes: class$_{i1,}$ ... class$_{i\,k}$;
- To find all explanation models AS (*R*, class$_{i1}$) and AS (*R*, class$_{i1}$).

  Constraints:

- All assertions of KB are concordant to findings and characteristics;
- R coordinated with domain ontology.

In diagnostics task, it is required to define possible cause-and-effect models of a system by its characteristics, observations and events, which were taken place. Here, a diagnosis is set of some internal processes of this system which aren't inherent to it.

The diagnosis task can be summarized as follows:
Givens:

- *R* (results of observations, characteristics and events);
- KB.

  Goals:

- To fit case into possible diagnoses $\Delta i1$, ..., Δik; the diagnosis is the subset of system internal processes;
- To find all possible cause-and-effect models AS (*R*, Δ_{i1}) and AS (*R*, $\Delta_{i\,k}$) for explanation of diagnoses Δ_{i1}, ..., Δ_{ik}.

  Constraints:

- All assertions of KB are concordant to findings, characteristics and events;
- All cause-and-effect models are concordant to KB

Classification of tasks with above-mentioned genetic relations gives the chance to choose suitable tasks, to consider their mathematical statements and formulate statements for the tasks of intelligent activity being automated, concretizing concepts and abstractions in the chosen statements.

3 IACPAAS Services Development Technologies with Reuse of Knowledge and Other Components

Just as the majority of cloud platforms, the IACPaaS has development tools for all components of applied services. IACPaaS has means of declarative submission of knowledge bases and other stored information in domain terms in the form of hierarchical semantic networks. The development of services processing declarative knowledge means creation (programming) of specialized solvers.

Each service is being designed with by a multi-component specialized solver and information stored in Fund. The technology is focused not so much on creation of services from scratch how on assembly of services from components—new, ready or modified copies of the existing components.

The main stages of service's development (by universal IACPaaS technology) are:

- Creation of input information resources—databases and knowledge bases;
- Creation of task solver as sets of interacting program units (PU).

The IACPaaS "Knowledge Base on Diagnosis of Diseases" was formed, representing a variety of diseases (Fig. 2).

To specify and construct the input information resource (IR), it is necessary to take care of the description of its structure (meta-information). If the structure was described earlier, it being reused, otherwise it must be formed explicitly by meta-language. Only if the structure (meta-information) of processed IR is available, it is possible to start designing of PU which processes the IR.

It is necessary to notice that structure of output information as rule is unique for every intelligent service. That is why the creation of structure of output information resources is important stage of software service's designing (in universal IACPaaS technology).

Software components of specialized solvers have an obligatory declarative presentation as IR. Each such IR is used by the IACPaaS solver processor in the course of running of service for dynamic connection of components. The common

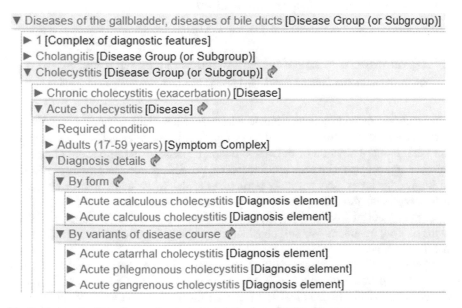

Fig. 2 Fragment of the disease diagnosis knowledge base (IACPaaS)

meta-information was formed for constructing these IR declarations. This meta-information is always reused.

The main stage of creation of the integrated solver is compound "work"; it includes creation of software logic components, creation of GUI and creation of meta-information for input and output IR. This work can be executed in two ways: (1) passing consistently all main stages of life cycle, beginning from requirements model, or (2) starting at once to creation of solver design.

The first way assumes "the planning work," including creation of model of requirements. Declarative representation (declaring) of a solver in the form of information resources provides a possibility of the structural description of model of requirements. This model will be projected onto program units being included in solver architecture. In the second way, a solver is created without these models. The first way facilitates maintenance process, and the second accelerates obtaining the first version of service.

Declarative representation of a solver contains formal parameters setting formats of the processed information resources (e.g., meta-information of used knowledge bases). Declarative representation of a solver contains name of the root agent and can specify set of all other agents who have to realize logic of the solution of a task. There are two levels of architecture designing of a solver. One can specify a set of all agents at one level or specify "control flow graph" which register concrete communications of agents with each other. So the declarative representation of a solver is its design model.

The development requires also: to declare a solver, to declare every PU (an « agent »), to execute design of every PU, to create their communication model, to execute coding and to execute loading and testing of PU.

4   Development of Medical Software Services with Reuse of Knowledge Bases and Solvers

The medicine remains subject domain where decision-making support by software services still isn't provided up to the mark [15, 16]. Some reasons of this situation are labor input of creation of the real knowledge bases taking into account a development of processes with time and a technical possibility of collective improvement of knowledge bases. The ontology (structure) of this knowledge, first of all, has to reflect representation of experts and be presented in their terms.

One of medical IACPaaS services being created is intended for support of decisions on drug treatment. We will consider the possibilities of a reuse at creation of a medical service.

One of medical IACPaaS services being created is intended for support of decisions on drug treatment. We will consider the possibilities of a reuse at creation of a medical service. This medical service is being created in conditions when other medical services are already created and placed on a platform. Therefore, the reuse of components is expected. First of all, the structure (meta-information) of case history (CH) will be reused. CH already created for service of medical diagnostics.

It is required in the task of drug treatment (planning of impacts on a system or a situation) to define such set of events and their time points at which findings (as functions of time in cause-and-effect models of a system or a situation) will reach desirable values, if characteristics, diagnosis and desirable findings of system "patient" are known.

The task of drug treatment of a patient with the established diagnosis (task of "management" of a patient health) can be summarized as follows:

Givens:

- R_O (patient's findings, observed symptoms and characteristics);
- Δ (diagnosis);
- R_{cond} (desirable findings);
- KB.

Goals:

- R_{ev} at which new findings R'_{ex} will meet R_{cond};
- Explanation AS ($R_{ex} \cup R_O \cup R_{ev}$, Δ, $R_{cond} \subseteq R'_{ex}$).

Constraints:

- R'_{ex} are functions of time which concordant to KB.

Since in practice doctors need consultation in this task, but not solution of a task instead of them, it is necessary to form full explanation for R_{ev}.

The services are constructed from two KB (the treatment scheme of concrete disease and The Index of medicaments—description of pharmacological means), a database of patient's case histories and a solver generating an explanation. Such composition of service provides (at the current stage of development) fast assembly of similar services for different medical profiles. In practice, the analysis of contraindications of tens or hundreds of known medicaments (drugs) with the necessary pharmacological influence is connected with essential expenses of time. The doctor, as rule, needs information on what concrete drug shouldn't be appointed for this patient with his "features." The ontology (i.e., structure and a meta-information) explanations of a solver of treatment are formed, taking into consideration these circumstances.

KB of medicaments is common and reusable; it is developed and is followed for all domains. There are a number of KB of schemes of treatment; every KB is created for a separate medical profile according to the standards approved by the Ministry of Health (e.g., the standard of "medical care by the patient with acute pancreatitis" is dated November, 2007).

The input data for service is one IB or the set of IB from which the next IB for decision-making will be chosen. Such set of IB is being formed not so for a separate medical profile how for "place" of application of service (for the doctor or concrete medical establishment).

Let us consider the possibilities of a reuse at creation of software logic components of the service. The nearest task from the point of view of formal description and problem-solving method (from task classification [14]) is the control planning task.

It is required in the task (planning of control) to define such set of events and their time points at which factors (as functions of time in cause-and-effect models of system) will reach desirable values, if characteristics, diagnosis and conditions on factors' values are known.

The task of planning of control can be summarized as follows:

Givens:

- R_O (system' characteristics, properties or factors);
- Δ (diagnosis);
- Cond (conditions on values of characteristics),
- KB.

Goals:

- R_{ev} (set of events and time points) at which new characteristics or factors R'_{ex} will meet Cond.

Constraints:

- R_{ev} are functions of in cause-and-effect models of system, concordant to KB.

In the presence of a ready solver of control planning task together with ontology of the processed information in the Fund, the reuse of this ontology and of method of decision-making for a solver of task of drug treatment of the patient with the established diagnosis is possible. Ontology of the processed information is the meta-information which sets terminology and associations between concepts. It has to be detailed: Characteristics and attributes of system are concretized to the patient's characteristics.

The ontology of knowledge for planning of control presents such entities and associations as impacts on a system and reactions to impacts. It has to be detailed for medicine domain. As a result, the ontology for task of planning of control in medicine concretizes such associations: therapeutic impacts (actions, drug intakes) and reactions to therapeutic impacts.

The solver components which implemented algorithm of information processing and decision-making for control planning task can be adapted (through detailing) for medicine domain. The solving process consists of some steps. The first step consists in obtaining a medicaments list recommended to patients with the specified diagnosis. Further, all medicaments which aren't suitable to specific patient (with specified age and other "features" of an organism) are being eliminated from this list. It is supposed that a doctor will want to choose several medicaments from the final list of potentially suitable ones. The scheme of application is usually standard; if necessary, it can be proposed by solver. In case of a joint disease (poly-pathology), the solving process analyzes a set of the recommended medicaments in terms of their compatibility in simultaneous reception.

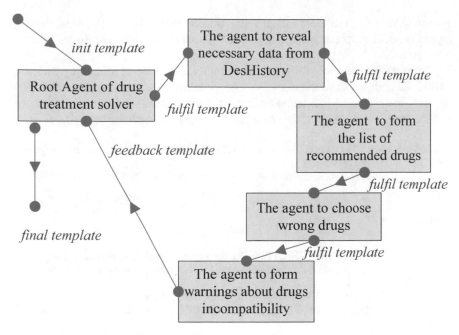

Fig. 3 Example of treatment solver architecture

Declarative representation of the solver contains reference to format of the treatment scheme of concrete disease, reference to format of The Index of medicaments (description of pharmacological means) and reference to format of stored set of CH, a name of root agent. Besides, depending on way of designing of a solver, the declaration can contain or a set of names of other program units or a model of messages transfer for those PU. These PU (agents) have to implement a logic of solution of a task; the IACPaaS communication system has to register concrete communications of PU agents with each other (Fig. 3). A program units of a solver have an obligatory declarative representation in the form of IR too. One of the versions of the architecture is shown in Fig. 3 (in graphical view).

This technology allows forming reuse knowledge bases for a medical knowledge portal, and designing and developing medical systems for supporting solutions. The knowledge about diagnostics of diseases is placed on the platform (Fig. 2).

5   Conclusion

The paper contains the methodology IACPaaS for intellectual services development and their knowledge bases formation. The IACPaaS cloud technology facilitates development of the intellectual services having knowledge bases in network representation and specialized solvers processing knowledge bases and other stored information in form of hierarchical semantic networks. The development technology of services supporting usage of premade decisions and the realized components is presented.

The platform has tools for support of cloud services development by means of various technologies and the tools for network knowledge bases and databases. The support for development, integration and maintenance of all components of services is necessary to receive the viable intellectual services using knowledge bases.

Acknowledgement The work is carried out with the partial financial support of the RFBR (projects nos. 17-07-00956 and 18-07-01079).

References

1. Akyol, K., Gültepe, Y.: A study on liver disease diagnosis based on assessing the importance of attributes. Int. J. Intell. Syst. Appl. (IJISA) **9**(11), 1–9 (2017). https://doi.org/10.5815/ijisa. 2017.11.01
2. Choubey, D.K., Paul, S.: GA_MLP NN: a hybrid intelligent system for diabetes disease diagnosis. Int. J. Intell. Syst. Appl. (IJISA) **8**(1), 49–59 (2016). https://doi.org/10.5815/ijisa.2016. 01.06
3. Perova, I., Pliss, I.: Deep hybrid system of computational intelligence with architecture adaptation for medical fuzzy diagnostics. Int. J. Intell. Syst. Appl. (IJISA) **9**(7), 12–21 (2017). https:// doi.org/10.5815/ijisa.2017.07.02
4. Bukhanovsky, A.V., et al.: CLAVIRE: perspective technology of cloud computing of the second generation. J. Instrum. Eng. (Izvestiya vysshikh uchebnykh zavedeniy. Priborostroenie) **10**, 7–14 (2011). (in Russian)
5. Sun, L., Dong, H., Hussain, F.K., Hussain, O.K.: Chang, E: cloud service selection: state-of-the-art and future research directions. J. Netw. Comput. Appl. **45**, 134–150 (2014)
6. Zhang, Q., Cheng, L., Boutaba. R.: Cloud computing: state-of-the-art and research challenges. J. Internet Serv. Appl.. **1**, 7–18 (2010)
7. Tudorache, T., Nyulas, C., Noy, N.F., Musen, M.A.: WebProtégé: a collaborative ontology editor and knowledge acquisition tool for the web. Semant. Web J. **4**, 89–99 (2013)
8. Zagorulko, Y.A.: Semantic technology for development of intellectual systems oriented to experts in subject domain. Ontol. Des **2**, 30–46 (2015). (in Russian)
9. OntoWiki: A tool providing support for agile, distributed knowledge engineering scenarios. [Electronic resource]. http://aksw.org/Projects/OntoWiki.html. Last accessed 2018/03/20
10. The Protégé Ontology Editor and Knowledge Acquisition System. [Electronic resource]. http://protege.stanford.edu/
11. Gribova, V.V., Kleshchev, A.S., Krylov, D.A., Moskalenko, F.M., Timchenko, V.A., Shalfeeva, E.A.: Basic technology of development of intellectual services on the cloudy IACPaaS platform. Part 1. Development of the knowledge base and solver of tasks. Program Eng. **12**, 3–11 (2015). (in Russian)
12. Sonar, R.M.: An enterprise intelligent system development and solution framework. Int. J. Appl. Sci. Eng. Technol. **1**, 34–39 (2007)
13. Puppe, F.: Knowledge reuse among diagnostic problem solving. Methods in the Shell-Kit D3. Int. J. Hum.-Comput. Stud. **49**, 627–649 (1998)
14. Kleschev, A.S., Shalfeeva, E.A.: An ontology of intellectual activity tasks. Ontol. Des. **2**, 179–205 (2015). (In Russian)

15. Soltan, R.A., Rashad, M.Z., El-Desouky, B.: Diagnosis of some diseases in medicine via computerized experts system. Int. J. Comput. Sci. Inf. Technol. **5**, 79–90 (2013)
16. Denekamp, Y., Peleg, M.: TiMeDDx—a multi-phase anchor-based diagnostic decision-support model. J. Biomed. Inform. **43**, 111–124 (2010)

A Life Ecosystem Management with DNA Base Complementarity

Moon Ho Lee, Sung Kook Lee and K. M. Cho

Abstract We present a DNA-RNA genetic code with their base complementarity, i.e., A = T (or U in RNA) = 30%, C = G = 20%, and C + U = G + A stemming from the fact that Erwin Chargaff discovered from his experimental results in 1950. These results strongly hinted that A and T are complementary base pairs, and C and G are also complementary base pairs, although he had no idea of them at that time. However, it has not been proved in a mathematically analytical point of view yet. In this paper, we have straightforward proof of this problem based on life ecosystem management by applying the doubly stochastic matrix to the Shannon entropy of the symmetric channel capacity.

Keywords Life ecosystem management · Chargaff rules · RNA entropy · Double stochastic matrix · Symmetric channel capacity

1 Introduction

In 1950, Erwin Chargaff pronounced upon two main rules in his lifetime which were appropriately named Chargaff's rules. The first and best known achievement was to show that in natural DNA, the number of guanine (G) units equals the number of cytosine (C) units and the number of adenine (A) units equals the number of thymine (T) units. For example, human DNA, consists of four bases as follows: A = 30.9%, T = 29.4%, G = 19.9%, and C = 19.8%. The Watson and Crick laboratory team deduced the double helical structure of DNA in [1].

M. H. Lee (✉) · K. M. Cho
Department of Electronic Engineering, Chonbuk National University, Jeonju, Republic of Korea
e-mail: moonho@jbnu.ac.kr

K. M. Cho
e-mail: sjjgz2011@jbnu.ac.kr

S. K. Lee
School of Economics, Indiana University Bloomington, Bloomington, IN, USA
e-mail: lee950@umail.iu.edu

© Springer Nature Switzerland AG 2020   15
Z. Hu et al. (eds.), *Advances in Artificial Systems for Medicine and Education II*,
Advances in Intelligent Systems and Computing 902,
https://doi.org/10.1007/978-3-030-12082-5_2

Fig. 1 Markov chain state

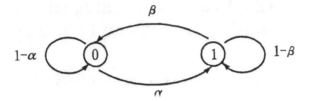

DNA sequences and DNA methylation were introduced by their novel approach in the recently published papers [2–4]. However, all of them haven't shown any mathematical proof and perfectly solved yet.

The RNA base [C U; A G] is presented to generate a sequence of the genetic code-based stochastic matrices. In connection with these code-based doubly stochastic matrices, we make use of a binary symmetric capacity based on the Markov process. It demonstrates the symmetrical relation between Shannon and the entries of RNA stochastic 2×2 transition matrix [C U; **AG**]. We recall some basic definitions of a stochastic matrix. A square matrix of $P = (p_{ij})$ is a stochastic matrix if all entries of the matrix are nonnegative and the sum of the elements in each row or column is unity or a constant. If the sum of the elements in each row and column is unity or the same, the matrix is called double stochastic. It represents a time-invariant binary communication channel, where $\mathbf{x}_n$ denotes the input and $\mathbf{x}_{n+1}$ the output. The input and the output of each process have two states, e_0 and e_1, which represent the two binary symbols "0" and "1", respectively. The channel delivers the input symbol to the output with a certain error probability that depends on the symbol being transmitted. Let $\alpha < 0.5$ and $\beta < 0.5$ represent the two kinds of channel error probabilities. In a time-variant channel, these error probabilities remain constant over various transmitted symbols

$$P\{\mathbf{x}_{n+1} = 1 | \mathbf{x}_n = 0\} = p_{01} = \alpha \quad P\{\mathbf{x}_{n+1} = 0 | \mathbf{x}_n = 1\} = p_{10} = \beta. \qquad (1)$$

and the corresponding Markov chain is homogeneous. The 2×2 homogeneous probability transition matrix P is given by

$$P = \begin{bmatrix} p_{00} & p_{01} \\ p_{10} & p_{11} \end{bmatrix} = \begin{bmatrix} 1-\alpha & \alpha \\ \beta & 1-\beta \end{bmatrix} = \begin{bmatrix} 1-p & p \\ p & 1-p \end{bmatrix}_{p=0.5} = \frac{1}{2}\begin{bmatrix} 1 & 1 \\ 1 & 1 \end{bmatrix}. \qquad (2)$$

In a binary symmetric channel, the two kinds of error probabilities [5, 6] are identical such that $\alpha = \beta = p$ as shown in Fig. 1.

This paper is organized as follows. First, we presented the RNA stochastic entropy in Sect. 2. The symmetric channel of RNA complementary bases [C U; A G] with noise immunity is described in Sect. 3. The conclusions are reached in Sect. 4.

2 RNA Stochastic Entropy Analysis

The genetic code of RNA stochastic complementary bases is given by Chargaff et al. [1] and He and Petoukhov [7]. If $C = G = 20\%$ and $A = U = 30\%$ (Table 1), then the transition channel matrix is

$$P = \begin{bmatrix} C & U \\ A & G \end{bmatrix} = \begin{bmatrix} 0.2 & 0.3 \\ 0.3 & 0.2 \end{bmatrix} \tag{3}$$

Let the RNA complementary bases (C U; A G) be the Markov process (2) with the two independent source probabilities, $0.2p$ and $0.3p$, maximizing its entropy, then, we have

$$P = \begin{bmatrix} 0.2p & 1-0.2p \\ 1-0.2p & 0.2p \end{bmatrix} = \begin{bmatrix} 0.5 & 1-0.5 \\ 1-0.5 & 0.5 \end{bmatrix} = \begin{bmatrix} 0.5 & 0.5 \\ 0.5 & 0.5 \end{bmatrix}, \quad \text{where } p \text{ is } 2.5 \tag{4}$$

From (12), we have

$$0.2p = 1 - 0.2p, \quad p = 2.5. \tag{5}$$

Table 1 Biochemical base composition ratio

Organism	Taxon	%A	%G	%C	%T	A/T	G/C	%GC	%AT
Chicken	Gallus	28.0	22.0	21.6	28.4	0.99	1.02	43.6	56.4
E. Coli	Escherichia	24.7	26.0	25.7	23.6	1.05	1.01	51.7	48.3
Grasshopper	Orthoptera	29.3	20.5	20.7	29.3	1.00	0.99	41.2	58.6
Human	Homo	29.3	20.7	20.0	30.0	0.98	1.04	40.7	59.3
Maize	Zea	26.8	22.8	23.2	27.2	0.99	0.98	46.1	54.0
Octopus	Octopus	33.2	17.6	17.6	31.6	1.05	1.00	35.2	64.8
Rat	Rattus	28.6	21.4	21.5	28.4	1.01	1.00	42.9	57.0
Sea Urchin	Echinacea	32.8	17.7	17.3	32.1	1.02	1.02	35.0	64.9
Wheat	Triticum	27.3	22.7	22.8	27.1	1.01	1.00	45.5	54.4
Yeast	Saccharomyces	31.3	18.7	17.1	32.9	0.95	1.09	35.8	64.4
$\varphi \times 174$	Phi × 174	24.0	23.3	21.5	31.2	0.77	1.08	44.8	55.2
Satsuma Mandarin	Citrus	20.57	32.71	30.00	16.71	1.23	1.09	62.7	37.3
Coffea arabica	Coffea	20.66	31.76	30.87	16.71	1.24	1.02	62.6	37.4
Trifoliate Orange	Poncirus	18.18	34.02	31.96	15.84	1.15	1.06	66.0	34.0

Fig. 2 Shannon and RNA
entropy with respect to
probability of RNA
complementary bases [C U;
A G]

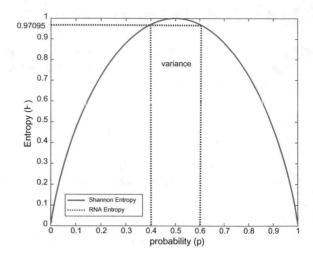

Also, in a similar fashion as (4)

$$P = \begin{bmatrix} 0.3p & 1-0.3p \\ 1-0.3p & 0.3p \end{bmatrix} = \begin{bmatrix} 0.498 & 1-0.498 \\ 1-0.498 & 0.498 \end{bmatrix} = \begin{bmatrix} 0.498 & 0.502 \\ 0.502 & 0.498 \end{bmatrix}$$
$$\approx \begin{bmatrix} 0.5 & 0.5 \\ 0.5 & 0.5 \end{bmatrix}, \tag{6}$$

where $0.3p = 1 - 0.3p, p = 1.6$.

Therefore, a sum of (4) and (6) is

$$2P = \begin{bmatrix} 0.5 & 0.5 \\ 0.5 & 0.5 \end{bmatrix} + \begin{bmatrix} 0.5 & 0.5 \\ 0.5 & 0.5 \end{bmatrix} = \begin{bmatrix} 1 & 1 \\ 1 & 1 \end{bmatrix}. \tag{7}$$

Finally, (3) is

$$2P = 2\begin{bmatrix} C & U \\ A & G \end{bmatrix} = 2\begin{bmatrix} 0.2 & 0.3 \\ 0.3 & 0.2 \end{bmatrix} = \begin{bmatrix} 0.4 & 0.6 \\ 0.6 & 0.4 \end{bmatrix}. \tag{8}$$

If p is the source probability of the first symbol event, then the entropy function
is

$$H_2(p) = p \log_2\left(\frac{1}{p}\right) - (1-p)\log_2\left(\frac{1}{1-p}\right), \tag{9}$$

and is tabulated in the last column of the Table 2. The graph of this entropy function
of Shannon and RNA are described in Fig. 2. Note that when $p = 0$ and $p = 1$, it has
a vertical tangent, because

Table 2 Entropy of RNA bases [C U; A G]

p	$\log_2\left(\frac{1}{p}\right)$	$p\log_2\left(\frac{1}{p}\right)$	$H(p)$
0.40	1.32193	0.52877	0.97095
0.41	1.28630	0.52738	0.97650
0.42	1.25154	0.52565	0.98145
0.43	1.21754	0.52356	0.98582
0.44	1.18442	0.52115	0.98959
0.45	1.15200	0.51840	0.99277
0.46	1.12029	0.51534	0.99538
0.47	1.08927	0.51596	0.99740
0.48	1.05889	0.50827	0.99885
0.49	1.02915	0.50428	0.99971
0.50	1.00000	0.50000	1.00000
0.51	0.97143	0.49543	0.99971
0.52	0.94342	0.49058	0.99885
0.53	0.91594	0.48545	0.99740
0.54	0.88897	0.48004	0.99538
0.55	0.86250	0.47437	0.99277
0.56	0.83650	0.46844	0.98959
0.57	0.81087	0.46225	0.98582
0.58	0.78588	0.45581	0.98145
0.59	0.76121	0.44912	0.97650
0.60	0.73697	0.44218	0.97095

$$\frac{d}{dp}\left[p\log_2\left(\frac{1}{p}\right)+(1-p)\log_2\left(\frac{1}{1-p}\right)\right]$$
$$=\left[\log_2\left(\frac{1}{p}\right)-1-\log_2\left(\frac{1}{1-p}\right)+1\right]\log_2 e$$
$$=\log_2\left(\frac{1}{p}\right)-\log_2\left(\frac{1}{1-p}\right)=0. \tag{10}$$

(9) comes to its maximum when $p=1/2$ where its derivative is zero as (10). Therefore,

$$\log_2\left(\frac{1}{p}\right)-\log_2\left(\frac{1}{1-p}\right)=0$$
$$\Rightarrow\left(\frac{1}{p}\right)-\left(\frac{1}{1-p}\right)=0 \tag{11}$$

Then, we have

$$p = 1 - p \Rightarrow p = \frac{1}{2}. \tag{12}$$

The symmetric entropy of RNA complementary bases (C U; A G) is given by

$$H_2(p)_{\text{RNA}} = p \log_? \left(\frac{1}{p}\right) - (1 - p) \log_? \left(\frac{1}{1 - p}\right),$$
$$= 0.97095, \quad \text{where } p \text{ is } 0.4 \text{ or } 0.6 \tag{13}$$

and the Shannon entropy is denoted by

$$H_2(p)_{\text{Shannon}} = p \log_2 \left(\frac{1}{p}\right) - (1 - p) \log_2 \left(\frac{1}{1 - p}\right) = 1, \quad \text{where } p \text{ is } 0.5. \tag{14}$$

The entropy of Shannon and RNA is shown in Table 2 and Fig. 2.

3 Symmetric Channel of RNA Complementary Bases [C U; A G] Without Noise Immunity

If the RNA genetic code has the complementary bases C $=$ U and A $=$ G, C + U $= p(Y_1)$ and G + A $= p(Y_2)$ over the symmetric channel. We have conditional probability $P(b_j|a_i) = P_{i,j}$ which defines a channel in (2) and (3), whose maximum amount of information of RNA complementary bases [C U; A G] that we can send through the binary symmetric channel with noise immunity is determined by its mutual information. If C and G are the probabilities of correct transmission, A and U are interference signals. Therefore, we have the channel matrix for a binary symmetric channel. Then, we have

$$\left[p(X_1) \ p(X_2) \right]_{1 \times 2} [P]_{2 \times 2} = \left[\alpha \ 1 - \alpha \right] \begin{bmatrix} C & U \\ A & G \end{bmatrix} == \left[p(Y_1) \ p(Y_2) \right]_{1 \times 2}. \tag{15}$$

Let p be the probability of choosing $\alpha = 0$, then $1 - p$ is the probability of choosing $\alpha = 1$, and the mutual information of this uniform channel is given by

$$I(X; Y) = H(Y) - H(Y|X). \tag{16}$$

Using (15), we can obtain,

Fig. 3 Binary symmetric channel (BSC) of RNA complementary bases (C U; A G)

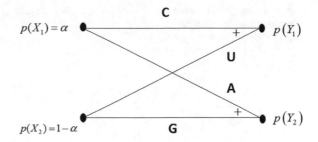

$$\begin{bmatrix} \alpha & 1-\alpha \end{bmatrix} \underbrace{\begin{bmatrix} -C\log_2 C & -U\log_2 U \\ -A\log_2 A & -G\log_2 G \end{bmatrix}}_{\text{Left hand-side}} = \begin{bmatrix} \alpha & 1-\alpha \end{bmatrix} \underbrace{\begin{bmatrix} -U\log_2 U & -C\log_2 C \\ G\log_2 G & -A\log_2 A \end{bmatrix}}_{\text{Right hand-side}},$$

$$(17)$$

where

$$\begin{aligned}
H(Y|X) &= \alpha C\log_2 C - (1-\alpha)A\log_2 A - \alpha U\log_2 U - (1-\alpha)G\log_2 G \\
&= -\alpha C\log_2 C - A\log_2 A + \alpha A\log_2 A - \alpha U\log_2 U \\
&\quad - G\log_2 G + \alpha G\log_2 G \\
&= -C\log_2 C - A\log_2 A = C\log_2 \frac{1}{C} + A\log_2 \frac{1}{A} = 0.9855. \quad (18)
\end{aligned}$$

where C is 20% and A is 30% (Fig. 3).

Note that

$$\begin{aligned}
C_{\text{RNA}} &= \max I(X;Y) \\
&= H(Y) - H(Y|X) = 1 - 0.9855 = 0.0145, \quad (19)
\end{aligned}$$

where $H(Y) = 1$, p is 0.4 or 0.6.

Also,

$$\begin{aligned}
C_{\text{Shannon}} &= \max I(X;Y) \\
&= H(Y) - H(Y|X) = 1 - 1 = 0 \quad (20)
\end{aligned}$$

where $H(Y) = 1$, p is 0.5.

Fig. 4 shows the comparison between Shannon capacity and RNA capacity for probability as well as the variance between them.

Fig. 4 Comparison between
Shannon capacity and RNA
capacity for probability

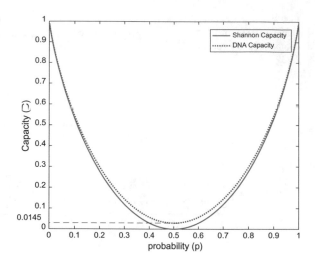

4   Conclusions

It is perfectly proved that Chargaff's experimental result of C = G = 20% and A = T (or U in RNA) = 30% in DNA and RNA is the same as that of the ecosystem management by the doubly stochastic matrix from a mathematically analytical point of view. The RNA entropy is 0.97095 given that the source probability is 0.4 or 0.6 for [C U; A G]. If C = U and A = G, then the channel capacity is 0.0145 with noise immunity. This is compared to Shannon entropy which arrives at its maximum and its corresponding capacity equals 0 when its source probability is given as 0.5. This enables optimal economic code to be designed when the source probability of the symmetric RNA bases [C U; A G] is set as 0.4 or 0.6, whose variance is equal to $\sqrt{0.1}$. Furthermore, Chargaff's second rule is able to be explained by the symmetric RNA bases [C U; A G], i.e., C + U and G + A.

Acknowledgements We are very thankful to AMIEE2018-editors, Z. B. Hu, Sergey Petoukhov, and Matthew He for their valuable comments.

References

1. Chargaff, E., Zamenhof, S., Green, C.: Composition of human desoxypentose nucleic acid. Nature **165**(4202), 756–757 (1950)
2. Khalil, M.I.: A new heuristic approach for DNA sequences alignment. Int. J. Image Graph. Signal Process. (IJIGSP) **7**(12), 18–23 (2015). https://doi.org/10.5815/ijigsp.2015.12.03
3. Zhang, W., Linghu, C., Zhang, J.: DbDMAD: a database of DNA methylation in human age-related disease. Int. J. Eng. Manuf. (IJEM) 2(5), 8–13 (2012). https://doi.org/10.5815/ijem.2012.05.02

4. Abo-Zahhad, M., Ahmed, S.M., Abd-Elrahman, S.A.; A novel circular mapping technique for spectral classification of exons and introns in human DNA sequences. Int. J. Inf. Technol. Comput. Sci. (IJITCS) **6**(4), 19–29 (2014). https://doi.org/10.5815/ijitcs.2014.04.02
5. Papoulis, A., Unnikrishna Pillai, S.: Probability, Random Variables and Stochastic Process, 4th edn. McGraw-Hill, Europe (2002)
6. Shannon, C.E.: A mathematical theory of communication. Bell Syst. Tech. J. **27**, 31–423 and 623–656 (1948)
7. He, M., Petoukhov, S.: Mathematics of Bioinformatics. Wiley, New York (2011)

Standard Genetic Code and Golden Ratio Cubes

Matthew He, Z. B. Hu and Sergey V. Petoukhov

Abstract The well-known golden ratio and knowledge about genetic coding systems are useful for artificial intelligence systems, engineering, and education. The golden ratio has been recently suggested for an expansion to the field of probabilistic and artificial intelligence in Tanackov et al. (Tehni ki vjesnikč 18:641–647, 2011) [1], in brain research (Conte et al. in Chaos, Solitons Fractals 41:2790–2800, 2009; Pletzer et al. in Brain Res. 1335:91–102) [2, 3], and in human facial proportion (Mizumoto et al. in Am. J. Orthod. Dentofac. Orthop. 136:168–74; Petoukhov in Binary sub-alphabets of genetic language and problem of unification bases of biological languages. Dubna, Russia, p. 191, 2002) [4, 5]. In this paper, we consider the number of hydrogen bonds of genetic code matrix and establish a relation with golden ratio matrix. We construct three-dimensional cubes from two-dimensional matrices. Furthermore, we revealed some numerical patterns of golden ratios under basic addition operation. The relationship data about the genetic stochastic matrices/cubes associated with the genetic codes play important roles on our new understanding of genetic code systems and can lead to new effective algorithms of information processing for modeling mutual communication among different parts of the genetic ensemble.

Keywords Genetic code · Hydrogen bonds · Golden ratio in artificial intelligence

M. He (✉)
Halmos College of Natural Sciences and Oceanography, Nova Southeastern University, Ft. Lauderdale, FL 33314, USA
e-mail: hem@nova.edu

Z. B. Hu
School of Educational Information Technology, Central China Normal University, Wuhan, China
e-mail: hzb@mail.ccnu.edu.cn

S. V. Petoukhov
Mechanical Engineering Research Institute, Russian Academy of Sciences, 101830 Moscow, Russia
e-mail: spetoukhov@gmail.com

© Springer Nature Switzerland AG 2020
Z. Hu et al. (eds.), *Advances in Artificial Systems for Medicine and Education II*,
Advances in Intelligent Systems and Computing 902,
https://doi.org/10.1007/978-3-030-12082-5_3

25

1 The Standard Genetic Codes

In biological systems, the texts from three-alphabetic words—*codons or triplets* encode the inheritable information. Each codon is compounded of four nitrogen alphabet bases: A (adenine), C (cytosine), G (guanine), and T (thiamine). The standard genetic code gives a complete list of $4^3 = 64$ codons, which correspond to 20 amino acids with three (one) letter code and stop codons (see Table 1).

The triplet locations of these 64 triplets or codons have huge number of variants approximately 10^{89} or 64! There are many formal characterizations of the particular structure of the code, which provide justification from physicochemical and/or evolutionary points of view [6]. Swanson [7] proposed a Gray code representation of the genetic code. In [8], it proposed a representation of the genetic code as a six-dimensional Boolean hypercube. Yang [9] applied a topological approach to rearrange the Hamiltonian-type graph of the codon map into a polyhedron model.

Table 1 Standard genetic code

		Second Position of Codon					
		U	C	A	G		
F i r s t	U	UUU Phe [F] UUC Phe [F] UUA Leu [L] UUG Leu [L]	UCU Ser [S] UCC Ser [S] UCA Ser [S] UCG Ser [S]	UAU Tyr [Y] UAC Tyr [Y] UAA *Ter* [end] UAG *Ter* [end]	UGU Cys [C] UGC Cys [C] UGA *Ter* [end] UGG Trp [W]	U C A G	T h i r d
P o s i t i o n	C	CUU Leu [L] CUC Leu [L] CUA Leu [L] CUG Leu [L]	CCU Pro [P] CCC Pro [P] CCA Pro [P] CCG Pro [P]	CAU His [H] CAC His [H] CAA Gln [Q] CAG Gln [Q]	CGU Arg [R] CGC Arg [R] CGA Arg [R] CGG Arg [R]	U C A G	P o s i t i o n
	A	AUU Ile [I] AUC Ile [I] AUA Ile [I] AUG Met [M]	ACU Thr [T] ACC Thr [T] ACA Thr [T] ACG Thr [T]	AAU Asn [N] AAC Asn [N] AAA Lys [K] AAG Lys [K]	AGU Ser [S] AGC Ser [S] AGA Arg [R] AGG Arg [R]	U C A G	
	G	GUU Val [V] GUC Val [V] GUA Val [V] GUG Val [V]	GCU Ala [A] GCC Ala [A] GCA Ala [A] GCG Ala [A]	GAU Asp [D] GAC Asp [D] GAA Glu [E] GAG Glu [E]	GGU Gly [G] GGC Gly [G] GGA Gly [G] GGG Gly [G]	U C A G	

By means of the nucleotide base representation on the square with vertices U or T = 00, C = 01, G = 10 and A = 11, Štambuk [10] defined universal metric properties of the genetic code. Petoukhov introduced the "Biperiodic table of genetic code" [5, 11, 12] shown (Table 2).

He and Petoukhov further investigated the symmetric characteristic of the biperiodic table and symmetries in structure of genetic code in [13–17].

In biology, the self-reproduction of biological organisms is embedded in the genetic system in their generations. In mathematics, the "golden ratio" (also called the "divine proportion") and its properties were the self-reproduction of a mathematical symbol of from the Renaissance, studied by Leonardo da Vinci, J. Kepler and many other prominent thinkers (see details in the Web site "Museum of Harmony and Golden Ratio" by A. Stakhov, www.goldenmuseum.com). For each full cycle of DNA double helix spiral, it measures 34 angstroms long by 21 angstroms wide. 34 and 21 are numbers in the Fibonacci sequence and their ratio, 1.6190476 closely approximates 1.6180339, it shows that the DNA molecules are based on the golden ratio. The a double-stranded helix B-DNA in the cell has a two grooves in its spirals with the major groove to the minor groove or roughly 21 angstroms to 13 angstroms, 21 and 13 are numbers in the Fibonacci sequence and their ratio 1.61538462 also closely approximates 1.6180339. It was shown in [18] that codon populations in single-stranded whole human genome DNA are fractal and fine-tuned by the golden ratio 1.618. It has been reported that a cross-sectional view from the top of the DNA double helix forms a decagon comprised with two pentagons, with one rotated by 36° from the other, so each spiral of the double helix must trace out the shape of a pentagon. The ratio of the diagonal of a pentagon to its side is φ to 1. So, no matter which way we look at it, even in its smallest element, DNA, and life, is constructed using the golden ratio $\varphi = (1 + 5^{0.5})/2 = 1.618...$ The Watson-Crick rules [19] of base pairing (or nucleotide pairing) are:

- **A** with **T**: the purine **adenine** (A) always pairs with the pyrimidine **thymine** (T)
- **C** with **G**: the pyrimidine cytosine (C) always pairs with the purine **guanine** (G)

This is because only with A and T and with C and G are there opportunities to establish **hydrogen bonds** between them (two between A and T; three between C and

Table 2 Biperiodic table of genetic code by Petoukhov

CCC	CCA	CAC	CAA	ACC	ACA	AAC	AAA
CCU	CCG	CAU	CAG	ACU	ACG	AAU	AAG
CUC	CUA	CGC	CGA	AUC	AUA	AGC	AGA
CUU	CUG	CGU	CGG	AUU	AUG	AGU	AGG
UCC	UCA	UAC	UAA	GCC	GCA	GAC	GAA
UCU	UCG	UAU	UAG	GCU	GCG	GAU	GAG
UUC	UUA	UGC	UGA	GUC	GUA	GGC	GGA
UUU	UUG	UGU	UGG	GUU	GUG	GGU	GGG

G). The rules of base pairing imply that if we can "read" the sequence of nucleotides on one strand of DNA, we can immediately deduce the complementary sequence on the other strand. The rules of base pairing also explain the phenomenon (called Chargaff's rule) that the amount of adenine (A) in the DNA of an organism equals the amount of thymine (T). Similarly, the amount of guanine (G) equals the amount of cytosine (C). The function of the golden ratio was recently established in human brain research in [2, 3] and human facial proportions in [4, 5].

In this paper, we establish a relation between the standard genetic codes and golden ratio cubes based on the fact that the complementary letters C and G have three hydrogen bonds (C = G = 3) and the complementary letters A and U have two hydrogen bonds (A = U = 2). This relationship provides a new connection between the number of hydrogen bonds and the golden ratio via a scheme of Hardmard product process, which is a type of iteration that has been frequently used in the signal processing and pattern recognition. This matrix relationship illustrates that the golden ratio plays an important role in DNA shapes and structures. Furthermore, we determine some numerical patterns of golden ratios under basic addition operation.

2 Hydrogen Bonds-based (2, 3) Equivalence Matrix and Golden Ratio Matrix

To introduce golden ratio matrices, we begin with a sequence of genetic matrices constructed by Hardmard product process from a 2×2 to 4×4 and then from 4×4 to 8×8 matrices as illustrated in Table 3.

The 8×8 matrix represents the Bi-periodic Table of Genetic Code by Petoukhov (Table 3). The complementary letters C and G have 3 hydrogen bonds (C = G = 3) and the complementary letters A and U have 2 hydrogen bonds (A = U = 2). A hydrogen bonds (2, 3) based equivalence matrices are:

One can show that all matrices $H^{(n)}$ are nonsingular. They are symmetrical relative to both diagonals. All rows and columns of this matrix are differentiated from each other by the sequences of their numbers. But the sums of entries of each row and of

Table 3 Genetic code matrices

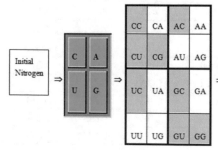

$$H^{(1)} = \begin{vmatrix} 3 & 2 \\ 2 & 3 \end{vmatrix} \; ; H^{(2)} = \begin{vmatrix} 9 & 6 & 6 & 4 \\ 6 & 9 & 4 & 6 \\ 6 & 4 & 9 & 6 \\ 4 & 6 & 6 & 9 \end{vmatrix} \; ; H^{(3)} = \begin{vmatrix} 27 & 18 & 18 & 12 & 18 & 12 & 12 & 8 \\ 18 & 27 & 12 & 18 & 12 & 18 & 8 & 12 \\ 18 & 12 & 27 & 18 & 12 & 8 & 18 & 27 \\ 12 & 18 & 18 & 27 & 8 & 12 & 12 & 18 \\ 18 & 12 & 12 & 8 & 27 & 18 & 18 & 12 \\ 12 & 18 & 8 & 12 & 18 & 12 & 27 & 18 \\ 12 & 8 & 18 & 12 & 18 & 12 & 27 & 18 \\ 8 & 12 & 12 & 18 & 12 & 18 & 18 & 27 \end{vmatrix}$$

Fig. 1 $H^{(n)} = [3\ 2;\ 2\ 3]^{(n)}$ ($n = 1, 2, 3$) based on product of hydrogen bonds (C = G = 3, A = U = 2)

each column in any matrix $H^{(n)}$ are identical. For example, in the case of the matrix $H^{(3)}$, these sums are equal to $125 = 5^3$ and the total sum of numbers inside the matrix equals 1000. A rank of this matrix is 8. Its determinant equals to 5^{12}. Eigenvalues of $H^{(3)}$ are 1, 5, 5, 5, 5^2, 5^2, 5^2, and 5^3 (Fig. 1).

We raise the power ½ (i.e., we take the square root) of the simplest genetic matrix $H^{(1)}$ in ordinary sense, the result matrix is the bi-symmetric matrix $\Phi = (H^{(1)})^{1/2}$, the matrix elements of which are equal to the golden ratio and to its inverse value. In general, if any other genomatrix $H^{(n)} = [3\ 2;\ 2\ 3]^{(n)}$ is raised to the power ½, the resulting matrix is the bi-symmetric matrix $\Phi^{(n)} = (H^{(n)})^{1/2}$, the matrix elements of which are equal to the golden ratio in various integer powers with elements of symmetry among these powers (Fig. 2). For instance, the matrix $\Phi^{(3)} = (H^{(n)})^{1/2}$ ($n = 1, 2, 3$) has only two pairs of inverse numbers: φ^1 and φ^{-1}, φ^3 and φ^{-3} (Fig. 2). Matrices with matrix elements, all of which are equal to golden ratio φ in different powers only, can be referred to as "golden matrices." Furthermore, product of all numbers in any row and in any column of these golden matrixes is equal to 1.

The molecular system of the genetic alphabet is constructed by the nature in such manner that other genetic matrices play role quint matrices and golden matrices for other parameters. The square root from such numeric matrices is connected with the golden ratio matrices.

3 Golden Ratio Cubes

In this section, we construct golden ratio cubes by using slice of golden ratio matric. In each stage of this process ($\Phi^{(1)} \Rightarrow \Phi^{(2)} \Rightarrow \Phi^{(3)}$), it generates a series of doubly stochastic matrices. A cube is triply stochastic if each (frontal, horizontal, and lateral)

$$(H^{(1)})^{1/2} = \\ = \Phi^{(1)} \quad \begin{vmatrix} \varphi & \varphi^{-1} \\ \varphi^{-1} & \varphi \end{vmatrix} \quad ; \quad (H^{(2)})^{1/2} = \\ = \Phi^{(2)} = \quad \begin{vmatrix} \varphi^2 & \varphi^0 & \varphi^0 & \varphi^{-2} \\ \varphi^0 & \varphi^2 & \varphi^{-2} & \varphi^0 \\ \varphi^0 & \varphi^{-2} & \varphi^2 & \varphi^0 \\ \varphi^{-2} & \varphi^0 & \varphi^0 & \varphi^2 \end{vmatrix}$$

$$(H^{(3)})^{1/2} = \\ = \Phi^{(3)} =$$

φ^3	φ^1	φ^1	φ^{-1}	φ^1	φ^{-1}	φ^{-1}	φ^{-3}
φ^1	φ^3	φ^{-1}	φ^1	φ^{-1}	φ^1	φ^{-3}	φ^{-1}
φ^1	φ^{-1}	φ^3	φ^1	φ^{-1}	φ^{-3}	φ^1	φ^{-1}
φ^{-1}	φ^1	φ^1	φ^3	φ^{-3}	φ^{-1}	φ^{-1}	φ^1
φ^1	φ^{-1}	φ^{-1}	φ^{-3}	φ^3	φ^1	φ^1	φ^{-1}
φ^{-1}	φ^1	φ^{-3}	φ^{-1}	φ^1	φ^3	φ^{-1}	φ^1
φ^{-1}	φ^{-3}	φ^1	φ^{-1}	φ^1	φ^{-1}	φ^3	φ^1
φ^{-3}	φ^{-1}	φ^{-1}	φ^1	φ^{-1}	φ^1	φ^1	φ^3

Fig. 2 The Kronecker family of the golden matrices $\Phi^{(n)} = (H^{(n)})^{1/2}$, where $\varphi = (1 + 5^{0.5})/2 = 1, 618\ldots$ is the golden ratio

sliced section is doubly stochastic matrix. From the matrix $\Phi^{(1)}$, we can construct a $2 \times 2 \times 2$ cube $\mathbf{C_2}$ with the following two frontal/lateral/horizontal slices of doubly stochastic matrices;

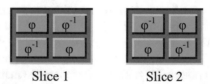

Slice 1   Slice 2

Take the sum of this $2 \times 2 \times 2$ cube from the frontal, lateral, and horizontal directions, respectively, each gives the common sum of $\sqrt{5}$. This cube C_2 is a triply stochastic cube. As we move on the next stages of this process, we'll see a sequence of triply stochastic cubes will emerge. From the matrix $\Phi^{(2)}$, we start with the central 2×2 submatrix to form a $2 \times 2 \times 2$ cube $\mathbf{C_{y4}^{(2 \times 2 \times 2)}}$ with 2 slices of matrices:

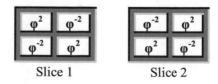

Slice 1   Slice 2

Take the sum of this $2 \times 2 \times 2$ cube from the frontal, lateral, and horizontal directions, respectively, each gives the common sum of **3**. This cube $\mathbf{C_4^{(2 \times 2 \times 2)}}$ is a triply stochastic cube.

Next, we expand the central matrix to all directions (left, right, up, and down) by one matrix dimension to form a $4 \times 4 \times 4$ cube $C_4^{(4 \times 4 \times 4)}$ with 4 slices of 4×4 matrices through a permutation of 4 columns.

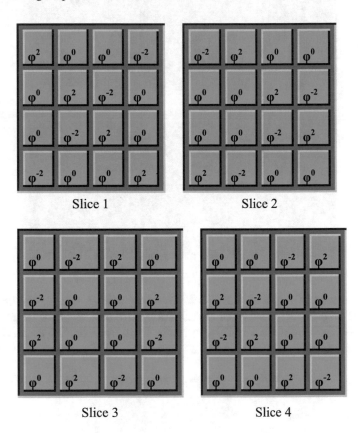

Slice 1   Slice 2

Slice 3   Slice 4

Take the sum of this $4 \times 4 \times 4$ cube from the frontal, lateral, and horizontal directions, respectively, each gives the common sum of **5**. This cube $C_4^{(4 \times 4 \times 4)}$ is a triply stochastic cube.

Next, we apply the same process to the matrix of $\Phi^{(3)}$ and form a series of $2 \times 2 \times 2$, $4 \times 4 \times 4$, $6 \times 6 \times 6$, and $8 \times 8 \times 8$ cubes as follows:

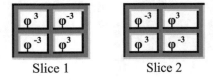

Slice 1   Slice 2

Take the sum of this $2 \times 2 \times 2$ cube from the frontal, lateral, and horizontal directions, respectively, each gives the common sum of $2\sqrt{5}$.

Next, we generate $4 \times 4 \times 4$ cube as follows:

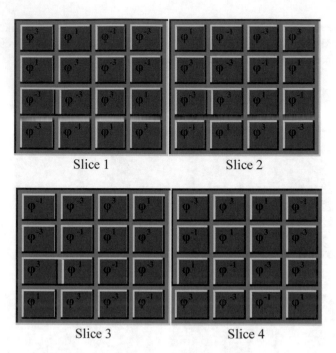

Slice 1   Slice 2

Slice 3   Slice 4

Take the sum of this $4 \times 4 \times 4$ cube from the frontal, lateral, and horizontal directions, respectively, each gives the common sum of $3\sqrt{5}$. Next, we generate **6 × 6 × 6** as follows, for the simplicity; we only display one of 6 slices as other slice can be easily generated by standard column permutations.

Slice 1

Take the sum of this $6 \times 6 \times 6$ cube from the frontal, lateral, and horizontal directions, respectively, each gives the common sum of $4\sqrt{5}$. Finally, we generate **8 × 8 × 8** cube as follows similar to $6 \times 6 \times 6$ cubes displayed above.

Table 4 Number of cubes and common sums of cubic slides

(2, 3) equivalence	No. of cubes	Common sums of cube slices
Initial stage (2^0)	0	0
1st stage (2^1)	1	$\sqrt{5}$
2nd stage (2^2)	2	3, 5
3rd stage (2^3)	4	$2\sqrt{5}, 3\sqrt{5}, 4\sqrt{5}, 5\sqrt{5}$
4th stage (2^4)	8	7, 10, 12, 15,17, 20, 22, 25

φ^3	φ^1	φ^1	φ^{-1}	φ^1	φ^{-1}	φ^{-1}	φ^{-3}
φ^1	φ^3	φ^{-1}	φ^1	φ^{-1}	φ^1	φ^{-3}	φ^{-1}
φ^1	φ^{-1}	φ^3	φ^1	φ^{-1}	φ^{-3}	φ^1	φ^{-1}
φ^{-1}	φ^1	φ^1	φ^3	φ^{-3}	φ^{-1}	φ^{-1}	φ^1
φ^1	φ^{-1}	φ^{-1}	φ^{-3}	φ^3	φ^1	φ^1	φ^{-1}
φ^{-1}	φ^1	φ^{-3}	φ^{-1}	φ^1	φ^3	φ^{-1}	φ^1
φ^{-1}	φ^{-3}	φ^1	φ^{-1}	φ^1	φ^{-1}	φ^3	φ^1
φ^{-3}	φ^{-1}	φ^{-1}	φ^1	φ^{-1}	φ^1	φ^1	φ^3

Slice 1

Take the sum of this $8 \times 8 \times 8$ cube from the frontal, lateral, and horizontal directions, respectively, each gives the common sum of $5\sqrt{5}$.

Here, we summarize all the cubes generated by Table 4.

Next, we apply addition operation (**+**) to $\Phi^{(1)}$ denoted by $\Phi_+^{(1)}$ (where the superscript (1) denotes the 1st matrix) to generate 4×4 matrix $\Phi_+^{(2)}$ (where the superscript (2) denotes the 2nd phase of iteration) and 8×8 matrix $\Phi_+^{(3)}$ (where the superscript (3) denotes the 3rd phase of iteration) as shown below:

$$\Phi_+^{(1)} = \begin{vmatrix} \varphi & \varphi^{-1} \\ \varphi^{-1} & \varphi \end{vmatrix} \quad ; \qquad \Phi_+^{(2)} = \begin{vmatrix} 2\varphi & \varphi+\varphi^{-1} & \varphi^{-1}+\varphi & 2\varphi^{-1} \\ \varphi+\varphi^{-1} & 2\varphi & 2\varphi^{-1} & \varphi^{-1}+\varphi \\ \varphi^{-1}+\varphi & 2\varphi^{-1} & 2\varphi & \varphi+\varphi^{-1} \\ 2\varphi^{-1} & \varphi^{-1}+\varphi & \varphi+\varphi^{-1} & 2\varphi \end{vmatrix}$$

$$\Phi_+^{(3)} = \begin{vmatrix}
3\varphi & 2\varphi+\varphi^{-1} & \varphi^{-1}+2\varphi & \varphi+2\varphi^{-1} & \varphi^{-1}+2\varphi & \varphi+2\varphi^{-1} & 2\varphi^{-1}+\varphi & 3\varphi^{-1} \\
2\varphi+\varphi^{-1} & 3\varphi & \varphi+2\varphi^{-1} & \varphi^{-1}+2\varphi & \varphi+2\varphi^{-1} & \varphi^{-1}+2\varphi & 3\varphi^{-1} & 2\varphi^{-1}+\varphi \\
\varphi^{-1}+2\varphi & \varphi+2\varphi^{-1} & 3\varphi & 2\varphi+\varphi^{-1} & 2\varphi^{-1}+\varphi & 3\varphi^{-1} & \varphi^{-1}+2\varphi & \varphi+2\varphi^{-1} \\
\varphi+2\varphi^{-1} & \varphi^{-1}+2\varphi & 2\varphi+\varphi^{-1} & 3\varphi & 3\varphi^{-1} & 2\varphi^{-1}+\varphi & \varphi+2\varphi^{-1} & \varphi^{-1}+2\varphi \\
\varphi^{-1}+2\varphi & \varphi+2\varphi^{-1} & 2\varphi^{-1}+\varphi & 3\varphi^{-1} & 3\varphi & 2\varphi+\varphi^{-1} & \varphi^{-1}+2\varphi & \varphi+2\varphi^{-1} \\
\varphi+2\varphi^{-1} & \varphi^{-1}+2\varphi & 3\varphi^{-1} & 2\varphi^{-1}+\varphi & 2\varphi+\varphi^{-1} & 3\varphi & \varphi+2\varphi^{-1} & \varphi^{-1}+2\varphi \\
2\varphi^{-1}+\varphi & 3\varphi^{-1} & \varphi^{-1}+2\varphi & \varphi+2\varphi^{-1} & \varphi^{-1}+2\varphi & \varphi+2\varphi^{-1} & 3\varphi & 2\varphi+\varphi^{-1} \\
3\varphi^{-1} & 2\varphi^{-1}+\varphi & \varphi+2\varphi^{-1} & \varphi^{-1}+2\varphi & \varphi+2\varphi^{-1} & \varphi^{-1}+2\varphi & 2\varphi+\varphi^{-1} & 3\varphi
\end{vmatrix}$$

From $\Phi_+^{(1)}$, we can generate a $2 \times 2 \times 2$ stochastic cube with a common sum of $\sqrt{5}$.

From $\Phi_+^{(2)}$, we can generate a $2 \times 2 \times 2$ and a $4 \times 4 \times 4$ stochastic cube with a common sum of $2\sqrt{5}$, $4\sqrt{5}$, respectively.

From $\Phi_+^{(3)}$, similarly, we can generate a $2 \times 2 \times 2$, $4 \times 4 \times 4$, $6 \times 6 \times 6$, and $8 \times 8 \times 8$ stochastic cube with a common sum of $3\sqrt{5}$, $6\sqrt{5}$, $9\sqrt{5}$, and $12\sqrt{5}$, respectively.

At the Nth stage ($N = 1, 2, 3, 4, \ldots$), the common sums of $(2 \times 2 \times 2)$, $(4 \times 4 \times 4)$, $(6 \times 6 \times 6)$, ... $(2k \times 2k \times 2k)$, ... $(2^N \times 2^N \times 2^N)$ stochastic cubes form an arithmetic sequence with the common difference $d = 5N$,

$$N\sqrt{5}, 2N\sqrt{5}, 3N\sqrt{5}, \ldots kN\sqrt{5}, \ldots, N2^{N-1}\sqrt{5}.$$

The sum of this arithmetic sequence is

$$S(\varphi, \varphi^{-1}) = \sqrt{5}N(1+2+3+\ldots+2^{N-1}) = \sqrt{5}N(1+2^{N-1})2^{N-1}/2 = \sqrt{5}N2^{N-2}(1+2^{N-1}), N = 1, 2, 3, \ldots$$

This shows that golden ratio related hydrogen bonds expansion is triply stochastic and its accumulation at each stage is governed by this arithmetic sequence with a common difference $\sqrt{5}N$.

4 Conclusions and Perspectives

By using genetic hydrogen bonds of (3, 2) associated with the genetic code, our study introduced a sequence of doubly stochastic cubes associated with golden ratios. We established a new relationship between the number of hydrogen bonds and the golden ratio via a scheme of Hardmard product process. Furthermore, we revealed some numerical patterns of golden ratios under basic addition operation. This matrix relationship illustrates that the golden ratio plays an important role in DNA shapes and structures. Biological organisms are based on a repertoire of structured and interrelated molecular building blocks. The same and related molecular structures and mechanisms show up repeatedly in the genome of a single species and across a very wide spectrum of divergent species. The matrices/cubes appear in various dimensions with different shapes. Stochastic matrices/cubes motivated by language of probability show up repeatedly in the nature. Many literatures on mathematics and biological systems have merged in recent years [20–22] to further advance our understanding of life and its evolutions [23]. DNA-Genetic Encryption Technique has been developed in [24]. DNA 3D Self-assembly algorithmic model has been recently developed to solve maximum clique problem in [25]. The golden ratio stochastic matrices/cubes could be served as a storage for data compression and encryption by using Dynamic Look up Table and Modified Huffman Techniques [26]. It is hoped that these relationships between hydrogen bonds of (3, 2) and golden ratio cubes will help us further explore our understanding of structure and function of genetic systems.

References

1. Tanackov, I., Jovan, T., Kostelacepi'c, M.: The golden ratio in probabilistic and artificial intelligence. Tehni ki vjesnikč **18**(4), 641–647 (2011)
2. Conte, E., Khrennikov, A., Federici, A., Zbilut, J. P.: Fractal fluctuations and quantum-like chaos in the brain by analysis of variability of brain waves: a new method based on a fractal variance function and random matrix theory: a link with El Naschie fractal Cantorian space–time and V. Weiss and H. Weiss golden ratio in brain. Chaos, Solitons Fractals **41**, 2790–2800 (2009)
3. Pletzer, B., Kerschbaum, H., Klimesch, W.: When frequencies never synchronize: the golden mean and the resting EEG. Brain Res. **1335**, 91–102 (2010)
4. Mizumoto, Y., Deguchi, T., Kelvin, W.C.F.: Assessment of facial golden proportions among young Japanese women. Am. J. Orthod. Dentofac. Orthop. **136**, 168–174 (2009)
5. Petoukhov, S. V.: Binary sub-alphabets of genetic language and problem of unification bases of biological languages. In: IX International Conference on Mathematics, Computer, Education, p. 191. Russia, Dubna, 28–31 Jan 2002 (in Russian)
6. Knight, R.D., Freeland, S.J., Landweber, L.F.: Selection, history and chemistry: the three faces of the genetic code. TIBS **24**, 241–247 (1999)
7. Swanson, R.: A unifying concept for the amino acid code. Bull. Math. Biol. **46**(2), 187–203 (1984)
8. Jimenéz-Montaño, M.A., Mora-Basáñez, C.R., Pöschel, T.: On the hypercube structure of the genetic code. In: Lim, H.A., Cantor, C.A. (eds.) Proceedings of 3rd International Conference on Bioinformatics and Genome Research, p. 445. World Scientific (1994)

9. Yang, C.M.: The naturally designed spherical symmetry in the genetic code (2003). http://arxiv.org/abs/q-bio.BM/0309014
10. Štambuk, N.: Universal metric properties of the genetic code. Croat. Chem. Acta **73**(4), 1123–1139 (2000)
11. Petoukhov, S.V.: The Bi-periodic Table of Genetic Code and Number of Protons, Foreword of K. V. Frolov, p. 258. Moscow (in Russian) (2001)
12. Petoukhov, S.V.: Genetic code and the ancient chinese book of changes. Symmetry: Cult. Sci. **10**(3–4), 211–226 (1999)
13. He, M.: Double helical sequences and doubly stochastic matrices. In: Symmetry: Culture and Science, Symmetries in Genetic Information, International Symmetry Foundation, Budapest, 2004, pp. 307–330 (2003a)
14. He, M.: Symmetry in structure of genetic code. In: Proceedings of the Third All-Russian Interdisciplinary Scientific Conference on Ethics and the Science of Future. Unity in Diversity, Moscow, 12–14 Feb 2003b
15. He, M.: Genetic code, attributive mappings and stochastic matrices. Bull. Math. Biol. **66**(5), 965–973 (2004)
16. He, M., Petoukhov, S.V.: Mathematics of Bioinformatics: Theory, Practice, and Applications, Wiley (2011)
17. He, M., Petoukhov, S.V., Ricci, P.E.: Genetic code, Hamming distance and stochastic matrices. Bull. Math. Biol. **66**, 1405–1421 (2004)
18. Perez, J.C.: Codon populations in single-stranded whole human genome DNA are fractal and fine-tuned by the Golden ratio 1.618. Interdiscip. Sci. **2**(3), 228–240 (2010)
19. Watson, J.: The Double Helix. Atheneum, New York (1968)
20. Higgins, D., Taylor, W.: Bioinformatics. Oxford University Press, Oxford (2000)
21. Percus, J.: Mathematics of Genome Analysis. Cambridge University Press Press, New York (2002)
22. Pevzner, P.: Computational Molecular Biology. MIT Press, Cambridge (2000)
23. Kay, L.: Who Wrote the Book of Life? A History of the Genetic Code. Stanford University Press, Stanford (2000)
24. Mousa, H.M.: DNA-genetic encryption technique. Int. J. Comput. Netw. Inf. Secur. (IJCNIS) **8**(7): 1–9 (2016). https://doi.org/10.5815/ijcnis.2016.07.01
25. Ma, J., Li, J., Dong, Y.: DNA 3D self-assembly algorithmic model to solve maximum Clique problem. Int. J. Image, Graph. Sign. Process. (IJIGSP) **3**(3), 41–48 (2011). https://doi.org/10.5815/ijigsp.2011.03.06
26. Hossein, S.M., Roy, S.: A compression & encryption algorithm on dna sequences using dynamic look up table and modified huffman techniques. Int. J. Inf. Technol. Comput. Sci. (IJITCS) **5**(10), 39–61 (2013). https://doi.org/10.5815/ijitcs.2013.10.05
27. Schmid, K., Marx, D., Samal, A.: Computation of a face attractiveness index based on neo-classical canons, symmetry, and golden ratios. Pattern Recogn. **41**, 2710–2717 (2008)

The Technique for Data Parallelism in Neural Processing Units

Vitaliy A. Romanchuk and **Ruslan I. Bazhenov**

Abstract In this paper, the authors propose a technique for efficient data parallelism in neural processing units through different dimensional data subsets and redistribution of similar operations between code segments that are executed in parallel. The authors observe a combined approach to optimize a solution of the one-dimensional optimization problem. The authors also consider a category of the neural processor bit depth, based on dynamic programming methods. Empirical study proves that the application of the method offered can improve significantly overall program instruction per second by 5–14%, depending on a complexity class of decision problem and the degree of operation homogeneity.

Keywords Data parallelism · Neural processing unit (NPU) · Executable compression · Parallelization

1 Introduction

There are two paradigms of parallel programming: data parallelism and problem parallelism. Mixed data and problem parallelism is the best kind of parallelism as it can be implemented simultaneously.

Data parallelism requires different elements of such an array to be processed/executed on a vector processor or different processors of a parallel computing system. Vectorization or parallelization within this approach is performed during transformation. Thus, when developing a parallel application, a programmer needs to

V. A. Romanchuk
Ryazan State University named for S. Yesenin, Ryazan, Russia
e-mail: v.romanchuk@rsu.edu.ru

R. I. Bazhenov (✉)
Sholom-Aleichem Priamursky State University, Birobidzhan, Russia
e-mail: r-i-bazhenov@yandex.ru

© Springer Nature Switzerland AG 2020
Z. Hu et al. (eds.), *Advances in Artificial Systems for Medicine and Education II*,
Advances in Intelligent Systems and Computing 902,
https://doi.org/10.1007/978-3-030-12082-5_4

- Set options to the translator those are going to be optimized either on vector or in parallel;
- Set the directives or pragmas for parallel compilation;
- Use formal parallel computing languages, libraries, or collections of computer software specially designed for specific computer hardware and optimized or modified regarding this or that specific kind of computing.

When performing programming by data parallelism, programmers generally follow the conventions of programming languages or supporting tools for language extension.

Data parallelism implementation model requires parallelism support at the translator level. Such support can be provided by

- Preprocessors using available direct translators and specialized libraries to implement parallel algorithmic structures;
- Pre-translators that deal with preliminary examination of a logical structure of a program, dependencies analysis, and selected parallel optimization;
- Parallelizing translators that detect parallelism in the source code of a program and transform it into parallel structures.

The basic principles of data parallelism are reviewed in scientific researches by Boyer and Pawley, Singh et al., Hillis and Steele, Flanders et al., Ebeling et al., Pan and Reif, Lim, Kong and Lieserson [1–8].

The computing industry has a lot of neural processing units nowadays. Some of them operate according to the principles of connectionism and provide an execution of various operations for emulating neurons. It seems better to make maximum use of data parallelism opportunities for those processing units with a lot of similar transformations without being programmed with transitions and other operations that prevent parallelism [9].

Many surveys are devoted to data parallelism in the GPU and neural processing units. They are those by Chen and Rabaey, Bottou, Noel and Osindero, Retcht et al., Jia, Boyd et al. [10–16].

In the papers mentioned, the authors point out data parallelism methods based on equal redistribution of operations with the same bitwise of operands, similar to the MMX (Multimedia extension, Multiple Math extension, or Matrix Math extension) technology for traditional processors. At the same time, if the original task should be performed online cyclically, then it would be efficient to minimize bitwise operations and redistribute similar operations to reduce downtime of hardware resource usage. For example, if there is an artificial neural network emulation task, where the input data are considered to be data patterns with different digits (even 10–10–20–2–20 bits), then it does not make any sense to level out a bitwise of operands to the maximum value since it is more productive to "pack data" into the 64-bitwise pattern.

The aim of this research is to develop a method for data parallelism in neural processing units by solving optimization problem (effective data and expressions packing) to increase neurocomputer program performance.

2 Methods of Investigation

In most scientific researches, the division into data sets is supposed to be sequential, i.e., a set is split into equal subsets. But one can increase the instruction per second by redistributing effectively operations of similar type between the code elements and changing the dimensional data subsets. The paper proposes a combinatorial optimization approach to solve the one-dimensional optimization problem taking into account bit depth features of a neural processing unit, using dynamic programming methods.

The efficiency of the method offered is achieved by parallel execution of operations with low-bit operands. Hardware data packing is possible in some models of neural processing units.

3 Methods of Solving the Problem of Effective Data Parallelism in Neural Processing Units

Each class of subtasks $Z^{(j)}$ presented as an artificial neural network (ANN) has a set of characteristics.

For instance, $NetX_{R_i}$—is the bit depth of i-type ANN input $Net_X = \{net_{x1}, \ldots, net_{xi}, \ldots, net_{xn}\}$, defined by its extreme value:

$$NetX_{R_i} = \log_2(\max(net_{xi}));$$

$NetW_{R_i}$ is a bit depth of i-type weight coefficient ANN $W = \{w_{w1}, \ldots, w_{wi}, \ldots, w_{wn}\}$ derived from its extreme value;

$$NetW_{R_i} = \log_2(\max(w_{xi}));$$

$NetY_{R_i}$ is bit depth of i-type ANN output $Net_Y = \{net_{y1}, \ldots, net_{yi}, \ldots, net_{yn}\}$, defined by its extreme value:

$$NetY_{R_i} = \log_2(\max(net_{yi}));$$

H_{Pr} is the length of the neural processing unit.

Then, when performing a neuron emulator, a certain number of bits must be reserved in the neural processing unit. It is supposed to be equal to the extreme bit capacity of the result. It is calculated as the sum of a bit depth of the input data and one bit for the transfer:

$$R = \max(NetX_{R_i}) + \max(NetW_{R_i}) + 1.$$

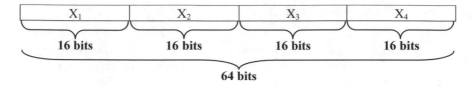

Fig. 1 Example of data packing

In that case, the number of parallel data sets P_t by data packing is defined as:

$$P_t = \frac{H_{Pr}}{\max(\mathrm{Net}X_{R_i}) + \max(\mathrm{Net}W_{R_i}) + 1}.$$

An example of data packing for the NM640x family of a neural processing unit is shown in Fig. 1.

A sum of digits of integer and fractional parts is calculated separately for real numbers of input values and weight coefficients.

To use K_{MAC} coefficient for implementing the observed method of information processing is impossible, since its value in most cases will be more than 1 (unit). Therefore, to evaluate the effectiveness of parallelization at this stage, the authors request a factor that implies the core workload rate of the neural processing unit during the execution of u-operation:

$$K_{Rju} = \frac{1}{Pt}.$$

The total parameters of the core load of the neural processing unit for the fragment SO_j:

$$K_{Rj} = 1 - \frac{\sum_{j=1}^{T} \frac{1}{Pt_i}}{T}.$$

For a microneuroprogram as a whole:

$$K_R = 1 - \frac{\sum_{i=1}^{p} \sum_{j=1}^{T} \frac{1}{Pt_i}}{p * T}.$$

The method outlined above can be used to simplify data packing, but does not provide the possibility of bit redistribution of the operation. So, the authors suggest another way for a programmer which is more time-consuming.

The purpose of information processing system is to redistribute a set of operations and consequently a set of microinstructions, so that the maximum number of operations would be performed in one time cycle by data packing.

For each tuple of operations $SO_i = \langle O_{i1}, O_{i2}, \ldots, O_{iinc} \rangle$, there a tuple of computing digits for each j operation $\langle R_{i1}, R_{i2}, \ldots, R_{iinc} \rangle$ can be defined as follows:

$$R_i = \text{Net}X_{R_i} + \text{Net}W_{R_i} + 1.$$

It means that it is necessary to redistribute a set of operations into tuples of operations, so that the difference between the capacity of the neural processing unit and the total number of calculations would be minimal:

$$SO_i = \langle O_{i1}, O_{i2}, \ldots, O_{iinc} \rangle = SO_{i1} \cup SO_{i2} \cup \ldots \cup SO_{iT},$$

$$SO_{ij} = \langle O_{ij1}, O_{ij2}, \ldots, O_{ijinc} \rangle, \quad j = \overline{1, T},$$

$$\left(H_{\text{Pr}} - \sum_{k=1}^{ijinc} R_{ijk} \right) \to \min.$$

where T is the number of neural processing unit cycles to implement ANNs.

Each operation O_{ijk} has an integer feature of the computational capacity equal to R_{ijk}.

Then, the operation tuple SO_{ij} can be performed in one time cycle of the neural processing unit, providing $ijinc$ neurons emulator.

The task of data packing in a neuroprogram can be formulated by mathematical proof in the following way:

$$\forall SO_i, i = \overline{1, p}; \sum_{j=1}^{T} \left(Pr - \sum_{k=1}^{ijinc} R_{ijk} \right) \to \min.$$

Thus, there is an NP-complete optimization problem, which is concerning the substance and reducible to the problem of integer linear programming (LP, also called linear optimization). In terms of computational complexity, the problem can be formulated as a cutting-stock problem. At the beginning of scientific investigation with regard to decreasing "costs" in objective functions, most of optimization methods used to apply linear models and linear programming methods.

The foundations of these methods were put in the writings by Gilmore, Gomory [17]. Also, accurate methods based on a branch-and-bound algorithm were viewed in theoretical studies by Martello, Toth [18]. They also developed them, as well as their further studies that they conducted without exceeding the value of all possible candidates for the solution (generate and test search) [19–24]. To solve these problems, simple one-pass and multi-pass heuristics are often used, as well as metaheuristic algorithms, for example, a genetic algorithm or a method of random tabu search. In this paper, the authors propose a combinatorial optimization approach to solve the problem of one-dimensional optimization, taking into account the characteristics of the neural processing unit capacity, based on dynamic programming methods.

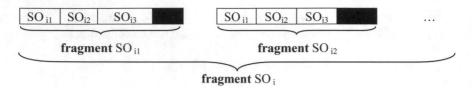

fragment SO $_{i1}$ **fragment SO $_{i2}$**

fragment SO $_i$

Fig. 2 Data compression

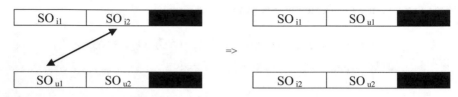

Fig. 3 Pairing operation exchange of the tuples

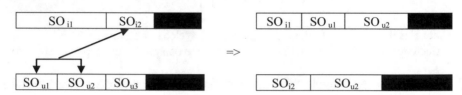

Fig. 4 Triple operation exchange of tuples

The authors include the notion of "the complement of the bit depth" that differentiates the characteristic of the neural processing unit capacity and the total digitization of calculations for all operations already entering the tuple

$$Ost_j = Pr - \sum_{k=1}^{ijinc} R_{ijk}.$$

First of all, all the tuples are primarily posted according to the following rule: the operation with the highest computational depth is put in the tuple having the largest remaining bit depth (Fig. 2) [12].

Secondly, operations in tuples are sorted into pairs, if this results in reducing the complement in the most completed tuple (Fig. 3). With each exchange of operations, all the tuples are sorted according to the remainder, as well as the operations in the tuple from most to least [21].

In order to minimize items wasted in step 2, in addition to pair wise exchange of details, three operations in tuples are exchanged, if it results in reducing the complement in the most completed tuple (Fig. 4). This addition provides an opportunity to reduce the total balance significantly, but increasing the algorithm execution time, depending on the number of operations in the conditions of the problem [21].

In this example:

$$K_{Rju} = 1 - \frac{Pr - \sum_{k=1}^{ujinc} R_{ujk}}{Pr}.$$

The total parameters of the core load of the neural processing unit for the fragment SO_j:

$$Ost_j = 1 - \frac{\sum_{j=1}^{T} \left(Pr - \sum_{k=1}^{ijinc} R_{ijk} \right)}{T * Pr}.$$

For a microneuroprogram as a whole:

$$K_R = 1 - \frac{\sum_{i=1}^{p} \sum_{j=1}^{T} \left(Pr - \sum_{k=1}^{iujinc} R_{iujk} \right)}{p * T * Pr}.$$

It should be emphasized that the accumulated mathematical methodology is effective only for low-bit input, output data, and weight coefficients. If there are low-bit only input data or weight coefficients, there will be delays, and that will reduce the efficiency of the information processing arrangement. Further increasing efficiency of information processing can be achieved via semantic analysis of the problem class $Z^{(j)}$.

The authors carried out experimental studies of the current scientific and practical findings. That is why they dealt with test activities for emulating artificial neural networks a layered perceptron with 10 layers and 1000–2000 neurons in each layer (MLP2), a layered perceptron with 50 layers and 500–2000 neurons in each layer (MLP3), an artificial neural network (ANN) of an adaptive resonance theory (ART), a counter propagation network (CPN), a convolutional neural network (CNN), and the control problem with a high degree of operations presented in a neural network logical basis. The scientists also did mathematical computations and practical experiments for neuroprocessor sites (models) including one-to-eight neuroprocessor devices. The findings of the experiment prove the reliability, adequacy, and efficiency of methods, models, algorithms, and software resulted in the research.

For the neural network emulation problem, the proposed method let software uptime in emulation mode reduce by on 9.089% on average; to reduce the equipment downtime on average by 4.874%; to increase the training speed on average by 6.835%. For the hexapod control problem, the productivity method of information processing made it possible to reduce software uptime by on average by 7.84%, and reduce the equipment downtime on average by 5.63%.

4   Conclusion

It is seen from the foregoing that it is efficient to use redistribution of operations to implement data parallelism for neural processing units with a high degree of operation homogeneity.

The paper touches upon a method of increasing productivity of neurocomputer programs by data and results compression (packing), i.e., a practical planning of streaming computing (calculations).

Applied studies based on the developed software have proved that the application of the proposed method improves significantly overall program instruction per second by 5–14%, depending on a complexity class of solution problem and the degree of operation homogeneity. The instruction per second increases because of "data and results compression" by the most effective method. It is possible to do since there are problems with a high degree of operation homogeneity for neural processors.

References

1. Boyer, L.L., Pawley, G.S.: Molecular dynamics of clusters of particles interacting with pairwise forces using a massively parallel computer. J. Comput. Phys. **78**(2), 405–423 (1988). https://doi.org/10.1016/0021-9991(88)90057-5
2. Singh, H., Lee, M.H., Lu, G., Kurdahi, F.J., Bagherzadeh, N., Filho, E.M.C.: MorphoSys: an integrated reconfigurable system for data-parallel and computation-intensive applications. IEEE Trans. Comput. **49**(5), 465–481 (2000). https://doi.org/10.1109/12.859540
3. Hillis, W.D., Steele Jr., G.L.: Data parallel algorithms. Commun. ACM **29**(12), 1170–1183 (1986)
4. Flanders, P.M., Hunt, D.J., Reddaway, S.F., Parkinson, D.: Efficient high speed computing with the distributed array processor. In: High Speed Computer and Algorithm Organization, pp. 113–128 (1977)
5. Ebeling, C., Cronquist, D.C., Franklin, P.: Configurable computing: the catalyst for high-performance architectures. In: IEEE International Conference on Application-Specific Systems, Architectures and Processors, pp. 364–372 (1997). https://doi.org/10.1109/asap.1997.606841
6. Pan, V., Reif, J.: Efficient parallel solution of linear systems. In: Proceedings of the Seventeenth Annual ACM Symposium on Theory of Computing, pp. 143–152 (1985)
7. Lim, W (ed.): Fast Algorithms for Labeling Connected Components in 2-D arrays. Thinking Machines Corporation (1987)
8. Kong, H.T., Lieserson, C.E.: Algorithms for VLSI processor arrays. In: Introduction to VLSI Systems, pp 271–292. Addison-Wesley, New York (1980)
9. Romanchuk, V.A.: The method of optimization of neuro-based concurrent operations in neurocomputers. In: IOP Conference Series: Materials Science and Engineering, vol. 177, no. 1, p. 012033 (2017). https://doi.org/10.1088/1757-899x/177/1/012033
10. Chen, D.C., Rabaey, J.M.: A reconfigurable multiprocessor IC for rapid prototyping of algorithmic-specific high-speed DSP data paths. IEEE J. Solid-State Circuits **27**(12), 1895–1904 (1992). https://doi.org/10.1109/4.173120
11. Bottou, L.: Large-scale machine learning with stochastic gradient descent. In: Proceedings of COMPSTAT'2010, pp. 177–186 (2010)
12. Bottou, L.: Stochastic gradient descent tricks. In: Neural Networks: Tricks of the Trade, pp. 421–436 (2012)

13. Noel, C., Osindero, S.: Dogwild!-distributed hogwild for CPU & GPU. In: NIPS Workshop on Distributed Machine Learning and Matrix Computations (2014)
14. Recht, B., Re, C., Wright, S., Niu, F.: Hogwild: a lock-free approach to parallelizing stochastic gradient descent. In: Advances in Neural Information Processing Systems, pp. 693–701 (2011)
15. Jia, Y.Q.C.: An open source convolutional architecture for fast feature embedding. In: Proceedings of the 22nd ACM International Conference on Multimedia, pp. 675–678 (2013)
16. Boyd, S., Parikh, N., Chu, E., Peleato, B., Eckstein, J.: Distributed optimization and statistical learning via the alternating direction method of multipliers. In: Foundations and Trends® in Machine Learning, vol. 3, no. 1, pp. 1–122 (2011). https://doi.org/10.1561/2200000016
17. Gilmore, P.C., Gomory, R.E.: The theory and computation of knapsack functions. Oper. Res. **14**(6), 1045–1074 (1966). https://doi.org/10.1287/opre.14.6.1045
18. Martello, S., Toth, P.: Knapsacks problems: algorithms and computer implementations. Wiley, Chichester (1990)
19. Kryuchkovsky, V.V., Usov, A.V.: Determinization of the multifactorial evaluation model for various types of uncertainty in setting parameters. In: Proceedings of Odessa National Polytechnic University, vol. 2, pp. 154–160 (2009)
20. Berezovsky, B.A., Baryshnikov, Y.M., Bozenko, V.I., Kempner, L.M.: Multicriteria optimization: mathematical aspects. Nauka, Moscow (1989)
21. Vasin, A.Y., Zadorozhny, V.N.: Solution of the production-related issue of one-dimensional cutting materials. Omsk Sci. Bull. **2**, 267–270 (2012)
22. Goswami, S., Chakraborty, S., Saha, H.N.: An univariate feature elimination strategy for clustering based on metafeatures. Int. J. Intell. Syst. Appl. **9**(10), 20–30 (2017). https://doi.org/10.5815/ijisa.2017.10.03
23. Barabash, O., Kravchenko, Y., Mukhin, V., Kornaga, Y., Leshchenko, O.: Optimization of parameters at SDN technologie networks. Int. J. Intell. Syst. Appl. **9**(9), 1–9 (2017). https://doi.org/10.5815/ijisa.2017.09.01
24. Yakkali, R.T., Raghava, N.S.: Neural network synchronous binary counter using hybrid algorithm training. Int. J. Image Graphics Sign. Process. **9**(10), 38–49 (2017). https://doi.org/10.5815/ijigsp.2017.10.05

Algebra for Transforming Genetic Codes Based on Matrices Applied in Quantum Field Theories

György Darvas◉ and Sergey V. Petoukhov◉

Abstract Genetic algebra shows 2 + 2 type character defined in a purine–pyrimidine field, and 1 + 3 type character defined in an RNA–DNA field. Its 1 + 3 type character has not been studied with algebraic methods. Physical matrices, among others in quantum electrodynamics (QED), demonstrate 1 + 3 character too. The latter is expressed with the help of the hypersymmetry and its algebra (called tau). The similitude between the two descriptions allowed showing that similar matrices can demonstrate the 1 + 3 character in genetic algebra like those that govern the matrices elaborated for QED. We show that certain algebraic methods and transformation formula developed in the theory of QED fields appear also in the transformation properties of certain genetic matrices. This paper is a minor contribution to demonstrate that combined transformations applied together in the convolution of the mentioned two fields may explain the structure change between RNA and DNA. This study shows that there are some general regularities in nature that appear in different domains of our knowledge about nature (here concretely, genetics and physical field theory), which have been considered distant and not overlapping in their methods and laws.

Keywords Genetic code · Genetic algebra · Hypersymmetry · Transformation of fields

1 Introduction

Based on previous studies of the two authors [2–5, 7–12] in the fields of physical field theories and genetic algebra, respectively, the paper makes an attempt to demonstrate

G. Darvas (✉)
Symmetrion, Budapest 1067, Hungary
e-mail: symmetrion@symmetry.hu

S. V. Petoukhov
Mechanical Engineering Research Institute of Russian Academy of Sciences,
Moscow 101990, Russia
e-mail: spetoukhov@gmail.com

© Springer Nature Switzerland AG 2020
Z. Hu et al. (eds.), *Advances in Artificial Systems for Medicine and Education II*,
Advances in Intelligent Systems and Computing 902,
https://doi.org/10.1007/978-3-030-12082-5_5

that certain similitudes in descriptions between the two disciplines allow to define those similar matrices, which can demonstrate the $1 + 3$ character in genetic algebra and in QED. The paper derives that certain algebraic methods and transformation formula developed in the theory of QED fields by the application of hypersymmetry appear also in the transformation properties of certain genetic matrices.

2 Application of Some Algebraic Methods Elaborated in the Theory of Quantum Electrodynamics (QED) to the Genetic Algebra

There are several forms of matrix representation of the genetic information that are coded in the DNA (RNA) molecules.

As a first step, let us decompose S. Petoukhov's Rademacher matrix R_4 [7, 9, 12] in the following way:

$$
\begin{bmatrix} 1 & 1 & 1 & -1 \\ -1 & 1 & -1 & -1 \\ 1 & -1 & 1 & 1 \\ -1 & -1 & -1 & 1 \end{bmatrix} = \begin{bmatrix} 1 & 0 & 0 & 0 \\ 0 & 1 & 0 & 0 \\ 0 & 0 & 1 & 0 \\ 0 & 0 & 0 & 1 \end{bmatrix} + \begin{bmatrix} 0 & 0 & 0 & -1 \\ -1 & 0 & 0 & 0 \\ 0 & 0 & 0 & 1 \\ 0 & 0 & -1 & 0 \end{bmatrix} + \begin{bmatrix} 0 & 1 & 0 & 0 \\ 0 & 0 & -1 & 0 \\ 0 & -1 & 0 & 0 \\ -1 & 0 & 0 & 0 \end{bmatrix} + \begin{bmatrix} 0 & 0 & 1 & 0 \\ 0 & 0 & 0 & -1 \\ 1 & 0 & 0 & 0 \\ 0 & -1 & 0 & 0 \end{bmatrix}
$$

This is the same form, like

$$
R_4 = \begin{bmatrix} I & \\ & I \end{bmatrix} + \begin{bmatrix} 0 & -\sigma_1 \\ -\sigma_1 & 0 \end{bmatrix} + \begin{bmatrix} i\sigma_2 & \\ & i\sigma_2 \end{bmatrix} + \begin{bmatrix} 0 & \sigma_3 \\ \sigma_3 & 0 \end{bmatrix} = I + \Sigma_1 + \Sigma_2 + \Sigma_3
$$

where σ_i are the [2x2] Pauli matrices, and Σ_i are Dirac-like bispinors composed of them.

Properties of Σ_i:
It is easy to check that $\{\Sigma_j, \Sigma_k\} = 0$ if $j \neq k$.

$$
\Sigma_1 \Sigma_2 = \Sigma_3, \quad \Sigma_2 \Sigma_1 = -\Sigma_3, \quad \Sigma_2 \Sigma_3 = \Sigma_1, \quad \Sigma_3 \Sigma_2 = -\Sigma_1,
$$
$$
\Sigma_3 \Sigma_1 = -\Sigma_2, \quad \Sigma_1 \Sigma_3 = \Sigma_2
$$

in short notation: $\Sigma_i \Sigma_j = \varepsilon_{ij}^k \Sigma_k$ where ε_{ij}^k is the Levi-Civita tensor ($\varepsilon_{ij}^k = 1$ at 1,2,3 cyclic permutation of the indices, and $= -1$ at opposite permutation.)
$\Sigma_1 \Sigma_1 = I$, $\Sigma_2 \Sigma_2 = -I$, $\Sigma_3 \Sigma_3 = I$ or we can denote $I = \Sigma_0$.

If we compose four-column vectors of σ_i and Σ_i, the former can be obtained from the latter by the following transformation:

$$\begin{bmatrix} \sigma_0 \\ \sigma_1 \\ \sigma_2 \\ \sigma_3 \end{bmatrix} = \begin{bmatrix} 1 & 0 & 0 & 0 \\ 0 & -\rho_1 & 0 & 0 \\ 0 & 0 & -i\rho_1 & 0 \\ 0 & 0 & 0 & 1 \end{bmatrix} \begin{bmatrix} \Sigma_0 \\ \Sigma_1 \\ \Sigma_2 \\ \Sigma_3 \end{bmatrix}$$

where ρ_1 is Dirac's $\rho_1 = \begin{bmatrix} 0 & 0 & 1 & 0 \\ 0 & 0 & 0 & 1 \\ 1 & 0 & 0 & 0 \\ 0 & 1 & 0 & 0 \end{bmatrix}$ matrix that has two important properties:

(1) Dirac defined his basic α_i matrices with the help of ρ_1: $\alpha_i = \rho_1 \sigma_i$ ($i = 1, 2, 3$), what he introduced in the wave equation of the electron to create QED; and
(2) Its role is to change the unit values of the 1st and 2nd columns of these matrices for the 3rd and 4th columns (and vice versa), what makes a correspondence between Petoukhov's U-complex numbers [10]. This is an important link [5] between the genetic matrices [12] and the Dirac algebra [2].

In the following, we show that the above introduced Σ-s allow a similar transformation like the QED Dirac matrices. We show that the Σ_i are subject of a Lorentz transformation, similar to the set of σ_i.

σ_i are generators of a 16-element group ($I, -I, iI, -iI, \sigma_i, -\sigma_i, i\sigma_i, -i\sigma_i$).

Σ_i are generators of an 8-element group: ($I, -I, \Sigma_i, -\Sigma_i$). The difference is due that all Σ_i are real.

Their multiplication table forms an 8x8 matrix, with a mosaic of $+$ and $-$ signature elements that can be arranged like R_8.

Σ_i are composed of the same elements like Dirac's 4x4 σ matrices. $I, -\sigma_1, i\sigma_2$ and σ_3 are elements of the same 16-element group.

Let it be $I = \Sigma_0 = \sigma_0$. Let us introduce the following notations: $\pi_0 = \sigma_0, \pi_1 = -\sigma_1, \pi_2 = i\sigma_2, \pi_3 = \sigma_3$.

We show that π_i can represent the same transformations like σ_i.

Let $A(x)$ be the matrix of a Lorentz transformation of the coordinates x^μ.

$$A(x) = \begin{bmatrix} x^0 + x^3 & x^1 - ix^2 \\ x^1 + ix^2 & x^0 - x^3 \end{bmatrix} = x^\mu \sigma_\mu.$$

Let us apply the same transformations for coordinates y^μ, with the matrices π_i.

$$B(y) = y^\mu \pi_\mu = \begin{bmatrix} y^0 & 0 \\ 0 & y^0 \end{bmatrix} + \begin{bmatrix} 0 & -y^1 \\ -y^1 & 0 \end{bmatrix} + \begin{bmatrix} 0 & -iy^2 \\ iy^2 & 0 \end{bmatrix} + \begin{bmatrix} y^3 & 0 \\ 0 & -y^3 \end{bmatrix}$$

$$= \begin{bmatrix} y^0 + y^3 & -(y^1 + iy^2) \\ -(y^1 - iy^2) & y^0 - y^3 \end{bmatrix}.$$

The two matrices differ in the sign of y^1, but their determinants coincide:

$$\det(A) = \left(x^0\right)^2 - \left(x^1\right)^2 - \left(x^2\right)^2 - \left(x^3\right)^2; \quad \det(B) = \left(y^0\right)^2 - \left(y^1\right)^2 - \left(y^2\right)^2 - \left(y^3\right)^2.$$

The Lorentz transformation is unitary, i.e., it keeps the vector('s unit) length. If x^μ and y^μ are coordinates of points in a Minkowski space, both define the length square of a vector arrowing from the origin to the respective points. In an inertial system in a Minkowski space, both must be equal to 0. This means, the **x** and **y** vectors can be transformed into each other by a rotation of the coordinate axes.

Let us define an abstract field over the Minkowski space (that may be even complex). Let us interpret the systems A and B in that field. One can imagine system B so that its basis axes coincide with those of A, only y^1 has an opposite direction. Changing its directedness to the opposite will cause changing permutation in the indexing of the axes. (This property is considered in the permutation rules of the multiplication of the Σ_i among themselves.)

Let us now identify the abstract field with a genetic one, whose basis is defined by the four units of the genetic alphabet. Let the basis axes of the genetic field stretched by spinors. Using the following notation:

$$\begin{bmatrix} C & U \\ A & G \end{bmatrix}$$

and indexing them according to [C, U; A, G]. The axis with index 1 will coincide with the dimension of the Uracil, the only member of the genetic alphabet, that changes (for T—Thymin) in RNA–DNA recombination. (On the DNA encryption technique, see also [13]). One may interpret such an index change, and change of the basis unit direction, as a transformation from an RNA basis to a DNA or *vice versa*. Note, upper row denotes pyrimidines, the lower row denotes purines; and we know that the 0-indexed nucleotide may pair with the 3-indexed one, and the 1-indexed nucleotide with the 2-indexed one.

In this sense, one can transform the basis of the abstract genetic field from the $B(y)$ reference frame system to $A(x)$ by a rotation in the latter's frame. Note, there is no translation in an inertial system, so a rotation around the x^1 axis in the x^2x^3 plane (we are free to choose the direction of the axes) can be described with the following Lorentz transformation:

$$\Lambda(\omega) = \begin{bmatrix} 1 & 0 & 0 & 0 \\ 0 & 1 & 0 & 0 \\ 0 & 0 & \cos\omega & -\sin\omega \\ 0 & 0 & \sin\omega & \cos\omega \end{bmatrix}.$$

We saw that vectors in the system A are subject of Lorentz transformation, we saw that the systems A and B can be Lorentz transformed into each other, and we know that the result of two consecutively executed Lorentz transformations is also a Lorentz transformation; thus, B must also represent a Lorentz transformation. In other words, σ_μ and π_μ represent Lorentz transformations and so does their product. (Since all elements of π_μ are real, it coincides with its complex conjugate.)

This correspondence allows two *conclusions*:

(a) The genetic matrices can be built from the same matrices that govern the quantum electrodynamic (QED) interactions;
(b) The interactions between the nucleotides appearing in the genetic molecules are subject to similar transformations like those in the physical processes.

3 Application of the Presented Algebra Combined with Hypersymmetry to the Transformation of Genetic Matrices in the Purine–Pyrimidine and RNA–DNA Fields

The golden matrices by S. Petoukhov are derived not directly from the Rademacher matrices that contain only unitary elements. They are derived from his matrices P that include the amount of hydrogenous bonds coupling the nucleotides (and other proportional properties of the nucleotides) (Cf., [8]). The basic matrix is

$$\mathbf{P}^{(1)}_{\mathrm{MULT}} = \begin{array}{|c|c|} \hline 3 & 2 \\ \hline 2 & 3 \\ \hline \end{array}$$

and its tensorial powers. The golden matrices are defined as the matrix square roots of the consecutive power genetic matrices $\mathbf{P}^{(n)}_{\mathrm{MULT}}$. (On the DNA sequence numerical representation, see also [1]). It is easy to check that all elements of these square root matrices are expressed in powers of the golden ratio $\varphi = (1 + \sqrt{5})/2$. For example:

$$(\mathbf{P}_{\mathrm{MULT}})^{1/2} = \mathbf{\Phi}_{\mathrm{MULT}} = \begin{vmatrix} \varphi & \varphi^{-1} \\ \varphi^{-1} & \varphi \end{vmatrix};$$

$$(\mathbf{P}^{(2)}_{\mathrm{MULT}})^{1/2} = \mathbf{\Phi}_{\mathrm{MULT}} = \begin{vmatrix} \varphi^2 & \varphi^0 & \varphi^0 & \varphi^{-2} \\ \varphi^0 & \varphi^2 & \varphi^{-2} & \varphi^0 \\ \varphi^0 & \varphi^{-2} & \varphi^2 & \varphi^0 \\ \varphi^{-2} & \varphi^0 & \varphi^0 & \varphi^2 \end{vmatrix}$$

and so on.

The matrix $\Phi^{(2)}_{\mathrm{MULT}}$ can be decomposed, similar to the matrix R_4 [6].

$$\Phi^{(2)}_{\mathrm{MULT}} = \begin{bmatrix} \varphi^2 & \varphi^0 & \varphi^0 & \varphi^{-2} \\ \varphi^0 & \varphi^2 & \varphi^{-2} & \varphi^0 \\ \varphi^0 & \varphi^{-2} & \varphi^2 & \varphi^0 \\ \varphi^{-2} & \varphi^0 & \varphi^0 & \varphi^2 \end{bmatrix}$$

$$= \varphi^2 \begin{bmatrix} 1&0&0&0 \\ 0&1&0&0 \\ 0&0&1&0 \\ 0&0&0&1 \end{bmatrix} + \varphi^0 \begin{bmatrix} 0&0&1&0 \\ 0&0&0&1 \\ 1&0&0&0 \\ 0&1&0&0 \end{bmatrix}$$

$$+ \varphi^{-2} \begin{bmatrix} 0 & 0 & 0 & 1 \\ 0 & 0 & 1 & 0 \\ 0 & 1 & 0 & 0 \\ 1 & 0 & 0 & 0 \end{bmatrix} + \varphi^{0} \begin{bmatrix} 0 & 1 & 0 & 0 \\ 1 & 0 & 0 & 0 \\ 0 & 0 & 0 & 1 \\ 0 & 0 & 1 & 0 \end{bmatrix}.$$

Similar to the matrix $P_{MULT}^{(2)}$, the matrix $\Phi_{MULT}^{(2)}$ is also weighted (with the powers of φ), so they are not unitary. (On the golden matrices, see also [14].) In their representation in a field, the respective vectors do not conserve their length under transformation. Consequently, they will not subject of any form of Lorentz transformation. At the same time, they show considerable symmetries and similarities in their transformations. The most important is, probably, that they can be composed of a part of the elements of the same spinor matrices which we learned in the QED [2], as well as applied for the genetic matrices in the Sect. 2. Moreover, they are all real and include less spinor elements than the genetic matrices do: this is the price that we earned in response to the loss of the unitarity.

At first approximation, the matrix $\Phi_{MULT}^{(2)}$ can be constructed with the help of the [2x2] Pauli σ_0 and σ_1 matrices.

$$\Phi_{MULT}^{(2)} = \varphi^2 \begin{bmatrix} \sigma_0 & 0 \\ 0 & \sigma_0 \end{bmatrix} + \varphi^0 \begin{bmatrix} 0 & \sigma_0 \\ \sigma_0 & 0 \end{bmatrix} + \varphi^{-2} \begin{bmatrix} 0 & \sigma_1 \\ \sigma_1 & 0 \end{bmatrix} + \varphi^0 \begin{bmatrix} \sigma_1 & 0 \\ 0 & \sigma_1 \end{bmatrix}.$$

The symmetry is obvious. However, the middle two matrices are not members of Dirac's [4x4] σ matrices. They are defined with the help of Dirac's matrix ρ_1 that we met in the previous Section (see also: [2]). It is the matrix that may change the values of the 1st and 2nd columns of the matrices for the 3rd and 4th columns (and vice versa), what makes correspondence between Petoukhov's U-complex numbers [10, 11], and marks a strong relation to the genetic matrices [8].

$$\begin{bmatrix} \sigma_0 & 0 \\ 0 & \sigma_0 \end{bmatrix} = \sigma_0; \quad \begin{bmatrix} 0 & \sigma_0 \\ \sigma_0 & 0 \end{bmatrix} = \rho_1; \quad \begin{bmatrix} 0 & \sigma_1 \\ \sigma_1 & 0 \end{bmatrix} = \rho_1\sigma_1; \quad \begin{bmatrix} \sigma_1 & 0 \\ 0 & \sigma_1 \end{bmatrix} = \sigma_1.$$

This connection between the genetic matrices and the golden matrices is an additional justification for the theory of genetic harmony [4, 8].

One can recognise in the form of the matrix $\Phi_{MULT}^{(2)}$ that it will not transform according to the Lorentz transformation. However, the above derived symmetric form with the help of the σ and ρ_1 matrices (ρ_1 formed by Dirac from σ_1) deserves interest in order to present the transformation rule of $\Phi_{MULT}^{(2)}$.

As it was mentioned above, the genetic matrices appear simultaneously in the (2 + 2) dimensional purine–pyrimidine field, and the (1 + 3) dimensional RNA–DNA field. The latter means that while the Uracil (U) transforms into Thymin (T) during the transformation from RNA to DNA, the other three letters of the genetic alphabet do not. In order to describe the combined transformation in the two genetic fields, one should define the matrices of the combined transformation.

First, let us define again an abstract field over the Minkowski space, with vectors z in it. Transformations of the vectors z can be composed by the following similar (but not Lorentzian) transformation $F(z^\mu)$:

$$F(z^\mu) = z^\mu \zeta_\mu$$

$$F^{(1)}(z) = z^i \begin{bmatrix} \varphi & \varphi^{-1} \\ \varphi^{-1} & \varphi \end{bmatrix} = z^0 \varphi[\sigma_0] + z^1 \varphi^{-1}[\sigma_1]$$

$$F^{(2)}(z^\mu) = z^0 \varphi^2[\sigma_0] + z^1 \varphi^0[\rho_1] + z^2 \varphi^{-2}[\rho_1\sigma_1] + z^3 \varphi^0[\sigma_1]$$

$$= \begin{bmatrix} \varphi^2 z^0 + z^3 & z^1 + \varphi^{-2} z^2 \\ z^1 + \varphi^{-2} z^2 & \varphi^2 z^0 + z^3 \end{bmatrix}$$

where $\zeta_0 = \varphi^2 \sigma_0$; $\zeta_1 = \varphi^0 \rho_1$; $\zeta_2 = \varphi^{-2}\rho_1\sigma_1$; $\zeta_3 = \varphi^0\sigma_1$.

This is easy to see that the sign of the z^1 components in the side diagonal of the matrix $F^{(2)}(z^\mu)$ differs from that of y^1 in $B(y)$. However, we have still considered only a transformation in the purine–pyrimidine field. In order to find the proper transformation that may change the sign of z^1 but leaves the sign of the other three genetic letters intact, we should combine it with a transformation in the RNA–DNA field. Since it has a $(1 + 3)$ character, we must turn to the tau (τ) algebra of hypersymmetry. It was described in [2, 3]. According to the theory of hypersymmetry [3], applied in $(-, + ,+, +)$ signature and in τ_3 representation, there are two matrices that govern the transformation in a $(1 + 3)$ field:

$$E = \begin{bmatrix} 1 & 0 & 0 & 0 \\ 0 & 0 & 0 & 1 \\ 0 & 0 & 0 & 1 \\ 0 & 0 & 0 & 1 \end{bmatrix} \quad \text{and} \quad \tau_3 = \begin{bmatrix} -1 & 0 & 0 & 0 \\ 0 & 0 & 0 & 1 \\ 0 & 0 & 0 & 1 \\ 0 & 0 & 0 & 1 \end{bmatrix}.$$

For the purpose of the combined transformation, we introduce a combined transformation matrix, denoted by ξ. Notice that similar to the role of σ_0 in the set of the Pauli matrices, E coincides also with the unit matrix in the hypersymmetry. They and their combined application leave the subjected matrices intact. Let us concentrate our attention to the role of τ_3. When U is in the position 1, τ_3 should be combined with σ_1. (When a permutation in the matrix [C, U; A, G] places U to another position, σ_1 should be replaced by ρ_1 and $\rho_1\sigma_1$, respectively.) So the matrix of the combined transformation will be the following:

$$\xi = \sigma_1\tau_3 = \begin{bmatrix} 0 & 1 & 0 \\ 1 & 0 & 0 & 0 \\ 0 & 0 & 0 & 1 \\ 0 & 0 & 1 & 0 \end{bmatrix} \begin{bmatrix} -1 & 0 & 0 & 0 \\ 0 & 0 & 0 & 1 \\ 0 & 0 & 0 & 1 \\ 0 & 0 & 0 & 1 \end{bmatrix} = \begin{bmatrix} 0 & 0 & 0 & 1 \\ -1 & 0 & 0 & 0 \\ 0 & 0 & 0 & 1 \\ 0 & 0 & 0 & 1 \end{bmatrix}.$$

The application of this transformation allows to change the sign of U and, in the transformation of $F^{(2)}(z^\mu)$, the sign of z^1. Now, the following transformations should be applied instead of those that appeared in ζ_μ above:

$$\zeta_0' = \varphi^2 \xi \sigma_0; \quad \zeta_1' = \varphi^0 \xi \rho_1; \quad \zeta_2' = \varphi^{-2} \xi \rho_1 \sigma_1; \quad \zeta_3' = \varphi^0 \xi \sigma_1.$$

where

$$\xi\sigma_0 = \begin{bmatrix} 0 & 0 & 0 & 1 \\ -1 & 0 & 0 & 0 \\ 0 & 0 & 0 & 1 \\ 0 & 0 & 0 & 1 \end{bmatrix}; \quad \xi\rho_1 = \begin{bmatrix} 0 & 1 & 0 & 0 \\ 0 & 0 & -1 & 0 \\ 0 & 1 & 0 & 0 \\ 0 & 1 & 0 & 0 \end{bmatrix};$$

$$\xi\rho_1\sigma_1 = \begin{bmatrix} 1 & 0 & 0 & 0 \\ 0 & 0 & 0 & -1 \\ 1 & 0 & 0 & 0 \\ 1 & 0 & 0 & 0 \end{bmatrix}; \quad \xi\sigma_1 = \begin{bmatrix} 0 & 0 & 1 & 0 \\ 0 & -1 & 0 & 0 \\ 0 & 0 & 1 & 0 \\ 0 & 0 & 1 & 0 \end{bmatrix}.$$

Although, in a general case, this is not a length-conserving transformation, and also not a proper Lorentz transformation in its classical sense, but formally similar, we gained, on the other hand, other advantages:

(a) It shows more (other kind) symmetry, and demonstrates the relation both to the quantum electrodynamics' algebra and the genetic algebra.

The last derived transformation demonstrates that

(b) The proper transformation between RNA and DNA can algebraically described by a combined transformation applied in the purine–pyrimidine and the RNA–DNA fields simultaneously. (We emphasize that his result was not shown earlier.)

The statements (a) and (b) are what we wanted to demonstrate. They are based on analogue algebras that govern physical field theory and the theory of the genetic codes. This pragmatic similitude is a nice example that certain symmetry methods, expressed in algebraic form, can be productive in apparently 'distant' disciplines.

4 Conclusions

The above paper contributed to demonstrate that combined transformations applied together in the convolution of two fields (2 + 2 and 1 + 3 dimensional) may explain the structure change between RNA and DNA. For this reason, mathematical analogies were borrowed from hypersymmetry applied to quantum electrodynamic (QED) field theory. This study demonstrated that there are some general regularities in nature that appear in different domains of our knowledge about nature (here concretely, genetics and physical field theory), which were considered distant and not overlapping in their methods and laws until now. The derived regularities may hopefully contribute to genetic surgery technologies in the future. Four important conclusions were specified at the end of Sect. 2 under (a) and (b), and at the end of Sect. 3 under (a) and (b).

Acknowledgments This paper is part of the results of a project in the framework of a long-term bilateral cooperation between the Russian and Hungarian Academies of Sciences under the item 5 in the working plan entitled "Non-linear models and symmetrologic analysis in biomechanics, bioinformatics, and the theory of self-organizing systems".

References

1. Abo-Zahhad, M., Ahmed, S.M., Abd-Elrahman, S.A.: Integrated model of DNA sequence numerical representation and artificial neural network for human donor and acceptor sites prediction. Int. J. Inf. Technol. Comput. Sci. (IJITCS) 6(8), 51–57 (2014). https://doi.org/10.5815/ijitcs.2014.08.07
2. Darvas, G.: Quaternion/vector dual space algebras applied to the Dirac equation and its extensions, Bull. Transilvania Univ. Brasov 8(57), 1; Series III: Math., Inf., Phys. 27–42 (2015)
3. Darvas, G.: Algebra of Hypersymmetry Applied to State Transformations in Strongly Relativistic Interactions, 20 p, (submitted, 2018)
4. Darvas, G., Koblyakov, A.A., Petoukhov, S.V., Stepanyan, I.: Symmetries in molecular-genetic systems and musical harmony. In: Petoukhov, S. (ed) Symmetry: Culture and Science, special issue "Symmetries in genetic information and algebraic biology, vol. 23, issues 3–4, pp. 343–375 (2012). https://doi.org/10.26830/symmetry_2012_3-4_343
5. Darvas, G., Petoukhov, S.V.: Algebra that demonstrates similitude between transformation matrices of genetic codes and quantum electrodynamics. Poster presentation to the workshop organised 27.11.2017 at the "Competence center for algorithmic and mathematical methods in biology, biotechnology and medicine" of the Hochschule Mannheim, 7 p (2017)
6. Hu, Z., Petoukhov, S.V., Petukhova, E.S.: I-Ching, dyadic groups of binary numbers and the geno-logic coding in living bodies. Prog. Biophys. Mol. Biol. 131, 354–368 (2017)
7. Petoukhov, S.: Genetic code and the Ancient Chinese "Book of Changes", Symmetry Cult. Sci. 10(3–4), 211–226 (1999). https://doi.org/10.26830/symmetry_1999_3-4_211
8. Petoukhov, S.: Bioinformatics: matrix genetics, algebras of the genetic code and biological harmony. Symmetry Cult. Sci. 17(1–4), 253–291 (2006). https://doi.org/10.26830/symmetry_2006_1-4_253
9. Petoukhov, S.: Matrix Genetics, Algebras of the Genetic Code, Noise Immunity, [in Russian], Moscow-Izhevsk: Regulyarnaya i khaoticheskaya dynamika, 316 p (2008)
10. Petoukhov, S.: Genetic coding and united-hypercomplex systems in the models of algebraic biology. Biosystems 158, 31–46 (2017). https://doi.org/10.1016/j.biosystems.2017.05.002
11. Petoukhov, S.: The Genetic Coding System and Unitary Matrices. Preprints (2018), 2018040131. https://doi.org/10.20944/preprints201804.0131.v1
12. Petoukhov, S., He, M.: Symmetrical Analysis Techniques for Genetic Systems and Bioinformatics: Advanced Patterns and Applications. IGI Global, Hershey, USA (2009)
13. Mousa, H.M.: DNA-genetic encryption technique. Int. J. Comput. Netw. Inf. Secur. (IJCNIS) 8(7), 1–9 (2016). https://doi.org/10.5815/ijcnis.2016.07.01
14. Wani, A.A., Badshah, V.H.: On the relations between lucas sequence and fibonacci-like sequence by matrix methods. Int. J. Math. Sci. Comput. (IJMSC) 3(4), 20–36 (2017). https://doi.org/10.5815/ijmsc.2017.04.03

Development of Methods for Solving Ill-Posed Inverse Problems in Processing Multidimensional Signals in Problems of Artificial Intelligence, Radiolocation and Medical Diagnostics

Alexander A. Potapov, Andrey A. Pakhomov and Vladimir I. Grachev

Abstract The present work develops the mathematical base of ill-posed inverse problems in optics, as well as a method of processing and reconstructing scenes from incomplete information of their Fourier spectra. The Hilbert equations for the components of the Fourier spectrum are generalized to two-dimensional continuous and discrete events. Also, the condition for an unambiguous analytical solution of a number of ill-posed problems is disclosed. The a priori information uses the finiteness and positivity of the unknown image, as well as the approximation of the transfer function by a Gaussian with an unknown variance. In this case, not a continuous case is considered, but discrete—post-detection processing of the image. Moreover, all the factors of the spatial spectra are two-dimensional discrete polynomials of finite degree. Taking into account their features and the Lebesgue measure apparatus, the problem reduces to Tikhonov regularizing filters, which contain, in addition to the regularizing factor, several unknown stabilizing parameters. The problem is solved with the use of set theory and, in particular, using projection operators on sets of functions with given properties, as well as methods for the cepstral processing of signals and images. This problem is part of the general problem of the eliminating the influence atmospheric and optical distortion when finding and registering an object at large distances under bad atmospheric conditions. The designed programmed methods are also used to process medical images.

Keywords Inverse problems · Convex sets · Cepstrum · Image processing · Location · Medical engineering

A. A. Potapov (✉) · A. A. Pakhomov · V. I. Grachev
V.A. Kotelnikov Institute of Radio Engineering and Electronics, Russian Academy of Sciences, Moscow, Russia
e-mail: potapov@cplire.ru

A. A. Potapov
Joint-Laboratory of JNU-IREE RAS, Jinan University, Guangzhou, China

© Springer Nature Switzerland AG 2020   57
Z. Hu et al. (eds.), *Advances in Artificial Systems for Medicine and Education II*,
Advances in Intelligent Systems and Computing 902,
https://doi.org/10.1007/978-3-030-12082-5_6

1 Introduction

Typical optical scenes tend to blur and defocus over large distances, as well as by the atmospheric distortions, fog, twilight, ascending atmospheric flows and the turbulent atmosphere of the Earth. Also of great interest are weak astronomical scenes from a remote cosmic object, which are actually a stream of photons, where in one frame pass from ten to one hundred photons. The final stage of the processing problem is the automatic recognition of the scene or object under study. In the present work, the so-called vector signs or the most important areas of the scene are distinguished for recognition. Practically, this work is the development of the well-known A. N. Tikhonov regularization method for solving nonlinear ill-posed problems [1] in which the number of unknown parameters exceeds the number of measured values and which have a set of solutions. The essence of the approach and the method for processing distorted images are set out in paragraphs 2–5. Successful digital processing of astronomical images, objects in the atmosphere and medical images was performed. As an example, the present work presents the results of processing medical images.

2 Statement of the Problem

From a mathematical point of view, the majority of such problems are a system of folding type equations. Summarizing Tikhonov's ideas at the modern level to solve such problems, we used a new approach based on the subsequent improvement of scene-assessment by coordinating the scene with available a priori information about an undistorted scene [2–12]. Coordination with a priori information is mathematically written in the form of a projection operator onto the corresponding ensemble.

2.1 Algorithms of the Reconstruction

The undistorted scenes are a sequence operator to projections and if prominent in relation to the ensemble, then similar procedures will reach a true decision. Of greatest interest is the determination of the analytical relationship between component Fourier spectrum, for instance, module and phase, both in univariate, and in two-dimensional events. Such an intercoupling, known as Hilbert transformations, was earlier received only for the univariate event. Generalization of the Hilbert equations is organized in our work on two-dimensional unceasing and discrete events. Also the condition of the unambiguous analytical decision of the row of the incorrect problems is revealed.

3   Mathematical Formulation in Information Processing Problems

When registration of image located at a considerable distance from the photodetector, the distorting influence of the medium of the propagation of the received radiation inevitably affects. Therefore, the general formulation of the problem of processing or restoring an image frame reduces to solving the convolution equation with two unknowns, after partial or complete elimination of the additive background [5, 6]:

$$i(\vec{x}) = \int o(\vec{r})h(\vec{x} - \vec{r})d\vec{r}. \tag{1}$$

Here $i(\vec{x})$ is the recorded image, $o(\vec{r})$ is the original unknown image of the unknown object and $h(\vec{x}, \vec{r})$ is the unknown impulse response of the optical system forming the image. Turning to spatial spectra, we arrive at the expression:

$$I(\vec{\omega}) = O(\vec{\omega})H(\vec{\omega}). \tag{2}$$

It is clear that without the use of additional information this task is unsolvable. As additional information, the authors used the finiteness and positivity of the unknown image $o(\vec{r})$. And also we approximated the transfer function by a Gaussian curve with an unknown variance. Herewith, not the continuous (2) but the discrete case was considered, i.e., the post-detect image processing. Herewith, all the factors (2) were two-dimensional polynomials of finite degree.

Taking into account, the singularities of two-dimensional discrete polynomials of finite degree and the Lebesgue measure apparatus one can prove that the convolution equation in the two-dimensional discrete case is almost always (in the sense of Lebesgue measure) uniquely can be solved [5]. The practical solution of this problem, even with an exact knowledge of the transfer function $H(\vec{\omega})$, where $\vec{\omega}$ are spatial frequencies, is also an uneasy task. It reduces either to Wiener filtration with an unknown regularizing factor or to regularizing Tikhonov filters that contain, in addition to the regularizing factor, several unknown stabilizing parameters [5, 8].

3.1   Methods of Solution

Thus, there is a need to use guaranteed and monotonically converging methods for solving the task. The answer to this question is provided by the application of set theory and, in particular, the use of projection operators on sets of functions with given properties. In solving this problem, we can use a set of finite and a set of positive functions, since it is assumed that the images have finite dimensions and represent the intensity distributions. Herewith, it is obvious that these sets are closed.

In general, the procedure for solving such ill-posed inverse problems reduces to choosing an initial estimate (image), designing it for a set of finite functions (accounting for boundedness with respect to linear dimensions), projecting for a set of positive functions (nulling negative values) and designing for a set of functions with given information on the Fourier spectrum.

According to the well-known theorem from functional analysis (the point theorem), if all a priori constraints correspond to convex sets, then such a procedure monotonically converges to the unique true solution [5]. In practice, when creating processing algorithms due to unavoidable rounding errors, image size determination, noise presence, etc., the restoration procedure converges to a certain neighborhood, covering the selected point and lying on the intersection of convex sets. From the point of view of image reconstruction when applying this approach, in the presence of noise, in the end, a lot of images will be obtained that are very similar to the true ones, but differ from it by small intensity fluctuations within the noise dispersion range. The external contour of the image is almost no different from the true one, i.e., is restored much more accurately than the distribution of the intensity of the image.

3.2 Transition to Convex Sets Based on a Cepstrum

Returning to the original problem (2) from the point of view of the theory described, it is easy to show that the set of positive and finite functions is convex [5]. At the same time, the set of functions (2) with a given product of spectra is a hyperbola, i.e., a nonconvex and nonclosed sets. At first glance, it seems that the theory of projections is inapplicable in this case. However, the described difficulty is quite manageable if we recall the methods of cepstral processing of signals and images that reduce to a transition from the complex spectra to their logarithms. The Fourier spectra can be written in the form:

$$I(\vec{\omega}) = |I(\vec{\omega})| \exp\{i \arg I(\vec{\omega})\} = |I(\vec{\omega})| \exp\{i\varphi_I(\vec{\omega})\}.$$

We take logarithms from both parts (2) and equate separately the real and imaginary parts:

$$Ln I(\vec{\omega}) = Ln|O(\vec{\omega})| + Ln|H(\vec{\omega})|,$$
$$\varphi_I(\vec{\omega}) = \varphi_O(\vec{\omega}) + \varphi_H(\vec{\omega}). \tag{3}$$

Thus, it is clear from (3) that after finding the logarithm, the set with a given product has become a set with a given sum. It is well-known that a set with a given sum is a line that is the limiting case of a convex set. If we introduce the notation: $K_I(\vec{\omega}) = Ln|I(\vec{\omega})| + i\varphi_I(\vec{\omega})$, where $K_I(\vec{\omega})$ is the cepstrum of the image, then in the language of cepstrums Eq. (2) is written not as a product but as a sum:

$$K_I(\vec{\omega}) = K_O(\vec{\omega}) + K_H(\vec{\omega}). \tag{4}$$

Thus, the possibility of translating a nonconvex set (2) into a convex set (4) becomes obvious. Convergence of the method for solving the convolution equation using cepstral spectra or cepstral methods also becomes apparent. It is worth noting that in modeling this procedure, the authors used modified cepsters associated with the specifics of the program (FFT). Herewith, the modified cepstrum looks like this:

$$K_I^M(\vec{\omega}) = Ln\{|I(\vec{\omega})| + c\} \exp i\varphi_I(\vec{\omega}) \tag{5}$$

where $c = 0.00001$ is a constant, the necessity of which is caused by the uncertainty of the value of the logarithm for small values of the argument.

It is easy to see that for modified cepstres the linearity of a set with a given sum is preserved, that is, the convergence of the method is guaranteed. The authors approximated the unknown transfer function $H(\vec{\omega})$ by a Gaussian curve, as well as its impulse response, so this function was real and positive and did not have a complex component, which led to an acceleration of convergence.

4 The Method Triple Correlation

The Knox-Thompson and triple-correlation methods are most often used at present for the processing of scenes distorted by the atmosphere [5], allowing for the restoration of the phase Fourier spectrum of the scene accurate to the remaining aberration of the telescope. Each k registered scene $i_k(\vec{x})$, represents a folding of the true scene $o(\vec{x})$ and pulsed response $h_k(\vec{x})$ of the system atmosphere-telescope:

$$i_k(\vec{x}) = \int o(\vec{r})h_k(\vec{x} + \vec{r})d\vec{r}. \tag{6}$$

Method triple correlation (TC) is concluded in calculation averaged Fourier spectrum on closed frequencies, i.e., under $\vec{f} = \vec{f_1} + \vec{f_2}$:

$$\left\langle J_k(\vec{f_1}) J_k(\vec{f_2}) J_k^*(\vec{f_1} + \vec{f_2}) \right\rangle$$
$$= o(\vec{f_1})o(\vec{f_2})o^*(\vec{f_1} + \vec{f_2})\left\langle H_k(\vec{f_1}) H_k(\vec{f_2}) H_k^*(\vec{f_1} + \vec{f_2}) \right\rangle \tag{7}$$

where transmission function is of the form:

$$H_k(\vec{f}) = k_0 \int d\vec{r} w(\vec{r}) w(\vec{r} - \lambda\vec{f})$$
$$\times \exp i\{\psi(\vec{r}) - \psi(\vec{r} - \lambda\vec{f})$$

$$\times \exp i\{\varphi_k(\vec{r}) - \varphi_k\left(\vec{r} - \lambda\vec{f}\right) \tag{8}$$

where k_0 is the normalizing factor.

Fourier spectrum of the k distorted scenes of the point source, $w(\vec{r})$ is the function of an eye pupil of the telescope, equal 1 within aperture and 0 outside of its, $\psi(\vec{r})$ is steady-state aberrations of the telescope, $\varphi_k(\vec{r})$ is phase atmospheric distortion. The last three multipliers in (7), which depend on the atmosphere, on residual aberration and optometry of the telescope, are named the transmission function of the method. In the event of uniform and isotropic atmosphere and Gaussian statistics of the atmospheric distortion, the transmission function of the method is a positive function, the type of which is investigated in multiple works of study of this method [2, 3, 5–10].

Leveling phases in (7) get:

$$\arg T_J = \arg T_O + \arg T_H, \tag{9}$$

where $\arg T_J$ is phase of the spectrum of the right part (7), $\arg T_O$ is phase of the first multiplier in left part, $\arg T_H$ is phase of the transmission function. Since in the absence of aberration the last summand in (9) is a zero, this allows restoring the undistorted phases of the spectrum of the scene on closed frequencies. Another obvious particularity of this method is its insensitivity (9) to linear shift of the processed scenes, hereby eliminating the need to center the images before processing. Thereby, from (9) follows:

$$\varphi\left(\vec{f_1}\right) + \varphi\left(\vec{f_2}\right) = \varphi\left(\vec{f_1} + \vec{f_2}\right).. \tag{10}$$

where $\varphi\left(\vec{f_1}\right) = \arg O\left(\vec{f}\right)$ is the unknown phase distribution Fourier spectrum of undistorted scene $o(\vec{x})$. The decision of the linear system (5) relatively $\varphi\left(\vec{f}\right)$ is the main task of the triple-correlation method.

5   Digital Distorted Scene Processing

5.1   A Scene Estimate in Presence Additive Noise

The problem is solved using the source system of Eq. (6). To obtain a scene estimate, if transmission functions are known, used square-law criterion of type:

$$\int \sum_{n=1}^{N} |I_n(\vec{\omega}) - O(\vec{\omega})H_n(\vec{\omega})|^2 d\vec{\omega} = \min. \tag{11}$$

Minimum to search for the whole number: $H_n(\vec{\omega})\ O(\vec{\omega})$ We shall find the minimum (6) on, rejecting dependency from frequency $\vec{\omega}$ and considering independent function O, H_n. After differentiating the integrand by O^*, we equate the result to zero. Hence, we get estimation of the scene O. In the manner of:

$$O = \sum_{n=1}^{N} I_n H_n^* / \sum_{n=1}^{N} |H_n|^2. \tag{12}$$

Substituting the estimation of the scene obtained from (12) into the source functional (11), we arrive at the following expression:

$$\begin{aligned} F &= \sum_{n=1}^{N} |I_n|^2 - \frac{\left| \sum_{n=1}^{N} I_n H_n^* \right|^2}{\sum_{n=1}^{N} |H_n|^2} \\ &= \sum_{n=1}^{N} |I_n|^2 - \left(\sum_{n=1}^{N} I_n^* H_n \right) \frac{\sum_{n=1}^{N} I_n H_n^*}{\sum_{n=1}^{N} |H_n|^2}. \end{aligned} \tag{13}$$

For finding estimations H_m, it is necessary to differentiate (13) on H_n and equaling the result to zero.

After a number of transformations, we get the expression (14) for optimum estimation the transmission function:

$$H_m = I_m \sum_{n=1}^{N} I_n^* H_n / \sum_{n=1}^{N} |I_n|^2. \tag{14}$$

The idea of the proposed method is to further improve the evaluation of the transmission function by including a priori information about the transfer function.

Thanks to the cyclic procedure of using a priori information, a strictly and quickly true solution is obtained, and with it, it is possible to restore the estimates of the transmission functions.

Then after inverse Fourier transformation, the undistorted scene is restored.

This procedure can be written in the language of the projection operator onto the corresponding sets [2–10]:

$$O_{k+1} = \hat{T}_1 \hat{T}_2 O_k \tag{15}$$

where k is number of the iterations, $\hat{T}_1$, $\hat{T}_2$ are the generalized operators of projections on positive finite function of the ensemble and on set of function type (14). Here,

$$\hat{T}_i = \hat{1} + \lambda_i \left(\hat{P}_i - \hat{1} \right), \tag{16}$$

where $\hat{P}_i$ is projection operator; $\hat{1}$ is single operator, not changing the value of the function; $\lambda_i \in (0, 1)$ is the relaxation multiplier.

5.2   Processing Images Distorted by Amplitude Grease

In a number of optical problems, the recorded image is often distorted by greases, when the observed object is displaced during exposure, and also by the random influence of an isotropic propagation medium. The original recorded image is a convolution (6), and the Fourier spectrum has the form (7). When observing cosmic objects with a long exposure, the transfer function is close to the Gaussian distribution and is a real and positive function. If the observation occurs not only with a long exposure, but also in white light, i.e., the additional averaging over the wavelengths takes place, then the transfer function can be regarded as a circular Gaussian with an unknown variance σ and zero mean. To reconstruct the image from incomplete information, a modification of the method of cyclic improvement of the image estimation in accordance with the available a priori information is used [2–11].

6   Medical Images Processing

Let the perimeter of the rectangular X-ray frame L is element of the resolution. Then the intensity J of the scene in frame is the sum of undistorted intensity I and intensity n of the background noise. The average value $\hat{n}$ of the unknown background is calculated along the perimeter of the rectangular frame, since it is expected that the investigated scene is in the center of the frame:

$$\hat{n} = \frac{\sum_{i=1}^{L} n_i}{L}, \tag{17}$$

where n_i is the value of the intensity of the background in I—resolution element of the perimeter frame L. Then we form the estimation J of the filtering scenes by the following rule:

$$J_k = \begin{cases} J_{k-1} - \hat{n}_{k-1} & \text{if } J_{k-1} > \hat{n}_{k-1} > 0, \\ 0, & \text{if } J_{k-1} < \hat{n}_{k-1} < 0. \end{cases} \tag{18}$$

By applying the given filter to the frame fewer than three times, the whole additive background is practically reduced to zero.

Filtering scenes J from the background described above by method (17) and (18) the scene is converted on Fourier and is realized in area of the spatial frequencies $F(\vec{\omega})$. It is realized by increasing the high spatial frequencies by erection amplitude

Fig. 1. Image processing of the patient's hand brush: the original image (**a**), the processed image by the background elimination method (**b**)

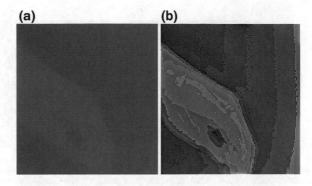

Fig. 2. Processing the face-scene structure of the person: source scene (**a**), processed scene by multifunction method of the eliminating the background and frequency correction (**b**)

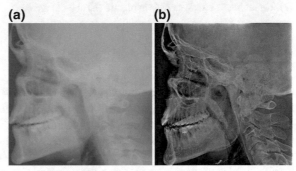

of the sharing the spectrum in degree, which factor α lies within the range of (0–1). It is possible to write the specified procedure as follows:

$$F(\vec{\omega}) = |F(\vec{\omega})|^{\alpha} \exp\{i\,\varphi(\vec{\omega})\}, \tag{19}$$

where are the phase distributions of the spatial spectrum of filtering scenes J. Then, inverse Fourier transformation is performed from, and the scene estimation is obtained with sharp detail. The operation (19) reminds one of partly fractal processing of the scenes [2–12]. Figures 1, 2 and 3 illustrate the results of the using of those methods in medicine [5].

(a)   (b)   (c)

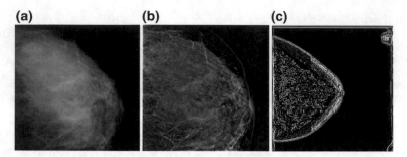

Fig. 3. Processing the scene of the female mammary gland at diagnostics of the primary appearance metastasis: source scene (**a**), scene processed by multifunction method (**b**) and fractal processing (**c**). Under intent visual analysis in picture (**b**) and (**c**) are seen small multiple compactions

7 Conclusion

In this paper, we present a method for processing and reconstructing a scene with incomplete information about its Fourier spectrum within the framework of the theory of ill-posed inverse problems in optics and radar. The Hubert equations for the components of Fourier spectra of images onto two-dimensional continuous and discrete events are generalized. The method of processing signals and images with modified cepstrums was used, the method of triple correlation for calculating the averaged Fourier spectrum forming the transfer function of the method. To reconstruct a distorted image, a modification of the method of cyclical improvement of the image estimation in accordance with available a priori information is used. It is shown that the defect of the described methods may be improved due to the transition to the analysis of fractal traits. As an example of the use of the developed method, the results of processing medical images are given.

Acknowledgements The authors prepared this article while working on the project "Leading Talents of Guangdong Province", № 00201502 (2016–2020) in the Jinan University (China, Guangzhou).

References

1. Tikhonov, A.N.: Nonlinear Ill-Posed Problems. Springer, Netherlands (1998)
2. Potapov, A.A.: Fractals in Radiophysics and Radiolocation: Topology of the Sample, 2nd edn, p. 848. University book Publ, Moscow (2005)
3. Potapov, A.A., Gulyaev, Y.V., Nikitov, S.A., Pakhomov, A.A., German, V.A.: The modern methods of the processing scenes. In: Potapov A.A. (ed.) 496 p. Fizmatlit Publication, Moscow (2008)
4. Mahajan, A., Gill, P.: 2D convolution operation with partial buffering implementation on FPGA. Int. J. Image, Graph. Signal Process. (IJIGSP) **8**(12), 55–61 (2016). https://doi.org/10.5815/ijigsp.2016.12.07

5. Pakhomov, A.A., Potapov, A.A.: The development of scaling effects in fractal theory and iterative procedures for processing of the distorted images. In: Proceedings of International Conference on Pattern Recognition and Information Processing (PRIP 2007), Minsk, Belarus, May 2007, 2, pp. 66–70
6. Bouhali, A., Berkani, D.: Combination of spatial filtering and adaptive wavelet thresholding for image denoising. Int. J. Image, Graph. Signal Process. (IJIGSP) **9**(5), 9–19 (2017). https://doi.org/10.5815/ijigsp.2017.05.02
7. Pakhomov, A.A., Potapov, A.A.: The improvement of the quality of images observed by the turbulent earth atmosphere. In: Proceedings of Second European Conference on Antennas and Propagation. Edinburgh, UK, Nov. 2007, TuPP.012. pdf. pp. 1–4, http://www.eucap2007.org/programme1.htm
8. Balovsyak, S.V., Odaiska, K.S.: Automatic highly accurate estimation of gaussian noise level in digital images using filtration and edges detection methods. Int. J. Image, Graph. Signal Process. (IJIGSP) **9**(12), 1–11 (2017). https://doi.org/10.5815/ijigsp.2017.12.01
9. Potapov, A.A., Pakhomov, A.A.: The methods of the digital processing of small-contrast scenes for the real-time recognizing system. In: Book Abstracts 2nd International Conference on Chaotic Modeling, Simulation and Applications (CHAOS' 2009), Chania, Crete, Greece, June 2009, p. 62
10. Potapov, A.A., German, V.A., Pakhomov, A.A.: Processing of images obtained from unmanned aerial vehicles in the regime of flight over inhomogeneous terrain with fractal-scaling and integral method. In: Proceedings of CIE International Conference on Radar 'Radar 2016', Guangzhou, China, Oct 2016, pp. 585–587
11. Khemis, K., Lazzouni, S.A., Messadi, M., Loudjedi, S., Bessaid, A.: New algorithm for fractal dimension estimation based on texture measurements: application on breast tissue characterization. Int. J. Image, Graph. Signal Process. (IJIGSP) **8**(4), 9–15 (2016). https://doi.org/10.5815/ijigsp.2016.04.02
12. Potapov, A.A.: Thematic course: statistical theory of fractal radar. In: Book of Abstracts of the 11th International Conference on Chaotic Modeling and Simulation' CHAOS 2018', Rome, Italy, June 2018, pp. 91–92

Numerical Approach to Solving the Problem of Choosing the Rational Values of Parameters of Dynamic Systems

I. N. Statnikov and G. I. Firsov

Abstract Discusses the use for research of problems of multicriteria synthesis of dynamic system method of planned LP search, which not only allows on the basis of the simulation model experiments to revise the parameter space within specified ranges of their change, but also through special randomized nature of the planning of these experiments to apply a quantitative statistical evaluation of influence of change in varied parameters and their pairwise combinations to analyze properties of the dynamic system.

Keywords Heuristic optimization methods · Monte Carlo methods · Planning of simulation experiment

1 Introduction

Designing and creating dynamic systems are essentially a process of man's creation of an artificial nature that helps him survive in a natural environment. Hence, the constant task of man (mankind) is to increase the effectiveness of systems of artificial nature from the point of view of its survival. For millennia, this problem was solved, mainly, by trial and error. With the advent of theoretical sciences and, first of all, natural sciences (physics, mathematics, theoretical mechanics, etc.), the roll in the solution of the above-mentioned problems shifts toward calculations with mathematical descriptions of the systems being developed mathematical models (MM).

The growth of the population of the Earth and the climatic diversity of living conditions constantly lead to the complication of mathematical models—the need for a more detailed mathematical description of the functioning of the projected system and, as a consequence, the emergence of increasingly sophisticated analytical methods for solving problems, approaching the possible limit of man's intellectual efforts. But the second half of the twentieth century pushed aside this limit somewhere: Elec-

I. N. Statnikov · G. I. Firsov (✉)
Blagonravov Mechanical Engineering Research Institute, Moscow, Russia
e-mail: firsovgi@mail.ru

© Springer Nature Switzerland AG 2020
Z. Hu et al. (eds.), *Advances in Artificial Systems for Medicine and Education II*,
Advances in Intelligent Systems and Computing 902,
https://doi.org/10.1007/978-3-030-12082-5_7

69

tronic computers (computers) have emerged, which are constantly improving up to now on two important parameters (the amount of computer memory and its speed).

In this paper, by solving the task of designing a dynamic system, we mean a set of recommendations for selecting values for design parameters of the system and conditions for controlling the operation of the system, ensuring its maximum efficiency in terms of technical, economic, and environmental quality criteria. Such an understanding clearly demonstrates that the tasks of designing dynamic systems are multiparametric and multicriteria.

From the theoretical point of view, the effectiveness of the application of one or another method of optimization, understood broadly, essentially depends on the degree of adequacy of the mathematical model used to the real dynamic processes occurring in the device being created or improved. Of course, in the narrow sense, when using the same mathematical model, there is always a competition of various optimization methods (in terms of accuracy, in the convergence rate of calculation results, in the clarity of the interpretation of these results). But even with the number of criteria $K \geq 2$ and the number of analyzed parameters, the parameter $J \geq 3$ became practically meaningless to talk about optimization of the required solutions in the narrow sense, but it can only be about finding rational solutions of the problem, corresponds to the search for compromise solutions. But in this case, the very effectiveness of the application of one or another method becomes a hostage to the volume and quality of the a priori information available at the time the solution to the applied optimization problem begins.

One of the ways to solve the problem can be the application of various heuristic methods of reducing the parameter space in which the best solutions are sought. Here, it is useful to rely on the cognitive rule derived by Fitts [1]: The time to reach the goal is inversely proportional to its size and distance to it. If the volume of the original search area is denoted by D, and the volume of the domain containing the preferred solutions is S, then the number of computational experiments can be determined by the formula: $N = a + b \log_2[(D/S) + 1]$, where a and b are some constants.

Therefore, it seems obvious that the most attractive are such methods of searching for rational solutions that, in the presence of an adequate mathematical model, require a minimum of a priori information about the problem being solved; moreover, it is possible to obtain such information easily and simply in the course of the decision. Such methods, of course, are called universal. We will refer to them a family of Monte Carlo methods and their various modifications [2]. The principles of using these methods are based on the principles of random search for the solution of a problem and statistical processing of the results obtained, which makes this approach universal. But the pay for such universality is a certain "blindness," and this leads to huge volumes of calculations, even for modern computers, especially as there is an increase in the dimension of the solved problems (the number of phase coordinates, the number of design parameters J, the number of quality criteria K characterizing the system (an object)). And huge volumes of the received information at carrying out of computing experiments naturally complicate its interpretation. There was a need to combine the universality of the Monte Carlo method with elements of a more intelligent analysis of the results of numerical experiments than a simple statement of statistical estimates, that is, improvements in the technology of conducting mathematical experiments. We illustrate this idea with the following quotation from [3]: "Historically, the problems of numerical modeling (in this concept we include the actual mathematical modeling associated with a numerical experiment), being significantly advanced in the" domashinny period "and developing at advanced rates in subsequent periods, have proved to be the most conservative component of modern mathematical technology solutions Tasks on a computer." We believe that to a significant extent, this need is met by the method of the planned LP search (PLP search).

2 Theoretical Aspects of a Research

We consider the method of PLP search (the method of planning LP_τ sequences [4–6]) and the formalized formulation of the problem being solved in its use. Note that the success of the application of PLP search is due to the fact that this method combines the ideas of a discrete survey of the space of analyzed parameters and the theory of planning of mathematical experiments [7], it is intended, mainly, for application in the preliminary stage of solving a problem when the information obtained allows us to make a decision about the use of other methods (but much more efficiently) or about the end of the solution (this is also possible).

At the base of the method is randomization of the location in the region of vectors calculated by LP_τ grids [8], and this is possible due to the fact that the entire computational experiment is performed in series. In PLP search, for today, it is possible to vary simultaneously the values up to the 51st parameters ($J = 51$). For randomization (random mixing of levels of variable parameters α_{ijk}) of a discrete survey $G(\vec{\alpha})$, many existing tables that are uniformly distributed in probability of integers can be used. In order to save the computer's memory in the PLP search, the randomization algorithm is constructed using a pseudo-random number sensor q ($0 < q < 1$) from [8]. Randomization consists in the fact that for each h series of experiments ($h = 1$, ..., H (i,j)), where H (i,j) is the sample size of the elements for one criterion Φ_{ijk}, its random number vector $\vec{j}$ $(j_{1h}, \ldots, j_{\beta h})$ in the table of directing numerators (TDN) according to the formula:

$$j_{\beta h} = [R * q] + 1, \tag{1}$$

and the values α_{ij} in the hth series are calculated using a linear transformation

$$\alpha_{ijh} = \alpha_{j*} + q_{ihj\beta h} \times \Delta\alpha_j, \tag{2}$$

where $\Delta\alpha_j = \alpha_{j**} - \alpha_{j*}$, α_{j**} and α_{j*}, respectively, the upper and lower boundaries of the region $G(\vec{\alpha})$; $\beta = 1, \ldots, J$; R—any integer number (in the PLP search $R = 51$); j—fixed number of the variable; $i = 1, ..., M$ (j) is the number of the level of the th parameter in the hth series; M (j) is the number of levels into which the parameter is divided; in the general case, $j_{\beta h} \neq j$ (which is one of the goals of randomization). It was proved with the help of the Romanovsky criterion [9] that the numbers $j_{\beta h}$, generated by the formula (1), turn out to be a set of numbers uniformly distributed in probability. Note that M (j) is the number of experiments implemented in one series. And if M $(j) = M = $ const and $H(i, j) = H = $ const, then in this case, the parameters NO, M, and H are related by a simple relation:

$$NO = M \times H, \tag{3}$$

where NO is the total number of computational experiments (CE), while the sample length Φ_{ijk} is exactly equal H. But in the general case, when $M(j) = $ var, then and $H(i, j) = $ var, and then formula (3) for one criterion takes the form:

$$NO = \sum_{i=1}^{M(j)} H(i, j)$$

Using the formulas (1) and (2) in PLP search, the following variants of the matrices of the planned experiments (MPE) are realized:

I. $M = $ const; NO is considered by formula (3); in this case, it is possible to construct MPE for such cases:

   (a) $\varepsilon = 0$; the exact values of the boundaries of the region $G(\vec{\alpha})$ are taken into account, but in this case it is necessary to increase the number of NO experiments, so the frequency of appearance of the boundary values α_j is two times lower than the frequency of appearance of internal values of this parameter;
   (b) $0 < \varepsilon \ll 1$; the boundaries of the jth parameter change form an interval $\left(\alpha_{j*} + \varepsilon; \alpha_{j**} - \varepsilon \right)$, and then calculation by formula (2) is used;
   (c) $0 < \varepsilon \ll 1$; the boundaries of the jth parameter change form the interval $\left(\alpha_{j*} - \varepsilon; \alpha_{j**} + \varepsilon \right)$, and then calculation by formula (2) is used;

II. $M_j = $ var; in this case, there are also three possible options for constructing an MPE, but for each jth parameter, one takes its own ε_j, wherein $0 < \varepsilon_j \ll 1$.

The implementation of the MPE options considered significantly extends the choice of rational values of the variable parameters of the system, allowing the boundaries of the variable parameters to be varied without changing the mathematical model.

To perform a single-factor analysis of variance [10] for all parameters, for each criterion, the results of the computations obtained in the calculation of the points of the matrix of the planned experiments (MPE) are sorted. As a result of sorting for one criterion, J matrices consisting of elements Φ_{ijk} will be obtained, and for K matrices—$J \times K$ matrices consisting of elements Φ_{ijhk}, where k is the number of the criterion. This analysis allows to accept (or reject) with the required probability

$P \geq 1 - \beta$, where β is a given level of significance, the following null hypothesis: The mean values $\bar{\Phi}_{ijk}$ are not significantly (randomly) different from the general average value of the criterion $\bar{\Phi}_{0k}$. If a positive answer is accepted (the hypothesis is accepted), then it is allowed at the next stage of the solution of the problem to neglect the insignificantly influencing parameter α_j and to fix one of its values, for example, $\alpha_j = \alpha_{ij}$, for the one i with the best value in the sense of the desired extremum.

Now, we describe a typical formalized formulation of the problem, for which it will be useful to use PLP search. Let the mathematical model of the system in question be given in the form

$$L(\bar{y}(\bar{\alpha}, t), (\bar{\alpha})) = 0, \bar{\varphi}(\bar{\alpha}) \geq 0, \tag{4}$$

where L is the operator acting on the system of Eqs. (4) (linear or nonlinear), $\bar{y}(\bar{\alpha}, t)$ is the vector of the system's phase coordinates, $\bar{\phi}(\bar{\alpha})$ is the vector of functional constraints on the parameters and behavior of the system (4), $\bar{\alpha} = (\alpha_1, \ .., \ \alpha_J)$ is the vector of the coefficients of the equations. The initial range $G(\bar{\alpha})$ of the coefficients is given in the form of inequalities $(\alpha_{j*} < \alpha_j < \alpha_{j**})$. And finally, we set the system of quality criteria for the device (in an explicit or implicit form) $\Phi_k = \Phi_k(\bar{\alpha}), (\bar{\alpha}) \in G(\bar{\alpha}), k = \overline{1, K}$.

On the basis of the numerical analysis performed, it is possible:

(a) Identify relevant parameters $\alpha_m(m \leq J)$ in order to reveal their influence on the values of each criterion, in other words, to evaluate statistically the changes in the derivatives $\partial \Phi_k(\bar{\alpha})/\partial \alpha_j$ in the region $G(\bar{\alpha})$;

(b) Determine the areas of concentration of the best solutions for each criterion. For this we need, based on the given metric $\rho(\Phi_k(\bar{\alpha}), \Phi_k^+)(\Phi_k^+$ is the experimental value of the kth quality criterion, known or determined in the course of numerical experiments), to find the area $G_k(\bar{\alpha})$ in which simultaneously execute the two conditions: $P[\rho(\Phi_k(\bar{\alpha}_u), \Phi_k^+) \leq \varepsilon_k, \quad u = \overline{1, n}; \quad n \leq N] \geq P_3$ and $n/N \geq 1 - \delta$, where $\alpha_u \in G_k(\bar{\alpha}), 0 < \varepsilon_k, \delta << 1$, P_3 is the given probability; N—total number of conducted numerical experiments;

(c) On the basis of determining the relevant parameters α_m and of concentration areas $G_k(\bar{\alpha})$, construct regression dependencies $\hat{\Phi} = \Psi_k(\alpha_1, \ldots, \alpha_m)$;

(d) To select in the K-dimensional space of criteria, the set of Pareto points (or, if possible, construct the Pareto surface), in the case of setting any compromise scheme, allocates an area containing compromise solutions [11, 12].

Implementation of pp. "b" and "d" are implemented algorithmically and with the help of the following graphs (Fig. 1).

In this case, the following heuristic rules for selecting the desired subregions were formulated:

(a) If the confidence (theoretical) probability P of the influence of the parameter on the values of the analyzed function is less than the preset P_3 and, for instance $P_3 < 0.95$, then the initial range of the change in this parameter does not change;

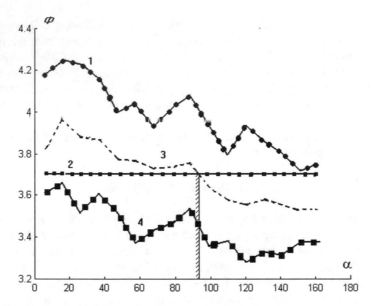

Fig. 1 Graphical analysis of the results of computational experiments

(b) If $P \geq P_3$, then we proceed as follows: If the general average curve $\overline{\Phi}_{0k}$ crosses the line of the curve of the means $\Phi_{ijk}(\alpha_j)$ one or two times, then we select a new subband from the source in accordance with the mathematical meaning of the sought extremum (min or max); if the number of intersections of the curve $\overline{\Phi}_{0k}$ of the mean curve is greater $\Phi_{ijk}(\alpha_j)$ than two, then in spite of the values of P, the original parameter range is not changed (although the decision in this case, like the choice of the value of P_3, remains with the researcher, since the decision itself affects the volume of the experiments).

1—The curve of the sampled values of the maxima $\Phi_{ijk}(\alpha_j)$; 2—the general mean curve $\overline{\Phi}_{0k}$; 3—average values of the criterion $\overline{\Phi}_{ijk}$ for the parameter α_j; 4—the curve of the sampled values of the minima $\Phi_{ijk}(\alpha_j)$.

3 Parametric Identification of System Parameters Arm Rotation of Industrial Robot

This approach was used to solve a number of specific problems of research, optimization, and identification of various mechanical and controllable systems. In particular, with the help of randomization of the parameter change area, the parameters of the robot arm rotation system with electrohydraulic drive and the position control system installed in the technological chain of the flexible production system have been identified.

High requirements to the accuracy of work and technological reliability of the robot require taking into account the dynamic properties of the robot control system when choosing the speed characteristics of the forward and rotational movements of the robot's arm. To reliably calculate the dynamic properties of the robot drive at the design stage, it is necessary to refine the mathematical model of the dynamic system on the basis of experimental studies. Such a refinement is essentially a parametric identification of the structure of the model proposed hypothetically on the basis of experimental data [8].

Based on the experimental studies carried out, the robot arm positioning mechanism is represented as a three-mass system with elastic and damping properties. The system is covered by negative position feedback. The amplitude–frequency characteristics of the servo valves used in this robot design showed that they can be described with sufficient accuracy by second-order differential equations. In general, the mathematical model of the robot is described by a system of nonlinear differential equations of the 10th order. The task of parametric identification is as follows.

In the space of variable parameters $\sigma(\bar{\alpha})$ (these parameters included inertial rigid elements, geometric parameters, control parameters, damping elements) to find a region $\sigma_1(\bar{\alpha})$ such that for any vector $\bar{\alpha} \in \sigma_1(\bar{\alpha})$, the criteria for the closeness of the calculated characteristics obtained on the model and the experimental ones were best satisfied. Two force characteristics of the pressure $P_1(t)$ and $P_2(t)$ in the cavities of the hydraulic cylinder were chosen as compared, and one kinematic pressure was the capture velocity $\dot{x}_3(t)$.

For the measure of closeness of the calculated and experimental data, the sums of root-mean-square deviations were chosen:

$$\Phi_1(\bar{\alpha}) = \frac{1}{n} \sum_{i=1}^{n} \left(\frac{P_{1\epsilon} - P_{1T}}{P_{1\epsilon}} \right)^2 ; \Phi_2(\bar{\alpha}) = \frac{1}{n} \sum_{i=1}^{n} \left(\frac{P_{2\epsilon} - P_{2T}}{P_{2\epsilon}} \right)^2 ;$$

$$\Phi_3(\bar{\alpha}) = \frac{1}{n} \sum_{i=1}^{n} \left(\frac{\dot{x}_{3\epsilon} - \dot{x}_{3T}}{\dot{x}_{3\epsilon}} \right)^2 ; \tag{5}$$

Further, in the first approximation, having no reason to consider the functions as unequal, we have chosen the convolution of these functions $\Phi_4(\bar{\alpha}) = \frac{1}{3n} \sum_{k=1}^{3} \Phi_k(\bar{\alpha})$. Naturally, the exact solution of the problem is to achieve an absolute minimum $\Phi_4(\bar{\alpha})$, which is practically impracticable. In solving the problem of parametric identification, taking into account the accepted equality of all the proximity criteria under consideration, the task was to find a region $\sigma_2(\bar{\alpha})$ such that for any vector $\bar{\alpha} \in \sigma_2(\bar{\alpha})$ an approximate equality

$$|\Delta\Phi_1/\Phi_1| \approx |\Delta\Phi_2/\Phi_2| \approx |\Delta\Phi_3/\Phi_3|. \tag{6}$$

Table 1 Utility function averaged over ten series of experiments

№ series	b_1	b_2	c_1	c_2	φ_1	φ_2	f_0
1	0.321	0.362	0.146	0.276	0.238	0.180	0.930
2	0.269	0.300	0.184	0.365	0.342	0.351	0.662
3	0.286	0.237	0.523	0.312	0.270	0.259	0.276
4	0.200	0.195	0.259	0.208	0.307	0.318	0.121
5	0.380	0.378	0.221	0.452	0.353	0.197	0.090
6	0.356	0.410	0.418	0.218	0.394	0.227	0.090
7	0.286	0.304	0.295	0.323	0.594	0.500	0.100
8	0.320	0.257	0.393	0.259	0.148	0.410	0.108

The physical meaning of equality (6) is obvious: We would like to identify the model in such a way that the best values $\Phi_4(\bar{\alpha})$ are achieved without discrimination of one of the any other criteria.

To solve this problem, we analyzed a function of the form $\Phi_5(\bar{\alpha}) = \left[\sum_{k=1}^{3} (\lambda_k(\bar{\alpha}) - 0,5)^2\right]^{1/2}$, $\lambda_k(\bar{\alpha}) = \Phi_{k\max} - \Phi_k(\bar{\alpha})/\Phi_{k\max} - \Phi_{k\min}$. Taking into account the fact that in the ideal case $\Phi_{k\min} = 0$, we obtain $\lambda_k(\bar{\alpha}) = 1 - \Phi_k(\bar{\alpha})/\Phi_{k\max}$, where $\Phi_{k\max}$ is the maximum error in formulas (5). It is clear that the ideal (unattainable) version of the solution of the whole problem corresponds to the case when $\lambda_k(\bar{\alpha}) = 1$.

The task of finding regions $\sigma_1(\bar{\alpha})$ and $\sigma_2(\bar{\alpha})$ was solved by performing mathematical experiments on a computer using the PLP search method in the MATLAB environment [6]. Based on the results of the experiments, the ranges of the following parameters were assigned: the stiffnesses of the system c_1 and c_2, the damping coefficients b_1 and b_2, the linear and quadratic losses φ_1 и φ_2, and the area of the slit spacing f_0. The following initial ranges of parameters were adopted with respect to the initial values:

$$0.2c_i \leq c_i^n \leq 5c_i, 0.5b_i \leq b_i^n \leq 3b_i, 0.2\varphi_0 \leq \varphi_0^{n} \leq 5\varphi_0.$$
$$0.2c_i \leq c_i^n \leq 5c_i, 0.5b_i \leq b_i^n \leq 3b_i, 0.2\varphi_0 \leq \varphi_0^{n} \leq 5\varphi_0.$$

The index "n" denotes the initial values of the variable parameters. At the first stage, 80 computer experiments were calculated. As a result, Tables 1 and 2 are constructed, where the average value of this proximity criterion corresponding to a particular value of the parameter is placed at the intersection of the row and column.

The variance analysis of Tables 1 and 2 made it possible to establish that the values $\Phi_4(\bar{\alpha})$ on the average are significantly affected by the parameters f_0, c_1, φ_1. Note that the effect of f_0 was almost 100% (proof that this is a control parameter). The criterion $\Phi_5(\bar{\alpha})$ was practically influenced by all the variable parameters.

At the second stage, the best parameters were selected for $\Phi_4(\bar{\alpha})$ and an additional series was performed (16 experiments), in which only the moment of inertia J_3 and

Table 2 Distance function

№ series	b_1	b_2	c_1	c_2	φ_1	φ_2	f_0
1	0.321	0.362	0.146	0.276	0.238	0.180	0.930
2	0.269	0.300	0.184	0.365	0.342	0.351	0.662
3	0.286	0.237	0.523	0.312	0.270	0.259	0.276
4	0.200	0.195	0.259	0.208	0.307	0.318	0.121
5	0.380	0.378	0.221	0.452	0.353	0.197	0.090
6	0.356	0.410	0.418	0.218	0.394	0.227	0.090
7	0.286	0.304	0.295	0.323	0.594	0.00	0.100
8	0.320	0.257	0.393	0.259	0.148	0.410	0.108

the feedback coefficient $k_{o.c}$ varied. In result we attain average value $\sigma_0 = 0{,}03$ and the average value of $\Phi_{40} = 0.0535$. And, finally, at the third stage, taking into account the results of stages I and II, the area in which the control experiments were conducted was determined.

The analysis of the comparative results for all three identification stages given in Table 3 shows that in the III stage, an area $\sigma_0(\bar{\alpha})$ is found that completely satisfies the criteria $\Phi_4(\bar{\alpha})$ and $\Phi_5(\bar{\alpha})$. Comparison of the corresponding calculated and experimental pressure change curves P_1 (t) in the head and P_2 (t) in the drain cavities of the hydraulic cylinder and the speed $\dot{x}_3(t)$ of the robot arm showed that the proposed mathematical model with acceptable accuracy describes the dynamics of the hand rotation mechanism. In addition, in area $\sigma_0(\bar{\alpha})$, compromise is really achieved for $\Phi_4(\bar{\alpha})$.

Thus, the constructed mathematical model with allowance for the region $\sigma_0(\bar{\alpha})$ allows: to make a more careful calculation of the dynamics of this mechanism already at the design stage; optimally select the parameters of the system to obtain the required characteristics; choose the law of braking the robot's arm in order to improve its speed and positioning accuracy.

The obtained model can serve as a basis for the development of diagnostic models of the robot.

Table 3 Comparative results for three stages of identification

Stage	$\bar{\Phi}_{40}$	σ_{40}	$\bar{\Phi}_{50}$	σ_{50}
I	0.2956	0.3070	0.5843	0.1457
II	0.0535	0.0300	0.7220	0.0154
III	0.0448	0.0155	0.7269	0.0085

4 Conclusions

Thus, the PLP search method not only makes it possible to perform a quasi-uniform scanning of the parameter space in given ranges of their changes on the basis of simulation simulations, but also as a result of a special, randomized nature of the planning of these experiments, to use quantitative statistical estimates of the effect of variation of the variable parameters and their pairwise combinations on analyzed properties of the dynamical system under consideration [13, 14].

At the same time, by constructing approximation models of criteria depending on the variable parameters, it is possible to estimate the sensitivity of the criteria on average for these parameters. The effectiveness of the plans for experiments in the PLP search is determined not only by the possibility of using them in dispersion analysis. These plans prove to be effective both in constructing regression dependencies and generally in regression analysis, both in the computational aspect and in terms of a number of optimality criteria for these plans. Proving the effectiveness of PLP search in comparison with the "blind" way of searching for extremes, it is not suggested to discard the latter. However, it is obvious that PLP search not only can help speedy search for extremes if it is required, but also provides information on the influence of variable parameters and, what is also important, contributes to issues related to the study of the space of variable parameters (in particular, the results of the PLP search can help to choose an effective compromise scheme). So, it can be argued that if the decision is made to investigate the formulated problem first in a discrete way (which is useful even in the case when it is possible to obtain analytical dependencies, but very complex ones), PLP search seems to be a very effective method of computer technologies in the sense of the previously mentioned second component. We also note that when using PLP search, it is possible to vary up to 51 parameters simultaneously. This possibility is determined by the time of calculation of one variant, which already with modern (and future) computers does not represent large obstacles.

References

1. Fitts, P.M.: The information capacity of the human motor system in controlling the amplitude of movement. J. Exp. Psychol. **47**(6), 381–391 (1954)
2. Sobol, I.M.: A Primer for the Monte Carlo Method, p. 128. CRC Press, Boka Raton (1994)
3. Belotserkovsk, O.M.: Rational numerical modeling in nonlinear mechanics. Moscow: Nauk. 224 p (1990). (In Russian)
4. Statnikov N., Firsov. G.I.: Combination use of discrete optimization methods in problems of research and modeling of dynamic systems of machines. South-Siberian Sci Herald. **2**, 74–78. (in Russian)
5. Statnikov, I.N., Firsov, G.I.: Interactive structuring of parameter space in the design of dynamic systems. Bull. Tambov State Tech. Univ. **1**, 36–41. (in Russian). https://doi.org/10.17277/vestnik.2015.01.pp.036-041

6. Statnikov, I.N., Andreenkov, E.V.: PLP-Search-Heuristic Method for Solving Mathematical Programming Problems, 140 p. Moscow State University of Technology and Design, Moscow (2006). (in Russian)
7. Montgomery, D.C.: Design and Analysis of Experiments, p. 684. Wiley, New York (2001)
8. Sobol I.M.: Multidimensional quadrature formulas and Haar functions, 288 p. Nauka, Moscow (1969). (in Russian)
9. Mitropolsky, A.K.: The technique of statistical computations, 576 p. Nauka, ГРФМЛ, Moscow (1971). (in Russian)
10. Scheffe, G.: Analysis of Variance, p. 477. Wiley, New York (1959)
11. Wang, L., Chen, Y.: Diversity based on entropy: a novel evaluation criterion in multi-objective optimization algorithm. Int. J. Intell. Syst. Appl. (IJISA), 4(10), 113–124 (2012). https://doi.org/10.5815/ijisa.2012.10.12
12. Hou, N., Wang, Z.: A growing evolutionary algorithm and its application for data mining. Int. J. Intell. Syst. Appl. (IJISA) 3(4), 8–16 (2011). https://doi.org/10.5815/ijisa.2011.04.02
13. Statnikov, I.N., Firsov, G.I.: Using sobol sequences for planning experiments. J. Phys. Conf. Ser. 937(1–3), 012050 (2017). https://doi.org/10.1088/1742-6596/937/1/012050
14. Roy, C., Rautaray, S.S., Pandey, M.: Big data optimization techniques: a survey. Int. J. Inf. Eng. Electron. Bus. (IJIEEB) 10(4), 41–48 (2018). https://doi.org/10.5815/ijieeb.2018.04.06

Algorithms for Topological Analysis of Spatial Data

Sergey Eremeev and Ekaterina Seltsova

Abstract The analysis of spatial relations and the search for abstract structures are some of the main tasks of geoinformatics. Huge arrays of spatial information contain hidden abstract structures. Existing methods of clustering do not take into account topological features of spatial objects. It is necessary to create new algorithms for finding regularities in the set of objects on the map. Algorithms for selecting abstract spatial structures based on topological data analysis are developed in the article. The developed algorithms allow us to find common structures in a set of unrelated spatial data. The algorithms for collecting spatial information from point and polygonal objects are also developed in the article. These algorithms allow us to present the initial information for processing by methods of persistent homology. An algorithm for constructing a barcode using simplicial complexes is shown. The study includes topological analysis of point and contour objects. An original approach for searching for repeating structures in a set of randomly distributed spatial objects is shown. Information about the repeating structures is calculated from the barcode. Computer modeling to analyze the structure of a sequence of objects with deformation is demonstrated. Examples for determining the similarity of natural spatial objects based on information from the barcode are given.

Keywords Topological data analysis · Spatial objects · GIS

1 Introduction

A topology in geographical information systems is a complex representation of data that describes the spatial relationships of neighboring spatial objects [1, 2]. Spatial objects are specific objects or phenomena defined by specific content and described

S. Eremeev (✉) · E. Seltsova
Vladimir State University, Vladimir, Russia
e-mail: sv-eremeev@yandex.ru

E. Seltsova
e-mail: catherine13nov@gmail.com

© Springer Nature Switzerland AG 2020

Z. Hu et al. (eds.), *Advances in Artificial Systems for Medicine and Education II*,
Advances in Intelligent Systems and Computing 902,
https://doi.org/10.1007/978-3-030-12082-5_8

in the form of data sets [3]. Unconnected objects are some abstract structures. The analysis of spatial relations and the search for abstract structures are some of the main tasks of geoinformatics and require the automation.

Huge arrays of spatial information include various hidden structures such as repeating, cyclical, symmetric and other abstract patterns. Traditional geometric approaches require a lot of time to search for these structures. Repeating structures are especially interesting for us. Searching for the same map areas can be based on repeating structures. It is necessary to create new and fast algorithms for selecting such structures on the map. The structures found on the map will allow us to better understand the nature of the spatial scene from unrelated geographic objects.

The aim of the work is to develop and study algorithms for searching abstract and repeating structures on a map from a set of non-structured spatial information.

2   Related Work

2.1   Method of Clustering

Clustering is the task of splitting a set of objects into groups. Within each group, the most similar objects should appear. Objects of different groups should be more distinct. Clustering is used to solve different tasks [4, 5].

The clustering method allows us to define the integration of spatial objects into clusters. Cluster analysis mainly considers only point data.

The similarity of objects determines the distance between them. There are several distances for calculating the similarity between objects: Euclidean distance, Manhattan distance, Chebyshev distance and others.

There are a lot of methods for clustering data. Widely used methods are clustering methods based on fuzzy logic [6, 7], algorithms based on graph theory [8], k-averages algorithms for large data [9] and others.

However, clustering only considers the distances between objects and does not take into account the topological features of objects and their structures.

2.2   Method of Persistent Homology

Another way to analyze spatial objects and structures is persistent homology. Persistent homology is a computational method developed to determine topologically significant features from a set of data points. A simplex is a convex hull of $n + 1$ points of an affine space of dimension n, i.e., assumed to be affinely independent and they do not lie in a subspace of dimension $n - 1$. These points are called the vertices of the simplex $\{v_0, v_1, \ldots, v_k\}$. For n equal to zero, the simplex is a point, for $n = 1$ it is a segment, for $n = 2$ it is a triangle and for $n = 3$ it is a tetrahedron (Fig. 1) [10, 11].

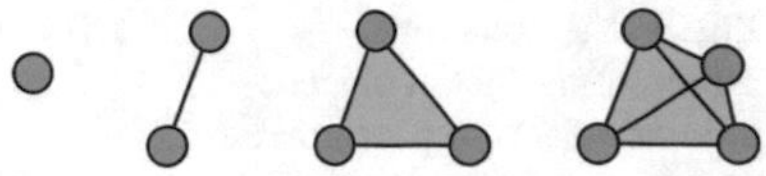

Fig. 1 Simplexes of
different dimensions

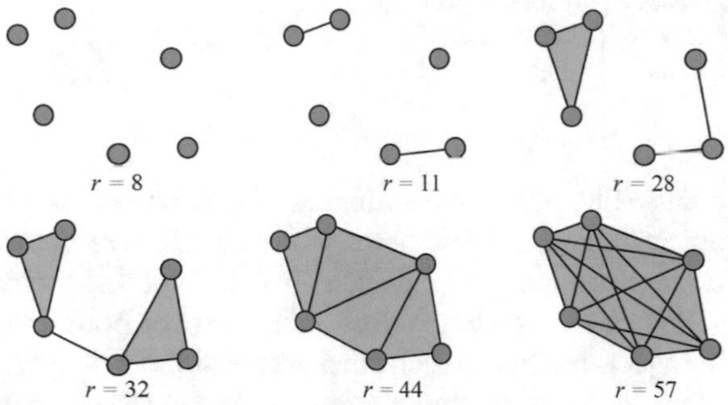

Fig. 2 The forming
structures with increasing
distance r

$r = 8$ $r = 11$ $r = 28$

$r = 32$ $r = 44$ $r = 57$

A simplicial complex is a finite set of simplexes. If two simplexes are in a simplicial complex then their intersection is either an edge of both simplexes or an empty.

The persistent homology method uses either structures from simplexes or holes. In the article, we consider structures from simplexes that form simplicial complexes.

The analysis is performed on the constantly increasing radius of finding neighboring objects r. If the distance between points is less than or equal to r, then a connection is established between them. In the case of combining three points, we can conclude that they compile a simplex. Simplexes make up structures that combine with each other and at the end of the analysis make up one single structure (Fig. 2) [12].

The result of the analysis is a set of selected topological features and noise. Purification of the received features from errors and inaccuracies is realized according to the following rule. Those features that manifest themselves over large ranges of distances are stable, and those features that are manifested at relatively small ranges of distances are noise and are not considered [13, 14].

3 Algorithms of Persistent Homology for Processing of Spatial Objects

3.1 Collecting Information About Map Objects

Depending on the task, each object on the map can be represented as a point or a set of points lying on the contour of the object.

When analyzing point, objects such as trees, intersections of roads, stops and metro stations, their coordinates are considered (Fig. 3a). If, in the analysis of area objects, their shape, size and rotation are not important, they are represented in the form of

Fig. 3 Methods for representing spatial objects: **a** point objects, **b** polygonal objects and their centers of gravity, **c** polygonal object and set of points on the contour

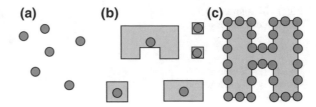

points (Fig. 3b). The coordinates of the points depend directly on the contour of the object. The arithmetic mean of the coordinates of the contour points is calculated. Examples of polygonal objects are land, buildings, parking lots, parks and reservoirs.

When representing objects in the form of points, only the features of the location of objects relative to each other are considered, namely the distance between centers. There are no restrictions on reading data. At the beginning of the analysis, simplicial complexes do not exist.

When analyzing contour objects, the points lying on the contour of the polygon are considered (Fig. 3c). All contour points or some points with a certain step can be taken into account. Analysis of the contours of objects considers not only their location on the map, but also their size, shape and rotation relative to each other.

In contrast to the analysis of points in the analysis of contours, there is a restriction on reading data from the map. The points are divided into groups, depending on their belonging to the object. Thus, at the beginning of the analysis, we have structures of points.

3.2 Splitting the Initial Set of Points Into Simplexes

Persistent homology considers structures composed of simplexes. It is necessary to collect triangles from the points found on the map. Both intersecting and disjoint triangles are considered (Fig. 4).

To solve the problem of partitioning, a set of points into triangles with allowance for intersections, the longest side of the triangle obtained must be considered. The algorithm of partitioning from a set of points forms combinations of three points, and the combinations should not be repeated. This approach more accurately represents the form of the obtained homogeneous regions. But result contains a large number of simplexes which can lead to a noticeable increase in the analysis time. Before

Fig. 4 **a** Non-intersecting and **b** intersecting triangles

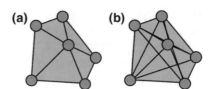

composing the structures, the list of obtained simplexes should be sorted by increasing the longest side of the simplex.

The partitioning of the initial set of points without intersecting triangles assumes the use of Delaunay triangulation algorithms.

3.3   Barcode Formation

The result of the analysis is a list of all compiled structures each of which has its own sequence number, a list of triangles, the initial and final distances and a list of parent structures. Using these data, the result of applying the method of persistent homology can be represented in the form of a tree where the root node is a finite structure. Internal nodes are structures that have parents. Terminal nodes are the structures that appeared at the time of the beginning of the analysis and do not have parents (Fig. 5). Also nodes can contain the initial or final distance of each structure.

In most cases, the results of applying the persistence homology method are presented as a barcode. A barcode is a diagram where each structure is designated as a bar [15].

The resulting diagram can be extended by adding the edges of the tree to it. These edges show which structures were combined (Fig. 6).

Fig. 5 Representation of results in the form of **a** diagram and **b** tree, h is the number of the structure, r is the distance of the length of the triangle

Fig. 6 Example of an extended barcode

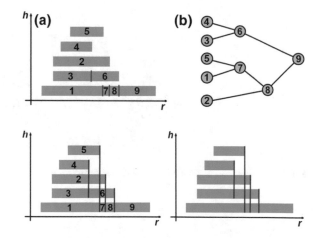

4   Results

4.1   Topological Analysis of Point Features

When analyzing spatial objects, each of them is represented in the form of a point located at the center of gravity of the figure. The program input is a raster map on which the contours of objects are searched. Then from all points of the contour of each object the center of gravity of the figure is calculated. As a result, we have a set of points for constructing triangles. Then the structures are compiled.

Figure 7 shows several point objects. With an increase in the search radius r, two simplicial complexes $S1$ and $S2$ were formed. Then a single simplicial complex $S3$ was formed which combined all the triangles found in it. The barcode shows three structures two of which appeared simultaneously and soon merged into one (Fig. 8).

Barcodes obtained by both methods are practically identical (Fig. 9).

On sufficiently large data, the errors of the method of disjoint triangles are almost invisible (Fig. 10). The use of all connections generates a huge number of triangles. For example, 252 map objects generate 2,635,500 triangles. The number of disjoint triangles is only 487.

The research has shown that when analyzing point objects only disjoint triangles should be considered. This method is faster in execution time and uses less memory.

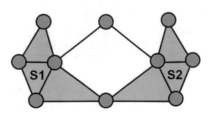

Fig. 7   The location of point objects and the composition of their structures

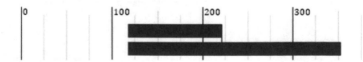

Fig. 8   The barcode for example in Fig. 7

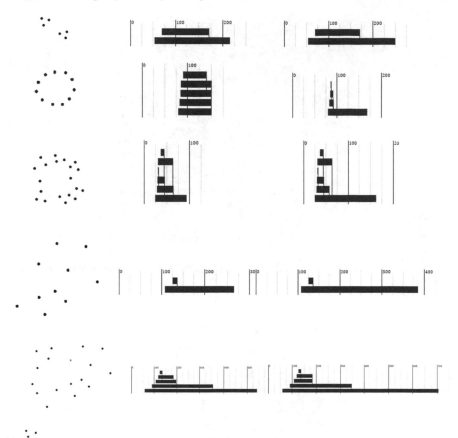

Fig. 9 Examples of sets of spatial data and the results of both ways of dividing the set into triangles: left diagrams are with non-intersecting triangles, right diagrams are with intersecting triangles

4.2 Topological Analysis of Contour Objects

When analyzing the contours of objects, either all points of the contour are used or a certain step between points is taken into account. As well as with point objects, connections between all points are established. The main difference is that the points belonging to one object are already at the beginning of the analysis within the structures corresponding to the objects. Accordingly, the initial distance of the existing structures will be zero.

Figure 11 shows the results of the analysis with and without intersections. As in the previous paragraph, the results are similar.

The most optimal way is to consider only disjoint triangles.

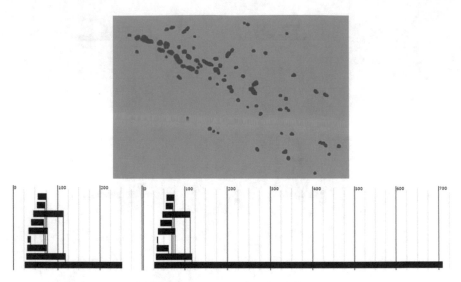

Fig. 10 A map of islands and barcodes

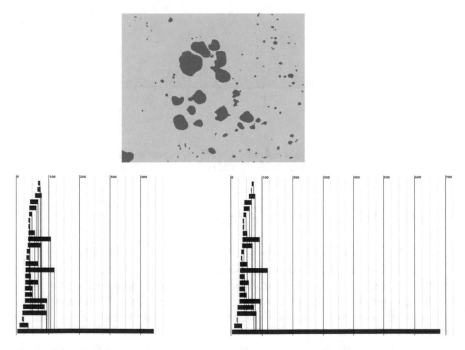

Fig. 11 Map of the lakes and the results of the analysis

4.3 Identification of Repeating Structures

The study has shown the following relationship. If the map has several identical sections, then the barcode also contains similar structures. This regularity manifests itself both in the analysis of point and contour objects.

Figure 12 shows the set of points and the resulting barcode. In this case, we do not observe any regularities on the map and barcode.

However, when duplicating the map, the following result is obtained (Fig. 13).

We see a repetition of bars. Each structure occurs twice. If we put the same bars on each other, we get the barcode of the map shown in Fig. 12.

This dependence is also observed in the case when the map consists of four symmetrical blocks (Fig. 14).

Next, we investigated the case where the structure contains objects of different shapes (Fig. 15).

In this case, there are repeating structures too. In addition to the four large lines, there are four bars of the same length. After sorting the received bars, we clearly see duplication of identical ones.

Thus, from an enormous number of points of unstructured data, the algorithm allows us to determine the number of identical or similar structures.

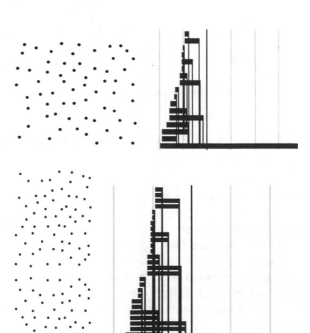

Fig. 12 A set of points and a barcode

Fig. 13 A map consisting of two identical blocks and its barcode

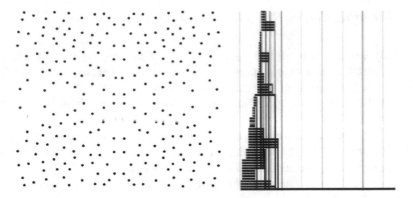

Fig. 14 A map consisting of four symmetrical blocks and its barcode

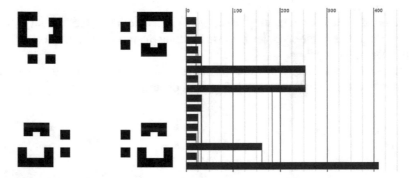

Fig. 15 A map of contour objects with different shapes and barcode

4.4 Search for a Common Structure for Similar Spatial Objects

Perform computer modeling for topological analysis of similar contour objects. Consider a map of the islands and various variants of its deformation, such as stretching, compression and small changes in the boundaries of islands (Fig. 16).

The first image is the original map of the islands and its barcode. Each barcode line corresponds to one of the islands. The second two images are the results of the map deformation. We can see on the barcode some noise that appeared after deformation of the sharpest corner of the upper island. However, in each case we can find a common structure for all barcodes, which is similar to the barcode of the original map.

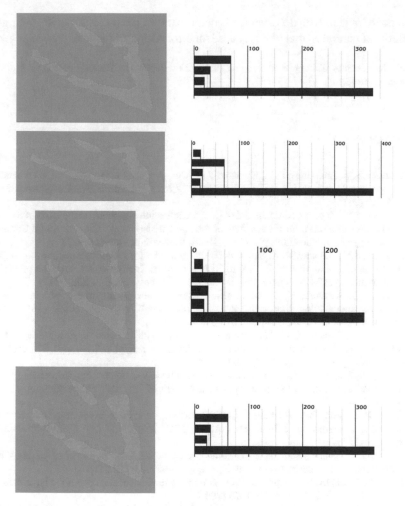

Fig. 16 Computer modeling of different deformations of the island map (left) and topological analysis results (right)

5   Conclusions

Algorithms for collecting information from a map, dividing the space into triangles, composing structures and forming a barcode were developed in the article. Any area object on the map can be represented as a point or a contour and can also be considered separately. The approach for selecting abstract and repeating structures on the map is shown. The study was conducted for a set of point and polygon objects. The analysis of repeating structures is based on information from the barcode. Barcode is built using persistent homology methods. Computer modeling to analyze the structure of a sequence of objects with deformation is demonstrated. Examples for determining the similarity of natural spatial objects based on information from the barcode are given.

Acknowledgements The reported study was funded by RFBR and Vladimir region according to the research project no. 17-47-330387.

References

1. Eremeev, S.V., Andrianov, D.E., Komkov, V.A.: Comparison of urban areas based on database of topological relationships in geoinformational systems. Pattern Recogn. Image Anal. **25**(2), 314–320 (2015)
2. Eremeev, S.V., Kuptsov, K.V.: Spatial objects classification algorithm on the basis of topological features of a form. In: Proceedings of the 2nd Russian-Pacific Conference on Computer Technology and Applications, pp. 44–48. IEEE, Vladivostok, Russia (2017)
3. Eremeev, S.V., Kuptsov, K.V., Seltsova, E.A.: Algorithm for selecting homogeneous regions from a set of spatial objects. In: XX IEEE International Conference on Soft Computing and Measurements (SCM 2017), pp. 109–113. IEEE, Saint-Petersburg, Russia (2017)
4. Saidala, R.K., Devarakonda, N.: Multi-swarm whale optimization algorithm for data clustering problems using multiple cooperative strategies. Int. J. Intell. Syst. Appl. (IJISA) **10**(8), 36–53 (2018). https://doi.org/10.5815/ijisa.2018.08.04
5. Anuradha, J., Tripathy, B.K.: Hierarchical clustering algorithm based on attribute dependency for attention deficit hyperactive disorder. Int. J. Intell. Syst. Appl. (IJISA) **6**(6), 37–45 (2014). https://doi.org/10.5815/ijisa.2014.06.04
6. Hu, Z., Bodyanskiy, Y.V., Tyshchenko, O.K., Samitova, V.O.: Fuzzy clustering data given in the ordinal scale. Int. J. Intell. Syst. Appl. (IJISA) **9**(1), 67–74 (2017). https://doi.org/10.5815/ijisa.2017.01.07
7. Hu, Z., Bodyanskiy, Y.V., Tyshchenko, O.K., Tkachov, V.M.: Fuzzy clustering data arrays with omitted observations. Int. J. Intell. Syst. Appl. (IJISA) **9**(6), 24–32 (2017). https://doi.org/10.5815/ijisa.2017.06.03
8. Al-Hasan, M., Salem, S., Zaki, M.J.: SimClus: an effective algorithm for clustering with a lower bound on similarity. Knowl. Inf. Syst. **28**(3), 665–685 (2011)
9. Pham, D.T., Dimov, S.S., Nguyen, C.D.: A two-phase K-means algorithm for large datasets. Mech. Eng. Sci. J. **218**(10), 1269–1273 (2004)
10. De Almeida, J., Morley, J., Dowman, I.: A graph theory in higher order topological analysis of urban scenes. Comput. Environ. Urban Syst. **31**(4), 426–440 (2007)
11. Edelsbrunner, H., Harer, J.: Computational Topology: An Introduction. American Mathematical Society (2009)
12. Carlsson, G.: Topology and data. Bull. Am. Math. Soc. **46**, 255–308 (2009)
13. Kelin, X., Zhixiong, Z.: Multiresolution persistent homology for excessively large biomolecular datasets. J. Chem. Phys. **143**, 5–13 (2015)
14. Kurlin, V.: A one-dimensional homologically persistent skeleton of an unstructured point cloud in any metric space. Comput. Graph. Forum **34**(5), 253–262 (2015)
15. Kalisnik, S.: Tropical coordinates on the space of persistence barcodes. Max Planck Inst. Math. Sci. **3**, 12–28 (2017)

Code Biology and Kolmogorov Complexity

Sergei V. Kozyrev

Abstract We discuss Code Biology in relation to approach "Kolmogorov complexity as energy" by Yuri Manin. Scaling laws in genomics are discussed in relation to Zipf's law. It is shown that Zipf's law implies scaling for sizes of families of orthologous genes. Relation of approach "complexity as energy" and regularization by complexity minimization in machine learning is discussed.

Keywords Kolmogorov complexity · Complexity as energy · Third evolutionary synthesis

1 Introduction

In this paper, we discuss the approach "Kolmogorov complexity as energy" by Manin [1] in application to genomics. We follow in general the construction of [2].

Code Biology approach [3] investigates multiplicity of codes in biological systems. These codes include genetic code: amino acid code, genomes; epigenetic codes: histone code, methylation code, splicing codes, chromatin codes; codes in multicellular organisms and bacterial colonies; and codes in neural sciences. Discussion of the Code Biology and different biological codes can be found in different papers in the issue of BioSystems [3]. Error correcting codes were discussed in relation to neural sciences in [4]. Something was discussed in [5–7].

The following properties of codes were discussed in Code Biology:

A code is a mapping between objects of two worlds implemented by objects of a third world.

A code is a pattern in a sequence which corresponds to some specific biological function.

The set of maps (codes) should be rich enough to generate a rich set of biological functions.

S. V. Kozyrev (✉)
Steklov Mathematical Institute of Russian Academy of Sciences, Moscow, Russia
e-mail: kozyrev@mi.ras.ru

© Springer Nature Switzerland AG 2020   93
Z. Hu et al. (eds.), *Advances in Artificial Systems for Medicine and Education II*,
Advances in Intelligent Systems and Computing 902,
https://doi.org/10.1007/978-3-030-12082-5_9

Codes are generated in the process of evolution—more complex organisms use more codes.

Code Biology is related to approach "Genome as a Program". In particular, in this approach, genetic code (sequence of nucleotides in a genome) is an algorithm and epigenetic code describes gene regulation, i.e., which part of a genome can be processed and which genes can be expressed. There are several forms of epigenetic code—methylation code, histone code, and chromatin codes which encode the information about genetic regulation. In particular, for a model of algorithm by a Turing machine, i.e., a strip of tape with symbols and a head with set of states, head performs movements, reading, and modification of tape, and the genomic analogue is as follows. Genome is a strip of tape with symbols (nucleotides), and the head of a Turing machine (state and position of the head) is described by epigenetic code. Since several genes can be expressed simultaneously, one can consider genome as a Turing machine with several heads.

We will use the notion of constructive world as a mathematical model of Code Biology.

Constructive Worlds An (infinite) constructive world is a countable set X given together with a class of structural numberings: computable $u\colon \mathbb{Z}_+ \to X$. Natural maps between constructive worlds should be given by computable functions. One can also consider structural numberings by (finite) sequences of bits.

Kolmogorov Complexity and Kolmogorov Order Let us recall the construction by Kolmogorov [8, 9].

Let us consider a set X (a constructive world) generated by "programs" (sequences of bits).

For $x, y \in X$, we consider the conditional entropy (complexity) as the minimal length of program p (in bits) satisfying

$$K_A(x|y) = \min_{A(p,y)=x} l(p). \tag{1}$$

i.e., program p computes x starting from y. Here, A is a "way of programming."

Unconditional complexity is given by application of the above definition to some "initial" object y_0

$$K_A(x) = K_A(x|y_0).$$

Logarithmic Kolmogorov complexity of x is the length (in bits) of the shortest program which generates x. In particular, there exists such way of programming A that for each other (semi)-computable B, some constant $c_{AB} > 0$, and all $x \in X$, one has

$$K_A(x) \le K_B(x) + c_{AB}.$$

Here, $K_A(x)$ is the logarithmic Kolmogorov complexity of x. Way of programming A is called optimal Kolmogorov numbering.

A Kolmogorov order in a constructive world X is a bijection $X \to \mathbb{Z}_+$ arranging elements of X in the increasing order of their complexities K_A.

Any optimal numbering is only partial function, and its definition domain is not decidable.

Kolmogorov complexity K_A is not computable. It is the lower bound of a sequence of computable functions. Kolmogorov order is not computable as well.

Zipf's Law and Kolmogorov Order Zipf's law describes frequencies of words of a natural language in texts. If all words w_k of a language are ranked according to decreasing frequency of their appearance in a corpus of texts, then the frequency p_k of w_k is approximately inversely proportional to its rank k: $p_k \sim k^{-1}$.

Zipf: This distribution "minimizes effort."

It is easy to see that Gibbs distribution with energy proportional to $\log k$ gives a power law.

A mathematical model of Zipf's law is based upon two postulates:

A.   Rank ordering coincides with the Kolmogorov ordering.
B.   The probability distribution producing Zipf's law is $\exp(-\beta K(w))$ with the inverse temperature $\beta = 1$.

For majority of natural numbers, Kolmogorov complexity of n is close to $\log n$. Hence, $\beta = 1$ is the point of phase transition (it separates divergent and convergent statistical sums with energy equal to logarithmic Kolmogorov complexity).

2   Zipf's Law and Scaling in Genomics

In the present section, we discuss a model of biological evolution based on constructive statistical mechanics and relation of scaling in genomics and Manin's approach to Zipf's law described in the previous section.

The following universal statistical properties were observed in genomics [10, 11]:

(1)   log-normal distribution of the evolutionary rates between orthologous genes;
(2)   power law—like distributions of membership in paralogous gene families;
(3)   scaling of functional classes of genes with genome size; and
(4)   Some of these observations can be compared to Zipf's law.

In [10] Eugene, Koonin proposed to develop a statistical mechanical model of biological evolution based on gene ensembles. This hypothetical model was based on the idea about interacting gas of genes and called "the third evolutionary synthesis" (the first is Darwinism, the second is Darwinism plus genetics, and the third should generalize Darwinism with genomics data).

In [2], a model of statistical mechanics in genomics was proposed as possible model of the third evolutionary synthesis. Instead of interaction of genes, the scaling in genomics was discussed as a result of the presence of contribution of complexity in energy (the evolutionary effort to generate a sequence).

Weighted Logarithmic Complexity as Evolutionary Effort Let us consider a structure of constructive world in the set of genomic sequences.

We start with a finite set S of sequences (genes, regulatory sequences, etc.) and a finite set O of genome editing operations which contain operations of gluing together sequences and operations similar to typical evolutionary transformations (point mutations, insertions, deletions, duplications of parts of a sequence, etc.).

To elements $s_i \in S$ and $o_j \in O$, we put in correspondence positive numbers $w(s_i)$ and $w(o_j)$, called scores (or weights), and to any sequence s obtained from elements in S by applications of operations in O, we put in correspondence score $w(s)$ equal to a sum of scores of elements $s_i \in S$ and operations $o_j \in O$: Composition of o_j generates the sequence s starting from s_i.

A sequence s can be obtained in this way nonuniquely—in particular, one can insert a symbol at a given position and then delete this symbol leaving s unchanged.

Let us define complexity of s as a minimum over possible compositions of operations $o_j \in O$ and elementary sequences $s_i \in S$ giving s

$$K_{SOW}(s) = \min_{A(s_1,\dots,s_n)=s} \left[\sum_i w(s_i) + \sum_j w(o_j) \right] \tag{2}$$

where A is a (finite) composition of o_j applied to sequences $s_1, \dots, s_n$.

Complexity $K_{SOW}(s)$ can be considered as a weighted number of genes and edit operations generating sequence s. Logarithmic complexity can be (approximately) discussed as a number of edit operations generating element of a constructive world, the above complexity (2) gives weighted version of estimate from above for logarithmic Kolmogorov complexity (in particular, this estimate from above should be computable).

Conditional version $K_{SOW}(s'|s)$ of weighted complexity (2) has the form

$$K_{SOW}(s'|s) = \min_{A(s_1,\dots,s_n)[s]=s'} \left[\sum_i w(s_i) + \sum_j w(o_j) \right] \tag{3}$$

where we generate sequence s' starting from sequence s. Here, $A(s_1, \dots, s_n)[s]$ is a combination of genome editing operations containing sequences $s_1, \dots, s_n$ applied to s.

Complexity as Energy in Biological Evolution Let us consider a statistical mechanical system where states s are sequences, generated as above, and the statistical sum is equal to

$$Z = \sum_s e^{-\beta H(s)}, \quad H = H_F + H_K \tag{4}$$

where β is the inverse temperature, and the Hamiltonian contains two contributions, the contribution $H_F(s)$ describes biological fitness of sequence s and the contribution $H_K(s)$ describes complexity of s (in particular, one can take $H_K = K_{SOW}$). Here, good fitness corresponds to low $H_F(s)$ (potential wells on the fitness landscape). The symbol K in H_K is for Kolmogorov (complexity).

This statistical sum describes a model of constructive statistical mechanics (statistical mechanics in the constructive world of generated sequences).

The contribution $H_K(s)$ in the energy describes the evolutionary effort to generate sequence s (sequences with less evolutionary effort are more advantageous).

We consider this constructive statistical system as a model of the third evolutionary synthesis.

Remark Using (2) and (4), the scaling in genomics, in particular power law—like distributions of membership in paralogous gene families, can be discussed as follows. Let us recall that paralogous genes are genes in the same genome generated by duplication events.

Let us assume that if the genome contains a paralogous family of genes with N elements, the Kolmogorov rank of this genome should be proportional to N (since the complexity (2) in this case will contain N contributions $w(s_i)$ for some gene s_i). Then, by Zipf's law, contribution of this genome to the statistical sum will be proportional to a (negative) degree of N which gives the power law.

3 Conditional Complexity and Alignment of Sequences

Actually, some methods of constructive mathematics are already applied in bioinformatics. In particular, alignment of sequences is a particular case of application of the conditional complexity.

Alignment of sequences. Let us consider the set X of sequences (genetic information). Let us consider the weighted conditional complexity (3) where the way of programming is given by combinations of insertions, deletions, and symbol substitutions. In this case, (3) will give a weight of alignment and the program $A(s_1, \ldots, s_n)[s]$ in (3) is alignment.

The standard definition of alignment is as follows [12].

Let A be a k-letter alphabet, $A' = A \bigcup \{-\}$ be expanded alphabet where $\{-\}$ is a space symbol.

Let V, W be finite sequences of symbols in A.

Alignment of two sequences $V = v_1 \ldots v_n$ and $W = w_1 \ldots w_m$ is a matrix with two lines of equal length $l \geq n, m$, the first line is a sequence $\tilde{V} = \tilde{v}_1 \ldots \tilde{v}_l$ obtained from V by insertion of $l - n$ spaces and the second line is a sequence $\tilde{W} = \tilde{w}_1 \ldots \tilde{w}_l$ obtained from W by insertion of $l - m$ spaces. Columns with two spaces are forbidden.

Columns with spaces in the first line are called insertions, and columns with spaces in the second line are called deletions.

Columns with equal symbols in both lines are called matches, and columns with two different symbols are called mismatches.

Score of a column is a real number (depends on symbols in the column). Score of alignment is a sum of scores of columns

$$\delta(\tilde{V}, \tilde{W}) = \sum_{i=1}^{l} \delta(\tilde{v}_i, \tilde{w}_i).$$

Example Scores $\delta(x, x) = 0$ for matches, $\delta(x, y) = \mu > 0$ for mismatches, and $\delta(x, -) = \delta(-, x) = \sigma > 0$ for insertions and deletions:

$$\text{score (alignment)} = \mu\#(\text{mismatches}) + \sigma\#(\text{indels})$$

(lower score—better alignment).

Alignment $(\tilde{V}, \tilde{W})$ of sequences V, W corresponds to a combination of editing operations of sequence V which converts V–W. Editing operations correspond to columns of the alignment matrix $(\tilde{V}, \tilde{W})$ and can be performed in arbitrary order. Insertion operation is the insertion of space in sequence V (at the position corresponding to the column), deletion operation is the deletion of the corresponding symbol in V, and mismatch operation is the substitution of a symbol in V by the corresponding symbol in W.

4   Evolution and Machine Learning

Coding Biology studies different codes in biological systems. In particular, to generate an animal, a multiplicity of codes should be used: genetic code, epigenetic codes, splicing code at the cell level, codes at the level of multicellular organism, and neural codes. One can compare this with much simpler situation—generative models in deep learning. Deep generative model allows to generate, for example, a picture of dog using a set of primitives. This fits in the framework of coding theory, of course a picture of dog is much simpler than the dog itself, and the corresponding model of coding theory will be much simpler as well.

Application of ideas of machine learning in biology is a natural approach. A problem of biological evolution can be considered as a problem of learning where genomes learn in the process of natural selection. In particular, the contribution $H_F(s)$ in (4) (biological fitness of sequence s) can be modeled by the functional of empirical risk (number of errors on a training set)

$$H_F(s) = R_{\text{emp}}(s) = \frac{1}{l} \sum_{j=1}^{l} (y_j - f(v_j, s))^2, \quad (y_1, v_1), \ldots, (y_l, v_l), \quad y_j \in 0, 1.$$

Here, genome s generates a classifier $f(v, s)$ which models biological function—it recognizes the situation v and classifies it (gives 0 or 1 for $f(v, s)$). Here, $(y_1, v_1), \ldots, (y_l, v_l)$ is the training set.

An important subject in theory of machine learning is Vapnik–Chervonenkis theory (or VC theory) [13] which states that a classifier can be taught if the family of classifiers has sufficiently low VC entropy (which is some kind of complexity). Minimization of entropy of classifiers in machine learning is performed by addition to the functional of empirical risk of some contribution which describes some kind of complexity of a classifier. In particular, for the support vector machine, this contribution reduces to a weighted sum of squares of coefficients of separating hyperplane (margin maximization).

The most fundamental approach to complexity is the definition of Kolmogorov complexity. In this sense, minimization of Kolmogorov complexity can be considered as the ultimate regularizer in machine learning. Therefore, the presence of the complexity contribution H_K in Hamiltonian (4) of evolution (where this contribution describes evolutionary effort) can be considered as a regularization by low Kolmogorov complexity in the model of learning by evolution. Let us note that Kolmogorov complexity (and considered in (2) estimates from above for Kolmogorov complexity) is different from VC entropy and gives an alternative regularization suitable for generative models in constructive worlds.

Minimization of complexity could also be important for deep learning where deep learning networks are considerably simpler (contain less nodes) in comparison with shallow learning networks.

5 Conclusion

In the present paper, we show that Zipf's law implies scaling for sizes of families of orthologous genes in genomes. Zipf's law is considered as a result of statistical mechanical model (Gibbs distribution where Hamiltonian contains contribution given by estimate for Kolmogorov complexity).

There are several point of views on complexity in biology. The first point of view was discussed by Schroedinger [14]: At molecular level, living systems have comparably low entropy (and therefore are simple in comparison with disordered systems). E. Schroedinger used the term "aperiodic crystal." The opposite point of view is common and states that complexity of living systems is large compared to more standard examples in physics.

Approach to biological evolution from point of view of physics was discussed by Koonin [10, 11], and it was proposed to explain scaling laws observed in genomics by some model of statistical mechanics which uses "interacting gas of genes." This hypothetical model was called "the third evolutionary synthesis."

Yu. I. Manin proposed to apply model of statistical mechanics, where logarithmic Kolmogorov (or algorithmic) complexity plays a role of energy [1], to explain Zipf's scaling law observed in linguistics.

In this paper, we have discussed a statistical mechanical model of biological evolution where Hamiltonian contains a contribution given by weighted estimate from above for Kolmogorov complexity of a genome, generated from a set of genes by editing operations. In some sense, this is a combination of Manin's and Koonin's approaches. In this approach, scaling in genomics is explained by the same mechanism as Zipf's law. Minimization of complexity resembles Schroedinger's statement.

Another useful point of view to biological evolution is related to theory of machine learning. In a typical problem of learning, one has to minimize a functional which contains two contributions: The first contribution is a functional of empirical risk (number of errors of classifier on a training set) and the second contribution controls complexity of a family of classifiers to avoid overfitting. In application to evolution, empirical risk describes biological fitness (lower the risk better the fitness), and the regularization (estimate for Kolmogorov complexity of genomic sequence) describes evolutionary effort to generate a sequence. Simultaneous presence of both contributions makes biological evolution a problem of learning.

Acknowledgments This work is supported by the Russian Science Foundation under grant 17-71-20154.

References

1. Manin, Y.I.: Complexity vs energy: theory of computation and theoretical physics. J. Phys. Conf. Ser. 532. Paper: 012018 (2014). arXiv:1302.6695
2. Kozyrev, S.V.: Biology as a constructive physics. **2018**, 1–10 (2018) arXiv:1804.10518
3. Barbieri, M.: What is code biology? BioSystems **164**, 1–10 (2018)
4. Manin, Yuri I.: Error-correcting codes and neural networks. Sel. Math. New Ser. **24**, 521–530 (2018)
5. Zhilyakova, L.Y.: Graph dynamic threshold model resource network: key features. Int. J. Math. Sci. Comput. (IJMSC) **3**(3), 28–38 (2017). https://doi.org/10.5815/ijmsc.2017.03.03
6. Katunin, A.: Construction of fractals based on catalan solids. Int. J. Math. Sci. Comput. (IJMSC) **3**(4), 1–7 (2017). https://doi.org/10.5815/ijmsc.2017.04.01
7. Rather, N.N., Patel, C.O., Khan, S.A.: Using deep learning towards biomedical knowledge discovery. Int. J. Math. Sci. Comput. (IJMSC) **3**(2), 1–10 (2017). https://doi.org/10.5815/ijmsc.2017.02.01
8. Kolmogorov, A.N.: Three approaches to the definition of the concept "quantity of information". Probl. Peredachi Inf. **1**(1), 3–11 (1965). (in Russian)
9. Kolmogoroff, A.: On the logical foundations of information theory and probability theory. Probl. Peredachi Inf. **5**(3), 3–7 (1969, in Russian); Translation: Probl. Inf. Trans. **5**(3), 1–4 (1969)
10. Koonin, E.V.: The logic of chance: the nature and origin of biological evolution. FT Press, USA (2012)

11. Koonin, E.V.: Are there laws of genome evolution?. PLoS Comput. Biol. **7**(8). Paper: e1002173 (2011)
12. Jones, N.C., Pevzner, P.A.: An introduction to bioinformatics algorithms. MIT Press, USA (2004)
13. Vapnik, V.N.: The nature of statistical learning theory. Springer, Germany (1995)
14. Schroedinger, E.: What Is Life? The Physical Aspect of the Living Cell. Cambridge University Press, United Kingdom (1944)

On Genetic Unitary Matrices and Quantum-Algorithmic Genetics

Sergey V. Petoukhov and Elena S. Petukhova

Abstract The article is devoted to parallelisms between the structural organization of the molecular genetic system and mathematics of quantum computers. This mathematics is perspective for developing systems of artificial intelligence. Using this mathematics to study, the genetic system can reveal bioinformational patents of the living nature for scientific and technologic progress. Unitary matrices are the basis of all calculations in quantum computers. Our described results show that structured systems of molecular genetic alphabets in their matrix forms of representations can be considered as sets of sparse unitary matrices related with phenomenologic features of the degeneracy of the genetic code. These sparse unitary matrices have orthogonal systems of functions in their rows and columns. A complementarity exists among some unitary genetic matrices in relation to each other. Tensor (or Kronecker) families of unitary genetic matrices with their fractal-like properties are also considered. These new results are interesting for development of quantum-algorithmic genetics and for revealing fundamental principles of bioinformatics.

Keywords Genetic code · Unitary matrix · Quantum computing · Algorithms

1 Introduction

The information molecules of DNA and RNA of the genetic system belong to the world of molecules, in which the principles of quantum mechanics manage. Correspondingly, study of informational molecular genetic structures can reveal their connection with algorithms of quantum computers, which are perspective for systems of artificial intelligence [1] and for many applications in different fields, including medical engineering and education. All calculations in quantum computers are based on using unitary (or orthogonal) operators, which are represented by unitary

S. V. Petoukhov (✉) · E. S. Petukhova
Mechanical Engineering Research Institute, Russian Academy of Sciences,
M. Kharitonievsky Pereulok, 4, Moscow, Russia
e-mail: spetoukhov@gmail.com

© Springer Nature Switzerland AG 2020
Z. Hu et al. (eds.), *Advances in Artificial Systems for Medicine and Education II*,
Advances in Intelligent Systems and Computing 902,
https://doi.org/10.1007/978-3-030-12082-5_10

(or orthogonal) matrices and which serve in quantum computers as quantum logic gates. Each of unitary matrices can be used as a logic gate in quantum computers [2, p. 18]. Taking this into account, the following results described in this article are interesting: Structured alphabets of DNA and RNA texts in their matrix representations have close connections with unitary matrices. Here, one should remind that matrix forms of representations for genetic alphabets are used in matrix genetics in line with our concept of resonance genetics, which assumes the key role of resonances in organization of genetic systems; the mathematical theory of resonances of vibration systems is based on using matrices [3, 4]. The goal of our research is a revealing new analogies between the structural organization of molecular genetic systems and mathematics of quantum computing for the development of quantum-algorithmic genetics, elements of which have been described in [5–7].

2   Matrix Representations of DNA Alphabets and Genetic Binary Oppositions

Science does not know why the basic alphabet of DNA has been created by Nature from just four letters (adenine A, thymine T, cytosine C and guanine G), and why just these very simple molecules were chosen for the DNA alphabet (out of millions of possible molecules). But science knows [8–12] that these four molecules are interrelated due to their symmetrical peculiarities into the united molecular ensemble with its three pairs of binary-oppositional traits or indicators:

(1)   Two letters are purines (A and G), and other two are pyrimidine's (C and T). From the standpoint of these traits, one can denote $C = T = 0$, $A = G = 1$;
(2)   Two complementary letters (A and T) have two hydrogen bonds, and other complementary letters (C and G) have three bonds. From the standpoint of these traits $A = T = 0$, $C = G = 1$;
(3)   Two letters are keto (G and T), and two other letters are amino (A and C). From the standpoint of these traits $A = C = 0$, $G = T = 1$.

One can check that each of DNA letters A, C, G and T is uniquely determined by any two of these three indicators. It is convenient to represent DNA alphabets of 4 nucleotides, 16 doublets, 64 triplets, etc., in a form of appropriate square tables (Fig. 1), which rows and columns are numerated by binary symbols in line with the following principle. Entries of each column are numerated by binary symbols in line with the binary-oppositional indicators "pyrimidine-or-purine" (for example, the triplet CAG and all other triplets in the same column are the combination "pyrimidine-purin-purin" and so this column is numerated 011). By contrast, entries of each of rows are numerated by binary numbers in line with the second set of indicators (for example, the same triplet CAG and all other triplets in the same row are the combination "amino-amino-keto" and so this row is numerated 001). In such tables (Fig. 1), each of 4 letters, 16 doublets, 64 triplets, etc., takes automatically its own individual place and all components of the alphabets are arranged in a strict order.

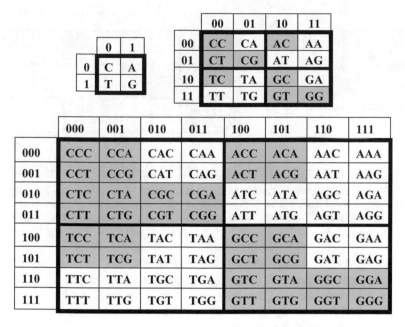

Fig. 1 The square tables of DNA alphabets of 4 nucleotides, 16 doublets and 64 triplets with a strict arrangement of all components in line with binary-oppositional traits of the nitrogenous bases (see explanations in the text)

It is essential that these 3 separate genetic tables form the joint tensor family of matrices since they are interrelated by the known operation of the tensor (or Kronecker) product of matrices [11, 12]. So, they are not simple tables but matrices. The second tensor power of the (2 * 2)-matrix [C, A; T, G] gives automatically the (4 * 4)-matrix of 16 doublets; the third tensor power of the same (2 * 2)-matrix gives the (8 * 8)-matrix of 64 triplets with the same strict arrangement of entries as in Fig. 1. In this tensor construction of the tensor family of genetic matrices, data about binary-oppositional traits of genetic letters C, A, T and G are not used at all. So, the structural organization of the system of DNA alphabets is connected with the algebraic operation of the tensor product. It is important since the operation of the tensor product is well known in mathematics, physics and informatics, where it gives a way of putting vector spaces together to form larger vector spaces. The following quotation speaks about the crucial meaning of the tensor product: *This construction is crucial to understanding the quantum mechanics of multiparticle systems* [2, p. 71]. For us, the most interesting is that the tensor product is one of basic instruments in quantum informatics.

As is known, the degeneracy of the genetic code has the important specificity: The entire set of 64 triplets is divided by Nature into 2 equal binary-opposition subsets [13]: 32 triplets with "strong roots" (black colors in Fig. 1), i.e., with 8 "strong" doublets AC, CC, CG, CT, GC, GG, GT, TC; 32 triplets with "weak roots" (white

V =

1	-1	1	-1
1	1	-1	-1
1	-1	1	-1
-1	-1	1	1

W =

1	1	-1	-1	1	1	-1	-1
1	1	-1	-1	1	1	-1	-1
1	1	1	1	-1	-1	-1	-1
1	1	1	1	-1	-1	-1	-1
1	1	-1	-1	1	1	1	1
1	1	-1	-1	1	1	-1	-1
-1	-1	-1	-1	1	1	1	1
-1	-1	-1	-1	1	1	1	1

Fig. 2 Numeric representations V and W of the genetic matrices of 16 doublets and 64 triplets from Fig. 1. Matrix cells with number $+1$ (-1) correspond cells with black (white) doublets and triplets in Fig. 1. Each of rows of the numeric matrices represents one of Rademacher–Walsh functions

colors in Fig. 1), i.e., with 8 "weak" doublets CA, AA, AG, AT, GA, TA, TG, TT. Code meanings of triplets with strong roots do not depend on the letters on their third position; code meanings of triplets with weak roots depend on their third letter. What are locations of these strong (black) and weak (white) members of DNA alphabets in the genetic matrices shown in Fig. 1?

The unexpected phenomenological fact is a symmetrical location (Fig. 1) of all black and white entries in the genetic matrices of 16 doublets and 64 triplets, which were constructed very formally without any mention about amino acids and the degeneracy of the genetic code. Symmetrical properties of mosaics in the genetic matrices in Fig. 1 are the following:

(1) The left and right halves of the matrix mosaic are mirror anti-symmetric each to other in its colors;
(2) The block mosaic of the matrix has the cruciform character: Both quadrants along each diagonals are identical each other from the standpoint of their mosaic;
(3) Mosaic of each of rows has the meander character identical to known Rademacher functions, which are particular cases of Walsh functions and contain only values $+1$ and -1.

Using this analogy with Rademacher–Walsh functions, one can represent the symbolic genetic matrices in Fig. 1 in forms of numeric matrices V and W with their entries $+1$ and -1 in Fig. 2 where numbers $+1$ (-1) represent black (white) doublets and triplets correspondingly. Taking into account that meander-like mosaics of rows of matrices V and W correspond Rademacher functions, we conditionally called these matrices "Rademacher matrices" in all our publications beginning from the book [11] (although Rademacher himself never worked with such matrices).

It should be noted that a huge quantity $64! \approx 10^{89}$ of variants exists for dispositions of 64 triplets in a separate (8 * 8)-matrix. For comparison, the modern physics estimates time of existence of the Universe in 10^{17} s. It is obvious that an accidental disposition of black and white triplets in a separate (8 * 8)-matrix will give almost never any symmetry. But in our approach, this matrix of 64 triplets is not a separate

$$
\begin{vmatrix} 1 & -1 & 1 & -1 \\ 1 & 1 & -1 & -1 \\ 1 & -1 & 1 & -1 \\ -1 & -1 & 1 & 1 \end{vmatrix}
=
\begin{vmatrix} 1&0&0&0 \\ 0&1&0&0 \\ 0&0&1&0 \\ 0&0&0&1 \end{vmatrix}
+
\begin{vmatrix} 0&-1&0&0 \\ 1&0&0&0 \\ 0&0&0&-1 \\ 0&0&1&0 \end{vmatrix}
+
\begin{vmatrix} 0&0&0&-1 \\ 0&0&-1&0 \\ 0&-1&0&0 \\ -1&0&0&0 \end{vmatrix}
+
\begin{vmatrix} 0&0&1&0 \\ 0&0&0&-1 \\ 1&0&0&0 \\ 0&-1&0&0 \end{vmatrix}
$$

Fig. 3 The dyadic-shift decomposition of the matrix V (Fig. 2) into the sum of 4 sparse matrices: $V = V_0 + V_1 + V_2 + V_3$, where V_0 is the identity matrix

matrix, but it is one of members of the family of matrices of genetic alphabets interrelated by means of binary-oppositional traits of nitrogenous bases A, T, C, G (and also it is one of members of the tensor family of matrices [C, A; T, G]^(n) of interrelated alphabets of DNA).

These numeric matrices V and W with their mosaics (Fig. 2) represent the phenomenological peculiarities of the degeneracy of the genetic code. The exponentiation of these genetic matrices in the second power leads to their doubling and quadrupling: $V^2 = 2 * V$ and $W^2 = 4 * W$. This resembles the doubling and quadrupling the genetic material under mitosis and meiosis of biological cells. Let us analyze algebraic properties of these genetic matrices V and W more deeply.

3 The Genetic Matrices and Sparse Unitary Matrices

We begin the algebraic analysis of the (4 * 4)-matrix V in Fig. 2. Figure 3 shows the decomposition of this matrix into a sum of 4 sparse matrices: $V = V_0 + V_1 + V_2 + V_3$ (this decomposition is not arbitrary but constructed on the principle of dyadic-shift decompositions known in technology of digital signal processing [5, 6, 11]).

By definition, a complex square matrix U is unitary if its conjugate transpose $U^\dagger$ is also its inverse: $UU^\dagger = I$, where I is the identity matrix (its conjugate transpose $U^\dagger$ is also its inverse matrix U^{-1}). The real analogue of a unitary matrix is an orthogonal matrix, for which the conjugate transposition $U^\dagger$ is identical to the ordinary transposition: $UU^T = I$. In this article, we consider only the case of real square matrix. Unitary matrices have significant importance in quantum mechanics because they preserve norms, and thus, probability amplitudes. The tensor product of two unitary matrices always generates a unitary matrix [14, p. 38].

It is interesting that each of sparse matrices V_0, V_1, V_2 and V_3 are unitary (or orthogonal since their entries are real):

$$V_0 * V_0^T = I, \quad V_1 * V_1^T = I, \quad V_2 * V_2^T = I, \quad V_3 * V_3^T = I. \tag{1}$$

This fact testifies that the degeneracy of the genetic code is connected with sparse unitary matrices, which can serve as logic quantum gates in supposed biological quantum computers.

Fig. 4 The multiplication
table of the unitary genetic
matrices V_0, V_1, V_2 and V_3
from Fig. 2

*	V0	V1	V2	V3
V0	V0	V1	V2	V3
V1	V1	- V0	V3	- V2
V2	V2	- V3	V0	- V1
V3	V3	V2	V1	V0

In molecular genetic systems, relations of complementarity play the very important role. The book [15, Chap. 1] notes that the proof of the complementary structure
of the bases in DNA has led to the most fundamental discoveries in modern biology:
This complementarity provides the most important properties of DNA as a carrier
of genetic information, including DNA replication in the course of cell division and
also all mechanisms of manifestation of genetic information. But one can note that
the set of unitary genetic matrices V_0, V_1, V_2 and V_3 (Fig. 2) contains the following
algebraic complementarities in their pairs: unitary matrices V_0 and V_2 form the first
pair of the algebraic complementarity since they are transformed into each other by
the mirror reflection relative to the average vertical line with simultaneous inversion
of signs of their non-zero entries (the mirror anti-symmetry). The same is true for
unitary matrices V_1 and V_3, which form the second pair with the similar algebraic
complementarity of the mirror anti-symmetric type. The degeneracy of the genetic
code is connected with such algebraic complementarities in the set of unitary genetic
matrices.

Determinants of all the unitary matrices V_0, V_1, V_2 and V_3 are equal to 1; by this
reason, these matrices belong to the type of so-called special unitary matrices. The
special unitary matrices are closed under multiplication and the inverse operation,
and therefore form a matrix group called the special unitary group [16].

The table of multiplication of the closed set of genetic unitary matrices V_0, V_1,
V_2 and V_3 is shown in Fig. 4. It coincides with the known multiplication table of the
algebra of split-quaternions by J. Cockle, which are used in mathematics and physics
including the Poincare disk model of hyperbolic geometry of Lobachevsky [17, 18].
The matrices V_2 and V_3 are additionally matrices of reflection transformations since
$(V_2)^2$ and $(V_3)^2$ are equal to the identity matrix.

One can mention that the similar dyadic-shift decomposition of the the (8 * 8)-matrix
W (Fig. 2) gives the set of 8 sparse unitary matrices, which is also closed under
multiplication and whose multiplication table coincides with the multiplication table
of bi-split-quaternions of Cockle (see details in the preprint [6]).

Why are DNA and RNA alphabets in their matrix representations connected with
split-quaternions of Cockle? In the whole, it is an open question now. However,
one can note the interesting fact about a connection of split-quaternions with the
non-Euclidean geometry of the space of visual perception. The split-quaternions
are connected with Poincare's disk model (or conformal model) of hyperbolic (or
Lobachevsky) geometry [19]. In accordance with the pioneer work of Luneburg
[20], the space of binocular visual perception is described by hyperbolic geometry

of Lobachevsky. These findings were followed by many papers in various countries, where the idea of a non-Euclidean space of visual perception was extended and refined. The Luneburg approach was thoroughly tested by Kienle [21]. In the main series of his experiments, where about 200 observers were involved, Kienle obtained about 1300 visual patterns of various kinds. The experiments confirmed not only that the space of visual perception is described by hyperbolic geometry but also that the Poincare disk (or conformal) model was an adequate model of that geometry. Kienle concluded his paper by writing: "Poincare's model of hyperbolic space, applied for the first time for a mapping of the visual space, shows a reasonably good agreement with experimental results" [21, p. 399].

The mentioned works of Luneburg and Kienle that the visual space of a human being is characterized by Lobachevsky geometry has attracted special attention of mathematicians and physicists. For example, their results were used in the article with the characteristic title: *Is the Brain a "Clifford Algebra Quantum Computer"?* [22]. This article was devoted to showing that hypercomplex algebras could be used for modeling visual perception in animal organisms to solve the problem of biological *pattern recognition in a multispectral environment in a natural and effective manner. … it is conceivable that nature has through evolution also learned to utilize useful properties of hypercomplex numbers. Thus, the visual cortex might have the ability to function as a Clifford algebra quantum device.* Additional materials to such modeling approach on the basis of hypercomplex algebras, where the visual cortex of a primate brain is considered as a "Fast Clifford algebra quantum computer," are presented in [23] and some other articles of these authors. The works [24, 25] introduced a non-Euclidean metric to describe color perception and constructed a perceptual theory of relativity by analogy with the notion of a space-time manifold in the special theory of relativity. One can mention here also some thoughts from [26, p. 115]: *artificial intelligence could be brought closer to mathematical thinking if it were possible to realize the metrical properties of the human mind. … the consciousness itself is structured geometrically: any person in his existential aspects is geometric. … in our minds, when constructing texts through which we perceive the World, something very similar to what happens in morphogenesis occurs. We are ready to see in the depths of consciousness the same geometric images that are revealed in morphogenesis.*

Our data about the hidden connection of the genetic system with split-quaternions of Cockle support the significance of such algebraic-geometric approaches to understand inherited properties of sensory systems and biological informatics in the whole. One can recall here that many devices of artificial intelligence are created on the basis of digital signal processing, where signals are represented in a form of a sequence of the numeric values of their amplitude in reference points. The theory of signal processing is based on an interpretation of discrete signals as a form of vector of multi-dimensional spaces. In each tact time, a signal value is interpreted as the corresponding value of one of coordinates of a multi-dimensional vector space of signals. In this way, the theory of discrete signals turns out to be the science of geometries of multi-dimensional spaces. The number of dimensions of such a space is equal to the quantity of referent points for the signal. Metric notions and all other necessary things are introduced in these multi-dimensional vector spaces for problems of maintenance of reliability, speed and economy of signal information.

The rows of each of the unitary genetic matrices V_0, V_1, V_2 and V_3 form a complete orthogonal system of functions. The action of each of these matrices on an arbitrary 4-dimensional vector $X = [x_0, x_1, x_2, x_3]$ transforms it into a new vector Y, which can be considered as a spectral representation of the vector X on the basis of the orthogonal system of functions in the rows of the given matrix. The action of the same unitary matrix, taken in its transposed form, on this vector Y restores the original vector X. The exponentiation of each of the matrices V_0, V_1, V_2 and V_3 in a tensor power generates a new unitary matrix with an orthogonal system of functions in its rows and columns.

One can add that each of (2 * 2)-matrices in 4 quadrants of the genetic matrix V is the sum of 2 unitary matrices. Really, the matrix $[1, -1; 1, 1]$, which is repeated in both quadrants along the main diagonal, is the sum of two unitary matrices $U_0 = [1, 0; 0, 1]$ and $U_1 = [0, -1; 1, 0]$; the matrix $[1, -1; -1, -1]$, which is repeated in both quadrants along the secondary diagonal, is also the sum of two unitary matrices $U_2 = [1, 0; 0, -1]$ and $U_3 = [0, -1; -1, 0]$. One of these unitary matrices is the quantum gate $Z = [1, 0; 0, -1]$ well known in quantum computers.

Unitary matrices are used in quantum informatics as quantum logic elements (quantum gates) for performing quantum computations on their basis. In the case of multi-qubit systems, the operation of the tensor product of matrices is of the key importance in connection with the postulate of quantum mechanics: The state space of a composite system is the tensor product of the state spaces of its components. In light of this, it is especially interesting that the entire genetic (4 × 4)-matrix V (Fig. 3) is constructed as the sum of the tensor products of the mentioned 4 unitary (2 * 2)-matrices U_0, U_1, U_2 and U_3 in line with the following expression (2):

$$V = U_0 \otimes U_0 + U_0 \otimes U_1 + U_3 \otimes (-U_2) + U_3 \otimes (-U_3). \qquad (2)$$

The set of these four (2 * 2)-matrices U_0, U_1, $-U_2$ and U_3 is also closed under multiplication; their multiplication table coincides with the multiplication table of split-quaternions by J. Cockle by analogy with the case of unitary (4 * 4)-matrices in Fig. 4. It should be noted that unitary matrices U_0, U_1, U_2 and U_3 have relations with quantum gates used widely in quantum computing [2, p. XXX].

Exponentiation of unitary matrices U_1, U_2 and U_3 into ordinary integer powers $n = 2, 3, 4,\dots$ gives cyclic groups of matrices with the following periods: $U_1^n = U_1^{n+4}$, $U_2^n = U_2^{n+2}$, $U_3^n = U_3^{n+2}$. In this article, we specially note a connection of cyclic groups with algebraic properties of genetic unitary matrices since such cyclic groups can be useful for modeling many inherited cyclic processes in physiology.

Exponentiation of each of unitary matrices U_0, U_1, U_2 and U_3 into tensor (or Kronecker) powers $k = 2, 3, 4, \dots$ generates corresponding tensor families of unitary matrices: $U_0^{(k)}$, $U_1^{(k)}$, $U_2^{(k)}$ and $U_3^{(k)}$ where (k) means the tensor power.

The preprint [6] contains additional data about interesting relations between amino acids and triplets and also about the cruciform principle in inherited sensory systems from the standpoint of unitary genetic matrices.

4   Some Concluding Remarks

The described results about connections of the genetic coding system with unitary matrices attract attention to opportunity of the development of quantum-algorithmic approaches for studying the genetic systems and bioinformatics in the whole. They are interesting by the following main reasons.

Firstly, they give new approaches to model some genetic structures and phenomena on the basis of mathematical formalisms of quantum mechanics and quantum informatics where unitary operators have a key meaning. Correspondingly—from this standpoint—a hidden logic organization of the genetic system should be considered in light of notions of quantum logic. Our results show that, from this modeling standpoint, the genetic system is a whole hierarchical system of interconnected unitary matrices of different orders woven together and formed tensor families of unitary matrices, which can serve as logic gates in biological quantum computers. One should add that quantum-information aspects of life are actively discussed in modern science, for example, in the book [27]; in articles about a possible meaning of the quantum algorithm of Grower in genetic information [28–30].

We suppose that unitary genetic operators (unitary matrices) are the basis for calculations in genetics by some analogy with calculations in quantum informatics. In the frame of our model approach, we put forward the working hypothesis that DNA and RNA sequences of n-plets (of doublets, triplets, etc.) serve to define unitary operators for quantum calculations in genetics by analogy with quantum-logical calculations in quantum computing. From this standpoint, DNA and RNA sequences are instruments to define systems of interconnected unitary operators for quantum calculations by means of the quantum logic. The presented materials about connections of genetic systems with quantum informatics (see additionally [5–7]) can lead to new studies of analogies between quantum physics and matrix representations of the genetic code. Here, one should note that the Hungarian scientist Gyorgy Darvas was the first who—in his study of quantum electrodynamics—paid attention on connections of the genetic numeric matrices with Pauli's matrices [31]. It is additional interesting that cyclic shifts of positions in doublets and triplets transform the mosaic matrices in Fig. 2 into new mosaic matrices [11, 12], which are connected with new systems of unitary genetic matrices.

Secondly, described unitary genetic matrices contain complete orthogonal systems of functions in their rows or columns. But it is known the following: *after Fourier it was found that for some problems, harmonic sinusoids rather than other systems of orthogonal functions, for example, the Legendre polynomials, are better suited. In fact, any particular problem needs its own system of orthogonal functions. This was most clearly manifested in the course of the development of quantum mechanics* [32].

Correspondingly, one can think that the genetic systems have their own orthogonal systems of functions, which should be used in physiology for appropriate spectral decompositions to study genetically inherited processes and structures (including genetic sequences, information processes in neuronal systems and cardio-vascular processes). The received results can be used in many genetic researches, for example [33–35].

Thirdly, described fractal features of the mentioned tensor families of unitary genetic matrices give additional materials to the wide topic of inherited fractal-like structures in biological bodies, including symmetries in long texts of single-stranded DNA [5, 7, 36] and facts about connections of fractals with cancer [37–41]. Fractal patterns are related with the theory of dynamic chaos, which has many applications in sciences and technology (see, for example, [42, 43]). The bridge between knowledges about fractals in information technologies and in bio-information systems can lead to a mutual enrichment of both these fields.

The authors hope that the further usage in genetics the concepts and formalisms of quantum informatics, which was undertaken in this article in the connection with unitary genetic matrices, will lead to the development of substantial quantum-algorithmic genetics. Consideration of biological phenomena, including the phenomena of inheritance of the intellectual abilities of biological bodies, from the standpoint of the theory of quantum computers, gives many valuable opportunities for their comprehension and also for development of artificial intelligence systems [44–48] and of genetic analysis methods [49, 50] (the work [1] contains a review about quantum computing and problems of artificial intelligence). For example, an adult human organism has around 10 trillion (10^{14}) human cells and each of cells contains an identical complect of DNA, whose genetic information is used for physiological functioning organism as the holistic system of cells. How such huge numbers of cells can reliably functioning as a cooperative whole? Associations with quantum computing can help to model and understand such holistic biological systems with their ability of computing complex tasks and transferring genetic information from one generation to another. Quantum-algorithmic approaches allow modeling complex biological systems without using data and hypotheses about interactions between adjacent molecules or between separate biological cells each with other; all of such separate elements are parts of a holistic organism as a quantum-algorithmic essence.

Acknowledgements Some results of this paper have been possible due to a long-term cooperation between Russian and Hungarian Academies of Sciences on the topic "Non-linear models and symmetrologic analysis in biomechanics, bioinformatics, and the theory of self-organizing systems," where S. V. Petoukhov was a scientific chief from the Russian Academy of Sciences. The authors are grateful to G. Darvas, E. Fimmel, M. He, Z. B. Hu, I. Stepanyan and V. Svirin for their collaboration. Special thanks to the German Academic Exchange Service (DAAD) for providing the very useful internship for S. V. Petoukhov in autumn 2017 at the Institute of Mathematical Biology of the Mannheim University of Applied Sciences (Germany) where his host was Prof. E. Fimmel.

References

1. Biamonte, J., Wittek, P., Pancotti, N., Rebentrost, P., Wiebe, N., Lloud, S.: Quantum machine learning. Nature **549**, 195–202 (14 Sept 2017). https://doi.org/10.1038/nature23474
2. Nielsen, M.A., Chuang, I.L.: Quantum Computation and Quantum Information. Cambridge University Press, New York (2010)
3. Balonin, N.A.: New Course on the Theory of Motion Control (Novyi kurs teorii upravleniia dvizheniem). Saint Petersburg State University, Saint Petersburg (2000, in Russian)
4. Gladwell, G.M.L.: Inverse Problems in Vibration. Kluwer Academic Publishers, London, 452 p (2004)
5. Petoukhov, S.V.: The Rules of Long DNA-Sequences and Tetra-Groups of Oligonucleo-Tides. https://arxiv.org/abs/1709.04943 (2017)
6. Petoukhov, S.V.: The Genetic Coding System and Unitary Matrices. Preprints, 2018040131 (2018). https://doi.org/10.20944/preprints201804.0131.v1
7. Petoukhov, S.V., Svirin, V.I.: The new wide class of symmetries in long DNA-texts. Elements of Quantum-Information Genetics. Biologia Serbica **40**(1), 51 (2018). Special Edition, ISSN 2334-6590, UDK 57(051). Book of Abstracts, Belgrade Bioinformatics Conference 2018, 18–22 June, 2018, Belgrade
8. Fimmel, E., Danielli, A., Strüngmann, L.: On dichotomic classes and bijections of the genetic code. J. Theor. Biol. **336**, 221–230 (2013)
9. Hu, Z.B., Petoukhov, S.V.: Generalized crystallography, the genetic system and biochemical esthetics. Struct. Chem. **28**(1), 239–247 (2017). https://doi.org/10.1007/s11224-016-0880-0
10. Hu, Z.B., Petoukhov, S.V., Petukhova, E.S.: I-Ching, dyadic groups of binary numbers and the geno-logic coding in living bodies. Prog. Biophys. Mol. Biol. **131**, 354–368 (Dec 2017). https://doi.org/10.1016/j.pbiomolbio.2017.08.018
11. Petoukhov, S.V.: Matrix Genetics, Algebras of the Genetic Code, Noise Immunity. Moscow, RCD, Russia, 316 p. (2008, in Russian)
12. Petoukhov, S.V., He, M.: Symmetrical Analysis Techniques for Genetic Systems and Bioinformatics: Advanced Patterns and Applications. IGI Global, Hershey, USA (2010)
13. Rumer, Y.B. Systematization of the codons of the genetic code. Dokl. Akad. Nauk SSSR **183**(1), 225–226 (1968, in Russian)
14. Rumer, Y.B., Fet, A.I.: The Theory of Unitary Symmetry (Teoria unitarnoi sim-metrii). Nauka, Moscow (1970, in Russian)
15. Chapeville, F., Haenni, A.-L.: Biosynthese des Proteines. Herman Collection, Paris Methodes (1974)
16. Special unitary matrix. http://mathworld.wolfram.com/SpecialUnitaryMatrix.html
17. Split-quaternion. http://en.wikipedia.org/wiki/Split-quaternion
18. Frenkel, I., Libine, M.: Split quaternionic analysis and separation of the series for SL(2,R) and SL(2,C)/SL(2,R). Adv. Math. **228**, 678–763 (2011)
19. Poincare Disk Model. https://en.wikipedia.org/wiki/Poincar%C3%A9_disk_model
20. Luneburg, R.: The metric of binocular visual space. J. Opt. Soc. Am. **40**, 627–642 (1950)
21. Kienle, G.: Experiments concerning the non-Euclidean structure of visual space, pp. 386–400. In: Bioastronautics. Pergamon Press, New York (1964)
22. Labunets, V., Rundblad, E., Astola, J.: Is the brain a 'Clifford Algebra Quantum Computer'? In: Dorst, L., Doran, C., Lasenby, J. (eds.) Applications of Geometric Algebra in Computer Science and Engineering. Birkhäuser, Boston, MA (2002)
23. Labunets, V., Rundblad, E., Astola, J.T.: Fast calculation algorithms of invariants for color and multispectral image recognition. In: Proceedings of the Conference "Algebraic Frames for the Perception-Action Cycle", AFPAC 2000, Kiel, Germany, Sept 10–11, 2000. https://doi.org/10.1007/10722492_5
24. von Schelling, H.: Die Geometrie des beideaugigen Sehens. Optik, Bd. 17, H. 7, S. 345–364 (1960)
25. von Schelling, H.: Experienced space and time. In: Bioastronautics, pp. 361–385. Pergamon Press, N. Y., L. (1964)

26. Nalimov, V.V.: I Am Scattering Thoughts (in Russian: Razbrasyvaiu mysli). Center for Humanitarian Initiatives; Moscow (2015). ISBN 978-5-98712-521-2
27. Abbott, D., Davies, P.C.W., Pati, A.K. (eds.): Quantum Aspects of Life. Foreword by Sir Roger Penrose (2008). ISBN-13: 978-1-84816-253-2
28. Patel, A.: Quantum algorithms and the genetic code. Pramana J. Phys. **56**(2–3), 367–381 (2001). arXiv:quant-ph/0002037
29. Patel, A.: Testing quantum dynamics in genetic information processing. J. Genet. **80**(1), 39–43 (2001)
30. Patel, A.: Why genetic information processing could have a quantum basis. J. Biosci. **26**(2), 145–151 (2001)
31. Darvas, G., Petoukhov, S.V.: Algebra that demonstrates similitude between transformation matrices of genetic codes and quantum electrodynamics. Presentation to the workshop organised 27.11.2017 at the Hochschule Mannheim (2017). https://www.cammbio.hs-mannheim.de/fileadmin/user_upload/projekte/cammbio/events/20171127-kickoff/EP-Petoukhov.pdf
32. Soroko, L.M.: Walsh's functions in physics and technics. Successes Phys. Sci. (Uspehi Fizicheskih Nauk) **111**(3) (1973)
33. Mousa, H.M: DNA-genetic encryption technique. Int. J. Comput. Netw. Inf. Secur. (IJCNIS) **8**(7), 1–9 (2016). https://doi.org/10.5815/ijcnis.2016.07.01
34. Abo-Zahhad, M., Ahmed, S.M., Abd-Elrahman, S.A.: Genomic analysis and classification of exon and intron sequences using DNA numerical mapping techniques. Int. J. Inf. Technol. Comput. Sci. (IJITCS) **4**(8), 22–36 (2012). https://doi.org/10.5815/ijitcs.2012.08.03
35. Srivastava, P.C., Agrawal, A., Mishra, K.N., Ojha, P.K., Garg, R.: Fingerprints, iris and DNA features based multimodal systems: a review. Int. J. Inf. Technol. Comput. Sci. (IJITCS) **5**(2), 88–111 (2013). https://doi.org/10.5815/ijitcs.2013.02.10
36. Petoukhov, S., Petukhova, E., Svirin, V.: New symmetries and fractal-like structures in the genetic coding system. In: Hu, Z., Petoukhov, S., Dychka, I., He, M. (eds.) Advances in Computer Science for Engineering and Education. ICCSEEA 2018. Advances in Intelligent Systems and Computing, vol. 754. Springer, Cham (2019). https://doi.org/10.1007/978-3-319-91008-6_59
37. Baish, J.W., Jain, R.K.: Fractals and cancer. Cancer Res. **60**, 3683–3688 (July 15, 2000)
38. Bizzarri, M., Giuliani, A., Cucina, A., Anselmi, F.D., Soto, A.M., Sonnenschein, C.: Fractal analysis in a systems biology approach to cancer. Semin. Cancer Biol. 21(3), 175–182 (June 2011). https://doi.org/10.1016/j.semcancer.2011.04.002
39. Dokukin, M.E., Guz, N.V., Woodworth, C.D., Sokolov, I.: Emergence of fractal geometry on the surface of human cervical epithelial cells during progression towards cancer. New J. Phys. **17**(3). pii: 033019 (Mar 10, 2015)
40. Lennon, F.E., Cianci, G.C., Cipriani, N.A., Hensing, T.A., Zhang, H.J., Chen, C.-T., Murgu, S.D., Vokes, E.E., Vannier, M.W., Salgia, R.: Lung cancer—a fractal viewpoint. Nat. Rev. Clin. Oncol. **12**(11), 664–675 (Nov 2015). https://doi.org/10.1038/nrclinonc.2015.108
41. Perez, J.C.: Sapiens mitochondrial DNA genome circular long range numerical meta structures are highly correlated with cancers and genetic diseases mtDNA mutations. J. Cancer Sci. Ther. **9**, 6 (2017). https://doi.org/10.4172/1948-5956.1000469
42. Dmitriev, A.A.: Design of message-carrying chaotic sequences. Nonlinear Phenom. Complex Syst. **5**(1), 78 (2002)
43. Potapov, A.A.: Chapter 12: Chaos theory, fractals and scaling in the radar: a look from 2015. In: Skiadas, C. (ed.) The Foundations of Chaos Revisited: From Poincaré to Recent Advancements, pp. 195–218. Springer International Publisher, Switzerland, Basel (2016). (ISBN 978-3-319-29701-9)
44. Petoukhov, S.V.: The system-resonance approach in modeling genetic structures. Biosystems **139**, 1–11 (Jan 2016)
45. Petoukhov, S.V.: Symmetries of the genetic code, Walsh functions and the theory of genetic logical holography. Symmetry Cult Sci **27**(2), 95–98 (2016)
46. Petoukhov, S.V., Petukhova, E.S.: Symmetries in genetic systems and the concept of genological coding. Information **8**(1), 2 (2017). https://doi.org/10.3390/info8010002, http://www.mdpi.com/2078-2489/8/1/2/htm

47. Petoukhov, S., Petukhova, E., Hazina, L., Stepanyan, I., Svirin, V., Silova, T.: Geno-logical coding, united-hypercomplex numbers and systems of artificial intelligence. In: Hu, Z., Petoukhov, S., He, M. (eds.) Advances in Artificial Systems for Medicine and Education. AIMEE 2017. Advances in Intelligent Systems and Computing, vol. 658, pp. 2–13. Springer, Cham (2018). https://doi.org/10.1007/978-3-319-67349-3_1
48. Darvas, G.: Petoukhov's rules on symmetries in long DNA-texts. Symmetry Cult. Sci. **29**(2), 318–320 (2018). https://doi.org/10.26830/symmetry_2018_2_318
49. Asir Antony Gnana Singh, D., Jebamalar Leavline, E., Priyanka, R., Padma Priya, P.: Dimensionality reduction using genetic algorithm for improving accuracy in medical diagnosis. Int. J. Intell. Syst. Appl. (IJISA) **8**(1), 67–73 (2016). https://doi.org/10.5815/ijisa.2016.01.08
50. Chawda, B.V., Patel, J.M.: Investigating performance of various natural computing algorithms. Int. J. Intell. Syst. Appl. (IJISA) **9**(1), 46–59 (2017). https://doi.org/10.5815/ijisa.2017.01.05

Genetic Code Modelling from the Perspective of Quantum Informatics

Elena Fimmel and Sergey V. Petoukhov

Abstract This paper's aim is to show the possibilities of modelling the information content carried by quantum mechanical DNA molecules by means of the formalism used in quantum informatics. Such modelling would open new options to reveal nature's information patents and to use them, for instance, in quantum computing and artificial intelligence (A.I.). Moreover, it would give an opportunity of understanding the ways of managing information in living organisms. As an empirical base, the open accessible data from GenBank which contains hundreds of millions of long DNA texts collected from thousands of organisms can be used.

Keywords Genetic code · Quantum informatics · Binary oppositions · Qubits

1 Introduction

The discovery of the double helix structure of the DNA by Watson and Crick in 1953 [1] was without doubt a breakthrough in the exploration of life's origin. The understanding that the genetic heritage of all living organisms is encoded by sequences of only four nucleotide bases, *adenine (A), cytosine (C), guanine (G) and thymine (uracil) (T/U)*, pointed to a natural connection between biology (genetics) and mathematics (e.g. coding theory). The biochemical enigma of genetic coding could be traced back to the abstract problem of manipulating symbols. All living organisms share the same principles of encoding their heritable information based on DNA- and RNA-molecules. The heritable information encoded in the DNA molecules of

E. Fimmel (✉)
Mannheim University of Applied Sciences, Paul-Wittsack-Strasse 10,
68163 Mannheim, Germany
e-mail: e.fimmel@hs-mannheim.de

S. V. Petoukhov
Mechanical Engineering Research Institute, Russian Academy of Sciences,
M. Kharitonievsky Pereulok, 4, Moscow, Russia
e-mail: spetoukhov@gmail.com

© Springer Nature Switzerland AG 2020   117
Z. Hu et al. (eds.), *Advances in Artificial Systems for Medicine and Education II*,
Advances in Intelligent Systems and Computing 902,
https://doi.org/10.1007/978-3-030-12082-5_11

an organism determines its physiological features is transmitted to descendants and has the property of the noise immunity, i.e. the ability of error detection and correction. All this is a result of a long process of evolution and allows, in particular, living organisms to inherit complex skills. To give some examples, a spider can weave a complex cobweb using up to seven different threads; bees build honeycombs, etc.

The questions that have arisen in the molecular biological context have led to a new interpretation of the role of mathematical modelling: "*Life is a partnership between genes and mathematics*" [2]. But what kind of mathematics is a partner of genes? And what mathematical methods are appropriate to create A.I. systems to reproduce the hereditary ability of living organisms to engage in complex activities?

DNA molecules are objects of the microworld ruled by quantum mechanical principles. Given that, on the one hand, the idea of exploring the possibilities of modelling the information content carried by DNA molecules is self-evident. On the other hand, this area is practically unchartered.

The role of quantum information in biology has begun to interest the scientific community only recently. For instance, in 2000, a very interesting article was published which emphasises the role of quantum information in biology: "*The role of quantum information in biology is intriguing. Rather, biology is quite unique in tailoring quantum mechanics for its own sake.... Biology is not about applying quantum mechanics as it is already known through the experiences of traditional physics, but rather about an attempt to extend quantum mechanics in the manner that the physicists have not tried.*" [3, p. 45]. However, the article [3] contains no clear suggestions of how to handle the outlined challenge. In [4], the first quantum mechanical model for the information exchange between the DNA and proteins is proposed. But the author ascertains at the end of the article: "*Many questions remain open and the model presented here should be further expanded and elaborated, but this is a first step towards the construction of a complete model for information processing in biosystems, which will be based on quantum information and quantum communication*". A few works published recently, as for example [5], leave many important questions unanswered.

In the present work, we will outline a way to establish models of the genetic code in terms of quantum informatics based on the authors' earlier works.

2 Basic Concept for a Quantum Informatical Modelling of the Genetic Code

It is still unclear why nature uses only four comparatively simple molecules, *adenine (A), cytosine (C), guanine (G) and thymine (uracil) (T/U)*, as letters for an DNA alphabet. However, the authors and their co-authors detected independently in their publications [compare 6–8] that due to their complementary chemical properties these four molecules build a uniform ensemble with three pairs of binary-oppositional characteristics (Fig. 1):

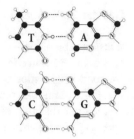

№	Binary Symbols	C	A	G	T/U
1	0_1 — pyrimidines 1_1 — purines	0_1	1_1	1_1	0_1
2	0_2 — amino 1_2 — keto	0_2	0_2	1_2	1_2
3	0_3 – strong bases 1_3 – weak bases	0_3	1_3	0_3	1_3

Fig. 1 Left: nitrogenous bases of DNA. Right: the system of their binary-oppositional traits

(1) Two letters are purines (A and G), and the other two are pyrimidines (C and T). If we assign 1 as an indicator for purines and 0 for pyrimidines ($C = T = 0$, $A = G = 1$) we can write, for example, the sequence GCATGAAGT as 101011110;

(2) Two letters are amino-molecules (A and C) and the other two are keto-molecules (G and T). If we assign 1 as an indicator for keto-molecules and 0 for amino-molecules ($A = C = 0$, $G = T = 1$) we obtain, correspondingly, for the same sequence, GCATGAAGT, the binary representation 100110011;

(3) The pairs of complementary letters, A-T (weak bases) and C-G (strong bases), are linked by 2 and 3 hydrogen bonds, respectively. If we assign 1 as an indicator for weak bases and 0 for strong bases ($C = G = 0$, $A = T = 1$) we obtain, correspondingly, for the same sequence, GCATGAAGT, the binary representation 001101101.

Accordingly, each of the DNA sequences carries three parallel messages on three different binary languages. At the same time, these three types of binary representations form a common logic set on the base of logic operation of modulo-2 addition denoted by the symbol $\oplus$: modulo-2 addition of any two such binary representations of the DNA sequence coincides with the third binary representation of the same DNA sequence: for example, $101011110 \oplus 100110011 = 001101101$. In these binary representations, each of DNA alphabets (16 dinucleotides, 64 trinucleotides, ..., 4^n n-nucleotides) corresponds its own group of n-bit binary numbers.

These phenomenological facts are important for the following reasons:

(1) Classical and quantum computers work on the base of Boolean algebra of binary numbers and their dyadic groups;

(2) The logic operation of modulo-2 addition is actively used in classical and quantum computers; for example, it is used by the quantum operator CNOT, which allows you to entangle two qubits $|x>$ and $|y>$;

(3) Algorithms of quantum informatics use Boolean functions; for example, within the Grover search algorithm such a function is called an oracle;

(4) Boolean algebra has a key meaning for construction of systems of artificial intelligence.

In [7], a new family of models of the genetic code, based on "binary questions" addressed to nucleotide bases, was presented. This type of models was implemented, among other things, in the open-source software package Genetic Code Analysis Toolkit (compare www.gcat.bio … [9, 10]).

A further important building stone, which seems to be necessary for modelling the genetic code from the viewpoint of quantum informatics, is a close connection between the structure of n-nucleotide DNA alphabets and the algebraic operation of the tensor product is explored in detail in [8, 11–13]. Let us recall the main features of how this connection can be established. First of all, we remember that two of three binary indicators determine a nucleotide base in a unique way. For example, speaking about a weak purine, we indicate the nucleotide base A uniquely.

Now, we represent the DNA alphabets of tupels consisting of n-nucleotide letters, n-nucleotides, (4 mononucleotides, 16 dinucleotides, 64 trinucleotides, … 4^n n-nucleotides) in a form of $2^n * 2^n$ square matrices (Fig. 2), whose rows and columns are tagged by binary symbols, for example, according to the following principles:

1. Entries of each column are labelled by binary symbols according to the first set of binary-oppositional indicators in Fig. 1 (purines/pyrimidines). For example, the trinucleotide CAG and all other trinucleotides in the same column are the combination "pyrimidine-purin-purin" and, thus, this column is correspondingly labelled 011

   while

2. Entries of each row are labelled by binary numbers according to the second set of indicators (amino/keto). For example, the same trinucleotide CAG and all other trinucleotides in the same row are the combination "amino-amino-keto" and, thus, this row is correspondingly labelled 001.

Fig. 2 The square tables of DNA alphabets of 4 nucleotides, 16 dinucleotides and 64 trinucleotides

	0	1
0	C	A
1	T	G

	00	01	10	11
00	CC	CA	AC	AA
01	CT	CG	AT	AG
00	TC	TA	GC	GA
01	TT	TG	GT	GG

	000	001	010	011	100	101	110	111
000	CCC	CCA	CAC	CAA	ACC	ACA	AAC	AAA
001	CCT	CCG	CAT	CAG	ACT	ACG	AAT	AAG
010	CTC	CTA	CGC	CGA	ATC	ATA	AGC	AGA
011	CTT	CTG	CGT	CGG	ATT	ATG	AGT	AGG
100	TCC	TCA	TAC	TAA	GCC	GCA	GAC	GAA
101	TCT	TCG	TAT	TAG	GCT	GCG	GAT	GAG
110	TTC	TTA	TGC	TGA	GTC	GTA	GGC	GGA
111	TTT	TTG	TGT	TGG	GTT	GTG	GGT	GGG

From the viewpoint of quantum informatics, it is important that for all $n = 1, 2,$ 3, ... these genetic tables are the n-th Kronecker tensor powers of the first matrix of nucleotides [C, A; T, G]. Let us recall the definition of the Kronecker tensor multiplication of two matrices: the Kronecker product, denoted by $\otimes$, is an operation on two matrices of arbitrary size resulting in a block matrix the way that each of the entries of the first matrix is multiplied with the whole of the second matrix. Besides, we define the multiplication in this context as a concatenation of symbols in the given order. Figure 3 shows that the second tensor power of the (2 * 2)-matrix [C, A; T, G] of 4 DNA-letters automatically gives the matrix of 16 dinucleotides; the third tensor power of the same matrix of 4 DNA-letters gives the matrix of 64 trinucleotides with the same strict arrangement of entries as in tables in Fig. 2.

Thus, the structural organisation of the system of interrelated DNA alphabets can be represented with the help of the algebraic operation of the tensor product. But the operation of the tensor product is well known in quantum mechanics and quantum informatics, where it presents a way of merging Hilbert spaces together to form larger Hilbert spaces. The following quotation emphasises the importance of the tensor product for the quantum mechanics: *"This construction is crucial to understanding the quantum mechanics of multiparticle systems .... The state space of a composite physical systems is the tensor product of the state spaces of the component physical systems"* [14, p. 71, 102].

In our model approach, we interpret DNA texts as quantum systems of qubits, where each of qubits is represented on the base of different pairs of binary-oppositional indicators of adenine A, guanine G, cytosine C and thymine T (Fig. 1). As we noted, each of these DNA bases can be uniquely defined by the first kind of indicators ("pyrimidine or purine") jointly with the second kind of indicators ("amino or keto"). On the basis of each of these pairs of binary-oppositional indicators, a cor-

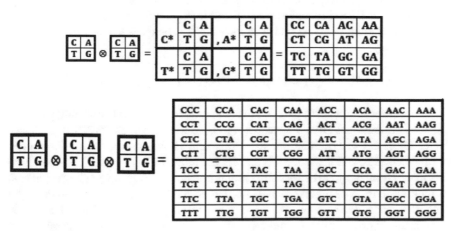

Fig. 3 The tensor family of genetic matrices [C, A; T, G]$^{(n)}$ (here tensor power $n = 1, 2, 3$) of DNA alphabets of 4 nucleotides, 16 dinucleotides and 64 trinucleotides. The symbol $\otimes$ denotes the tensor product

responding two-level quantum system can be formally introduced with a definition of its appropriate qubit. For example, in the case of a genetic qubit as a two-level quantum systems on the basis of the indicators "pyrimidine or purine", one level corresponds to the indicator "pyrimidine" and the second level to the oppositional indicator "purine". The state of such qubit is a vector in a two-dimensional Hilbert space H_1:

$$|\psi_1 >= \alpha_0|0 > +\alpha_1|1 >, \quad \alpha_0^2 + \alpha_1^2 = 1 \tag{1}$$

where α_0 and α_1 are correspondingly amplitudes of probabilities of pyrimidines and purines in the considered DNA text as a quantum system.

By analogy, for the case of the indicators "amino or keto" , another genetic qubit is defined with its state in another two-dimensional Hilbert space H_2:

$$|\psi_2 >= \beta_0|0 > +\beta_1|1 >, \beta_0^2 + \beta_1^2 = 1 \tag{2}$$

where β_0 and β_1 are correspondingly amplitudes of probabilities of amino-molecules and keto-molecules in the considered DNA text as a quantum system.

The tensor product of the two-dimensional Hilbert space $H_1 \otimes H_2$ gives one four-dimensional Hilbert space with the following separable pure state of a quantum 2-qubit system:

$$|\psi_{12} >= |\psi_1 > \otimes|\psi_2 >= (\alpha_0|0 > +\alpha_1|1 >) \otimes (\beta_0|0 > +\beta_1|1 >)$$
$$= \alpha_0\beta_0|00 > +\alpha_0\beta_1|01 > +\alpha_1\beta_0|10 > +\alpha_1\beta_1|11 > . \tag{3}$$

In the expression (3), cytosine C corresponds to the computational basis state |00>, thymine T corresponds to |01>, guanine G corresponds to |11>, adenine A corresponds to |10>. Probabilities P(C), P(T), P(G), P(A) of computational basis states of nucleotides C, T, G and A in long DNA sequences can be obtained in the following way:

$$P(C) = (\alpha_0\beta_0)^2, \quad P(T) = (\alpha_0\beta_1)^2, \quad P(G) = (\alpha_1\beta_1)^2, \quad P(A) = (\alpha_1\beta_0)^2. \tag{4}$$

To consider similarly the case of 16 dinucleotides, we should accordingly expand the four-dimensional Hilbert space $H_1 \otimes H_2$ to the 16-dimensional Hilbert space $H_1 \otimes H_2 \otimes H_3 \otimes H_4$. Here the spaces H_1 and H_2 are related to the first letters of dinucleotides, and the spaces H_3 and H_4 are related to the second letters of dinucleotides. By analogy, in the general case of n-nucleotides, a corresponding expansion of multidimensional Hilbert spaces is needed on the basis of the tensor product of n Hilbert spaces of separate nucleotides. This model approach is useful to explain some phenomenological features of long DNA texts as quantum systems in separable pure states [13, 15] and is also useful to construct biological-like systems of artificial intelligence since the problems of effective systems of A.I. can be closely connected with quantum computing [16]. In addition, it can lead to new methods of genomic analysis including compression and encryption algorithms on DNA sequences [17–19].

3 Prospects

In the previous part of this work, we have shown how an arbitrary DNA sequence can be modelled by means of a quantum informatical formalism. Of course, we acknowledge that the genetic code is a far more complicated subject than the basic model we have proposed in this work. Hence, further investigations in this field should be undertaken. To achieve this, we suggest taking the following first steps:

- With the last years having seen the evolution of the theory of so-called circular codes which seem to maintain the right reading frame during the process of protein building and, thus, play an important role for the ability of noise immunity of the genetic code (compare [9, 20–24]), it is important to establish a link between quantum informatics models of the genetic code and circular codes in order to understand the nature of noise immunity;
- An analysis of the adaptability of the quantum informatics Grover- and Shor-algorithms to model genetic code;
- A study of the utility of the Hadamard operators and Walsh functions, which play an important role in quantum informatics and signal processing, in modelling of the genetic information;
- Creating and implementing genetic databases and software tools satisfying the requirements of quantum informatics modelling [10, 25, 26]
- Understanding the results of quantum informatics modelling of the genetic code in terms of reversible logic of quantum computing using the Boolean reversible quantum gates such as quantum Peres and Toffoli gates.

4 Conclusions

This paper presents a basic concept of modelling the genetic code by means of the quantum informatics formalism. We have shown that the four nitrogenous bases, which are chosen by nature as letters for the genetic alphabet, as well as oligonucleotides can be represented in a natural way using their biochemical properties as quantum systems of qubits. The model described uses an important construction of a tensor product in a natural way to recursively expand the model from the level of single nucleotides to oligonucleotides. In the present work, we have also outlined possible next steps for developing the approach suggested.

We sincerely hope that, with a quantum informatics approach for modelling the genetic code, providing better explanations for other more complex phenomena of life including inherited intelligence properties of living organisms will be possible.

References

1. Watson, J.D., Crick, F.H.C.: Molecular structure of nucleic acids: a structure for deoxyribose nucleic acid. Nature. Band **171**, S. 737–738, 25 (April 1953). https://doi.org/10.1038/171737a0
2. Stewart, I.: Life's Other Secret: The New Mathematics of the Living World. Penguin, New York (1999)
3. Matsuno, K., Paton, R.C.: Is there a biology of quantum information? BioSystems **55**, 39–46 (2000)
4. Karafyllidis, I.G.: Quantum mechanical model for information transfer from DNA to protein. Biosystems **93**(3), 191–8 (2008)
5. Patel, A.: Quantum Algorithms and the Genetic Code. arXiv:quant-ph/0002037 (2001)
6. Fimmel, E., Danielli, A., Strüngmann, L.: On dichotomic classes and bijections of the genetic code. J. Theor. Biol. **336**, 221–230 (2013)
7. Gumbel, M., Fimmel, E., Danielli, A., Strüngmann, L.: On models of the genetic code generated by binary dichotomic algorithms. BioSystems **128**, 9–18 (2015)
8. Petoukhov, S.V., He, M.: Symmetrical Analysis Techniques for Genetic Systems and Bioinformatics: Advanced Patterns and Applications. IGI Global, Hershey, USA (2009)
9. Fimmel, E., Strüngmann, L.: Mathematical Fundamentals for the noise immunity of the genetic code. BioSystems **164**, 186–198 (2018)
10. Fimmel, E., Gumbel, M., Strüngmann, L.: Exploring structure and evolution of the genetic code with the software tool GCAT. In: AIMEE 2017: Advances in Artificial Systems for Medicine and Education, 658, pp. 14–22. Springer, Berlin (2018). https://doi.org/10.1007/978-3-319-67349-3-2
11. Petoukhov, S.V.: The system-resonance approach in modeling genetic structures. Biosystems **139**, 1–11 (2016)
12. Petoukhov, S.V.: Genetic coding and united-hypercomplex systems in the models of algebraic biology. Biosystems **158**, 31–46 (2017)
13. Petoukhov, S.V.: The rules of long DNA-sequences and tetra-groups of oligonucleotides. https://arxiv.org/abs/1709.04943, the 4th version (25.12.2017)
14. Nielsen, M.A., Chuang, I.L.: Quantum Computation and Quantum Information. Cambridge University Press, New York (2010)
15. Petoukhov, S.V., Petukhova, E.S., Svirin, V.I.: New symmetries and fractal-like structures in the genetic coding system. In: Hu, Z., Petoukhov, S., Dychka, I., He, M. (eds.) Advances in Computer Science for Engineering and Education. ICCSEEA 2018. Advances in Intelligent Systems and Computing, vol. 754. Springer, Cham (2019). https://doi.org/10.1007/978-3-319-91008-6_59
16. Biamonte, J., Wittek, P., Pancotti, N., Rebentrost, P., Wiebe, N., Lloud, S. Quantum machine learning. Nature **549**, 195–202 (14 Sept 2017). https://doi.org/10.1038/nature23474
17. Abo-Zahhad, M., Ahmed, S.M., Abd-Elrahman, S.A.: Genomic analysis and classification of exon and intron sequences using DNA numerical mapping techniques. Int. J. Inf. Technol. Comput. Sci. (IJITCS) **4**(8), 22–36 (2012). https://doi.org/10.5815/ijitcs.2012.08.03
18. Hossein, S.M., Roy, S.: A compression & encryption algorithm on DNA sequences using dynamic look up table and modified Huffman techniques. Int. J. Inf. Technol. Comput. Sci. (IJITCS) **5**(10), 39–61 (2013). https://doi.org/10.5815/ijitcs.2013.10.05
19. Meher, J.K., Panigrahi, M.R., Dash, G.N., Meher, P.K.: Wavelet based lossless DNA sequence compression for faster detection of eukaryotic protein coding regions. Int. J. Image Graph. Signal Process. (IJIGSP) **4**(7), 47–53 (2012). https://doi.org/10.5815/ijigsp.2012.07.05
20. Fimmel, E., Michel, ChJ, Strüngmann, L.: Diletter circular codes over finite alphabets. Math. Biosci. **10**(294), 120–129 (2017). https://doi.org/10.1016/j.mbs.2017.10.001
21. Fimmel, E., Michel, ChJ, Strüngmann, L.: Strong comma-free codes in genetic information. Bull. Math. Biol. **79**(8), 1796–1819 (2017). https://doi.org/10.1007/s11538-017-0307-0
22. Fimmel, E., Strüngmann, L.: Codon distribution in error-detecting circular codes. Life **6**(1), 14 (2016). https://doi.org/10.3390/life6010014

23. Fimmel, E., Michel, ChJ, Strüngmann, L.: n-nucleotide circular codes in graph theory. Phil. Trans. R. Soc. A **374**, 20150058 (2016)
24. Fimmel, E., Giannerini, S., Gonzalez, D., Strüngmann, L.: Circular codes, symmetries and transformations. J. Math. Biol. **70**(7), 1623–44 (2014)
25. Fickett, J.W., Burks, C.: Development of a database for nucleotide sequences. In: Waterman, M.S. (ed.) Mathematical Methods for DNA Sequences, pp. 1–34. CRC Press, Inc., Florida (1989)
26. Kraljic, K., Strüngmann, L., Fimmel, E., Gumbel, M.: Genetic code analysis toolkit: a novel tool to explore the coding properties of the genetic code and DNA sequences. SoftwareX **7**, 1214, (January–June 2018), https://doi.org/10.1016/j.softx.2017.10.008

Focal Model in the Pattern Recognition Problem

T. Rakcheeva

Abstract The pattern recognition problem in the focal paradigm is considered, which makes it possible to apply the focal model both at the learning stage and at the identification stage. The theoretical substantiation and methods of algorithmic realization of the description of classes and their boundaries with the help of the focal model of representation of smooth curves and surfaces by multifocal lemniscates are given. As a result, at the learning stage for each class in the feature space a multipolar space with a focal distance is formed, in which the class boundary is described by a lemniscate and is an isometric surface. An important advantage is that the focal classification space also forms a continuous affiliation function. At the decision-making stage, the advantage of the focal model is that identification is an elementary operation of calculating the focal distance of a given point and comparing its value with the boundary parameter. Thus, the multidimensional classification problem reduces to optimization and decision-making in one-dimensional space. The complexity of the traditional solution, resulting in a fragmented description of class boundaries, is compensated by a high level of computer power that allow successfully cope with practical classification tasks. At the same time, a person has, in addition to a strong applied rationale, the desire for adequacy and organicity of the approach, the result of which is simplicity and sensibility. Such a solution of the classical recognition problem in the focal paradigm is proposed in this paper.

Keywords Pattern recognition · Decision-making · Focal model · Lemniscate · Curves · Surfaces · Affiliation function

T. Rakcheeva (✉)
Mechanical Engineering Research Institute of the Russian Academy of Sciences 4, Maly Kharitonievskiy Pereulok, Moscow 101990, Russia
e-mail: rakcheeva.tanja@yandex.ru

© Springer Nature Switzerland AG 2020
Z. Hu et al. (eds.), *Advances in Artificial Systems for Medicine and Education II*,
Advances in Intelligent Systems and Computing 902,
https://doi.org/10.1007/978-3-030-12082-5_12

127

1 Introduction

One of the fundamental problems of artificial intelligence is the traditional problem of pattern recognition. The automation of the decision-making process in the recognition problem continues to pose considerable difficulties especially in the multidimensional feature space. To describe the boundaries of classes of recognizable objects, as a rule, the dichotomy procedure is used. As a result, the boundaries receive a piecewise description by the hyperplanes, which grows exponentially with the increasing dimension. The description is complicated by the fact that in most problems class boundaries are not convex and the regions themselves are not simply connected. At the identification stage, the implementation of the decision-making procedure in conditions of fragmented boundaries requires, as you know, the comparison of a large number of logical constructions. The convex–concave character of the boundaries and their discontinuity for multiply connected regions significantly complicate the decision-making procedures.

The complexity of the traditional solution, resulting in a fragmented description of class boundaries, is compensated by a high level of computer power. At the same time, an artificial intelligence is not only technology, but also new approaches to the known problems. Man in addition to a strong applied rationale is characterized by a desire for adequacy and naturalness of a solution. This is the solution of the classical recognition problem in the focus paradigm proposed in this paper. The adequacy of the focus paradigm in the problem of classifying is provided by modeling the curves and surfaces by multifocal lemniscates, proposed by Hilbert and developed by the author [1–7]. The principal features of the focus model give advantages to its application for both the describing class boundaries at the learning stage and the decision-making at the identification stage. The class boundaries in the feature space are continuous and smooth. The focal method is not critical to the convexity of the described regions and also to their connectivity. The boundaries of the multiply connected regions of one class are described not by several curves or surfaces, united by logical conditions according to belonging to one class, but by one general analytic curve or surface. At the learning stage, boundaries are formed in the form of multifocus lemniscates. Together with them, affiliation function which essentially simplifies the identification of the newly presented object at the decision-making stage is also formed. The multidimensional classification problem reduces to optimization and decision-making in the one-dimensional space [5].

Thus, since the traditional solutions are purely technological taking little account of the specificity of the recognition problem the goal of this study is to develop a method based on a model adequate to the problem of recognition the manifestation of which would be the simplicity and the organic nature of the solution.

2 Focal Space of Multifocal Lemniscates

The classical *lemniscate* L [1] is a smooth closed curve on the plane defined by its invariant through the system m of focal points $\{f_1, \ldots, f_m\}$ inside as the locus of points the product of distances from which to all focuses is constant and equal to its focal radius R_m (see Fig. 1a). The invariant and the equation of the multifocal lemniscate have the form:

$$\prod_{j=1}^{m} r_j = R^m \tag{1}$$

where r_j is the Euclidean distance from an arbitrary point of the lemniscate to the j-th focus ($j = 1, \ldots, m$). We take the following notation: the mf-system is the set of m focuses of the lemniscate, the mf-lemniscate is a lemniscate with m focuses, and mf-radius is the radius R of its lemniscate.

Classical definition of a lemniscate on the plane (1) is automatically generalized to an N dimensional space by a generalization to a multidimensional space of the Euclidean distance [4]:

$$r_j = \left(\sum_{i=1}^{N} (x_i - f_{ij})^2 \right)^{1/2} \tag{2}$$

where x_i are the coordinates of the point x, f_{ij} are the coordinates of the j-th focus f_j. This circumstance has got a fundamental importance for the classification problem since the number of informative characteristics of the description of classification objects, as a rule, is not limited to two and the feature space is not limited to the plane [8–14]. In the future, two-dimensional models will be used only for illustrations.

For any fixed focal system, *confocal* lemniscates with different radii form a *family* of nested curves (surfaces) (see Fig. 1a): Lemniscates with a larger radius surround lemniscates with a smaller radius without intersecting. For small values of the radius of the lemniscate, it is m small circles (spheres) around each of the m focuses which increase and deform with increasing R merging in a simply connected curve (surface) of complex shape, and in the limit, the lemniscate degenerates into one circle (hyper-

Fig. 1 Multifocal lemniscates: **a** focal distances for the point (x, y) forming the lemniscate invariant; family of confocal $3f$-lemniscates; **b** focal approximation of the empirical curve

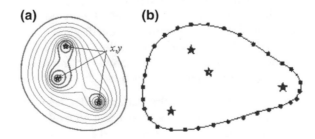

sphere) with the center inside the mf-system (the arithmetic mean of the coordinates of the focuses).

We introduce the concept of the *focal distance* from an arbitrary point x to an mf-system by an expression analogous to the radius of the lemniscate (1):

$$\rho(x, f)^m =_{\text{def}} \prod_{j=1}^{m} r_j \qquad (3)$$

This is the radius of the lemniscate to which the point x belongs in this focal mf-system. The focal distance ρ defined in this way exists at every point of the space has a value in the range from 0 to $\{L\}$. Focal distance takes the value $\rho = 0$ only at the focal points of the given mf-system [6]. By definition, all the points of the mf-lemniscate L as well as all the points of the lemniscates $\{L\}$ that are confocal with it have equal distances to their mf-system. Therefore, any mf-system is a multifocal center (mf-center), similar to the center of the polar coordinate system with the $1f$-system. Thus, (1) and (2) for each mf-system define a continuous mf-factorization of the space by multifocal lemniscates with respect to the focal radius.

Let us consider successively both stages of solving the classification problem: the stage of learning, which consists in the formation of the *classification space*, and the decision-making stage, which consists in constructing a *decisive rule* in this space.

3 Formation of Classification Space

The initial information for the classification problem is, as a rule, the sets of typical samples of each class. The problem of forming a classification space can be formulated as follows:

> Suppose that the sets of typical samples of mf_k-classes are given in the N dimensional space of features by a finite number of m_k points ($k = 1, ..., K$) representing typical samples of classes in their sufficient representation. It is required to form boundaries that describe or divide class areas.

Here, it is necessary to distinguish two variants of the description of boundaries that affect the formation of a classification space:

(1) The boundaries describing the class regions include only the elements of the class without extraneous elements—the entire feature space is conditionally divided into "one's own" and "another's";

(2) The boundaries separating the class regions include not only the elements of the class, but also the elements that do not represent any class—the entire feature space is conditionally divided into "spheres of influence."

Let's first consider the first variant, call it "one's own and another's." The region of each class must contain all elements of the set of typical samples and a free space representing the remaining elements of the class. The application of the focal model with the invariant (1) for the description of classes can be realized in this case by constructing for each class a boundary lemniscate in the format of a *direct focal problem*.

> Each of the points representing samples of the learning set is taken as a focus, and the entire learning set of the class forms a focal mf-system for the boundary lemniscate of the given class.

By definition, a lemniscate of any radius contains within itself all its own focuses, which in this case are the *mf*-system of the learning set of the class. Therefore, the *mf*-system of the learning sample generates a family of confocal lemniscates each of which can be the class boundary.

> The learning samples of K classes consisting of m_k samples each create in the feature space K of focal mf_k-systems with their families of confocal potentially boundary lemniscates $\{L_k\}$.

As follows from the above, every mf_k-lemniscate from the family of confocal lemniscates $\{L_k\}$ with increasing radius surrounds all the learning samples of the given mf_k-system and all lemniscates with smaller radius, monotonically increasing the free space. To complete the formation of the classification space for the "one's own and another's" variant, it is required to select from the family $\{L_k\}$ a single boundary lemniscate L_k by fixing the value of the focal radius $R_L(mf_k)$ in each mf_k-class. The choice depends on the additional information regarding the internal and external properties of each class of the classification space generated by the original data. In the absence of such information, the requirement of "sufficient representation" of the learning set contained in the statement of the problem allows us to use the statistical characteristics of the mf_k-systems of each class to determine the boundary lemniscate radius. Thus, the problem of forming the description of identification classes in the format of learning samples is reduced to the determining in each class the radius of the boundary lemniscate $R_L(mf_k)$ in accordance with the criterion for the required compromise. The specification of the criterion depends on the conditions of the practical problem, which determines both the degree of variability of the initial data and the parameters of the acceptable risk.

4 Decision-Making and Affiliation Function

The invariant of the boundary of the mf_k-lemniscate and the monotonicity with respect to the radius of its confocal family determine the partition of the classification space for each class into an inner part containing the mf_k-system and an exterior one.

All points belonging to a region bounded by a lemniscate with a focal radius R_L have values ρ of mf-radius less than $R_L : \rho < R_L$, all points belonging to the lemniscate have the same radius value equal to $R_L : \rho = R_L$, and all points of space that do not belong to the region of a given lemniscate have radius values greater than $R_L : \rho > R_L$. This property of the lemniscate, which is fundamental for the classification problem, includes the case when the region bounded by a single lemniscate can be multiply connected. In accordance with this, the problem of identification in the focal *KL*-polycentric space is extremely simple. To identify a new object represented by a point in this space, it is required to calculate the values of the focal distance ρ_0 from this point to the focal mf_k-systems of each class and compare with the focal radii $R_L(mf_k)$ of their boundary lemniscates. The newly presented object must be identified with the k-th of the K regions if the following condition is satisfied:

$$\rho_0(mf_k) \leq R_L(mf_k), \quad k = 1, \ldots, K. \tag{4}$$

Criterion (4) is used in the absence of intersections of class regions. If the focal distance from the presented object ρ_0 to the focal systems of different classes satisfies the criterion (4) more than for one class, then the belonging is identified by the minimum value of these focal distances:

$$\min\{\rho_0(mf_k)|\rho_0(mf_k) \leq R_L(mf_k)\}, \quad k = 1, \ldots, K. \tag{5}$$

In the case of class intersections, it is preferable to use the other variant of class boundaries discussed below which divide space into "spheres of influence."

The focal KL-polycentric space forms not only continuous class boundaries, but also the continuous affiliation function F_P.

For each k-class, the affiliation function is identical to the focal distance $\rho_k(x)$ from an arbitrary point x to the mf_k-system. Like the focal distance, this function is nonnegative equal to 0 at the points of the mf_k-system and only in them, continuous, monotonic, unboundedly increasing, having a minimum in the mf_k-center of the class. The affiliation function exists on the whole space and indicates the degree of proximity in quantitative terms of the newly presented identification object to the focal mf_k-center of the class.

For a problem of K classes, the *affiliation function* F_P is defined at the point x as a minimum from the values at this point of the affiliation functions of all K classes:

$$F_P(x) = \min(\rho_k(x)), \quad k = 1, \ldots, K. \tag{6}$$

As an extremum of continuous functions, the F_P inherits the existence, continuity, and nonnegativity on the whole space and also the equality 0 at the points of all mf_k-systems ($k = 1, \ldots, K$). The property of monotonicity has in this case a local character bounded by the region of each class. The F_P increases from the mf_k-center either to ∞ or to the point x in which the distances $\rho_k(x) = \rho_l(x), \quad l \neq k(l = 1, \ldots, K)$. Thus, in addition to the minima in the focal centers of classes, F_P has extremes (maximums) forming the boundaries separating classes.

It is natural to use the affiliation function F_P in the classification problem to determine the separating boundaries between classes. The boundaries separating the "spheres of influence" of the classes are obtained in a natural way as an *extremal manifold of the affiliation function*. At the same time, F_P can testify not only to the belonging or nonbelonging of the presented object to a given class but also about the *degree of identification proximity* in quantitative terms.

It should be noted that the decision-making procedure in a space with traditional *fragmentary* boundaries is a *step function* with a binary response that can only indicate belonging to one or another region but does not say anything about the degree of proximity quantitatively. With this approach, the risks of identification errors increase for border areas of space. Focal F_P allows us to construct an index function F_P, whose value at each point is the class number. The focal index space will also have a step structure that marks each point of the space as belonging to a class with a certain number, but it also preserves the quantitative evaluation of the degree of this belonging.

5 Model Computer Experiment 1

Let's consider an example of an identification task for one class in the format "one's own and another's." In this case $K = 1$, so the identification space is formed by one focus mf-system with its family of confocal lemniscates $\{L\}$. It is required to define a boundary lemniscate dividing the space into "one's own" and "another's." As an assessment of the internal variability of the set of typical samples of class, we use basic statistical characteristics, such as the average value and variance of interfocal distances. Figure 2 shows the results of modeling a feature space with a set of 16 elements marked with the symbol "*.". The elements of the set are fairly compact, so that the region of the class can be considered simply connected (see Fig. 2a).

The class boundary is represented by a lemniscate corresponding to a focal system identified with all 16 points of the learning set. The radius of this lemniscate R'_L can initially be defined as the minimum value at which the lemniscate takes a simply connected form, encompassing the entire set as a whole (see Fig. 2a, inner contour). Then, the class region is increased with the appropriate radius correction. The correction to the radius is determined statistically by the average (or largest)

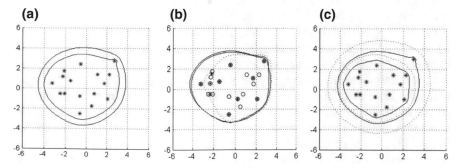

Fig. 2 Focal representation of the class: sampling elements—focuses (description in the text)

minimum interfocal distance. For each point of the focal system, distances to the nearest focus are calculated which are then averaged, and according to the obtained value of $\Delta f_{\min}$, the focal radius R_L' is recalculated. As a result, the $16f$-system, together with the lemniscate family and the boundary lemniscate L of radius R_L taking into account the intra-class variability (see Fig. 2a, outer contour), completes the formation of the identification space.

6 Optimization of Initial Information

An example of the considered problem (see Fig. 2a) suggests the redundancy of the focal system for describing the class boundary by means of lemniscate. Eliminating redundancy compressing the initial information is an important component of the formation of the identification space.

The peculiarity of recognition problems with respect to the shape of class boundaries is that, as a rule, these problems do not require excessive detailing of the shape. The description of the boundaries without significant loss of information can be compressed by eliminating the redundancy of the focal system representing the learning samples.

> Thus, the problem arises to minimize the focal system under the condition of equivalence of the boundary lemniscate from the point of view of identification.

To solve this problem, a procedure has been developed for reducing the number of focuses while preserving the shape of the boundary in the sense that its deformation is not essential for making a decision. The procedure implements the principle of "annihilation" when a group of neighboring focuses is replaced by a single focus as long as the reduced focal system maintains the original shape of the boundary, possibly with a new value of the focal radius. An iterative procedure at each step of "annihilation" generates a lemniscate whose shape with radius correction is insignificantly different from the original form of the boundary lemniscate (the proximity criterion in the Hausdorff metric is used to estimate the distance between boundaries). The degree of proximity of focuses subjecting to "annihilation" is also determined statistically through the interfocal distance.

Figure 2b shows the results of the described procedure for compressing the learning sample of the previous problem. The focal samples of the initial set ($m = 16$) representing the $16f$-system are marked with the symbol "o." The internal solid contour represents the corresponding $16f$-lemniscate of radius $16f\text{-}R_L$. The "*" symbol marks focal points in the number $m' = 10$ of a new $10f$-focal system, obtained as a result of "annihilation," with some focuses remaining the same (double marking). As a result, the focal system was reduced from 16 to 10 focuses with insignificant changes in the boundary of the class region with a new focal radius of $10f - R_L$ (two close contours) that were not significant for identification. The third contour (internal dotted line) corresponds to a lemniscate with a $7f$-focus system that does not support the original shape of the boundary of learning set. Thus, the "annihilation" procedure allows to optimize the focal system, optimizing the initial information.

Figure 2c shows the transformation of the learning set boundary in connection with yet another traditional problem of preliminary processing of initial data—the problem of eliminating artifacts. One element is at a considerable distance from the main set of sample data $(f_{i1} \gg \Delta f_{min})$. The boundary lemniscate for the full set, as can be seen from the Figure, is redundant in coverage of the class region; in addition, the remote point significantly shifts the average value of the interfocal distance which introduces redundancy also in correction of the radius. The outer contour of the shortened set in spite of the remote "artifact" almost completely repeats the lemniscate by the original $16f$-focal system (see Fig. 2a).

7 Fractional–Rational Invariant

The case $K = 2$ is a binary problem called "yes-no" or "one-zero." The solution is realized in the above format for the direct problem, i.e., all points of each of the two learning sets are mf_k-systems ($k = 1, 2$) of class regions. In contrast to the invariant functional (1), in the given binary problem we use another, rational–fractional functional of the form:

$$\left(\prod_{j=1}^{mc} r\left(l_i, fc_j\right) \right) / \left(\prod_{j=1}^{mp} r\left(l_i, fp_j\right) \right) = R^m \tag{7}$$

where the focal radius $R_L = R^m$ is optimized under the condition of minimizing errors and maximizing the distances from each set (fc_j, fp_j) to the boundary lemniscate $L(l_i)$. Figure 3 shows the result of the algorithm work for two model samples.

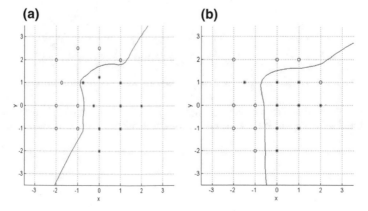

(a)   **(b)**

Fig. 3 The interclass boundary in the binary problem "zero-one": o—"zero," •—"one"

In Fig. 3a, the separating lemniscate passes without errors between class sets and in Fig. 3b with the assumption of errors (presumably, artifacts). To minimize errors on the control samples, the combined distances from the points of the learning samples to the boundary lemniscate were maximizing the radius is chosen in accordance with the above criterion.

8 Model Computer Experiment 2

We illustrated the case of two equivalent classes. The forming of each region is performed in accordance with the above solution to the problem "one's own and another's" without mutual intersection (see Fig. 4a). The construction of the class boundary of one sample that formed the $16f$-system marked with "*"was discussed above (see Fig. 2). Elements of another set in the number $m = 20$, marked "o," form a $20f$-system generating respectively $20f$-lemniscate. The border of this sample differs strongly pronounced convex–concave shape; therefore, in this case it is more effective to use the fractional–rational criterion (6). The product in the numerator represents the focal distance of the entire $20f$-system. One additional focus representing the denominator is located outside this region and is marked by the double symbol "o" and "●"(see Fig. 4a). Figure 4b shows the results of "annihilation" of focuses.

It should be noted that in the case of a traditional piecewise fragmented representation of convex–concave boundaries or discontinuous boundaries, the complexity of logical constructions for localizing a point relative to such boundaries is increasing radically.

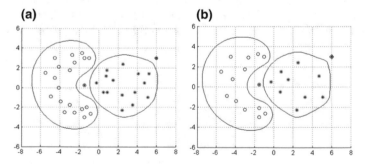

Fig. 4 Formation of domains in a multiclass problem: **a** construction of boundaries by focus systems identified with samples; **b** class boundaries with "annihilation"

9 Conclusion

A significant difference between the focal paradigm and traditional recognition approaches is that the decision rule is based on the integral perception of the set of learning samples in which the influence of all elements of the sample organized in a specific structure is taken into account and not just the boundary ones.

The multidimensional classification problem in the focal paradigm as a whole is reduced to the formation of a classification space in the form of focal systems associated with set of learning samples and the construction of continuous class boundaries in the form of lemniscates ensuring the simplicity of the decision-making procedure. The focal polycentric identification space factorized by the lemniscates on the radius also forms a continuous affiliation function, indicating not only the belonging or nonbelonging of the presented object to a given class, but also the degree of identification proximity in quantitative terms. Moreover, the boundaries separating the "spheres of influence" of classes are obtained in a natural way as an extreme manifold of the affiliation function. The simplicity of making a decision in the focal paradigm is also preserved in the case of the multivalued character of the recognize image. An example is the task of recognizing letters or any symbols, where there are different inscriptions for semantically the same sign, identification a new sample with such a multivalued area does not require additional calculations [15].

The creation of class boundaries can also be implemented in an interactive format of computer power. A distinctive feature of the method of focal modeling is a unified representation space for the system of focal degrees of freedom and a simulated geometric shape [2, 3, 7]. This allows natural driving of the focal freedoms providing the possibility of solving the direct problem of modeling the shape of the boundaries of identification classes in the search-interactive mode of visual control.

References

1. Hilbert, D.: Gessamelte Abhandlungen, 435 p. Springer, Berlin (1935). Bd. 3
2. Rakcheeva, T.A.: Multifocal lemniscates: approximation of curves. Comput. Math. Math. Phys. **50**(11), 1956–1967 (2010)
3. Rakcheeva, T.A.: Focal approximation in the complex plane. Comput. Math. Math. Phys. **51**(11), 1847–1855 (2011)
4. Rakcheeva, T.A.: Approximation of surfaces by multifocal lemniscates. In: XX International Conference on Mathematics, Economy, and Education. Rostov n/D, p. 136 (2012)
5. Rakcheeva, T.A.: Focal approximation in the recognition problem. Math. Comput., Educ. Sat. **18**(Part 1), 141 (2011)
6. Rakcheeva, T.A.: Polypolar lemniscate coordinate system. Comput. Res. Model. **1**(3), 256–263 (2009)
7. Rakcheeva, T.A.: Managing multifocal degrees of freedom in the problem of shape formation. In: Parallel computations and control tasks. Proceedings of the International Science Conference on M. IPP RAS (2001)

8. Murali Krishna, N., Sirisha Devi, J., Yarramalle, S.: A novel approach for effective emotion recognition using double truncated gaussian mixture model and EEG. Int. J. Intell. Syst. Appl. (IJISA), 9(6), 33–42 (2017). https://doi.org/10.5815/ijisa.2017.06.04
9. Hamd, M.H., Ahmed, S.K.: Biometric system design for iris recognition using intelligent algorithms. Int. J. Mod. Educ. Comput. Sci. (IJMECS) 10(3), 9–16 (2018). https://doi.org/10.5815/ijmecs.2018.03.02
10. Onifade, O.F.W., Akinyemi, J.D., Adebimpe, O.S.: A recursive binary tree method for age classification of child faces. Int. J. Mod. Educ. Comput. Sci. (IJMECS) 8(10), 56 66 (2016). https://doi.org/10.5815/ijmecs.2016.10.08
11. Gupta, S.L., Baghel, A.S.: Efficient feature extraction in sentiment classification for contrastive sentences. Int. J. Mod. Educ. Comput. Sci. (IJMECS) 10(5), 54–62 (2018). https://doi.org/10.5815/ijmecs.2018.05.07
12. Nandi, D., Saif, A.F.M.S., Prottoy, P., Zubair, K.M., Shubho, S.A.: Traffic sign detection based on color segmentation of obscure image candidates: a comprehensive study. Int. J. Mod. Educ. Comput. Sci. (IJMECS) 10(6), 35–46 (2018). https://doi.org/10.5815/ijmecs.2018.06.05
13. John, A.K., Williams, O.O., Adewale, M.A.: Evaluating the effect of JPEG and JPEG2000 on selected face recognition algorithms. Int. J. Mod. Educ. Comput. Sci. (IJMECS) 6(1), 41–52 (2014). https://doi.org/10.5815/ijmecs.2014.01.05
14. Ullah, Z., Fayaz, M., Iqbal, A.: Critical analysis of data mining techniques on medical data. Int. J. Mod. Educ. Comput. Sci. (IJMECS), 8(2), 42–48 (2016). https://doi.org/10.5815/ijmecs.2016.02.05
15. Benafia, A., Mazouzi, S., Sara, B.: Handwritten character recognition on focused on the segmentation of character prototypes in small strips. Int. J. Intell. Syst. Appl. (IJISA) 9(12), 29–45 (2017). https://doi.org/10.5815/ijisa.2017.12.04

Experimental Detection of the Parallel Organization of Mental Calculations of a Person on the Basis of Two Algebras Having Different Associativity

A. V. Koganov and T. A. Rakcheeva

Abstract In this work, we research a human's ability to use the parallel calculations for improving and accelerating of an information processing. We measured the amount of information that a person processes per unit of time. In the described experiment, the ability of a person to parallel pattern recognition in a large array of graphic information was determined. A person receives a series of tasks, the solution of which requires the processing of a certain amount of information. Computer recorded the time and correctness of the decision and calculated the dependence of the average solution time on the amount of information in the task for correctly solved problems. In accordance with the earlier proposed method (Koganov and Rakcheeva in Advances in Intelligent Systems and Computing. Springer, Warsaw, Poland, pp. 68–78, 2017; Koganov and Rakcheeva in The Computer Research and Modeling. The Computer Research Institute (UGU), pp. 621–638, 2017) [1, 2], the problems contain calculations of expression in two algebras, one of which is associative and the other is nonassociative. To facilitate the work of the subjects, we used in the experiment the figurative graphic images for representation of an algebraic elements. The computations in associative algebra allow the parallel counting. In the algebra where the associativity is absenting, the computations may be only sequential. Therefore, the analysis of the solution time in a series of problems allows to discover the computing strategy: uniformly sequential or accelerated sequential or parallel computing. The experiment was showed that all subjects used a uniform sequential strategy in nonassociative algebra, and parallel computations for associative problems.

Keywords Parallel counting · Engineering psychology · Testing · Algebra · Associativity · Recognition of visual images

A. V. Koganov (✉)
FGU FNC NIISI RAN, 36 Nakhimovsky st., corp. 1, Moscow 117218, Russia
e-mail: akoganov@yandex.ru

T. A. Rakcheeva
IMash RAN, 4, Bardina st., 117334 Moscow, Russia

© Springer Nature Switzerland AG 2020   139
Z. Hu et al. (eds.), *Advances in Artificial Systems for Medicine and Education II*,
Advances in Intelligent Systems and Computing 902,
https://doi.org/10.1007/978-3-030-12082-5_13

1 Introduction

Pattern recognition is an urgent task of artificial intelligence development [3–8]. We are investigating the organization of this information processing in the human brain. In the described experiment, the ability of a person to parallel pattern recognition in a large array of graphic information was determined. In the articles [1, 2], we developed theoretically the method of human testing in order to determine the strategy of his brain in adapting to the increasing complexity of computational problems. This technique was based on the use of two different types of calculations: associative and nonassociative. To do this, it was proposed to teach the subject two different formal algebras with associative and nonassociative operations. However, subsequent research has shown that it is more natural for a person to work in a game environment where the calculations are disguised as work with recognizable images. In previously our work, the experiments were carried out with the tasks of finding the maximum in the numerical table and calculating the trajectory of movement by cells of the numerical table [8–12]. The task of finding the maximum is associative, and the task of calculating the trajectory is nonassociative. It turned out that working in a formal environment caused some people disorientation and reduced concentration. This led to unreliable results, which did not allow a identifying of the adaptation strategy for about 25% of the subjects. In a new series of experiments, a person was presented with a problem in a game-shaped form that hid the calculations in the desired formal algebra. The results were much more stable and reliable.

The research method is based on the difference in possible ways of accelerating associative and nonassociative calculations. In the associative case, you can split a string into segments and perform simultaneous calculations on each segment, and then perform calculations on the line of this particle results. This is a parallel strategy, which gives acceleration approximately as many times as there were segments. In the nonassociative case, only sequential calculations from the first to the last operand in the string are allowed. In this case, the way of speeding up the work of those areas of the brain that are engaged in calculations is only to speed up the calculation process. However, this method gives acceleration in the associative case also. Therefore, if a decrease in the specific time per one operation, when the numbers of operations in the task increase, is observed in both types of tasks, then most likely, we observe a strategy of sequential solution with acceleration. If in the associative problem, the specific time decreases with the increase in the number of operations, and in the nonassociative case, the specific time remains almost constant, then we can talk about the registration of a parallel account strategy. In the case of constant specific time in both cases, a uniform sequential solution strategy is observed without acceleration when the complexity of the calculations increases. The remaining variants of dependence of the specific time and a task complexity correspond to inaccurate information about the adaptation strategy. Most often, this is due to the low concentration of the subject, usually due to low involvement in the solution of problems.

In fact, we measure the time of the task solution for different volumes of initial dates in the problems. If the time is a linear increasing of the date volume, then we register uniform sequential strategy of adaptation. If the time function has plateau in nonassociativity calculation, then we register the acceleration sequential strategy. If the time for nonassociative tasks is linear function of the volume initial date and the time for associative tasks has a plateau, then we register the parallel strategy of adaptation in the associative task series. In other variants, we register uncertainty. In new experiments, the obtained results do not contain an uncertainty cases.

2 Mathematical Basis of the Experiment

The full description of general mathematical principles of the specified research contained in the articles [1, 2]. From the point of view of mathematics, two algebras of a universal class were used in the implemented experiment.

2.1 Algebra of Cyclic Order (ACO)

The finite set of elements (carrier) is cyclically ordered:

$$x_1 < x_2 < \ldots < x_n < x_1. \tag{1}$$

A binary operation is specified on the carrier

$$(x_i, x_j) = \max\{x_i, x_j\}.$$

The correct expression in this algebra is an arbitrary finite string consisting of carrier elements. The calculation is performed by successive binary operations on the left pair of elements, and, further, on the last result and the next element of the line from left to right to the last right element.

$$z = y_1 y_2 y_3 \ldots y_{m-1} y_m = (((\ldots((y_1, y_2), y_3), \ldots), y_{m-1}), y_m). \tag{2}$$

Or in expanded form

$$z_1 = (y_1, y_2);$$
$$z_2 = (z_1, y_3);$$
$$z_3 = (z_2, y_4);$$
$$\ldots$$
$$z_{m-2} = (z_{m-3}, y_{m-1});$$
$$z = z_{m-1} = (z_{m-2}, y_m). \tag{3}$$

Here $y_1, y_2, y_3, \ldots, y_{m-1}, y_m \in \{x_1; \ldots; x_n\}$.
The algebra of cyclic ordering is nonassociative.

$$(x_n x_1) x_2 = x_2;$$
$$x_n (x_1 x_2) = x_n. \tag{4}$$

Therefore, in ACO, the evaluation of expression (2) requires a sequential procedure (3). In [1, 2], it is shown that the parallel counting in such problems generates a very low coefficient of use for computational elements (processors). Psychologically, this corresponds to large inefficient energy costs.

2.2 Algebra of Recognition (AR)

A finite set of elements (carrier) is specified

$$\{x_1; x_2; \ldots; x_n; 0; 1\}.$$

The other elements have a form of a string

$$y_1 y_2 y_3 \cdots y_{m-1} y_m;$$
$$y_1, y_2, y_3, \ldots, y_{m-1}, y_m \in \{x_1; \ldots; x_n\}. \tag{5}$$

The elements $\{x_1; \ldots; x_n\}$ are called the main carrier.
In AR, the two-place operations $f_1, \ldots, f_n$ on $\{x_1; \ldots; x_n\}$ are defined.

$$f_i(a, b) = \begin{cases} 1 \Leftarrow \{a; b\} \cap \{1; x_i\} \neq \emptyset; \\ 0 \Leftarrow \{a; b\} \cap \{1; x_i\} = \emptyset. \end{cases} \tag{6}$$

This operation recognizes the specified element x_i of main carrier among the arguments by the result value "1," or records the fact that it was recognized earlier, and this is recorded in at least one of the arguments. From the point of view of psychology, it simulates the elementary act of recognizing a given object.

The characteristic of the expression (5) is defined as a set of those elements of the main carrier, which absent in this line. If the string (5) contains all elements of main carrier, then the characteristic is defined as 0.

The problem of calculation of characteristics is reduced to the calculation of the value of the sequential transactions of recognition in relation to the entire row (5).

$$z = f_i(y_1 y_2 y_3 \cdots y_{m-1} y_m) =_{\text{def:}}$$
$$= f_i(f_i(f_i(\ldots f_i(f_i(y_1, y_2), y_3), \ldots), y_{m-1}), y_m). \tag{7}$$

However, decomposition into a sequence similar to (3) is not necessary in this case, and parallel computation of recognition on different parts of the sequence is possible. This is due to the associativity of the binary recognition operation.

$$f_i(f_i(y_1, y_2), y_3) = f_i(y_1, f_i(y_2, y_3)). \tag{8}$$

For example, expression (7) can be written in equivalent form with parallelism 2 and 3.

$$
\begin{aligned}
z &= f_i(y_1 y_2 y_3 \ldots y_{m-1} y_m) \\
&= f_i(f_i(y_1 \ldots y_k), f_i(y_{k+1} \ldots y_m)) \\
&= f_i(f_i(y_1 \ldots y_k) f_i(y_{k+1} \ldots y_{k+l}) f_i(y_{k+l+1} \ldots y_m)).
\end{aligned} \tag{9}
$$

It is also possible to calculate all the recognitions in a row at the same time. Therefore, the calculation of the string characteristic is possible in parallel with high and efficient load of parallel blocks. This allows you to reduce the time of calculations for giving volume of the input data information. Note that if the characteristic is not zero, then the calculation requires processing all the elements of the string. This is important for the assessing of the real processed information.

As shown in [1, 2], with the help of the described algebras, ACO and AP can be realized an experiment to identify the ability of a person to move to parallel computing, when it is effective.

3 Experiment Description

In the experiment, we proposed a series of computation problems in the above algebras to the test subject. In each series, the lengths of the sequences of the input date were the same. The subject solved several series of problems with different lengths of string, which grew from series to series. Initially, the main task was at ACO (nonassociative), and then, it was the task of characteristic calculation in AR (associative). In the each algebra before beginning of the work, the subject was trained on a series with a maximum length of the string. Training stopped at the request of the subject. The working series contained 10 tasks of each length.

Technically, the experiment was implemented on a computer screen in a special window in MATLAB. As elements of the algebra carrier, a set of simple pictures depicting well-known objects with clear names was proposed. This is important for recognition. Pictures were placed in standard small squares. The task string was formed by these squares with the transfer of the left, as in the letter. The length of the graphic line was 7 squares. A table of several lines appeared on the screen, which had to be interpreted as one long line. The test showed that this does not cause difficulties in a person who can read.

In the same tables, there were several buttons and windows, which the subject or experimenter operated with a computer mouse and click, or keyboard. The experimenter sets the length of the string in the current series. The subject started the next task, stopped the countdown when he calculated the answer (the line disappeared), and recorded the result. After that, he could rest and run the next task. The number of the current task in the series was displayed in a special window. When the last contractual (tenth) problem was solved, the subject could ask for the next series. If an external interference occurs when solving problems, the series could contain additional tasks to replace the corrupted ones.

The parameters of the experiment.

3.1 In the Problem of Cyclic Ordering, the Alphabet of Three Pictures "Stone," "Scissors," and "Paper" Was Used

Their ordering corresponded to the well-known game: scissors cut paper, paper wraps stone, and stone breaks scissors (Fig. 1a). We had to determine the "winner" in a random line of these pictures at the action from left to right. The pictures on each item occurred equiprobable likely and statistically independent; it is possible a repeat of neighboring the images (Fig. 2). The answer was given in the form of a picture number of the winner. The alphabet of pictures with numbers has always been on the screen in the field of view under the task line. Further, this is the task of "rock-paper-scissors" or the RPS.

Number of task series 4.

The length of the lines in the series: 8, 12, 16, 20.

On-screen view: table of figures with line length 7. The last line could contain empty cells on the right.

Number of tasks in one series is 10.

3.2 In the Problem of Calculating the Characteristics, It Is Used Alphabet of Six Images "Ship," "Locomotive," "Car," "House," "Tree," and "Horse" (Fig. 1b)

The subject was warned that in the line either there is no exactly one picture, or there are all pictures. The problems were generated by the following probabilistic algorithm. Initially, one "discarded" picture of alphabet whose chose equiprobable. Then, a line was generated in which all the pictures, except for the discarded one, appeared on each position equally probable and statistically independent. In the second stage, all the pictures were entered in a row in the initial positions of the line "forcibly," except for the discarded one, and random permutations of each of these first positions with a randomly selected position were made. The choice of the

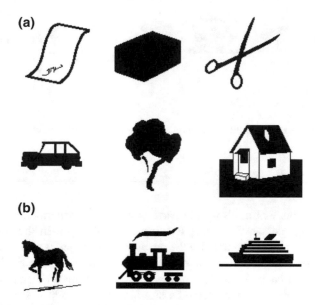

Fig. 1 The pictures which were using in the test: **a** the problem of cyclic ordering; **b** the problem of recognition. In real display, the pictures were color

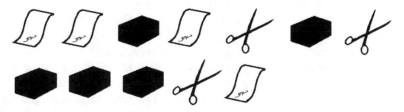

Fig. 2 The example of a string for a cyclic ordering problem of length 12 (the result is "the paper")

position of the exchange was carried out uniformly across the row and is statistically independent of the other random selections. Before the second stage, with a given probability (0.1 was used), it was determined that there should be no missing picture in the line. Then, the discarded picture was entered into the line at the second stage before mixing. This method of generation guaranteed that the line does not have exactly one picture, or the entire alphabet. The subject gave the answer in the form of the number of the missing picture, or 0, if all were present. The alphabet of pictures with numbers has always been on the screen in the field of view under the task line. Next, this is the "recognition problem" (Fig. 3).

Number of task series 4.

The length of the lines in the series: 8, 12, 16, 20.

On-screen view: table of figures with line length 7. The last line could contain empty cells on the right.

Number of tasks in one series 10.

Fig. 3 The example string for the recognition problem of length 8 (the result is "the tree")

4 Experimental Result

At the time of this writing, 8 people participated in the experiment. The duration of testing of one person, including training, is 30–40 min. In the resulting data processing, the average time to solve the problem for each series was calculated, and the standard deviation of this average was estimated from the empirical variance of the solve time for each series. The average time of solving problems in two series of the same algebra was considered as significantly different if their intervals of plus or minus standard deviation did not intersect. Data processing were carried out individually for each test subject, because the average for people in this case does not reflect individual strategies for solving the problem.

In all experiments, no errors in the solution of problems were recorded. This allows you to evaluate the processing speed of the input information directly as the ratio of the string length to the average solution time in a series with a given string length. All subjects showed a significant increase in the average time of solving the problem of RPS depending on the length of the string. The graph of average times in all subjects is close to the linear function. The deviation of the experimental points from the linear approximation does not exceed the standard deviation. The slope of the approximating straight line was different in different subjects. This allows us to speak confidently about a sequential and uniform method of solving the nonassociative problem of RPS in all subjects. However, the speed of solution varies from person to person (Fig. 4).

In the recognition problem, all subjects showed unreliable oscillations of the average solution time around a certain level independent on the length of the string in the series. This suggests a parallel mechanism of recognition for humans. However, some subjects showed a small and unreliable tendency to reduce the solution time as the length of the string increased (Fig. 5). Perhaps this indicates an increase the concentration of humans with increasing complexity of the task. In addition, two subjects had a significant small increase in the average solution time at a length of 20 (Fig. 4 right). This may indicate a shift to a partially sequential solution. This is probably the result of additional checks of the already obtained solution.

In the experimental work [8], it was shown that in the problem of calculating the trajectory from a numerical table, which does not allow an effective parallel solution, some test subjects used a sequential account with acceleration as the complexity increased. In the task of the RPS, this was not observed. Perhaps this is due to the

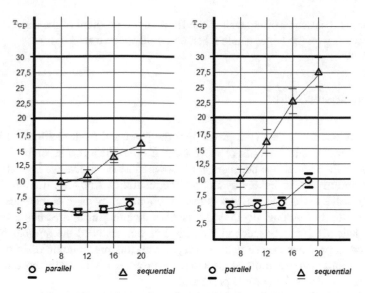

Fig. 4 The examples of characteristic type of the solution time dependence on the date string long. Circles designate the points of recognition problem, and triangles designate RPS problem

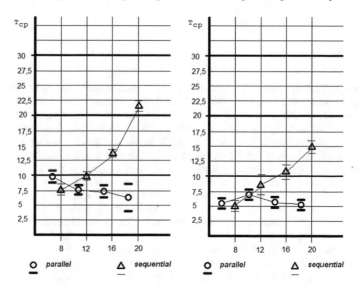

Fig. 5 Examples with a no reliable acceleration of recognition

purely logical structure of the task of the RPS. In the problem of the trajectory used visual tracking on the table, which a person could speed up.

5   Conclusion

The results obtained allow us to assert that the proposed method allows to identify sequential and parallel methods of processing of symbolic information by the human brain. Specifically in this experiment, it is shown that nonassociative logical calculations are intuitively carried out by a person sequentially. And after training a person begins to solve the associative problems of recognition by parallel methods.

The method of coding algebraic signs with meaningful figures has given good results. This significantly reduced training time and reduced the number of errors compared to previous experiments. Perhaps, with this presentation of information, any additional brain structures connect to solve problems, but usually those structures are not involved to solving logic problems.

The work is done by the government specify of science research, the theme of 0065-2018-0004.

References

1. Koganov, A.V., Rakcheeva, T.A.: Tests of parallel information processing on the basis of algebra and formal automata. In: Kacprzyk, J. (ed.) Advances in Intelligent Systems and Computing; Advances in Artificial Systems for Medicine and Education, vol. 658, pp. 68–78. Polish Academy of Sciences, Springer, Warsaw, Poland (2017). ISBN 978-3-319-67348-6
2. Koganov, A.V., Rakcheeva, T.A.: Tests for checking the parallel organization of the logical calculation on the basis of algebra and automata. In: The Computer Research and Modeling, pp. 621–638 (T. 9, N 4). The Computer Research Institute (UGU), The Machine research institute at RAN in the name of A.A. Blagonravov (2017). ISSN 2076-7633, Russian
3. Fischer, R., Plessow, F.: Efficient multitasking: parallel versus serial processing of multiple tasks. Front. Psychol. **6**, 1366 (2015). https://doi.org/10.3389/fpsyg.2015.01366
4. Chabbah Sekma, N., Elleuch, A., Dridi, N.: Automated forecasting approach minimizing prediction errors of cpu availability in distributed computing systems. Int. J. Intel. Syst. Appl. (IJISA) **8**(9), 8–21 (2016). https://doi.org/10.5815/ijisa.2016.09.02
5. Zoubida, L., Adjoudj, R.: Integrating face and the both irises for personal authentication. Int. J. Intel. Syst. Appl. (IJISA) **9**(3), 8–17 (2017). https://doi.org/10.5815/ijisa.2017.03.02
6. Maurya, M., Singh, A.: An approach to parallel sorting using ternary search. Int. J. Mod. Educ. Comput. Sci. (IJMECS) **10**(4), 35–42 (2018). https://doi.org/10.5815/ijmecs.2018.04.05
7. Nandi, D., Saifuddin Saif, A.F.M., Prottoy, P., Zubair, K.M., Shubho, S.A.: Traffic sign detection based on color segmentation of obscure image candidates: a comprehensive study. Int. J. Mod. Educ. Comput. Sci. (IJMECS) **10**(6), 35–46 (2018). https://doi.org/10.5815/ijmecs.2018.06.05
8. Popov, G., Mastorakis, N., Mladenov, V.: Calculation of the acceleration of parallel programs as a function of the number of threads. https://www.researchgate.net/publication/228569958. Article Jan 2010

9. Koganov, A.V., Pyatecky-Shapiro, I.I., Fejgenberg, I.M.: The depending of the speed of the solution from the complexity and the coding method of initial dates. In: Compendium the Equations of the Experimental Investigation of the Reacting Speed. Tartu (1971)

10. Koganov, A.V.: The growing inductor spaces and the analysis of the parallel algorithms. The programming products and systems, the application to the international journal "The problems of the theory and practicum of the control", pp. 33–38, N 2 (2010)

11. Koganov, A.V., Zlobin, A.I., Rakcheeva, T.A.: The research of the possibility of the parallel processing of the information by man in the series of the tasks of the high complexity. In: The Computer Research and Modeling, pp. 845–861 (t. 5, N 5). The Computer Research Institute (UGU), The Machine research institute at RAN in the name of A.A. Blagonravov (2013)

12. Koganov, A.V., Zlobin, A.I., Rakcheeva, T.A.: The task of the calculation of the trajectory with homogenous the distribution of solutions. In: The Computer Research and Modeling, pp. 803–828 (t. 6, N 5). The Computer Research Institute (UGU), The Machine research institute at RAN in the name of A.A. Blagonravov (2014)

Determinant Optimization Method

Nikolay A. Balonin and Mikhail B. Sergeev

Abstract This article is devoted to matrices which play a significant role in digital signal processing, artificial intelligence systems, and mathematical natural sciences in the whole. The study of the world of matrices is going on intensively all over the world and constantly brings useful and unexpected results. This paper discusses the determinant optimization method for finding so-called quasi-orthogonal matrices. It is shown that the area of its application is extended due to matching of quasi-orthogonal matrices, which are locally optimal by their determinants (Cretan matrices), and non-orthogonal matrices of determinant maximum (D-matrices). All necessary definitions of matrices and their properties, used in the optimization algorithm, are provided. Two realizations of the algorithm are given—the basic and the reversed one. Examples of matrices calculated using the proposed method are given.

Keywords Hadamard matrices · Determinant · Determinant maximum · D-matrices · Quasi-orthogonal matrices · Cretan matrices · C-matrices · Cyclic matrices

1 Introduction

Many digital technologies of artificial intelligence and noise-immune coding of information are created on the basis of Hadamard matrices [1], which applied in bioinformatics to study noise-immune properties of the genetic code systems [2, 3]. Different kinds of matrices are used in biometric person identification systems, web video object mining, and many other modern practical tasks [4–6].

N. A. Balonin (✉) · M. B. Sergeev
Saint Petersburg State University of Aerospace Instrumentation, 67, B. Morskaia St., 190000 St. Petersburg, Russian Federation
e-mail: korbendfs@mail.ru

M. B. Sergeev
e-mail: mbse@mail.ru

© Springer Nature Switzerland AG 2020
Z. Hu et al. (eds.), *Advances in Artificial Systems for Medicine and Education II*,
Advances in Intelligent Systems and Computing 902,
https://doi.org/10.1007/978-3-030-12082-5_14

The special work of Balonin and Seberry is concentrated on remarks on extreme and maximum determinant matrices with moduli of real entries ≤ 1 giving an overview of related works [7]. Matrix determinant optimization is a relatively new issue, which is not widely studied yet. It is significantly more complicated compared to the simple determinant calculation.

For example, the totality of matrices with entries $\{1, -1\}$ consists of maximum determinant Hadamard matrices ($\mathbf{H}$) of orders 1, 2, and $n = 4t$, where t is a natural number, $\mathbf{H}^T\mathbf{H} = n\mathbf{I}$, $\mathbf{I}$ is the $n \times n$ identity matrix, and "T" stands for transposition [8].

The columns and rows of a matrix, which is extreme by determinant, are strictly orthogonal.

A real square matrix $\mathbf{A}$ of order n is called quasi-orthogonal if it satisfies $\mathbf{A}^T\mathbf{A} = \omega\mathbf{I}$, where ω is a constant real number.

In this and future works, we will only use quasi-orthogonal to refer to matrices with real elements; a least one entry in each row and column must be 1. Mentioned Hadamard matrices are the best known among these matrices with entries from the unit disk.

The class of quasi-orthogonal matrices with maximal determinant and entries from the unit disk may have a very large set of solutions. Different solutions may give the same maximal determinant. Symmetric conference matrices, a particularly important class of $0, \pm 1$ matrices, are the most well known [7]. A symmetric conference matrix, $\mathbf{C}$, is an $n \times n$ matrix with elements $0, +1$, and -1, satisfying $\mathbf{C}^T\mathbf{C} = (n-1)\mathbf{I}$.

Conference matrices can only exist if the number $n - 1$ is the sum of two squares. Similar to symmetric conference matrices are quasi-orthogonal matrices $\mathbf{W}(n, n - m)$ of order n with elements $0, +1$, and -1, satisfying $\mathbf{W}^T\mathbf{W} = (n-m)\mathbf{I}$.

The values of the entries of the quasi-orthogonal matrix, $\mathbf{A}$, are called *levels*, so Hadamard matrices are two-level matrices, and symmetric conference matrices and weighing matrices are three-level matrices. Quasi-orthogonal matrices with maximal determinant of odd orders have been discovered to have a larger number of levels [1].

The Balonin–Mironovsky matrix (2005), $\mathbf{A}$, of order n, is a quasi-orthogonal matrix of maximal determinant. In this paper, they are called BM matrices. Conjecture (Balonin [1, 9, 10]): There are only 5 Balonin–Mironovsky matrices of orders 3, 5, 7, 9, and 11 with $(n + 1)/2 \pm m$ levels, $m \leq 1$. Order 13 was unresolved. During 2006–2014, Balonin and Sergeev carried out many computer experiments to find the absolute maximum of the determinant of $\mathbf{A}$, order 13. It was speculated [9] that 13 is a critical order for matrices of odd orders with maximal determinant. Starting from this odd order, the number of levels $k >> (n + 1)/2$ (level explosion).

The absence of a solution with a low number of levels for $n \geq 13$ led Balonin and Sergeev to search and classify quasi-orthogonal matrices with other properties [9–11]. A quasi-orthogonal matrix with extreme or fixed properties, global or local extremum of the determinant, saddle points, minimum number of levels, or matrices with fixed numbers of levels, is called a Balonin–Sergeev matrix. They are called also BSM matrices [7]. A Balonin–Mironovsky matrix is a Balonin–Sergeev matrix with the absolute maximum determinant. Balonin–Sergeev matrices with fixed numbers of levels were first mentioned during a conference in Crete, so we will call them Cretan matrices (CM-matrix). A Cretan matrix $\mathbf{A}$, $\mathbf{A}^{\mathrm{T}}\mathbf{A} = \omega \mathbf{I}$, of order n, which has k indeterminate entries, $a = 1$, b, c, d, ..., is said to have k levels [1, 7, 12–17].

To get Cretan matrices, we need an instrument to search it: determinant optimization method. The optimization of determinants of matrices that are orthogonal by columns with finding local determinant maximums is possible, in general, for any odd and even orders, including the orders of Hadamard matrices.

Figure 1 shows the images of sub-optimal quasi-orthogonal matrices of orders 17 and 14, the variety of entry values of which is represented by the shades of gray.

Rounding the entries of these matrices to integers $\{1, -1\}$ gives so-called **D**-matrices that are not orthogonal, but strictly optimal by matrix determinant. The rule, demonstrated by the pictures, is common. The absolute and relative maximums of determinants of matrices that are non-orthogonal and orthogonal by their columns (quasi-orthogonal) match by corresponding matrix structure.

The structures of optimal matrices are similar. The entry values that are described by parametric dependencies allow for getting matrices of both types, in particular by way of rounding. However, the absolute determinant maximum of non-orthogonal matrices corresponds to not biggest optimum in the class of orthogonal matrices, and vice versa.

So, an extreme quasi-orthogonal matrix with a small number of different elements is called *Cretan matrices* [7]. The interrelation of these extreme tasks allows for searching both Cretan and D-matrices using the same numerical method.

The peculiarity of Cretan matrices is that only a part of them is binary (including entries with only two values), like Hadamard matrices.

It is worth mentioning that for binary matrices, indirect optimization methods are developed, which hide the extreme task behind the solution of an equation.

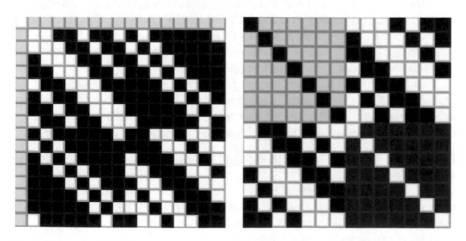

Fig. 1 Sub-optimal BSM matrices of orders 17 and 14

For Hadamard matrices, this equation is $\mathbf{H}^T\mathbf{H} = n\mathbf{I}$. The Galois field arithmetic allows for finding both blocks of the $\mathbf{H}$-matrix, and Cretan matrices of odd orders. However, the totality of such solutions is limited to the area of application of the theory of finite fields to matrix orders that are prime integers or their degrees.

This paper discusses the numerical method of determinant optimization, which does not impose artificial restrictions mentioned above. The number of values of entries of the required Cretan matrices can be more than two, and their orders can be not only prime numbers.

In a broad sense, we will call *quasi-orthogonal* all matrices that are orthogonal by their columns (and rows) $\mathbf{A}^T\mathbf{A} = \omega(n)\mathbf{I}$. In this case, they will include orthogonal matrices with weight $\omega = 1$ and maximal entry $m < 1$ in modulus.

Let us observe more precious and detailed definitions.

2   Terms and Definitions

Definition 1 A *quasi-orthogonal* matrix [1, 7, 9] is a square matrix $\mathbf{A}$, order n, with the maximum of absolute values of its entries reduced to a unit, and satisfying the quadratic constraint condition $\mathbf{A}^T\mathbf{A} = \omega(n)\mathbf{I}$, where $\omega(n)$ is some *weight function* that determines the matrix type, and $\mathbf{I}$ is the identity matrix.

This paper discusses matrices that are extreme by determinant when the entry values are limited to ≤ 1 [1, 7]. It is obvious that after being multiplied by $1/m$, any orthogonal matrix with determinant equal to 1 is reduced to quasi-orthogonal, and its determinant is increased by a factor of $1/m^n$. Further increase in the determinant by way of scaling is impossible, since this will lead to breaking the entry value limit.

Definition 2 Maximal entry m of an orthogonal matrix, associated with corresponding quasi-orthogonal matrix, is called their m-norm [9].

The m-norm parameter of extreme solutions is minimal.

Actually, $|\det(\mathbf{A})| = 1/m^n$, and the lesser is m, the greater the determinant value is. It follows from $\det(A)^2 = \omega^n$ that weight $\omega = 1/m^2$.

It is notable that the extreme points of matrices by determinants can be global or local, but *saddle points* should also be taken into consideration. Strictly extreme quasi-orthogonal matrices have corresponding minimax matrices—orthogonal matrices of given orders with minimal value of maximal (in modulus) entries.

Definition 3 Hadamard norm h (h-norm) [1] of a quasi-orthogonal matrix is $h = m\sqrt{n} \geq 1$.

Hadamard norm of a quasi-orthogonal matrix is more convenient than the m-norm when controlling the completion of calculation of Hadamard matrices, since for them (and only for them) its value is 1. This is the invariant of Hadamard matrices. All other quasi-orthogonal matrices have $h > 1$.

Definition 4 The *reduced* determinant D_n (h-determinant) of a quasi-orthogonal matrix is $D_n = 1/h^2 \leq 1$.

It follows from $\det(\mathbf{A})^2 = \omega^n$ and $\omega = 1/m^2 = n/h^2$ that $\det(\mathbf{A})^2 = n^n D_n^n$. Matrix determinant is a fairly large number, but the value inverse to h-norm is always equal to 1 for all Hadamard matrices ($\det(\mathbf{H}) = n^{n/2}$) and is less than 1 for all other matrices.

For example, the maximal entries (m-norms) of sub-optimal orthogonal matrices of orders 17 and 14, shown in Fig. 1, are 0.2941 and 0.3035. The reduced determinants of these matrices are 0.6801 and 0.7755.

Definition 5 *The Balonin Constant* $B = D_{13}$ [9] is the reduced determinant of the absolute maximum determinant matrix of order 13.

For matrix of order 13, $m \cong 0.31$, $h \cong m\sqrt{13}$ and $D_{13} \cong 0.8$. The value of this limit is due to the structural rebuilding of matrices—at order 13, the number of entry values of a strictly optimal matrix grows, and the disperse state of elements becomes optimal [1]. If the algorithm of determinant optimization gets cycled on the disperse structure, then regardless of the order $n \geq 13$, the value of the reduced determinant is close to 0.8.

If the determinant can be increased to $D_n > 0.8$, which can be achieved by increasing the size of matrix, it is a sure sign of getting closer to the ordered structure.

Definition 6 *Entropy* of a quasi-orthogonal matrix $E = \lg(1/D_n)$.

It is a yet another useful metric to control the quality of optimization. The entropy of Hadamard matrices is always 0. For all other matrices, it is above zero, and the entropy of chaotic matrices approximately is equal to 1.

Definition 7 *Cretan matrices* [7] are extreme quasi-orthogonal matrices and matrices of saddle points with small numbers of values of their entries, defined by the order functions.

Matrices of different families and corresponding functions of values of their entries are given in papers [1, 10, 13]. In the general case, the number of entry values of strictly optimal and sub-optimal (locally optimal) quasi-orthogonal matrices can be significantly greater than two [7].

3 Minimal Calculation Method of Optimization

The simplified form of the iteration scheme of calculations for optimizing of a quasi-orthogonal matrix determinant is shown in Fig. 2.

Any non-orthogonal matrix $\mathbf{M}$ can be orthogonalized, for instance, using the Gram–Schmidt algorithm [9]. After that, to ensure the growth of the matrix determinant, which is inversely proportional to the m degree, the matrix's maximal entry is decreased by the way of multiplication of entries by a saturation function.

In the simplest case, all entries that are by absolute value bigger than the threshold $p_k m$, where $p_k \leq 1$, k—iteration index, are reduced to the threshold (saturation).

Let us call the initial value p_0 of the scale multiplier *the initial compression index*. Its value at the start is 0.5 and it changes by calculation of $p_{k+1} = a p_k + b$, where $b = 1 - a$ and $a < 1$. Therefore, the compression is decreased with the increase in the number of iterations, since the *compression index* p_k tends to 1.

Fig. 2 Iteration scheme of
determinant optimization

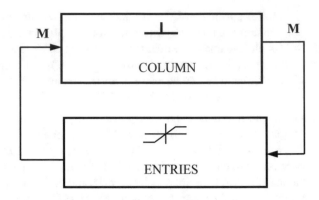

4 Calculations with Reversed Sequence and Initial Selection

The effectiveness of the minimal scheme of calculation, shown above, is significantly expanded by two additions:

- Reversed sequence of the calculation that was first used in the search for a rare artifact—minimax six-level matrix of the 22 order [14];
- Selection of a matrix of initial approximation that is consistent with the reversed sequence of calculation and taking into account their explicit or hidden symmetry [16, 17].

At the initial stage, the matrix generator forms a lot of **A**-matrices of a certain form, for instance, circulant, partly symmetric. The symmetry of the solution is known in advance or can be found by analyzing the results of calculation of low-order matrices.

Quadratic residual $\xi = \left\| \mathbf{A}^\mathrm{T}\mathbf{A} - \omega(n)\mathbf{I} \right\|$ is calculated for each matrix. From these matrices, only one should be selected with the smallest value of ξ as a starting approximation of $\mathbf{A}_0$, and norms of its columns are reduced to 1 by way of scaling. The initial vector $\mathbf{X}_0 = [0, 1, 2, ..., n - 1]$ of the numbers of future reversed column permutation is set.

Procedure 1 Permutation of columns of an iterated matrix $\mathbf{A}_k$ and elements of the vector of reversed permutation $\mathbf{X}_k$ in the descending order of absolute values of maximal entries in columns.

Procedure 2 Matrix compression. Absolute values of matrix entries are set in the way not to exceed $p_k m$. The compression is big at the start, but the coefficient changes further as $p_{k+1} = a p_k + b$, where $b = 1 - a$, and usually $a = 0.995$.

Procedure 3 Orthogonalization of a compressed matrix using the Gram–Schmidt method. Permutation of columns creates effective "linking" to a vector that was maximally changed in the desired way.

Upon completion of these three procedures, a reduced determinant D_n should be calculated. Otherwise, the matrix entropy E_n should be calculated, which together with the quadratic residual gives an indication of the quality of calculations and their closeness to completion.

If the calculations are unsuccessful, the generation of a new matrix of initial approximation is needed. Otherwise, the calculations end with a reversed permutation of columns using vector $\mathbf{X}_k$ to rebuild the initial matrix structure.

The algorithm ends after reducing the compression and calculating of the final Cretan matrix by way of dividing all entries by the maximal entry m or by rounding, if the objects of search were Hadamard matrices or **D**-matrices that include them.

5 Examples of Calculated Matrices

Figure 3 shows quasi-orthogonal BM matrices of absolute maximum of determinants $\mathbf{A}_3$, $\mathbf{A}_5$, $\mathbf{A}_7$, $\mathbf{A}_9$, $\mathbf{A}_{11}$, $\mathbf{A}_{13}$ found and systematized by possible structures. As can be seen on the histograms, the number of different entries by modulus grows in optimal matrices of odd orders.

At order 13, the qualitative rebuild can be seen, when weakly dispersed composition of entries becomes non-optimal.

All optimal matrices with orders over 13 have complex structures. It makes the search for locally optimal Cretan BSM matrices with *two or three* values of modules of their entries very important.

It is the matrices of local maximums that, being rounded, correspond to Hadamard matrices [8], Belevitch **C**-matrices, and weighing **W**- and **D**-matrices [9]. The universal determinant optimization algorithm allows for studying the families and the resulting chains of calculated Cretan matrices [1, 18–20].

The shown optimization method brought many new ideas, and scientific and practical results. The first of all, it established that besides famous Hadamard matrix conjecture (Hadamard matrices of all orders $4t$, t integer, exist), there is so named

Fig. 3 Images of matrices $\mathbf{A}_3$, $\mathbf{A}_5$, $\mathbf{A}_7$, $\mathbf{A}_9$, $\mathbf{A}_{11}$, $\mathbf{A}_{13}$ with histograms of their entries

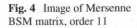

Fig. 4 Image of Mersenne
BSM matrix, order 11

Balonin conjecture [9, 11] accordingly two-level Mersenne matrices of odd orders
$4t - 1$, Fig. 4. They exist for every $4t - 1$.

6 Conclusion

The method for optimization of determinants of matrices of even and odd combined
orders, described in this paper, is effective where other methods, including the method
using the arithmetic of Galois fields, cannot be successfully applied.

The effectiveness of the approach and its independence from the simplicity of
the order of the sought matrix open the door to revising the idea about the non-
demonstrability of the Hadamard hypothesis, formed due to the use of the combinatorial approach to solving the problem discussed in the article.

In different to integer Hadamard matrices, the two-level odd-order matrices,
named Mersenne matrices, can have rational and irrational levels and defined less
hard.

From existence of these matrices for all odd orders follows existence of Hadamard
matrices, since any Hadamard matrix is the Mersenne matrix with rounded levels
and the border consisting of -1. Instead of a task of combinatorial mathematics, we
have a typical task of extreme search, not bounded by integer reasons.

This conclusion has a big practical meaning for technical tasks. Cretan matrices,
simple in structure, exist on all orders, odd and even, that is important for new methods
of image compression and fault detection. There is a big set of them described in [1]
and other papers of the authors and of other scientific workers [7, 16] etc.

Acknowledgements The authors wish to sincerely thank Tamara Balonina for converting this
paper into printing format. The authors also would like to acknowledge the great help of Professor

Jennifer Seberry. The research leading to these results has received funding from the Ministry of Education and Science of the Russian Federation according to the project part of the state funding assignment 2.2200.2017/4.6.

References

1. Balonin, N.A., Sergeev, M.B.: Quasi-orthogonal local maximum determinant matrices. Appl. Math. Sci. **9**(6), 285–293 (2015). https://doi.org/10.12988/ams.2015.4111000
2. Petoukhov, S.V.: The genetic code, algebra of projection operators and problems of inherited biological ensembles, pp. 1–93. http://arxiv.org/abs/1307.7882, 8th version of the article from 3 May 2017
3. Petoukhov, S.V., He, M.: Symmetrical Analysis Techniques for Genetic Systems and Bioinformatics: Advanced Patterns and Applications. IGI Global, Hershey, USA (2010)
4. Angadi, S.A., Hatture, S.M.: Biometric person identification system: a multimodal approach employing spectral graph characteristics of hand geometry and palmprint. Int. J. Intell. Syst. Appl. (IJISA) **8**(3), 48–58 (2016). https://doi.org/10.5815/ijisa.2016.03.06
5. Sahana, S.K., Mohammad, A.L.F., Mahanti, P.K.: Application of modified ant colony optimization (MACO) for multicast routing problem. Int. J. Intell. Syst. Appl. (IJISA) **8**(4), 43–48 (2016). https://doi.org/10.5815/ijisa.2016.04.05
6. Algur, S.P., Bhat, P.: Web video object mining: a novel approach for knowledge discovery. Int. J. Intell. Syst. Appl. (IJISA) **8**(4), 67–75 (2016). https://doi.org/10.5815/ijisa.2016.04.08
7. Balonin, N.A., Seberry, J.: Remarks on extremal and maximum determinant matrices with real entries ≤ 1. Informatsionno-upravliaiushchie sistemy [Inf. Control Syst.] **5**(71), 2–4 (2014)
8. Hadamard, J.: Resolution d'une question relative aux determinants. Bull. Sci. Math. **17**, 240–246 (1893)
9. Balonin, N.A., Sergeev, M.B.: Local maximum determinant matrices. Informatsionno-upravliaiushchie sistemy [Inf. Control Syst.] **1**(68), 2–15 (2014). (In Russian)
10. Balonin, N.A., Sergeev, M.B.: Mersenne and Hadamard matrices. Informatsionno-upravliaiushchie sistemy [Inf. Control Syst.] **1**(80), 92–94 (2016). https://doi.org/10.15217/issn1684-8853.2016.1.2. (In Russian)
11. Sergeev, A.M.: Generalized Mersenne matrices and Balonin's conjecture. Autom. Control Comput. Sci. **48**(4), 214–220 (2014). https://doi.org/10.3103/S0146411614040063
12. Balonin, N.A., Sergeev, M.B.: The generalized Hadamard matrix norms. Vestn. St.-Petersb. Univ. Appl. Math. Comput. Sci. Control Process. **2**, 5–11 (2014). (In Russian)
13. Balonin, N.A., Djokovic, D.Z., Seberry, J., Sergeev, M.B.: Hadamard-type matrices. Available at: http://mathscinet.ru/catalogue/index.php. Accessed 16 Aug 2018
14. Balonin, N.A., Sergeev, M.B.: M-matrix of the 22nd order. Informatsionno-upravliaiushchie sistemy [Inf. Control Syst.] **5**, 87–90 (2011). (In Russian)
15. Balonin, N.A., Sergeev, M.B.: Initial approximation matrices in search for generalized weighted matrices of global or local maximum determinant. Informatsionno-upravliaiushchie sistemy [Inf. Control Syst.] **6**(79), 2–9 (2015). https://doi.org/10.15217/issn1684-8853.2015.6.2. (In Russian)
16. Balonin, N.A., Djokovic, D.Z.: Symmetry of two circulant Hadamard matrices and periodic Golay pairs. Informatsionno-upravliaiushchie sistemy [Inf. Control Syst.] **3**(76), 2–16 (2015). https://doi.org/10.15217/issn1684-8853.2015.3.2. (In Russian)
17. Seberry, J., Yamada, M.: Hadamard matrices, sequences, and block designs. In: Dinitz, J.H., Stinson, D.R. (eds.) Contemporary Design Theory: A Collection of Surveys, pp. 431–560. Wiley, New York (1992)
18. Balonin, N.A., Sergeev, M.B., Mironovsky, L.A.: Calculation of Hadamard-Mersenne matrices. Informatsionno-upravliaiushchie sistemy [Inf. Control Syst.] **5**(60), 92–94 (2012). (In Russian)

19. Balonin, N.A., Vostrikov, A.A., Sergeev, M.B.: On two predictors of calculable chains of quasi-orthogonal matrices. Autom. Control Comput. Sci. **49**(3), 153–158 (2015). https://doi.org/10.3103/S0146411615030025

20. Balonin, N.A., Djokovic, D.Z.: Negaperiodic Golay pairs and Hadamard matrices. Informatsionno-upravliaiushchie sistemy [Inf. Control Syst.] **5**(78), 2–17 (2015). https://doi.org/10.15217/issn1684-8853.2015.5.2. (In Russian)

Analysis of Oscillator Ensemble with Dynamic Couplings

M. M. Gourary and S. G. Rusakov

Abstract Challenges to analyze networks of weakly coupled oscillators are considered. The model of the oscillator ensemble with the frequency-dependent couplings has been developed. The construction of the model is based on special locking functions obtained by the analysis of synchronized oscillator under external excitation. It is shown that the set of locking functions completely define coupling functions of well-known Kuramoto model. The ensemble interconnections are presented by linear time-invariant dynamical systems that correspond to the frequency-dependent transfer functions included in the model. The order of the resulting system is equal to the number of oscillators, and it does not depend on orders of transfer functions. Two approaches to take into account the frequency dependence of couplings are proposed—by the average frequency and by instantaneous frequencies. The simplified form of the model for weakly nonlinear sinusoidal oscillators is derived. Presented numerical examples illustrate the usage of the model and demonstrate some new effects arising in the oscillator ensemble with the dynamic couplings.

Keywords Coupled oscillators · Kuramoto model · Phase macromodel · Synchronization · Dynamical system · Transfer function

1 Introduction

Analysis of the coupled oscillator networks is one of the main problems in nonlinear dynamics. It plays an important role in various fields of research including biology, chemistry, physics, and electronics. [1, 2]. The analysis of oscillatory neural networks [3, 4] is of particular interest. The simulation of the complete set of the ensemble equations is suitable for a small number of simple oscillators [5, 6], but such evaluation of high-dimensional networks requires inadmissible computational efforts. This

M. M. Gourary (✉) · S. G. Rusakov
Institute for Design Problems in Microelectronics, Russian Academy of
Sciences (IPPM RAS), Moscow, Russia
e-mail: ippm@ippm.ru

© Springer Nature Switzerland AG 2020   161
Z. Hu et al. (eds.), *Advances in Artificial Systems for Medicine and Education II*,
Advances in Intelligent Systems and Computing 902,
https://doi.org/10.1007/978-3-030-12082-5_15

leads to the need for simplified techniques. The most widespread simplification is the Kuramoto model (KM) [1, 3], where each ordinary differential equation (ODE) defines the phase of the corresponding oscillator. KM presents oscillator interconnections by static transfer factors that can result in the incorrect solution. So, to consider the dynamic properties of interconnects, several model modifications were proposed.

The simplest coupling dynamics determines the time delay of the signal passage through the interconnection. This transforms KM to the system of delayed differential equations [1, 7]. A special form of dynamic couplings appears in many biological systems due to the diffusion of a chemical substance. Additional ODE for each coupling in the KM provides the correct description of such ensemble [7]. A similar approach with another form of the additional ODE is referred to as an adaptive coupling method [8, 9]. The KM supplemented by second-order phase derivatives [1, 10] provides the simulation of the inertial dynamics of power grids with rotating generators and motors.

However, these modifications do not cover dynamic effects in the propagation of electric signals along interconnects (e.g., neural axons in biological networks). Electrical interconnects in the form of networks containing traditional linear components (resistances, capacitances, inductances) represent linear time-invariant (LTI) dynamical system defined by complex-valued frequency-dependent transfer functions (TF). Known publications on KM-based methods do not consider general LTI interconnects, unlike the alternative to KM approach based on the nonlinear phase macromodel [2, 11, 12]. This approach presents the solution of the oscillator ODE system under small perturbation as the unperturbed signal with the time-varying delay. The delay satisfies nonlinear ODE representing an oscillator macromodel. The ensemble model can also include the ODE system of LTI interconnects. Examples of the analysis of oscillators with resistive, capacitive, inductive, and transmission line couplings are given in [13]. Shortcomings of this approach are associated with highly oscillatory terms in the macromodel solution [14]. This leads to the significant increase in the number of steps when ODE solving, large errors in the instantaneous frequency, difficulties in determining synchronization conditions. Besides, highly dimensional interconnect equations essentially increase the size of the total system.

The goal of this research is the proposal of an approach to analyzing the oscillator ensemble with arbitrary LTI-defined couplings. The approach defined by the extended KM considers dynamic characteristics of both oscillators and couplings.

2 Mathematical Description of Known Models

KM is presented in general form by the following system of equations [1]:

$$d\theta_m/dt = \omega_m + \sum_{n=1}^{N} h_{mn}(\theta_m - \theta_n). \tag{1}$$

Here, θ_m, ω_m are the phase and the fundamental (intrinsic frequency) of mth oscillator, N is a number of oscillators, $h_{mn}(\Delta\theta)$ are coupling functions.

Usually, the special case of weakly nonlinear harmonic oscillators is considered—$h_{mn}(\Delta\theta) = K_{mn}\sin(\Delta\theta)$ where K_{mn} determine coupling strength from nth to mth oscillator. When strengths are sufficiently weak, the oscillators run incoherently [1], otherwise the oscillators become synchronized. In the fully synchronized state, all oscillators have a common frequency $\omega = d\theta_m/dt$ (for all m) where instantaneous phases $\theta_m = \omega t + \varphi_i$ after substituting into (1) transform it to the algebraic system

$$\omega - \omega_m = \sum_{n=1}^{N} h_{mn}(\varphi_n - \varphi_m). \tag{2}$$

The set of locked oscillators produces free-running oscillations with an arbitrary phase shift that can be fixed by setting the phase of an oscillator to zero, e.g., $\varphi_N = 0$. Then, (2) defines the system of N algebraic equations with N variables: $\omega, \varphi_1, \ldots, \varphi_{N-1}$.

If the ensemble depends on external stimuli with frequency ω_{N+1}, then a natural extension of KM considers also an external periodic force [1, 15]. It can be included in (1, 2) as $(N + 1)$th term $h_{m,N+1}(\theta_{N+1} - \theta_m)$ added to the sum in right-hand side (RHS) of each equation. The number of equations of the extended system is saved, and its variables are N phases $\varphi_1, \ldots, \varphi_N$ because the common frequency of externally synchronized ensemble is known ($\omega = \omega_{N+1}$), and the excitation phase φ_{N+1} is also known.

Coupling functions h_{mn} in KM (1) are not defined through characteristics of oscillators and interconnects. This shortcoming of KM is eliminated in the nonlinear phase macromodel [5] defined as follows. Let $x(t)$ be any periodic steady-state (PSS) solution of the free-running oscillator equations, and $b(t)$ be its small perturbations. Then, perturbed solution presented as $x_p(t) = x(t+\alpha(t))$ with the time-varying phase shift $\alpha(t)$ defined by the nonlinear ODE [11, 12]:

$$d\alpha(t)/dt = v^T(t + \alpha(t)) \cdot b(t), \quad \alpha(0) = 0 \tag{3}$$

Here, periodic vector-function $v(t)$ is the perturbation projection vector (PPV) which is computed directly from the oscillator ODE system [12]. PPV defines the phase sensitivity of the oscillator. The important advantage of (3) in comparison with the KM (1) is the separation of the coupling strength into the oscillator sensitivity $v(t)$ and the perturbation intensity $b(t)$. The sensitivity (PPV) is the property of the oscillator only, and the perturbation intensity is defined by both PSS of other oscillators and parameters of interconnects. The approach was applied [2, 13] to solve problems of the oscillator network by the extension of (3) to N coupled oscillators

$$d\alpha_m(t)/dt = v_m^T(t + \alpha_m(t)) \cdot \sum_{n=1}^{N} \gamma_{mn}(t), \quad m = 1, \ldots, N. \tag{4}$$

Here, $\gamma_{mn}(t)$ is the perturbation of mth oscillator excited by PSS vector $x_n(t)$ of nth oscillator through the coupling (mn interconnect) defined by LTI system [16]

$$C_{mn}\mathrm{d}z_{mn}(t)/\mathrm{d}t + G_{mn}z_{mn} = F_{mn}x_n(t + \alpha_n(t)),$$
$$\gamma_{mn}(t) = D_{mn}z_{mn}, \quad m, n = 1, \ldots, N. \tag{5}$$

Here, z_{mn} represents internal states of the mn interconnect. Matrices F_{mn} define projections of jth oscillator waveforms onto mn interconnect inputs, and D_{mn} define projections of mn interconnect outputs onto perturbations of nth oscillator. Matrices G_{mn} and C_{mn} contain coefficients of linear ODE system for mn interconnect that correspond to conductance and capacitance matrices in the electrical network. Thus, joint solving of ODE systems (4, 5) determines the behavior of coupled oscillators.

It has been pointed out that high-frequency oscillations in the solution of (4, 5) lead to numerical difficulties [14] in simulations by the macromodel. Another shortcoming of the model is the impossibility to define synchronization conditions like (2).

3 Synchronization Model of Weakly Coupled Oscillators

To perform the analysis of synchronized oscillators, we firstly present basic expressions derived in [17] for the single oscillator synchronized by the external perturbation.

Any oscillator is characterized by its fundamental ω_0, and the PSS time-domain vector $x(t)$, comprising waveforms of the oscillator ODE state variables. PSS solution $x(t)$ can be also represented in the frequency domain by harmonic components X_k. Then the double-sided Fourier series defines the PSS waveform (j is the imaginary unit)

$$x(t) = \sum_{k=-K}^{K} X_k \exp(\mathrm{j}k\omega_0 t). \tag{6}$$

The behavior of the oscillator under small periodic excitation is characterized by PPV $v(t)$ and perturbation $b(t)$ vectors or by their Fourier harmonics V_k, B_k.

PSS of the free-running oscillator (6) possesses an arbitrary phase shift. But if the perturbation frequency $\omega = \omega_0 + \Delta\omega$ is sufficiently close to ω_0, then the oscillator is locked (synchronized) and locking phase φ is unambiguously obtained [17] by:

$$W(B, \varphi) = \Delta\omega/\omega_0. \tag{7}$$

where the locking function $W(B, \phi)$ is 2π-periodical dependence on φ

$$W(B, \varphi) = \sum_{k=-K}^{K} B_k V_k \exp(-jk\varphi). \tag{8}$$

Usually, perturbations B are induced by the excitation waveform $x^{\mathrm{ex}}(t)$ like (6) through an interconnect between the waveform source and perturbed oscillator. We assume that the interconnect is represented by LTI system like (5) which is defined in the frequency domain by TF $H^{\mathrm{ex}}(\omega)$. Hence entries of B^{ex} are $B_k^{\mathrm{ex}}(\omega) = H^{\mathrm{ex}}(k\omega)X_k^{\mathrm{ex}}$, where X_k^{ex} is kth entry of frequency-domain vector X^{ex} (or kth harmonic of $x^{\mathrm{ex}}(t)$).

The case of N oscillators with close fundamentals ω_n and known PPV vectors V^i is considered in [18]. The effect of the waveforms X^n of nth oscillator on the perturbation of the mth one is defined by the TF $H^{mn}(\omega)$ that for the system like (5) has the form

$$H^{mn}(\omega) = D^{mn}(G^{mn} + j\omega C^{mn})F^{mn}, \quad m, n = 1, \ldots, N. \tag{9}$$

If all oscillators are synchronized with the common frequency ω and phases φ_m, then we can represent (7) for mth oscillator by

$$(\omega - \omega_m)/\omega = W^m(B^m(\omega), \varphi_m). \tag{10}$$

Here, $B^m(\omega)$ is the perturbation vector of mth oscillator that contains components induced by waveforms of all oscillators. Let φ_n be the phase of nth oscillator. Then

$$B^m(\omega) = \sum_{n=i}^{N} \sum_{k=-K}^{K} B_k^{mn}(\omega) \exp(jk\varphi_n). \tag{11}$$

Here, $B_k^{mn}(\omega) = H^{mn}(k\omega)X_k^n$ with H^{mn} obtained by (9). After substituting (8, 11) into (10), the RHS of (10) takes the form

$$\begin{aligned}
W^m(B^m(\omega), \varphi_m) &= \sum_{k=-K}^{K} \exp(-jk\varphi_m) \cdot V_k^m \sum_{n=1}^{N} B_k^{mn}(\omega) \exp(jk\varphi_m) \\
&= \sum_{n=1}^{N} \left(\sum_{k=-K}^{K} V_k^m B_k^{mn}(\omega) \exp jk(\varphi_m - \varphi_n) \right) \\
&= \sum_{n=1}^{N} u^{mn}(\omega, \varphi_m - \varphi_n) \tag{12}
\end{aligned}$$

where functions $u^{mn}(\omega, \Delta\varphi)$ are defined by Fourier series with respect to $\Delta\varphi$

$$u^{mn}(\omega, \Delta\varphi) = \sum_{k=-K}^{K} U_k^{mn}(\omega) \exp(jk\Delta\varphi), \quad U_k^{mn}(\omega) = V_k^m H^{mn}(k\omega)X_k^m. \tag{13}$$

After substituting (12) into (7), one can write the system of algebraic equations

$$(\omega - \omega_m)/\omega = \sum_{n=1}^{N} u^{mn}(\omega, \varphi_m - \varphi_n), \quad m = 1, 2, \ldots, N. \tag{14}$$

System (14) can be simplified by considering the accuracy order of the locking Eq. (7) that is obtained in [17] by the linearization of full oscillator equations. Its error is evaluated as the second-order value:

$$\Delta\omega/\omega_0 - W(B, \varphi) = O(\|B\|^2) = O(\Delta\omega^2). \tag{15}$$

Therefore, we can perform transformations of (10, 14) within the accuracy order (15). Particularly, the denominator ω in the left-hand side of (10, 14) can be replaced by ω_m because $\Delta\omega/\omega_m \approx \Delta\omega/\omega + \Delta\omega^2/2$. Thus, (14) can also take the form

$$(\omega - \omega_m)/\omega_m = \sum_{n=1}^{N} u^{mn}(\omega, \varphi_m - \varphi_n), \quad m = 1, 2, \ldots, N. \tag{16}$$

Similarly, considering $O(B) = O(\Delta\omega)$, RHS of (10) can be evaluated under frequency $\omega + O(\Delta\omega)$ instead of ω due to the linearity of (8) with respect to B. So, we have

$$B(\omega + O(\Delta\omega)) - B(\omega) = O(\partial B/\partial \omega)O(\Delta\omega) = O(B)O(\Delta\omega) = O(\Delta\omega^2). \tag{17}$$

Thus, one can replace ω in RHS of (16) by a common value $\omega' = \omega + O(\Delta\omega_i)$ (e.g., $\omega' = \sum_{n=1}^{N} \omega_n/N$) and obtain a linear system with respect to ω

$$(\omega - \omega_m)/\omega_m = \sum_{n=1}^{N} u^{mn}(\omega', \varphi_m - \varphi_n), \quad m = 1, 2, \ldots, N. \tag{18}$$

System (18) coincides with the algebraic system (2) derived from KM (1), if

$$h_{mn}(\phi_m - \phi_n) = \omega_m \cdot u^{mn}(\omega', \phi_m - \phi_n), \quad m, n = 1, 2, \ldots, N. \tag{19}$$

Thus, (13) provides the evaluation of KM parameters.

Equations (18) are linear with respect to the synchronization frequency ω. Hence, we can eliminate ω and derive the system with respect to $n - 1$ phases φ_m. Specifically, (18) for two oscillators has the form ($\omega' = (\omega_1 + \omega_2/2)$)

$$(\omega - \omega_1)/\omega_1 = u^{1,2}(\omega', \varphi_1), \quad (\omega - \omega_2)/\omega_2 = u^{2,1}(\omega', -\varphi_1). \tag{20}$$

After eliminating ω from (20), the equation with respect to $\varphi = \varphi_1$ is obtained

$$(\omega - \omega_1)/\omega_1 = u^{1,2}(\omega', \varphi), \quad (\omega - \omega_2)/\omega_2 = u^{2,1}(\omega', -\varphi). \tag{21}$$

Equation (21) can be numerically solved by sweeping $0 \le \varphi < 2\pi$. After calculating φ, the locking frequency can be obtained from (20) as $\omega = \omega_1(1 + u^{1,2}(\varphi))$.

4 Time-Variant Model of Coupled Oscillators

The evaluation of the synchronized ensemble with more than two oscillators by solving (18) is a much more difficult problem than the case of the oscillator pair (20, 21). RHS of Eq. (18) comprises periodic functions u^{mn} (19) that can lead to multiple solutions including both stable and unstable solutions. In addition, the solution does not exist under weak couplings or/and large discrepancies between fundamentals. Thus, the simplest way to obtain the solution of (18) is to define the ODE system similar to KM (1) whose steady-state solution is the synchronization mode of the ensemble. Another purpose of ODE system formulation is the analysis of transients under synchronization approaching and the evaluation of the unsynchronized behavior of oscillators.

To develop the ODE system for coupled oscillators with frequency-dependent interconnects, we can substitute (19) into the original KM (1) and obtain a system

$$\dot{\theta}_m = \omega_m \left(1 + \sum_{n=1}^{N} u^{mn}(\omega', \theta_m - \theta_n) \right), \quad m = 1, 2, \ldots, N. \tag{22}$$

Here, $\dot{\theta}_m = d\theta_m/dt$. System (22) uses average oscillator frequency ω' due to the accuracy order estimate. However, the second-order error in (17) can essentially surpass the error in (15) of the same order for highly nonlinear frequency dependence $B(\omega)$.

This error can be responsible for the main part of the error in (22). Better accuracy is achieved by replacing average frequency ω' in the nth RHS term u^{mn} in (22) by the instantaneous frequency of nth oscillator $\omega_n^{\text{inst}} = \dot{\theta}_n$. Thus, we form an ODE system

$$\dot{\theta}_m = \omega_m \left(1 + \sum_{n=1}^{N} u^{mn}(\dot{\theta}_n, \theta_m - \theta_n) \right), \quad m = 1, 2, \ldots, N. \tag{23}$$

The steady-state solution ($\dot{\theta}_m = \omega$ for all m) of (23) is equal to the synchronization state obtained by solving (16) while (22) converges to the less accurate solution of (18).

An important special case of the application of (22, 23) is an ensemble of sinusoidal oscillators with waveforms $x^n(t) = |X_1^n| \cos\theta_n(t)$. PPV of the sinusoidal oscillator based on LC tank is also a sinusoid with phase shift $\pi/2$ with respect to the oscillator waveform [19]—$v^m(t) = |V_1^n| \sin\theta_m(t)$. All other PPV and waveform harmonics are zeroes: $V_k^m = X_k^n = 0$ for $|k| \neq 1$. Then taking into account complex conjugate properties: $V_1^m = \overline{V}_{-1}^m$, $X_1^m = \overline{X}_{-1}^m$, $H^{mn}(\omega) = \overline{H}^{mn}(-\omega)$, one can obtain from (13)

$$u^{mn}(\omega, \Delta\phi) = K^{mn}|H^{mn}(\omega)| \sin(\Delta\phi + \psi^{mn}(\omega)), \quad K^{mn} = 4|V_1^m X_1^n|. \quad (24)$$

where $\psi^{mn}(\omega)$ is the phase of TF $H^{mn}(\omega) = |H^{mn}(\omega)| \exp(i\psi^{mn}(\omega))$. Using (24), we can present (22, 23) for sinusoidal ensembles as

$$\dot\theta_m = \omega_m(1 + K_{mn}|H^{mn}(\omega')| \sin(\theta_m - \theta_n + \psi^{mn}(\omega'))), \quad m = 1, 2, \ldots, N. \quad (25)$$

$$\dot\theta_m = \omega_m(1 + K_{mn}|H^{mn}(\dot\theta_n)| \sin(\theta_m - \theta_n + \psi^{mn}(\dot\theta_n))), \quad m = 1, 2, \ldots, N. \quad (26)$$

For frequency-independent couplings, $H^{mn} = $ const, $\psi^{mn} = 0$ (initial KM).

5   Numerical Experiments

The numerical analysis of two coupled sinusoidal oscillators is presented below. Both oscillators and interconnects are identical. The average frequency of oscillators was defined as $f' = (f_1 + f_2)/2 = 100$ Hz. Time dependences of the oscillators' instantaneous frequencies ($\omega_1^{inst} = \dot\theta_1$, $\omega_2^{inst} = \dot\theta_2$) were obtained by solving (25) or (26). The results are presented in Figs. 1, 2, 3, and 4. Initially, disabled couplings are switched on at $t = 0.2$ s.

In the first series of experiments, we considered couplings with fixed phase shift ψ that used in the analysis of chimera states combining coherent and incoherent

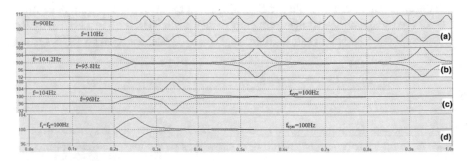

Fig. 1 Static couplings: **a** $\Delta f = 20$ Hz, **b** $\Delta f = 8.4$ Hz, **c** $\Delta f = 8$ Hz, **d** $\Delta f = 0$

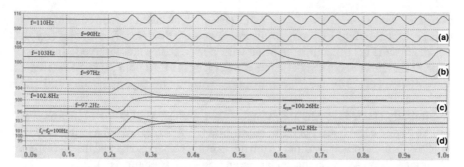

Fig. 2 Phase shift $\psi = \pi/4$ in couplings: **a** $\Delta f = 20$ Hz, **b** $\Delta f = 6$ Hz, **c** $\Delta f = 5.6$ Hz, **d** $\Delta f = 0$

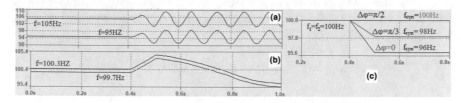

Fig. 3 Phase shift $\psi = \pi/4$ in couplings: **a** $\Delta f = 10$ Hz, **b** $\Delta f = 0.6$ Hz, **c** $\Delta f = 0$: different synchronization frequencies for different initial phases of oscillators $\Delta\varphi = \theta_2(0) - \theta_1(0)$

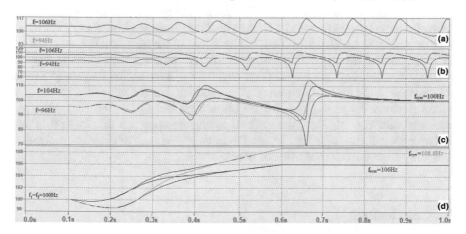

Fig. 4 RLC couplings. I. Beats mode ($\Delta f = 12$ Hz) analyzed **a** by (25), **b** by (26). II. Synchronization mode analyzed by both versions of TF evaluations: **c** $\Delta f = 8$ Hz, **d** $\Delta f = 0$

oscillations [20]—$u^{1,2} = u^{2,1} = K \sin(\Delta\phi + \psi)$. The case of static couplings ($\psi = 0$) is presented in Fig. 1 for various discrepancies of oscillators frequencies (Δf). At $\Delta f = 20$ Hz (Fig. 1a), antiphase beats with period 0.05 s are observed. The period of beats increases to 0.4 s under $\Delta f = 8.4$ Hz (Fig. 1b). At both $\Delta f = 8$ Hz (Fig. 1c) and $\Delta f = 0$ (Fig. 1d), the synchronization with the average frequency 100 Hz is achieved.

Similarly obtained results are presented for the phase shift $\psi - \pi/4$ in Figs. 2a–d. Unlike Figs. 1a and b, the curves in beat mode (Figs. 2a, b) are not in antiphase. The synchronization frequency differs from the average one (Fig. 2c), and the discrepancy essentially increases under the equal frequencies of oscillators (Fig. 2d).

The behavior of oscillators under couplings with the phase shift $\psi = \pi/2$ has unique features. In beats mode (Figs. 3a and b), the waveforms are common-mode signals and under any $\Delta f \neq 0$, the synchronization of oscillators does not occur due to the incompatibility of Eq. (20):

$$u^{1,2}(\phi_1) = K \cos(\phi_1) = K \cos(-\phi_1) = u^{2,1}(-\phi_1).$$

At $\Delta f = 0$, (20) can possess any number of solutions. In Fig. 3c, the existence of different synchronization states dependent on the ODE initial conditions is demonstrated.

Comparisons of models based on TF evaluations by (25) and by (26) were performed using couplings through RLC tanks with the resonance frequency $f_p = 110$ Hz and the quality factor $Q = 5.28$ (Fig. 4). For $\Delta f = 12$ Hz, curves in Fig. 4a obtained by (25) have the same beat frequencies as slightly different waveforms in Fig. 4b obtained by (26). For $\Delta f = 8$ Hz (Fig. 4c), the synchronization frequency almost coincides with the average frequency $f' = 100$ Hz, but for $\Delta f = 0$ (Fig. 4d), we obtain different values $f_{syn} = 108.8$ Hz under (25) and $f_{syn} = 106$ Hz under (26) similarly with Fig. 2d. Thus, the error of averaging approach (25) is 2.8%. Note that here we can also, as in Fig. 2d, see the notable deviation of the synchronization frequency from the common value of fundamentals.

6 Conclusions

The developed extended KM has the following features and advantages in comparison with the traditional one

– The coupling strength separately represents sensitivity/intensity factors.
– The model parameters relate to characteristics of full ODE systems.
– Dynamic couplings defined by any LTI ODE system are allowed in the model.
– Two frequency-dependent forms (average/instantaneous) can be used.
– The simplified model is developed for the ensemble of sinusoidal oscillators.

Computational experiments showed the following effects in the symmetrical pair of the dynamically coupled oscillators

- Unlike, the case of static couplings beat curves is not in antiphase.
- The synchronization frequency differs from the average value of oscillators fundamentals; the maximal difference is achieved at coinciding fundamentals.
- For TF with fixed phase shift $\pi/2$ beat curves are common-mode signals and under any nonzero discrepancy of fundamentals the synchronization never occurs otherwise the set of the synchronization modes exists.
- The more expensive model based on the instantaneous frequencies provides somewhat better accuracy than the model with the frequencies averaging.

Thus, proposals presented in this paper can essentially extend the range of the possibilities of analyzing oscillator ensembles.

References

1. Acebrón, J.A., et al.: The Kuramoto model: a simple paradigm for synchronization phenomena. Rev. Mod. Phys. **77**(1), 137–185 (2005)
2. Bhansali, P., Roychowdhury, J.: Injection locking analysis and simulation of weakly coupled oscillator networks. In: Li, P., et al. (eds.) Simulation and Verification of Electronic and Biological Systems, pp. 71–93. Springer, New York (2011)
3. Ashwin, P., Coombes, S., Nicks, R.J.: Mathematical frameworks for oscillatory network dynamics in neuroscience. J. Math. Neurosci. **6**(2), 1–92 (2016)
4. Ziabari, M.T., Sahab, A.R., Fakhari, S.N.S.: Synchronization new 3D chaotic system using brain emotional learning based intelligent controller. Int. J. Inf. Technol. Comput. Sci. **7**(2), 80–87 (2015). https://doi.org/10.5815/ijitcs.2015.02.10
5. Lytvyn, V., Vysotska, V., Peleshchak, I., Rishnyak, I., Peleshchak, R.: Time dependence of the output signal morphology for nonlinear oscillator neuron based on Van der Pol model. Int. J. Intell. Syst. Appl. **10**(4), 8–17 (2018). https://doi.org/10.5815/ijisa.2018.04.02
6. Alain, K.S.T., Bertrand, F.H.: A secure communication scheme using generalized modified projective synchronization of coupled Colpitts oscillators. Int. J. Math. Sci. Comput. **4**(1), 56–70 (2018). https://doi.org/10.5815/ijmsc.2018.01.04
7. Jörg, D.J., et al.: Synchronization dynamics in the presence of coupling delays and phase shifts. Phys. Rev. Lett. **112**(17), 174101 (2014)
8. Batistaa, C.A.S., et al.: Synchronization of phase oscillators with coupling mediated by a diffusing substance. Physics A **470**, 236–248 (2017)
9. Gushchin, A., et al.: Phase-coupled oscillators with plastic coupling: synchronization and stability. IEEE Trans. Netw. Sci. Eng. **3**(4) (2016)
10. Grzybowski, J.M.V., et al.: On synchronization in power-grids modelled as networks of second-order Kuramoto oscillators. Chaos **26**(11), 113113 (2016)
11. Demir, A., Mehrotra, A., Roychowdhury, J.: Phase noise in oscillators: a unifying theory and numerical methods for characterization. IEEE Trans. CAS I. **47**(5), 655 (2000)
12. Demir, A., Roychowdhury, J.: A reliable and efficient procedure for oscillator PPV computation, with phase noise macromodeling applications. IEEE Trans. Comput. Aided Des. Integr. Circ. Syst. **22**(2), 188–197 (2003)
13. Harutyunyan, D., et al.: Simulation of mutually coupled oscillators using nonlinear phase macromodels and model order reduction techniques. In: Gunter, M. (ed.) Coupled Multiscale Simulation and Optimization in Nanoelectronics, Mathematics in Industry, vol. 21, pp. 398–432. Springer, Berlin, Heidelberg (2015)
14. Gourary, M.M., Rusakov, S.G., et al.: Smoothed form of nonlinear phase macromodel for oscillators. In: Proceedings of the IEEE/ACM International Conference on Computer-Aided Design, pp. 807–814 (2008)

15. Moreira, C.A., de Aguiar, M.A.: Global synchronization of partially forced Kuramoto oscillators on networks. arXiv:1802.07691v2 [nlin.AO] (2018)
16. Karabutov, Nikolay: Methods of estimation Lyapunov exponents linear dynamic system. Int. J. Intell. Syst. Appl. **7**(10), 1–11 (2015). https://doi.org/10.5815/ijisa.2015.10.01
17. Gourary, M.M., Rusakov, S.G., et al.: Injection locking conditions under small periodic excitations. In: 2008 IEEE International Symposium on Circuits and Systems, pp. 544–547 (2008)
18. Gourary, M.M., Rusakov, S.G., et al.: Mutual injection locking of oscillators under parasitic couplings. In: Michielsen, D., Poirier, J. R. (eds.) Scientific Computing in Electrical Engineering (SCEE) 2010, pp. 303–312. Springer, Berlin, Heidelberg (2012)
19. Lai, X., Roychowdhury, J.: Analytical equations for predicting injection locking in LC and ring oscillators. In: IEEE Custom Integrated Circuits Conference, pp. 461–464 (2005)
20. Panaggio, M.J., Abrams, D.M., Ashwin, P., Laing, C.R.: Chimera states in networks of phase oscillators: the case of two small populations. Phys. Rev. E **93**(1), 012218 (2016)

The Genetic Language: Natural Algorithms, Developmental Patterns, and Instinctive Behavior

Nikita E. Shklovskiy-Kordi, Victor K. Finn, Lev I. Ehrlich
and Abir U. Igamberdiev

Abstract The genetic system of biological organisms possesses a structure which corresponds to the general principles of linguistics and can be defined as the genetic language. In this study, we suggest to analyze the mechanisms for interpretation of genetic texts based on the universal model of operation of the programs in electronic computers as initially suggested by Efim Liberman. Ontogenetic development is realized at the level of reading of genetic texts by the structure named by Liberman as a molecular computer of the cell (MCC), which includes DNA, RNA, and the corresponding enzymes that work with molecular addresses. The main feature of the biological computer is the search for addresses using the thermal Brownian motion and the complex formation of weak bonds without the cost of free energy. The implementation of genetic programs takes place not only in the course of individual development, characterized by the encoding of the sequences of reading proteins, but also in the execution of instinctive behavior. The description of external reality occurs in terms of the genetic language in all living beings. In addition, the reality is universally described in the natural (human) language. In both cases, the description is implemented in the form of using models, the calculation of which allows prediction of the future of the simulated reality and its management. The success of such control depends on the choice of model and the correct scale, which determines the energy and time spent on the calculation. This quantity, equal to the production of energy and time, is quantized and is related to Planck's constant. An attempt has been made to construct a semantic system of the genetic language, for which a deliberately narrowed but instrumental definition of "text" and "meaning" is given.

N. E. Shklovskiy-Kordi
National Research Center for Hematology, Moscow, Russia

V. K. Finn
Federal Research Centre of Computer Sciences, Management of RAS, Moscow, Russia

L. I. Ehrlich
Research Computing Center, Moscow State University, Moscow, Russia

A. U. Igamberdiev (✉)
Department of Biology, Memorial University of Newfoundland, St. John's, NL, Canada
e-mail: igamberdiev@mun.ca

© Springer Nature Switzerland AG 2020
Z. Hu et al. (eds.), *Advances in Artificial Systems for Medicine and Education II*,
Advances in Intelligent Systems and Computing 902,
https://doi.org/10.1007/978-3-030-12082-5_16

Keywords Biological computer · Encoding · Genetic language · Natural algorithm · Noosphere

1 Introduction Genetic Language and Its Functioning

Until recently, all the texts known to Humanity were written by a human person. But in 1953, it was finally established that the genetic information in the cell is contained in the form of text based on the molecular alphabet and a linear sequence presented as a combination of four nucleotides. Many terms describing the new language called the "DNA code" were borrowed from linguistics. Information in DNA, as in the "natural" language, is written ultimately as the order of the nucleotide letters acting as signs. Obviously, like in books where the information includes pictures, binding etc., the secondary, tertiary, and quaternary structures of DNA are important, but, as binding for most readers of books, they are secondary in comparison with the primary linear sequence of the text.

The genetic language in its basis possesses all fundamental characteristics of the linguistic semiotic system formulated by de Saussure [1]. First of all, it is the arbitrariness of the sign and the linear character of the signifier, which completely corresponds to the linearity of writing in the human language. The sign is always material, in the sense that it is somehow made, and in this respect, each sign represents a separate substance that is not identical to the others. However, for a reader all instances (inscriptions) of the same sign are identical, like the same letters in the natural alphabet. The letters of the DNA code (nucleotides) are signs of the genetic language. It is distinctive that the correspondence of the triplets of genetic code to amino acids is not of a physical nature.

2 The Molecular Cell Computer is a Reader of Texts of the Genetic Language

The genetic code is the basis of the genetic language, but for the functioning of this language, a device that reads it and a subject (self) who would perceive it are necessary [2, 3]. For the first time, the concept of the operation of this device in living cell was proposed by Liberman [4–6]. According to this concept, DNA is rather not a set of genes, but a set of molecular text programs for molecular cell computer. This molecular computer works with the system of DNA, RNA, and targeted proteins (Fig. 1). Its operators cut and glue molecular text at certain places according to a program written on DNA. The enzymatic activity required for this function is associated with both the corresponding enzyme proteins and the enzymatic activity of RNA (Fig. 2). The molecular computer uses the thermal Brownian motion of molecular structures in the calculation process. Searches for the address are due to

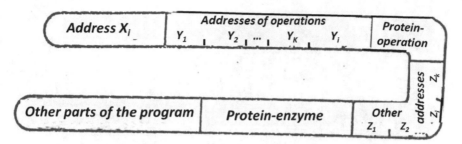

Fig. 1 Schematic representation of the structure of the molecule–word. Three-dimensional structure of the molecule is represented by the pattern of recognizable sites–addresses. The active sites possessing enzymatic activity are designated as operations with molecules–words. The scheme assumes that, depending on the context, the molecule–word has many modes of its reading (polysemantic property). The original scheme of Liberman [4] is modified and adapted here

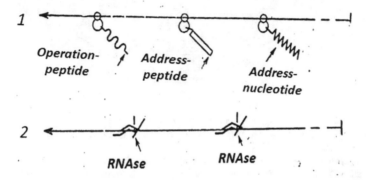

Fig. 2 Top panel: schematic representation of the readout of peptides and nucleotides from the word program by the ribosomes containing ribonuclease and RNA-dependent RNA and DNA polymerases. Ribosome function, according to this representation, consists in not only the translation of the nucleotide code into the protein sequence but also the formation of polysemantic structure of protein molecule, which, according to Liberman, takes place not only during post-translational enzymatic modification but also in the direct interaction with the text of controlling program. Bottom panel: Operations of specific RNases by which the word program is cut out into pieces according to specific addresses. The process was predicted by Liberman [4] before the discovery of splicing in 1977 (Nobel prize in 1993) and developed from Liberman [4]

thermal motion [4]. The definiteness of the interpretation of the sequence and its irreversible readout over time are determined, according to Liberman [7, 8], by the irreversible loss of energy spent on the calculation, i.e., by the price of the action which determines the physical limitation of the operation (Fig. 3).

Gamkrelidze [9] examined the structural isomorphism between the genetic and linguistic codes and analyzed two approaches to this problem corresponding to the views of Jacob [10] and Jakobson [11]. Jacob claimed that the isomorphism between genetic and linguistic originated as a result of the structural coincidence of the two systems performing similar information functions, whereas Jakobson believed that

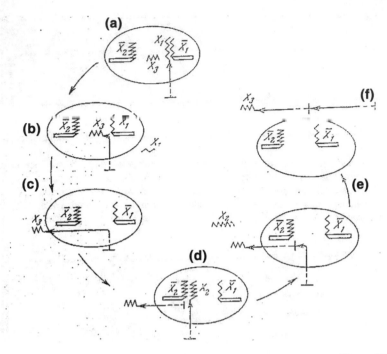

Fig. 3 Operation of specific ligase: Combination of the words with the addresses $\overline{X}_1$ and $\overline{X}_2$ into a new word with the nucleotide X_3. Ligase has two complementary addresses $\overline{X}_1$ and $\overline{X}_2$ that can be used by corresponding peptides.

a—$\overline{X}_1$ connects to the complementary nucleotide X_1, which is bound to ligase by the use of peptide;
b—The nucleotide X_1 is cut off by nuclease and becomes substituted by the nucleotide X_3 located on ligase; nucleotide X_1 is removed or destroyed;
c—The word is pulled across the enzyme so its second end becomes located near the active site of ligase;
d—The address $\overline{X}_2$ of the second word becomes complementary attached to the nucleotide X_2;
e—The nucleotide X_2 is cut off and becomes removed or destroyed while the beginning of the second word becomes complementary attached to the address $\overline{X}_2$;
f—As a result of the described operation, a new word with new address is obtained
The whole scheme shows the operation of the structure called spliceosome, which consists of the molecules of RNA and proteins and realizes the removal of non-coding sequences (introns) from the precursor RNA.
The scheme is modified from Liberman [4]. All figures represent an attempt to show the isomorphism between the natural language, the genetic language, and the computer language

this isomorphism is a consequence of the phylogenetic construction of the language code according to the structural principles of the genetic code. This debate remains actual in the understanding of the origin and operation of the genetic and human languages and requires further analysis and thorough exploration.

The whole scheme shows the operation of the structure called spliceosome, which consists of the molecules of RNA and proteins and realizes the removal of non-coding sequences (introns) from the precursor RNA.

3 Generativity and Control of the Behavior of Living Beings

The ambiguity and complexity of the genetic language [12] are realized in somewhat different ways than in the human language. Popov et al. [13] noted that the genetic text can be read in different ways, each time using an alternative choice of signs of one text and ignoring other signs. This is possible, e.g., through the use of overlapping sequences, which makes genetic texts often more structurally complex than the texts of the human language, whose ambiguity and dependence on the reader were considered by many writers and scientists, including Eco [2], see also the works of Finn [3]. Collado-Vides [14, 15] examined in detail the generative principles of the genetic and human language, defined the linguistic principles of molecular biology, and formulated the rules for the transformation of the genetic language. The genetic syntactic structure is considered as the equivalent of the sentence [16]. In the transformation processes that provide generative properties of the genetic language, an important role is played by non-coding DNA sequences that are responsible for the redundancy of the genetic texts necessary for realization of the transformational capabilities of the genetic system and for its ability to rebuild and evolve [17]. There are reasons to explore the detection of the non-coding part of the genome texts similar to the algorithms and programs for controlling behavior of the organism [8].

Ji [18] analyzed in detail the isomorphism between genetic and human languages. Ten out of thirteen properties of the human language specified by Hockett [19] were also inherent in the genetic language. According to Ji [18], the structural genes correspond to lexicon, and the "space-time" genes correspond to the grammar of the genetic language. This grammar was identified as a mapping of the nucleotide sequences of DNA into its four-dimensional folding patterns that control the expression of genes. This mapping is considered as the second genetic code.

Later, Ji [20] developed a model of the cellular computer, which essentially repeats the model of Liberman [4, 7, 21]. Conrad [22] actually followed the concept of Liberman in understanding the genetic language, but made it more reductionist, without the idea of an internal quantum regulator acting as a subject of these operations. Recognizing the patterns of proteins and nucleic acids based on the complementarity of forms leads, according to Conrad, to the multiplicity of solutions that have different levels of dependence on the context. The "self-organizing" method of calculation using DNA molecules as an example was analyzed by Conrad and Zauner [23].

It should be noted that the pioneering nature of the work of Adleman [24] that revealed the computational properties of DNA is somewhat exaggerated: This concept was put forward by Liberman in 1972, and in the following year, the equivalence of the molecular cell computer to a universal computer (Markov's algorithms) was proven [25].

Generativity (corresponding to the "logos heauton auxon" (self-growing logos)) of Heraclitus can be viewed as a consequence of the ability of the text to generate hypertext, which the text itself stands for. In this case, some assertions of the system acquire the property of mapping the system as a whole. In simple words, it defines the *behavior* of a living being. The control systems of the genome contain algorithms that allow the body to read the parameters of the external environment, substitute them in appropriate models that describe reality, predict the future and generate a response aimed at changing the situation in the optimal direction from the point of view of an integrated system.

According to Darwin, the generativity in evolution is a consequence of errors; however, the error itself must be internalized in a complex formal system. Von Neumann [26] proved that a sufficiently high level of complexity is needed to reproduce finite state machines. This level of complexity may correspond to the ability of the formal system to generate Gödel numbers (in linguistics corresponding to hypertextual statements), in other words, to be generative and to build a sufficiently advanced hypertext that allows reproducing the existing system and making it more complex, e.g., to be able to complexify in the course of evolution.

The alternative implementations for reading appearing as read options, and the processes such as processing, splicing, the role of "short" RNAs as "exclamations," i.e., the shortest and most frequently used words, the interpretation of which depends entirely on the context, were predicted by Liberman [4] several years before their discovery. The system of alternative implementations laid in the system of language is carried out in the process of individual development and individual adaptation of the organism, and the possibility of new realizations not envisaged by the system and arising from the incompleteness of the formal system opens the way to a different level of generativity and determines biological evolution. The "generative grammar" of genetic system is determined, in particular, by the rules for the recombination of genetic material in accordance with molecular addresses.

If molecular computer works with molecular addresses and has a certain set of possible realizations, then the question arises, what determines the choice of these realizations? The answer to this question is given by Liberman based on the existence of a kind of "self" in the biosystem, namely, treating the living cell as "a system with a purely internal point of view" [27–30]. Since the MCC works at the quantum level, it is possible to find out that "it thinks" only by "talking" with it, i.e., having received the "regular" report of the MCC, provided by its structure [31, 32]. Otherwise, interference in the work of the IMM uncontrollably (unpredictably) changes the quantum mechanical characteristics of the system. The cell's self is defined by Liberman [28] as "the quantum regulator," and later, it was suggested that it is a coherent "internal quantum state" [33, 34], sending decoherence commands, implemented by what is called MCC by Liberman. The initial structure of the genetic language (genetic code) is more stringent than that of the human language, and consistent with the analysis that uses the unique rules of formal logic. But at the level of hypertexts, a large field for freedom opens up, which is responsible for the individual development and adaptation of organisms, and generates potentially infinite possibilities for evolution.

Considering that brain operation is determined not by the neural network but by the network consisting of powerful molecular computers using the bios of the genetic language [21], it is possible to trace the analogy between the hypertexts of the genetic and natural languages. The hypertext organization of the genetic language, revealed in the molecular biology studies, became much more understandable for us than the processes taking place during operation of the natural language in human brain. We do not know the material substrate for the natural language, and the neural network hypotheses such as the synaptic facilitation cannot be taken seriously due to the price of action (time × energy) which should be spent for the elementary operation of calculation. The suggestions of Vyacheslav V. Ivanov (personal communication) on the application in linguistics of the laws of operation of genetic texts are becoming more and more actual.

The genome is a system that has an internal complementarity between linear texts and their superposition. This addition means that text and hypertext cannot be read at the same time: they must be separated by a time interval, representing the separation of the system and its embedding. Overlapping genes, sequences resulting from alternative splicing, editing of DNA and RNA, introns, recombination in accordance with molecular addresses all these phenomena are important features of the hypertext, generating potentially infinite number of language games [35, 36]. The genome as a complete linguistic structure exists as a complementary set of its alternative combinations, which inevitably includes the logical paradoxes that determine its temporal dynamics [37]. This superposition is the basis of ontogeny, adaptation, and evolution. Thus, the complete structure of the genome is a superposition of alternative implementations that generate one implementation at a particular time. Combinatorial capabilities of the genome are multiplied many times by means of a pool of mobile genetic elements, including viruses, transposons, and other agents of the horizontal transfer of genetic information.

4 The Role of RNA in the Establishment of Genericity of the Genetic Language

In the dual gene–protein system, the recursion is impossible or strongly limited. Generativity in this case is limited to the Darwinian (in a narrow sense) evolution through mutations and their selection. Liberman [7] calculated that if an accidental search of DNA was to take place in the entire Universe, densely packed with cells, and mutations occur every 10–12 s, then for the entire time of the existence of the Universe, it would be possible to obtain a DNA text sensing for a specific reading device of only 400 nucleotides in the length. However, in a system with an intermediate component (RNA), flexibility appears, and evolution can occur as a complex language game in which RNA becomes part of a temporal component mediating incomplete identification.

Liberman [4, 7] pointed to the primary role of RNA in the functioning of the molecular reader and its evolution. Even the role of randomness increases significantly in a system that includes RNA. A random enumeration of nucleotides in the additional provision of RNA generation may, in particular, be provided by the enzyme polynucleotide phosphorylase, operating without the expenditure of energy. A new word arising from an equilibrium molecular enumeration, e.g., from the RNA rearrangement by polynucleotide phosphorylase, may prove to be appropriate for intracellular tasks which are solved in small compartments. The work of the MCC, according to Liberman, is not limited to operations on DNA and RNA. Post-translational modification of proteins represents the level of genetic language that works not with linear but with the three-dimensional objects. A significant role belongs to the allosteric regulation of the function of protein operators.

5   Conclusion

In living cells, new DNA texts arise in a special cell division with the formation of sex gametes (meiosis), when homologous texts of parents are mixed. Also, horizontal transfer of genes is essential. All diversity of living organisms is mediated by these molecular mechanisms of processing of the genetic text. Humans have acquired the opportunity to create independently of the genetic information, new texts that are, like the DNA text, the sequences of signs. Signs that people use are the most diverse: phonemes, letters of various alphabets, and since the twentieth-century molecular letters of DNA. The latter, together with the use of nuclear energy, makes our time absolutely exceptional in determination of the future of Humanity.

The phenomenon of life exists in the only known to us "natural" form which is a living cell with a molecular computer, operating on a DNA program, written not by a human being. The statement that the cell only occurs from the cell and the idea of a single ancestor of all cells are generally accepted in modern science. Effective for the scientific study of the life model is an electronic computer that operates on silicon crystals and metallic conductors according to a program written by a person.

The phenomenon of man can be defined as "an animal that can be taught the human language." This language gives a freedom to operate with signs, i.e., the ability to create new texts. The concept of noosphere introduced by Édouard Le Roy, Pierre Teilhard de Chardin, and Vladimir Vernadsky also acquires a clear meaning. Noosphere is a collection of texts written by humanity.

All that a person knows about the Universe comes either in the form of genetic texts or in the form of a code of nerve impulses from the different types of external and internal receptors. All man's creations, even if they are not accompanied by a verbal description, are the product of the work of muscles, which are controlled by the efferent fluxes of the same linear sequences of nerve impulses. We propose to introduce the term "Descriptiveness of the Universe," which, from the point of view of the semiotic system, is the most common characteristic of the natural language. The introduction of genetic and neural algorithms in the system of human knowledge brings a new power and advanced solutions for optimization of technical and social systems [38–43]. Future analysis should determine the nature and origin of isomorphism of the genetic and natural languages.

In this analysis, the following instrumental definitions could be considered. The "TEXT" is a sequence of signs that can be interpreted by presenting a living cell, a cell system (e.g., human brain), or a computer. The "MEANING" is an interpretation of the text in a cell, a system of cells, or computing devices.

References

1. Saussure, F. de: Course in General Linguistics, 1911 pp. McGraw-Hill Humanities (1965)
2. Eco, U.: The Role of the Reader Explorations in the Semiotics of Texts. University of Indiana Press, Bloomington (1979)
3. Finn, V.K.: Toward logical-semantic issues in the theory of text comprehension. Autom. Doc. Math. Linguist. **44**(5), 235–245 (2010)
4. Liberman, E.A.: Cell as a molecular computer (MCC). Biofizika **17**, 932–943 (1972)
5. Liberman, E.A.: Molecular computer—biological physics and physics of real world. Biofizika **23**, 1118–1121 (1978)
6. Liberman, E.A.: Analog–digital molecular cell computer. BioSystems **11**, 111–124 (1979)
7. Liberman, E.A.: Molecular quantum computers. Biofizika **34**, 913–925 (1989)
8. Liberman, E.A.: Biophysical and mathematical principles and biological information. Biofizika **42**, 988–991 (1997)
9. Gamkrelidze, T.V.: The unconscious and the problem of isomorphism between the genetic code and semiotic systems. Folia Linguist. **23**, 1–5 (1989). Hockett, 1960
10. Jacob, F.: The linguistic model in biology. In: van Schooneveld, C.H., Armstrong, D. (eds.) Roman Jakobson. Echoes of His Scholarship. Peter de Ridder, Lisse (1977)
11. Jakobson, R.: Selected Writings, vol. 2. Mouton, Hague (1971)
12. Trifonov, E.N.: The multiple codes of nucleotide sequences. Bull. Math. Biol. **51**, 417–432 (1989)
13. Popov, O., Segal, D.M., Trifonov, E.N.: Linguistic complexity of protein sequences as compared to texts of human languages. BioSystems **38**, 65–74 (1996)
14. Collado-Vides, J.: A transformational-grammar approach to the study of the regulation of gene expression. J. Theor. Biol. **136**, 403–425 (1989)
15. Collado-Vides, J.: Grammatical model of the regulation of gene expression. Proc. Natl. Acad. Sci. USA **89**, 9405–9409 (1992)
16. Monod, J., Jacob, F.: Teleonomic mechanisms in cellular metabolism, growth, and differentiation. Cold Spring Harb. Symp. Quant. Biol. **26**, 389–401 (1961)
17. Mantegna, R.N., Buldyrev, S.V., Goldberger, A.L., Havlin, S., Peng, C.K., Simons, M., Stanley, H.E.: Linguistic features of noncoding DNA sequences. Phys. Rev. Lett. **3**, 3169–3172 (1994)
18. Ji, S.: Isomorphism between cell and human languages: molecular biological, bioinformatic and linguistic implications. BioSystems **44**, 17–39 (1997)
19. Hockett, C.F.: The origin of speech. Sci. Am. **203**, 89–96 (1960)
20. Ji, S.: The cell as the smallest DNA-based molecular computer. BioSystems **52**, 123–133 (1999)
21. Liberman, E.A., Minina, S.V., Shklovsky-Kordi, N.E.: Quantum molecular computer model of the neuron and a pathway to the union of the sciences. BioSystems **22**, 135–154 (1989)
22. Conrad, M.: Molecular and evolutionary computation: the tug of war between context freedom and context sensitivity. BioSystems **52**, 99–110 (1999)
23. Conrad, M., Zauner, K.P.: DNA as a vehicle for the self-assembly model of computing. BioSystems **45**, 59–66 (1998)
24. Adleman, L.M.: Molecular computation of solutions to combinatorial problems. Science **266**, 1021–1024 (1994)
25. Weinzweig, M.N., Liberman, E.A.: Formal description of cell molecular computer. Biofizika **18**, 939–942 (1973)
26. Von Neumann, J.: Theory of Self-Reproducing Automata. University of Illinois Press, Urbana (1966)

27. Conrad, M., Liberman, E.A.: Molecular computing as a link between biological and physical theory. J. Theor. Biol. **98**, 239–252 (1982)
28. Liberman, E.A.: Extremal molecular quantum regulator. Biofizika **28**, 183–185 (1983)
29. Minina, S.V., Liberman, E.A.: Input and output ionic channels of quantum biocomputer. Biofizika **35**, 132–136 (1990)
30. Liberman, E.A., Minina, S.V.: Cell molecular computers and biological information as the foundation of nature's laws. BioSystems **38**, 173–177 (1996)
31. Liberman, E.A., Minina, S.V., Shklovskiy Kordi, N.E.: Biological information and laws of nature. BioSystems **46**, 103–106 (1998)
32. Liberman, E.A., Minina, S.V., Shklovsky-Kordi, N.E.: Problems combining biology, physics, and mathematics. Ideas of the new science. Biofizika **46**, 765–767 (2001)
33. Igamberdiev, A.U.: Quantum computation, non-demolition measurements, and reflective control in living systems. BioSystems **77**, 47–56 (2004)
34. Igamberdiev, A.U.: Physical limits of computation and emergence of life. BioSystems **90**, 340–349 (2007)
35. Igamberdiev, A.U., Shklovskiy-Kordi, N.E.: Computational power and generative capacity of genetic systems. BioSystems **142–143**, 1–8 (2016)
36. Igamberdiev, A.U., Shklovskiy-Kordi, N.E.: The quantum basis of spatiotemporality in perception and consciousness. Prog. Biophys. Mol. Biol. **130**, 15–25 (2017)
37. Isalan, M.: Gene networks and liar paradoxes. BioEssays **31**, 1110–1115 (2009)
38. Kaur, S., Verma, A.: An efficient approach to genetic algorithm for task scheduling in cloud computing environment. Int. J. Inf. Technol. Comput. Sci. (IJITCS) **4**(10), 74–79 (2012). https://doi.org/10.5815/ijitcs.2012.10.09
39. Kumar, K., Thakur, G.S.M.: Advanced applications of neural networks and artificial intelligence: a review. Int. J. Inf. Technol. Comput. Sci. (IJITCS) **4**(6), 57–68 (2012). https://doi.org/10.5815/ijitcs.2012.06.08
40. Ebrahimzadeh, R., Jampour, M.: Chaotic genetic algorithm based on Lorenz chaotic system for optimization problems. Int. J. Intell. Syst. Appl. (IJISA) **5**(5), 19–24 (2013). https://doi.org/10.5815/ijisa.2013.05.03
41. Lytvyn, V., Vysotska, V., Peleshchak, I., Rishnyak, I., Peleshchak, R.: Time dependence of the output signal morphology for nonlinear oscillator neuron based on Van der Pol model. Int. J. Intell. Syst. Appl. (IJISA) **10**(4), 8–17 (2018). https://doi.org/10.5815/ijisa.2018.04.02
42. Babichev, S., Korobchynskyi, M., Mieshkov, S., Korchomnyi, O.: An effectiveness evaluation of information technology of gene expression profiles processing for gene networks reconstruction. Int. J. Intell. Syst. Appl. (IJISA) **10**(7), 1–10 (2018). https://doi.org/10.5815/ijisa.2018.07.01
43. Bilgaiyan, S., Aditya, K., Mishra, S., Das, M.: A swarm intelligence based chaotic morphological approach for software development cost estimation. Int. J. Intell. Syst. Appl. (IJISA) **10**(9), 13–22 (2018). https://doi.org/10.5815/ijisa.2018.09.02

Part II
Advances in Medical Approaches

Planck–Shannon Classifier: A Novel Method to Discriminate Between Sonified Raman Signals from Cancer and Healthy Cells

Sungchul Ji, Beum Jun Park and John Stuart Reid

Abstract The Planckian distribution equation (PDE), also called blackbody radiation-like equation, BRE, was derived from the Planck radiation formula by replacing its universal constants and temperature with free parameters, A, B, and C, resulting in $y = A/(x + B)^5/(e^{C/(x+B)} - 1)$, where x is bin variable and y is frequency. PDE has been found to fit many long-tailed asymmetric histograms (LAHs) reported in various fields, including atomic physics, protein folding, single-molecule enzymology, whole-cell metabolism, brain neurophysiology, electrophysiology, decision-making psychophysics, glottometrics (quantitative study of words and texts), sociology, econometrics, and cosmology (http://www.conformon.net/wp-content/uploads/2016/09/PDE_Vienna_2015.pdf). The apparent universality of PDE is postulated to be due to the principle of wave-particle duality embodied in PDE that applies not only to quantum mechanics but also to macrophysics regardless of scales. In this paper, the new classification method referred to as the Planck–Shannon classifier (PSC) or the Planck–Shannon plot (PSP) is formulated based on the two functions, i.e., (i) the Planckian information of the second kind, I_{PS}, and (ii) the Shannon entropy, H, that can be computed from PDE. PSC has been shown to successfully distinguish between the digital CymaScopic images generated from the sonified Raman signals measured from normal and cancer cells in human brain tissues. PSC is a general purpose classifier and can be applied to classifying long-tailed asymmetric histograms generated by many physical, chemical, biological, physiological, psychological, and socioeconomical processes called Planckian processes, i.e., those processes that generate long-tailed asymmetric histograms fitting PDE.

S. Ji (✉)
Department of Pharmacology and Toxicology, Ernest Mario School of Pharmacy, Rutgers University, Piscataway, NJ, USA
e-mail: sji.conformon@gmail.com

B. J. Park
The Doctor of Pharmacy Program, Ernest Mario School of Pharmacy, Rutgers University, Piscataway, NJ, USA

J. S. Reid
Sonic Age, Inc., St. John's-in-the-Vale, Cumbria, England, UK

© Springer Nature Switzerland AG 2020
Z. Hu et al. (eds.), *Advances in Artificial Systems for Medicine and Education II*,
Advances in Intelligent Systems and Computing 902,
https://doi.org/10.1007/978-3-030-12082-5_17

Keywords Digital CymaScope · Planck–Shannon classifier · Planckian distribution equation · Planckian information of the second kind · Shannon entropy · Sonified Raman spectral features of cancer cells

1 Introduction

There are many natural and human phenomena that generate long-tailed asymmetric histograms (LAHs) fitting the so-called Planckian distribution equation (PDE) discovered in 2008 [1, 2]. Some examples of LAHs that fit PDE are provided in Fig. 1. PDE has been found to fit almost all long-tailed asymmetric histograms generated in a wide range of scientific disciplines, including atomic physics, protein folding, single-molecule enzyme catalysis (see Fig. 1b), whole-cell metabolism (Fig. 1c), fMRI (functional magnetic resonance imaging), linguistics, psychophysics, econometrics, and cosmology [2, 3, Fig. 8.6]. The natural and human-mediated processes that produce the LAHs fitting PDE have been defined as the *Planckian processes*. It was demonstrated in [3, pp. 356–357] that the *Planckian process* involves various selection mechanisms that separate subsets of events out of all possible randomly accessible events. The main purpose of this paper was to describe how PDE can be utilized to classify LAHs into functionally distinct groups such as those derived from cancer or healthy cells (see Tables 1 and 2).

2 Data Acquisition Using the Digital CymaScope

The CymaScope (cyma- = wave; -scope = viewing device) was invented by John Stuart Reid in 2002 [4, 5]. This instrument may be currently the most sensitive device of its kind that can transform sound vibrations (as low as 10 Hz) into exquisitely detailed images of the standing waves formed in water that are visible via direct ocular viewing and by camera with macro lens. There are many devices that can convert sound into visible images, such as spectrograms, but the CymaScope is unique in its use of *water* as the *sound-detecting molecular sensor*, which is sensitive to sounds as low as LAFmin 25 (A-weighted, fast time-weighted, minimum sound level 25 dB). It should be noted that the sound-induced water wave images obtained by the CymaScope are much more detailed than those reported by Jenny [6], mainly because the CymaScope utilizes the reflection optics and the 32 LEDs (light-emitting diodes) arranged in a circle as the light source in contrast to the Schlieren optics with one light source employed by Jenny. Reid has an extensive gallery of the CymaScope images (called *CymaGlyphs, cymaglyphs, water waves, water wave patterns, aquawaves, aquaresonances*, etc. [3, Sect. 4.21]) produced by a variety of sound vibrations including music [4, 5, 7]. All these CymaGlyphs that he has accumulated until early 2017 are *analogical* and popularized mainly among artists and sonic therapists. In early 2017, the *analog* CymaScope was transformed into the *digital* CymaScope by replacing

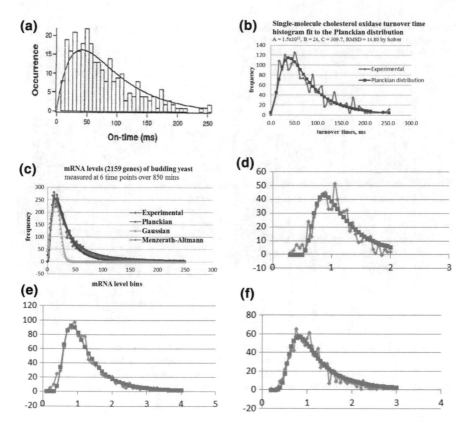

Fig. 1 Some examples of experimental long-tailed histograms (see blue curves) that fit the Planck-ian Distribution Equation (see red curves). **a** The histogram of the on-times of single-molecule cholesterol oxidase measured and simulated by a double exponential function of Lu et al. [17]. **b** The histogram in (**a**) was hand-digitized (blue curve) and simulated using the Planckian distribution equation, Eq. (3) (red curve). **c** The fitting to PDE of the genomic mRNA level data measured from the human breast tissues [18]. **d** Fitting to PDE of the mRNA level data of the MAPK (mitogen-activated protein kinase) pathway in the human breast tissues [18]. **e** Fitting to PDE of the mRNA level data of the CGI (kinase binding protein) pathway [18]. **f** Fitting to PDE of the mRNA level data of the ZFP (zinc-finger protein) pathway. [18]

the conventional camera with a Blackmagic Cinema camera. The digital CymaScope so constructed can transform sounds into visual images of water wave patterns (see Table 1) and outputs the corresponding digital data series which can be displayed as long-tailed asymmetric histograms (LAHs) using the histogram software in Excel (see Fig. 1). The LAHs can then be fitted to PDE (also called BRE, blackbody radiation-like equation) as exemplified in Fig. 1.

Table 1 The Planck–Shannon Classifier-based analysis of the CymaScopic images of the sonified Raman signals from cancer and healthy cells

Time	Cancer cells						Healthy cells					
	1		2		3		5		6		7	
	#	I_{PS}/H	#	I_{PS}/H	#	I_{PS}/H	#	I_{PS}/H	#	I_{PS}/H	#	I_{PS}/H
1	80	6.095 3.017	80	6.024 3.448	80	6.302 2.782	57	7.372 1.616	27	7.345 1.636	88	7.339 1.676
2	81	6.035 3.147	81	6.163 3.313	81	6.102 3.076	58	7.328 1.614	28	7.423 1.667	89	6.417 2.328

I_{PS} (Planck information of the second kind) and H (Shannon entropy) were calculated based on Eqs. (8) and (9), respectively. Each image took 45 ms to be captured by the Blackmagic Cinema camera installed in the CymaScope. The first two sets of the 10 measurements made at 10 time points. The symbol # refers to the frame number of an image

3   The Planckian Distribution Equation (PDE)

The Planckian distribution equation (PDE) was derived from the blackbody radiation equation, Eq. (1), discovered by M. Planck (1858–1947) in 1900 [8] by replacing its universal constants and temperature with free parameters, A, B, and C, resulting in Eq. (3) (see Fig. 2).

As already alluded to, PDE has been found to fit almost any LAHs regardless of their origin. One way to account for the apparent universality of PDE is to postulate that the *principle of wave-particle duality* discovered in quantum mechanics is not confined to micro-physics but applies also to macrophysics, including living systems, the human society, and the Universe itself (see Fig. 8 in [9] and Fig. 1 in [10]). This possibility is suggested by the fact that the first and the second terms appearing in the blackbody radiation equation, Eq. (1), are related to the *number of standing waves* in the system under consideration and the *average energy of the standing waves*, respectively [8]. Since PDE is isomorphic with Planck's blackbody radiation equation, Eq. (1), in mathematical form, it seems reasonable to assume that the first term, $(A/(x + B)^5$, and the second term, $(e^{C/(x+B)} - 1)$, of PDE are also related, respectively, to the number of standing waves and the average energy of the standing waves.

Table 2 The Planck–Shannon classifier-based (or the Planck–Shannon plot-based) analysis of the CymaScopic images of the sonified Raman signals from healthy and cancer single cells

Cells	Cancer cells			Healthy cells		
1	Planck-Shannon plot, Cancer Cell 1, 10 images analyzed $y = -0.7366x + 8.3379$ $R^2 = 0.8781$			Planck-Shannon plot, Healthy Cell 1, 13 images analyzed $y = -1.3748x + 9.6091$ $R^2 = 0.8647$		
2	Planck-Shannon plot, Cancer Cell 2, 10 images analyzed $y = -0.5803x + 8.0816$ $R^2 = 0.9397$			Planck-Shannon plot, Healthy Cell 2, 10 images analzyed $y = -0.8235x + 8.6815$ $R^2 = 0.5492$		
3	Planck-Shannon plot, Cancer cell 3, 10 images, analyzed $y = -0.4909x + 7.6252$ $R^2 = 0.5742$			Planck-Shannon plot, Healthy Cell 3, 10 images analyzed $y = -1.4231x + 9.8223$ $R^2 = 0.9761$		
Cells	I_{PS}	H	Slope (Corr. coeff.)	I_{PS}	H	Slope (Corr. coeff.)
1	6.131	2.996	−0.737 (0.878)	7.285	1.690	−1.375 (0.865)
2	6.268	3.124	−0.580 (0.940)	7.299	1.679	−0.824 (0.549)
3	6.139	3.029	−0.491 (0.574)	7.421	1.87	−1.423 (0.976)
Average	6.179	3.050	−0.603	7.335	1.685	−1.207
STD	0.004	0.066	0.094	0.075	0.006	0.333
p–value	0.001 for H	0.004 for I_{PS}	0.094 for slopes	0.001 for H	0.004 for I_{PS}	0.094 for slopes

Red color indicates the Plamckian informaion of the second kind, I_PS, while the blue color indicates Shannon entropy, H
Data from [11]

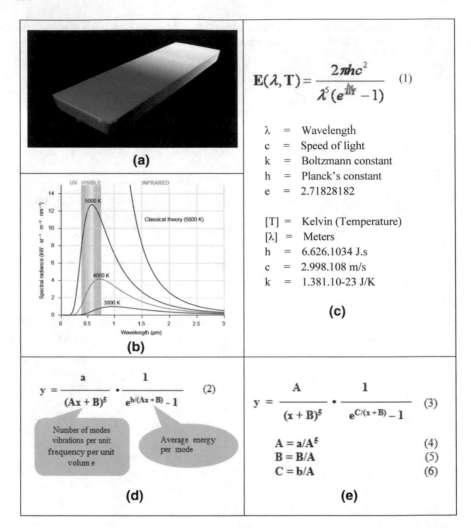

Fig. 2 a Blackbody radiation. **b** The blackbody radiation spectra. Both (**a**) and (**b**) were retrieved from https://en.wikipedia.org/wiki/Black-body_radiation on 01/05/2016. **c** Planck radiation equation (PRE). Reproduced from [8]. **d** The blackbody radiation-like equation or BRE [2, Chaps. 11 and 12], also called the generalized Planck equation (GPE) or the Planckian distribution equation (PDE). In Eqs. (2) and (3), x is the bin variable and y is the frequency. The interpretation of the two terms given in Eq. (2) assumes that a similar interpretation given to PRE in [8] can be justifiably transferred to BRE due to the similarity in their mathematical forms. **e** The 3-parameter (colored red) version of BRE/GPE/PDE. The relations between the 4- and 3-parameter versions of BRE/GPE/PDE are given in Eqs. (4–6)

4 Sonified Raman Spectroscopic Signals from Cancer and Healthy Cells in Human Brain Tissues

Surgically removing tumor cells (tumorectomy) from the brain is one of the best methods to decrease the recurrence rate of brain cancer [11]. In order to increase clinical outcome of tumorectomy, it is important to have minimal damage on normal tissues. To help surgeons to distinguish between cancer and healthy cells based on *sounds* so that their eyes can remain focused on the surgical target during surgery, M. Baker and his coworkers at the University of Strathclyde in Glasgow developed a sonified Raman spectroscopic method [12] to distinguish between cancer and healthy cells in brain tissues [11]. Here, we report that, by converting the audio files that they generated to the visual images using the digital CymaScope and then analyzing the digital images based on a PDE-based algorithm, we have been able to distinguish between cancer and healthy cells with p-values far less than 0.05 (see the last row in Table 2). Thus, in the present manuscript, we have

(i) Constructed the digital version of the CymaScope to successfully produce the first *digital CymaScopic images* of the cancer and healthy cells from their audio files (see Table 1) and

(ii) Demonstrated that the PDE and its derived functions (i.e., Eqs. (8) and (9) below) can distinguish between cancer and healthy cell CymaGlyphs (see Table 2).

5 The Planck–Shannon Plot

Once a long-tailed histogram is fitted into PDE, two numbers, I_{PS} and H, can be generated from the resulting PDE (see Fig. 3):

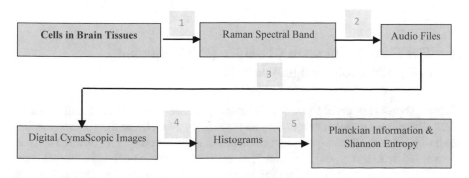

Fig. 3 The 5 key steps underlying the *PDE-based Digital Raman CymaScopy* (PDRC) for distinguishing between cancer and healthy cells in human brain tissues

(i)   The *Planckian Information of the First Kind* (I_{PF}) defined as the binary logarithm of the ratio of the area under the curve of PDE to that of Gaussian-like symmetric curves, Eq. (7).

$$y = Ae^{-(x-\mu)^{\wedge 2}/(2\sigma^{\wedge 2})} \tag{7}$$

where A is a free parameter, μ is the mean, and σ is the standard deviation [10, 13].

(ii) The *Planckian Information of the Second Kind* (I_{PS}) defined as the negative binary logarithm of the *skewedness* of the long-tailed histogram [10, 13].

$$I_{PS} = -\log_2(|\mu - \text{mode}|/\sigma) \tag{8}$$

The *Shannon entropy* (H) calculated based on Eqs. (9) and (10):

$$H = -\sum p_i \log_2 p_i \tag{9}$$

where p_i is the probability of observing the ith data point calculated as

$$p_i = y_i / \sum y_i \tag{10}$$

where y_i is the frequency of the ith data point and the index i runs from 1 to n, the total number of the data point.

It is often more convenient and more reproducible to calculate I_{PS} from PDE than to calculate I_{PF}. Thus, the information encoded in the shape of a long-tailed asymmetric histogram can be visualized as a point in what is here called the *Planck–Shannon space* (or *plane, plot*) constructed by plotting H on the x-axis and I_{PS} on the y-axis. Some examples of the Planck–Shannon plots of the mRNA levels measured from the human breast cancer tissues are shown in Table 2.

6   The Planck–Shannon Classifier (PSC) Applied to CymaScopic Image Analysis

Table 1 shows two sets of CymaScopic images, labeled "Cancer Cells" and "Healthy Cells," that were visualized from their audio files at two different time points (the first two of a total of 10 time-point measurements), each set consisting of three subsets numbered 1, 2, and 3, and 5, 6, and 7, respectively. The CymaScopic images were transformed into their corresponding histograms which were then reduced to two numbers, I_{PS} (in red) and H (in blue), by fitting to PDE, as described in Sect. 5. These numbers allowed each histogram to be represented as a point on the Planck–Shannon plot (see the graphs in Table 2). When the 10 CymaScopic images belonging to a given cell type are plotted on the Planck–Shannon plane, a straight line was obtained with the correlation coefficient ranging from 0.549 to 0.976 (see the lower portion of Table 2).

As evident in the last row of Table 2, the Planck–Shannon classifier (PSC) or the Planck–Shannon plot (PSP) can successfully distinguish between cancer cells and healthy cells with the p-values of 0.001 when comparing them in terms of H and with the p-values of 0.004 when comparing them in terms of I_{PS}. But when these two cell types are compared in terms of the slopes of the Planck–Shannon plots, the p-value is found to be 0.094, which is larger than the p-values based on I_{PS} or H alone. Since the slope of the line on a Planck–Shannon involves both I_{PS} and H, we thought it would be a better discriminator than I_{PS} or H individually, but apparently it is not. One possible reason for this finding may be that some information is lost when I_{PS} and H are combined into a ratio, I_{PS}/H. Perhaps some essential information common to I_{PS} and H got cancelled out. In other words, the information content of the ratio I_{PS}/H seems to be less than that of either I_{PS} or H alone.

When the two sets of the CymaGlyphs of cancer and healthy cells shown in Table 1 are examined visually (i.e., *qualitatively*), it is almost possible to distinguish them in words or verbally due to the extreme variability of individual images regardless of cell types. However, when PSC is used to compare them *quantitatively*, a clear distinction could be made (see the last row in Table 2). Thus, Table 2 provides the first concrete experimental evidence for *the power of PSC to classify CymaScopic images or CymaGlyphs unambiguously.*

7 Conclusions

This paper represents the first application of the digital CymaScope (DCS) constructed in 2017 and the Planck–Shannon classifier (PSC) formulated in 2018 to distinguishing between cancer and healthy cells in the human brain tissues. Two improvements were made over the previous attempt to accomplish the same goal by Stables et al. [11]. First, the audio files generated from the Raman spectral signals from cancer and healthy cells from brain tissues were transformed into digital CymaScopic images, rather than being analyzed directly via the human ear. Second, these digital CymaScopic images were then *analyzed quantitatively* utilizing the Planck–Shannon classifier (PSC) or the Planck–Shannon plot (PSP). Both these novel features, i.e., the digital CymaScopy and the PSC-based AI, may improve the success rate of distinguishing cancer and healthy cells based on their sonified Raman spectral features, relative to the original procedures that depended on the HI (human intelligence), i.e., surgeon's ability to differentiate between the sounds generated from cancer and healthy cells. Although the PSC has been applied to analyzing the digital CymaScopic images in this paper, it was also successfully applied to analyzing human breast cancer transcriptomics, fMRI signals from the human brain before and after psilocybin administration, etc. [3], indicating that PSC can be used as a *general purpose classifier*. It would be a great challenge to compare in the future

PSC's performance with other classifiers such as those of Hu et al. [14], Shi et al. [15], and Kalaiselvi et al. [16].

Acknowledgements We thank Drs. R. Stables and M. Baker for providing us with the audio files of the sonified Raman spectral signals of the cancer and healthy human brain cells and Sayer Ji, Founder of GreenMedInfo, Bonita Springs, FL, USA, for the financial support enabling us to convert the analog CymaScope to the digital CymaScope in 2017. Sungchul Ji also thanks Dr. Sergey Petoukhov for helpful suggestions and informative discussions over the years.

References

1. Ji, S.: Modeling the single-molecule enzyme kinetics of cholesterol oxidase based on Planck's radiation formula and the principle of enthalpy-entropy compensation. In: Short talk abstracts. The 100th Statistical Mechanics Conference, Rutgers University, Piscataway, 13–16 Dec, 2008
2. Ji, S.: Molecular Theory of the Living Cell: Concepts, Molecular Mechanisms, and Biomedical Applications. Springer, New York (2012)
3. Ji, S.: The Cell Language Theory: Connecting Mind and Matter. World Scientific Publishing, New Jersey (2018)
4. Reid, J.S. The Science of the CymaScope. https://www.cymascope.com/cymascope.html (2017)
5. The Science of the CymaScope. http://www.cymascope.com/cymascope.html
6. Jenny, H.: Cymatics: A Study of Water Phenomena and Vibration. MACROmedia Publishing, ME (2001)
7. Socolofsky, S., Reid, J.S.: CymaScope video projected in Trinity Episcopal Cathedral. https://www.youtube.com/watch?v=n_e-vhGbqFQ (2015)
8. Blackbody Radiation. http://hyperphysics.phy-astr.gsu.edu/hbase/mod6.html
9. Ji, S.: Planckian distributions in molecular machines, living cells, and brains: the wave-particle duality in biomedical sciences. In: Proceedings of the International Conference on Biology and Biomedical Engineering, Vienna, Mar 15–17, 2015, pp. 115–137, 2015. PDF at http://www.conformon.net/wp-content/uploads/2016/09/PDE_Vienna_2015.pdf
10. Ji, S.: Waves as the symmetry principle underlying cosmic, cell, and human languages. Information **8**(1), 24 (2017). https://doi.org/10.3390/info8010024. PDF at http://www.mdpi.com/2078-2489/8/1/24
11. Stables, R., Clemens, G., Butler, H.J., et al.: Feature driven classification of Raman spectra for real-time spectral brain tumor diagnosis using sound. Analyst **142**, 98–109 (2017)
12. Cheng, J.-X., Xie, X.S.: Vibrational spectroscopic imaging of living systems: an emerging platform for biology and medicine. Science **350**(6264) aaa8870-1–9 (2015)
13. Ji, S.: RASER model of single-molecule enzyme catalysis and its application to the ribosome structure and function. Arch. Mole. Med. Genet. **1**(1), 31–39 (2018)
14. Hu, Z., Mashtalir, S.V., Tyshchenko, O.K., Stolbovyi, M.I.: Video shots' matching via various length of multidimensional time sequences. Int. J. Intell. Syst. Appl. (IJISA) **9**(11), 10–16 (2017). https://doi.org/10.5815/ijisa.2017.11.02
15. Shi, Y.F., He, L.H., Chen, J.: Fuzzy pattern recognition based on symmetric fuzzy relative entropy. Int. J. Intell. Syst. Appl. (IJISA) **1**(1), 68–75 (2009). https://doi.org/10.5815/ijisa.2009.01.08
16. Kalaiselvi, T., Kalaichelvi, N., Sriramakrishnan, P.: Automatic brain tissues segmentation based on self initializing K-means clustering technique. Int. J. Intell. Syst. Appl. (IJISA) **9**(11), 52–61 (2017). https://doi.org/10.5815/ijisa.2017.11.07

17. Lu, H.P., Xun, L., Xie, X.S.: Single-Molecule Enzymatic Dynamics. Science **282**, 1877–1882 (1998)
18. Perou, C.M., Sorlie, T., Eisen, M.B., et al.: Molecular portraits of human breast Tumors. Nature **406**(6797), 747–752 (2000)

ECG Signal Spectral Analysis Approaches for High-Resolution Electrocardiography

A. V. Baldin, S. I. Dosko, K. V. Kucherov, Liu Bin, A. Yu. Spasenov, V. M. Utenkov and D. M. Zhuk

Abstract This article considers the possibility of application of high-resolution electrocardiography spectral methods to the analysis of the variability of cardiac cycle spectral analysis results, which will significantly improve the diagnostic capabilities of these methods. Implementation of the proposed approaches will move from a one-dimensional (the duration of the cardiac cycle) analysis of heart rate variability to a multidimensional (parameters of the form and location of the peaks of the cardiac cycle) analysis of the cardiac parameters.

Keywords Electrocardiograph · Electrocardiosignals · Cardiac cycle · Spectral analysis · HRV analysis · High-resolution electrocardiography · Mathematical processing

1 Introduction

One of the main diagnostic tools of the functional state of the patient and the activities of his cardiovascular system in current medical practice is the analysis of heart rate variability (HRV) [1]. It should be noted that traditional approaches to the HRV analysis are not reliable for diagnosing a particular disease, although some researchers are trying to use it this way [2]. Substantially, HRV analysis allows to evaluate general state of the subject at the current time, and the change of HRV parameters allows to assess the dynamics of this state.

The analysis of HRV for short periods of approximately 2–10 min is quite well developed by now. The following groups are well-known methods of HRV analysis:

- Time domain methods, statistical method;
- Geometric methods;

A. V. Baldin · S. I. Dosko · K. V. Kucherov · A. Yu. Spasenov (✉) · V. M. Utenkov · D. M. Zhuk
Bauman Moscow State Technical University, Moscow 105005, Russia
e-mail: a.spasenov@bmstu.ru; 15988233333@163.com

L. Bin
Panther Healthcare, Beijing 102209, China

© Springer Nature Switzerland AG 2020
Z. Hu et al. (eds.), *Advances in Artificial Systems for Medicine and Education II*,
Advances in Intelligent Systems and Computing 902,
https://doi.org/10.1007/978-3-030-12082-5_18

- Frequency domain methods, methods of spectral analysis;
- Nonlinear methods;
- Correlation rhythmography.

In this article it is assumed to consider the features of HRV analysis methods in the frequency domain, which include spectral methods. Spectral methods of HRV analysis are very common in the diagnosis of subject's functional state now. Modern functional diagnostics offers a wide variety of methods, among which the most common is the method of recording and processing the electrocardiogram, including high-resolution electrocardiography.

The main goal of the study is to identify the signal robust parameters for the analysis of HRV in order to improve the methods of diagnosing the disease. The Prony method allowing spectral analysis of the components of the ECG signal with the required accuracy is investigated.

2 Methods of Spectral Analysis of HRV

Analysis of the fluctuations' spectral power density gives information about the power distribution depending on the fluctuation frequency. Applying of spectral analysis allows to quantify various frequency components of heart rate fluctuations and visualize graphically ratios of various components of heart rate, reflecting the activity of certain regulatory mechanism units [1, 3–5].

The advantages and disadvantages of HRV analysis spectral methods are discussed in sufficient detail in many publications (e.g., in [6]). Heart rate is regulated primarily by vegetative nervous system and the most effective method of its analysis today is the evaluation of HRV. Perennial explore of HRV analysis practical applications lead to international guidelines for research methodology development [7]. It is recommended at least 4 frequency ranges suitable for assessing the functional state of the subject:

- (0.15–0.4) Hz—tidal range, vagal tone (HF);
- (0.05–0.15) Hz—10 second rhythm reflecting pressosensitive regulation (LF);
- (0.04–0.004) Hz—range of very slow waves (VLF);
- Ultra Low Frequency (ULF) range (less than 0.004 Hz), whose physiological significance of the least studied.

Research on the origin of VLF range have been published in [8]. In the clinical use of HRV VLF range proved prognostically significant, i.e. it was revealed a large variety of body functions which are reflected in the amplitude-frequency spectrum of the range [9–11]. Such a variety of functions requires a more detailed study of this and other ranges. This possibility has appeared thanks to the use of improved techniques of fast Fourier transform (FFT) and new methods—continuous wavelet analysis and Hilbert–Huang transform [12].

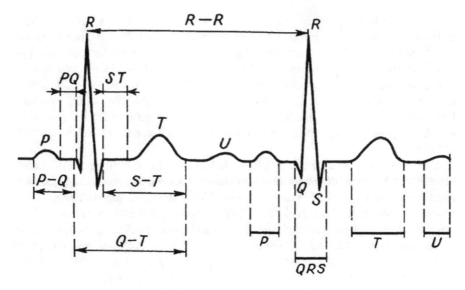

Fig. 1 Electrocardiocycle parameters

Generally, all methods consider rhythmograms based on calculated sequences of the ECG R-R intervals. The R-R interval includes: half-wave R, peak S, segment S-T, peak T, segment T-P of current cycle and segment P-Q, peak Q, half of the R-wave of the next cycle (see Fig. 1).

Thus R-R interval variability consists of variabilities of two adjacent partial cycles and does not fully meet the individual variability of the cardiac cycle. In the analysis of HRV one dimensional variability of only parameter—the duration of the cardiac cycle—is investigated. This limits the amount of diagnostic information. More informative to raise diagnostic information is to study the variability of the difference between spectral analysis relevant results (e.g., the same name harmonics) of two adjacent cardiocycles. Such an investigation may be based on spectral techniques used in high-resolution electrocardiography [13, 14].

3   High Resolution Electrocardiography

This method is based on the stringent requirements to the parameters of the electrocardiograph. An example of such requirements is given in Chap. 7 of [15], though the current level of microelectronics development allows for higher values of the basic parameters of the high-resolution electrocardiograph. E.g., electrocardiograph designed in NPO "Geophysics" provides bandwidth from 0 to 500 Hz, the sampling frequency up to 4 kHz, 24 ADC bit capacity and a very low own noise level. Besides, it is necessary to use high quality connecting wires with silver chloride electrodes.

The main objectives of this method is averaging and filtering electrocardiogram different sections with their subsequent mathematical processing. All this allows to isolate and analyze the low amplitude signals which are not available for analysis using conventional ECG detection methods. Meanwhile these signals may contain important diagnostic information. Using the methods of high-resolution electrocardiography spectral analysis at the level of the cardiac cycle may allow to obtain information about shape parameters and position of the peak of the cardiac cycle, which will greatly complement the diagnostic information obtained in the analysis of HRV.

High-resolution electrocardiography is mainly used for measuring the potentials of delayed myocardial depolarization, so-called ventricular late potentials (VLP) and atrial late potentials (ALP). When analyzing VLP three basic approaches are used: temporal, spectral, and spectral-time ECG analysis techniques. Spectral analysis estimates changes in the amplitude-frequency characteristics of the particular portion of the ECG signal. Often a FFT is used for spectral analysis method, which imposes a requirement of continuity and the periodicity of the signal. To meet this condition, the "sliding window" is applied [7, 16]. Choosing the size of the "window" is a compromise between the desired accuracy of the localization of low amplitude signals within the QRS-complex (this means the selection of a short period), and preserving the spectral resolution (the ability to distinguish the spectral components of the two or more signals) since spectral resolution deteriorates with a decrease of "window" width.

A significant drawback of the spectral analysis using FFT is the inability to determine the exact location of the desired high-frequency components in the analyzed segment. To overcome this drawback a spectral-time mapping method has been developed, the essence of which is in the calculation of the spectrum on the time axis of a moving "window" in the final part of the QRS-complex and S-T segment.

A further development of these methods was a series of non-stationary signal analysis techniques (ECG). The most famous was the wavelet transform (Wavelet-transform). It is the expansion of the signal by a set of basic functions that are defined on an interval shorter than the duration of cardio. Moreover, all functions of the set are generated by using two-parameter conversion (time axis shift and zoom) of one of the basic functions. Large scale parameter values correspond to the original signal applied to the low pass filter, low values—applied to the high-pass filter. Unlike Fourier transform, wavelet transform contains signal to "window" multiplication in basic function wherein "window" can adapt to zoom.

However, listed spectral methods of high-resolution electrocardiography do not have sufficient robustness of the results and require interactive support for customization of the method parameters, which leads to the need to develop new methods and approaches for the automatic spectral analysis of high-resolution electrocardiography at the cardiac level.

One way to increase the resolution is the use of parametric methods of digital spectral analysis, which belong to the class of superresolution. The most effective, in terms of feasibility and potential resolution, according to many authors, include the MUSIC (Multiple Signal Classification) and ESPRIT (Estimation of Signal Parameters via Rotational Invariance Techniques), as well as their variations [17–20]. It should be noted that these methods allow several times to increase the resolution, but their application in practice is constrained by increased requirements for analyzers electronic components and high computational load in signal processing.

If we consider the patient-electrocardiograph system, it can be assumed that the ECG signal is determined by the kind of control inputs and signal paths from them to the system output signal. Signal paths can be considered as a kind of distributed system, the immutability of the own frequency spectra in which is an important criterion of constancy of its physical properties and integrity.

In our opinion the possibility of use Prony spectral analysis [21, 22, 23], as one of the high-resolution methods for determining the characteristics of the temporal evolution of dynamic processes, is of considerable interest.

Consider the time series

$$v[k] = v[x_i, y_i, (k-1)\Delta t], \quad k = [1 : N], \tag{1}$$

which represents a fragment of the time dependence $v(x_i, y_i, t)$ of transient oscillations of fixed point (x_i, y_i) belongs to distributed system for $t \in [0, t_N], t_N = N\Delta t$ (where Δt is time sampling interval). Let the time series contains noise due to measurement inaccuracies and environmental influences.

Prony spectral decomposition of the time series (1) has the form:

$$v[k] = \sum_{j=1}^{p} r_l(z_l)^{\Delta t(k-1)} + n(\Delta tk), \quad k = [1 : N], \tag{2}$$

where: p is the number of oscillations of the poles; $z_l = \exp(\delta_l + j\omega_l), l = [1 : p]$ is poles, which are determined on the basis of the analyzed time series $\{v[k], k = [1 : N]\}$; $r_l = \alpha_l \cdot \exp(j\varphi_l), l = [1 : p]$ is the residues at these poles; $n(\Delta tk)$ is noise. Thus, the oscillation poles determine discrete spectra of decrements and frequencies $\{\delta_l \cdot \omega_l, l = [1 : p]\}$, and the residues at these poles—the corresponding discrete spectra of amplitudes and phases $\{\alpha_l, \varphi_l, l = [1; p]\}$ for a considered time window $t \ni [0, t_N]$ [1].

After determining the poles $z_l, l = [1 : p]$ residues $l = [1 : p]r_l$ at these poles are determined by the least squares method [12].

It should be noted [16], that the main advantages of the Prony method are: the ability to allocate the principal energy components of oscillation and the ability to recover digitized oscillation by the formula (2), and on the basis of this—possibility of estimating the accuracy of the spectral analysis.

Evaluation of the time dependence of discrete decrements and frequencies spectra $\{\delta_l, \omega_l, l = [1 : p]\}$ and the corresponding discrete spectra of amplitudes and phases $\{\alpha_l, \varphi_l, l = [1 : p]\}$ is performed by successive shift of the time window of fixed length or segment identification in signal according to certain rules. In our case the varied width of the window may be in the range from 0.1 s, which characterizes the duration of QRS complex, to 0.7 s.

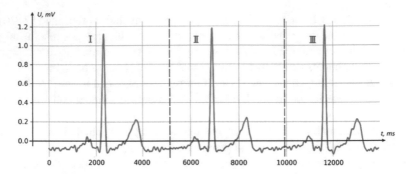

Fig. 2 The original signal

Fig. 3 The first segment

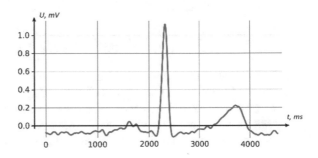

Fig. 4 The second segment

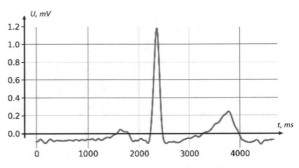

For example, consider ECG signal picked using electrocardiograph with a sampling rate of 4000 Hz. Cardiocycles parameter variability analysis will be performed on consecutive portions of the electrocardiogram recording. To reduce the computational load decimation procedure was used. The decimation ratio is 10, however in all graphs frequency axis is 10 times stretched. Figures 2, 3, 4 and 5 show successive fragments of ECG record.

During record processing the approximation of each fragment with the required accuracy is executed. Figures 6 and 7 show a first approximation of a signal segment and the percentage of the approximation error as a function of the number of function approximation modes.

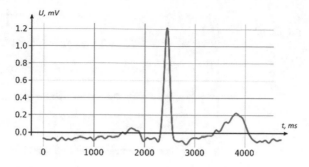

Fig. 5 The third segment

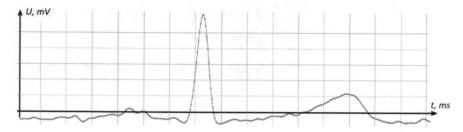

Fig. 6 Approximations of the first ECG segment

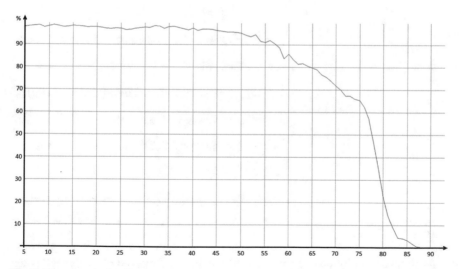

Fig. 7 The error obtained by the approximation of the record first fragment with different numbers of modes

Fig. 8 Modal
decomposition of signal of
the first segment

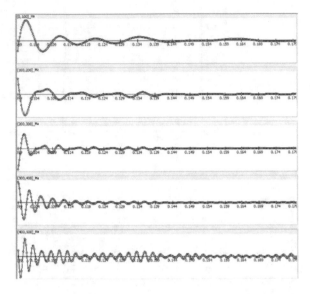

Fig. 9 Modal
decomposition of signal of
the second segment

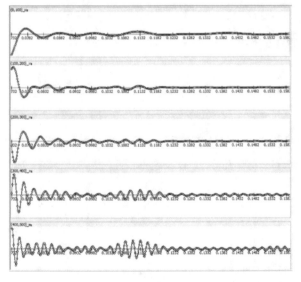

Analytical description of each fragment obtained is followed by analytical modal decomposition (AMD) of signal and calculating its analytical Fourier spectra (AFS). Figures 8, 9 and 10 show modal decomposition of the signal segments, Figs. 11, 12 and 13 show spectra of Prony amplitudes and Figs. 14, 15 and 16 show analytical Fourier amplitude spectra.

Fig. 10 Modal
decomposition of signal of
the third segment

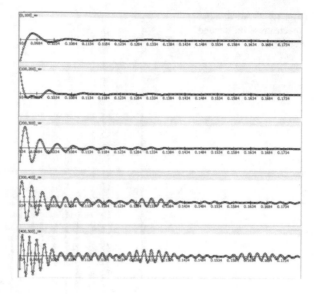

Fig. 11 Prony amplitude
spectrum of the first segment

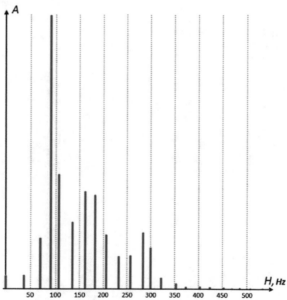

Fig. 12 Prony amplitude
spectrum of the second
segment

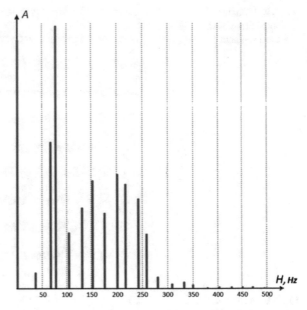

Fig. 13 Prony amplitude
spectrum of the third
segment

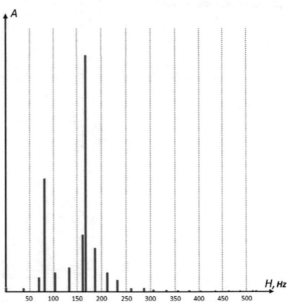

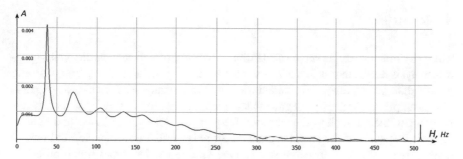

Fig. 14 Analytical Fourier amplitude spectrum of the first segment

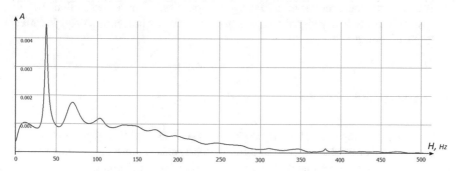

Fig. 15 Analytical Fourier amplitude spectrum of the second segment

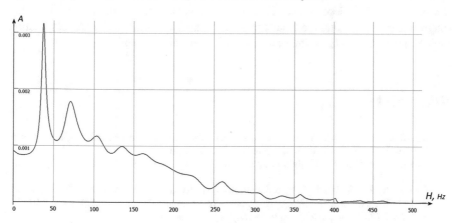

Fig. 16 Analytic Fourier amplitude spectrum of the third segment

4 Results

The experimental results show substantial variability of adjacent cardiocycles spectral analysis results that allows to analyze the variability of the cardiac parameters and greatly enhance its diagnostic capabilities in comparison with the analysis of HRV. The example of three ECG segments shows sufficient sensitivity of several information parameters such as the vibration frequencies, Prony amplitudes and Fourier analytical spectra of the signal segments. Signals' modal decomposition allows to dissect signal of frequency bands wherein the first step of decomposition is proposed to use empirical modes by Huang method [12], and the second step—to use decomposition by analytical modes of the Prony method. Studies have shown that such an analysis of the frequency signal can give good results for processing each fragment of the ECG record.

5 Conclusions

The prospect of using time frequency Prony analysis to assess HRV appears attractive. Since the frequency, decrements, amplitudes and phases are a set of independent variables, they and functions of some of them (e.g., squared amplitudes) are most appropriate information parameters. In this regard, the Fourier spectra do not possess the properties of independence. Future work involves the formation of additional diagnostic criteria, using the results of spectral analysis of the ECG and conducting their experimental verification to assess their diagnostic capabilities.

References

1. Baevsky, R.M.: Analysis of heart rate variability using different electrocardiographic systems. Bull. Arrhythmol. **1**, 66–85. Ryabykina G.V., Moscow (2001)
2. Sobolev, A.V.: Holter ECG Monitoring and Bifunctional and Blood Pressure. Medpraktika-M, Moscow (2010)
3. Darwan, Hartati, S., Wardoyo, R., Setianto, B.Y.: The feature extraction to determine the wave's peaks in the electrocardiogram graphic image. Int. J. Image Graph. Signal Process. (IJIGSP) **9**(6), 1–13 (2017). https://doi.org/10.5815/ijigsp.2017.06.01
4. Ahmad, A.A., Kuta, A.I., Loko, A.Z.: Analysis of abdominal ECG signal for fetal heart rate estimation using adaptive filtering technique. Int. J. Image Graph. Signal Process. (IJIGSP) **9**(2), 19–26 (2017). https://doi.org/10.5815/ijigsp.2017.02.03
5. Choubey, D.K., Paul, S.: GA_MLP NN: a hybrid intelligent system for diabetes disease diagnosis. Int. J. Intell. Syst. Appl. (IJISA) **8**(1), 49–59 (2016). https://doi.org/10.5815/ijisa.2016.01.06
6. Mikhailov, V.M.: Heart Rate Variability. Practical Experience Method. Ivanovo State Medical Academy, Ivanovo (2000)
7. Task Force of the European Society of Cardiology the North American Society of Pacing Electrophysiology: Heart rate variability. Standards of measurement, physiological interpretation, and clinical use. Circulation **95**, 1043–1065 (1996)

8. Taylor, J.A., Carr, D.L., Myers, C.W., Eckberg, D.L.: Mechanisms underlying very-low-frequency RR-interval oscillations in humans. Circulation **6**(98), 547–555 (1998)
9. Akselrod, S., Gordon, D., Madwed, J.B.: Hemodynamic regulation: investigation by spectral analysis. Am. J. Physiol. **249**, 867–875 (1985)
10. Kleiger, R.E., Miller, J.P., Bigger, J.T., Moss, A.J.: Multicenter. Decreased heart rate variability and its association with increased mortality after acute myocardial infarction. Am. J. Physiol. **59**, 256–262 (1987)
11. Gerus, A.Yu., Fleishman, A.N.: Heart rate variability in patients with diabetes mellitus. Biol. Klin. Med. **1**(8), 96–100. Novosibirsk (2010)
12. Li, H., Kwong, S., Yang, L., Huang, D., Xiao, D.: Hilbert-Huang transform for analysis of heart rate variability in cardiac health. IEEE/ACM Trans. Comput. Biol. Bioinf. **8**, 1557–1567. Pubmed (2011)
13. Ivanov, G.G., Gracheva, S.V., Syrkina, A.L.: High-Resolution Electrocardiography. Triada-H, Moscow (2003)
14. Grishaev, S.L.: High-Resolution Electrocardiography. Russian Military Medical Academy, St. Petersburg (2004). (in Russia)
15. Ardashev, A.V.: Clinical Arrhythmology. Medpraktika-M, Moscow (2009)
16. Ardashev, A.V., Loskutov, A.Yu.: Practical Aspects of Modern Methods of Analysis of Heart Rate Variability. Medpraktika-M, Moscow (2010)
17. Pesavento, M., Gershman, A.B., Haardt, M.: Unitary root-MUSIC with a real-valued eigendecomposition: a theoretical and experimental performance study. IEEE Trans. Signal Process. **48**, 1306–1314 (2000)
18. Roy, R., Kailath, T.: ESPRIT-estimation of signal parameters via rotational invariance techniques. IEEE Trans. Acoust. Speech Signal Process. **37**, 984–995 (1989)
19. Lau, C.K.E., Adve, R.S., Sarkar, T.K.: Mutual coupling compensation based on the minimum norm with applications in direction of arrival estimation. IEEE Trans. Antennas Propag. **52**, 2034–2041 (2004)
20. Van Trees, H.L.: Detection, Estimation, and Modulation Theory, Optimum Array Processing. Wiley, New York (2004)
21. Schmidt, R.O., Frank, R.E.: Multiple source DF signal processing: an experimental system. IEEE Trans. Antennas Propag. **34**, 276–280 (1986)
22. Pariyal, P.S., Koyani, D.M., Gandhi, D.M., Yadav, S.F., Shah, D.J., Adesara, A.: Comparison based analysis of different FFT architectures. Int. J. Image Graph. Signal Process. (IJIGSP) **8**(6), 41–47 (2016). https://doi.org/10.5815/ijigsp.2016.06.05
23. Marple, S.L.: Digital Spectral Analysis with Applications. Prentice Hall, Englewood Cliffs, New York (1987)

Intelligent System for Health Saving

V. N. Krut'ko, V. I. Dontsov and A. M. Markova

Abstract Multifactorial nature of human health and need in personifying the approach to each person in the health saving programs should use modern information and cognitive technologies for the tasks of health assessment and control. The article presents a concept, basic methods, and a structure of intelligent system of health saving (InSyHS), created by the authors to solve these tasks. This system implements the intelligent Internet technology based on modern cognitive methods and information about health, considering all possible health-determining essential factors (nutrition, physical activity, lifestyle, social and nature environment), and doing people to form an active relation to health with the possibility of self-diagnostics (physical and mental reserves, stress, psycho-emotional characteristics), optimization and personalization of personal health saving programs.

Keywords Health · Health Saving · Big data · Healthy lifestyle · Internet technology · Preventive medicine · Personalized medicine

1 Introduction

Medical and demographic situation in many countries is one of the main obstacles for an effective socioeconomic development.

Modern informational and cognitive technologies allow considering many personal health and environmental factors for personalized optimization of health saving (HS) programs and prolongation of active life for 10–15 years. They give people many opportunities in learning [1–3], health assessment [4, 5], choosing a healthy lifestyle (HLS), and personal HS [6–8].

V. N. Krut'ko (✉) · V. I. Dontsov · A. M. Markova
Federal Research Center "Computer Science and Control" of RAS,
60-let Oktyabria str. 9, 117312 Moscow, Russia
e-mail: krutkovn@mail.ru

V. N. Krut'ko
Sechenov First MSMU, Moscow, Russia

© Springer Nature Switzerland AG 2020
Z. Hu et al. (eds.), *Advances in Artificial Systems for Medicine and Education II*,
Advances in Intelligent Systems and Computing 902,
https://doi.org/10.1007/978-3-030-12082-5_19

211

This article explores scientific and technical solutions and a new intelligent Internet technology for personalized support of health saving processes, based on the analysis of big data of health and its key factors. The article presents a concept, basic methods, and a structure of intelligent system of health saving (InSyHS), created by the authors to solve these tasks.

This system implements the intelligent Internet technology based on modern cognitive methods and information about health, considering all possible health-determining essential factors (nutrition, physical activity, lifestyle, social and nature environment), and doing people to form an active relation to health with the possibility of self-diagnostics (physical and mental reserves, stress, psycho-emotional characteristics), optimization and personalization of personal health saving programs.

2 Basic Principles of Intelligent System for Health Saving (InSyHS)

Effective health control is the most complex interdisciplinary and interdepartmental (more correctly—overdepartmental) task and the <<function of health building>> is implemented via so-called <<so-called healthy life style>> (HLS).

HLS control is carried out via mild methods of individual information impacts, proposing modern scientifically substantiated HS technologies of HLS considering the individual habits and preferences of the person. Internet may be an effective means for these impacts, and our project implements just this approach. The main principle of InSyHS is creation of HS information medium with using the Internet technology for personal informing people about the modern HS methods, added motivation for HLS and personal information support in solving the problem of optimizing the range of methods for applying these technologies, considering the personal characteristics and preferences of the person.

The key principle is personification, i.e., an individual approach to each person, considering his/her gender, age, habits and form of activity, together with genetic, psycho-physiological and medical characteristics.

The principle of continuous evolutionary development of the system while maintaining the optimum balance between its conservative and revolutionary elements is important that is caused by continuous emergence of new knowledge about health and methods of its correction.

The principle of standardization stipulates that the existing standards of preventive medicine should be used (on the analogy of medical treatment standards). New standards should also be developed.

The principle of health saving motivation and psychological support determines the interest of people, its attraction and retention in the HS sphere.

Other essential principles of InSyHS include: completeness, systematicity, bio-psycho-social concept of health. This principles require flexible, open, modular system design that allows a qualitative and quantitative developing, supplementing the bases with new information and the system with new algorithms and methods, connecting new external services to the system.

3 Informatics Methods in InSyHS

Based on the literature analysis and using the results obtained earlier, the most promising methods of work in the information space were identified and allowed analyzing big data of health and its key factors and providing personalized support for HS processes.

When implementing the pilot sample of the Internet system for personalized HS support, the following programming languages were used: C++ (ISO/IEC 14882: 2011), HTML (version 5.0), XML, Python, PHP, Java, operating systems Windows and Linux.

The task of developing a method for assessing the importance of key health characteristics and factors was mathematically formalized. The most currently effective mathematical methods of choosing the ideal set of characteristics in machine learning tasks, as well as their application to the modern problems of public health and medicine, were considered. The main method consists of a combined model of several filter methods, their comparison with the results of wrapping methods and nested models, testing with cross-validation on an available data set, and comparison of the results of the developed method for evaluating the efficiency of machine learning HS tasks.

To show the correlation of a person's health status with the characteristics of his lifestyle and other health saving factors such methods of intelligent data analysis as methods of classification, search for associations, forecasting, including neural network are used.

Fuzzy logic control was useful in our case, since often the system under consideration is not defined or cannot be determined mathematically, but the linguistic description of the system can be obtained from the expert on the basis of his experience, intuition, or heuristic approach: Mamdani's fuzzy logic conclusion and Takagi–Sugeno fuzzy logic conclusion. To find the fuzzy rules and functions of referring to the linguistic terms available in fuzzy rules, fuzzy clustering, such as *fuzzy c-means algorithm*, was used; unlike hard clustering methods, fuzzy clustering methods can assign each observation to more than one cluster with different degrees of membership.

Application of the decision trees was considered to be useful for the Project, and this was one of the most widely known and applied methods of machine learning, used in a great number of real applications. The k-nearest neighbor algorithm is actively used in the problems of diagnostics and computer analysis of human state, and it can also be useful in the Project in the section on diagnosing the patient's state. Logistic regression method is useful for choosing the therapeutic drugs and determination their interaction according to the analysis of records in the online social networks on medical topics: this is one of the most widely used methods for solving binary classification problems, which appears to be a statistical model predicting the probability of occurrence of an event by matching logistic curve with the data.

The developed method of data obtaining and entering to the integrated health passport from the various sources involves the creation of a number of loosely coupled functional units (components) designed to solve the subtasks of data collection. These components include: social network crawlers, thematically focused Web-crawler, user data loader from the information networks, Rosstat data loader, and component for primary data analysis, which allows extracting fragments of useful information, as well as Web links from the documents in xml, html, and csv formats. These blocks interact with each other via the JSON-RPC protocol and together give the solution of the task of obtaining data from heterogeneous sources and entering them to the integrated health passport.

To reveal important psychological and motivational characteristics of the users, psycholinguistics methods are used to analyze their posts in social networks, which, in particular, allow determining the risk of depression, as well as the attitude of the users to certain HS technologies and methods.

Information about all characteristics of the user's health and its key factors are contained in the system database, and it virtually presents an integrated health passport, which collects and stores information from various sources: social networks; medical records; statistical databases; information given by the user and the data of the remote diagnostics with the help of InSyHS tests. One of the effective approaches is application of our methods of smart analysis of medical texts and messages of social networks, health-related problems, which makes it possible to get new knowledge about the impact of various factors on human health [1, 9–11] and to use this knowledge at personal programs developing HS.

4 Contents and Structure of InSyHS

In the sense of the creators of InSyHS technology, the health means not the absence of certain diseases, but the level of working capacity, physical and mental capabilities, the level of opportunities to live actively and to solve all life problems. InSyHS provides the user with a range of information services for monitoring and managing his/her own health. The information support and the user motivation are achieved by several modules:

- Specially arranged arrays of text data, containing a scientific evidence base of efficiency and safety of the proposed HS tools and methods.
- Specially arranged and/or selected texts and visual materials (e.g., video clips or their online links), made according to modern laws of effective advertising.
- Feedback module of the system of dynamic monitoring of the HS program effects.
- The scientific and analytical block of the Internet system allows analyzing the obtained data, improving the work during the development of the system.

In the section "All about health," the user can get up-to-date, scientifically grounded information about HS methods: official recommendations of the World Health Organization; the Ministry of Health of Russia and other countries; links to the most interesting modern Web resources in this area. There is a service for searching and analyzing texts about health, which unlike traditional search engines offers advanced capabilities for a comprehensive solution of search and analysis tasks in HS.

InSyHS includes algorithms for: automatic generation of personalized recommendations for nutrition, geroprophylactic measures and correction of psycho-emotional problems and a choice of individualized physical activities.

In the Personal account, the user can comprehensively assess his/her health and lifestyle and get automated personalized advice on HS. The user can assess his/her physical and mental potential, namely:

- Assess the "General Health Status" by determining the level of stress in 12 main vital body systems and the level of environmental stress;
- Assess his/her physical working capacity, group health, indices of functional status, recommended exercise load;
- Assess his/her mental working capacity and its key factors, such as health, activity, mood, and anxiety;
- Assess his/her biological age in general and by a number of essential functional systems (aging profile), get a forecast of life expectancy.

In the section "Nutrition and quality of life," the user can assess the quality of his/her nutrition and lifestyle characteristics: the approach used to assess nutrition allows the user to compare his/her diet with the standard comfortably and without a wearily calorie count—the so-called Nutrition pyramid—and get personal recommendations on the correction of the diet and weight.

The mode of life is described with the help of the so-called Rose of the life quality, which clearly shows the discrepancy between the levels of priority for the user of various spheres of his/her life and the real-life conditions.

"Problem health zones" service allows the user to get information about the latest results of those tests which revealed health problems, requiring correction or special attention of the user.

With "Health Monitoring" service, the user can assess the dynamics of his/her health state and quality of life and thus do self-assessment of the effectiveness of his/her Personal HS Program. In the section "My personal recommendations," the user can get automated personal recommendations of InSyHS for the personal HS program, and the recommendations of the doctor and/or fitness trainer.

In general, the modular structure of InSyHS is shown in Fig. 1.

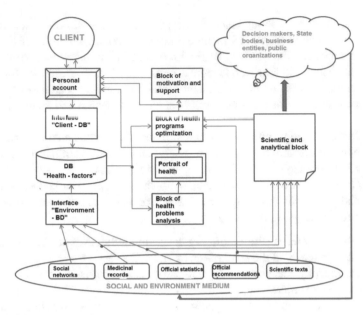

Fig. 1 Modular structure of InSyHS

5 Samples of InSyHS Operation

The communication of the client with a computer system that provides the implementation of HS technology is carried out through his/her Personal account on the portal that implements the Internet technology of personalized HS support. The interface of the Client-DB account includes three main blocks:

- System of electronic questionnaire;
- System of remote self-testing;
- System of data check and correction.

The known to the client medical information about his/her health is structured and standardized for further automated data processing.

The operation of the system is shown on the example of the "Nutrition and health" service. The revolutionary concept of nutrition diagnosis and optimization is important achievement (Fig. 2).

Using the proposed structural model and previously developed desktop system "Nutrition for Health and Longevity" [12], the system of online support for the choice of nutrition is offered. It includes service and content blocks. The most important and interesting feature of the second level of the user data formation is a possibility of using an interactive nutrition optimization procedure:

- Algorithms of formation of personal standards, considering personal data about psycho-physical load, lifestyle, and environment.
- Subsystem of evaluation of daily individual nutrition, including:

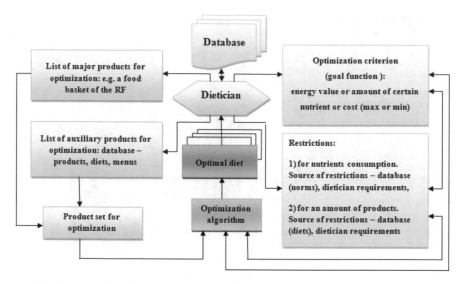

Fig. 2 Subsystem of nutrition monitoring and optimization in InSyHS

- express evaluation based on a comparison with the Nutrition pyramid;
- detailed evaluation based on a comparison with the individual norms including about 10–30 nutrients.

– Subsystem of monitoring, correction planning, and optimization of nutrition according to the individual norms.
– Databases and knowledge bases.

The results of analysis of the client's nutrition structure in comparison with the norms (Fig. 3).

The client's lifestyle is described in the system with the help of the so-called Rose of Life Quality, which clearly shows the discrepancy between the priority levels for the user of various spheres of his/her life and their real levels (Fig. 4).

6 Conclusion

Developed intelligent Internet technology of personalized HS (InSyHS) allows getting a conceptually new system result, i.e., a new, patentable, effective, Internet technology of personalized HS having non-direct analogs and allowing to convey personally to each person a reliable information about modern HS technologies, to carry out more additional human motivation to a healthy mode of life and to carry out personal information support in solving the problem of optimization for the spectrum and methods of applying these technology according to the characteristics and preferences of the person.

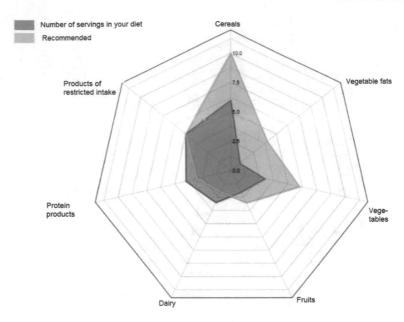

Fig. 3 Results of analysis of the client's nutrition structure in comparison with the norms

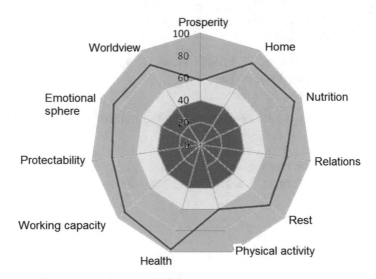

Fig. 4 Results of the user's life quality analysis: "Rose of Life Quality"

The expected results of HS practice are the following: improving the quality of life, improving the demographic parameters, increasing of social activity of the people, eliminating the risk of depopulation, ensuring social progress and sustainable progressive development of the country as the basis for everyone's well-being, and forming a healthy society. The effect of competent use of HLS can be expressed in 10–15 more years of active healthy life, and the assessments carried out in many countries show that the increase of life expectancy for one year may cause an increase of the country's GDP for about 4–5%.

References

1. Mestadi, W., Nafil, K., Touahni, R., Messoussi, R.: Knowledge representation by analogy for the design of learning and assessment strategies. Int. J. Mod. Educ. Comput. Sci. (IJMECS) 9(6), 9–16 (2017). https://doi.org/10.5815/ijmecs.2017.06.02
2. Kotevski, Z., Tasevska, I.: Evaluating the potentials of educational systems to advance implementing multimedia technologies. Int. J. Mod. Educ. Comput. Sci. (IJMECS) 9(1), 26–35 (2017). https://doi.org/10.5815/ijmecs.2017.01.03
3. Satyanarayana Murthy, T., Gopalan, N.P., Alla, D.S.K.: The power of anonymization and sensitive knowledge hiding using sanitization approach. Int. J. Mod. Educ. Comput. Sci. (IJMECS) 10(9), 26–32 (2018). https://doi.org/10.5815/ijmecs.2018.09.04
4. Mazurov, M.: Intelligent recognition of electrocardiograms using selective neuron networks and deep learning. Adv. Intel. Syst. Comput. 658, 182–197 (2017)
5. Rustembekova, S.A., Gorshkov, V.V., Sharipova, M.M., Khazova, A.S.: Detection of hidden mineral imbalance in the human body by testing chemical composition of hair or nails. Adv. Intel. Syst. Comput. 658, 215–228 (2017)
6. Takizawa, K., Takesako, K., Kawamura, M., Sakamaki, T.: Development of medical communication support system "health life passport". Stud. Health Technol. Inform. 192, 1027 (2013)
7. Hsieh, S.H., Hsieh, S.L., Cheng, P.H., Lai, F.: E–Health and health saving enterprise information system leveraging service oriented architecture. Telemed. e–Health 18(3), 205–212 (2012)
8. Krut'ko, V.N., Bolshakov, A.M., Dontsov, V.I., Mamikonova, O.A., Markova, A.M., Molodchenkov, A.I., Potemkina, N.S., Smirnov, I.V.: Intelligent internet technology for personalized health saving support. Adv. Intel. Syst. Comput. 658, 157–165 (2017)
9. Osipov, G.S., Smirnov, I.V., Tikhomirov, I.A.: Relational-situational method for text search and analysis and its applications. Sci. Tech. Inf. Process. 37(6), 432–437 (2010)
10. Shelmanov, A.O., Smirnov, I.V., Vishneva, E.A.: Information extraction from clinical texts in Russian. In: Computational Linguistics and Intellectual Technologies: Papers from the Annual International Conference "Dialogue", vol. 14, no. 21, pp. 537–549 (2015)
11. Panda, M.: Developing an efficient text pre-processing method with sparse generative Naive Bayes for text mining. Int. J. Mod. Educ. Comput. Sci. (IJMECS) 10(9), 11–19 (2018). https://doi.org/10.5815/ijmecs.2018.09.02
12. Krut'ko, V.N., Potemkina, N.S., Mamikonova, O.A., Markova, A.M.: Individual optimization of nutrition on the basis of big data analysis in human-computer dialogue. In: Data Analytics and Management in Data Intensive Domains. Collection of Scientific Papers of the XIX International Conference DAMDID/RCDL' 2017, pp. 486–487. FRC CSC RAS, Moscow, Russia, Oct 10–13 2017

Novel Approach to a Creation of Probe to Stop Bleeding from Esophageal Varicose Veins Based on Spiral-Compression Method

Sergey N. Sayapin, Pavel M. Shkapov, Firuz G. Nazyrov
and Andrey V. Devyatov

Abstract This article relates to an area of a clinical engineering, and it presents a novel approach to a creation of effective mechanical methods and devices for treatment of bleeding esophageal varices. Currently, numerous endoscopic methods are widely used in the diagnostics and the treatment of bleeding esophageal varices. However, in the case of acute bleeding from esophageal varices such methods will be ineffective because the blood will drabble the optics of the endoscope and take away sclerosing agents and/or glues from an injury. Therefore, the endoscopic methods are used after a stopping of acute bleeding by balloon tamponade, etc. The paper presents a new mechanical method for the treatment of acute bleeding from esophageal varices by a transformable probe with a superelastic spiral. This method versus balloon tamponade permits not only to arrest the bleeding but also to give a free access toward the injury for endoscopic and other methods. In the present paper, we will describe the new concept of the transformable probe for effective stop bleeding from esophageal varicose veins based spiral-compression method.

Keywords Clinical engineering for stop bleeding from esophageal ·
Spiral-Compression method · Superelastic spiral · Transformable probe

S. N. Sayapin (✉)
Blagonravov Institute of Machines Science, Russian Academy of Sciences,
Moscow 101990, Russia
e-mail: S.Sayapin@rambler.ru

S. N. Sayapin · P. M. Shkapov
Bauman Moscow State Technical University, Moscow 105005, Russia

F. G. Nazyrov · A. V. Devyatov
Republican Specialized Center of Surgery Named After Academician V. Vakhidov,
Tashkent, Uzbekistan

© Springer Nature Switzerland AG 2020　　　　　　　　　　　　　221
Z. Hu et al. (eds.), *Advances in Artificial Systems for Medicine and Education II*,
Advances in Intelligent Systems and Computing 902,
https://doi.org/10.1007/978-3-030-12082-5_20

1 Introduction

The treatment of acute bleeding from esophageal varices remains a difficult problem [1, 2]. Currently, numerous endoscopic methods are widely used in the area of clinical engineering for the diagnostics and the treatment of bleeding esophageal varices. These methods include injection sclerotherapy and glue injection; thermal methods and mechanical methods. The classification of the base mechanical methods and the devices, including the mechanical endoscopic methods, is shown in Fig. 1 [1–8]. However, in the case of acute bleeding from esophageal varices the optics of the endoscope will get dirty by the blood. Furthermore, the blood will be able to take away the agents and/or glues from an injury. Therefore, the endoscopic methods, for example, injection sclerotherapy and glue injection (Fig. 2a) [4, 5], are used after a stopping of the acute bleeding by balloon tamponade (Fig. 2b, c) [6, 7], etc. But the use of balloon tamponade has a high risk of rebleeding after deflation and of a risk major complications. Furthermore, there is not any access toward a bleeding place during all period of the bleeding's stopping. Therefore, there is a growing interest to new safer and less traumatic methods. Some approaches to a realization of scientific and technical ideas and knowledge in practice, and to learning and the structural analysis of a content of their transformation processes into innovation in technoparks are shown in [9].

A spiral-compression method [2, 10] (Figs. 1 and 3), including installation self-expanding stent [10], versus balloon tamponade [5, 6] permits not only to arrest the bleeding but also to keep on the necessary access toward the bleeding place during all period of the treatment. These possibilities give to the spiral-compression method the indubitable advantages versus other compression methods used for the treatment of bleeding esophageal varices [2, 10]. This research is a part of a bigger research, which relates to the area mechanical methods and devices of the clinical engineering for the effective treatment of the bleeding esophageal varices. The paper presents a new concept of the spiral-compression method for the treatment of acute bleeding from esophageal varices by a transformable probe with so-called a superelastic spiral [2, 11]. It includes the introduction with the description of the problem of the stop bleeding from the esophageal varicose by the known mechanical methods and devices; the description of the transformable probe, and the conclusions.

2 Description of Transformable Probe

2.1 Comparison of Spiral-Compression Methods

As in Fig. 1, for example, the spiral-compression methods are divided into material of the spiral and the methods of the placement (extraction) into (out of) esophagus. The spirals are divided into an elastic spiral, for example, the spiral made of steel wire; a superelastic spiral, for example, the spiral made of monocrystal Cu–Al–Ni

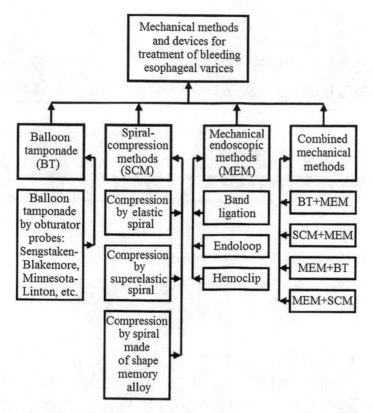

Fig. 1 The classification of mechanical methods and devices for treatment of bleeding esophageal varices

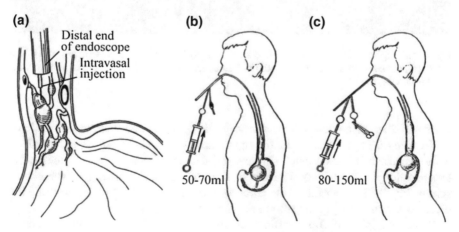

Fig. 2 Endoscopic sclerotherapy of esophageal varices: intravasal injection (**a**) or/and paravasal injection; installation and insufflations of Sengstaken-Blakemore balloon, including the gastric (**b**) and esophageal (**c**) balloons

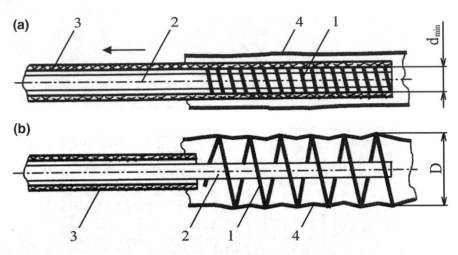

Fig. 3 The schematic view of placement of some spiral into esophagus

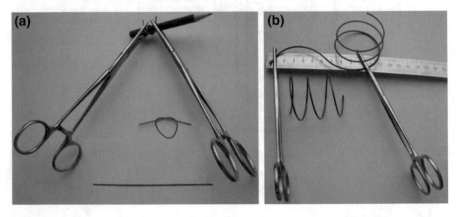

Fig. 4 The superelastic rod with 1.0 mm diameter monochrystal Cu–Al–Ni wire and 100.0 mm length (**a**); the superelastic spiral with 50.0 mm free diameter and 1.5 mm diameter monocrystal Cu–Al–Ni wire (**b**)

wire, and a spiral made of shape memory alloy, for example, the spiral made of Ni–Ti wire [2, 11, 12]. Herewith, the superelastic spiral can be coiled and unfolded without permanent set on much slimmer rod than the elastic spiral. Therefore, the superelastic spiral has a biggest transformation coefficient ($k = D/d_{min}$) versus elastic spirals (Fig. 3). This permits to decrease some traumatism of the patients during the placement (extraction) of the spiral into (out of) esophagus.

Some examples of superelastic properties of the rod with 1.0 mm diameter monocrystal Cu–Al–Ni wire and 100.0 mm length, and the spiral with 50.0 mm free diameter and 1.5 mm diameter monocrystal Cu–Al–Ni wire are shown in Fig. 4.

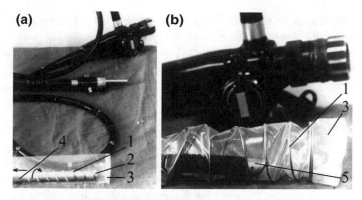

Fig. 5 Installation the superelastic spiral 1 into the model of esophagus 3 by rod 2 and tube 4 (**a**); the distal end of endoscope 5 is positioned inside the superelastic spiral 1 which was self-unfolded inside the model of esophagus 3

The spiral made of shape memory alloy can be also coiled and unfolded on much slimmer rod than the elastic spiral. However, there is necessity in heating such coiled spiral and thereafter, the spiral can have incorrect form in esophagus.

Reference [10] shows a method of placement (extraction) of the spirals into (out of) esophagus. At first, the spiral 1 (Fig. 3) is coiled round the distal end 2 of endoscope and it is fixated by means of a sliding cuff 3. Thereafter, the endoscope with the coiled spiral 1 is inserted in esophagus 4(a) and the spiral 1 is unfixed on level of bleeding site by means of an axial movement of the sliding cuff 3(b). The spiral self-unfolds and after spiral coils compress the bleeding esophageal varices, the bleeding will be arrested.

Herewith, there are free accesses between the spiral coils for endoscopic manipulations, for example, sclerosing therapy, a visual control. After all, manipulations are finalized and the bleeding is stopped fully, the spiral will be extracted from esophagus by means of biopsy forceps through biopsy channel of endoscope. However, after the steel elastic spiral or the spiral made of shape memory alloy are extracted, they will have need of some reconstructive operations for second use [2, 11]. A structure identification of such dynamic system with elastic hysteresis in the conditions of uncertainty may be realized by the method in [13]. The superelastic spiral versus other spirals can be used many times without any reconstructive operations.

The superelastic spiral 1 made of monocrystal Cu–Al–Ni wire was self-unfolded inside a polyethylene tube which was used as a simplified model of the esophagus 3 by rod 2 and tube 4 (Fig. 5). As shown in Fig. 5, the superelastic spiral 1 permits not only to compress the esophagus but also to afford the free accesses between the spiral coils for endoscopic manipulations by the distal end of endoscope 5.

However, this method of the spiral extraction can traumatize the patients and it has a high risk of rebleeding from esophageal varices [2]. Furthermore, there are uncontrolled movements of the spiral during of the placement (extraction) of the spiral into (out of) esophagus. Therefore, there is a need of a new effective method for the controlled placement (extraction) of the spiral into (out of) esophagus which would be safer and less traumatic for the patients than the extant method.

A new spiral-compression method for the treatment of bleeding esophageal varices and a transformable probe for its implementation are presented below [11].

2.2 Description Transformable Probe

A general view of the transformable probe (a) and that of the placement of the superelastic spiral 1 into esophagus 2(b) are shown in Fig. 6. The new method includes: compression of esophageal varices by the superelastic spiral 1 which should be introduced into esophagus 2 in its folded preset state to be then unfolded at the level of bleeding vessels into working state with the help the transformable probe (b); extraction of the transformable probe without superelastic spiral 1 from esophagus 2; control of bleeding's stopping and carrying out curative manipulations with the help of an endoscope and other means, with using a spiral's distal end 3; extraction of the superelastic spiral through mouth cavity by the transformable probe with using distal 3 and proximal 4 spiral ends as guides.

The transformable probe (a) comprises the exchangeable superelastic spiral 1 which is mounted on a transformable device of its placement (extraction) into (out of) esophagus. This device consists of a body 5 with a control lever 6 and a flexible slim tube 7 with a rotating end 8. The body 5 contains a screw gear 9 which is connected with an output shaft 10 and the control lever 6 and a revolution counter 11. There is a window 12 for an observation. The output shaft 10 is connected with the rotating end 8 by a rotating flexible shaft which is placed inside the flexible slim tube 7. There also are two through channels. First, channel 13 comes through the output shaft 10, the rotating flexible shaft, and the rotating end 8. This channel is not axial. Second, channel 14 comes through the body 5 and the flexible slim tube 7.

2.3 Principle of Work of Transformable Probe

The use of the transformable probe consists of four phases. The first phase includes some preparation of the transformable probe for the installation of the superelastic spiral into esophagus. At first, the necessary superelastic spiral 1 (for child or adult) is mounted at the transformable device of the transformable probe here. Herewith, the distal and proximal ends 3, 4 are inserted into the channels 13 and 14. Thereafter, the superelastic spiral 1 is folded around the rotating end 8 by a turn of the control lever 6 (Fig. 6b). The superelastic spiral 1 cannot self-unfold because the screw gear 9 is a self-braking gear. This number of turns one can see through the window 12 and it is memorized.

The second phase includes the placement of the superelastic spiral 1 into esophagus (Fig. 6b). The rotating end 8 with the folded superelastic spiral 1 is inserted

Fig. 6 The general view of
the transformable probe
(**a**) and the placement of the
superelastic spiral into
esophagus (**b**)

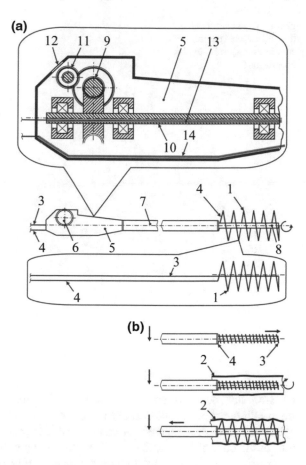

into esophagus 2 on level of bleeding site (Fig. 6b) and the superelastic spiral 1 is
unfolded by the turn of the control lever 6. The unfolded superelastic spiral 1 com-
presses the bleeding esophageal varices and the bleeding is arrested. Further, the
transformable device is extracted through mouth cavity out of esophagus, using the
distal and proximal ends 3, 4 as the guides (Fig. 6). The unfolded superelastic spiral
1 remains into esophagus 2, and the distal and proximal ends 3, 4 are fixated on the
outside, for example, with the help of the sticking-plaster 5 (Fig. 7).

The third phase includes carrying out curative manipulations with the help of the
endoscope and other means with using distal end 3 as guide to deliver the endoscope's
distal end and other means for the treatment of bleeding esophageal varices. There
is a possibility of delivery of necessary nutritive substances directly into stomach 6
of patient bypassing the affected esophageal sector.

The fourth phase is final one. After all manipulations are finalized and the bleeding
is stopped fully, the spiral will be extracted out of esophagus by the transformable
device with using the distal and proximal spiral ends as guides. Herewith, the distal
and proximal ends 3, 4 are passed through the channels 13, 14 (Fig. 4). Further, the

Fig. 7 The general view of
the patient with the unfolded
superelastic spiral into his
esophagus

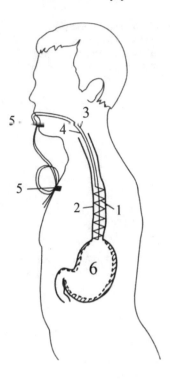

transformable device is moved with the help of the distal and proximal ends 3, 4 until
the rotating end 8 and the end of the flexible slim tube 7 will abut to bases of the
spiral ends. After the superelastic spiral is folded around the rotating end 8(b) by the
turn of the control lever 6, the transformable probe will be extracted. Herewith, the
turn of the control lever 6 has to have the same number of the turns as in first phase.
After the transformable probe and its parts are cleaned and ones are disinfected, it
will able to reuse.

3   Conclusions

The superelastic spiral can be coiled and unfolded without permanent set on much
slimmer rod than the elastic spiral. This permits to decrease some traumatism of the
patients during the placement (extraction) of the spiral into (out of) esophagus.

The superelastic spiral versus other spirals can be used many times without any
reconstructive operations.

The new effective method for the controlled placement (extraction) of the superelastic spiral into (out of) esophagus by the transformable probe is shown here. This
method is safer and less traumatic for the patients than the extant one and it has better
functional possibilities than others.

The unfolded and fixated superelastic spiral permits to keep to the patient some mobility in the hospital and it gives a comfort and reduces terms of convalescence.

References

1. Nazyrov, F.G., Devyatov, A.V., Babadjanov, A.H., Djumaniyazov, D.A., Baybekov, R.R.: Efficacy of endoscopic intervention in the prevention of portal genesis bleedings. Vestnik Exp. Clin. Surg. **10**(3) (2017, in Russian)
2. Saiapin, S.N., Nazyrov, F.G., Deviatov, A.V, Sokolova, A.S.: Treatment of esophageal varices bleeding with the use of transformable probes. Khirurgiia (Mosk) **12**, 58–64 (2010, in Russian)
3. Kovach, T.O.G., Jensen, D.M.: Recent advances in the endoscopic diagnosis and therapy of upper gastrointestinal, small intestinal, and colonic bleeding. Med. Clin. N. Am. **86**, 1319–1356 (2002)
4. Caldwell, S.H., Shami, V.M., Hespenheide, E.E.: Injection therapy of esophageal and gastric varices: sclerosis and cyanoacrylate. In: Drossman, D.A (Senior ed.) Handbook of Gastroenterologic Procedures, 4th edn., pp. 103–111. Lippincott Williams & Wilkins, Philadelphia (2008)
5. Carvalho, E., Nita, M.H., Paiva, L.M.A., Silva, A.A.R.: Gastrointestinal bleeding. J. Pediatr. **76**(2), 135–146 (2000)
6. Isaacs, K.L.: Balloon tamponade. In: Drossman, D.A. (Senior ed.) Handbook of Gastroenterologic Procedures, 4th edn., pp. 119–126. Lippincott Williams & Wilkins, Philadelphia (2008)
7. Shertsinger, A.G.: Treatment of haemorrhages from oesophageal varices with the aid of an obturator probe. Khirurgiya (Mosk) **3**, 94–98 (1980, in Russian)
8. Zehetner, J., Shamiyeh, A., Wayand, W., Hubmann, R.: Results of a new method to stop acute bleeding from esophageal varices: implantation of a self-expanding stent. Surg Endosc **22**, 2149–2152 (2008)
9. Aliyev, Alovsat Garaja, Shahverdiyeva, Roza Ordukhan: Conceptual bases of intellectual management system of innovative Technoparks. Int. J. Educ. Manag. Eng. (IJEME) **7**(2), 1–7 (2017). https://doi.org/10.5815/ijeme.2017.02.01
10. Nazyrov, F.G., Devjatov, A.V.: Method for stopping hemorrhage out of esophageal varicose veins. U.S.S.R. Patent 1362462, 12 Dec 1987
11. Sajapin, S.N., Nazyrov, F.G., Devjatov, A.V., Sajapina, A.S.: Method for stopping hemorrhage out of esophageal varicose veins and transformable probe for its implementation. R.F. Patent 2316272, 10 Feb 2008
12. Komarov, S.M., Lukyanychev, S.Y.: NiTi is flexible and one all remembers. J. Chem. Life **3**, 40–45 (1998). (in Russian)
13. Karabutov, Nikolay: Structural identification of dynamic systems with hysteresis. Int. J. Intell. Syst. Appl. (IJISA) **8**(7), 1–13 (2016). https://doi.org/10.5815/ijisa.2016.07.01

Parallel Mechanisms in Layout of Human Musculoskeletal System

A. V. Kirichenko and P. A. Laryushkin

Abstract The hypothesis about the structure of the musculoskeletal system of a human, as a system of coordinated parallel mechanisms, first expressed in the article. Clinical and technical examples demonstrate the similarity of the behavior of body segments and mechanical devices of a parallel structure in similar modes of operation. An explanation offered for the quantitative composition of the skeleton as a permanently inherited trait. It is known that parallel mechanisms are characterized by strict requirements to the proportions of links and special sizes and shapes of working and dead zones. It is concluded that the logic of a multi-link musculoskeletal system, as a logical mechanical machine, can be determined by the geometric-kinematic characteristics of parallel mechanisms with a given (inherited) number and configuration of links. Reasonings were given by generalizing the elements of the theory of mechanisms and machines in biomechanics. The work is the initial stage of identification of the inertial mass model of the musculoskeletal system of vertebrates for the needs of technology, medicine, and sports.

Keywords Locomotor system · Model identification · Parallel mechanism · Biomechanics · Logical machine · Theory of machines and mechanisms

A. V. Kirichenko (✉)
Medical Center "Mediccity", 2, Poltavskaya St., 127220 Moscow, Russian Federation
e-mail: bio.symmetry@yandex.ru

Mechanical Engineering Research Institute, Russian Academy of Sciences, 4, Malyi Kharitonievsky Pereulok, 101990 Moscow, Russian Federation

P. A. Laryushkin
Bauman Moscow State Technical University, 5, 2 Baumanskaya St., 105005 Moscow, Russian Federation
e-mail: pav.and.lar@gmail.com

© Springer Nature Switzerland AG 2020   231
Z. Hu et al. (eds.), *Advances in Artificial Systems for Medicine and Education II*,
Advances in Intelligent Systems and Computing 902,
https://doi.org/10.1007/978-3-030-12082-5_21

1 Introduction

In connection with the development of technology, the particular importance attached
to the model identification of live systems. In particular, the study of the locomotor
system opens the way to the use of engineering solutions that evolved in nature for
use in engineering and medicine due to the high degree of specialization and system
integration of living systems.

The purpose of the article is a theoretical justification of the musculoskeletal system structure as "logical machine" which logic is defined by the geometric-kinematic
patterns of functioning and design of parallel mechanisms.

To the evidence-based description of the human body, people aspired in antiquity.
Pythagoras, Polyclute, and Vitruvius, for example, described the proportions of the
body, as manifestations of the harmony of development. To a greater or lesser extent, it
was clear that the human body, which is different in each person, has constant features
of its structure and geometrical proportions. There are works by Andreas Vesalius
"De Humanis Corporis Fabrica," even according to modern standards, authentically
describing the structure of man, Agrippa's drawing of the fifteenth century, which
gave the proportions of the human body a mystical meaning and many works of other
artists and scientists. To a greater or lesser extent, the structure of the human body
seemed to be a regular and unambiguous phenomenon.

Until some time, researchers of antiquity and the Middle Ages had only aesthetic
significance. By the middle of the twentieth century, by the time of development of
orthopedic surgery, the works of antiquity were in demand due to the work of G. A.
Ilizarov, who developed methods for the treatment of chondrodysplasia, a congenital
disease that disrupts the growth and shaping of bones, which required the knowledge
of ancient and medieval predecessor scientists. Accurate comprehension of proportions became necessary for determining the regimes of elongation of shortened and
deformed limb segments.

Since the beginning of the twenty-first century, intensive development is undergoing bioengineering and regenerative medicine, which solve the overriding task of
transforming genetic information through the embryonic phase of development to the
cultivation of mature organs and tissues. This formulation of the problem involves the
processing of significantly larger amounts of data related to significant dependencies.
And the weighty importance, together with chemical, metabolic, electromagnetic,
and other factors, given to the influence on the tissue growth and differentiation of
the mechanical environment of morphogenesis is locomotor activity, closely related,
in turn, with the properties of the mechanisms that realize it [1–3].

Empirical description of human body structure, which is largely characteristic of
traditional medicine, inevitably gives way to evidence-based methods of research,
an important place among which is the interdisciplinary approach, which consists
in the generalization of a number of mathematical and physical theories in natural
science [4].

2 ToMM in Biomechanics

Forming in recent years, the concept of the embodiment of intelligence, as a result of the interdependent development of the object and the control system [5–7], allows us to consider the features of the kinematic component of the human body as a logical machine. Thus, the subject of the article is the logic of the mechanical structure of the musculoskeletal system of vertebrates and humans, which provides necessary and sufficient opportunities for implementing the locomotor design—a complex system of control signals emanating from higher nervous centers. A tool that makes it possible to prove conclusively the description of the kinematic component of locomotion is the Theory of Mechanisms and Machines, which is surprisingly rare and, to a limited extent, generalized in biomechanics.

Structural analysis and synthesis of mechanisms under conditions of uncertainty is traditionally the most difficult and creative element in the work of the research engineer. The authors of this article attempt to generalize ToMM in the description of the macroscopic structure of the hierarchically organized human musculoskeletal system with the aim of developing a method for the systematic study of biological mechanisms. For the study, parallel-structure mechanisms chosen that were most appropriate for the use of limbs as manipulators.

We use the terminology of ToMM and consider the musculoskeletal technical system as a hierarchically organized *machine aggregate*—a complex machine consisting of several *mechanisms* connected in series or in parallel. In our model, the support segments of the limbs, the bones covered with cartilage, serve as links, as elastically deformable solids. Links are divided into input (leading) and output (slave)—here, we should make a reservation, since in living systems the drives, that is, the muscles, are located crosswise and, thus, the links, depending on the character of the locomotion act, alternately and consistently play the role of drive link and driven link.

Next, consider the kinematic pair. Kinematic pairs were created in evolution in the form of synovial joints, characterized by low friction. Here, strictly speaking, one should ask the question: Does each locomotion act require so low a friction? At the time of free transfer of the lower limb when walking, for example, it is required. However, at the moment of support and push, when it is necessary to securely fix the achieved position of the limb? Small friction is harmful, like a slippery shoe on ice. People suffering from chronic synovitis walk cautiously and slowly, despite the fact that there is no strong pain and the ligamentous apparatus of joints is intact.

An explanation of this state of affairs provided by the study of the mechanisms of energy transfer: The synovial fluid is two-phase and has a low viscosity due to its pseudoplasticity. When the limb is loaded, the fluid becomes elastic, thus ensuring the transfer of mechanical energy from the link to the link [8]. Apparently, the concept of an articular joint, as a hinge, describes only the free transfer of the limb—half the cycle of mutual movement of the links during walking.

The elasticity of the synovial fluid that arises in the joints under the load provides a dynamic gearing in the joint [9], which gives it the properties of a self-locking transmission mechanism [10]. Ligaments and curvature of the articular surfaces also give the joint the properties of the directing mechanism. As an evolutionary adaptation, the joint mechanism is a method of effective directed energy transfer. The hinge, however, can only provide energy dissipation to friction, which seems to be an unjustifiable expenditure of it under the pressure of natural selection. Thus, apparently, the joint should be considered as a special case of a completely developed mechanism: a kinematic pair with active friction and direction control.

Kinematic pairs unite links into closed or non-closed kinematic chains that represent separate segments of the body—limbs—mechanisms that have their own drives (extrafusal muscles), control links (intrafusal muscles), and working organ—a hand or foot, in turn having links, drives, feedback, and working organs—the surfaces and tips of the fingers.

Bilateral and pentameric symmetry, characteristic of living beings, makes us think about the completely regular, evolutionarily conditioned necessity of the subordinate parallel arrangement of living mechanical systems, the inherited metric relations of which are reflected in the configuration canon—the subject of the search for many generations of scientists.

3   Parallel Mechanisms as a Universal Evolutionary Adaptation

Since earlier information on parallel mechanisms, as a universal evolutionary adaptation of living mechanical systems, has not been found in the literature, let us touch upon the definition of PM in the general case.

Parallel mechanisms are technical devices containing more than one closed kinematic chain. This is a general definition because there are a very large variety of parallel mechanisms associated with different technical tasks. The properties of such mechanisms are complex and diverse, and their rigorous description is a non-trivial mathematical problem [11].

The interest of engineers and researchers in this class of mechanisms is due, first, to their characteristics. Parallel mechanisms are characterized by high positioning accuracy, better stiffness characteristics, and load-carrying capacity relative to the weight of the moving parts of the mechanism [12].

In the wild nature, such general trends in the continuously evolving evolutionary process of locomotor adaptation seem quite valid and are present in an explicit form: in the late Devonian period, marine animals began to emerge on dry land, which required significant improvements in the skeleton in the form of the development of the belts of multi-link limbs.

The accuracy of movements, as a characteristic of the grasping function of arboreal primates, was undoubtedly a factor of survival and led, in turn, to the formation of binocular vision, much contributing to further evolutionary progress. Binocular vision, which is necessary for an accurate estimate of the distance, is also realized, in

Fig. 1 Normal wrist (left) and wrist in SLAC

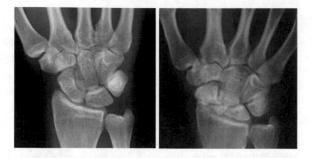

essence, thanks to the parallel mechanism control system, since formally the output link can be even a point—the accommodation point in this case.

The functional hand, which appeared during the evolution, has anatomical and biomechanical traits of a parallel structure: 1–3 and 4–5 rays (the radial and ulnar halves of the wrist) supported by the bones of the distal wrist row on the semilunar bone connected to the medial column of the radius bone. Radial fractures type "C" in the area of the support of the semilunar bone, as well as the fractures of the semilunar bone itself, lead to permanent losses of the wrist joint function and treated surgically with the aim of the most accurate restoration of the segment length. It can be assumed that the semilunar bone provides a kinematic decoupling, and the radius is the output link and their damage leads to a gross disruption of the parallel structure, which is sensitive to changing geometry.

The attention of orthopedists to the scaphoid bone of the wrist is well known. Its unhealed lesions reduce the mobility of the hand and severely limit the grasping force. The scaphoid encloses the links of three kinematic chains of 1–3 rays, which provide 70% of the wrist function. A catastrophe for the wrist joint is the simultaneous defeat of the scaphoid and semilunar bones, called SLAC syndrome—scapholunate advanced collapse. Movement in the wrist joint with SLAC is sharply limited, painful; the hand is of little use. When other bones of the wrist damaged, such large loss of function not observed (Fig. 1).

In functional terms, the accuracy of the hand as a parallel mechanism most fully revealed in the work, for example, of a microsurgeon. In the work area—a surgical field with a volume much less than a cubic centimeter—the surgeon holds the micro-tool with the first three fingers, turning it around the longitudinal axis, the fourth and fifth fingers accurately moving the radial half of the hand within a fraction of a millimeter.

A forearm reveals a similarity to the layout of the Stewart platform drive units: in the distal part, the radial and ulnar bones are connected by one joint, and in the proximal part, these bones are connected to the humerus by separate joints in such a way that the ray bone makes a helical movement around the immovable ulnar, providing pronation and supination hands. The upper extremity belt is in an obvious way two conjugate kinematic chains. The output link is the sternum, elastically connected with the axial skeleton by means of the ribs.

Foot and lower limb belt arranged similarly. Parallel location of the external and internal parts of the foot is clearly noticeable on the osseous–ligamentous preparation: 1–3 rays through the wedge-shaped and navicular bone attached to the talus bone. The fourth and fifth rays through the cuboid bone connected with the heel. The talus, which has two saddle articular surfaces in the subtalar and ankle joints, thus, plays the role of a link providing a kinematic decoupling, similar to the crosspiece of Hook's hinge, and the output link is the supporting surface of the distal epiphysis of the tibia.

Counting the links in the kinematic chains of hands, feet, and belts of the limbs shows that their number surprisingly turns out to be constant: The output link is sure to be the seventh one. A simple calculation of the vertebrae as a link in an isolated kinematic chain shows that the seventh links are the twelfth and sixth thoracic vertebrae,[1] located in the zones of maximum mobility of the spine: The twelfth vertebra Th12 occurs in the transitional thoracolumbar zone, and the sixth vertebra Th6 is the geometric center of the thoracic kyphosis.

The continuation of calculation of the links in the cranial direction determines the seventh cervical vertebra C7, as the next output link, which allows the calculation to be completed: on the first cervical vertebra is located the head.

How such a quantitative ratio of the links is natural, is unknown, but the number of bones and vertebrae in the norm is inherited by a constant. The change in the number of vertebrae resulting from, for example, sacralization or lumbarization often accompanied by low mobility and persistent low back pain. The fusion of the first cervical vertebra with the base of the skull, the assimilation of the atlas, leads to symptoms that are extremely onerous for the patient and, in some cases, life-threatening conditions.

There are known manifestations of secondary instability in the joints of the lower limbs, resulting from the acquired shortening of one lower limb or decompensation of the asymmetry of limb growth in childhood. The constant pain and the growing tension of the capsule and ligaments in the joints of the longer leg lead eventually to the formation of contractures—the outcome of chronic overload arthropathy associated with the difference in the length of the links (Fig. 2).

The study of properties of parallel mechanisms, constructed with the discrepancy between real structural and geometric parameters by calculation, shows the appearance of a "parasitic" mobility of links, leading to their rapid deterioration and destruction. Such an observation, in essence, describes the pathogenetic process of disruption of compensation in people who have a significant difference in the length of the lower limbs.

One of the main topics—the tasks of the work area and the special positions to the mechanisms of the parallel structure—should be touched. The generalization of these problems in biomechanics provided the validity of the hypothesis about PM, as a universal evolutionary adaptation, reveals the keys to the demonstrative description of a highly integrated hierarchically organized locomotor system. Now problems

---

[1]The order of counting organized in such a way that the underlying "mechanism" considered first in the following kinematic chain. For example, the wrist is the first link, followed by the forearm, shoulder, scapula, clavicle, sternum handle, and the seventh link—the sternum proper.

Fig. 2 Posterior tibial subluxation and degenerative lesion of posterior cruciate ligament in female patient 1.55 m tall with contralateral leg shortening up to 1.5 cm

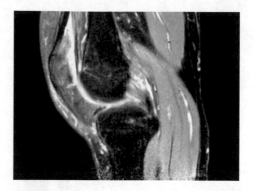

successively and successfully solved in ToMM, in biomechanics often solved by the method of "trial and error." In support of this inference, we present the work of two scientific groups.

For many years, Ian Loram's working group examines the issues of locomotion control in children suffering from various palsies. The authors point out the need to take into account the movements of each segment of the body in the whole model to describe the observed working and dead zones. The study of movements of isolated segments turned out to be of little informative due to the complexity of the dynamic interaction with neighboring and distant segments of the body [13].

In the second case, locomotion researchers constructed a biped robot and, after fourteen configurations, found out that the special lengths of the segments, the position of the hip, and knee joint axes of the robot, matching the rhythm of the gait, allowing for the inertial properties, saved 29% of the energy consumed.

The authors conclude hip and knee are coupled and biarticular. They indicate that the efficiency of movement will grow with the inclusion of the robot coordinated hand movement. The calculation of the design was successful in the form of a parallel mechanism. The patterns of gait, which is remarkable, were calculated by genetic algorithms [14] (Fig. 3).

In essence, the authors of these studies have faced the task of calculating the PM work areas and found that this task solved within the framework of the general paradigm of analyzing and synthesizing parallel structures, eventually successfully completing the research.

In the PM design, special zones are necessarily calculated, the achievement of which by the output link can lead to loss of control and mobility of the mechanism. Based on the paper [11], we give the correlation of known anatomical and physiological data on the operation of the musculoskeletal system and technical information on the special positions of the parallel mechanisms.

In general, PM "special position" is understood such a configuration of the mechanism when it is observed: loss of one or more degrees of freedom by the output link of the mechanism, the emergence of uncontrolled mobility of the output link of the mechanism, and the emergence of uncontrolled mobility of the intermediate links of the mechanism. The actual and very non-trivial task of avoiding special positions is

Fig. 3 CAD file and schematic of SAFFiR lower body [14]

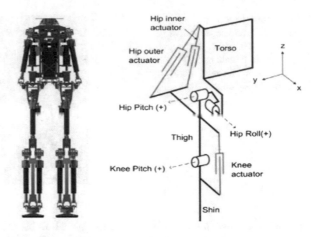

solving at the stage of the mechanism synthesis due to combinations of selection of the number of links, their length, setting the number of degrees of freedom, imposing links, reserving mobility, drives, sensory elements, and so on.

With respect to the musculoskeletal system, it can be concluded that the achievement of such positions by the body segments leads to their traumatic injuries and evolutionarily there must be a way of avoiding them, for example, a control algorithm that prevents such development of events.

This control algorithm, apparently, is the awareness of a person about discomfort when approaching special positions and the pain that arises when they reach and overcome them. In addition to the control algorithm that prevents the achievement of traumatic positions, there must also exist evolutionary mechanical solutions that determine the range of effective work of the output link. It can be assumed that these design solutions realized due to the guiding properties of the joints due to the curvature of the surfaces and the presence of a ligamentous apparatus. At the present time, such adaptive control algorithms already exist as software products and seem necessary in the near future robotics and synthetic biology [15].

One of the solutions to the problem of the number of links, their length, and the arrangement of kinematic chains in the construction of the human body is given in the work [16]. In this work, the author forms the total number of independent configurational links when the scheme of the human body is inscribed in the pentagram and gives a comparative analysis of the number of invariants obtained because of fitting into simplexes of different orders from three to eight. It turned out that the layout of the human body, inscribed in the pentagram, gives the maximum number of invariants. In other words, this arrangement of the links of a given number and length is the most compact and allows the largest number of segment arrangements to be achieved.

The review tasks do not include a strong evidence of the parallel mechanisms of identity to the human body structure. We suggested that it allows us to consider the intentions of living nature in somewhat more detail. Based on fundamental theoretical knowledge, it seems to us that it is possible to see the whole picture and to collect in small parts the details of an evolutionary plan—a reliable and accurate live locomotor system.

4 Conclusions

The multi-link musculoskeletal system of a human has the traits of parallel mechanisms. Information about parallel mechanisms and requirements for their design can be used to solve the following tasks:

- Development of technical conditions for the design of joint endoprostheses, characterized by functional invariance of the living joint structure and reflecting uniquely determined parameters of personification.
- Diagnostics and prediction of the physiological condition and disturbances of locomotor function in medicine and sports.
- Development of technical conditions for designing automated control systems and optimizing the locomotor function in medicine and sport.
- Development of technical conditions for the design of bioreactors intended for the creation and cultivation of tissue structures for the purpose of bioengineering replacement of joint and other complex defects.
- Identification of the inertial mass model of the human body for use in engineering in view of the effectively controlled energy transfer observed in living mechanical systems.

It is necessary and planned additional kinematic and dynamic studies and computer modeling of human locomotion, which will allow the parametric analysis of the obtained data within the framework of the ToMM, based on the assumption of a parallel arrangement of mechanisms phenomenologically observed in multi-tiered, hierarchically organized mechanical systems.

References

1. Mouthuy, P.-A., Carr, A.: Growing tissue grafts on humanoid robots: a future strategy in regenerative medicine? Sci. Robot. **2**, 1–3 (2017)
2. Patwari, P., Lee, R.T.: Mechanical control of tissue morphogenesis. Circ. Res. NIH Publ. Access **103**, 234–243 (2008)
3. Tennakoon, H., Paranamana, C., Weerasinghe, M., Sandaruwan, D., Mahindaratne, K.: A novel musculoskeletal imbalance identification mechanism for lower body analyzing gait cycle by motion tracking. Int. J. Inf. Technol. Comput. Sci. (IJITCS) **10**(3), 27–34 (2018). https://doi.org/10.5815/ijitcs.2018.03.04
4. Kaur, Sukhraj, Malhotra, Jyoteesh: On statistical behavioral investigations of body movements of human body area channel. Int. J. Comput. Netw. Inf. Secur. (IJCNIS) **8**(10), 29–36 (2016). https://doi.org/10.5815/ijcnis.2016.10.04

240   A. V. Kirichenko and P. A. Laryushkin

5. D'Avella, A., Giese, M., Ivanenko, Y.P., Schack, T., Flash, T.: Editorial: modularity in motor control: from muscle synergies to cognitive action representation. Front. Comput. Neurosci. Front. **9**, 1–6 (2015)
6. Hara, F, Pfeifer, R.: In: Hara, F., Pfeifer, R., (eds.) Morpho-Functional Machines: The New Species. Springer, Tokyo (2003)
7. Lara, B., Astorga, D., Mendoza-Bock, E., Pardo, M., Escobar, E., Ciria, A., Embodied cognitive robotics and the learning of sensorimotor schemes. Adapt. Behav (2018). 105971231878067
8. Bird, R.B., Stewart, W.E., Lightfoot, E.N., Transport Phenomena, 2nd edn. Accreditation and Quality Assurance. Wiley, New York (2002)
9. Carrier, D.R., Gregersen, C.S., Silverton, N.A.: Dynamic gearing in running dogs. J. Exp. Biol. **201**, 3185–3195 (1998)
10. Veitz, V.L., Kolchin, N.I., Martynenko, A.M.: Some questions of the dynamics of self-locking mechanisms. J Mech. **4**, 93–104 (1969)
11. Erastova, K.G., Laryushkin, P.A.: Working zones of mechanisms of a parallel structure and methods for determining their shape and measurement. Izv. VUZov. Mashinostroenie **8**, 78–87 (2017). (in Russian)
12. Glazunov, V.A., Aleshin, A.K., Kovaleva, N.L., Skvortsov, S.A., Rashoyan, G.V.: Prospects for the development of mechanisms of a parallel structure. Stankoinstrument. Robototehnika i mehatronika 86–88 (2016, in Russian)
13. Nema, S., Kowalczyk, P., Loram, I.: Complexity and dynamics of switched human balance control during quiet standing. Biol. Cybern. **109**, 469–478 (2015)
14. Lahr, D., Yi, H., Hong, D.: Minimizing the energy loss of the bi-articular actuation in bipedal robots. In: 39th Mechanisms and Robotics Conference, ASME, vol. 5B, pp. 1–6 (2015)
15. Krishnamurthy, E.V.: Agent-based models in synthetic biology: tools for simulation and prospects. Int. J. Intell. Syst. Appl. (IJISA) **4**(2), 58–65 (2012). https://doi.org/10.5815/ijisa. 2012.02.07
16. Rakcheeva, T.A.: Anthropometric configurational pentacanon. In: VI International Congress "Weak and superweak fields and radiation in biology and medicine" 0207-0607, St. Petersburg, pp. 1–5 (2012, in Russian)

Methods of Mutual Analysis of Random Oscillations in the Problems of Research of the Vertical Human Posture

G. I. Firsov and I. N. Statnikov

Abstract The application of the maximum correlation coefficient, the dispersion ratio, and the mutual dispersion function for revealing the nonlinear relationship of the oscillations of the human body in the sagittal and frontal directions is considered. For some neurological diseases, a high degree of coherence of oscillations in the sagittal and frontal directions is noted in sufficient wide frequency bands, which indicates the presence of a strong linear inertial statistical coupling of the oscillations.

Keywords Posture · Computer stabilography · The dispersion relation · Mutual dispersive function

1 Introduction

The use of well-known and well-tested electrophysiological methods (electroencephalography, electromyography, evoked potentials, and rheography) in neurological practice does not always make it possible to sufficiently accurately assess the functional state of individual body systems for more accurate and adequate diagnosis of various pathological conditions. One of these states in the medical, and particularly neurological practice, is imbalance, that is the condition and the basis for normal interaction with the surrounding environment and human activities therein [1–4].

Maintaining balance and posture in a gravitational field is achieved by a coordinated activity of a large complex of analyzers and systems of the brain (the sensitive, visual, vestibular, cerebellar, and extrapyramidal system), an important role in the play and the parts of the brain that are responsible for formulating the strategy of maintaining posture, as a special case of human behavior (frontal parts of the brain). The study of the processes of regulation of the human pose as an ability to maintain equilibrium occupies an important place in the general problem of analyzing the regulation of various functions of the organism when interacting with the envi-

G. I. Firsov (✉) · I. N. Statnikov
Blagonravov Mechanical Engineering Research Institute, Moscow, Russia
e-mail: firsovgi@mail.ru

© Springer Nature Switzerland AG 2020
Z. Hu et al. (eds.), *Advances in Artificial Systems for Medicine and Education II*,
Advances in Intelligent Systems and Computing 902,
https://doi.org/10.1007/978-3-030-12082-5_22

241

ronment. At the same time, researchers are attracted to both the biomechanical and physiological aspects of the problem. On the one hand, the conservation predetermined vertical position may be considered as a special case of the motor control device of the person as a system with many degrees of freedom, on the other hand, disequilibrium observed in various neurological diseases can carry a large amount of diagnostic information.

It should be noted that the process of maintaining a pose is, like any homeostasis, dynamic and a person can never stand completely still and constantly makes oscillatory movements. And the challenge is to choose the most appropriate method for recording these movements. Currently, one of these methods is stabilography—registration of the total pressure of the center of the plane through the platform on the pickoff sensors (stabilograph). The method was developed at the turn of the 40–50 s and is widely used in medical practice around the world [1–3, 5].

2   Methods of Stabilographic Analysis

The basic form of studying the regulation of the human body in norm and pathology is a passive experiment. In this case, the subject is in a free vertical posture on the stabilograph platform. Possible effects on the part of the experimenter are limited to individual changes in the condition of the subject (closing and opening of the eyes), the performance of simple tasks in the mind (counting, singing, and reading poems), using the so-called substitution technique, which consists in following the movement of a luminous point on the oscilloscope screen, reflecting the position of the center of pressure on the brake stabilograph platform (the subject was instructed to watch the screen and store the point within the circle in the center, this test can be regarded as a simple non-invasive method of changing the dynamic system of a person standing for the purpose of extracting additional information, a strong visual feedback was included in the regulation of the posture, in addition to proprioceptive mechanisms.

In our experiments conducted on the basis of the nervous diseases clinic of the First Moscow State Medical University named after I. M. Sechenov with the help of the Force platform Ela of the French company l'Elecctronique Applique, the subject was placed on a free-standing platform in such a way that the projection of the center of gravity was located on 3–4 cm behind the center of the platform. The study was conducted in different modes: open eyes, closed eyes, and open eyes. The experimental realizations were sampled at a frequency of 25 Hz. Each record contains two data sets, each of 4096 points (about 160 s long). More than 750 stabilograms were obtained in the study of practically healthy patients with different neurological pathologies.

The obtained stabilograms of healthy people look like polymorphic, of different amplitude and frequency of the wave and visually characterized by low- and medium-amplitude oscillations. In the frequency spectrum, one could note oscillations with a frequency of 0.1–3 Hz. When the eyes were closed, in most cases the amplitude of the oscillations increased somewhat while maintaining their frequency. After opening the eyes, the character of the stabilograms returned to the original one.

In the literature, various indicators characterizing a person's ability to maintain equilibrium are described [1–3]. These indicators, which are highly integrative, do not allow us to distinguish the subtle characteristics of the trajectory of the human center of gravity movement, and their use in solving diagnostic problems is very difficult. Below we consider the possibility of analyzing stabilograms as realizations of random processes on the basis of probability-theoretic methods for describing them, since in general the stabilogram can be regarded as a two-dimensional time process reflecting the features of the regulation of the body's posture and carrying information on the state of the musculoskeletal system and the human nervous system. Therefore, to analyze the stabilogram, it is advisable to use the mathematical apparatus of time series analysis, which makes it possible to evaluate the nature, the level and frequency composition of the oscillations, and the projections of the center of gravity of the body in mutually perpendicular planes—sagittal (forward–backward) and frontal (right–left).

Currently, there are two basic approaches to the analysis of stabilograms—either separately in the sagittal (forward–backward) and frontal (right–left) directions, or together. The first approach is quite traditional, it is widely used in assessing the nature of the stabilographic curve and its changes in various pathologies or functional loads. As indicators characterizing the ability of a person to maintain equilibrium, the number of oscillations in 1 min, the mean and maximum amplitude, the area of the vector of the stabilogram, and the "Romberg coefficient" are usually used—the ratio of the average amplitude of body oscillations with closed eyes to the average amplitude of vibrations with open eyes; "total amplitude of oscillations" Yu. V. Terekhov (mm); "coefficient of stability" I. I. Rosen, equal to the ratio of the sum of all deviations of the common center of gravity of the body in one of the planes or their resultant Y and some constant T to the latter; "coefficient of mobility" I. I. Rosen, equal to the ratio of the total amplitude of the oscillations of the common center of gravity of the body in one of the planes Y and the projection of the curve on the X-axis to the latter. These figures having very integrative character, not possible to distinguish the subtle characteristics of the path of movement of center of gravity, and their use in solving diagnostic problems, as well as modeling of regulatory processes is very difficult. The simplest and therefore widespread method of investigating trajectories is the direct image of the trajectory in the plane (statokinesiogram) in the selected coordinates with subsequent visual control. For simple laws of motion, the trajectories obtained are sufficiently visual, allowing them to qualitatively evaluate their location on the plane (frequency of lines) in different areas of the plane and some other features. In practice, especially with the random nature of the oscillations inherent in stabilograms, such a visual qualitative analysis becomes ineffective. In this case, it is advisable to focus on the method of figurative analysis of experimental data, or on analysis of the joint density of the distribution of two geometrically folded processes [6].

As is known, the joint distribution density of two processes $P(x, y)$ (two-dimensional histogram) determines the probability that instantaneous values of processes at an arbitrary moment of time will be enclosed simultaneously in two definite intervals. An important quantitative measure of motion is the maximum area of the histogram, defined as the reference area of the corresponding two-dimensional histogram: $S_{max} = \iint dxdy$. Estimation of the trajectory area (statokinesiogram) makes it possible to compare different objects of movements in terms of the size of the filled part of the plane and the extent of the boundaries of the limiting motion. We note that the problem of estimating the area of the statokinesiogram is solved both with the help of a rather crude method of counting the number of elementary squares on the reference plane, where the elementary "quantum" along the coordinate axes is determined from the number of intervals of the histogram, and by constructing scattering ellipses or constructing convex hulls [7]. It is of interest to use the methods of contour analysis and R-functions to determine the area of the statokinesiogram [8, 9]. However, the estimation of the area of the trajectory does not contain information on the degree of filling the trajectory of this area. Such information lies in the features of the lateral surface shape of a two-dimensional histogram that can be detected using its plane sections parallel to the coordinate plane, determining the area of each section S and the relative probability F of the trajectory falling into the region bounded by the corresponding section $F(S) = \iint P(x, y)dxdy$. As a result, we obtain the dependence of the probability of filling the given area on the value of this area, in other words, the probability distribution function of the area coverage. The specific form of the obtained distribution function, naturally, will be determined by the form of the original two-dimensional histogram. In order to obtain a quantitative estimate of the uniformity of the area of the trajectory, we can use a power-law approximation of the distribution function: $F(S) = S_{OTH}^c$, where c is the uniformity of area coverage, $S_{OTH} = S/S_{max}$ is the relative trajectory area ($0 < S_{OTH} < 1$), and $F(S)$ is the probability of filling a given area ($0 < F(S) < 1$). Large values of the exponent c indicate a more uniform filling of the trajectory area. If, for example, the trajectory is unevenly distributed in density around the center of its mole, as can be seen from the high sharp peak and the steep receding lateral surfaces of the histogram, then the function in question quickly reaches large values even with a relatively small area. If the trajectory more uniformly fills the occupied area, i.e., the histogram has a flat top or horizontal areas (which is typical, for example, for atactic patients), then the function will slowly reach large values. The use of data on the area of the statokinesiogram and the values of the coefficient c makes it possible to evaluate the effectiveness of various mediocom—metotic correction of disturbances of the equilibrium function [10].

The software of domestic computer stabilographs often includes calculation of the stability indicators of the vertical posture, based on the determination of the speed of the projection of the common pressure center on the reference plane and its orthogonal components. It is believed that the average speed of movement (sometimes called the looseness index) characterizes the stability of the vertical posture—the higher the speed, the less stable the posture. Sometimes the so-called path is calculated as the product of the average speed for the duration of the survey. It is obvious that the correlation coefficient between the mean velocity and the path is practically equal to unity at close values of the experiment time. With the increase in the time of the experiment, the increase in fatigue, both purely physical and psychological, can significantly affect the obtained values of the mean velocity. We also note that the usually used formula for estimating the average speed uses neighboring values of the stabilograms for calculation, which can lead to significant errors due to a rather coarse discretization of the stabilogram. More accurate and reliable mean speeds can be obtained using approximate formulas to calculate the derivatives of three or five points of the curve, or by applying digital filtration algorithms, synthesized by the Remez method.

When describing the stabilogram as a function that varies with time, the speed of this change is important; Frequency of the process. With respect to the frequency composition, it is of interest to determine the spectral peaks at individual frequencies, and the slope (steepness) of the spectrum in the high- and low-frequency region. Spectral density is the main characteristic of the frequency composition of the process; it is usually interpreted as the distribution of its dispersion in frequency and has a dimension equal to the square of the dimension of the measured parameter divided by Hz. Stabilogram spectra of practically healthy people are smoothly falling from 0 to 12.5 Hz curves, sometimes with a peak in the range of 8–12 Hz, which reflects the presence of physiological tremor in the regulation of equilibrium. The presence of Parkinson's tremor is signaled by a wide peak in the 3–5 Hz region. In some cases, the registration of such a peak made it possible to reveal subclinical manifestations of Parkinsonian tremor [11].

The so-called nonparametric methods of spectral estimation are widely used in the practice of spectral analysis; the most frequently used approach is based on the direct Fourier analysis of a sequence of data samples using the fast Fourier transform (FFT). It is known that this approach, along with the high efficiency in the computational sense, has a number of basic limitations, manifested in the analysis of sufficiently short data realizations and related both to a lower resolution in frequency and to implicit weight processing of data in the calculation of the FFT. In recent years, various procedures for spectral estimation have been proposed that allow us to relax these limitations, which is especially important in the examination of neurological patients who find it difficult to stand on the stabiloplatform for more than 1 min. A characteristic feature of the new approach, called parametric, is the presence of a model of the spectral density of the time series in the form of a fractional–rational function, the parameters of which are estimated on the basis of observation of the corresponding process over a certain limited time interval [12, 13].

The use of such models can be very promising in the search for informative signs characterizing the slope of the spectrum (the rate of its decay) and the presence of bright spectral components. In the general case, this is the autoregressive-moving average model (ARMA), in which this fractional–rational function has both zeros and poles. In the absence of a fractional–rational function of zeros or poles, the autoregressive (AR) model and the moving average model (MA) are different. In this case, the AP-part model allows a sufficiently accurate description of the spectrum peaks, and the MA-part—depression. Characteristic properties of ARMA models is the ability to work with a fairly short implementations, flexibility and relative ease of calculation procedures, and a natural extension to some of the simplest nonlinear cases. To construct the model can use some variational principle, in particular entropy maximization process, which is known for the individual values of the correlation function. This approach leads to the construction of the AR model Mth order.

If, for classical spectral methods, the limiting resolution is achieved when the relation $\mathrm{Df} \bullet N = 1$, then for the maximum entropy method the empirical relation $\mathrm{Df} \bullet N = 0.4$ is known, where Df is the frequency resolution, N is the number of terms of the time series, that is, the resolving The frequency capability of the maximum entropy method, provided that the optimal number of autoregression terms is chosen, is at least 2.5 times greater than the resolving power of classical spectral methods. An important factor in the construction of the AP model is the correct choice of its order. If the order of the model is too low, a smoothed spectral density is obtained, and if it is too high, false peaks appear on the spectral density graph. The most popular method of choosing the order of the model is the information criterion Akaike [14].

The level of oscillations on stabilograms can be characterized as their peak value or amplitude, and their dispersion. The information on the variance of the oscillations in each of the orthogonal directions and its variation under various functional influences on the subject, expressed as the ratio of the variances of the stabilograms in the frontal and sagittal planes, or as the ratio of the variances of the same stabilograms when the state changes (e.g., when the eyes are closed) can serve as a diagnostic sign. In particular, the variance analysis was very informative in the study of some psychogenic diseases, and a sharp increase in dispersion when closing the eyes may indicate a violation of deep sensitivity [15].

3 Mutual Analysis of Stabiligraphic Curves

In the literature devoted to the analysis of the processes of vertical posture regulation, the opinion was repeatedly expressed that there is no correlation between the sagittal and frontal components of the stabilogram. This was usually justified by small values of the correlation coefficient r. Our experiments [16, 17] showed that such a conclusion is not always valid, the relationship between the orthogonal components of the stabilogram can be nonlinear, which cannot be established with the help of the correlation coefficient r. In addition, the relationship between frontal and sagittal stabilograms depends on the functional state of a man. Therefore, when processing stabilograms, in addition to calculating the values of the correlation coefficient r, the calculation of the $x(y)$ and the η cross-correlation function, the dispersion relation mutual dispersion function should be performed [18]. The variance (correlation) ratio of the random variable Y with respect to X is defined as

$$\eta^2_{y|x} = DM(X|Y)/DY,$$

$$DM(Y|Y) = M[M(Y|X) - M(Y)]^2 = \int_{-\infty}^{\infty} \left[\int_{-\infty}^{\infty} y\varphi(y|x)\mathrm{d}y - m_y \right]^2 \varphi(x)\mathrm{d}x$$

– variance of the conditional expectation, characterizing the fluctuations of the variable part of the Y, which is caused by the influence of variable the X, DY—the variance of the Y. The dispersion ratio $\eta_{x(y)}$ determines the extent to which the functional relationship between the variables x and y is observed. When $\eta_{x(y)} \approx 1$ practically functional dependence; closer $\eta_{x(y)}$ to zero, so it is more disturbed. In the case of the independence of X and Y, $\eta_{x(y)} = 0$. In the general case $0 \leq \eta_{x(y)} \leq 1$. The dispersion ratio $\eta_{x(y)}$ can be interpreted as a quantitative measure of certainty characteristic of the random variable y from the values of the random variable x. In this case, the dispersion ratio is always greater than or equal to the coefficient of correlation and is not symmetrical, i.e., $\eta_{x(y)} \neq \eta_{y(x)}$.

The $x(y)$ allows to reveal the existence of η use of the dispersion relation $\eta_{x(y)}$, a functional connection between body oscillations in the sagittal and frontal directions, especially in patients with organic lesions of the central nervous system. So, for the patient A.B. (diagnosis—multiple sclerosis), $r = 0.03$, $\eta_{x(y)} = 0.28$, was more than an order of magnitude r. This may indicate both the existence of delayed elements in the control system for maintaining the posture and the nonlinear nature of the relationship of oscillations in the orthogonal planes. At the same time, in some diseases, the difference between the value of $\eta_{x(y)}$ and r was relatively small, as in the case of the patient S.I. (parkinsonism) value $r = 0.72$, the value of $\eta_{x(y)}$ with respect to the sagittal and vice versa was approximately 0.73. Such large values of the correlation coefficient r and the $\eta_{x(y)}$ indicate a possible presence of a single powerful η dispersion relation source of oscillations along both planes.

The correlation coefficient r and values of the dispersion relation $\eta_{x(y)}$ quite fully describe the general form of the statistical inertial-free relationship of two random processes, implying no time shift (in phase) between the values of the two coupled processes. Consideration of the same dynamic system of regulation of the pose makes it expedient to resort to mutual variance analysis. The mutual dispersion function for each pair of values t_1, t_2 is equal to the variance of the conditional mathematical expectation of the cross section of one function x relative to the cross section of another function y shifted by the interval t_1, t_2

$$\theta_{yx}(t_1, t_2) = M\left[M(Y_{t_1}|X_{t_2}) - MY_{t_1}\right]^2$$

$$= \int_{-\infty}^{\infty} \left[\int_{-\infty}^{\infty} y_{t_1}\varphi(y_{t_1}; t_1|x_{t_2}; t_2)dy_{t_1} - \int_{-\infty}^{\infty} y_{t_1}\varphi_y(y_{t_1}; t_1)dy_{t_1} \right]^2 \varphi_x(x_{t_2}; t_2)dx_{t_2},$$

where $\varphi(y_{t_1}; t_1|x_{t_2}; t_2)$—the conditional density of probability of $Y(t_1)$ it is relative $X(t_2)$; $\varphi_y(y_{t_1}; t_1)$ and $\varphi_x(x_{t_2}; t_2)$—one-dimensional density of probability of stochastic functions of $Y(t_1)$ and $X(t_2)$.

This allows us to estimate, the mutual correlation function $\rho(t)$, $\eta_{x(y)}(t)$ the magnitude of the inertial statistical relationship of processes in time. The quantitative measure of the magnitude of this connection is the maximum correlation coefficients r_{max} (and, correspondingly, the maximum of the values of the mutual dispersion function $\eta_{\max}$) and mutual correlation or dispersion function $\rho(t)$ and $\eta_{x(y)}(t)$, respectively. These coefficients allow us to assess the degree of coherence of the two processes, even if there are phase shifts between them. So, for the case of functional left-sided hemiparesis in patient V.V. At the value of the correlation coefficient when standing with closed eyes 0.0677 the maximum correlation coefficient was 0.4448, which indicates the presence of a certain linear inertial statistical coupling of the oscillations in the frontal and sagittal planes. Thus, in healthy subjects the difference in the values of the dispersion ratio of the maximum dispersion ratio is very small, no more than 10–15%. The results of calculations of correlation and dispersion relation stabilograms show that the correlation coefficients, the maximum correlation coefficients, dispersion relations and high mutual dispersion function have a certain information value in terms of medical diagnosis, so it is advisable to enter into the software commercially available computerized stabilography and stabiloanalizator packages calculation programs and a graphic display of correlation–dispersion characteristic line provider stabilograms.

At some types of neurologic pathologies, initiation of the interconnected fluctuations of the center of gravity is observed, including the neuroses and hysteria. This can be evidenced by the high degree of coherence of oscillations observed on the stabilograms in the sagittal and frontal planes observed in fairly wide frequency ranges (from 4 to 8 Hz) [10, 16]. The large value of the coherence function indicates the presence of a strong linear inertial statistical coupling of oscillations in two planes and can be explained either by the existence of a single powerful oscillation source in the CNS or by the synchronization of oscillations in two planes [19]. The revealed phenomenon can be a manifestation of a violation of the programming of the regulation of the equilibrium of the vertical posture and the interest of the systems responsible for this programming, in particular of the frontal sections. In particular, the electric mechanism of the brain, which manifests itself in the theta rhythm of the EEG at a frequency of 4–7 Hz and in the amplitude of tens of μV, acts as a possible mechanism causing the emergence of areas of the coherence function, while the intensity of the theta rhythm depends on the degree of emotional, mental tension, background main activity, and age. The existing relationship between the activity of theta rhythm and mental stress, the effectiveness of activity is explained by the fact that the theta rhythm reflects the activity of the mid-stem brain formations and is an electrophysiological correlator of the mechanism that quantifies the flow of engrams extracted from memory. In general, the results of our experiments can serve as a confirmation of the well-known hypothesis [20] that the dynamics of a healthy physiological system must produce highly irregular and complex types of variability, while disease and aging are associated with the loss of complexity and greater regularity. To estimate the coupling of oscillations in the sagittal and frontal planes, it is very promising to use mutual generalized information, which was considered in detail in [16].

4 Conclusions

In conclusion, we note that to solve the problems of functional diagnostics of neurological diseases based on computer stabilography, it is necessary to apply the entire spectrum of methods for calculating the probabilistic characteristics of stabilograms, both individually for the sagittal and frontal directions, and their mutual characteristics.

References

1. Gagey, P.-M., Weber, B.: Posturologie: Regulation et dereglements de la station debout, p. 224. Masson, Paris (2004)
2. Gurfinkel, V.S., Kots, Y.M., Shik, M.L.: Regulation of the Person's Pose, p. 256. Nauka, Moscow (1965). (in Russian)
3. Skvortsov, D.V.: Diagnosis of Motor Pathology by Instrumental Methods: Gait Analysis, Stabilometry, p. 640. T.M. Andreeva, Moscow (2007). (in Russian)
4. Maatar, D., Lachiri, Z., Fournier, R., Nait-Ali, A.: Stabilogram mPCA decomposition and effects analysis of several entries on the postural stability. Int. J. Image Graph. Signal Process. **4**(5), 21–30 (2012). https://doi.org/10.5815/ijigsp.2012.05.03
5. Baron, J.: History of posturography. In: Igarashi, M., Black, F. (eds.) Vestibular and Visual Control of Posture and Locomotor Equilibrium, pp. 54–59. Karger, Basel (1983)
6. Vinarskaya, E.N., Kuuz, R.A., Ronkin, M.A., Firsov, G.I.: Topological aspects of afferent and efferent system synthesis in the problems of studying postural human activity. Inform. Control Syst. **4**(22), 44–46 (2009). (in Russian)
7. Belyaev, V.E., Kononov, A.F., Sliva, S.S.: Approaches to the Assessment of the Area of the Statokinesigram. Clinical Postural Studies, Posture and Bite, pp. 81–86. SPb.: OOO "Publishing House of SPbMAPO" (2004). (in Russian)
8. Furman, Y.A., Krevetskiy, A.V., Peredreyev, A.K.: Introduction to Contour Analysis and its Applications to the Processing of Images and Signals, p. 592. Fizmatlit, Moscow (2002). (in Russian)
9. Rvachev, V.L.: Theory of R-Functions and Some of its Applications, p. 552. Dumka, Kiev (1982). (in Russian)
10. Kuuz, R.A., Firsov, G.I.: Application of computer stabilometry methods for solving functional diagnostics problems in neurology. Biomed. Radioelectron. **5–6**, 24–33 (2001). (in Russian)
11. Kuuz, R.A., Magomedova, R.K., Rozenblyum, M.G., Suslov, V.N., Firsov, G.I.: Investigation of spectral features and non-linear dynamics of physiological, essential and parkinsonian tremor. Bull. Sci. Tech. Dev. **11**(39), 12–20 (2010). (in Russian)
12. Kuuz, R.A., Ronkin, M.A., Firsov, G.I.: Methods of parametric spectral analysis of stabilographic information in clinical neurology. Medico-ecological information technologies, pp. 11–14. Kursk: KSTU (2004). (in Russian)
13. Khashei, M., Montazeri, M.A., Bijari, M.: Comparison of four interval ARIMA-base time series methods for exchange rate forecasting. Int. J. Math. Sci. Comput. **1**(1), 21–34 (2015). https://doi.org/10.5815/ijmsc.2015.01.03
14. Akaike, H.: Canonical correlation analysis of time series and the use of an information criterion. In: System Identification: Advanced and Case Studies, pp. 27–96. Academic Press, New York (1976)
15. Diukova, G.M., Stoliajrova, A.V., Kuuz, R.A., Firsov, G.I., Vein, A.M.: Posturography in hysteria. In: International Symposium of Gait Disorders, Prague, Czech Republic, 4–6 Sept 1999. Book of Abstracts, Qualisis, Prague, pp. 122 (1999)

16. Rosenblum, M.G., Firsov, G. I., Kuuz, R.A., Pompe, B.: Human postural control: force plate experiments and modeling. In: Kantz, H., Kurths, J., Mayer-Kress, G. (eds.) Nonlinear Analysis of Physiological Data, pp. 283–306. Springer, Berlin (1998)
17. Rosenblum, M.G., Firsov, G.I., Kuus, R.A., Suslov, V.N.: Investigation of chaotic oscillations in a nonlinear control system for maintaining the vertical posture of the human body. Bull. Sci. Tech. Dev. **3**, 32–42 (2007). (in Russian)
18. Raybman, N.S., Kapitonenko, V.V., Ovsepyan, F.A., Varlakov, P.M.: Dispersion Identification, p. 336. Nauka, Moscow (1981). (in Russian)
19. Alain, K.S.T., Bertrand, F.H.: A secure communication scheme using generalized modified projective synchronization of coupled Colpitts oscillators. Int. J. Math. Sci. Comput. **4**(1), 56–70 (2018). https://doi.org/10.5815/ijmsc.2018.01.04
20. Ehlers, C.L.: Chaos and complexity: can it help us to understand the mood and behavior. Arch. Gen. Psych. **52**, 960–964 (1995)

Verification of Mathematical Model for Bioimpedance Diagnostics of the Blood Flow in Cerebral Vessels

Anna A. Kiseleva, Petr V. Luzhnov and Dmitry M. Shamaev

Abstract The electrical impedance method is considered as an evaluation of cerebral blood circulation. An electrode construction has been developed for recording the pulse blood filling of main large arteries of the brain: internal carotid artery, anterior cerebral artery, middle cerebral artery and ophthalmic artery. The electrical scheme for the replacement of cerebral vessels is verified.

Keywords Bioimpedance method · Blood flow · Cerebral vessels · Electrodes · Mathematical modeling

1 Introduction

At present, the methods of magnetic resonance imaging (MRI) [1–3], computed tomography (CT) [1], Doppler ultrasound [4] and ultrasound images [5] are used to diagnose cerebral circulation disorders. Possessing undeniable advantages, such as high spatial resolution, the possibility of obtaining three-dimensional images and a comprehensive examination of the required body area [6, 7], these methods have significantly increased the level of diagnostic quality. The method of electrical impedance tomography is one of the ways to solve a number of limitations [8–11]. It is widely used in various fields of research: in vitro diagnostics, in respiratory system studies [12, 13], diagnostics in ophthalmologic practice [14, 15], brain neural activity studies, tumor studies and cerebral circulation studies [16–18]. Losing in the spatial

A. A. Kiseleva · P. V. Luzhnov (✉) · D. M. Shamaev
Bauman Moscow State Technical University, 2-nd Baumanskaya St. 5, 105005 Moscow, Russian Federation
e-mail: petervl@yandex.ru

A. A. Kiseleva
e-mail: kiseleva.anna.a94@gmail.com

D. M. Shamaev
e-mail: shamaev.dmitry@yandex.ru

© Springer Nature Switzerland AG 2020
Z. Hu et al. (eds.), *Advances in Artificial Systems for Medicine and Education II*,
Advances in Intelligent Systems and Computing 902,
https://doi.org/10.1007/978-3-030-12082-5_23

251

resolution, this method significantly benefits in the duration of the study, the cost and mobility of using.

Thus, the purpose of this work is the development of an electrode construction for diagnosing cerebral circulation disorders using an electrical impedance method. The electrode system is intended for measuring pulse blood filling of the main large arteries of the brain.

The electrical impedance method is the technique of obtaining images in the sections of the impedance distribution body by means of non-invasive electrical sounding. The current, flowing through the bio-object, creates a volumetric distribution of the electrical potential. The potential decreases along the streamline as it moves away from the injection current electrode. The voltage drop per unit length is proportional to the magnitude of the current and the resistance of the medium. By measuring the voltage drop and knowing the value of the current, it is possible to calculate the resistance value. The reconstruction algorithm allows to use voltages measured only on the body surface to calculate the spatial distribution of the resistivity inside it [19].

In practice, this is done using an electrode structure located around the surveyed area. Currently, from 64 to 512 electrodes are used in laboratory studies [20] with the magnitude of the probing current having a frequency within the range of 20–100 kHz and a value of 1–5 mA. Thus, on signals, there are pulse oscillations of a blood flow. Verification of pulse oscillations is the basic problem of bioimpedance researches. For verification, we propose to use ultrasound signals.

2   Development of an Electrodes System

The minimum required number of electrodes is the problem in the study with the method of impedance tomography of the head. The 64-electrode systems cannot be precisely located at the necessary points. The inaccuracy of the location of the electrodes will lead to a significant decrease in the quality of the resulting image [21, 22]. To solve this problem, it has been proposed to use an electrode structure consisting of 16 electrodes arranged on a silicone tape. The location of the electrodes only in the front part of the head will allow to receive information on the blood supply to the anterior part of the brain, while maintaining the simplicity of the algorithm.

Figure 1 shows an image of an electrode structure with a fixed distance between the electrodes in the projection on the cross section of the head. The distance is chosen in such a way as to fix the blood supply of the main arteries: internal carotid artery, anterior cerebral artery, middle cerebral artery and ophthalmic artery [23]. It is known that the range of head circumference in adults varies in the range from 55.0 ± 5.5 to 58.0 ± 5.8 cm in women and 58.0 ± 5.8 to 60.0 ± 6.0 cm in men [22, 23]. Accordingly, on average, the spacing of the electrode will not exceed 3.0 cm, which is acceptable for the impedance method of investigation.

Also, a significant advantage of this arrangement is the possibility of analyzing the blood supply to the eye (recording the pulse blood filling of the ophthalmic artery). Existing research methods (color Doppler, magnetic resonance imaging and

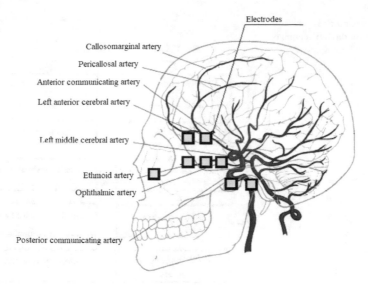

Fig. 1 The electrodes location on the person head with the anatomic localization of arteries

computed tomography) are complex and cumbersome in using [24]. The obtained analysis method of the eye vessels will be a continuation of the eye impedance researches, proposed in the works [25–27].

To verify the obtained electrode structure, an electrical circuit for replacing the cerebral vessels was developed, which allows us not only to confirm the adequacy of the results obtained, but also to reconstruct the signal of pulse blood filling in small vessels.

From the standpoint of functional significance, all vessels can be divided into 5 main groups: (1) elastically extensible (aorta, large arteries of the great circle of the circulation and pulmonary artery); (2) resistance vessels or resistive vessels (arterioles and precapillary sphincters); (3) exchange vessels (capillaries); (4) venules; and (5) capacitive vessels (veins). We took as a basis the work [28], where the separation of the vessels was represented as follows:

- Elastic-expandable vessels are represented as resistors;
- Resistive vessels in the form of two resistors and a capacitor;
- Exchange vessels in the form of two parallel-connected resistors;
- Venules in the form of two resistors and a capacitor;
- Capacitive vessels in the form of a capacitor.

To calculate the resistive components, we use Poiseuille's law, where the hemodynamic resistance of each vessel depends on its length, blood viscosity, but most on the radius of the vessel. To calculate the capacitive components, we use the mathematical model of the electrical capacitor.

Accordingly, for the quantitative calculation of the chain, two important parameters of the vessel are needed: length and diameter. Due to the fact that the data on

Fig. 2 Generalized block diagrams for the replacement of large- and medium-sized vessels

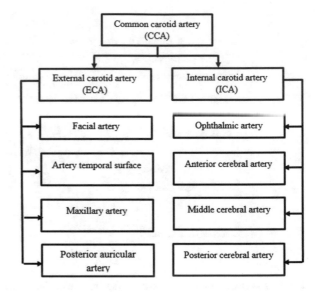

the characteristics of the vessels vary from ±10% in different works, it was decided to use the averaged values for each vessel. Figure 2 presents the generalized blocks for the replacement of large- and medium-sized vessels.

This structure is used as input for our model.

3   Verification of the Model

The next step is the verification of the obtained replacement circuit. The main stages of verification are shown in Fig. 3 and include the following steps: registration of pulsatile blood supply signals of the common carotid artery (CCA) and the middle cerebral artery (MCA) by the Doppler ultrasound; feed to the input of the circuitry of the registered CCA signal; simulation of the MCA signal; correlation analysis of registered and simulated signals; and conclusion on the efficiency of the substitution scheme.

Figure 4 shows the scheme of simultaneous recording of pulsatile blood flow signals of CCA and MCA by Doppler ultrasound method and electrical impedance method.

The study was conducted on 3 healthy persons aged 25 ± 3 years. There was a sequential recording of signals from the common carotid artery and the middle cerebral artery Doppler ultrasound. Then, in the Micro-Cap environment, the portion of the obtained chain was modeled, where the source was the signal of pulse blood filling of the common carotid artery (Fig. 5).

At the same time, a bioimpedance signal was recorded from the middle cerebral artery.

Fig. 3 The verification stages of the electrical circuit of substitution

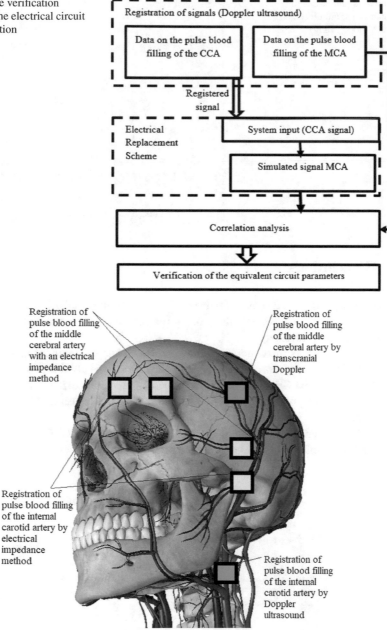

Fig. 4 The scheme of pulse signal recording by Doppler ultrasound (green) and electrical impedance method (yellow)

Fig. 5 Input data of the
substitution circuit (pulse
CCA blood filling)

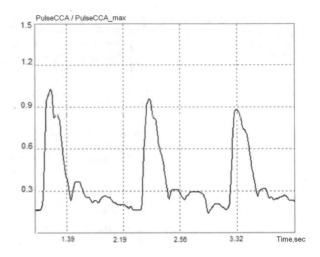

4 Conclusions

As a result, a simulated signal of pulse blood filling of the middle cerebral artery was
obtained. Figure 6 shows simultaneously the recorded and simulated signals from
the middle cerebral artery.

The last stage is the conduct of the correlation analysis and comparison of the
data obtained for the researches group. Table 1 shows the obtained results.

As a result, the SE value is on average 2.11, and the correlation coefficient is
0.87. It confirms possibility of the developed algorithm use for cerebral blood flow
studying.

In this work, an electric impedance method has been offered for assessing the state
of cerebral circulation. An electrode construction has been described that makes it

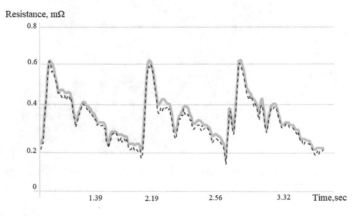

Fig. 6 Registered (solid green) and simulated (dotted black) pulse filling MCA signals

Table 1 The comparative data for the researches group

Volunteer	SE	Correlation coefficient
Volunteer 1	2.12	0.91
Volunteer 2	1.87	0.84
Volunteer 3	2.36	0.86
Average value	2.11	0.87

possible to obtain data on the state of main arteries (internal carotid artery, anterior cerebral artery, middle cerebral artery and ophthalmic artery). The main advantage is the reduction of electrodes necessary for registration (to 16). Also, an electrical circuit for replacing the cerebral vessels has been developed, which makes it possible to verify the results obtained with the help of an electrode tape. An algorithm for converting signals obtained from an electrode tape is developed on the basis of an electrical circuit of substitution. The scheme has checked on the signals of the common carotid and middle cerebral arteries, obtained by the method of Doppler ultrasound. Comparative data on the researches group showed high accuracy of the obtained simulated signals.

In the future, we plan to check the remaining parts of the layout: the internal carotid, the external carotid, the anterior cerebral and ophthalmic arteries, and compare the simulated signals with the signals obtained through the electrode system. The developed replacement scheme will provide data on the blood supply of medium and small vessels, analyze the state of the microcirculatory bloodstream. It will enable us to investigate the relationship between the paired arteries of the brain, which in the future will allow us to derive a quantitative assessment of the cerebral blood flow state.

Acknowledgements The paper was supported by a grant from RFBR (No. 18-08-01192).

References

1. Audebert, H., Fiebach, J.: Brain imaging in acute ischemic stroke—MRI or CT. Curr. Neurol. Neurosci. Rep. **15**(3) (2015). https://doi.org/10.1007/s11910-015-0526-4
2. Nageswara Reddy, P., Mohan Rao, C.P.V.N.J., Satyanarayana, C.: Optimal segmentation framework for detection of brain anomalies. Int. J. Eng. Manuf. (IJEM) **6**(6), 26–37 (2016). https://doi.org/10.5815/ijem.2016.06.03
3. Neela, R., Kalaimagal, R.: BRAINSEG—Brain structures segmentation pipeline using open source tools. Int. J. Math. Sci. Comput. (IJMSC) **1**(1), 1–10 (2015). https://doi.org/10.5815/ijmsc.2015.01.01
4. Kristensen, T., Hovind, P., Iversen, H., Andersen, U.: Screening with doppler ultrasound for carotid artery stenosis in patients with stroke or transient ischaemic attack. Clin. Physiol. Funct. Imaging (2017). https://doi.org/10.1111/cpf.12456
5. Gupta, N., Shukla, A.P., Agarwal, S.: Despeckling of medical ultrasound images: a technical review. Int. J. Inf. Eng. Electron. Bus. (IJIEEB) **8**(3), 11–19 (2016). https://doi.org/10.5815/ijieeb.2016.03.02

6. Pagán, R., Parikh, P., Mergo, P., Gerber, T., Mankad, R., Freeman, W., Shapiro, B.: Emerging role of cardiovascular CT and MRI in the evaluation of stroke. Am. J. Roentgenol. **204**(2), 269–280 (2015). https://doi.org/10.2214/ajr.14.13051

7. Naess, H., Tatlisumak, T., Kõrv, J.: Stroke in the Young. Stroke Res. Treat. **2011**, 1–2 (2011). https://doi.org/10.4061/2011/271979

8. Scholz, F., Weiler, N.: Electrical impedance tomography and its perspectives in intensive care medicine. Intensiv. Care Med. 437–447 (2006). https://doi.org/10.1007/0-387-35096-9_40

9. Adler, A., Gaburro, R., Lionheart, W.: Electrical impedance tomography. In: Handbook of Mathematical Methods in Imaging. Springer (2016)

10. Holder, D.: Electrical Impedance Tomography. Institute of Physics Publication, Bristol (2005)

11. Polydorides, L., Borsic, A.: The reconstruction problem. In: Electric Impedance Tomography: Methods, History and Applications. IOP Publishing, England (2004)

12. Reinius, H., Borges, J.B., Fredén, F., et al.: Real-time ventilation and perfusion distributions by electrical impedance tomography during one-lung ventilation with cannothorax. Acta Med. Scan **59**, 354–368 (2015). https://doi.org/10.1111/aas.12455

13. Luzhnov, P.V., Dyachenko, A.I., Semenov, Y.S.: Research of impedance characteristics with a negative pressure breathing using rheocardiographic and rheoencephalographic signals. IFMBE Proc. **68**(2), 937–940 (2018). https://doi.org/10.1007/978-981-10-9038-7_173

14. Luzhnov, P.V., Shamaev, D.M., Iomdina, E.N., et al.: Using quantitative parameters of ocular blood filling with transpalpebral rheoophthalmography. IFMBE Proc. **65**, 37–40 (2017). https://doi.org/10.1007/978-981-10-5122-7_10

15. Shamaev, D.M., Luzhnov, P.V., Iomdina, E.N.: Modeling of ocular and eyelid pulse blood filling in diagnosing using transpalpebral rheoophthalmography. IFMBE Proc. **65**, 1000–1003 (2017). https://doi.org/10.1007/978-981-10-5122-7_250

16. Akhtari-Zavare, M., Latiff, L.: Electrical impedance tomography as a primary screening technique for breast cancer detection. Asian Pac. J. Cancer Prev. **16**(14), 5595–5597 (2015). https://doi.org/10.7314/apjcp.2015.16.14.5595

17. Brown, B., Leathard, A., Sinton, A., McArdle, F., Smith, R., Barber, D.: Blood flow imaging using electrical impedance tomography. Clin. Phys. Physiol. Meas. **13**, 175–179 (1992). https://doi.org/10.1088/0143-0815/13/a/034

18. Shi, X., You, F., Fu, F., Liu, R., You Y, Dai M., Dong, X.: Preliminary research on monitoring of cerebral ischemia using electrical impedance tomography technique. In: 30th Annual International Conference of the IEEE Engineering in Medicine and Biology Society (2008). https://doi.org/10.1109/iembs.2008.4649375

19. Yorkey, T., Webster, J., Tompkins, W.: Comparing reconstruction algorithms for electrical impedance tomography. IEEE Trans. Biomed. Eng. **34**(11), 843–852 (1987). https://doi.org/10.1109/tbme.1987.326032

20. Dimas, C., Tsampas, P., Ouzounoglou, N., Sotiriadis, P.: Development of a modular 64-electrodes electrical impedance tomography system. In: 6th International Conference MOCAST (2017). https://doi.org/10.1109/mocast.2017.7937666

21. Dimas, V., Sotiriadis, P.: Conductivity distribution measurement at different low frequencies using a modular 64 electrode electrical impedance tomography system. In: Panhellenic Conference on Electronics and Telecommunications PACET (2017). https://doi.org/10.1109/pacet.2017.8259973

22. Hyvönen, N., Majander, H., Staboulis, S.: Compensation for geometric modeling errors by positioning of electrodes in electrical impedance tomography. Inverse Prob. **33**(3), 035 (2017). https://doi.org/10.1088/1361-6420/aa59d0

23. Fritsch, H., Kuehnel, W.: Color Atlas of Human Anatomy. Stuttgart (2014)

24. Fan, N., Wang, P., Tang, L., Liu, X.: Ocular blood flow and normal tension glaucoma. Biomed. Res. Int. **2015**, 1–7 (2015). https://doi.org/10.1155/2015/308505

25. Luzhnov, P.V., Shamaev, D.M., Iomdina, E.N., et al.: Transpalpebral tetrapolar reoophtalmography in the assessment of parameters of the eye blood circulatory system. Vestn. Ross. Akad. Med. Nauk **70**(3), 372–377 (2015). https://doi.org/10.15690/vramn.v70i3.1336

26. Shamaev, D.M., Luzhnov, P.V., Iomdina, E.N.: Mathematical modeling of ocular pulse blood filling in rheoophthalmography. IFMBE Proc. **68**(1), 495–498 (2018). https://doi.org/10.1007/978-981-10-9035-6_91
27. Luzhnov, P.V., Shamaev, D.M., Kiseleva, A.A., Iomdina, E.N., et al.: Analyzing rheoophthalmic signals in glaucoma by nonlinear dynamics methods. IFMBE Proc. **68**(2), 827–831 (2018). https://doi.org/10.1007/978-981-10-9038-7_152
28. Cassani, S.: Blood circulation and aqueous humor flow in the eye: multi-scale modeling and clinical applications. Purdue University, Ph.D. (2016)

Evaluation of Adhesive Bond Strength of Dental Fiber Posts by "Torque-Out" Test

Anna S. Bobrovskaia, Sergey S. Gavriushin and Alexander V. Mitronin

Abstract A new technique for evaluation of adhesive bond strength of fiber posts is presented. It called by the authors the "torque-out" test. Unlike the known test methods push-out test and pull-out test, in which a tensile or compressive load is applied to the sample, the load is produced by the torque. Samples for testing of a specific combination of post and cement can be manufactured chairside without using of special equipment. A compact installation developed by the authors designed to assess the adhesive bond strength of the connection by the value of the maximum torque is described. Using numerical modeling, the stress–strain state that occurs in the samples during push-out and torque-out tests was compared.

Keywords Fiber post · Adhesive bond strength · Postendodontical restoration · Torque-out test · Testing installation

1 Introduction

Restoration of teeth with coronal part destroyed by trauma or caries and its complications is one of the current tasks of modern dentistry. Usually, such treatment requires using of posts. The technique of tooth strengthening with fiber post has gained popu-

A. S. Bobrovskaia (✉) · A. V. Mitronin
Moscow State University of Medicine and Dentistry Named After A.I. Evdokimov, 20, Delegatskaia St., Moscow 127473, Russian Federation
e-mail: aggi@yandex.ru

A. V. Mitronin
e-mail: mitroninav@list.ru

S. S. Gavriushin
Mechanical Engineering Research Institute of the Russian Academy of Sciences, 4, Malyi Kharitonievsky Pereulok, 101990 Moscow, Russian Federation
e-mail: gss@bmstu.ru

Bauman Moscow State Technical University, 5, 2nd Baumanskaya St., 105005 Moscow, Russian Federation

© Springer Nature Switzerland AG 2020
Z. Hu et al. (eds.), *Advances in Artificial Systems for Medicine and Education II*,
Advances in Intelligent Systems and Computing 902,
https://doi.org/10.1007/978-3-030-12082-5_24

larity last years [1–3]. Fiber posts are luted adhesively in root canals of the teeth with composite cement. However, according to the published data, due to the weakness of the adhesive compound, more than 60% of the failures [4] are associated with construction's "debonding." A large number of foreign studies [4–7] are known, devoted to the search for rational post pretreatment and fixation methods. Studies which deal with the adhesive bond strength evaluation of fiber posts, as a rule, are carried out under laboratory conditions, and also require a complicated procedure and special equipment. Standardized evaluation conditions and direct measurement of adhesion between fiber posts and cements for their fixation have not been developed to date.

Thus, the actual practical task is the development of techniques that allow assessing the adhesive bond strength in clinical settings, and make a valid choice of fiber posts luting protocol that provides required adhesive strength.

There are two fundamentally different approaches to carrying out tests used to evaluate the adhesive bond strength in dentistry: microshear and microtensile tests [8]. When conducting microshear tests, one of the adhesion substrates is fixed immovably, and force acting in the direction of the axis parallel to the contact plane between the substrates is applied to the other. During the test, the maximal force that occurs when the sample is destroyed is determined. The tests of the first group include the so-called micro "push-out" and "pull-out" tests. When implementing a micro push-out test, the fragment of the post is extruded from the prepared sample obtained by horizontal saw cuts of the preparation. When carrying out a pull-out test, the pin previously locked in the prepared preparation is pulled.

Microshear tests are widespread today, but among their main drawbacks, it should be noted a significant frequency of cases of cohesive destruction of the substrate. This is due to the emergence of a complex load distribution scheme during the test and may lead to erroneous interpretation of the results [5].

The second group includes microtensile tests [9], in which force is applied along an axis directed perpendicular to the plane of the adhesive compound. The microtensile test was originally developed to evaluate the strength of tooth tissues, but was later used to measure the adhesive bond strength to enamel and dentin [10]. In this test, preparation of test specimens requires cutting the sample into "bars" containing a fragment of the post, the fixing cement, and the area of the adhesive bonding between them. Then, the stretching load is applied to the preparations until the moment of destruction. The small size of the samples provides a fairly even distribution of the load, which limits the possibility of cohesive fractures and helps to evaluate the adhesive bond strength directly [6]. In addition, from one sample it is possible to obtain several fragments for testing, which leads to a smaller scatter of data [5]. At the same time, it is noted in [10] that 46 (!) specimens out of 50 were subjected to premature failure in the preparation of samples for a microtensile test, which forced the authors to abandon the application of this method.

Thus, the main shortcomings of the described methods include the complexity and laboriousness of preparing the specimens used in the study. In addition, the implementation of cuts requires the use of special high-precision technological equipment. At the stage of preparation of samples, it is possible to change the physic-mechanical properties and structure of samples right up to its destruction due to the aggressive influence of the tool. It must also be taken into account that a very complex stress distribution law in the preparation may lead to erroneous interpretation of the results. Finally, the direct carrying out of the experiment requires the use of expensive stationary equipment (universal test machines, such as Instron [11]) and is practically inaccessible in clinical settings.

2 The Methodology Essence

The new technique for measuring the adhesive bond strength of luted fiber posts has been called the "torque-out" test. Unlike the known test methods described above, in which a tensile or compressive load is applied to the sample, the samples are loaded with torque.

Samples for the study are made using the required kind of posts, cements (and, if necessary, removed teeth) in clinical conditions without the use of technological

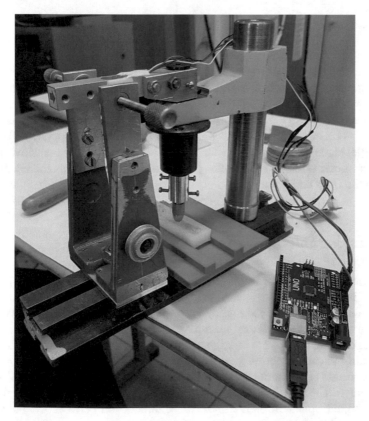

Fig. 1 Testing installation for torque-out test

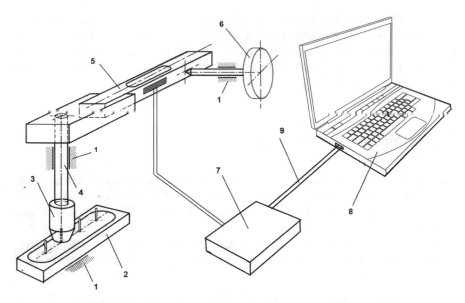

Fig. 2 Installation diagram

equipment for sawing. The developed compact installation (Fig. 1) for torque measurement makes it possible to evaluate the adhesive bond strength without the use of
stationary testing equipment.

The installation (Fig. 2) for measurement of the adhesive bond strength between
dental posts (2) and fixing cements consists of a rigid frame, a template for the posts
(1), a collet clamp (3), a torque transferring shaft (4), a tensometric sensitive element
(5), micrometer screw (6), controller (7) connected to the computer (9) by wires
through the USB connector (8).

3   Carrying Out the Test

Preparation of samples for torque-out test is carried out as follows: The posts are
installed in the prepared template, and the groove around the posts is filled with
cement, which polymerizes according to the manufacturer's instructions. When the
test is carried out, the post being examined is fixed by means of a collet clamp, the
strain gage sensor being set to its original position, controlled by a micrometer screw.
The loading is carried out by successive rotation of the micrometer screw. In this
case, the torque is transmitted through the strain gauge element, shaft, and collet to
the specimen. The law of variation of the magnitude of the torque in time is reflected
in the computer diagram (Fig. 3). When the connection between the pin and the
cement is broken, the maximum torque value is fixed.

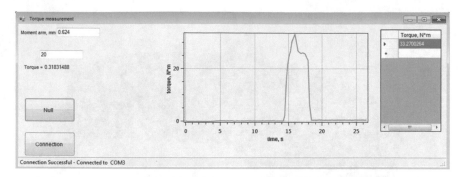

Fig. 3 Screenshot of torque-out measurement

4 Numerical Modeling

The numerical modeling of the proposed torque-out test and the known push-out test was carried out with finite element method [12–14] in the ANSYS software package (Figs. 4 and 5).

The Young modulus and the Poisson's ratio for the post were assumed to be equal to the characteristics of dentin [12, 13]: $E1 = 18.6$ GPa and $v1 = 0.31$. The corresponding parameters for cement were $E2 = 5.1$ GPa, and $v2 = 0.27$ [15, 16].

The materials were assumed to be isotropic.

To compare the results of the analysis of the stress–strain state, we used the averaged value of shear stresses. With respect to the push-out test, the averaged tangential stress level was estimated using the formula

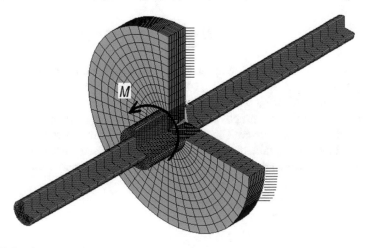

Fig. 4 Finite element model for torque-out test

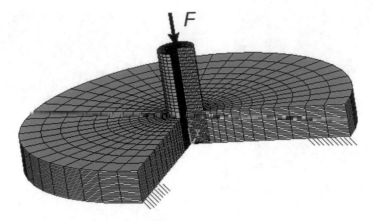

Fig. 5 Finite element model for push-out test

$$\tau_{yz} = \frac{F}{\pi D h},$$

where D is the diameter of the pin, h is the thickness of the cement layer.

With respect to the torque-out test, the average level of tangential stresses was estimated using the formula

$$\tau_{xz} = \frac{2M}{\pi D^2 h}.$$

Equating the average level of the tangential stresses of the tests under consideration, we can establish an approximate correspondence between the load parameters

$$M = \frac{FD}{2}.$$

The Figs. 6 and 7 show color graphic chars of the equivalent stresses distribution according to the Huber–Mises theory [14]. The samples were loaded with a torque of $M = 6$ N * mm and an ejection force of 10 N, respectively [17].

The Figs. 8 and 9 show graphs of the distribution of equivalent stresses along the height of the cement layer. The nature of the stress distribution allows us to conclude that there is a qualitative coincidence of these dependencies for torque-out and push-out tests, which makes it possible to compare the experimental data.

5   Conclusions

Thus, the developed technique, called the torque-out test, makes it possible to evaluate the adhesion strength of dental posts and fixing cements and to make an informed

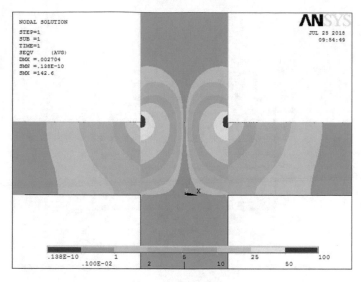

Fig. 6 Color graphic chars of the equivalent stresses distribution for torque-out test

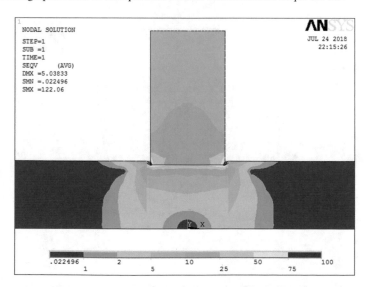

Fig. 7 Color graphic chars of the equivalent stresses distribution for push-out test

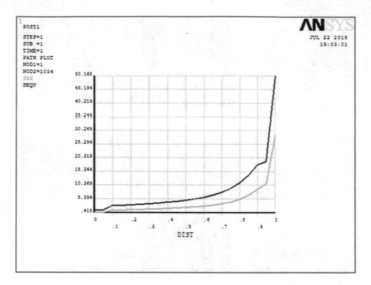

Fig. 8 Graph of the equivalent stresses distribution along the height of the cement layer for torque-out test

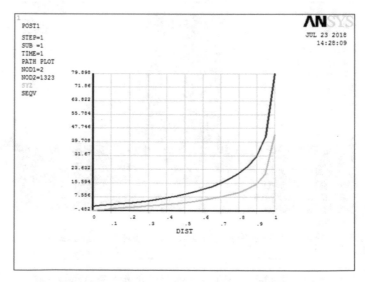

Fig. 9 Graph of the equivalent stresses distribution along the height of the cement layer for push-out test

choice of the method of restoration in a specific clinical case. The proposed installation is an alternative to the use of expensive stationary equipment and is available in clinical settings. The conservative technique of samples preparation is less laborious, since it does not require sawing, and thus excludes the possibility of changing the physical and mechanical properties and structure of the samples in preparation for the testing. The results of numerical simulation have shown that it is possible to compare the test-to-test data with other known techniques for determining the adhesive bond strength.

References

1. Duret, B., Reynaud, M., Duret, F.: A new concept of corono-radicular reconstruction, the Composipost (2). Le Chirurgien-dentiste de France. **60**(542), 69–77 (1990)
2. Bobrovskaia, A.S., Mitronin, A.V.: Evaluation of the effectiveness of postendodontic restoration by the improved method fiber post luting. Russ. Dent. **11**(2), 46–47 (2018). (In Russian)
3. Venkateshwar Reddy, C., Ramesh Babu, P., Ramnarayanan, R.: Effect of various filler materials on interlaminar shear strength (ILSS) of glass/epoxy composite materials. Int. J. Eng. Manuf. (IJEM) **6**(5), 22–29 (2016). https://doi.org/10.5815/ijem.2016.05.03
4. Mosharraf, R., Ranjbarian, P.: Effects of post surface conditioning before silanization on bond strength between fiber post and resin cement. J. Adv. Prosthodont. (5), 126–132 (2013). https://doi.org/10.4047/jap.2013.5.2.126
5. Cecchin, D., Farina, A.P., Vitti, R.P., Moraes, R.R., Bacchi, A., Spazzin, A.O.: Acid etching and surface coating of glass-fiber posts: bond strength and interface analysis. Braz. Dent. J. **27**(2), 228–233 (2016). https://doi.org/10.1590/0103-6440201600722
6. Sumitha, M., Kothandaraman, R., Sekar, M.: Evaluation of post-surface conditioning to improve interfacial adhesion in post-core restorations. J. Conserv. Dent. **14**, 28–31 (2011). https://doi.org/10.4103/0972-0707.80728
7. Lin, J.-L., Ay, C., Cheng, J.-Y., Young, C.-W.: Optimal measurement model for the assessment of cell adhesive force by using the dielectrophoresis force. Int. J. Eng. Manuf. (IJEM) **2**(5), 36–43 (2012). https://doi.org/10.5815/ijem.2012.05.06
8. Castellan, C.S., Santos-Filho, P.C.F., Soares, P.V., Soares, C.J., Cardoso, P.E.C.: Measuring bond strength between fiber post and root dentin: a comparison of different tests. J. Adhes. Dent., 12, 477–485 (2010). https://doi.org/10.3290/j.jad.a17856
9. Sano, H., Shono, T., Sonoda, H., Takatsu, T., Ciucchi, B., Carvalho, R., Pashley, D.H.: Relationship between surface area for adhesion and tensile bond strength–evaluation of a micro-tensile bond test. Dent. Mater. **10**(4), 236–240 (1994)
10. Goracci, C., Grandini, S., Bossu, M., Bertelli, E., Ferrari, M.: Laboratory assessment of the retentive potential of adhesive posts: a review. J. Dent. **35**, 827–835 (2007). https://doi.org/10.1016/j.jdent.2007.07.009
11. Soares, S.J., Santana, F.R., Castro, C.G., Santos-Filho, P.C.F., Soares, P.V., Qian, F., Armstrong, S.R.: Finite element analysis and bond strength of glass post to intraradicular dentin: comparison between microtensile and push-out tests. Dent. Mater. **24**, 1405–1411 (2008). https://doi.org/10.1016/j.dental.2008.03.004
12. Zienkiewicz, O.C.: The Finite Element Method, 3rd edn, p. 483. McGraw-Hill, London (1977)
13. Gavriushin, S.S., Baryshnikova, O.O., Boriskin, O.F.: Numerical Analysis of Structure's and Device's Elements. M, BMSTU Publishing (2014). 479 p. (in Russian)
14. Tong, J., He, Y.: Vane design which based on finite element analyzing method. Int. J. Educ. Manage. Eng. (IJEME) **2**(10), 72–81 (2012). https://doi.org/10.5815/ijeme.2012.10.12

15. Asmussen, E., Peutzfeldt, A., Sahafi, A.: Finite element analysis of stresses in endodontically treated, dowel-restored teeth. J. Prosthet. Dent. **21**, 709–715 (2005). https://doi.org/10.1016/j.prosdent.2005.07.003
16. Borodin, I.N., Seyedkavoosi, S., Zaitsev, D.V., Drach, B., Mikayelyan, K.N., Panfilov, P.E., Gutkin, M.Y., Sevostianov, I.: Viscoelasticity and mechanisms of plasticity of dentin of human teeth. Solid State Phys. **60**(1), 118–126 (2018). (in Russian)
17. Valishvili, N.V., Gavriushin, S.S.: Strength of Materials and Structures. M, Urait, (2017), 429 p. (in Russian)

Image Segmentation Method Based on Statistical Parameters of Homogeneous Data Set

Oksana Shkurat, Yevgeniya Sulema, Viktoriya Suschuk-Sliusarenko and Andrii Dychka

Abstract In this paper, we present a new automatic method of image segmentation, which uses statistical parameters of a homogeneous data set. The method can be applied to image sets obtained from CT, MRI, ultrasound, and histological investigations. The result of applying the proposed method is image components and their contours.

Keywords Image segmentation · Contour detection · Medical images analysis

1 Introduction

Existing information systems of biomedical [1, 2] and spatial data processing [3, 4], video surveillance [5, 6], pattern recognition systems [7, 8], and e-medicine [9] are usually based on image analysis methods. Medical image analysis methods are used in such medical intelligent systems as diagnostic and monitoring systems, decision support systems, training systems, etc. [10].

Image segmentation is one of the stages of image analysis. The segmentation procedure transforms an initial image into a set of segments which represent the image objects. It enables simplifying further image processing and solving data-compression task in a certain way.

O. Shkurat · Y. Sulema (✉) · V. Suschuk-Sliusarenko · A. Dychka
Igor Sikorsky Kyiv Polytechnic Institute, 37 Peremogy Ave, Kiev 03056, Ukraine
e-mail: sulema@pzks.fpm.kpi.ua

O. Shkurat
e-mail: shkurat.ksusha@gmail.com

V. Suschuk-Sliusarenko
e-mail: viss@pzks.fpm.kpi.ua

A. Dychka
e-mail: andriydychka@gmail.com

© Springer Nature Switzerland AG 2020
Z. Hu et al. (eds.), *Advances in Artificial Systems for Medicine and Education II*,
Advances in Intelligent Systems and Computing 902,
https://doi.org/10.1007/978-3-030-12082-5_25

271

The segmentation methods can be divided into thresholding and clustering methods, edge detection, region-growing methods, graphs partitioning methods, etc. The choice and application of segmentation methods depend on research objectives as well as on a required level of details of segmentation results. Excessive detailing leads to impossibility of interest objects detection and, thus, it complicates the image analysis. At the same time, insufficiently detalization leads to errors in further stages of the image processing. Also, most of the existing segmentation methods are focused on images of a specific type of medical investigation and/or a specific organ, e.g., computerized tomography (CT) of kidney, magnetic resonance imaging (MRI) of brain, X-ray investigation of bones, and ultrasound of thyroid gland. Such methods are not supposed to be used for other types of images, thus, it narrows their application area.

Thus, the research objective is to develop an accurate segmentation method which can be used for different types of images with both minimal time consumption and minimal computing resources requirements to solve complex tasks in image analysis.

2 Related Work

There are many methods for medical images segmentation. These methods are based on defining an image gradient [11, 12], threshold processing [13], mathematical morphology operations [14, 15], Canny and Deriche detectors [16, 17], etc.

Thus, in [18], the Norouzi et al. considered and analyzed four approaches to the medical image segmentation: region-based approach, clustering, classification, and hybrid approaches. The authors described advantages and disadvantages of these segmentation methods by testing images obtained as a result of CT and MRI.

In [19], Costin and Rotariu developed semi-automatic method of medical images segmentation. At first, this method defines and detects image contours by applying procedures of fuzzy logic. Then, these contours are used to creating a contour model based on a genetic algorithm and the active contour method. The method has been tested on CT images of brain.

In [20], the Bouchet et al. developed a method of angiographic images segmentation. This method is based on fuzzy sets theory and mathematical morphology procedures.

The Mozaffari and Lee presented in [21] a segmentation method for MRI of brain. This method combined Convergent Heterogeneous Particle Swarm Optimization algorithm and threshold processing methods (Otsu, Kapur).

Lakshmi and Ravi in [22] proposed a double-layered segmentation algorithm for cervical cell images based on Generalized Hierarchical Fuzzy C Means and Artificial Bee Colony. The performance of the proposed segmentation algorithm is analyzed in terms of accuracy, sensitivity, and specificity.

In [23], Isah et al. presented their approach to the medical image segmentation based on an improved Bat-Active Contour Method.

Besides, Alyahya, and Abu-Shareha in [24] presented accuracy evaluation of brain tumor detection using entropy-based image thresholding. Five entropies (Renyi, Maximum, Minimum, Tsallis, and Kapur) are evaluated.

Mwambela in [25] presented comparative performance evaluation of entropic thresholding algorithms based on Shannon, Renyi, and Tsallis entropy definitions for electrical capacitance tomography measurement systems.

Gourav et al. presented computational analysis of image segmentation algorithms in [26].

Thus, these segmentation methods are object-oriented, since their effectiveness depends on the medical image types. The method proposed in this paper can be applied for different image types (CT, MRI, ultrasound, and histological images). The proposed segmentation method enables automatic data processing.

3   Method Description

The input data of the proposed method is homogeneous images arrays. Images can be considered as homogeneous ones if they contain the same objects (organs, vessels, etc.) and obtained by using the same imaging tools (CT, MRI, ultrasound, etc.).

Elements of the initial array $A = \{I_1, I_2, \ldots, I_q, \ldots, I_Q\}$ are gray scale or color images of the size $M \times N$ which can be described by three-dimensional matrix $I_q(i, j, k) = a$ where a is a pixel color intensity of k-component with coordinates $(i, j), i = 1, 2, \ldots, M, j = 1, 2, \ldots, N; k$ is an order number of the color component of a color model, $k = 1, 2, 3; I_q$ is a q-image of the array $A, q = 1, 2, \ldots, Q; Q$ is an image number of the array A.

The proposed method of medical image segmentation includes the following steps:

– Color space conversion
– Image array statistical parameters computation
– Initial segment centers detection
– Detection of image elements belonging to segments
– Contours detection and analysis.

3.1   Color Space Conversion

In our research, we use images conversion from RGB model into *XYZ* model [27]:

$$
\begin{bmatrix} X \\ Y \\ Z \end{bmatrix} = \begin{bmatrix} 0.4124\ 0.3576\ 0.1805 \\ 0.2126\ 0.7152\ 0.0722 \\ 0.0193\ 0.1192\ 0.9505 \end{bmatrix} \times \begin{bmatrix} R^{(1)} \\ G^{(1)} \\ B^{(1)} \end{bmatrix}
\tag{1}
$$

where $R^{(1)}$, $G^{(1)}$, $B^{(1)}$ are calculated according to (2).

$$
\text{Component}^{(1)} = \begin{cases} \left(\dfrac{\text{Component}+0.055}{1.055} \right)^{2.4}, & \text{Component} > 0.04045 \\ \dfrac{\text{Component}}{12.92}, & \text{otherwise} \end{cases}
\tag{2}
$$

where component variable means *R*, *G*, and *B* components, $R, G, B \in [0; 1]$.

The input data for converting is Z-component data which can be represented by an array $A' = \{I'_1, I'_2, \ldots, I'_q, \ldots, I'_Q\}$. The elements of A' array are 2D matrices $I'_q(i, j) = a'$ where a' is q-image pixels intensity of Z-component, $a' \in [0; 1.089]$.

3.2   Image Array Statistical Parameters Computation

To detect the segments centers, the statistical parameters of images array are determined. These parameters are mathematical expectation (3) and dispersion (4).

$$
\mu = \frac{1}{Q} \times \sum_{q=1}^{Q} \left(\frac{1}{M \times N} \times \sum_{i=1}^{M} \sum_{j=1}^{N} I'_q(i, j) \right)
\tag{3}
$$

where I'_q is intensity values of q-image pixels, $q = 1, 2, \ldots, Q; (i, j)$ is the pixel coordinates, $i = 1, 2, \ldots, M, j = 1, 2, \ldots, N; Q$ is the elements number in A' array.

$$
d = \frac{1}{Q} \times \sum_{q=1}^{Q} \left(\frac{1}{M \times N} \times \sum_{i=1}^{M} \sum_{j=1}^{N} \left(I'_q(i, j) - \mu \right)^2 \right)
\tag{4}
$$

where μ is a mathematical expectation of pixel intensity values of images array A'.

3.3 Initial Segment Centers Detection

The formation of homogeneous areas in the proposed method is independent of spatial proximity of corresponding image pixels. Segment centers are pixels intensity values computed according to (5) and (6).

$$T_+(h_1) = \mu + h_1 \times d \tag{5}$$

$$T_-(h_2) = \mu - h_2 \times d \tag{6}$$

where $T_+(h_1)$ is a function which determines "rising" values of segments centers; $h_1 = 1, 2, \ldots, H_1$; $T_-(h_2)$ is a function which determines "falling" values of segments centers, $h_2 = 1, 2, \ldots, H_2$; H_1, H_2 are the maxima of h_1, h_2 variables accordingly; μ is a mathematical expectation of A' array pixels intensity; d is a dispersion.

Since the intensity value of A' array pixels belongs to the range [0; 1.089], the maximal quantity of segments can be determined by (7).

$$C = \text{int}\left(\frac{\mu}{d}\right) + \text{int}\left(\frac{1.089 - \mu}{d}\right) \tag{7}$$

where int means rounding.

That is, $H_1 = \text{int}\left(\frac{\mu}{d}\right)$ and $H_2 = \text{int}\left(\frac{1.089-\mu}{d}\right)$. Thus, elements of the array (8) are centers of segments; the size of this array is finite.

$$L = \{l_c | l_c \in (T_+ \cup T_-), \quad c \in [1; C]\}. \tag{8}$$

The array L can be supplemented by $l_{C+1} = 0, l_{C+2} = \mu, l_{C+3} = 1.089$ centers for more accuracy of the segmentation.

3.4 Detection of Image Elements Belonging to Segments

The quantitative value of pixels homogeneity degree of A' array images according to L centers can be determined by a distance function (9). To find a number of the segment, which includes a certain pixel, it is necessary to calculate a minimum of this function.

$$\text{Distance}\,(c) = \sqrt{\left(I_q'(i, j) - l_c\right)^2} \tag{9}$$

where Distance (c) is a belonging degree of I_q' image pixel to l_c centers, $I_q' \in A'$; l_c is a c-element value of L array, $c = 1, 2, \ldots, C$; $I_q'(i, j)$ is the pixel intensity of q-image, and $q = 1, 2, \ldots, Q$; (i, j) are the pixel coordinates, $i = 1, 2, \ldots, M$, $j = 1, 2, \ldots, N$.

Thus, the I_q' image pixel with (i, j) coordinates belongs to p-segment if $\min(\text{Distance}) = \text{Distance}(p)$, $p \in [1; C]$.

The values redistribution of A' array image pixels according to the segment numbers can be represented by the array of index images $A_s = \{S_1, S_2, \ldots, S_q, \ldots, S_Q\}$. The A_s array elements are described by 2D matrices $S_q(i, j) = c$ where c is the segment number to which I_q' image pixel with coordinates (i, j) belongs; $c \in [1; C]$.

The color palette is used for visual estimation of the obtained segments. The pixels values of S_q images are the palette indexes (10) and the palette size equals C in (7).

$$\text{map}(c) = (r_c, g_c, b_c) \tag{10}$$

where (r_c, g_c, b_c) is a color vector of RGB model, $r_c, g_c, b_c \in [0; 255]$.

3.5 Contours Detection and Analysis

The S_q image can be represented by the array of bilevel images $W = \{B_1, B_2, \ldots, B_c, \ldots, B_C\}$ whose elements are determined by (11).

$$B_c(i, j) = \begin{cases} 1, & S_q(i, j) = c \\ 0, & S_q(i, j) \neq c \end{cases} \tag{11}$$

where $B_c(i, j)$ is the pixel intensity of c-segment image, $c = 1, 2, \ldots, C, B_c \in W$; $S_q(i, j)$ is the pixel value of an index image, and $q = 1, 2, \ldots, Q, S_q \in A_s$; (i, j) are the pixel coordinates, $i = 1, 2, \ldots, M, j = 1, 2, \ldots, N$.

The B_c images represent background, objects, and their contours of I_q' image. To detect the contours of all obtained segments, the difference between B_c image and the result of applying the erosion is defined. The erosion procedure is determined by (12) for the structural element and an image fragment in Fig. 1.

$$E_c(i, j) = \begin{cases} 1, & \sum_{a=-1}^{1} \sum_{b=-1}^{1} B_c(i + a, j + b) = 9 \\ 0, & \text{otherwise} \end{cases} \tag{12}$$

(a)

$B_c(i-1, j-1)$	$B_c(i-1, j)$	$B_c(i-1, j+1)$
$B_c(i, j-1)$	$B_c(i, j)$	$B_c(i, j+1)$
$B_c(i+1, j-1)$	$B_c(i+1, j)$	$B_c(i+1, j+1)$

(b)

1	1	1
1	1	1
1	1	1

Fig. 1 An example of erosion procedure data: **a** image fragment and **b** structural element

where $E_c(i, j)$ is the pixel intensity of a resulted image for c-segment, $c = 1, 2, \ldots, C$.

The contours detection is performed according to (13):

$$K_c(i, j) = B_c(i, j) - E_c(i, j) = \begin{cases} 0, & B_c(i, j) = E_c(i, j) \\ B_c(i, j), & \text{otherwise} \end{cases} \tag{13}$$

where $K_c(i, j)$ is the pixel intensity of contour points image for c-segment, $c = 1, 2, \ldots, C$.

4 Practical Implementation

The MedPix medical images database [28] was used for testing the proposed method. This database is organized according to several parameters: disease location, pathology category, image classification, patient profile, etc. Thus, the test data satisfied the requirement on images homogeneity because each image array contains data of one type for one patient. The minimum quantity of such test data set is six images; the maximum quantity is 460 images.

An example of an image array and its segmentation result is shown in Fig. 2. The input array of CT images contains ten images, its mathematical expectation is 0.2949, the dispersion is 0.1550, and clusters quantity is 9. The segmentation result is the index images array (Fig. 2b).

The proposed segmentation method has been applied to different types of test images: CT, MRI, ultrasound, and histological images. The maximal segments quantity has reached 244. This is due to low contrast of input images (MRI and ultrasound images) and is expressed by non-zero difference of numbers order between values

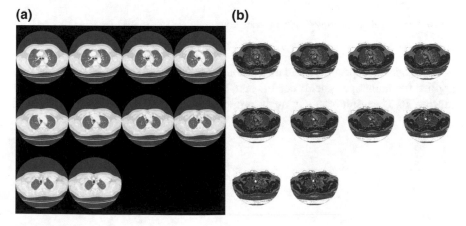

Fig. 2 Test image array: **a** input images and **b** segmentation result

Fig. 3 Segmentation result

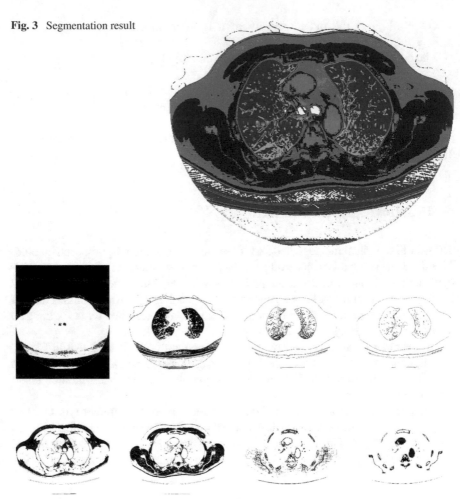

Fig. 4 Segmented image array (inverse visualization)

of mathematical expectation and dispersion. In such cases, the weighting factors are applied for defining segments centers and, thus, the maximal segments quantity has reached 26. The results of applying the proposed method are shown in Figs. 3, 4 and 5.

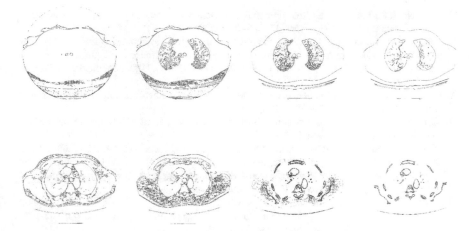

Fig. 5 Image segment contours array (inverse visualization)

5 Conclusion

The proposed segmentation method is automatic. The determination of image segments centers is performed by computing several statistical parameters of homogeneous images set. The formation of image segments is independent of location and spatial connections of image pixels.

The segmentation method has been applied to CT, MRI, ultrasound, and histological images. The result of the segmentation method is both image components (segments) and their contours. The proposed method is flexible in usage because the results can be presented as separate images. For one image, the average computation time is 5.8 s., and resource usage intensity is 14.5%. Thus, further research can be focused on parallel realization of the proposed method what can increase the overall efficiency of image analysis.

References

1. Fujitaa, H., Uchiyamaa, Y., Nakagawaa, T., Fukuoka, D., et al.: Computer-aided diagnosis: the emerging of three CAD systems induced by Japanese health care needs. Comput. Methods Programs Biomed. **92**, 238–248 (2008)
2. Doi, K.: Computer-aided diagnosis in medical imaging: historical review, current status and future potential. Comput. Med. Imaging Graph. **31**, 198–211 (2007). https://doi.org/10.1016/j.compmedimag.2007.02.002
3. Armenakis, C., Savopol, F.: Image processing and GIS tools for feature and change extraction. Int. Arch. Photogramm. Remote Sens. Spat. Inf. Sci. **35**, 611–616 (2004)
4. Klimesova, D., Oselikova, E.: GIS and image processing. Int. J. Math. Models Methods App. Sci. **5**, 915–922 (2011)

5. Micheloni, C., Foresti, G.L.: Real-time image processing for active monitoring of wide areas. J. Vis. Commun. Image Represent. **17**, 589–604 (2006)
6. Oh, S., Park, S., Lee, C.: A platform surveillance monitoring system using image processing for passenger safety in railway station. In: International Conference on Control, Automation and Systems (ICCAS), pp. 394–398. IEEE Press, Seoul (2007). https://doi.org/10.1109/iccas.2007.4406975
7. Lee, J.G., Jun, S., Cho, Y.W., Lee, H., et al.: Deep learning in medical imaging: general overview. Korean J. Radiol. **18**(1), 570–584 (2017)
8. Luculescu, M.C., Lache, S.: Computer-aided diagnosis system for retinal diseases in medical imaging. WSEAS Trans. Syst. **7**, 264–276 (2008)
9. Mironov, R., Kountchev, R.K.: Architecture for medical image processing. In: Kountchev, R.K., Iantovics, B. (eds.) Advances in Intelligent Analysis of Medical Data and Decision Support Systems Data and Decision Support Systems. SCI, vol. 473, pp. 225–234. Springer, Heidelberg (2013). https://doi.org/10.1007/978-3-319-00029-9_20
10. Dinevski, D., Bele, U., Sarenac, T., Rajkovic, U., Sustersic, O.: Clinical decision support systems. In: Graschew, G. (ed.) Telemedicine Techniques and Applications, pp. 185–210. InTech, Rijeka (2011). https://doi.org/10.5772/25399
11. Mehena, J., Adhikary, M.C.: Medical image edge detection based on soft computing approach. Int. J. Innov. Res. Comput. Commun. Eng. (IJIRCCE) **3**, 6801–6807 (2015). https://doi.org/10.15680/ijircce.2015.0307033
12. Saif, J.A.M., Hammad, M.H., Alqubati, I.A.A.: Gradient based image edge detection. Int. J. Eng. Technol. **8**, 153–156 (2016)
13. Aja-Fernandez, S., Vegas-Sanchez-Ferrero, G., Martin Fernandez, M.A.: Soft thresholding for medical image segmentation. In: Annual International Conference of the IEEE Engineering in Medicine and Biology Society, pp. 4752–4755. IEEE Press, Buenos Aires (2010). https://doi.org/10.1109/iembs.2010.5626376
14. Zhao, Y.-Q., Gui, W.-H., Chen, Z.-C., Tang, J.-T., Li, L.-Y.: Medical images edge detection based on mathematical morphology. In: 27th Annual International Conference of the IEEE Engineering in Medicine and Biology Society, pp. 6492–6495. IEEE Press, Shanghai (2005). https://doi.org/10.1109/iembs.2005.1615986
15. Gui, L., Lisowski, R., Faundez, T., Huppi, P.S., Lazeyras, F., Kocher, M.: Morphology-driven automatic segmentation of MR images of the neonatal brain. Med. Image Anal. **16**, 1565–1579 (2012). https://doi.org/10.1016/j.media.2012.07.006
16. Ludwiczak, A., Slosarz, P., Lisiak, D., Przybylak, A., et all: Different methods of image segmentation in the process of meat marbling evaluation. In: 7th International Conference on Digital Image Processing. SPIE Press, Los Angeles (2015). https://doi.org/10.1117/12.2197071
17. Laishram, R., Singh, W.K.K., Kumar, N.A., Robindro, K., Jimriff, S.: MRI brain edge detection using GAFCM segmentation and canny algorithm. Int. J. Adv. Electron. Eng. **2**, 168–171 (2012)
18. Norouzi, A., Rahim, M., Altameem, A., Saba, T., et all: Medical image segmentation methods, algorithms, and applications. IETE Techn. Rev. **31**, 199–213 (2014). https://doi.org/10.1080/02564602.2014.906861
19. Costin, H., Rotariu, C.: Medical image processing by using soft computing methods and information fusion. In: 11th WSEAS International Conference on Wavelet Analysis and Multirate Systems: Recent Researches in Computational Techniques, Non-Linear Systems and Control, pp. 182–191. WSEAS Press, Iasi (2011)
20. Bouchet, A., Pastore, J., Ballarin, V.: Segmentation of medical images using fuzzy mathematical morphology. J. Comput. Sci. Technol. **7**, 256–262 (2007)
21. Mozaffari, M.H., Lee, W.: Multilevel thresholding segmentation of T2 weighted brain MRI images using convergent heterogeneous particle swarm optimization. https://arxiv.org/pdf/1605.04806.pdf
22. G. Anna Lakshmi, S. Ravi: A double layered segmentation algorithm for cervical cell images based on GHFCM and ABC. Int. J. Image Graph. Signal Process. (IJIGSP) **9**(11), 39–47 (2017). https://doi.org/10.5815/ijigsp.2017.11.05

23. Isah, R.O., Usman, A.D., Tekanyi, A.M.S: Medical image segmentation through bat-active contour algorithm. Int. J. Intell. Syst. Appl. (IJISA) **9**(1), 30–36 (2017). https://doi.org/10.5815/ijisa.2017.01.03

24. Alyahya, A.Q., Abu-Shareha, A.A.: Accuracy evaluation of brain tumor detection using entropy-based image thresholding. Int. J. Inf. Technol. Comput. Sci. (IJITCS) **10**(3), 9–17 (2018). https://doi.org/10.5815/ijitcs.2018.03.02

25. Mwambela, A.J.: Comparative performance evaluation of entropic thresholding algorithms based on Shannon, Renyi and Tsallis entropy definitions for electrical capacitance tomography measurement systems. Int. J. Intell. Syst. Appl. (IJISA) **10**(4), 41–49 (2018). https://doi.org/10.5815/ijisa.2018.04.05

26. Gourav, Sharma, T., Singh, H.: Computational approach to image segmentation analysis. Int. J. Mod. Educ. Comput. Sci. (IJMECS) **9**(7), 30–37 (2017). https://doi.org/10.5815/ijmecs.2017.07.04

27. Pascale, D.: A comparison of four multimedia RGB spaces. http://www.babelcolor.com/index_htm_files/A%20comparison%20of%20four%20multimedia%20RGB%20spaces.pdf

28. MedPix. https://medpix.nlm.nih.gov/home

Steganographic Protection Method Based on Huffman Tree

Yevgen Radchenko, Ivan Dychka, Yevgeniya Sulema,
Viktoriya Suschuk-Sliusarenko and Oksana Shkurat

Abstract This paper presents an advanced method of graphical data protection. The novelty of the proposed method consists in the use of bits correspondence scheme, Huffman tree, and specific data encoding based on variable length of color hue code. The steganographic stability evaluation of the proposed method is presented and discussed in the paper. The application of the method to the technology of archive medical image processing is discussed as well.

Keywords Steganography · Data protection · Huffman tree

1 Introduction

The task of the information protection never loses its relevance. Nowadays, a significant amount of graphical data is being transmitted over the Internet. The majority of this data represents certain personal information which users prefer to be confident. A significant part of such data is of multimedia nature (images, video streams, audio records). In particular, there is huge amount of graphical data, which need to be protected. A special category of images is medical images (MRI, CT, X-ray

Y. Radchenko · I. Dychka · Y. Sulema (✉) · V. Suschuk-Sliusarenko · O. Shkurat
Igor Sikorsky Kyiv Polytechnic Institute, Peremogy 37, Kiev 03056, Ukraine
e-mail: sulema@pzks.fpm.kpi.ua

Y. Radchenko
e-mail: radchenko.zh@gmail.com

I. Dychka
e-mail: dychka@pzks.fpm.kpi.ua

V. Suschuk-Sliusarenko
e-mail: viss@pzks.fpm.kpi.ua

O. Shkurat
e-mail: shkurat.ksusha@gmail.com

© Springer Nature Switzerland AG 2020
Z. Hu et al. (eds.), *Advances in Artificial Systems for Medicine and Education II*,
Advances in Intelligent Systems and Computing 902,
https://doi.org/10.1007/978-3-030-12082-5_26

images, ultrasound, histological images, etc.). In most cases, patients prefer to hide their health problems and, thus, medical images require proper protection.

There are two main approaches to solving the data security problem: cryptography and steganography [1]. Cryptography secures confident data by encrypting it. At the same time, steganography protects confident data by hiding it in other data, which is open, in order to hide the fact itself of this confident data existence. Steganographic approach is optimal for medical images because it allows patients to hide the fact of illness itself. This approach can be useful for eHealth [2–4] and mHealth [5] applications.

Disadvantage of the classical steganographic technique is that if the fact of hidden data presence in certain open data is disclosed the secret data can be easily extracted. That is why the development of new advanced methods of steganographic protection is a topical task. Such new methods should be reliable and, at the same time, easy-to-use for daily protection of personal graphical data of end users.

One of the new approaches to steganographic data protection developed by the authors of this paper is based on the bits correspondence scheme [6, 7]. The objective of the research presented in this paper is to improve the steganographic approach based on bits correspondence scheme by applying an additional level of data protection—the data encoding based on variable length of color hue code. This new level allows us to arise the steganographic stability of the data protection procedure. We consider the method proposed in this paper within the context of medical images protection.

2 Related Work

The essence of the classical least significant bits (LSBs) method is to replace the least significant bits in the cover (graphical data, audio, and video) [1, 8] by the bits of the secret data. The difference between filled and empty covers should not be noticeable for human perception [9]. However, this method is quite weak if the fact of secret information presence is disclosed. That is why there is a necessity in more complex algorithms of data protection.

A modified pixel-value differencing image steganographic scheme with LSBs substitution method is presented in [10]. The authors propose to divide the cover image into the blocks of two consecutive pixels and calculate the absolute difference between the pixels of a block. If the difference is less than a particular threshold, then four bits of secret data are taken, and these bits are embedded onto the LSBs of the block pixels through least-significant-bit substitution method, otherwise the number of bits to be hidden is selected based on some characteristics of the block and hidden.

The authors, in [11], considered the particle swarm optimization algorithm to be applied to the spatial domain steganographic technique. The authors assert that the improved algorithm called the accelerated particle swarm optimization converges faster than the usual particle swarm optimization.

In [12], the authors proposed to use statistical image classification for image in steganographic techniques. In particular, two-level unsupervised image classification algorithm based on statistical characteristics of the image is developed. It helps a sender to make reasonable selection of cover image to enhance performance of steganographic method based on his specific purpose.

The image protection method based on palette encryption [13] uses replacement of color hues of secret image pixels by coordinates of the pixels of corresponding color hues in the key image (palette) where all color hues of the secret image are presented.

In [6], it is proposed to use a combined crypto-steganographic method based on a bit-value transformation according to both a certain Boolean function and a specific scheme of correspondence between most significant bits (MSBs) and LSBs. The scheme of correspondence is considered as a secret key.

A special case of information hiding is the use of watermarking. A digital watermark is a special label placed on digitalized image which remains invisible, but it can be recognized by specialized software. Existing methods for solving the problem of copyright protection can be divided into two groups: methods which hide information in the spatial domain of an image and methods which embed the watermark in the frequency domain [14].

3   Method Description

The proposed method is an advancement of the steganographic protection method based on bits correspondence scheme [6], which uses bit values transformation according to both a certain Boolean function (e.g., ternary exclusive disjunction) and a specific scheme of correspondence between MSBs and LSBs. The scheme of bits correspondence is considered as a secret key. This scheme (Fig. 1) sets relation between LSBs and MSBs of a bit sequence which represents color hue of an image pixel. Such pair "LSB–MSB" is used as the first and the second operands of the ternary operation for calculating a new value of the LSB to be used for the substitution of this LSB according to data hiding principle in LSB-steganography. The third operand is the secret graphical data bit:

$$x = a \oplus b \oplus c \tag{1}$$

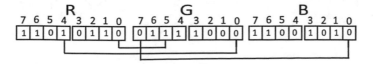

Fig. 1 Example of bits correspondence scheme

where

a   is a MSB of the cover image graphical data sequence,
b   is a LSB of the cover image graphical data sequence,
c   is a bit of secret image graphical data sequence,
x   is a new value of the LSB of the cover image graphical data sequence.

The difference of the method we propose in this paper and the method based on bits correspondence scheme is the following. In the method based on bits correspondence scheme, a secret image is converted into a bit sequence by placing bits of R, G, and B components [15, 16] of a pixel hue sequentially (pixel by pixel). At the same time, the secret image in the proposed method is converted into an array of bits using another image (key). This key can be used for multiple image transfers. The only requirement for the key is that the key must contain all possible values of a color component [0; 255]. If at least one integer value between 0 and 255 is missing, the image cannot be used as a key. The selection of the key is the first stage of the method.

At the second stage, we form a "dictionary of colors" which is a table of conversion between a color hue value and its binary code. To construct this dictionary, we use Huffman algorithm [17] which is based on the statistical features (frequencies of occurrence) of a data set and is resulted in the construction of Huffman tree. An example of the Huffman tree for values [251; 255] (i.e., a subset of the full-color values set [0; 255]) is depicted in Fig. 2 where values in circles are frequencies of occurrence of the color values from 251 to 255 in a certain picture chosen as a key. The dictionary for this binary tree is given in Table 1.

At the third stage, we convert a secret image into an array of bits using the created dictionary. As a result, we obtain a bit sequence ready for embedding it into a cover image according to the steganographic principle of data hiding.

Fig. 2 Example of a binary Huffman tree for the color values subset [251; 255]

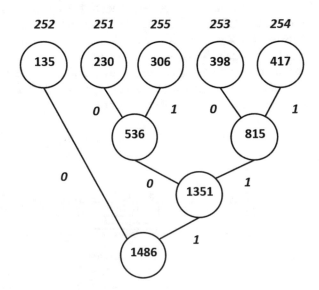

Table 1 Huffman's dictionary for five possible values of each component

Color value	Binary code
251	100
252	0
253	110
254	111
255	101

Since the cover image must exceed the secret image in several times, it is necessary to determine whether the size of the cover image is sufficient to integrate the secret data:

$$secret\_bit\_arr\_length = \sum_{0}^{255} cc \times bc \qquad (2)$$

where cc (component count) is a number of repetitions of this value (a value belongs to the interval [0; 255]); bc (bits count) is a number of bits in the dictionary, which represents the given value.

The procedure for the steganographic embedding of a secret image graphical data includes the following steps:

1. An image to be used as a key is randomly selected from an open database or chosen by a user. Let this image has the size 256 × 256 pixels or more.
2. The key image is checked for the presence of all values in the interval [0; 255]. If all values are present (i.e., the criterion called "available bytes completeness" [13] is satisfied), the image may act as a key; otherwise, another image must be selected.
3. The frequencies of color values occurrence in the key image are calculated and the array of color values [0; 255] is sorted accordingly (Fig. 2 shows an example of the color values array ascending sorting by the color values frequencies).
4. Based on the sorted array, a Huffman tree is constructed as well as the Huffman dictionary table is formed. In this table, each value [0; 255] will correspond to a certain binary code. The length of the code depends on the number of repetitions of this value. The higher occurrence frequency of the value is, the smaller code length is. It enables certain compression of embedded graphic data.
5. The secret image graphical data is converted into a bit array according to the Huffman dictionary table (e.g., Table 2).
6. Both the Boolean function and the bits correspondence scheme are selected. They serve as a compound private key. The Boolean function must be reversible operation to guarantee the recovery of encrypted graphical data. An example of such Boolean function is exclusive disjunction.

Table 2 Results of the methods comparison

Parameter	LSB	Method based on bits correspondence scheme	Proposed method
Resistance to breaking	$\frac{1}{4}$	$\frac{1}{24(3n)(24-3n)^3}$	$\frac{1}{24(3n)(24-3n)^3 256!}$
Minimal cover size (pixels count)	8·SIS	8·SIS	[4·SIS; 13·SIS]
Redundancy	8	8	[4; 13]
Complexity of the algorithm	Constant	Cubic	Factorial

SIS is the Secret Image Size (a number of pixels in the secret image)

7. The secret image is embedded into the cover image, which is an open image either randomly selected from an open database or chosen by a user, according to both the image key and the compound private key.

The size of the output data depends on a number of identical components in the secret image as well as on a size of the compound private key. For the correct decoding of graphic data, it is necessary to know the resolution of the hidden image. In this method, information about the resolution of the hidden image is embedded into the first (the width of the hidden image) and the last (the height of the hidden image) several pixels of the cover image. To calculate the exact quantity of these pixels, the following formula can be used:

$$maxHiddenPixels = imgOriginal.Width \times imgOriginal.Height/8$$
$$= 256 \times 256/8 = 8192 \qquad (3)$$

The following formula allows us to calculate the number of digits for recording the maximum resolution of the hidden image:

$$capacity = log_2(maxHiddenPixels) + 1 = 14 \qquad (4)$$

where $log_2(maxHiddenPixels)$ rounds the value down; $capacity$ is the number of bits necessary to hide height/width of graphic data.

4   Method Evaluation

In order to evaluate the steganographic stability of the proposed method, we compared it with the classical LSB method [1, 8, 18] and the method based on the bits correspondence scheme [6] according to the following approach.

Let event F consists in decryption of the secret image at the first attempt, and let this event occurs under condition that the following information is true [19]:

- The container includes the hidden data
- The size of the secret graphic data is known
- The secret image graphical data is embedded into the cover image in a consistent manner
- The method for embedding the secret image graphical data is known but the parameters of this method (keys) are unknown.

Let us analyze the steganographic stability of the classical LSB method which supposes that bits of the secret image are embedded into LSBs of the cover image without any additional transformations.

Since the classical LSB method enables the replacement of the last four LSBs and the choice of the number of bits that are changing, it is possible to decrypt the secret image from the first attempt:

$$P(F) = \frac{1}{4}. \tag{4}$$

Let us analyze the steganographic stability of the method based on bits correspondence scheme [6].

Let event H_1 consists in the fact that the Boolean function is chosen. There are six ways to form this Boolean function:

$$P_{H_1}(F) = \frac{1}{4} \times \frac{1}{C_4^2} = \frac{1}{4} \times \frac{1}{6} = \frac{1}{24}. \tag{5}$$

Probability of event F under condition that the correspondence scheme is selected (event H_2) can be estimated as follows:

$$P_{H_1 H_2}(F) = \frac{1}{24(3n)(24 - 3n)^3} \tag{6}$$

where $n = \{1, 2, 3, 4\}$ is a number of LSBs to be changed.

Let us analyze the steganographic stability of the proposed method.

Let event H_3 be the application a Huffman dictionary (unknown). Then the probability of event F depends on the number of possible permutations of 256 components in the dictionary:

$$P_{H_1 H_2 H_3}(F) = \frac{1}{24(3n)(24 - 3n)^3 256!}. \tag{7}$$

The results of the comparison of these three methods are shown in Table 2. As one can see, the proposed method under the same conditions shows higher steganographic stability than other two methods.

5 Method Application for Medical Image Protection

Medical image protection is an integral part of data processing and handling in any medical information system. In this section, we consider the technology of medical graphical data processing and analysis (Fig. 3) based on the information system for archival medical images automated processing, which was earlier presented by the authors [20]. It can be used for design and development of both a complex medical information system and its separate integral parts (modules).

The technological chain of the operations applied to a medical image depend of the image type.

If the image is non-digitized (old medical archive image), its basic processing includes the following stages: digitization; preprocessing (quality enhancement); classification and documentation; data protection; digital archiving.

If the image is obtained from medical imaging device in a digitalized form, its basic processing includes classification and documentation; data protection; digital archiving.

Thus, the main idea of the proposed informational system is to integrate data protection in the medical image processing procedure at the basic level of the technology in order to guarantee that only protected graphical data is circulating in the complex medical information system. It means that all medical images and their metadata are

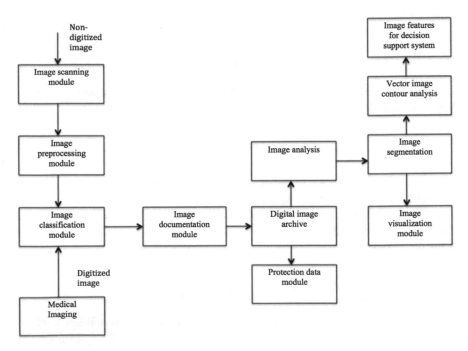

Fig. 3 General scheme of proposed medical archive information system

stored in the medical archive (database) in a steganographically protected form. In case of unauthorized access to this medical database, which can be distributed one, a violator will have access only to cover images. They should be also of medical content to eliminate any suspicion about their value.

The proposed medical archive information system can also include modules for the following procedures: image segmentation, vectorization, and analysis, including feature detection. These procedures are usually applied to the original image. However, the further research can be also focused on processing of hidden data without their evident extraction.

6 Conclusions

The method of graphical data steganographic protection presented in this paper includes an additional level of protection. This level is guaranteed by the procedure of secret data encoding based on variable length of color hue code. Such variable length of color hue code can be obtained from Huffman tree and the dictionary based on this binary tree. The result of data encoding is a bit sequence which is ready for embedding into a cover image.

Another specificity of the proposed method is the use of bits correspondence scheme and a Boolean operation which compose a complex private key. Another key used in the method is a full-color image which is the basis for the Huffman tree construction.

The steganographic stability evaluation of the proposed method has shown that the method can be used for graphical image protection. One of the possible areas of its application is healthcare, in particular, eHealth and mHealth.

References

1. Cummins, J., Diskin, P., Lau, S., Parlett, R.: Steganography and Digital Watermarking. School of Computer Science, The University of Birmingham, London (2004)
2. French-Baidoo, R., Asamoah, D., Oppong, S.O.: Achieving confidentiality in electronic health records using cloud systems. Int. J. Comput. Netw. Inform. Secur. (IJCNIS) **10**(1), 18–25 (2018). https://doi.org/10.5815/ijcnis.2018.01.03
3. Alfonse, M., Aref, M.M., Salem, A.-B.M.: An ontology-based system for cancer diseases knowledge management. Int. J. Inf. Eng. Electron. Bus. (IJIEEB) **6**(6), 55–63 (2014). https://doi.org/10.5815/ijieeb.2014.06.07
4. Adebayo, K.J., Ofoegbu, E.O.: Issues on E-health adoption in Nigeria. Int. J. Mod. Educ. Comput. Sci. (IJMECS) **6**(9), 36–46 (2014). https://doi.org/10.5815/ijmecs.2014.09.06
5. Mwammenywa, I.A., Kaijage, S.F.: Towards enhancing access of HIV/AIDS healthcare information in Tanzania: is a mobile application platform a way forward? Int. J. Inf. Technol. Comput. Sci. (IJITCS) **10**(7), 31–38 (2018). https://doi.org/10.5815/ijitcs.2018.07.04
6. Hu, Z., Dychka, I., Sulema, Y., Radchenko, Y.: Graphical data steganographic protection method based on bits correspondence scheme. Int. J. Intell. Syst. Appl. (IJISA) **9**(8), 34–40 (2017). https://doi.org/10.5815/ijisa.2017.08.04

7. Radchenko, Y., Sulema, Y.: Graphical data steganographic protection method based on bits correspondence scheme and cover visual properties analysis. In: The 6th International Conference on Methods and Means of Information Coding, Protection and Compression, pp. 51–53. Ukraine (2017)

8. Morkel, T., Eloff, J.H.P., Olivier, M.S.: An overview of image steganography. In: The Fifth Annual Information Security South Africa Conference (ISSA2005). Sandton, South Africa (2005)

9. Dumitrescu, S., Wu, X., Wang, Z.: Detection of LSB steganography via sample pair analysis. IEEE Trans. Signal Process. **51**, 1995–2007 (2003). https://doi.org/10.1109/tsp.2003.812753

10. Malik, A., Sikka, G., Verma, H.K.: A modified pixel-value differencing image steganographic scheme with least significant bit substitution method. Int. J. Image Graph. Signal Process. (IJIGSP) **7**(4), 68–74 (2015). https://doi.org/10.5815/ijigsp.2015.04.08

11. Divya, E., Kumar, P.R.: Steganographic data hiding using modified APSO. Int. J. Intell. Syst. Appl. (IJISA) **8**(7), 37–45 (2016). https://doi.org/10.5815/ijisa.2016.07.04

12. Seyyedi, S.A., Ivanov, N.: Statistical image classification for image steganographic techniques. Int. J. Image Graph. Signal Process. (IJIGSP) **6**(8), 19–24 (2014). https://doi.org/10.5815/ijigsp.2014.08.03

13. Sulema, Y., Shyrochyn, S.: Image protection method based on palette encryption. Her. Khmelnytskyi Natl. Univ. **3**, 114–119 (2014)

14. Yaremchuk, Y., Karpinets, V.: The digital watermarks usage for copyright protection in the images. Legal Norm. Metrol. Provis. Inf. Prot. Syst. Ukr. **2**, 63–70 (2006)

15. Pascale, D.: A Review of RGB Color Spaces … from xyY to R'G'B'. BabelColor Company, Montreal, Canada (2003)

16. Ibraheem, N.A., Hasan, M.M., Khan, R.Z., Mishra, P.K.: Understanding color models: a review. ARPN J. Sci. Technol. **2**, 265–275 (2012)

17. Huffman, D.A.: A method for the construction of minimum-redundancy codes. Inst. Radio Eng. **40**, 1098–1101 (1952)

18. Adak, C.: Robust steganography using LSB-XOR and image sharing. In: International Conference on Computation and Communication Advancement (IC3A), pp. 97–102. West Bengal, India (2013)

19. Korolyov, V., Polinovskyy, V., Gerasymenko, V.: RS-Seoanalysis. Principles of work, disadvantages and the concept of the method bypass. Her. Vinnytsky Polytech. Inst. **6**, 66–71 (2010)

20. Sulema, Y., Shkurat, O.: Information system for archival medical images automated processing. In: The 3rd International Conference Health Technology Management (HTM-2016), p. 72. Chisinau, Moldova (2016)

The Role of Hybrid Classifiers in Problems of Chest Roentgenogram Classification

Rimma Tomakova, Sergey Filist, Roman Veynberg, Alexey Brezhnev and Alexandra Brezhneva

Abstract The hybrid technology of chest roentgenogram classification, based on three-level hierarchic structure, has been suggested. "Weak classifiers," based on two ways of data analysis, are formed on the first level. The approach is to make a "weak" classifier using the first way which is based on the analysis of Fourier amplitude spectra in sliding window. An X-ray image is sequentially scanned by windows of various scales. A "weak" classifier is made based on Fourier amplitude spectrum, defined in each window. It refers to an image segment, which got into the sliding window, to a certain class. The second way of making a "weak" classifier is based on the descriptors, which were received because of intensity histogram approximation in the analysis window. The number of received "weak classifiers," based on the two ways of analysis, depends on the number of analysis window scales chosen. On the second hierarchic level, the solutions of "weak" classifiers within every way of the first hierarchic-level analysis are integrated. A final classifier makes the ultimate solution, which aggregates the solutions of two second hierarchical level classifiers.

Keywords Chest roentgenogram · Classifier · Analysis window · Intensity histogram · Fourier spectrum · Solution aggregator

R. Tomakova · S. Filist
Southwest State University, 94, 50th Anniversary of October St., 305040 Kursk, Russian Federation
e-mail: rtomakova@mail.ru

S. Filist
e-mail: SFilist@gmail.com

R. Veynberg (✉) · A. Brezhnev · A. Brezhneva
Plekhanov Russian University of Economics, 36, Stremyanny lane, 36, 117997 Moscow, Russian Federation
e-mail: veynberg.rr@rea.ru

A. Brezhnev
e-mail: brezhnev.av@rea.ru

A. Brezhneva
e-mail: a.brezhneva@hotmail.com

© Springer Nature Switzerland AG 2020
Z. Hu et al. (eds.), *Advances in Artificial Systems for Medicine and Education II*,
Advances in Intelligent Systems and Computing 902,
https://doi.org/10.1007/978-3-030-12082-5_27

1 Introduction

Processing of medical digital images has been a subject of close attention for researchers for the latest decades. The themes of many scientific works are connected with mathematical and program methods used in this sphere and hardware tool development. The researchers' interest in modern facilities of medical roentgenogram processing is determined by the increased requests to the quality and reliability of newly developed diagnostic systems [1–5].

The goal of the study is to develop a tool environment designed to automatically classify of X-ray images based on the Fourier descriptors. In accordance with the goal to realize the capabilities of the software, it was necessary to solve the following tasks: to develop algorithmic and mathematical support for the hybrid neural network structures and to identify software modules implementing spectral methods of the image processing.

2 Methodology and Experiment

The new software, enabling to receive a set of descriptors, which are meant for making image classifiers, has been created at the biomedical engineering department of SWSU in order to study digital biomedical images [6–8]. The original interface (the main window) of this software, which was developed in MATHLAB environment, is presented in Fig. 1.

Among various possibilities of this software, it is necessary to single out the programming modules, which realize the spectral methods of image processing. The purpose of the development of these modules is to receive an instrumental environment for classifying images on basis of Fourier descriptors. Fourier descriptors are formed from modules of window two-dimensional Fourier transformation counting on basis of the proposed concept of making chest roentgenogram classifier. According to this conception, a chest roentgenogram is scanned by windows of variable size. A suitable classifier, which refers to "weak classifiers," is made for the size of every window. Descriptors, defined by intensity histogram of Fourier amplitude two-dimensional spectrum, are used for making a "weak" classifier [9–12]. The two segments of chest roentgenogram, received in the windows of the same size, are represented in Fig. 2 as an example. The corresponding Fourier amplitude spectra are shown on the right.

In the course of the research of the chest roentgenogram window spectra, there were discovered the evolutions of Fourier amplitude spectra depending on the existence or absence of pathology. These evolutions can be seen with the help of "weak" classifiers made on basis of trainable neural networks. Image intensity histograms of Fourier window amplitude spectrum were used as informative signs. The examples of such histograms are shown in Fig. 3.

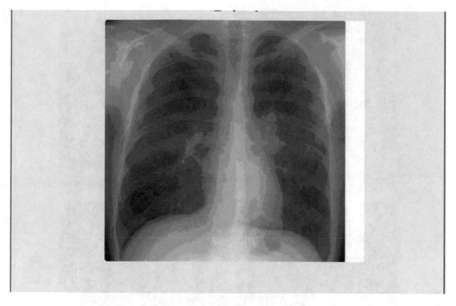

Fig. 1 Original interface of the software for processing biomedical images in MATHLAB environment

Intensity histograms of corresponding original window images are shown on the left for the comparison.

The analysis of the images, in which examples are represented in Figs. 2 and 3, has demonstrated that the form of window spectra intensity histogram has showed higher stability in comparison to corresponding forms, received from original images. This enables to generate descriptors on its basis using the method, suggested in [2, 3].

The second "weak" classifier is based on the descriptors received as a result of histogram approximation in the analysis window. For understanding this method of making a "weak" classifier, it needs to consider Fig. 4. The roentgenogram analysis window in the region of a lung without pathological changes is shown in Fig. 4a on the left. The intensity histogram in this window is shown in Fig. 4 on the right.

The analysis of analogous histograms of various roentgenograms has shown that histograms have a multimodal form on condition there are no pathological changes in the analysis window. The analysis window, simulating pneumonia, is shown in Fig. 4 on the left. Low-frequency filtration by means of Fourier two-dimensional transformation was used for pneumonia model in the photofluorogram analysis window of a healthy patient. The modality of the intensity histogram disappears with low-frequency filtration and the histogram gets a form close to a triangle [7, 8, 13, 14].

The analysis windows of the same photofluorogram, which has the segments, classified as pneumonia, are shown in Fig. 5.

The analysis if the represented images enables to make a conclusion that real pathological formations with pneumonia intensity histograms in analysis window

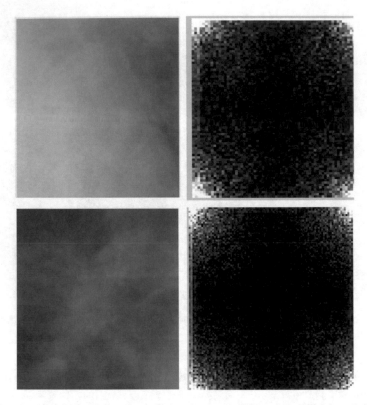

Fig. 2 Images of chest roentgenogram window segments (left) and their Fourier window two-dimensional transformations (right)

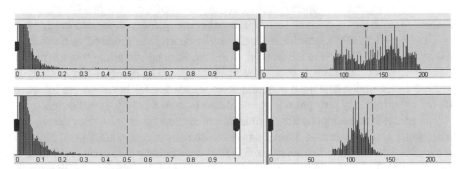

Fig. 3 Intensity histograms of chest roentgenogram window segments (right) and histograms of their Fourier window two-dimensional transformations (left)

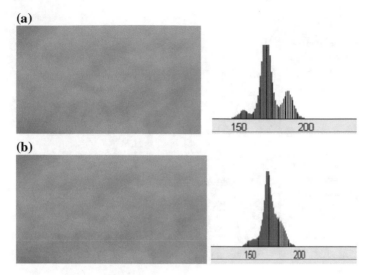

Fig. 4 Windows of analysis and roentgenogram intensity histograms: the window without pathology (**a**); the model of a window with pathology, which was received as a result of the window without pathology low-frequency filtration (**b**)

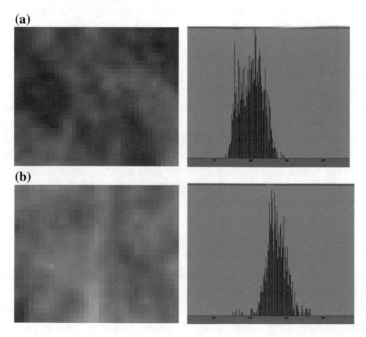

Fig. 5 Windows of analysis and roentgenogram intensity histograms: the window without pathology (**a**); the window with pneumonia (**b**)

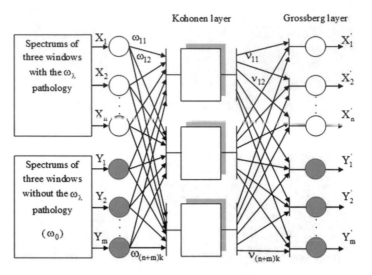

Fig. 6 Counter-propagation network architecture

get a triangle shape. This is the main reason for descriptors of the second level for forming as primitives, which approximate intensity histograms in the analysis window.

"Weak" classifiers are made for every size of the window. For integrating "weak" classifiers into "strong" ones, trainable neural networks of direct distribution are used [15–17]. The training is performed for every pathology and size of a window.

Following on from the suggested conception of building classifier of chest X-ray, analytical techniques for the purpose of discovering optimal values qualitative and quantitative measures were held. In the capacity of a target of the research on the purpose of image processing, we used chest X-rays of fifty patients of emergency care hospital of Kursk city which had pneumonia and which were healthy.

For developing a model of a morphological formation, a bilayer neural network of counter-propagation (CP) was implemented, which is shown in Fig. 6.

The neural network of CP consists of two layers: the Kohonen layer and the Grossberg layer of neurons. Vectors $\vec{X}$ and $\vec{Y}$ are input of the network. The first vector $\vec{X}$ contains spectral indexes of three windows of the morphological formation, which are relevant to a ω_ℓ disease. The second vector $\vec{Y}$ contains spectral indexes of X-ray segment of a patient, who does not have the ω_ℓ disease.

It should be noted that $\vec{X}$ and $\vec{Y}$ vectors dimensions coincide each other, as the same sizes of the image windows with the ω_ℓ disease and without were selected.

For scientific researches, a database was built to develop such networks; it contains examples of X-ray segments of morphological formations with the ω_ℓ disease and without.

For building the database of a learning sample, experts established clusters, which are relevant to that disease. Moreover, clusters can be distributed in the entire area of the X-ray or just in lung segments. The number of clusters is defined by experts or by the number of original inclusions of pathological mass on the X-ray, which are relevant to the ω_ℓ disease.

The neural network of CP functions in two modes: learning and using. In the first case, we input vectors $\vec{X}$ and $\vec{Y}$ and do correct weighting factor; in the second case, we input in already taught network $\vec{X}$ or $\vec{Y}$ and output $\vec{X}$ and $\vec{Y}$.

General method of the neural network's work: Do input the first image $\vec{Z}$. The image $\vec{Z}$ forms as a combination of vectors $\vec{X}$ and $\vec{Y}$:

$$\vec{Z} = \vec{X} \cup \vec{Y}, \tag{1}$$

In each neuron of the Kohonen layer, an activation is calculating

$$A_k = \sum_{i=1}^{n+m} w_{ik} z_i = W^\mathrm{T} Z, \tag{2}$$

Output of one of neurons of Kohonen's j layer equals unit, viz

$$\mathrm{OUT}_j^K = \begin{cases} 1, & \text{if } A_j = \max_K A_k, \\ 0, & \text{else.} \end{cases} \tag{3}$$

The next step is to correct weighting factors.

The Grossberg layer functions similar to the Kohonen layer. The activation of the neuron g and its layer is given as follows:

$$B_g = K \sum_{k=1} v_{kg} \mathrm{OUT}_k^K, \tag{4}$$

There are weights on the output of the Grossberg layer's neurons, which are relevant to "winning" neuron of the Kohonen layer $\mathrm{OUT}_j^G = B_j$. A result of looping of the procedure is that weighting factors of the Grossberg layer might be sufficiently close to input images or to coincide with them.

Scheme of the counter-propagation network includes the following steps.

Step 1. Do normalize each rate; for example, normalized analogue of the second element of the first row can be calculated using the following formula

$$z_2^1 = \frac{x_2^1}{\sqrt{\sum_{j=1}^n \left(x_j^1\right)^2 + \sum_{j=1}^m \left(y_j^1\right)^2}}, \tag{5}$$

Step 2. Do randomly generate weighting factors, and do normalize them, which help to lighten the process of the learning.

Step 3. Do input on the network a row of the matrix Z, and do calculate a dot product of it and the vectors of the weight, which are connected with each Kohonen's neurons.

Step 4. Do choose the largest product and do adjust weights of the required neuron according to formula $W_h = W_G + \alpha(Z - W_G)$ where W_G is preceding value of the weight, W_h is its new value, and α is index of learning, which is average out of 0.7 at the beginning and is slowly decreased during the process of learning. The output of the "winning" neuron equals unit and the others zero.

Step 5. Do correct all weights of the Grossberg layer according to the formula:

$$V_{ijh} = V_{ijG} + \beta \cdot \left(Z_j - V_{ijG}\right) \cdot \chi_i^K \tag{6}$$

where β is index of learning (equals 0.1 at the beginning), Z_j is jth constituent of the real vector of the output, V_{ijG} is preceding value of the weighting factor of the Grossberg layer, and V_{ijh} is a new value:

$$\chi_i^K = \begin{cases} 1, & \text{if } i\text{th neuron of the Kohonen layer "won",} \\ 0, & \text{else.} \end{cases} \tag{7}$$

Step 6. Do decrease the α and β values.

Step 7. Do repeat steps 3–7 until each input couple would generate equivalent couple on the output.

3   Results and Discussions

During the work, a valuation of a classification quality of the morphological formations on images of the chest X-rays was obtained. For this purpose, an index of the classification quality verification was formed, which is based on a ratio of a number of incorrectly classified pixels of input ω_ℓ to a total number of pixels of that class on the master image:

$$M_{\omega_\ell} = \frac{\sum_{i=1}^{n_{\omega_\ell}} C_{i\omega_\ell} - C_{i\omega_\ell}}{\sum_{i=1}^{n_{\omega_\ell}} C_{i\omega_\ell}} \times 100 \tag{8}$$

where n_{ω_ℓ} is the number of classes, C_{ω_ℓ} is the number of pixels correctly classified to ω_ℓ class, and $\sum_{i=1}^{n_{\omega_\ell}} C_{i\omega_\ell}$ is the number of pixels, which actually belong to ω_ℓ class.

Figure 7 shows diagrams, which reflect an allocation of that index on samples of the morphological formations of different classes, which were obtained from 12 X-rays.

Fig. 7 Results of M_{ω_ℓ} index, which were obtained from 12 chest X-rays

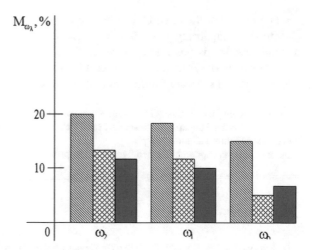

 is sample of active circuit; is sample of segmentation based on data preelaboration and spectrum analysis; is sample of segmentation based on data without preelaboration by multiwindow spectrum analysis.

Samples of the morphological formations, such as pneumonia (ω_1), and also oncology (ω_2), form on each of the films.

Thus, a comparative analysis of the quality indexes obtained by means of various X-ray segmentation methods has shown that the proposed method of multiwindow spectral transformation allows differential diagnosis of pneumonia and oncological morphological formations by M criterion below 15%.

4 Conclusion

Thus, a three-level classifier and software for analyzing and classifying X-ray images have been developed, allowing the analysis of amplitude Fourier spectra and brightness histograms in a sliding window. The X-ray image is successively scanned by sliding windows of different sizes. In each window, the amplitude Fourier spectrum and the histogram of the brightness of the original image are determined. On the basis of the analysis of the histograms of the amplitude Fourier spectrum and brightness histograms, descriptors are obtained in the window, which are used as a space of informative features for "weak" classifiers based on the trained neural networks. "Weak" classifiers, based on two methods of the analysis, receive as much as the scales of the analysis windows were selected. At the second hierarchical level, the solutions of the "weak" classifiers are combined within each method of analysis of the first hierarchical level. The final decision is made by the final classifier, which aggregates the decisions of the two classifiers of the second hierarchical level.

The results of the research can be used to build the intelligent decision support systems for diagnosing and predicting of socially significant diseases. In the future, further research is possible to develop similar systems for early and prenosological diagnosis of diseases, as well as related to the development of mathematical algorithms and software for the automated workplace of a radiologist.

Acknowledgements We thank Plekhanov Russian University of Economics, Southwest State University and Russian Fundamental Research Fund, who helped us in developing this research, believed in us, and supported all the way up.
The research was supported financially by RFBR, project No. 16-07-00164a.

References

1. Kudryavtsev, P., Kuzmin, A., Filist, S.: Developing the boosting technology for classification of the photofluorographies. Biomed. Radioeng. **9**, 10–14 (2016)
2. Tomakova, R., Emelyanov, S., Filist, S.: Intellectual Technologies of Segmentation and Classification of Biomedical Images, 222 p. South-West State University, Kursk (2012)
3. Filist, S., Dyudin, M.: Automatic classifiers of complex structured images based on multimethod technologies of multi-criteria choice. Questions of radio electronics. Series "Systems and means of information display and control of special equipment" (SOIU), No. 1, pp. 130–140 (2015)
4. Filist, S., Chernicov, K.: Miglioratadiagnosidelle prime fasi di sviluppo di tubercolosipolmonarebasatosusistemeautomatizzati di analisi espresso e classificazione bitmap photofluorogramm torace. Ital. Sci. Revier **2**(11), 54–57 (2014)
5. Babenko, A.: Neural codes for image retrieval. In: European Conference on Computer Vision. Springer International Publishing, pp. 584–599 (2014)
6. Tomakova, R., Filist, S., Pykhtin, A.: Comparative analysis of segmentation efficiency method the halftont image based on the selection of priority direction of machining segment boundaries. IJAER **11**(5), 3199–3206 (2016)
7. Tomakova, R., Filist, S., Pykhtin, A.: Intelligent medical decision support system based on internet-technology. In: International Multidisciplinary Scientific Geoconference SGEM 2016, Albena, 263–270
8. Candemir, S.: Lung segmentation in chest radiographs using anatomical atlases with nonrigid registration. IEEE Trans. Med. Imaging **33**(2), 577–590 (2014)
9. Umbaugh, S.: Digital Image Processing and Analysis: Human and Computer Vision Applications with CVIPtools, 2nd edn, 980 p. CRC Press, Boca Raton (2010)
10. Marr, D., Hildreth, E.: Theory of edge detection. Proc. R. Soc. **207**, 187–217 (1980). https://doi.org/10.1098/rspb.1980.0020
11. Chanda, B., Majumde, D.: Digital Image Processing and Analysis, 384 p. Prentice Hall of India, New Delhi (2004)
12. Kang, W.X., Yang, Q., Liang, R.: The comparative research on image segmentation algorithms. In: IEEE Conference on ETCS, 703–707 (2009)
13. Karch, P., Zolotova, I.: An experimental comparison of modern methods of segmentation. In: SAMI: 8th International Symposium on Applied Machine Intelligence and Informatics, Herlany, Slovakia, 247–252 (2010)
14. Singh, K., Singh, A.: A study of image segmentation algorithms for different types of images. Int. J. Comput. Sci. Issues **7**(5), 414–417 (2010)
15. Jyothirmayi, T., Srinivasa Rao, K., Srinivasa Rao, P., Satyanarayana, Ch.: Performance evaluation of image segmentation method based on doubly truncated generalized Laplace mixture

model and hierarchical clustering. Int. J. Image Graph. Signal Process. (IJIGSP) **9**(1), 41–49 (2017).https://doi.org/10.5815/ijigsp.2017.01.06

16. Bahy, R.M.: New automatic target recognition approach based on Hough transform and mutual information. Int. J. Image Graph. Signal Process. **10**(3), 18–24 (2018). https://doi.org/10.5815/ijigsp.2018.03.03

17. How to cite this paper: Hu, Z., Dychka, I., Sulema, Y., Valchuk, Y., Shkurat, O.: Method of medical images similarity estimation based on feature analysis. Int. J. Intell. Syst. Appl. (IJISA) **10**(5), 14–22 (2018). https://doi.org/10.5815/ijisa.2018.05.02

Patient-Specific Biomechanical Analysis in Computer Planning of Dentition Restoration with the Use of Dental Implants

I. N. Dashevskiy and D. A. Gribov

Abstract This study considers the technology of patient-specific computer planning of dentition restoration of edentulous mandible using dental implants. A model of the jaw and distribution of elastic modules by jaw's volume are reconstructed from a computer tomogram. The model is supplemented with virtual implants and a model of the prosthetic structure, and is passed on to the finite element suite, in which the loading and supporting conditions are specified. Biomechanical analysis and comparison of two implant placement schemes are carried out for two types of loading: one modeling biting and the other—chewing. The problem of automation of a patient-specific choice of the optimal implantation scheme is discussed taking into account the stress-strain state of the system "jaw–implants–prosthetic construction".

Keywords Edentulous mandible · Dental implants · Patient-specific planning · Implantation schemes · Biomechanical analysis

1 Introduction

Rehabilitation of edentulous patients is one of the most difficult, relevant and still not completely solved problems in dental implantology [1]. One of the main problems is that in the absence of a tooth (or rather, in the absence of mechanical loads), the alveolar bone tissue in its place undergoes resorption [2]. With that the most difficult is the placement of a large number of implants with a pronounced atrophy of the entire jaw and its low mineral density. Up to this date, there is still no universal method to achieve a guaranteed stability of the prosthesis on the edentulous mandible.

In 1988, Nobel BioCare offered a treatment concept called "All-on-Four" [3], in which a full-arch prosthesis (12–14 teeth) is fixed on an edentulous jaw, supported

I. N. Dashevskiy (✉) · D. A. Gribov
Ishlinsky Institute for Problems in Mechanics RAS, Moscow, Russia
e-mail: dash@ipmnet.ru

D. A. Gribov
e-mail: denis4@inbox.ru

© Springer Nature Switzerland AG 2020
Z. Hu et al. (eds.), *Advances in Artificial Systems for Medicine and Education II*,
Advances in Intelligent Systems and Computing 902,
https://doi.org/10.1007/978-3-030-12082-5_28

by 4 implants placed in the anterior section of the upper or lower jaw. To increase the placement base and, accordingly, to reduce the cantilever part of the prosthesis, the lateral implants are tilted at an angle of 30 or 45°. Such an approach has a number of advantages including the possibility of fixing a full prosthesis of the jaw even with a small amount of bone tissue in the chewing portion and some others. Today, this technology has become quite widespread in practice.

However, under such a scheme, some of the implants may have a greater load than in the classical prosthetic scheme with more implants arranged in parallel. Reducing the risk of stress concentration is provided by the choice of both the characteristics of the implants themselves and their placement schemes. In view of this, of particular importance is mathematical modeling and patient-specific quantitative analysis of the jaw stress–strain state (SSS), being performed for a specific patient and taking into account individual morphological features and variability in the mechanical characteristics of bone tissue by its volume. There have been published some papers on creation of jaw models and modeling their behavior under various implantation schemes, prosthetic configurations and materials, loading conditions, etc. [4–12].

The aim of this work was the development of an individualized biomechanical model of a mandible with implants and a prosthesis model as well as an assessment of the stress–strain state of the model components with two different implant placement options (including <<All-on-Four>>) and under different loading conditions.

2 Materials and Methods

2.1 Finite Element Model of the Lower Jaw

Computer planning of implantation with subsequent prosthetics was carried out on the basis of the cone-beam computed tomography (CBCT) shots of a 70-year-old female patient with a complete absence of teeth on the lower jaw.

Two schemes of implant placement on the lower jaw were simulated: scheme 1—all implants placed vertically; and scheme 2—the lateral implants placed at an angle of 45°. Implant parameters, as indicated in Fig. 1, were set according to [13, 14].

Conversion of raster CBCT images into a three-dimensional solid model of the mandible was done using the software suites Mimics 17.0 and 3-matic 6.1. In the first stage, automatic segmentation of CBCT images was carried out.

Then, based on the results of the segmentation, a surface model of the jaw was created that was supplemented with the surface implant models, after which, a three-dimensional mesh for model components was created based on four-node tetrahedra. Only the case of complete osseointegration of the implants was considered. The typical number of nodes of the model was 180,556, of elements—1,027,638.

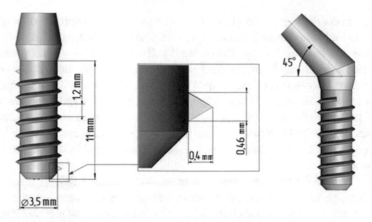

Fig. 1 Implants models with straight (left) and tilted (right) abutments used in modeling

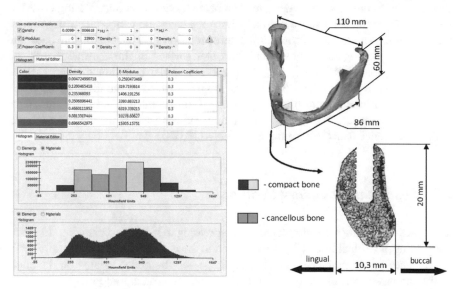

Fig. 2 Result of the mechanical properties assignments of mandible bone: Mimics dialog box with defined empirical expressions for density and elastic modulus (on the left); mandible model and its cross-section in the area of placement of one of the implants (on the right)

The material of the jaw and the implants was modelled as isotropic linearly elastic, while the modulus of elasticity of the LJ model was determined discretely for each element using the corresponding Mimics calculation module (Fig. 2). To this end, the program calculated the average HU value for each model tetrahedron used to determine the physical density. Density values, in turn, were used to calculate the Young's modulus from a given empirical relationship. Expression

$$\rho = (0.668 \cdot \mathrm{HU} + 9.84) \times 10^{-3} \ \mathrm{g/cm^3}, \tag{1}$$

which connects the X-ray and physical density, as well as empirical dependencies for determining the Young's modulus E were taken according to [5, 11]. As in [11], elements with an average X-ray density HU $\geq$ 1000 ($\rho \geq 0.67$ g/cm^3, $E \geq$ 4583 MPa) were assigned to compact bone tissue. Poisson's ratio of spongy and compact bone tissues was taken as equal to 0.3 [11, 15]. Figure 2 (left) presents the Mimics program interface, in which empirical dependencies were defined and the properties of each finite element were determined. According to the results of the operation, 1/3 (288,637 elements) of the total number of LJ elements were found to be compact bone tissue and this type of tissue predominated in the area of implant placement.

After the model was completed, it was exported to the ANSYS finite element suite. Since the models of LJ and implants with abutments consisted of four-node tetrahedra, the latter were associated with an element of the SOLID285 type with three translational degrees of freedom in each node. When planning prosthetics, it was planned to mount a beam denture. The beam model was created using ANSYS tools and approximated using a 2-node BEAM188 beam element with six degrees of freedom in each node (three translational, three rotational). The relationship between the nodes belonging to the beam and the outer surface of the abutments was set by imposing kinematic constraints using the multipoint coupling method (MPC-algorithm). With the parallel placement of implants, the average distance between them was 12 mm, and with tilted placement of the lateral implants—14 mm. The mechanical characteristics of titanium implants and beams were taken in accordance with the reference data [16]: Young modulus—110,000 MPa, Poisson's ratio—0.35.

2.2 Loads and Boundary Conditions

Two types of loading were simulated, corresponding to biting and chewing. In modeling biting, a vertical force (F_B) was applied to the nodes located in the central part of the beam, while in the case of chewing (F_C)—to the nodes located on the edge of the cantilever part of the beam. In both cases, force magnitude was assumed to be equal to the typical occlusal load for the first premolar and molars $F_B = F_C =$ 200 N [11, 17]. As boundary conditions, fixing all degrees of freedom was taken in the nodes belonging to the heads and coronoid processes of LJ (Fig. 3).

3 Results

An evaluation was made of the stress state of both the jawbone tissue in the implant placement area and the implants themselves. The results of determining the equivalent stresses are given in Tables 1 and 2. In all cases, the maximum stresses occurred in the zone of the first (cervical) turn of the implant thread. When biting off (Table 1), the stresses in the upper layer of LJ compact bone tissue and the implants themselves

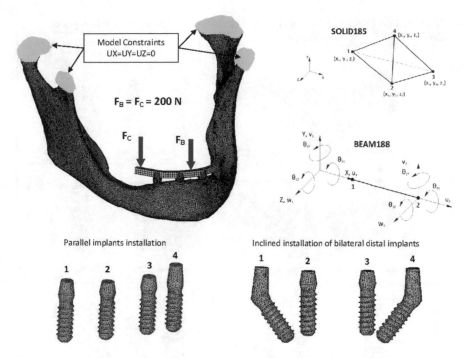

Fig. 3 Mandible finite element model in the biting and chewing processes modeling and model constraint conditions (top left); finite elements types used to create the model (top right); implants placement schemes (below)

were distributed more evenly than when chewing (Table 2). The maximum stresses in the bone tissue, when placing implants according to scheme 1, were slightly higher than in scheme 2 (difference no more than 9%), whereas for the implants the reverse situation was observed, with the difference in stresses reaching 47%. In the simulation of the chewing process, the lateral implant and the corresponding LJ region located on the side of the applied load were the most loaded. The stresses in this area of bone tissue for Scheme 1 were significantly lower than in Scheme 2 (the difference of 79%), while in the lateral implant, on the contrary, they were higher (32% difference). In relation to this area, the level of stress in the placement areas of other implants was insignificant, without pronounced dependence on the implant placement scheme (difference no more than 21%).

4 Discussion

As a result of the study, it was found that under given loading conditions, maximum stresses in the LJ bone tissue and implants arise, as a rule, in the area of the first (cervical) turn of the implant thread. Comparison of the calculations results for the

Table 1 Results of mandible bone and implants stress state calculation in modeling of biting

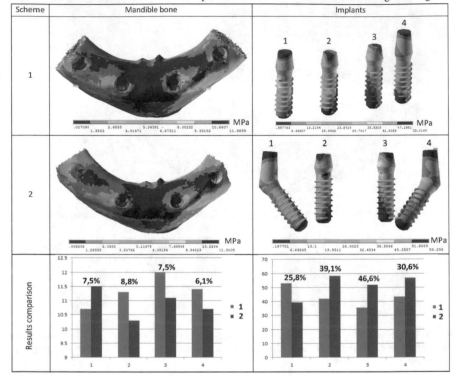

two schemes of implant placement shows that when biting off, the choice of the scheme is of no fundamental importance, since the stress level in the jaw bone is comparable and in the implant material—low enough. When chewing, the parallel placement proves to be preferable from the point of view of the stressed state of bone tissue. However, for such a scheme, maximum stresses in implants (at the first thread turn in the lateral implant) are significantly higher than in the analogous tilted implant and are comparable with the yield strength of titanium $\sigma_{0.2} \approx 440$–480 MPa [16].

In [12], two schemes in question were studied without regard to variability of bone mechanical characteristics by volume and with implants nonthreaded. Other than some specific thread effects, one can state general consistency of our results and [12].

The stresses obtained in LJ do not exceed the strength of compact bone tissue (according to [18] $\sigma_B = 130$ MPa). At the same time, the maximum values of equivalent deformations in bone tissue when loading the cantilever part of the beam significantly exceeded the bone damage threshold, which is about 4×10^{-3} according to H.M. Frost theory of Mechanostat [19].

Table 2 Results of mandible bone and implants stress state calculation in modeling of chewing

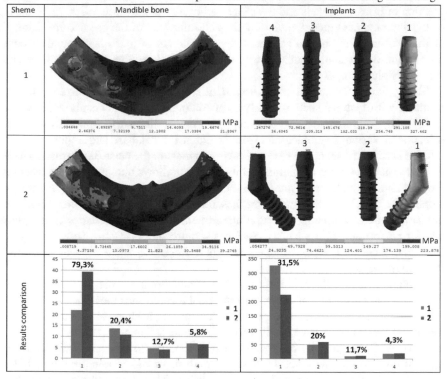

Let us note that grayscale values (GV) in CBCT images are always lower than Hounsfield Units (HU) in the spiral CT images. In the study [20], performed using three CBCT scanners and one CT scanner, a linear relationship was empirically established between CBCT-GV and HU. Therefore, when working with CBCT, the previous two-step scheme for determining the Young's modulus E from the tomograms: (1) HU $\rightarrow \rho$, (2) $\rho \rightarrow E$—must be supplemented by one more preliminary step (0) CBCT-GV $\rightarrow$ HU using formulas of the type found in [20].

The main advantage of the described approach is a deeper individualization of the dentition restoration planning, taking into account both the morphology of the jaw and the variability of the physico-mechanical characteristics by its volume. One of the main problems preventing its implementation in a wide range of practices is the high labor intensity and the need to use several expensive software products. First, regardless of the quality of the obtained tomography images, the results of automatic segmentation require manual modification of almost every image. Given the number of shots in the tomogram (300–600 shots), this process can take 3–8 h. Secondly, the quality of the surface model obtained after automatic optimization is usually unsatisfactory for creating a tetrahedral mesh. Due to the complex geometry of the jaw bone, a number of triangular faces of the model do not meet the standard quality criteria, and so require correction using interactive editing tools. Thirdly, the creation of implant models and their integration into the jaw model, the processing of the finite element model, including the determination of the physico-mechanical characteristics of the model elements, the assignment of loads and boundary conditions and the evaluation of the results obtained, all require significant time and specific engineering skills.

Today, when planning implantation, dentists use actively specialized software products with features such as: analysis of tomography images using a variety of tools; image segmentation and creation of corresponding surface models; implant selection of various manufacturers from an extensive database and importing them into the surface model of the jaw; and saving models in the basic 3D formats. These programs have a simple and intuitive interface that allows one to quickly master all their functionality. However, due to the above limitations, the resulting models are not suitable for creating finite element meshes, and the software systems themselves do not have modules neither for determining the mechanical characteristics of materials nor for performing SSS analysis. Surfaces require optimization of the quality of triangular faces and implant models are conditional and serve only to plan their location in the jaw.

Another serious problem is the need for systematization, justification, verification, validation and unification of numerous empirical formulas for the transition from X-ray to physical density and further to the values of the mechanical characteristics of biological tissues. This issue is still very far from its satisfactory solution.

Thus, it is not currently possible to fully automate the process of selection the optimal implant scheme based on the SSS analysis of the system, which includes bone tissue, implants and prosthetic construction. Moreover, although at the 8th World Congress of Biomechanics (Dublin, 8–12 July 2018), Materialize's representatives promised to release an update of the Mimics suite with full automation of CT images segmentation in 2019, specialists in biomechanics are much more restrained in their assessments, and some of them in general doubt the basic possibility of full automation of this kind.

Anyway, for significant advances in solving this problem, it is necessary to develop micromechanical models and experimental methods for determining the physico-mechanical characteristics of biological tissues and also medical diagnostic tools, software and computer technology.

An important contribution should make use of methods of intelligent information processing. As for the determination of the physico-mechanical characteristics of biological tissues from CT, big data and data mining can play an important role. To automate the creation of high-quality geometric and grid models—it must be new and more powerful methods and algorithms for pattern recognition, machine learning methods, including neural networks, deep learning, etc., which are increasingly used in various areas of biomedicine [21–27].

5 Conclusions

The presented technology makes it possible to take into account the individual geometric and mechanical characteristics of the bone structures and tissues of a particular patient in digital planning and comparing different implantation options on the edentulous jaw.

For the time being, the topic of the accuracy, reliability, conditions of applicability, and the possibilities of unifying a wide variety of formulas of an empirical nature for the transition from X-ray density to physical density and then to mechanical characteristics are poorly understood.

In order to speed up the automation of planning implant surgery interventions, it is vitally important to involve as much as possible intelligent data processing techniques and algorithms, in particular, methods of pattern recognition, machine learning and other approaches of artificial intelligence.

Acknowledgments The research was performed on the theme of the state assignment (state registration number AAAA-A17-117021310386-3) and with partial support of RFBR grants No. 17-08-01579 and No. 17-08-01312.

References

1. Chumachenko, E.N.: Prediction of possible complications in orthopedic dentistry on the basis of analysis of the stress–strain state of supporting tooth tissues. Vestn. Russ. Acad. Nat. Sci. **7**(3), 42–49 (2007). (in Russian)
2. Sukharsky I.I.: Optimization of the surgical stage of dental implantation on the basis of computer modeling. Author's thesis abstract MD, Moscow (2013). (in Russian)
3. Nobel Biocare. https://www.nobelbiocare.com/ru/ru/home/company/about-us/history.html
4. Liao, S.H., Tong, R.F., Dong, J.X.: Anisotropic finite element modeling for patient-specific mandible. Comput. Methods Programs Biomed. **88**(3), 197–209 (2007). https://doi.org/10.1016/j.cmpb.2007.09.009
5. Arahira, T., Todo, M., Matsushita, Y., Koyano, K.: Biomechanical analysis of implant treatment for fully edentulous maxillas. J. Biomech.Sci. Eng. **5**(5), 526–538 (2010). https://doi.org/10.1299/jbse.5.526
6. Chuiko, A.N., Shinchukovsky, I.A.: Biomechanics in dentistry. Fort, Kharkov (2010). (in Russian)
7. Chuiko, A.N., Kalinovsky, D.K., Levandovsky, R.A., Gribov, D.A.: Biomechanical support of operations in maxillofacial surgery using MIMICS and ANSYS programs. Orthop. Traumatol. Prosthet. **2**, 57–63 (2012). (in Russian)
8. Chuiko, A.N., Ugrin, M.M.: Biomechanics and computer technologies in maxillofacial orthopedics and dental implantology. Galent, Lviv (2014). (in Russian)
9. Polyakova, T.V., Chumachenko, E.N., Arutyunov, S.D.: Features of mathematical modeling of a segment of the dentoalveolar system from computed tomography data. Russ. Herald. Dent. Implantol. **29**(1), 7–13 (2014). (in Russian)
10. Polyakova, T.V., Gavryushin, S.S., Arutyunov, S.D.: Modeling the planning of placement of temporary implants for support of prototypes of bridge dentures for the period of osseointegration of two-stage dental implant. Eng. J. Sci. Innov. **12**, 1–18 (2016). (in Russian)

11. Horita, S., Sugiura, T., Yamamoto, K., Murakami, K., Imai, Y., Kirita, T.: Biomechanical analysis of immediately loaded implants according to the "All-on-Four" concept. J. Prosthodont. Res. **61**, 123–132 (2017). https://doi.org/10.1016/j.jpor

12. Bronstein, D.A.: Fixed prosthetics with complete absence of teeth using intraosseous implants in the frontal jaw (clinical, biomechanical and economic aspects). Author's thesis abstract Grand MD, Moscow (2018). (in Russian)

13. Lan, T.H., Du, J.K., Pan, C.Y., Lee, H.E., Chung, W.H.: Biomechanical analysis of alveolar bone stress around implants with different thread designs and pitches in the mandibular molar area. Clin. Oral Invest. **16**(2), 363–369 (2012). https://doi.org/10.1007/s00784-011-0517-z

14. Ryu, H.S., Namgung, C., Lee, J.H., Lim, Y.J.: The influence of thread geometry on implant osseointegration under immediate loading: a literature review. J. Adv. Prosthodont. **6**, 547–554 (2014). https://doi.org/10.4047/jap.2014.6.6.547

15. Lin, C.L., Wang, J.C., Ramp, L.C., Liu, P.R.: Biomechanical response of implant systems in various areas of angulation, bone density, and loading. Int. J. Oral Maxillofac. Implants **23**, 57–64 (2008)

16. Il'in, A.A., Kolachev, B.A., Polkin, I.S.: Titanium alloys. In: Composition, Structure, Properties. Handbook, VILS-MATI, Moscow (2009). (in Russian)

17. Mericske-Stern, R., Assal, P., Mericske, E., Bürgin, W.: Occlusal force and oral tactile sensibility measured in partially edentulous patients with ITI implants. Int. J. Oral Maxillofac. Implants **10**, 345–353 (1995)

18. Obraztsov, I.F. (ed.): Problems of Strength in Biomechanics. Higher School, Moscow (1988). (in Russian)

19. Frost, H.M.: Skeletal structural adaptations to mechanical usage (SATMU): 1. Redefining Wolff's law: the remodeling problem. Anat. Rec. **226**, 414–422 (1990). https://doi.org/10.1002/ar.1092260402

20. Razi, T., Niknami, M., Ghazani, F.A.: Relationship between Hounsfield unit in CT scan and gray scale in CBCT. J. Dent. Res. Dent. Clin. Dent. Prospects **8**, 107–110 (2014)

21. Sfeir, R.F., Julien, C.H.: Tomographic convex time-frequency analysis. Int. J. Image, Graph. Signal Process. (IJIGSP) **7**(7), 33–41 (2015). https://doi.org/10.5815/ijigsp.2015.07.05

22. Sharma, G.T., Singh, H.: Computational approach to image segmentation analysis. Int. J. Mod. Educ. Comput. Sci. (IJMECS) **9**(7), 30–37 (2017). https://doi.org/10.5815/ijmecs.2017.07.04

23. Bhima, K., Jagan, A.: An improved method for automatic segmentation and accurate detection of brain tumor in multimodal MRI. Int. J. Image, Graph. Signal Process. (IJIGSP) **9**(5), 1–8 (2017). https://doi.org/10.5815/ijigsp.2017.05.01

24. Fradi, M., Youssef, W.E., Lasaygues, P., Machhout, M.: Improved USCT of paired bones using wavelet-based image processing. Int. J. Image, Graph. Signal Process. (IJIGSP) **10**(9), 1–9 (2018). https://doi.org/10.5815/ijigsp.2018.09.01

25. Mazurov, M.: Intelligent recognition of electrocardiograms using selective neuron networks and deep learning. In: Hu, Z., Petoukhov, S., He, M. (eds.) Advances in Artificial Systems for Medicine and Education. AIMEE 2017. Advances in Intelligent Systems and Computing, vol. 658, pp. 182–197. Springer, Cham (2018). https://doi.org/10.1007/978-3-319-67349-3_17

26. Ivaniuk, N., Ponimash, Z., Karimov, V.: Art of recognition the electromyographic signals for control of the bionic artificial limb of the hand. In: Hu, Z., Petoukhov, S., He, M. (eds.) Advances in Artificial Systems for Medicine and Education. AIMEE 2017. Advances in Intelligent Systems and Computing, vol. 658, pp. 176–181. Springer, Cham (2018). https://doi.org/10.1007/978-3-319-67349-3_16

27. Izonin, I., Trostianchyn, A., Duriagina, Z., Tkachenko, R., Tepla, T., Lotoshynska, N.: The combined use of the wiener polynomial and SVM for material classification task in medical implants production. Int. J. Intell. Syst. Appl. (IJISA) **10**(9), 40–47 (2018). https://doi.org/10.5815/ijisa.2018.09.05

A Viscoelastic Model of the Long-Term Orthodontic Tooth Movement

Eduard B. Demishkevich and Sergey S. Gavriushin

Abstract The objective of this study was to propose an approach to simulate orthodontic tooth movement by modeling of periodontal ligament as a viscoelastic material. Although there are various models of tooth movement proposed by several authors, most of them are based on empirical remodeling laws without considering nature of biochemical processes occurring in the periodontal ligament during orthodontic tooth movement. The proposed approach allows to describe the process of long-term orthodontic tooth movement more accurately and use it to improve the efficiency of orthodontic treatment.

Keywords Dentofacial system · Orthodontic treatment · Finite element analysis · Tooth movement

1 Introduction

Orthodontic treatment is intended to correct or prevent teeth position disorder and malocclusion. Treatment process consists in the application of forces to teeth; these forces are generated by various orthodontic appliances. The biochemical process, occurring in dentition under the influence of orthodontic forces, results to movement of teeth through their supporting bone. The treatment success depends mainly on the accuracy of selected appliances. In complicated clinical cases, it is difficult to provide an efficient treatment plan without preliminary analysis of patient's dentition

E. B. Demishkevich (✉) · S. S. Gavriushin
Bauman Moscow State Technical University 5c1, 2nd Baumanskaya st.,
105005 Moscow, Russian Federation
e-mail: mail@edtech.ru

S. S. Gavriushin
e-mail: gss@bmstu.ru

E. B. Demishkevich · S. S. Gavriushin
Mechanical Engineering Research Institute of the Russian Academy of Sciences, 4, Malyi
Kharitonievsky Pereulok, 101990 Moscow, Russian Federation

© Springer Nature Switzerland AG 2020
Z. Hu et al. (eds.), *Advances in Artificial Systems for Medicine and Education II*,
Advances in Intelligent Systems and Computing 902,
https://doi.org/10.1007/978-3-030-12082-5_29

315

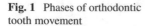

Fig. 1 Phases of orthodontic tooth movement

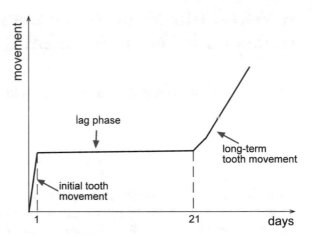

biomechanical state. So, the determination of optimal orthodontic forces is the actual task of modern prosthetic dentistry.

There are three phases of orthodontic tooth movement [1]. The first phase is initial tooth movement, and it is an immediate response of dentition to applied forces. No bone remodeling happens on this stage. Displacements of dentition during initial tooth movement are caused by deformations in periodontal ligament (PDL) [2], which is a thin connecting tissue located between cementum of a tooth root and alveolar bone. The second phase is usually called a lag phase, there is no significant tooth movement on this stage, and it is characterized by the hyalinization process in PDL where its structure is changing. The third phase is a long-term orthodontic tooth movement; it is represented by movement of teeth due to resorption and apposition of the bone tissues supporting them (Fig. 1).

Several approaches in investigations are used to model orthodontic tooth movement. Schneider et al. [3] used the finite elements method to estimate strains inside PDL of 2D modeled tooth and calculated an empirical stimuli function which is a key for bone remodeling. Soncini and Petrabissa [4] modeled the alveolar bone as a viscoelastic material and got accurate results of simulation compared to experimental data. But they couldn't determine integrated parameters of their model to use it for prediction of tooth movement. Mengoni and Ponthot [5] concentrated on bone behavior during remodeling and proposed a remodeling law based on microdamage in the alveolar bone.

The main goal of this research is to build more accurate models and use them for prediction of long-term orthodontic tooth movement. This article presents a brief description of the processes occurring in the PDL under the influence of forces because it is important to understand the underlying processes that cause bone remodeling. Also, a new approach is suggested to simulate long-term orthodontic tooth movement based on the internal processes in the PDL.

2 Biological Process in the PDL Under Orthodontic Appliances

As a result of external influence to dentition, two zones occur in the periodontal ligament: a zone of compression and a zone of tension. Thereby, a biological system leaves the state of physiological equilibrium. Shape and density of the bone tissue supporting dentition are changing to adjust system to the new equilibrium state.

Numerous investigations [6, 7] show that key stimuli to bone tissue remodeling are a deformed state in the PDL. Its main purpose apart from load transfer between tooth and jaw is blood and nutrients supply of surrounding tissues.

The main components of PDL are [8]:

- Function-oriented fibers and elastic fibers
- Connecting tissue containing cells typical for their location (fibroblasts, osteo-clasts, osteoblasts, cementoblasts, mastocytes, macrophages) and lymph
- Blood and lymphatic vessels
- Nerve fibers connected to pulp, tooth socket, and the gingival.

An important role of PDL is its active participation in remodeling the surrounding bone tissues [9]. Under influence of long-term forces, circulatory arrest is happening in zones where hydrostatic pressure is over than blood pressure. It stops the production of osteoblasts responsible for the formation of bone tissues. In the same time, compressed tissues continue to produce osteoclasts, and it results to bone resorption. And vice versa, production of osteoblasts predominates in areas of PDL with low hydrostatic pressure, causing bone formation. Using this principle, it is possible to create a phenomenological model of bone remodeling during orthodontic treatment. The model will describe the biomechanical state of the system "bone tissue–PDL–tooth" (displacements, stresses, and deformations) without defining changes on microstructure level.

It makes sense to note that orthodontic tooth movement caused by biochemical changes in PDL goes on much faster than during classic bone remodeling theories like Cowin's theory of adaptive elasticity [10], the Stanford model [11] of bone remodeling and so on.

3 Bone Remodeling Algorithm

Bone remodeling is considered as a step-by-step process of iterations, each lasting a defined amount of time Δt (usually one day, but it might differ depending on a specific patient). The process consists of three steps, represented in Fig. 2. On the first step, we apply forces to the biomechanical system and calculate deformations occurring in the PDL. Total deformation ε_Σ is composed of two parts:

$$\varepsilon_\Sigma = \varepsilon_e + \varepsilon_p.$$

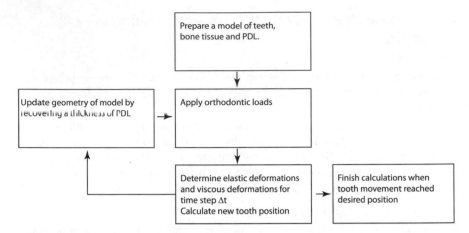

Fig. 2 Schematic representation of bone remodeling algorithm

ε_e is the elastic part of deformation, and it corresponds to initial tooth movement. ε_p is a component responsible for long-term tooth movement on a period Δt. To calculate the components of deformations, the PDL was considered as a viscoelastic material represented by Maxwell's model which is a series combination of elastic and dash-pot elements [12]. Elastic deformations correspond to ε_e and viscous deformations correspond to ε_p.

We consider the elastic mechanical characteristics of PDL as linear and isotropic, with $E_{PDL} = 1$ MPa, $\mu = 0.45$. Viscous behavior can be defined by η, the viscosity of the dashpot.

On the second step, we calculate the displacement of the tooth caused by long-term deformations ε_p. On the next step, we simulate the reconstruction of PDL thickness by remodeling biomechanical system. Then, iterative process is repeated with updated geometry for the following time span Δt_{i+1}.

Simulation process can be stopped once required tooth movement is reached.

4   Intuitive 1D-Model of Long-Term Orthodontic Tooth Movement

Let's use the model shown in Fig. 3. Rigid element (tooth) is linked to rigid structure (alveolar bone tissue) with two elastic elements and two viscous elements. PDL layer is considered as a Maxwell's viscoelastic element. Single force F is applied to the tooth. The metabolic process of bone remodeling that occurred in this system was investigated.

Reaction forces for the system are shown in Fig. 4.

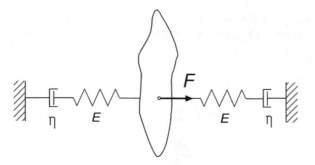

Fig. 3 Schematic representation of the investigated model

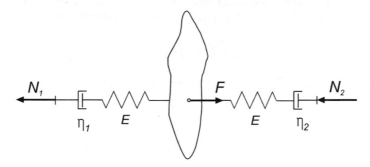

Fig. 4 Reaction forces in the system

The investigated model is symmetric relatively to the point of force application. Elastic characteristic of spring k corresponds to Young's module of PDL. According to investigations [13], expanded and compressed areas of PDL have different influence to bone resorption/apposition, so viscous characteristics of the system depend on whether deformations are negative or positive.

$$\eta = \begin{cases} \eta_1, & \varepsilon \geq 0 \\ \eta_2, & \varepsilon < 0 \end{cases}.$$

That's why it is not possible to consider this mechanical system as a statically determinate structure. To determine all the parameters of the system, we have to use the mechanical equilibrium equation:

$$F - N_1 - N_2 = 0 \tag{1}$$

and the displacements compatibility equation:

$$\delta_{p1} + \delta_{e1} + \delta_{e2} + \delta_{p2} = 0 \tag{2}$$

where δ_{e1}, δ_{e2} indicate displacements of elastic elements, δ_{p1}, δ_{p2} the displacements of viscous elements. Displacements of elastic elements are prorated to their internal stresses:

$$\delta_e = k_e \cdot N.$$

Displacement speeds of viscous elements are also prorated to stress factors:

$$\frac{d\delta_p}{dt} = \eta N.$$

Let's consider that orthodontic force is constant, it will allow us to determine displacement value of the elastic element by time integration:

$$\delta_p = \eta N t.$$

Therefore, compatibility Eq. (2) takes the following form:

$$\eta_1 N_1 t + k_1 N_1 - k_2 N_2 - \eta_2 N_2 t = 0. \tag{3}$$

It is possible to solve the system of Eqs. (1), (3) and determine bone tissue reactions:

$$N_1 = \frac{F(\eta_2 t + k)}{\eta_1 t + 2k + \eta_2 t}, \quad N_2 = \frac{F(\eta_1 t + k)}{\eta_1 t + 2k + \eta_2 t}.$$

Then, tooth displacement was determined:

$$\delta_\Sigma = \delta_{p1} + \delta_{e1} = \frac{F(k + \eta_1 t)(k + \eta_2 t)}{\eta_1 t + \eta_2 t + 2k},$$

Long-term tooth movement corresponds to displacement of viscous elements:

$$\delta_p = \delta_{p1} + \delta_{p2} = \frac{k(\eta_2 + \eta_1)F t}{\eta_1 t + 2k + \eta_2 t}.$$

δ_p is the amount of orthodontic movement, determined on the current step of the iteration process, as a result of influence of orthodontic load F to the system. However, updated tooth position is not a physiologic equilibrium state yet, because PDL still contains compressed and expanded fibers under the influence of the force. The fibers produce osteoblasts and osteoclasts causing bone resorption/apposition. In the long run, it will result in the recovery of PDL thickness, but in updated position of tooth. So, the model geometry is updated in the following way: Tooth is moved with δ_p displacement, PDL is remodeled with its initial thickness l, and bone tissue borders

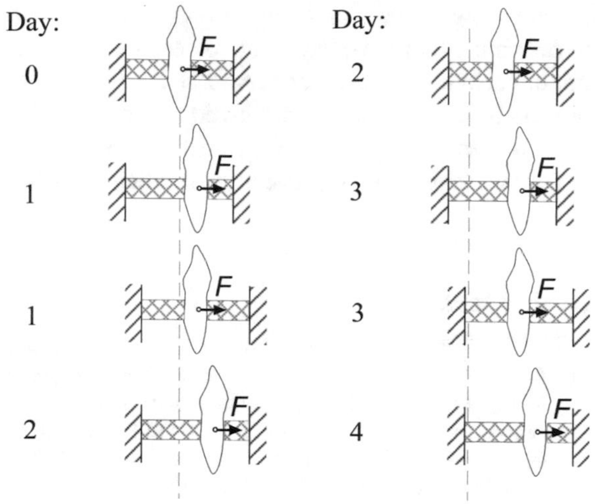

Fig. 5 Simulation results for the test model

Day: 0, 1, 1, 2

Day: 2, 3, 3, 4

move accordingly. The next iteration step is started from applying forces to the updated model.

5 Results

Simulation results of the model shown in Fig. 3 for the first days of the remodeling phase are presented in Fig. 5.

The tooth was loaded by the force of 0.5 N. The elastic and viscous parameters of the PDL are chosen to make the remodeling process stable and corresponding to reference values ($k = 1, \eta_1 = 0.5, \eta_2 = 0.3$). The daily amount of long-term remodeling is 0.14 mm.

According to the remodeling algorithm presented in this paper, there are two steps for each remodeling iteration: The first step is the movement of the tooth, and the second step corresponds to recovery of the PDL thickness.

6 Conclusion

A new approach allowing to describe the process of long-term orthodontic tooth movement accurately has been presented, since the proposed model is based on the knowledge of cellular changes occurring in the PDL under influence of orthodontic appliances. Further development of the model will allow to simulate orthodontic tooth movement for each patient individually. As a source of geometric data, it is possible to build 3D-models of dentition based on the patient's computer tomography images using contemporary algorithms [14, 15]. Finite elements method can be used to determine tooth displacements on each iteration of the described algorithm. It is important to note that it becomes possible to choose the parameters of the remodeling process by a preliminary examination of the patent's PDL properties. Prediction of tooth movement as a result of mechanical loading can help clinician to reduce risks of complications and improve the efficiency of orthodontic treatment.

References

1. Nanda, R.: Biomechanics and Esthetic Strategies in Clinical Orthodontics. Elsevier Saunders, St. Louis (2005). ISBN 978-0-7216-0196-0
2. Arutyunov, S.D., Gavryushin, S.S., Demishkevich, E.B.: Finite-Element Modeling of the Orthodontic Tooth Movement, No. 3 (54), pp. 108–120. Herald of the Bauman Moscow State Technical University (2014) (in Russian)
3. Schneider, J., Geiger, M., Sander, F.-G.: Effects of bone remodeling during tooth movement. Russ. J. Biomech. 4(3), 57–72 (2000)
4. Soncini, M., Pietrabissa, R.: Quantitative approach for the prediction of tooth movement during orthodontic treatment. Comput. Methods Biomech. Biomed. Eng. 5(5), 361–368 (2002). https://doi.org/10.1080/1025584021000016852
5. Mengoni, M., Ponthot, J.: A damage/repair model for alveolar bone remodelling. Paper presented at: CMBBE 2008. Proceedings of the 8th Computer Methods in Biomechanics and Biomedical Engineering, Porto, Portugal
6. Tanne, K., Sakuda, M., Burstone, C.J.: Three-dimensional finite element analysis for stress in the periodontal tissue by orthodontic forces. Am. J. Orthod. Dentofac. Orthop. 92(6), 499–505 (1987). https://doi.org/10.1016/0889-5406(87)90232-0
7. Chen, J., Li, W., Swain, M.V., Ali Darendeliler, M., Li, Q.: A periodontal ligament driven remodeling algorithm for orthodontic tooth movement. J. Biomech. 47(7), 1689–1695 (2014). https://doi.org/10.1016/j.jbiomech.2014.02.030
8. De Jong, T., Bakker, A.D., Everts, V., Smit, T.H.: The intricate anatomy of the periodontal ligament and its development: lessons for periodontal regeneration. J. Periodontal Res. 52(6), 965–974 (2017). https://doi.org/10.1111/jre.12477
9. Henneman, S., Von den Hoff, J.W., Maltha, J.C.: Mechanobiology of tooth movement. Eur. J. Orthod. 30(3), 299–306 (2008). https://doi.org/10.1093/ejo/cjn020
10. Cowin, S.C., Hegedus, D.H.: Bone remodeling I: theory of adaptive elasticity. J. Elast. 6(3), 313–326 (1976). https://doi.org/10.1007/bf00041724
11. Beaupré, G.S., Orr, T.E., Carter, D.R.: An approach for time-dependent bone modeling and remodeling-theoretical development. J. Orthop. Res. 8(5), 651–661 (1990). https://doi.org/10.1002/jor.1100080506
12. Sokolovskyy, Y., Levkovych, M.: Two-dimensional mathematical models of visco-elastic deformation using a fractional differentiation apparatus. Int. J. Mod. Edu. Comput. Sci. (IJMECS) 10(4), 1–9 (2018). https://doi.org/10.5815/ijmecs.2018.04.01
13. Su, M.-Z., Chang, H.-H., Chiang, Y.-C., Cheng, J.-H., Fuh, L.-J., Wang, C.-Y., Lin, C.-P.: Modeling viscoelastic behavior of periodontal ligament with nonlinear finite element analysis. J. Dental Sci. 8(2), 121–128 (2013). https://doi.org/10.1016/j.jds.2013.01.001
14. Sharma, G.: Performance analysis of image processing algorithms using matlab for biomedical applications. Int. J. Eng. Manuf. (IJEM) 7(3), 8–19 (2017). https://doi.org/10.5815/ijem.2017.03.02
15. Hamdi, M.A.: A comparative study in wavelets, curvelets and contourlets as denoising biomedical images. Int. J. Image Graph. Signal Process. (IJIGSP) 4(1), 44–50 (2012). https://doi.org/10.5815/ijigsp.2012.01.06

Metric Properties of the Visual Illusion of Intersection

T. Rakcheeva

Abstract The work is devoted to an experimental investigation of the well-known optical-geometric Poggendorff illusion (of the intersection illusion). With a visual perception of the intersection of the line with the opaque strip, an error arises due to inadequate extrapolation of the line continuation after passing through the strip. Experimental computer researches of factors influencing the existence and magnitude of illusion have been carried out. The main factors: the strip width, the strip orientation and the line inclination angle relative to the strip. After presenting a part of the line in front of the strip (of the "input fragment") on the computer screen, the subject had to draw a line after the passage through the strip ("output fragment") with the mouse. The following characteristics were measured: the displacement along the strip of the line "output fragment" and its direction. It is shown that the illusion exists irrespective of the spatial orientation of the screening strip, its color and texture. Quantitative estimates of the dependence of the illusion value on the strip width and the line inclination angle are obtained. Regression models of functional dependences of the illusion value from these factors are constructed. The effect of accumulating the illusion value at the intersection by the line of several parallel strips was studied also (the subjects had making subjective correction independently for each fragment of the line between the strips). The test revealed additive properties of Poggendorff's illusion. The conducted researches made it possible to form a hypothesis combining the manifestations of Poggendorff's illusion independently of the action of various factors, namely: there is a steady tendency to shorten the distance between the point of the "entrance" of the line into the strip area and the point of its "exit." The field of practical applications, where optical illusions should be given great attention, is engineering psychology, medical engineering and education.

Keywords Visual illusions · Optical-geometric illusions · Illusion of intersection · Poggendorff · Metric factors · Engineering psychology

T. Rakcheeva (✉)
Mechanical Engineering Research Institute of the Russian Academy of Sciences,
4 Maly Kharitonievskiy Pereulok, Moscow 101990, Russia
e-mail: rta_ra@list.ru

© Springer Nature Switzerland AG 2020
Z. Hu et al. (eds.), *Advances in Artificial Systems for Medicine and Education II*,
Advances in Intelligent Systems and Computing 902,
https://doi.org/10.1007/978-3-030-12082-5_30

1 Introduction

Optical illusions are objectively existing mechanism of human visual perception, known from ancient times [1]. They are of a steady, systematic nature. People, as a rule, cannot correct these errors of perception self-dependently, even if they have the objective information of the observed phenomenon [1–5].

For specialists in a number of professions, the influence of illusions on the result of their professional activity, based on visual estimates of metric characteristics, is very critical. At the same time, the character of the manifestation of visual illusions, as shown by scientific research, can indicate the features and pathologies of human development and the state of its health [6–8].

Over the past two hundred years, much attention has been paid to the study of the phenomenon of visual illusions by researchers of different fields of knowledge [9–15]. During this time, many images were created demonstrating the presence of significant systematic errors in the estimation of the sizes and shapes of geometric figures. All the results were mainly reduced to a classification of known illusions and hypotheses about the nature of their occurrence [2, 3, 11–13]. Most of the experimental studies are aimed at confirming the existence of visual illusions and identifying the groups of people most prone to these illusions [3–10]. Much less work contains studies of regularities that determine the quantitative characteristics of visual illusions and how to compensate them [9–16].

In this regard, it seems important to explore the nature of visual illusions, their variability relative to metric factors, as well as the possibility of compensating for these illusions, which provides a more adequate solution to professional problems.

At the heart of visual illusions is the fact that objects located in the neighborhood in the field of vision with the object under consideration always more or less influence the formation of the visible image. In other words, our visual impression of the size and shape of an object depends on the context in which it is viewed. This property of our vision was noticed long ago. In particular, it was found that the perceived length, curvature and orientation of the lines are greatly influenced by the dimensions of the figures in which they are included, as well as the presence of adjacent or intersecting lines. It was also discovered by Poggendorff that in the visual perception of an inclined line partially passing behind any opaque object, an error arises due to inadequate extrapolation of the continuation of the line after passing behind the object [2–4]. Japanese psychologist Shogu Imai experimented with various versions of Yastrov's illusion in 1960: measured angles, inner and outer radius, placed figures horizontally and vertically. It turned out that the biggest effect of illusion occurs when the figures are horizontal and the inner radius should be 60% of the outer radius.

Of particular interest to the researchers represented the causes of the appearance of visual illusions in general, and the illusions of Poggendorff, in particular. There are many reasons for explaining the nature of illusions: the spatial perspective, the visual estimation of acute angles, the context, etc. But until now, there are not enough convincing explanations for this illusion. At the same time, factors that significantly influence the manifestations of illusion are known, among them two factors [4]:

(1) *The width of the opaque strip covering the straight line.* The wider the closing strip, the more difficult it is for a person to continue the line correctly; with a decrease in the width of the strip, the manifestations of illusion decrease and disappear.

(2) *The inclination of the straight line with respect to the strip.* Illusion does not arise when the orientation of the line is orthogonal to the strip, but the smaller the angle between the straight line and the strip, the stronger the illusion of displacement, i.e. for the illusion of intersection it is important that the angle be acute.

This work is devoted to an experimental investigation of the metric factors of the well-known visual illusion of Poggendorff—a systematic error of perception that occurs when a line crosses an opaque strip.

2 Formulation of the Problem

When an inclined line passes behind an opaque strip, the systematic error of visual perception consists in the fact that the line before the intersection with the strip and its continuation after the strip do not form a united straight line (Fig. 1).

The fragment of the line to the intersection with the strip, which is traced as the initial one, is called the "input" fragment or the etalon one, and the fragment of the line after the intersection, which is traced as the final one, is called the "output" fragment or test one.

The following notations are also accepted:

I The value of illusion;
α The angle of inclination of the "input fragment" of the line relative to the strip;
β The slope angle of the "output fragment" of the line relative to the strip;
S The width of the strip;
A The point of intersection of the "input fragment" with the strip;

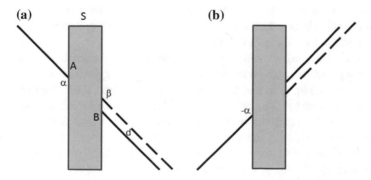

Fig. 1 The schemes of Poggendorff's illusion

B   The point of intersection of the "output fragment" with the strip in perception;
d   The displacement of the "output fragment" in perception relative to the "output fragment" in reality.

When passing the line of the opaque strip, its continuation shifts in visual perception along the strip by a significant amount d (Fig. 1). This shift is the metric magnitude of illusion I. In addition, the illusion manifests itself in the direction of the "output fragment."

The aim of the work was to study factors that could influence the appearance of the illusion and the degree of its expression, as well as in obtaining quantitative estimates of the value of illusion, depending on the parameters of these factors.

A number of questions follow from the formulation of the problem:

(1) Is there a visual illusion, i.e. whether the errors of perception have a certain, statistically significant reliability in magnitude and direction.
(2) Whether the illusion depends on the horizontal-vertical orientation of the shielding strip.
(3) Whether the illusion depends on the inversion of the intersecting line—whether the illusion is symmetric.
(4) What is the character of the dependence of the value of illusion on the width of the shielding strip.
(5) What is the character of the dependence of the value of illusion on the slope of the straight line.

The answers to these and other questions were obtained in computer experiment.

3   Description of the Experiment

To accomplish this work, computer software has been written that allows us to experimentally investigate quantitative estimates of a wide range of factors that determine the mechanism of Poggendorff's illusion [17].

During the experiment for each subject, the computer automatically performed all necessary records of the personal data, the results of their actions, calculations and comparisons, both length and direction, providing an opportunity for accurate quantification of the value of the illusion.

Each test of the experiment was as follows. On the screen of the monitor, the subject was presented with an incomplete scheme of Poggendorff's illusion, where there was a screening strip and an "input" fragment of the line for given values of the parameters S and α. The test subject, in accordance with the instruction, drew on the screen of the computer with the mouse the "output" fragment of the straight line. The initial position of the mouse was fixed on the "output" border of the strip. Drawing with the mouse was performed in the "straight line" mode and ended at an arbitrary accessible point on the screen. The testing time was not limited, so that the test subject could concentrate on the accuracy of the task making. The program was recording the start and end points of the drawn line fragment. From these data, both the metric values of the illusion and the orientational values were calculated.

The test subjects of both sexes, aged 20–30 years, who had computer skills, participated in the experiment, the total number of examinee was 25 people. On the day of the test, the participant twice participated in testing, which, as a rule, took about 20 min.

For each test subject, the complete experiment consisted of a series of tests with a grid of parameters for each factor studied. In order to neutralize the effect of random factors accompanying individual data, for each of the values of the experimental parameters, the test subjects performed a series of 4 tests, the data of which were averaged.

4 Results of the Experiment

Each experiment included a series of tests that make up the parameterization of a particular task. In the program, two arrays were specified: width strip values S (in pixels): 20, 40, 60, 80, 100, 120, 140, 160, 180, 200 and slope of the etalon line α (in degrees): 30, 40, 45, 50, 55, 60, 65, 70, 75, 80. A set of pair combinations of the values S and α constituting a grid of $10 \times 10 = 100$ tests was presented in random order. The results of the performed experiments are shown in Fig. 2.

In Fig. 2 shows the dependence of the value of the illusion of displacement I on the strip width (S) for a fixed angle of the slope of the straight line (α_0): $I = f(S, \alpha_0)$. Here, the strip width S and the value of the illusion I were measured in millimeters or pixels, and the value of the angle α_0 in degrees.

Figure 2a shows the dependence of illusion I on the width of the strip S in the range from 20 to 200 mm for a line with a slope of $45°$, the graph is provided with error bars. Figure 2b shows 3 similar graphs to lines with a slope of 30, 45, $60°$. From the data presented in the graphs, it follows that the value of illusion depends on the width of the strip and with satisfactory accuracy this dependence has a linear form.

The dependence of the illusion I on the angle of the straight line slope (α) for a fixed strip width (S_0): $I = f(S_0, \alpha)$ is shown in Fig. 3. Figure 3a shows the dependence of illusion I on the slope of the line α in the range from 30 to $80°$ for a strip width of 27 mm, the graph is provided with error bars. Figure 3b shows 3 similar graphs to strips with a width of 60, 100, 160 pixels (16.2, 27, 43.2 mm). From the data presented in the graphs, it follows that the dependence of the value of the illusion on the slope of the straight line is significantly non-linear.

The carried out experiments allow receiving answers to questions posed.

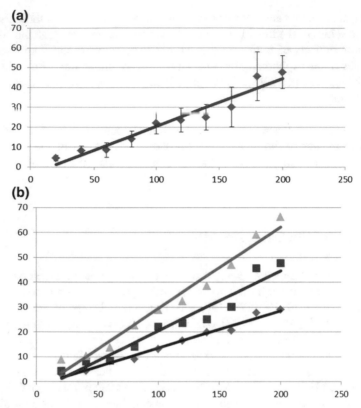

Fig. 2 Dependence of the illusion of displacement on the strip width for a fixed slope of the line: **a** graph of the illusion $I(S)$ for $\alpha_0 = 45°$ and for full range of strip width values S; **b** graphs of illusion $I(S)$ for different angles: $\alpha_0 = 30, 45, 60°$

The results showed that the Poggendorff's illusion in the form of a systematic bias error of the line was statistically reliably recorded for all test subjects in different experimental configurations. The illusion significantly does not depend on either the horizontal-vertical orientation of the strip or the vertical inversion of the line, but depends only on factors such as the strip width and the slope of the straight line. Regression models of functional dependencies of the value of illusion from these factors allow quantifying the required dependencies.

Thus, in the qualitative sense of the illusion of the intersection of Poggendorff is invariant with respect to orientation of the strip and symmetric with respect to the vertical inversion of the line. In the quantitative sense, the value of the illusion depends linearly on the strip width of S, the linearity coefficient depends on α_0, and depends nonlinearly on the line slope α, the "degree of nonlinearity" depends on the strip width of S_0.

Another result relates to the direction of the illusion of bias observed in all the experiments performed: the systematic error of the "output" fragment is always

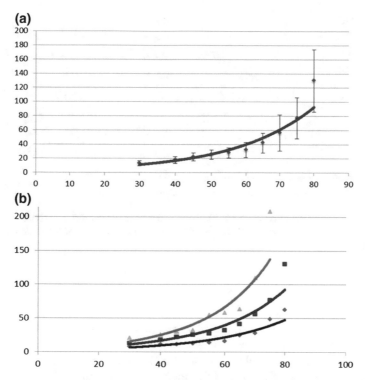

Fig. 3 Dependence of the illusion of displacement on the slope of the line for a fixed strip width: **a** the graph of the illusion $I(\alpha)$ for $S_0 = 27$ mm and for full range of angle values of the line α; **b** graphics of illusion $I(\alpha)$ for different strip widths: $S_0 = 16.2, 27, 43.2$ mm

directed toward the "input" fragment of the line [12]. This effect is illustrated in Fig. 4, where the objective and visually perceived intersection points are presented together for the dependence on the width of the strip (Fig. 4a) and the slope of the line (Fig. 4b). These figures combine graphs for the classical test configuration (Figs. 1a, 4a, b, above) and for configuration with vertical inversion (Figs. 1b, 4a, b, below). External lines represent real data, and internal lines—visually perceived. From these graphs, it is seen that the illusion of displacement is directed inwards.

The additional experiments show that the Poggendorff illusion has also the additive property—it accumulates at the intersection of several strips, preserving the character of the dependence for each of the strips and practically no significantly depends from the test color.

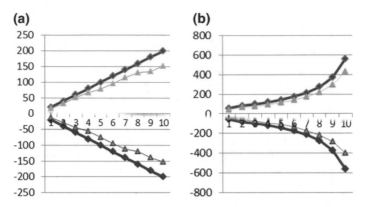

Fig. 4 Illustration of the direction of the illusion of displacement for different test configurations (Fig. 1a—above, Fig. 1b—bottom): **a** factor of the strip width, **b** factor of the slope angle of the straight line

5   Conclusions

The conducted studies also confirmed the author hypothesis that the illusion of Poggendorff has a certain tendency to shorten the distance between the "entrance" of the line to the end of the visible part of the first fragment and its "exit" to the beginning of the visible part of the second fragment, i.e. *subjects try to bring together the points of the "entrance" and "exit" of the line*. As for the reproduction of the given direction of the line interrupted by the strip, according to preliminary estimates, the subjects coped much better with this task, i.e. *the person basically keeps the direction well*. The trend is maintained under conditions of direct inversion, changing the orientation of the strip and color, and also the composition of the strips.

The carried out researches have confirmed the author's hypothesis that the illusion of Poggendorff has a certain tendency to shorten the distance between the "input" fragment of the line and the "output" fragment.

A person tends to reduce the distance between the points of the "input" and "output" of the line with respect to the strip.

The trend is persistently maintained for all parameters of the strip width and the slope of the line, regardless of the orientation of the strip, inversion of the straight line, changing the composition of the strips and the color.

The study of optical illusions occupies a large place in engineering psychology and medicine. The obtained results can have practical application, for example, for the correction of metric estimates in the work of the human-operator, where the accuracy of visual assessments is associated with high risks. The development of automation, the need for rapid assessments of the situation for various indications of numerous devices or test results make us pay attention to the possible danger associated with the accuracy of visual assessments, which medical technology should also take into account [18–21].

In addition, a significant role is played by the use of visual illusions in general and optic-geometric illusions in particular in social and medical psycho-diagnostics [8, 22–25].

References

1. Tit Lucretius, K.: On the Nature of Things, M., p. 698 (1947)
2. Gregory, R.: Seeing Through Illusions, p. 253. Oxford University Press, Oxford (2009)
3. Fress, P., Piaget, J.: Experimental Psychology, Issue 4, Progress, 344 p. Moscow (1978)
4. Tolansky S., F.R.S.: Optical Illusions, 172 p. Pergamon Press, Oxford, London (1964)
5. Milrud, R.P.: Dependence of visual illusions on the level of cognitive development. In: RP Mil'rudud. Questions of Psychology, pp. 114–120 (1997)
6. Bondarko, V.N., Semenov, L.A.: Estimation of size in the illusion of Ebbinghaus in adults and children of different ages. Physiol. Man **30**(1), 31–37 (2004)
7. Evershed, N.: Schizophrenics less fooled by visual illusions. Cosmos Online. No. 9 (April, 2009)
8. Medvedev, L.N., Shoshina, I.I.: Quantitative evaluation of the effect of sex and the type of interhemispheric asymmetry on the distortion of the visual perception of Poggendorff's figure in the modification of Jastrow. Physiol. Man **30**(5), 5–11 (2004)
9. Shoshina, I.I., Shelepin Y.E., Pronin, S.V.: The use of wavelet filtering of the input image to study the mechanisms of Muller-Layer visual illusion. Opt. J. 5, 70–75 (2011)
10. Hayashi, E., Kuroiwa, Y., Omoto, S., Kamitani, T., Li, M., Koyano, S.: Visual evoked potential changes related to illusory perception in- normal human subjects. Neurosci. Lett. **359**, 29–32 (2004)
11. Yarbus, A.L.: About some illusions in assessing the visible distances between the object edges. In: Studies on the Psychology of Perception, pp. 289–303. Institute of Philosophy, Sector of Psychology, M.-L.: Publishing House of the USSR Academy of Sciences (1948)
12. Howe, C.Q., Yang, Z., Purves, D.: The Poggendorff illusion explained by natural scene geometry. In: Proceedings of National Academy of Sciences, vol. 102. pp. 7707–7712. USA (2005)
13. Bulatov, A.N., Bertulis, A.V., Mickene, L.I.: Quantitative research of geometric illusions. In: Sensory Systems, T. 9, vol. 2–3, pp. 79–93. Nauka (1995)
14. Rakcheeva, T.A.: The influence of form on the perception of distance. In: Eighths Anniversary Kurdyumov Readings: Synergetics in the Natural Sciences, Proceeding of the International Interdisciplinary Conference, pp. 117–119 (2012)
15. Rakcheeva, T.A., Nikolaeva, E.C., et al.: Orthogonal illusion of visual perception. Mod. Med. Theory Pract. **1**, 42–51 (2004)
16. Rakcheeva, T.A.: Metric factors of optical-geometric illusion of intersection. In: Synergetic in the Social and Natural Sciences. Tenth Anniversary Kurdyumov Readings: Synergetic in the Natural Sciences, Proceedings of the International Interdisciplinary Conference, pp. 142–146 (2015)
17. Rakcheeva, T.A., Zholudev, E.P.: Investigation of factors of visual illusion of intersection. In: Mathematics Computer Education, XXII International Conference (MCE), vol. 22, p. 100 (2015)
18. Ajay, A., Singh, P.K.: Novel digital image water marking technique against geometric attacks. Int. J. Mod. Educ. Comput. Sci. (IJMECS) **7**(8), 61–68 (2015). https://doi.org/10.5815/ijmecs.2015.08.07
19. Mukherjee, S., Ganguly, D., Mukherjee, P., Mitra, P.: A novel technique for copyright protection of images using hybrid encryption model. Int. J. Mod. Educ. Comput. Sci. (IJMECS) **4**(5), 10–17 (2012). https://doi.org/10.5815/ijmecs.2012.05.02
20. Portugal, C., de Souza Couto, R.M.: Educational support for hypermedia design. Int. J. Mod. Educ. Comput. Sci. (IJMECS) **4**(6), 9–16 (2012). https://doi.org/10.5815/ijmecs.2012.06.02

21. Nandi, D., Saifuddin Saif, A.F.M., Paul, P., Zubair, K.M., Shubho, S.A.: Traffic sign detection based on color segmentation of obscure image candidates: a comprehensive study. Int. J. Mod. Educ. Comput. Sci. (IJMECS) **10**(6), 35–46 (2018). https://doi.org/10.5815/ijmecs.2018.06.05
22. Onifade, O.F.W., Akinyemi, D.J.: A review on the suitability of machine learning approaches to facial age estimation. Int. J. Mod. Educ. Comput. Sci. (IJMECS) **7**(12), 17–28 (2015). https://doi.org/10.5815/ijmecs.2015.12.03
23. Dominic, M., Francis, S.: An adaptable E-learning architecture based on learners' profiling. Int. J. Mod. Educ. Comput. Sci. (IJMECS) **7**(3), 26–31 (2015). https://doi.org/10.5815/ijmecs.2015.03.04
24. Kumar, M., Singh, A.J.: Evaluation of data mining techniques for predicting student's performance. Int. J. Mod. Educ. Comput. Sci. (IJMECS) **9**(8), 25–31 (2017). https://doi.org/10.5815/ijmecs.2017.08.04
25. Isong, B.: A methodology for teaching computer programming: first year students' perspective. Int. J. Mod. Educ. Comput. Sci. (IJMECS) **6**(9), 15–21 (2014). https://doi.org/10.5815/ijmecs.2014.09.03

Method of the Data Adequacy Determination of Personal Medical Profiles

Yuriy Syerov(iD), Natalia Shakhovska(iD) and Solomiia Fedushko(iD)

Abstract This paper presents the current problem of determining the level of data adequacy of personal medical profiles comparing to the online information tracks of the same persons. In this work, the analysis of data verification of personal medical profiles of several major medical information systems is conducted. The analysis of data adequacy level of personal medical profiles of major medical centers is carried out based on the algorithm for determining the personal data adequacy and model of personal medical profiles based on online information tracks. The results of the study in the information support system for making medical decisions are implemented. The proposed method of data verification of personal medical profiles is in demand and effective in detecting non-valid accounts and accounts with incorrect or stale data. The effectiveness of the verification system of personal medical profiles shows that the volume of checking personal data by medical workers is reduced by 3–5 times.

Keywords Online communities · Personal medical profile · Information track · Data adequacy · Medical information system · Verification

1 Introduction

The online communities collected immeasurable database of veracious and mendacious information about medical department, its clients, and medical specialists. Veracious data in medical information system profiles in most cases is incomplete and unreliable. Nowadays, new approaches to content of personal medical profiles, which

Y. Syerov · N. Shakhovska · S. Fedushko (✉)
Lviv Polytechnic National University, Lviv, Ukraine
e-mail: solomiia.s.fedushko@lpnu.ua

Y. Syerov
e-mail: yurii.o.sierov@lpnu.ua

N. Shakhovska
e-mail: nataliya.b.shakhovska@lpnu.ua

© Springer Nature Switzerland AG 2020
Z. Hu et al. (eds.), *Advances in Artificial Systems for Medicine and Education II*,
Advances in Intelligent Systems and Computing 902,
https://doi.org/10.1007/978-3-030-12082-5_31

consist in synchronizing the personal data of patients of medical institutions from communication social services (social networks, virtual communities, chat rooms, Web sites, etc.), resolved the problem of missing important information and data untruthfulness in personal medical profiles of medical decision-making information support system. Verifying the personal profile of medical support system and determining the level of data adequacy of personal medical profiles comparing to their online information tracks are the topical issues. The information profiles that consolidate all personal data with height level adequacy will increase productivity and reduce time spent of medical specialists, and improve diagnostic accuracy of patients based on reliable and complete data about patients.

Creating a database of verified personal data of patients will simplify diagnostics of patients in various medical institutions. Information medical profile of patient contains only verified personal data of patients, and it will be a component of the medical decision-making information support system.

The aim of research is to develop an effective method of determining the level of data adequacy of personal medical profiles. Proposed solution will help to regularly update the data in personal medical profiles of clients of medical decision-making information support system. The scientific novelty of the work is to develop new methods for collecting and verifying information from social services of communication. The consolidation of data in personal medical profiles of medical decision-making information support system is key issue of functioning e-health.

2 Review of Recent Works

An analysis of the online community activities is the subject of research, among which are clearly distinguished in three main areas: Web usage mining, Web structure mining, and Web content mining. One of the important problems of analyzing the content of online communities is the analysis of the Web-user personal data. Despite the significant importance for further development of this research area, no effective methods for analyzing personal medical information in the account of the online community users have yet been developed. Verifying and consolidating the personal data in Web are actual and important subjects of scientific research in such areas:

- Validation of online information (Abdiansah et al. [1], Amsbary, Metzger et al. [2]);
- Strategies for assessing the reliability and usefulness of data (Morse, Babichev et al. [3]);
- Definition and justification of validity and reliability content (Carmines, Sihare [4], Crocker), relevance of the content (Beck's, Siddikjon, Abduganiev [5]);
- Validation of crystallographic files used in a magazine article (Westrip [6]);
- Accuracy and reliability of Internet sources, educational and scientific Web content, highly specialized content; perception of Web users on the reliability of the content (Johnson and Kaye [7], Wright, Korzh et al. [8–11], Rao and others);

- Research of authoritativeness and reliability (Freeman, Mahadevaswamy [12], Goldsmith, Flanagin), personal data quality (Park, Rubio, Kumar et al. [13]);
- Detecting fake profiles in online social networks (Meligy et al. [14]);
- Modeling and developing the hospital information system (El Azami, Ouçamah [15], Winter, Schreiweis [16]),
- Research of cloud computing environment for e-health (Chen et al. [17]);
- Security and protection of health information (Malin [18], Zou, Dai);
- Decision-making support information system (Fradian, Manjafarashvili et al. [19]).

The latest research direction is insufficiently researched. In modeling and developing the hospital information system, the conducted researches to verifying the data of medical information system profiles and consolidated its data of online community users are dedicated. The results of scientific research in this area are in demand by a wide range of specialists in the organization and functioning of online communities, as those that should ensure their success and effectiveness.

Considering the above analysis of scientific works, among the well-known literary sources, there is a lack of thorough research on the verification of personal information of Internet users [20–22] and the study of the data reliability of medical information support system [23], in order to improve their functioning. This, in turn, generates the actual problem of developing new methods and tools to analyze the adequacy of the data of medical information system profiles which would have sufficient scientific justification, formalization, predictable efficiency, and versatility.

3 Models of Personal Medical Profiles Based on Online Information Tracks

Modeling the process of patient profiles based on information content verification is a complex task; its solution entirely depends on further areas and the application purpose of the developed models. The models, which have been proposed in this work, are adapted for computer-linguistic analysis of the informational track of the online community user and information content in the Web in general.

In this section are substantiated the forming principles of a system of linguistic and communicative indicators based on a training sample of users in online community to validate the patient data in online communities. The system of linguistic and communicative indicators is sets of linguistic and graphic indicative features that are inherent for the online communication of a particular user in the online community, which establish the membership participation of the Web community to profile data and actually determine the importance of personal characteristics. This system is the basis for formulating and solving a complex of validating personal data problems of online users by the method of computer-linguistic analysis of information content.

The main purpose of the system's operation is to validate the data of patient profiles based on online information tracks in the online communities. Also was done the models construction of computer-linguistic analysis for the reliability of personal characteristics in the online communities.

3.1 Model of the Information Track of Online Communities Users

The verification of personal data based on the results of computer-linguistic analysis of information content in online communities is part of the method of system verification personal data of the user in the online community. It is based on verified personal characteristics of the user in the online community using the created information systems to form the information profile of the online user. For both analysis and computer-linguistic analysis, it is necessary to create a database for personal data and data created by a user during his communicative activities in the online community. For the same reason, the notion of "information track" is introduced and outlines all the basic information that is necessary for the verification of the personal characteristics of a certain user in online communities and the construction of his information profile. The information track of the online community user is the set of all data of the user of the online community and the results of his communicative activity—the information content created by user. In turn, the information track in the online community introduced for the convenience of analyzing information content created within a single online community by a specific user or group of users. This analysis is carried out in order to identify the personal characteristics of the user in the online community and verify the authenticity of the personal data specified by the user.

The information track of the Web user P_i will be described as follows:

$$InfTrack(P_i) = \langle Content(P_i), PersonalData(P_i) \rangle \tag{1}$$

The components of an information track are: $Content(P_i)$ created by a member of the online community, and personal data—$PersonalData(P_i)$. The information content is determined by the tuple, namely subsets such as discussions, polls, and messages:

$$Content(P_i) = \langle Thread(P_i), Poll(P_i), Post(P_i) \rangle \tag{2}$$

where $Thread(P_i) = \{Thread_j(P_i)\}_{j=1}^{N_i^{UThead}}$ is set of discussions created by a member in the online community P_i; N_i^{UThead} is number of such discussions. $Poll(P_i) = \{Poll_j(P_i)\}_{j=1}^{N_i^{UPoll}}$ is set of polls created by a member of the online community P_i; N_i^{UPoll} is number of such polls. $Post(P_i) = \{Post_j(P_i)\}_{j=1}^{N_i^{UPost}}$ is set of online community posts P_i; N_i^{UPost} is number of posts by a member in online community P_i.

Computer-linguistic analysis is made only for the personal data that the user of the online community has specified in user account.

The most prioritized data for forming the data profile of the patients in the online community is the mandatory information about the online community user, less important—important data. However, an analysis of the additional personal data of such member is also required for a complete analysis. The amount of personal information, which the user provides in his personal account, is divided into blocks. This division is not monotonous in many cases and depends on the type of community. The personal data of a user is divided into blocks. The content of the blocks and the priority of personal data, which is provided by a member of the online community, are determined by the developers and administrators of this online community, focusing on the subject and purpose of this online community.

Formally, the distribution of the personal data of the online community member account to the blocks is as follows:

$$PersonalData(P_i) = \left\langle \begin{array}{l} BasicInfo(P_i), EduInfo(P_i), InterestsInfo(P_i), \\ WorkInfo(P_i), ContactInfo(P_i), FotoInfo(P_i) \end{array} \right\rangle$$

$$(3)$$

where $BasicInfo(P_i)$ is block of basic information; $EduInfo(P_i)$ is information block about education; $WorkInfo(Pi)$ is information block about the work of the online community user; $ContactInfo(P_i)$ is contact information block; $FotoInfo(P_i)$ is block graphic and media information; $InterestsInfo(P_i)$ is a block of information about the hobby and interests of the online community user.

Formal description of the online community member account:

$BasicInfo(P_i)$ is block of basic personal information of the online community user.

$$BasicInfo(P_i) = \left\langle \begin{array}{l} Name(P_i), NickName(P_i), Age(P_i), \\ Gender(P_i), Region(P_i), Lang(P_i) \end{array} \right\rangle \qquad (4)$$

where $Name(P_i)$ is full name; $NickName(P_i)$ is nick name; $Gender(P_i)$ is gender; $Age(P_i)$ is age; $Lang(P_i) = \{Lang_j(P_i)\}_{j=1}^{N_i^{Lang}}$ is plural of languages signed by a user; N_i^{Lang} is number of language; $Region(P_i) = \{Region_k(P_i)\}_{k=1}^{N_i^{Region}}$ is set of regions with which a user is associated; N_i^{Region} is set of regions.

$EduInfo(P_i)$ is information block about education.

$$EduInfo(P_i) = \langle EduLevel(P_i), Specialization(P_i) \rangle \qquad (5)$$

where $EduLevel(P_i)$ is level of education received, $Specialization(P_i)$ is specialty.

$WorkInfo(P_i)$ is block of data about the work of online community user.

$$WorkInfo(P_i) = \langle Company(P_i), Position(P_i) \rangle \qquad (6)$$

where $Company(P_i)$ is institution where works, $Position(P_i)$ is a position taken by a member of the online community in that institution.

$ContactInfo(P_i)$ is contact information for the member of the online community.

$$ContacInfo(P_i) = \langle Email(P_i), SocialNets(P_i), Website \rangle \qquad (7)$$

where $Email(P_i)$ is main email address, $SocialNets(P_i) = \left\{ SocialNets_j(P_i) \right\}_{j=1}^{N_i^{(Up)}}$ is plurality of pages in social networks, $N_i^{(Up)}$ is the number of pages in social networks; $Website$ is Web site.

$FotoInfo(P_i)$ is graphic information block.

$$FotoInfo(P_i) = \langle Avatar(P_i), Userbar(P_i), Foto(P_i) \rangle \qquad (8)$$

where $Avatar(P_i) = \left\{ Avatar_j(P_i) \right\}_{j=1}^{N_i^{Ava}}$ is set of avatars, N_i^{Ava} is number of avatars, N_i^{Foto} is a number of photographs, $Userbar(P_i) = \{ Userbar_k(P_i) \}_{k=1}^{N_i^{Userbar}}$ is set of graphic signatures, $N_i^{Userbar}$ is number of signatures, $Foto(P_i) = \{ Foto_m(P_i) \}_{m=1}^{N_i^{Foto}}$ is set of photographs.

$InterestsInfo(P_i)$ is a block of information about the hobby and interests of the online community user.

$$InerestsInfo(P_i) = \langle Byline(P_i), Activity(P_i), Quot(P_i), Biography(P_i) \rangle \qquad (9)$$

where $Byline(P_i) = \left\{ Byline_j(P_i) \right\}_{j=1}^{N_i^{Byline}}$ is number of signatures, N_i^{Byline} is number of signatures; $Activity(P_i) = \{ Activity_k(P_i) \}_{k=1}^{N_i^{Act}}$ is a set of favorite lessons and phrases, N_i^{Act} is number of lessons, phrases, $Quot(P_i) = \{ Quot_l(P_i) \}_{l=1}^{N_i^{Quot}}$ is a plurality of quotations, N_i^{Quot} is number of quotations, $Biography(P_i)$ is a biography.

Information about contacts and Web sites where the Web user displays communicative activity is placed in the $ContactInfo(P_i)$ block. Each account blocks contain information from three groups of personal data of the online user. Preferably, in the $BasicInfo(P_i)$ block, compulsory data is placed; without this data registration in the online community is not possible.

4   Method of Determining the Level of Data Adequacy of Personal Medical Profiles

The notion of the adequacy of the personal data of the user profile of the patient of the medical institution by the authors is introduced to compare the personal data of the informational profile in the online communities with the data of the medical

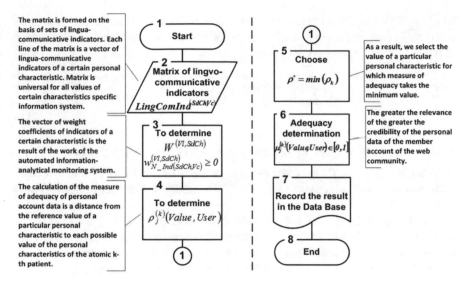

Fig. 1 Block diagram of the algorithm for determining the personal data adequacy of account

information system. The information profile in the online communities is generated by the method of computer-linguistic analysis of the information track of user.

The adequacy of personal profile data is a characteristic of the personal data of the online account, which indicates the degree of probability of verifying the personal characteristics of a particular patient to the personal data that he/she has specified in his own account in the virtual communities. That is, determining the truth of the specified personal information in the profile of the information medical system.

Determining the personal data adequacy of the account to the real information of system user consists in the implementation of the main stages of the algorithm of determining the of personal data adequacy of the account. The block diagram of this algorithm in Fig. 1 is presented.

The result of the step-by-step and correct implementation of the algorithm is the formation of information personal profiles of system users with verified personal characteristics. Moreover, the value of the measure of adequacy is directly proportional to the reliability of the user's personal data of the system.

The measure of the adequacy of the personal data of the account is the definite level of probability of the computer-linguistic analysis of the online community user account for the control (reference) account of the user of the online community, which is based on actual and up-to-date information about the patient of the medical institution. The difference between 1 and $\rho_j^{(k)}(Value, P)$ is the distance between the reference value of the personal characteristics, and the value of the personal characteristics of the atomic kth user is determined by the adequacy of the personal data of the kth user profile.

$$\mu_j^{(k)}(Value, P) = 1 - \rho_j^{(k)}(Value, P) \tag{10}$$

where $\rho_j^{(k)}(Value, P)$ is the distance from the reference value of the personal characteristic to each possible value of the personal characteristics of the atomic kth user of the online community according to (10):

$$\mu_j^{(k)}(Value, P) = 1 - \sqrt{\sum_{i=1}^{N\_Ind(PrCh,k)} \left(Ind_{i,j}^{(PrCh,Vc)} - Ind_{i,j}^{(PrCh,P)}\right)^2 * w_i^{(PrCh)}}$$
(11)

where $k \in 1 \ldots N\_Vl(PrCh, Vc)$. Moreover, $\mu_j^{(k)}(Value, P) \in [0, 1]$.

When $\mu_j^{(k)}(Value, P) \to$ max, then the degree of probability of personal characteristics of a particular user in the online community to this user personal data is high.

The proposed method of vectorization consists in transforming the data into a vector form, which will enable to determine the degree of similarity between the values of personal characteristics. The value of a similarity measure between the value of personal characteristics and the control vector indicates the importance of membership by the online community to a certain value characteristics.

5 Practical Implementation of Research Results

The results of analysis of data level adequacy of personal medical profiles of Ukrainian medical centers comparing to these patients online information tracks are shown in Fig. 2.

The indicator of the effectiveness of the developed methods of data verification of personal medical profiles is determined in Eq. (12).

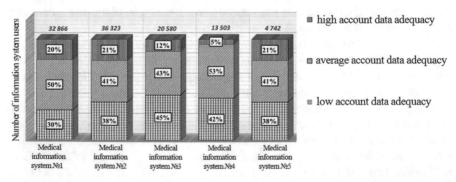

Fig. 2 Results of medical profiles data adequacy analysis of Ukrainian medical information systems

Method of the Data Adequacy Determination of Personal Medical …

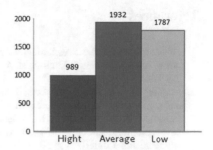

Fig. 3 Results of analysis of level of data verification of personal medical profiles

$$Efficiency = \frac{N^{VerPD}}{N^{VerPD} - N(LAdequacy)^{APD}}, \quad N^{VerPD} \neq N(LAdequacy)^{APD}$$

(12)

where $N(LAdequacy)^{APD}$ is number of personal medical profiles with low account data adequacy, N^{VerPD} is the total number of verified personal medical profiles.

Based on Eq. (12) the results of data verification level of personal medical profiles, investigated profiles are classified according to data verification of personal medical profiles (21% of all investigated accounts contained high level of data verification, 41% of all investigated accounts contained average level of data verification, and 38% of all investigated accounts contained low level of data verification). These results are presented graphically in Fig. 3.

The results show that 23% of the patients (total of 4708 person) provided reliable information in their accounts. 28% of members updated their credentials in the accounts. 4% of all personal medical accounts are blocked.

6 Conclusion

The results of this study will be applied in the decision-making information support systems, which will significantly enhance the effectiveness of the provision of medical services. The result of the research shows that the proposed method of data verification of personal medical profiles is in demand and effective in detecting non-valid accounts and accounts with incorrect or stale data. The relative quantity of mistakes in data verification systems on the test set of personal medical profiles is permissible for performance of assigned tasks, and it is actually a system error probability.

The effectiveness of developed methods shows that the volume of checking personal data on medical workers is reduced by 3–5 times. Also, the implementation of these methods significantly reduces the time and cost of system administration.

References

1. Abdiansah, A., Azhari, A., Sari, A.K.: Survey on answer validation for Indonesian question answering system. Int. J. Intell. Syst. Appl. **10**(4), 68–78 (2018). https://doi.org/10.5815/ijisa. 2018.04.08
2. Metzger, M.J.: Making sense of credibility on the web: models for evaluating online information and recommendations for future research. JASIST **58**(13), 2078–2091 (2007)
3. Babichev, S., Korobchynskyi, M., Mieshkov, S., Korchomnyi, O.: An effectiveness evaluation of information technology of gene expression profiles processing for gene networks reconstruction. Int. J. Intell. Syst. Appl. **10**(7), 1–10 (2018). https://doi.org/10.5815/ijisa.2018.07. 01
4. Sihare, S.R.: Roles of e-content for e-business: analysis. Int. J. Inform. Eng. Electron. Bus. **10**(1), 24–30 (2018). https://doi.org/10.5815/ijieeb.2018.01.04
5. Abduganiev, S.G.: Towards automated web accessibility evaluation: a comparative study. Int. J. Inform. Technol. Comput. Sci. **9**(9), 18–44 (2017). https://doi.org/10.5815/ijitcs.2017.09.03
6. Westrip, S.: publCIF: software for editing, validating and formatting crystallographic information files. J. Appl. Crystallogr. **43**(4), 920–925 (2010)
7. Johnson, T., Kaye, B.: Webelievability: a path model examining how convenience and reliance predict online credibility. JMCQ **79**, 619–642 (2002)
8. Korzh, R., Peleshchyshyn, A., Syerov, Y., Fedushko, S.: University's information image as a result of university web communities' activities. Adv. Intell. Syst. Comput. **512**, 115–127 (2017)
9. Korzh, R., Peleshchyshyn, A., Syerov, Y., Fedushko, S.: Principles of university's information image protection from aggression. In: 2016 XIth International Scientific and Technical Conference Computer Sciences and Information Technologies (CSIT), Lviv, pp. 77–79 (2016)
10. Korzh, R., Peleshchyshyn, A., Fedushko, S., Syerov, Y.: Protection of university information image from focused aggressive actions. In: Recent Advances in Systems, Control and Information Technology. SCIT 2016. Advances in Intelligent Systems and Computing, vol. 543, pp. 104–110. Springer, New York (2017)
11. Korzh, R., Fedushko, S., Trach, O., Shved, L., Bandrovskyi, H.: Detection of department with low information activity. In: 12th International Scientific and Technical Conference on Computer Sciences and Information Technologies (CSIT), Lviv, pp. 224–227 (2017)
12. Mahadevaswamy, M.U.B.: Wireless wearable smart healthcare monitoring using android. Int. J. Comput. Netw. Inform. Secur. **10**(2), 12–19 (2018). https://doi.org/10.5815/ijcnis.2018.02. 02
13. Kumar, V., Chaturvedi, A., Dave, M.: A solution to secure personal data when aadhaar is linked with digilocker. Int. J. Comput. Netw. Inform. Secur. **10**(5), 37–44 (2018). https://doi.org/10. 5815/ijcnis.2018.05.05
14. Meligy, A.M., Ibrahim, H.M., Torky, M.F.: Identity verification mechanism for detecting fake profiles in online social networks. Int. J. Comput. Netw. Inform. Secur. **9**(1), 31–39 (2017). https://doi.org/10.5815/ijcnis.2017.01.04
15. Azami, I., Ouçamah, M.: Integrating hospital information systems in healthcare institutions: a mediation architecture. J. Med. Syst. **36**, 3123–3134 (2012)
16. Schreiweis, B.: Modelling the hospital information system of the Karolinska University Hospital in Stockholm. University of Heidelberg, Heilbronn University and Karolinska Institutet (2010). http://ki.se/content/1/c6/10/46/20/Diplomarbeit_Bjoern_Schreiweis.pdf
17. Guo, L., Chen, F., Chen, L., Tang, X.: The building of cloud computing environment for e-health. In: E-Health Networking, Digital Ecosystems and Technologies (EDT), pp. 89–92 (2010)
18. Malin, B.: Guidance on de-identification of protected health information. Office for Civil Rights., U.S. Department of Health & Human Services (2010). www.hhs.gov/ocr/privacy/ hipaa/understanding/coveredentities/Deidentification/hhs_deid_guidance.pdf

19. Faradian, A., Manjafarashvili, T., Ivanauri, N.: Designing a decision making support information system for the operational control of industrial technological processes. Int. J. Inform. Technol. Comput. Sci. **7**(9), 1–7 (2015). https://doi.org/10.5815/ijitcs.2015.09.01
20. Korobiichuk, I., Fedushko, S., Juś, A., Syerov, Y.: Methods of determining information support of web community user personal data verification system. In: Automation 2017. Advances in Intelligent Systems and Computing, vol. 550, pp. 144–150. Springer, New York (2017)
21. Korzh, R., Fedushko, S., Peleschyshyn, A.: Methods for forming an informational image of a higher education institution. Webology **12**(2), Article 140 (2015). Available at: http://www.webology.org/2015/v12n2/a140.pdf
22. Syerov, Y., Fedushko, S., Loboda, Z.: Determination of development scenarios of the educational web forum. In: 2016 XIth International Scientific and Technical Conference Computer Sciences and Information Technologies (CSIT), Lviv, pp. 73–76 (2016)
23. Bilushchak, T., Komova, M., Peleshchyshyn, A.: Development of method of search and identification of historical information in the social environment of the internet. In: Proceedings of the 12th International Scientific and Technical Conference on Computer Sciences and Information Technologies, CSIT 2017, pp. 196–199 (2017)

Experimental Method for Biologically Active Frequencies Determination

Victor A. Panchelyuga, Victor L. Eventov and Maria S. Panchelyuga

Abstract The recent studies showed that frequencies related to spin subsystem of biological system and defined by values of their Zeeman splitting are biologically active. Theoretical calculation of these frequencies for real biosystems is practically impossible due to complexity of the systems. Therefore, the paper proposes an idea of the experimental method for biologically active frequencies determination and an experimental setup to realize the idea. The method is based on connection between Zeeman and Faraday effects. The difference of polarizations of two orthogonally polarized laser modes is used as a parameter under control. High sensitivity of the method is provided by usage of principles of intracavity laser polarimetry. The method presented in the paper was used for treatment of model, alloxan, and of a spontaneous diabetes mellitus in experimental animals. The treatment was successful in both cases.

Keywords Biologically active frequencies · Polarization · Time series · Primary targets

1 Introduction

Numerous research works performed mainly during the latter half of the twentieth century brought the reliable experimental evidences of the low-frequency (of kHz and sub-kHz frequency range) electromagnetic radiation effects on biological systems. A case of special interest is the effect of superweak (many orders lower than a typical value of thermal vibrations—kT) magnetic fields (SMF).

V. A. Panchelyuga (✉) · M. S. Panchelyuga
Institute of Theoretical and Experimental Biophysics, Russian Academy of Sciences, Institutskaya str., 3, 142290 Pushchino, Moscow Region, Russia
e-mail: panvic333@yahoo.com

V. L. Eventov
Petrovsky Russian Research Centre of Surgery, Abrikosovsky Side str., 2, 1 19991 Moscow, Russia

© Springer Nature Switzerland AG 2020
Z. Hu et al. (eds.), *Advances in Artificial Systems for Medicine and Education II*,
Advances in Intelligent Systems and Computing 902,
https://doi.org/10.1007/978-3-030-12082-5_32

The major step in understanding of the conditions of the SMF action is researches of Libov et al. [1–3] who used the combined magnetic fields (CMF):

$$\vec{B} = \vec{B}_0 + \vec{B}_\sim \cos \Omega t, \tag{1}$$

where $\vec{B}_0$ and $\vec{B}_\sim$ are direct and alternating magnetic fields (MF). It is important to note that (1) is a sufficiently general case that is always present in form of the geomagnetic field $\vec{B}_0$ and its fluctuations. The same works show that in case of $\vec{B}_0 || \vec{B}_\sim$ and $\Omega = n \Omega_c$ ($n = 1, 2, 3, \ldots$), where Ω_c is a cyclotron frequency of an ion with charge q and mass m:

$$\Omega_c = \frac{q}{m} B_0, \tag{2}$$

the fields (1) can affect various biological processes.

V. V. Lednev reached the deepest, to our opinion, insight into mechanisms of CMF biological action. His theoretical model [4, 5] and experimental researches of his group [4–7] showed that the primary targets of the CMF (1) include not only biologically important ions (Ca^{2+}, Mg^{2+}, and K^+) at the $n \Omega_c$ frequencies, but also a spin subsystem at the $\Omega = n \Omega_L$ frequencies:

$$\Omega_L = \gamma B_0, \tag{3}$$

where γ is a gyromagnetic ratio for a certain magnetic moment. The importance of the spin subsystem was experimentally proved in [7]. Note that a biological effect depends also on the amplitude of alternating magnetic field $\vec{B}_\sim$ along with (2) and (3). A biosystem response has, at that, a polyextremal character. Details of the amplitude conditions are presented in the review [4], and not considered in this paper.

The mentioned researches [1–7] imply, usually, CMF with single alternating frequency defined by (2) or (3). Nevertheless, recent experimental works show that poly-frequency effects are more favorable in the achievement of practically meaningful results. For example, the authors of [8] use a signal of the form $B_0 + B_1 \cos \Omega_1 + B_2 \cos \Omega_2 + B_3 \cos \Omega_3$, where $B_0 = 42$ μT, $B_1 = 300$ nT; $\Omega_1 = 1$ Hz, $B_2 = 100$ nT; $\Omega_2 = 4{,}4$ Hz, $B_3 = 150/300$ nT; $\Omega_3 = 16.5$ Hz. At that, the authors show that a signal containing a sum of the mentioned frequencies causes an anticancer activity in mice with the grafted Erlich carcinoma. An exposition of the carcinoma in such a field is followed by its complete degradation. The authors [8] emphasize that only the signal $B_0 + B_1 \cos \Omega_1 + B_2 \cos \Omega_2 + B_3 \cos \Omega_3$ is especially effective, while its separate mono-frequency components $B_1 \cos \Omega_1$, $B_2 \cos \Omega_2$, and $B_3 \cos \Omega_3$ are not effective. They suppose the presence of some set of receptors in the biosystem, which results in higher performance, when SMF are tuned to their frequencies. While the research [8] uses a superposition of several harmonic components, the authors of [9] use a set of several hundred frequencies of 100 Hz–21 kHz range to modulate a carrier frequency 27.12 MHz. The set demonstrated its effectiveness and inhibiting growth of hepatic tumor cells of two types. Another frequency set inhibited the growth of breast cancer cells. At that, the frequencies suitable for the therapy of a hepatic tumor did not affect the breast cancer cells and vice versa. The authors of [10] postulate that sensitivity of a biosystem to the external superweak influence is caused not by separate frequencies but by the general structure of their set.

These examples [8–10] show that the more complex is a spectrum of an affecting field, the more specific effects it causes. It seems clear that an ideal, in terms of the effectiveness and specificity, effect corresponds to the complete set of the resonance frequencies Ω_i of a biosystem under examination

$$\Omega_i = \gamma_i B_0, \quad i = \overline{1, \ N} \tag{4}$$

and is realized in a polyharmonic signal

$$\vec{B} = \vec{B}_0 + \sum_i^N \vec{B}_i \cos \Omega_i t. \tag{5}$$

The greatest challenge of the (5) implementation is that Ω_i cannot be calculated theoretically due to the great complexity of real biosystems. So, the experimental ways of frequencies set (4) search seem more effective and accurate.

The goal of the study is the development of a method and its technical realization allowing for experimental obtaining of a signal (5), which contains information on a frequency set (4) and can be used further for the specific affecting an appropriate biological subsystem.

The method is based on a connection between Zeeman and Faraday effects. The relation provides a possibility to obtain information on biologically active frequencies related to a value of Zeeman splitting, thus recording polarization-time pattern of $\Delta P(t)$ with the use of a special laser system described in Sects. 2 and 3. The special experiments used as examples in Sect. 4 demonstrated a high biological activity of the signal $\Delta P(t)$.

2 Background for an Experimental Method

For this purpose, we shall use a laser system with two orthogonally polarized modes P_1 and P_2 whose difference ΔP value is an output parameter:

$$\Delta P(t) = P_1(t) - P_2(t). \tag{6}$$

Values of P_1 and P_2 are set by the rotation of an angle of laser beams polarization at their pass through a biological sample under examination.

As well know, a polarization plane rotation is determined by the Faraday effect [11], the main idea of which is that a magnetized substance cannot be characterized only by a single refraction index n. Under a magnetic field B_0, the refraction indices n^+ and n^- for a circular polarized light become different. As a result, a plane of monochromatic light with wavelength λ passed the way L, in a medium rotates by an angle θ:

$$\theta = \frac{\pi L(n^+ - n^-)}{\lambda}. \tag{7}$$

In a case of not very strong magnetic fields ($B_0 < 1.5$ T), interesting for practical use, the difference $n^+ - n^-$ linearly depends on a magnetic field intensity B_0 and, in general, a Faraday rotation angle can be described by the expression:

$$\theta = VLB_0, \tag{8}$$

where V is a Verdet constant (specific magnetic rotation) depending on the light wavelength λ, and substance properties, including its temperature.

The Faraday effect is tightly bound with the Zeeman effect expressed in splitting of energy levels of atoms and molecules in the magnetic field. At the longitudinal, relative to the magnetic field, observation, the spectral components of the Zeeman splitting appear circular polarized. The corresponding circular anisotropy can be found in the spectral course of the refraction index of Zeeman transitions. So, in its simplest form, the Faraday effect is a consequence of the Zeeman splitting of dispersion curves of the refraction indices for two circular polarizations.

For a central symmetrical force field, a Zeeman splitting of an energy level relates to the Larmor precession of electron orbits and equals to

$$\Delta E = \frac{q}{m} B_0. \tag{9}$$

Lowering of symmetry complicates the Larmor precession, and the last expression includes a factor $\gamma < 1$:

$$\Delta E = \gamma \frac{q}{m} B_0. \tag{10}$$

The lower is the symmetry, the more it differs from the spherical one and the less is the Zeeman splitting; nevertheless, it exists in any case. Therefore, effects of Zeeman and Faraday should always be observed for molecules of any symmetry that can be proved by the experiments [12, 13].

Behavior of the polarization plane of laser radiation interacting with a biological object is analyzed in [14]. The analysis is based on the V.V. Lednev's model [4, 5]. The analysis [14] shows that a CMF in the case of parametric resonance can induce rotation of a polarization plane only at the certain ("resonance") values of an alternating field equal to a cyclotron frequency or its subharmonics.

The result tells about significance of the frequencies of (2) and (3) in relation to their effects on the rotation of a polarization plane.

3 System for Registration of the Difference Polarization, ΔP

A scheme of a system used for a signal (6) obtaining, is presented in Fig. 1.

An active element (AE) of the laser has a spectrum of output radiation in form of two orthogonally polarized modes P_1 and P_2 with 640 MHz difference. Powers of the modes change oppositely, when the length ΔL of the optical resonator (R) and frequency of laser radiation change. The changes take place with a period corresponding to an intermode laser interval $\lambda/2$ that is used to stabilize the radiation frequency. For this purpose, radiation from semitransparent mirror $M1$ of the active element, after spatial separation by polarizations with a birefringent crystal (BC), comes to the photodiodes $D1$ and $D2$. Photoinduced currents from $D1$ and $D2$ are subtracted in the differential amplifier DA and come into the automatic frequency control system (AFC), in which they, after the processing in the proportionally integro-differential (PID) regulator, are delivered to power amplifier (PA) and then to the heater (H), mounted on the lateral walls of the AE. The heater set such temperature of the active element AE, at which length of a resonator cavity R corresponds to the equality of powers of the orthogonally polarized radiation modes. At a laser heating by an external heat source, a heater current decreases, while at cooling it increases. This maintains a constant distance between mirrors $M1$ and $M2$ of the optical resonator and stabilizes a radiation frequency [15].

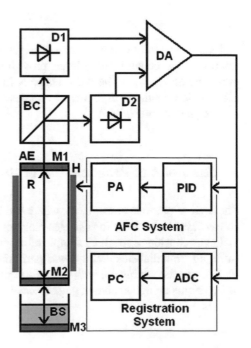

Fig. 1 System for registration of the difference polarization ΔP. Pleases, refer to the text for explanations

As one can see from Fig. 1, a signal obtained at the output of the differential amplifier DA corresponds to the desired signal ΔP (6). For its recording, we use a system consisting of two blocks labeled at Fig. 1 as "Registration system": of an analog to digital converter (ADC) and personal computer (PC) with the appropriate software. After an analog signal from the DA is transformed by the ADC into the digital form, it is recorded in the PC memory for the further use.

An important feature of the system shown at Fig. 1 is the presence of the third reflecting mirror $M3$ at the external part of the AE. The mirror is fixed on the pedestal allowing for its justification in a way that provides reflection of the laser beam back to the resonator R. A biological sample (BS) under examination is mounted over the $M3$, therefore, a beam reflected into a resonator carries information on its optical activity. A beam reflected by the $M1$ passes a substance under examination once more and is reflected by the $M3$. The process is reiterated. That is why we may consider the system presented at Fig. 1 as an intracavity laser polarimeter.

It is well-known [10] that a sign of an angle of a polarization plane rotation in the Faraday effect, contrary to the natural optical activity, does not depend on a direction of light propagation (along or against a magnetic field). Therefore, repeated light passages through a medium placed into a magnetic field (we mean, the Earth magnetic field, first of all) result in the increase of an angle of the polarization plane rotation by a correspondent number of times. This property makes it possible to increase a total angle of a polarization plane rotation at k-fold light passing through a magnetoactive substance. An angle value θ in such a case will be:

$$\theta = 2kVLB_0. \tag{11}$$

From (11) it follows that in a case $k \gg 1$, a value of an angle of a polarization plane rotation θ can be sufficiently large, providing for a high sensitivity of the system, presented at Fig. 1, to the weak changes in optical activity of the material under examination, realizing in such a way a possibility to obtain some information about its "frequency portrait" (4).

4 Examples of Practical Use of the Method

The section presents some cases, where we used a signal $\Delta P(t)$ obtained with the above-described system for diabetes mellitus treatment.

Let us consider a specific example of the treatment use. A seventeen-year-old cat suffering from diabetes mellitus was permanently treated with 10 U/day caninsulin (that is, special vet insulin with two action modes). The animal had a stable performance status though his blood glucose never dropped below 10–14 (healthy glucose level is 3.8–5.8 mmol/L).

We extracted donor beta-cells from the pancreas of a two-week-old kitten and, using the device described in the previous section, registered the $\Delta P(t)$ signal from these properly functioning beta-cells.

After that, we held three runs of the cat irradiating with radiowaves modulated with the $\Delta P(t)$ signals. Level-time pattern of the cat's blood glucose is presented at Fig. 2. Sixteen days after irradiation, the cat's blood glucose fell to 1.3 mmol/L. A caninsulin dose was decreased by 2 U. Twenty-five days later, a caninsulin dose was decreased by 2 U more, and 50 days later canisulin administration was discontinued. Glucose level in the cat's blood came to norm, the animal felt well, gained weight, his hair, dull earlier, became shining, urination became normal, and disposis stopped. He was followed-up for three years after treatment and was healthy and wealthy.

The similar results were observed at treatment of several other cats, we managed to cure eight animals in total.

However, no positive results were obtained at the treatment of dogs suffering from diabetes mellitus with records of the $\Delta P(t)$ of normally functioning kitten cells. To treat the dogs according to the same scheme, we used beta-cells of puppies. Figure 3 presents a level-time pattern glucose in the blood of a dog suffering from diabetes mellitus (eight-year-old) at its treatment with the $\Delta P(t)$ signals of a puppy beta-cells. We can see that the pattern observed in Fig. 3 is similar to that presented in Fig. 2.

Five of six treated dogs were completely healed, and one was partially rehabilitated.

The examples tell that the $\Delta P(t)$ signals are very specific, they affect only the same cells of a recipient: records of the kitten's beta-cells $\Delta P(t)$ never affect cells of other organs of cats, and beta-cells of dogs.

More details of the above-presented results are provided in [16].

A similar method was used in [17], where the method was successfully used for the treatment of a model (alloxan) diabetes mellitus in rats. The experiments were performed by different independent research teams [18, 19] and had the same positive results.

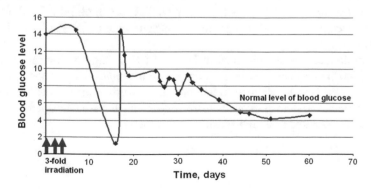

Fig. 2 Course of treatment of a diabetes mellitus cat

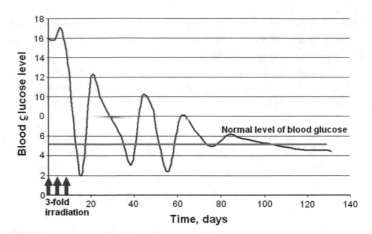

Fig. 3 Course of treatment of a diabetes mellitus dog

5 Conclusion

The researches [1–7] show that biological actions of superweak magnetic fields are intimately connected with the mechanism of Zeeman splitting of energy levels of atoms and molecules. At the same time, the researches [12, 13] show that Zeeman effect is closely connected with Faraday effect. Therefore, the obtained information on a value of Faraday rotation of a polarization plane ΔP can give us information on a value of Zeeman splitting and, further, on values of Ω_i.

However, as Faraday rotation angle of real biological samples has too low values, we used a special system to get information on a value of rotation of polarization plane ΔP. The system is based on a helium–neon laser with two orthogonally polarized radiation modes realizing the scheme of intracavity laser polarimetry: a laser beam, having passed a biological sample and reflected from an external mirror, repasses a sample and returns into the resonator cavity. The last thing provides the system with high sensitivity to small changes in optical activity of a biological sample under examination.

The preliminary results obtained in [16–19] tell on high biological activity of the signal $\Delta P(t)$. As can be seen from the above examples, the method along with [20–24] can be used for diabetes treatment. It is worth a special mentioning that the experimental system presented in the paper can be used independently as a spectrometer $\{\Omega_i\}$ and as a high-sensitive polarimeter.

References

1. Liboff, A.R.: Cyclotron resonance in membrane transport. In: Chiabrera, A., Nicolini, C., Schwan, H.P. (eds.) Interactions Between Electromagnetic Fields and Cells, pp. 281–296. Plenum, New York (1985)
2. Liboff, A.R.: Interaction mechanism of low-level electromagnetic fields and living systems. In: Norden, B., Ramel, C. (eds.), pp. 130–147 Oxford University Press, Oxford (1992)
3. Liboff, A.R., McLeod, B.R.: Kinetics of canalized membrane ions in magnetic fields. Bioelectromagnetics **9**(1), 39–51 (1988)
4. Belova, N.A., Panchelyuga, V.A.: Lednev's model: theory and experiment. Biophysics **55**(4), 661–674 (2010)
5. Lednev, V.V.: Bioeffects of weak combined, constant and variable magnetic fields. Biophysics **41**(1), 241–252 (1996)
6. Belova, N.A., Ermakov, A.M., Znobishcheva, A.V., Srebnitskaya, L.K., Lednev, V.V.: The influence of the extremely weak alternating magnetic fields on the regeneration of planarians and the gravitropic response of plants. Biophysics **55**(4), 704–709 (2010)
7. Belova, N., Ermakov, A., Lednev, V.: Effects of weak combined magnetic fields tuned resonance for nuclear spins on the regeneration of planaria. In: BioEM2013, Thessaloniki, Greece, 10–14 June 2013, PA-221
8. Novikov, V.V., Novikov, G.V., Fesenko, E.E.: Effect of weak combined static and extremely low-frequency alternating magnetic fields on tumor growth in mice inoculated with the ehrlich ascites carcinoma. Bioelectromagnetics **30**, 343–351 (2009)
9. Zimmerman, J.W., Pennison, M.J., Brezovich, I., et al.: Cancer cell proliferation is inhibited by specific modulation frequencies. Br. J. Cancer **106**, 307–313 (2012)
10. Zaguskin, S.L.: Bioresonans and biocontrol in course of laser therapy. Photonics **3**, 62–68 (2012)
11. Alexandrov, E.B., Zapasskiy, V.S.: Laser Magnetic Spectroscopy, p. 280. Nauka, Moscow (1986)
12. Verchozin, A.N.: May low-symmetry molecules have degenerate energy mode? Nat. Sci. Math. **11**, 5–7 (2007)
13. Verchozin, A.N.: Magnetostatics of Weak-Magnetic Molecular Systems. St. Petersburg, 286 p (2006)
14. Panchelyuga, V.A., Eventov, V.L.: Determination of biologically active frequencies and based on them method of therapy. Med. Phys. **3**(67), 48–54 (2015)
15. Vlasov, A., Hilov, S.: Frequency-stabilized He-Ne lasers for interferometry. Photonics **5**, 7–9 (2007)
16. Eventov, V.L., Tertyshny, G.G., Shydkov, I.L., Adrianova, MYu., Sitnichenko, N.B.: Wave correction of cells functioning. Bull. Russ. Acad. Nat. Sci. **11**(1), 22–25 (2011)
17. Eventov, V.L., Tertyshny, G.G., Shydkov, I.L., Sitnichenko, N.B., Adrianova, M.Y.: Wave control of organism cell functioning. In: Modern Information Technologies in Science, Education, and Practice, VII All-Russian conference, Orenburg, 2008, pp. 618–623
18. Kokaya, N.G., Kokaya, A.A., Mukhina, I.V.: The effect of corrective and preventive electromagnetic radiation modulated by biostructures on the course of acute insulin insufficiency in rats. STM **3**, 11–15 (2011)
19. Gariaev, P.P., Kokaya, N.G., Mukhina, I.V.: The effect of electromagnetic radiation modulated by biostructures on the course of acute insulin insufficiency in rats. Bull. Exp. Biol. Med. **2**, 155–158 (2007)
20. Akyol, K.: Assessing the importance of attributes for diagnosis of diabetes disease. Int. J. Inf. Eng. Electron. Bus. (IJIEEB) **9**(5), 1–9 (2017). https://doi.org/10.5815/ijieeb.2017.05.01
21. Priyanka Shetty, S.R., Joshi, S.: A tool for diabetes prediction and monitoring using data mining technique. Int. J. Inf. Technol. Comput. Sci. (IJITCS) **8**(11), 26–32 (2016). https://doi.org/10.5815/ijitcs.2016.11.04

22. Choubey, D.K., Paul, S.: GA_MLP NN: a hybrid intelligent system for diabetes disease diagnosis. Int. J. Intell. Syst. Appl. (IJISA) 8(1), 49–59 (2016). https://doi.org/10.5815/ijisa.2016.01.06
23. Jain, V., Raheja, S.: Improving the prediction rate of diabetes using fuzzy expert system. Int. J. Inf. Technol. Comput. Sci. (IJITCS) 7(10), 84–91 (2015). https://doi.org/10.5815/ijitcs.2015.10.10
24. Allam, F., Nossair, Z., Gomma, H., Ibrahim, I., Abdelsalam, M.: Evaluation of using a recurrent neural network (RNN) and a fuzzy logic controller (FLC) in closed loop system to regulate blood glucose for type-1 diabetic patients. Int. J. Intell, Syst. Appl. (IJISA) 4(10), 58–71 (2012). https://doi.org/10.5815/ijisa.2012.10.07

From the Golem to the Robot
and Beyond to the Smart Prostheses

Anatoly K. Skvorchevsky, Alexander M. Sergeev and Nikita S. Kovalev

Abstract Golem and the Robot are a historical example of artificially created humanoids—"programmable" organisms. The ideas, underlying the creation of these historical characters have influenced many areas of modern mechatronics. The concept of "smart", which is equally applicable in social and technological environments is considered separately. The article emphasizes the development of systems with electronic control, called "smart prosthesis".

Keywords Golem · Robot · Humanlike creatures · Mind · Anthropomorphic · Smart prosthesis

1 Introduction

It is a paradox, but in modern science (even in such an advanced branch as the radioelectronics) uses the terms which were applied in the past century and, moreover, in the Middle Ages. The viability of these terms is explained by the fact that they have a rich meaning. But that is why their use today and, especially in the future, every time should be based on the realization of the richness of these meanings. As an example of this one, we consider the Golem and the Robot concepts and connection of these concepts with the modern research and development in the field of the limb's prosthetics.

A. K. Skvorchevsky · A. M. Sergeev (✉) · N. S. Kovalev
Mechanical Engineering Research Institute of the Russian Academy of Sciences,
4, Malyi Kharitonievsky pereulok, 101990 Moscow, Russian Federation
e-mail: xemirc@yahoo.com

© Springer Nature Switzerland AG 2020   355
Z. Hu et al. (eds.), *Advances in Artificial Systems for Medicine and Education II*,
Advances in Intelligent Systems and Computing 902,
https://doi.org/10.1007/978-3-030-12082-5_33

2 Golem

According to Jewish tradition, a *Golem*, (from the Hebrew word meaning "unformed matter") is an artificial human moulded from clay, just as God had shaped Adam from the dust of the earth. But Golem, unlike Adams, is "unformed," because he lacks the soul and ability to speak, which God bestowed exclusively on people. However, a Golem can move and act.

Because God created the world and humankind using the power of speech, Jewish mystics believed they could create a humanoid in a similar way [1].

According to various myths, dating back to the eleventh century, Jewish mystics created Golems to serve as personal servants.

Legends say—not only clay was used. It is also known that not only male individuals were created. A Spanish poet and philosopher Solomon ibn Gabirol, as the story goes, created his own Golem. It was a woman. And she was made from the pile of wood.

The most famous stories about the Golem date back to the sixteenth century and are connected with the territory of Poland.

A German philologist, jurist and mythologist Jakob Grimm wrote about this. "Having created certain prayers and having sustained several days of fasting, Polish Jews create a man out of clay or mud, and when they pronounce over him the miraculous Shem ha-Mephorash (the God's Ineffable Name), he comes to life. True, he does not have the gift of speech, but is able to understand the speech and orders addressed to him. They call him a Golem and use as an assistant in the household; he can not leave the house. On the forehead of the Golem the word EMET (TRUTH) is written. Every day he grows and,—no matter how small he was at first,—gradually becomes taller and stronger than all the inhabitants of the house. Therefore,—when he begins to cause fear,—they erase the first letter, so that the word MET (DEAD) remains,—and then the Golem falls and crumbles,—turning into clay. Once a careless master allowed the Golem to grow so tall that he could no longer reach his forehead. Filled with fear of such a servant, the master ordered him to take off his boots,—hoping to reach the forehead of the Golem, when he bended down. So it happened. But as soon as the letter was erased, a mountain of clay hit the Jew and crushed him" [2, 3].

The literary tradition connects this legend with Rabbi Elijah Baal Shem from the Polish city Helm. This story emphasized that the man was created without speech and served the master as a slave (Fig. 1).

At the turn of the second half of the eighteenth century the legend of Rabbi Eliyahu Baal Shem from Helm reached Prague where it was associated with a much more famous figure Judah Loew ben Besalel or Maharal from Prague. In the Prague tradition of the nineteenth century, this legend was associated with a special custom in the liturgy of the Saturday evening. It is said that Rabbi Levy created a Golem who served him six days a week, doing all sorts of work.

The most common story is that the Rabbi used a Shem, a slip of paper (according to other sources—it was a sign) with God's name written on it, to keep the Golem

Fig. 1 The animation of the Golem

alive. The Shem, placed within the mouth of the Golem, gave it life. He extracted the Shem from the Golem's mouth every Friday afternoon, so as to let it rest on Sabbath. At that moment the Golem turned into a lifeless body. Once Rabbi Loew forgot to do this. Suddenly, the mighty Golem began to rage threatening to destroy everything around. People summoned Rabbi Loew to help. The sun had not set yet, and the Sabbath had not begun. Rabbi Loew rushed to the running Golem, snatched from him Shem ha-Mephorash and the Golem fell crumbling to dust. After this incident Rabbi Loew did not return the Golem to life at the end of the Sabbath, but "buried" its remains in the attic of the old synagogue, where they are still [2, 3].

Towards the end of the nineteenth century, many Jewish writers began to remake the Golem into a folk-hero, a symbol of renewed Jewish strength and heritage in the form of a man-made messiah.

We are obliged to "Polish Rabbi" Yudel Rosenberg and Hayim Bloch for wide dissemination of the legend about the creation of the Golem by Maharal from Prague. Rosenberg published in 1909 a book titled "The Golem and the Wondrous Deeds of the Maharal of Prague" allegedly written the famous Rabbi's son-in-law and containing a story about the struggle of Maharal with blood libels, his public dispute with a Christian priest, a description of the creation of the Golem, and a number stories how Maharal with the help of the Golem miraculously saved the Jews from oppression [4]. The book concludes that Maharal achieved the prohibition to examine blood libel in court and destroyed the Golem as superfluous.

Ultimately Golem became bodyguards to protect their people against devastating pogroms when prayer alone did not suffice.

Proceeding from the above, we can draw several conclusions:

- A Golem is an artificially created human being that is given life by supernatural means;
  A Golem was created as a dumb slave, at best, as a submissive servant, as creature that is completely subservient to the owner.
- The creation of a Golem is associated with danger. In this case the source of danger is not the Golem but the creator. Errors in the execution of prescriptions can lead to the death of not only the creature, but also its creator.
- The anthropomorphic nature of the Golem is determined by the desire to resemble God who created Adam, the first man.
- The phenomenon of the Golem is based on the belief that with "God's assistance" an anthropomorphic creature can perform dirty work instead of the person and better than the person.

The ability to create "living matter" from "lifeless matter" still attracts researchers.

This is confirmed by some projects being developed today. Among these projects there is an abstract framework called Golem for the design and modelling of behaviour of autonomous robotic systems [5]. Another project is the design of the robot based on a conceptual framework that is centred on the notion of dialogue models with the interaction-oriented cognitive architecture (IOCA) and its associated programming environment, SitLog [6].

3   Robot

By a strange coincidence, it was in the Czech Republic that the history of the Golem got its further development.

Back in 1908, the Chapek brothers (Karel—the writer and Josef—the artist) wrote the story "System", the hero of which—an American manufacturer—invented a working machine: operarius utilis—a useful worker. When Karel came up with the idea of a play about minded living machines, he turned to his brother: "I do not know (…) how I'll call these artificial workers. I would call them laborji (Chech. laboři from English "labour.") but it seems to me that it is too bookish. "So call them robots," suggested Joseph. In the Czech language there are two verbs for labor activity with similar synonymic meanings: pracovati—to work and robiti—to do. From the latter, a few words are formed with a derogatory connotation robotiti—to graft, to grub; robota—drudgery, corvee labor; rob—slave; robotnik—corvee, laborer; robotny—serf. Thus, the term *robot* is formed from these derogatory words and conveys their meaning [7].

The play was written and titled "R.U.R."—Rossumových Univerzálni Roboti (Fig. 2).

Fig. 2 The Czech playbill of
RUR

In the Prague newspaper Prager Tagblatt (23 September 1935) Chapek wrote "R.U.R. is in fact a transformation of the Golem legend into a modern form. However, I realized this only when the piece was done" [8].

Chapek's idea and his *humanoid automat* attracted popular attention: the play was translated into many languages and staged in many countries. Already in 1921, the abbreviation R.U.R. was deciphered in English like Rossum's Universal Robots without the translation of the last word.

In the German translation, made in 1922 by Otto Pick, the name of the inventors was "Werstand" probably from the German word verstand—the mind. For this reason, the play became known as WUR (Fig. 3).

In 1924 Chapek's play was translated from German into Russian with the name "VUR" (ВУР)—"Verstand's Universal Robots". In the Russian version Rossum turned into Verstand. Apparently, the translator did not use the Czech name Rossum (mind) which was in the original title of the play, but took it from the German translation, transforming Werstand into Verstand (mind). At the same time, the main idea of the play that the human mind with the help of science is able to create similar creatures was preserved [9] (Fig. 4).

In his play, Chapek described the heyday and death of the robots manufacturing company, whose brand included the names of two robot's "fathers"—professor Rozum and his nephew, engineer Rozum. With the aid of the words of his work's hero, he gave the main characteristics of both Rosum's humanlike creatures. Robot has a normal size and very high class human finish. It is the cheapest and its requirements are the smallest. As one of the protagonists of the play said: "In fact he (prof. Rozum) rejected man and made the Robot … the Robots are not people. Mechanically they

Fig. 3 Cover of the German
translation of the RUR

Fig. 4 Title page of the
Russian translation of the
RUR

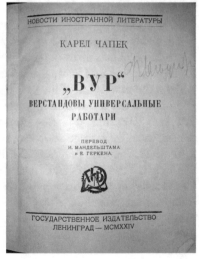

are more perfect than we are, they have an enormously developed intelligence, but they have no soul" [10, 11].

Chapek foresaw the need to create specialized robots. The play emphasizes that Rossum's Universal Robot factory does not produce a uniform brand of Robots, and they have Robots of finer and coarser grades. The play also notes that "the best will live about twenty years".

From the point of view of the robot design Chapek stressed that it should be supplied with tactile sensations including a sense of pain [10, 11].

Do different Golems and Chapeck's robots have *artificial intelligence*? It would be more correct to say that they were *smart*.

4 Smart: The Word for Two Worlds

In the English language the disticrion between the concepts "the natural mind" and "the mind in the technological sphere" has long been accepted. But there is a term equally referring to both the "natural" and the "artificial" world—this is the word "smart". So, we have a "smart vehicle", and at the same time, we find the expressions "a smart carpenter", "a smart servant" or "a smart driver".

The basis through which this word connects the two worlds is the quality of work. In other words, here we are dealing with functions performed "wisely" and, as a result flawlessly or perfectly, according to the criteria that correspond to the ideas of a person or community (Fig. 5).

Smart is a very capacious concept which has a wide range of applications in the field of technology. So, today we can talk not only about smart phones, but also about smart homes, smart factories and even smart cities [12–15].

Thus we can talk about a variety of objects that are capable of using various forms of assimilation and processing of information into special signals in an optimal way to carry out useful work. At the same time, the criterion of the work performed will be the satisfaction of the user's requirements. The concept of smart also assumes a certain degree of autonomy and implies such a solution to the problem, which can be

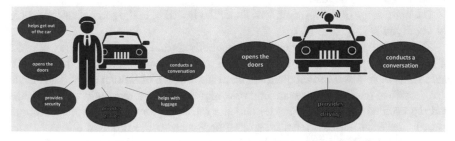

Fig. 5 This is the case when the person and the machine are "smart" each in their own way, but both are "smart"

called skillful, refined or elegant. With regard to the word smart, we can talk about a "organic" way solution to a problem.

In order to characterize somebody of something as "smart", we of course must take into account the perception. But there are objects in which perception is the most important factor. And here, we can not do without smart technologies. We are talking about prostheses—devices from the field of an "artificial" nature, which are connected with a living organism.

Today, UNYQ designs and manufactures smart 3D-printed prosthetics that are personalized for each individual and integrate embedded activity tracker sensors [16].

The research and development of the mechatronic system design of an instrumented lower-limb prosthetic leg is currently held. This is the semi-active prosthetic knee. In future, the prosthetic leg system will have the capability of auto-settings and tuning based on the amputees requirements [17].

The work is also carried out with the initial design of an EEG mind-controlled smart prosthetic arm. The arm is controlled by the brain commands, obtained from an electroencephalography (EEG) headset and equipped with a network of smart sensors and actuators that give the patient intelligent feedback about the surrounding environment and the object in contact. This network provides the arm with normal hand functionality, smart reflexes and smooth movements. Various types of sensors are used including temperature, pressure, ultrasonic proximity sensors, accelerometers, potentiometers, strain gauges and gyroscopes. The arm is completely 3D-printed built from various lightweight and high strength materials that can handle high impacts and fragile elements as well [18].

Another approach to the "smart prosthesis" is connected with a control scheme for hand prostheses implementing multiple pinches and grasps. The control signals for the hand are determined by myoelectric signals from the arm and volitionally generated signals by tongue through an inductive interface with a mouth piece [19].

At present the laboratory of "Dynamics of human-machine systems" of the Mechanical Engineering Research Institute of the Russian Academy of Sciences is working on a project called "Humanics". The result of our work will not be a "robotic arm", which includes the legacy of rebellious and threatening Golems of Jewish myths or the pseudo-people of Chapek, but a "smart arm" that will be maximally "friendly" to the user.

The main stage of the research was launched in 2012. In the same year the article by Vorobyov et al. [20] (Fig. 6).

The article described the scientific basis, methods and tools for the study of tasks of biomechanics in the view of activation of the human factor, namely the methods and means of removal of biopotentials from the surface of human body. The article also described hardware–software complex algorithm of human body and its organs and represented the physical model and the rationale for performance of motion, named "artificial muscles". An integrated approach to the problem allowed us to formulate and build a working model of anthropomorphic bioprosthesis of hands.

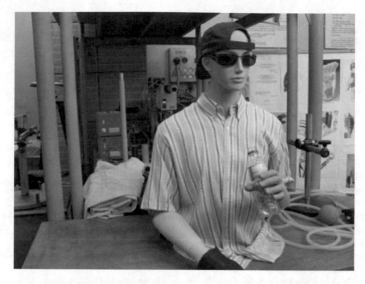

Fig. 6 Anthropomorphic bioprostheses of an arm, made in the laboratory of "Dynamics of human-machine systems" of Mechanical Engineering Research Institute of the Russian Academy of Sciences

In 2015, the studies went on and the main ideas were reflected an article by K.A. Skvorchevsky et al. [21].

In 2017, the theme of anthropomorphic bioprostheses was further developed. The result of the conducted studies was the article by A.K. Skvorchevsky et al. [22] and then one more article by the same authors [23].

Today in the laboratory, the main attention is paid to the design of the prosthesis with the aim of making it as close as possible to the anatomy of the human hand and its natural movements. Thus a transition is made from attempts to create an "anthropomorphic" prosthesis to an "anthropoessential" one based not on external similarity, but on the essence of the human organ itself.

5 Conclusions

Golem did not die. In any case, the name "Golem" today is reflected in the works on the creation of "anthropomorphic automata". As for the robots this theme dominates in scientific research and development all over the world. Being aware of the history of these creatures, we tried to use the term "smart" at the new stage of our project (aimed to create the most comfortable and effective prosthesis of arm).

Once Wiener, the father of cybernetics, described robot as a slave. Today the term "slave" is not so acute in the world of technology. A "slave" is just an executing mechanism in a particular system, for example, the part of surgery system DaVinci.

But undertaking the development of a prosthesis, we would not want to establish "master–slave" relationship between our device and the user. Let these relations be friendly.

References

1. Bertman, S.: The role of the Golem in the making of Frankenstein. Keats-Shelly Rev. **29**(1), 42–50 (2015). https://www.researchgate.net/publication/276165839_The_Role_of_the_Golem_in_the_Making_of_Frankenstein
2. Scholem, G.: The Idea of a Golem in Its Telluric and Magical Connections (In Russian). http://online-books.openu.ac.il/russian/germanhasidism/appendix/app16(2).html
3. Gershovich, U.: Golem: the Legend or Myth (In Russian). http://old.lechaim.ru/3519
4. Rosenberg, Y.: The Golem and the wondrous deeds of the Maharal of Prague. Yale University Press, New Haven/London (2007)
5. Messina, F., Pappalardo, G., Santoro, C.: Designing Autonomous Robots using GOLEM. http://ceur-ws.org/Vol-1260/paper15.pdf
6. Reyes, Y. M.: Sistema de monitoreo de signos vitales usando visión computacional para el robot Golem-II+, 20–21. http://golem.iimas.unam.mx/pubs/martinez15sistema.pdf
7. Shustov, A.I.: Once again about the Robot. Russian language. No. 6, pp. 96–99 (1993) (In Russian) http://russkayarech.ru/files/issues/1993/6/18-shustov.pdf
8. Horakova, J.: Robot: The modern age Golem. In: Beyond AI: Artificial Golem Intelligence. Proceedings of the International Conference Beyond AI 2013, pp. 5–6. Pilsen, Czech Republic, 12–14 Nov 2013. https://www.beyondai.zcu.cz/files/BAI2013_proceedings.pdf
9. Maslova, K.K.: On the "Theme Genesis" in sci-fi dramas by K. Čapek and A.N. Tolstoy. Vestnik slavianskikh kul'tur (In Russian language), vol. 44, pp. 117–125 (2017)
10. Chapek, K., R.U.R.: (In Russian). https://royallib.com/read/chapek_karel/RUR.html#0
11. Capek, K., R.U.R.: Rossum's Universal Robots. http://preprints.readingroo.ms/RUR/rur.pdf
12. Mannion, P.: Smart Vehicles Launch the Autonomous Moonshot. https://www.electronicproducts.com/Automotive/Smart/Smart_vehicles_launch_the_autonomous_moonshot.aspx
13. Teslyuk, V., Beregovskyi, V., Denysyuk, P., Teslyuk, T., Lozynskyi, A.: Development and implementation of the technical accident prevention subsystem for the smart home system. Int. J. Intell. Syst. Appl. **1**, 1–8 (2018). http://www.mecs-press.org/ijisa/ijisa-v10-n1/IJISA-V10-N1-1.pdf
14. Edvards, J.: Building a Smart Factory with AI and Robotics. https://www.roboticsbusinessreview.com/download/building-a-smart-factory-with-ai-robotics/
15. Faisal, H., Usman, S., Zahid, S.M.: In what ways smart cities will get assistance from internet of things (IOT). Int. J. Educ. Manag. Eng. **2**, 41–47 (2018). http://www.mecs-press.org/ijeme/ijeme-v8-n2/IJEME-V8-N2-5.pdf
16. UNYQ.: Personalized, Smart Prosthetics & Orthotics. http://unyq.com/us/
17. Awad, M.I., Abouhossein, A., Dehghani-Sanij, A.A., Richardson, R., et al.: Towards a Smart Semi-Active Prosthetic Leg: Preliminary Assessment and Testing. http://eprints.whiterose.ac.uk/102329/1/Mechatronics%20conference%202016%20ver.4.pdf
18. Beyrouthy, T., Al Kork, S., Korbane, J.A., Abouelela, M.: EEG Mind Controlled Smart Prosthetic Arm—A Comprehensive Study. https://www.astesj.com/publications/ASTESJ_0203111.pdf
19. Johansena, D., Popović, D. B., Struijkc, L., Sebeliusd, F., Jensene, S.: A novel hand prosthesis control scheme implementing a tongue control system. Int. J. Eng. Manuf. **5**, 14–21 (2012). http://www.mecs-press.org/ijem/ijem-v2-n5/IJEM-V2-N5-3.pdf

20. Vorobyov, E.I., Skvorchevsky, A.K., Sergeev, A.M.: The problems of creating the control algorithms of anthropomorphic bioprostheses of hands and feet. Med. High Technol. **1**, 7–12 (2012) (In Russian)
21. Skvorchevsky, K.A., Sergeev, A.M., Akentiev, A.A., Skvorchevsky, A.K., Gudushauri, E.G.: Medical robotics for compensation of lost functions of patients on quantum and radioelectronic elements implanted into bioprostheses. Med. High Technol. **4**, 45–54 (2015) (In Russian)
22. Skvorchevsky, A.K., Vorobiev, E.V., Sergeev, A.M., Kovalev, N.S.: Information models of anthropomorphic bioprostheses with artificial muscles based on the hydrolaser effect and control algorithms. Med. High Technol. **2**, 63–71 (2017) (In Russian)
23. Skvorchevsky, A.K., Vorobiev, E.V., Sergeev, A.M., Kovalev, N.S.: Algorithms for controlling the brush orientation for the anthropomorphic prosthetic hands with artificial muscles. Med. High Technol. **2**, 72–77 (2017) (In Russian)

Part III
Advances in Technological and Educational Approaches

Lifelong Education of Sports Media Professionals Based on System Theory

Ziye Wang, Qingying Zhang and Mengya Zhang

Abstract Lifelong education is a new idea of education and the demand of the new time. From a system perspective, the paper discusses the lifelong education and the significance for sports media professional's cultivation. Lifelong education is typically a comprehensive system, with the entirety and pertinence as the main features, and is composed of multiple elements. By probing the career demands of sports media professionals in three aspects, the paper lists various approaches which are available to accomplish lifelong education and learning. The paper also presents some combinations, i.e., a combination of university and continuing education, off-job and part-time training, specialty and technical skill cultivation, system education and qualification certification, theory study and practice drilling, and online learning and offline practical operation, which are important to lifelong education and training of sports media professionals.

Keywords Lifelong education · Sports media professionals · System theory

1 System View on Lifelong Education

Lifelong education refers to all kinds of educations a person accepted throughout his lifetime, including school and social, formal and informal. Propounding that everyone should get necessary knowledge and skill by the best means in times of need, lifelong education becomes the guideline of educative reform in many countries.

From the viewpoint of system theory, lifelong education is completely a system which is one part of the nation's education system, but also an independent or self-contained system operated in its own regulation and law.

Z. Wang (✉)
Wuhan Sports University, Wuhan 430079, People's Republic of China
e-mail: kathy8899@126.com

Q. Zhang · M. Zhang
Wuhan University of Technology, Wuhan 430063, People's Republic of China

© Springer Nature Switzerland AG 2020
Z. Hu et al. (eds.), *Advances in Artificial Systems for Medicine and Education II*,
Advances in Intelligent Systems and Computing 902,
https://doi.org/10.1007/978-3-030-12082-5_34

1.1 System Construction

According to the system theory originated from natural science and propounded by Ludwig Von Bertalanffy, a system is an organic and intricate whole composed of lots of factors in a given construction aiming at the definite target. The basic idea of system thought is to consider every issue of the world as a system, in which the factors affect each other while making the system to be functional and stable in construction. The elements of the system are not isolated, and each of them has its own position and role. By taking the object as a whole, analyzing it from three directions, i.e., the microlevel of the elements, the middle plane of the structure, and the macrolevel of the environment, and searching their relations and dynamics, it is capable to realize the structure and feature of the system and determine the correct path to solve the problems.

1.2 Entirety and Relatedness of a System

Entirety, the most prominent feature of the system, means the relation between the system and its factors. What decide the properties and functions of a system are the factors and their connection way rather than the simple superposition of the factors or elements. By connecting separate elements organically in different ways, specific structures are set up, which bring distinctive synergistic effects which are not equipped on those elements before being associated together.

Pertinence, or dependency, correlation, refers to the interacting, interrelation, interdependency, and mutual restriction between different factors, parts, and system. This correlation is between factors and factors, system and factors, system and system, system and environment and is about the construction, function, behavior, and their universal integration.

2 Issues About Lifelong Education

The terminology "lifelong education" was proposed by Paul Lengrand, director of adult education, UNESCO, 40 years ago. A popular acceptable concept is that lifelong education is the summation of all education and training a person received in his lifetime [1]. The fundamental objective of it is to keep and raise the quality of people in their social life [2], and to develop their personality and characters, discover their potential and play maximum to their talents.

2.1 Cause of Lifelong Education

Research shows that the aging rate of human's knowledge is increasingly accelerated. It is said that aging cycle of knowledge was 5–8 years 40 years ago, 3–5 years 20 years ago, while nowadays, the cycle becomes shorter and shorter, and outdated speed of knowledge is unimaginable high. This is why the concept of lifelong education has appeared to the public, attracted attention, and become widely acceptable.

Lifelong education is most often regarded as a supporting system for lifelong learning which encompasses all activities of an educational nature, with an ultimate goal of universalization of self-education and self-growth.

2.2 A New Idea of Education

Lifelong education is a new idea which affirms that people have the demand of self-education, self-training, and self-developing at any time and any place. School and university, as well as society education, should be available to serve it, and any form of barriers and obstacles must be eliminated.

Lifelong education system in China is roughly divided into three subsystems. One is the school or university education which is mainly about degree courses, named academic education [3]. This is the backbone of the educational system and the foundation of the lifelong education. The second is the industry education focused on professional qualification, especially when a person is ready to enter a new business which needs specific aptitude or planned to change his area of expertise into the given industry, where versatile and flexible education and training are indispensable. The third is social education system dealt with cultural, life, and continuing learning, which has become a characteristic system and gets expanding influence along with the development of computer, Internet, public media, and various advanced technologies [4].

2.3 Significance of Lifelong Education

To survive is to challenge. Lifelong education system is a modern mechanism which is sized up in a big angle of education, and an organic association of family education, school and university education as well as separate fields of social education, such as minor course, advanced study, specialized training, practice, and adult education. It goes beyond the transcendental stage and institutionalization, but with individuality in the value design of education target, with diversity in the implementation form of the education procedure, with the integration in the temporal and spatial variation of the category, while with the openness on the education resource support.

Some decades ago, public believes that a person could rise to the occasion at anytime of his life as long as he/she masters the amount of knowledge and skills [5]. This thought is absolutely outdated currently [6]. It is the process and progress rather than the knowledge which is helpful and powerful for people to improve themselves continuously.

It is said that education and training do not get ending when the school study is finished, while running through people's whole life. This is the true essence of lifetime education.

3 Approaches to Lifelong Education

The establishment of lifelong education system is the inevitable choice of the development of education which includes two contents mainly. One is from the society where the unabridged architecture and surrounding are to be built to offer the chance and condition to the civics. The other is from the citizen to promote and stimulate themselves to get their lifelong education.

Various approaches mentioned below are available for lifelong learning.

The first one is to change the idea of studying. Lifelong learning, study every time and everywhere, team learning, research learning, and reflective learning become a new trend and people's survival state instead of traditional classroom regular studying.

The second thing is to explore learning potential of the persons and integrate them as scarce resources.

The third issue is to wake up members' awareness of study by creating learning organizations and to foster their interesting of continual learning.

The fourth thing of lifelong education is to provide learning conditions to ensure it to be fruitful by guiding learners' behavior and correcting their mode. Improving learning network is also important to make the learner a big gain. Also, an effective carrier is vital to enlarge the learners' harvest.

The last but not least is to innovate learning pattern to inject more energy into lifelong education. Facing the demand of people's development to arrange and evaluate the learning effectiveness is considered to be valid to keep learning a survival precondition and progress motivation.

4 Career Demands of Sports Media Professionals

Systematically speaking, sports media professional (SMP) has some specific career demands. Aside from mastering issues about sports games and activities as comprehensive as possible, they need to have a wisdom mind, sharp perspective, quick and accurate judgment, precise and reasonable expression, warm and moderate interaction with the audience, and to be able to grasp new skills and techniques [7]. Along with the development of science and technology, new style of media forms and communication tools, mode of propagation emerge constantly [8]. Keeping pace with the times is definitely essential for the professionals of this field to progress and improve continually [9].

Figure 1 shows the career demands of the sports media professionals. Three aspects are personal quality, professional quality, and professional skills.

Personal qualities include:

(1) Social responsibility which is indispensable for a media worker who was usually viewed as a gatekeeper of public opinion [10];
(2) Humanistic quality, the internal character of a person with the core of humanity spirit of people-oriented or people-centered;
(3) New media thinking, which is vital for media professionals at the information era;
(4) Creative spirit, i.e., new viewpoint, new thinking, and new method, to help media staff changing mind to do work more effective and more valuable [11].
(5) As for the professional quality, knowledge of sports and game is essential definitely. Besides, language skill is the base, while commentary ability enables sports media professionals do their job more powerful.
(6) Professional skills include:
(7) Spot control ability: When the sports game is extremely fierce, audiences may lose their temper where the media personnel; e.g., commentator should try his best to lead the fans to chill out and calm down; thus
(8) Interacting with audience actively and initiatively are serviceable. Making the spot warm when it is cold and cheerless, while turning it down when it is too turbulent and tempestuous in a kind way is the duty of sports media professionals;
(9) To be able to accomplish those works such as interview, producing and broadcasting concurrently are the basic requirements of the sports media professionals of this time [12]; while,
(10) Proficient use of new media tools is obbligato.

Personal quality	Professional quality	Professional skill
Social responsibility	Sports knowledge	Spot control ability
Humanistic quality	Game knowledge	Interact with audience
New media thinking	Language skill	Interview, production and broadcasting
Creative spirit	Commentary ability	Use of new media tools

Career demand of sport media professionals

Fig. 1 Career demand of sports media professionals

5 Lifelong Learning and Education System of Sports Media Professionals

5.1 Lifelong Learning of SMP

As shown in Fig. 2, lifelong learning can be divided into two parts: conscious learning (presented as C) and unconscious learning (U).

Conscious learning could be guided by both guidance: external and internal (self-guidance), signed as A_1 and A_2.

External guidance includes adult education (B_{11}), professional training (B_{12}), and MOOC study (B_{13}). Self-study is marked as B_{21}.

Unconscious learning incorporates planed but aimless learning (A_3), accidental or random learning (A_4), and social learning (A_5).

What called planed but aimless learning could be by journey (B_{31}), by game watching (B_{32}), and by imitation (B_{33}).

These are useful ways for sports media professionals to improve themselves greatly. Adventives learning (B_{41}) is a kind of accidental learning (A_4).

As for the social learning, those items, such as professional training (B_{51}), regular study (B_{52}), practice learning (B_{53}), and by MOOC (B_{54}) are all available to SMP.

Figure 2 illustrates that conscious learning and social learning are necessary for SMP to get the qualification to sustain the career.

As seen from Fig. 2, the comprehensive effect of various indexes determines the quality of sports media talents.

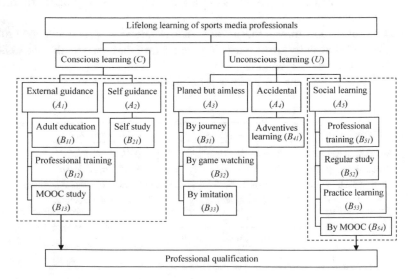

Fig. 2 Lifelong learning of sports media professionals

Assuming that the quality of sports media talents is Z, it is the sum or combination of conscious learning C and unconscious learning U.

$$Z = C + U \tag{1}$$

The model of the first stage index synthesis is written as:

$$C = \sum A_i R_i \tag{2}$$

$$U = \sum A_j R_j \tag{3}$$

where A_i represents the score of the first stage index item, and R_i and R_j, respectively, represent the weight coefficient of each index item. Among them, $i = 1, 2$ while $j = 3, 4, 5$.

The model of the second stage index synthesis is described as:

$$A = \sum B_{ik} R_{ik} + \sum B_{jk} R_{jk} \tag{4}$$

In formula (4), B_{ik} and B_{jk} represent the scores of the second stage index items, and R_{ik} and R_{jk}. respectively, represent the weight coefficient of each index item. Among them, $k = 1, 2, 3, 4$.

For a specific case, by substituting values of separate index and the corresponding coefficient into the formula, the evaluation result of a lifelong learning and professional qualification is capable to be attained concretely.

Massive Open Online Courses (MOOC), appeared twice in Fig. 2 (A_{13} and A_{54}), are viewed as a sonic storm of education and learning, which is a fenceless universities opened to the public of the whole world [13]. People get into the course anywhere with the Internet and electronic reading tools (computer or mobile) by watching a video of teacher's lecture and reading materials on the course Web site [14]. They interact with the teacher and other learners, so as to make their mind clear by discussion even contesting at the interactive zone [15]. For SMP, a special area well designed on the Web site of the course is capable to offer the learners to show their work and cause comment and criticism by each other.

5.2 Lifelong Education System for SMP

Lifelong education and training for sports media professionals is a complex system which deals with lots of elements and is affected by inner or outer factors.

The system is set on an environmental surrounding which puts various effects on the education system [16]. Based on the career demands of SMP, a lifelong education and training system supports and promotes professional to grow steadily and persistently.

As in Fig. 3, it is presented that some combinations are momentous, which are the combination of university and continuing education, off-job and part-time training, specialty and technical skill cultivation, system education and qualification certification, theory study and practice drilling, and online learning and offline practical operation.

Taking sports media professionals as the object of the system, career demand is the inner factor, while externality, i.e., the environment, refers to multifarious outer factors affected on the system [4]. Lots of contents build a lifelong education and training systems which support sports media professionals to grow unceasingly and improve continuously.

6 Conclusion

The objectives of research issues about lifelong education involve successful adjustment to life; all-round development of the person; and establishment of an equitable society. For sports media professionals, keeping themselves improving and growing persistently is the basic requirement to rise to the challenge of the society and the times. Lifelong learning supported by a perfect social education system is therefore vital.

In order to cultivate more and more excellent sports media professionals, a lifelong education and training system is necessary to be built, the curriculum system should be well designed, and the teaching methods should be reformed. All of those have

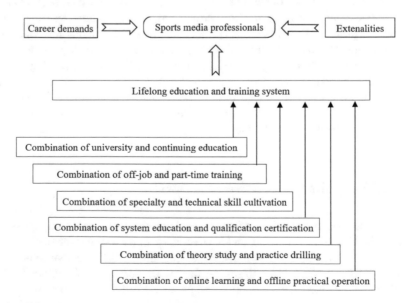

Fig. 3 Lifelong education and training systems for SMPs

to be done under the base of a combination of theory and practice, facing the career requirements of sports media professional.

References

1. Robles, A.C.M.O.: Blended learning for lifelong learning: an innovation for college education students. IJMECS **4**(6), 1–8 (2012). https://doi.org/10.5815/ijmecs.2012.06.01
2. Wu, Z., Guo, H., Zhao, H.: Retrospective analysis of Chinese lifelong education policy. Educ. Dev. Res. **17**, 16 (2012)
3. Alkhathlan, A.A., Al-Daraiseh, A.A.: An analytical study of the use of social networks for collaborative learning in higher education. Int. J. Mod. Educ. Comput. Sci. (IJMECS) **9**(2), 1–13 (2017). https://doi.org/10.5815/ijmecs.2017.02.01
4. Kotevski, Z., Tasevska, I.: Evaluating the potentials of educational systems to advance implementing multimedia technologies. Int. J. Mod. Educ. Comput. Sci. (IJMECS) **9**(1), 26–35 (2017). https://doi.org/10.5815/ijmecs.2017.01.03
5. Khan, A.A., Madden, J.: Speed learning: maximizing student learning and engagement in a limited amount of time. Int. J. Mod. Educ. Comput. Sci. (IJMECS) **8**(7), 22–30 (2016). https://doi.org/10.5815/ijmecs.2016.07.03
6. Cao, Y., Liu, Y.: Attainment of press-type host and cultivating under the background of convergence media. Sci. Technol. Commun. **6**(22), 208–209 (2014)
7. Zhang, K.: Assessment of media professionals with four dimensions. Youth Report. **25**, 68–69 (2015)
8. El Haji, E., Azmani, A., El Harzli, M.: Using AHP method for educational and vocational guidance. Int. J. Inf. Technol. Comput. Sci. (IJITCS) **9**(1), 9–17 (2017). https://doi.org/10.5815/ijitcs.2017.01.02
9. Papadimitriou, A., Gyftodimos, G.: The role of learner characteristics in the adaptive educational hypermedia systems: the case of the MATHEMA. Int. J. Mod. Educ. Comput. Sci. (IJMECS) **9**(10), 55–68 (2017). https://doi.org/10.5815/ijmecs.2017.10.07
10. Li, Y.: Cultivation of inter-disciplinary media talent under the background of convergence media. Beijing Educ. **02**, 17–18 (2015)
11. Lu, L., Wu, W.: Analysis of the coexistence between sport and the media in the modern society. In: 4th International Conference on Education, Management and Computing Technology (ICEMCT 2017), pp. 1289 1294
12. Shang, F., Fu, J.: MOOC and the transformation of Chinese adult education. Chin. Adult Educ. **11** (2015)
13. El Mhouti, A., Nasseh, A., Erradi, M.: Stimulate engagement and motivation in MOOCs using an ontologies based multi-agents system. Int. J. Intell. Syst. Appl. (IJISA) **8**(4), 33–42 (2016). https://doi.org/10.5815/ijisa.2016.04.04
14. Jun, X., Hui, L.: Opportunity and challenge: adult education in the era of MOOC. J. Educ. Coll. Hebei Univ. **3**, 2 (2014)
15. Nandi, D., Hamilton, M., Harland, J.: What factors impact student—content interaction in fully online courses. Int. J. Mod. Educ. Comput. Sci. (IJMECS) **7**(7), 28–35 (2015). https://doi.org/10.5815/ijmecs.2015.07.04
16. Li, D.: Lifelong education system and its implementation patterns. Mod. Educ. Sci. **6**, 117–118 (2016)

Fuzzy Classification on the Base of Convolutional Neural Networks

A. Puchkov, M. Dli and M. Kireyenkova

Abstract The paper deals with the algorithm of object classification based on the method of fuzzy logic and the application of artificial convolutional neural networks. Every object can be characterized by a set of data presented in the numerical form and in the form of images (photographs in different parts of the light spectrum). In this case, one object can be matched with a few images associated with it; they can be received by different methods and from different sources. In the algorithm, this generalized totality of images is recognized by convolutional neural networks. A separate neural network is formed for every channel of data receiving. Then, the network outputs are combined for processing in the system of classification on the basis of fuzzy logic output. The normalized outputs of convolutional neural networks are used as values of a membership function to terms of outputs variables when a fuzzy classifier works. For the first adjustment of the convolution neural network hyperparameters, the gradient method is applied. The algorithm is realized in Python language with the use of Keras deep learning library and Tensor Flow library of parallel computation with CUDA technology from NVIDIA company. This paper presents the results of practical application of the developed neuro-fuzzy classifier to forecast the problem of working time losses.

Keywords Fuzzy logic · Classification · Image recognition · Convolutional neural networks

A. Puchkov (✉) · M. Dli · M. Kireyenkova
National Research University "Moscow Power Engineering Institute" (Branch)
in Smolensk, Energetichesky Proyezd 1, g., Smolensk 2014013, Russia
e-mail: putchkov63@mail.ru

M. Dli
e-mail: midli@mail.ru

M. Kireyenkova
e-mail: bitser1@mail.ru

© Springer Nature Switzerland AG 2020   379
Z. Hu et al. (eds.), *Advances in Artificial Systems for Medicine and Education II*,
Advances in Intelligent Systems and Computing 902,
https://doi.org/10.1007/978-3-030-12082-5_35

1   Introduction

The problem of object distribution according to different groups is the base of the algorithmic ware of most automated systems of information processing. The necessity of the initial distribution or classification is justified by the presence of different approaches to the processing and usage of data coming from the objects of different classes. The diagnostic systems of technical equipment failure, the search for potentially interested customers in the organization services according to their profiles in social networks (targeted advertising) can be the examples of such systems [1]. Marketing in social networks, Social Media Marketing (SMM), is actively developing now. So, the development and improvement of the support methods of SMM analysis, including the classification as a base of the market segmentation, presents an actual problem [2].

2   Related Work

The aim of the investigation is to develop the architecture of the multichannel hybrid classifier based on the convolutional neural networks and to estimate its validity on the practical experiment. The class is characterized by a certain combination of characteristics (attributes); according to their values, the splitting of objects into classes is done [3]. Though in the case of profile analysis in social networks, there are sources of feature information presented in different formats: a text, photos, video content, and sound massages. To distinguish the feature space in such a set is not easy as these features are not evidently detected and understood, this can appear in their quality. The methods of splitting into classes, basing on the technology of deep learning, give good results in this situation. First of all these are the methods of fuzzy clusterization [4–6] and neural network approaches [7–9].

The recognition methods of images should be specially mentioned; they present algorithms of classification basing essentially on the processing of visual information about objects. In this direction, there are many approaches toward the solving of recognition problem [10–13]. Some of them offer to form triplet features of a new class recognition on the base of stochastic geometry and functional analysis. The source of triplet features formation is the geometrical conversion connected with the scanning according to the complex trajectories [11]. The others, vice versa, do not go deep into details of images structure and apply algorithms similar to the processes of forming the images in human brains.

Deep neural networks such as convolutional neural networks [14–16] and capsule neural networks, announced at the end of 2017 [17], deserve special attention. In capsule neural networks, the groups of virtual neurons (capsules) are added allowing to take into consideration not only some details of images but their mutual orientation as well. British scientist in the field of neuron networks G. Hinton was the first who introduced the idea of these networks in 1976, but this idea became more or less concrete only in 2011, and the first publication was appeared in October 2017.

The above-mentioned interest to the creation of the new architectures of neuron networks for realization of computer vision systems is explained by their being in great demand in the modern world and the presence of technical opportunities to perform parallel computing allowing to carry out recognition in real time [18]. Besides, the appearance of such algorithms and technological opportunities makes the preconditions to apply them in the fields considered, until recently, to be unpromising from the point of view of these approach implementations into them (e.g., because of too high development cost compared with the expected effect). The situation has changed. So, the problem of using the methods of computer recognition of images for those applications where the necessity to use them is not obvious considers to be actual. The example of such problems can be the estimation of an organization staff on the base of totality of some images connected with the concrete employer. In this case, such images can be already in the database of Human Resources Department (photos, diagrams, the results of graphical psychological tests) or can be taken in some artificial way. In particular, the signal recorded in the time domain can be presented by Fourier frequency spectrum and the spectrum itself can be transformed into cepstrum and visualized on the base of mel–coefficient by the bitmap [19].

In spite of the convolutional neuron network success at image recognition, the network does not give the absolutely definite answer. The membership to the class is given by the level of signal on the appropriate output of the network at the normalized range [0, 1], the closer the value to one on this output there is more confidence that the object under research belongs to this class.

The authors propose the algorithm of object classification, and the information about them is represented in different format (numbers, a text, images) for which the part of feature space is not identified. The basis of the algorithm is a system of fuzzy logic output. Its base of rules works on the basis of values of linguistic variables forming by including the results of the extraction of the object with hidden features from the image by convolutional neural networks.

3   Fuzzy Classification Problem Statement

Let the disjoint classes of objects be defined and presented as a finite set of $A = \{a_1, a_2, ..., a_k\}$. The space of objects X and the set of binary vectors of dimension k: $Y = \{0; 1\}^k$ are given. Every object is connected only with one class from A. At the number of classes $k = 2$, there is a binary classification, and at $k > 2$, we have a multiclass classification [20].

The object $x \in X$ is described by the vector of features $x = \{x_1, x_2, ..., x_p\}$ which are the linguistic variables. They are the quintuple in the form of $x_i = <n\_x_i, T(x_i), X_i, \Pi, M>$ where n_x_i are the names of a feature, $T(x_i)$ is a term set of values of linguistic variable x_i, every of which is a fuzzy variable on the set X_i, Π is a syntactic rule for forming new names, M is a semantic procedure of converting the new name into the fuzzy variable [21].

Suppose that the training set of data $S = (x_i, y_i)$, $1 \leq i \leq n$ is given, it consists of n objects ($x_i \in X, y_i \in Y$) and taken from the unknown distribution D. The application of the linguistic variables makes possible the presence among features $\{x_1, x_2, ..., x_p\}$ both numerical values and qualitative values which can be received including though on the basis of image recognition connected with the object by the convolutional neural networks. It is supposed that each group of images is received by the distinct method and has some characteristics of object $x_i \in X$ which can be detected from the image by its recognition. It is required to build a classifier $h: X \rightarrow Y$ minimizing the given function of losses [22]. The application of fuzzy output and neuron networks in this case leads to algorithm h gives some real-value function $f: X \times Y \rightarrow R$ instead of binary classification for each class. As a result, the evaluation vector of membership $G = \{g_1, g_2, ..., g_k\}$ requires the application of one more algorithm which transforms it into binary vector $\{y_1, y_2, ..., y_k\} \in Y$ [20].

4 The Methods of Fuzzy Classification Problems Solving

4.1 The Preparation of Associated Images

The offered method of fuzzy classification problems solving presents a superposition of a recognizing operator (forms the vector of evaluations of the membership to classes) and decisive rule (transforms the evaluation of membership into binary vector).

The recognizing operator contains the performing of a few stages:

- The preparation of learning sample of object images;
- The design of convolutional neuron network architecture and its training on the learning sample;
- The development of knowledge base for the system of fuzzy output with regard to the data of image recognition by the convolutional networks;
- The application of fuzzy output system on the working sample.

The Preparation of Associated Images

The preparation of the learning sample of object images is the detection of the data sources on the base of which the images, these images normalization and splitting according to classes $\{a_1, a_2, ..., a_k\}$ are formed. The sources of images can be as already existed images of the object and the images visualizing not graphic information. Thus, one object can be matched with k_s groups of images which will be named associated with the object or just associated images. The methods of the image visualization can be various; the main thing is they definitely match the image with the object.

One of the most universal methods of receiving associated images considers to be the approach implementing for the speech recognition by convolutional networks. It is based on Fourier transformation of the speech signal and mel–frequency of cepstral coefficients [23]. Cepstrum is the result of discrete cosine transformation from the algorithm of amplitude spectrum signal. The speech signal considers as some function changing in time. The procedure of receiving of such time dependence characterizing the object is not considered, as it does not have the principle character in this context, it is enough that it can be received by anyway from the available content (site or some other sources).

The main idea of receiving associated images is in the following. Let us define through $s[t]$ some of the process or the dependence characterizing the object behavior in time. The splitting of $s[t]$ into K shot on N discrete intersecting on $N/2$ of its width: $s[t] \rightarrow S_n[t]$, $n = 1, \ldots K$. In each shot, discrete Fourier transformation is performed; the spectrum density of power is calculated [24]. Different frequencies can take part in further transformations with different significance which is given by different sensitivity. It is realized with the help of mel–scale which gives the required sensitivity to frequencies. Mel–frequency cepstral coefficients are the values of cepstrum distributed along the mel–scale with the usage of filters bank. Then, the matrix of cepstral coefficients is calculated and the image (bitmap) is matched it. This image is used for recognition by the convolutional network as it is supposed that the image is unique [23].

4.2 The Applicable Method of Associated Images Recognition

The received associated images are matched the object of classification by no means, and, as it is expected, can contain features according to which the classification can be performed. Visualization of cepstral coefficients does not give an opportunity for the researcher to see and point to that feature on the image of a bitmap which can be used when classifying. At the same time, the usage of the technology of the deep learning on the base of convolutional neural networks makes automatic detection of the hidden features from the images and carrying out the classification of objects on their base possible [25].

The general structure of the multichannel classifier is shown in Fig. 1, and the recognizer of associated images on the base of ensemble of convolutional networks is in the rectangle fragment.

Primarily, the images can be presented in any graphic format so with the aim of the further certainty of processing these formats are transformed to format *jpg* using additive color model *RGB*. In this format, each image contains three color channels (red, green, and blue) presented by matrixes of $n \times m$ size where n is a number of pixels along the vertical, m is a number of pixels along the horizontal. As a result, every image is a matrix of $n \times m \times 3$ size.

The classifier on the basis of neural networks is a part of general algorithm on the basis of fuzzy logic output.

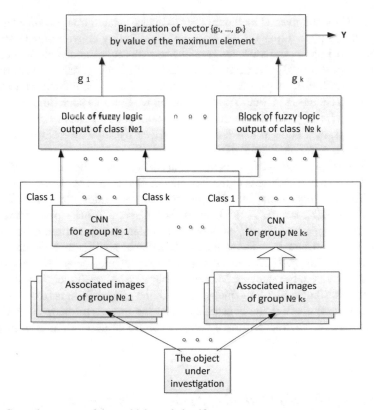

Fig. 1 General structure of the multichannel classifier

4.3 The Structure of the Fuzzy Logic Output System

The recognition of every associated image is performed by a separate convolutional network each of which has the number of outputs of the last fully connected layer matching the number of classes. As the result of i convolutional network working, the evaluation vector of object membership to every from $a_1, a_2, \ldots, a_k$ classes on the base of recognition of i associated image: $\boldsymbol{gc}_i = \{gc_{i,1}, gc_{i,2}, \ldots, gc_{i,k}\}, i = 1, \ldots, k_s$ is formed.

Separate blocks of fuzzy logic output (see Fig. 1) are created according to the numbers of classes, besides every j-block collects vectors $\boldsymbol{gc}_i$ elements from convolutional networks corresponding j-class on its input. The applied ontology of the subject area under research serves as a base for fuzzy systems creation, let us consider this area to be developed (this process deserves special attention but does not have theoretical importance in the context of the presented material). The blocks of fuzzy output contain the sets of rules in the form of:

$$If \bigcup_{i=1}^{kp_1} (gc_{1,1} = T_1 \wedge / \vee gc_{2,1} = T_1 \wedge / \vee \ldots \wedge / \vee gc_{k_s,1} = T_1 \ TO \ g_1 = TG_1)$$

$$If \bigcup_{i=1}^{kp_2} (gc_{1,2} = T_2 \wedge / \vee gc_{2,2} = T_2 \wedge / \vee \ldots \wedge / \vee gc_{k_s,2} = T_2 \ TO \ g_2 = TG_2),$$

$$\ldots$$

$$If \bigcup_{i=1}^{kp_k} (gc_{1,k} = T_k \wedge / \vee gc_{2,k} = T_k \wedge / \vee \ldots \wedge / \vee gc_{k_s,k} = T_k \ TO \ g_k = TG_k)$$

where $kp_1, \ldots, kp_k$ is the number of rules for each class; $T_i, \ldots, T_k$ are the terms for the corresponding components of the evaluation vector membership to classes $\{gc_{i,1}, gc_{i,2}, \ldots, gc_{i,k}\}$, $i = 1, \ldots, k_s$ $TG_j, j = 1, \ldots, k$ are the term membership to classes of components of evaluation vector $\{g_1, g_2, \ldots, g_k\}$.

The sets of rules, their composition, and the content depend on the ontology of the subject area.

5 The Realization and Results of Application

5.1 Software Architecture

The program, realizing the described method of fuzzy classification on the base of convolutional network, is written in Python 3.6 language in Spyder environment from Python Anaconda assembly, operating system Linux. For the opportunity to apply deep neural networks, the library of machine learning TensorFlow was additionally installed [26]. It should be noted that to learn CNN is a resource–intensive process as thousands of synaptic weights adjust on the great number of images and the usage of the central processor (CPU) leads to the long process of learning. So, it is reasonable to transfer calculations on the graphic processor containing thousands of more simple than CPU microprocessors but being able to perform many not difficult operations paralleling the process of the network training. The redirection of library calculations TensorFlow to GPU is so far possible only for video maps of Nvidia company with the support of CUDA—software–hardware architecture of parallel calculations [27].

While the program of binary classification working the graphic processor, GeForce GTX 1060 was used containing 1280 cores of Maxwell/GP106 released in 2016. The transfer of calculations on the notebook Asus FX502VM from CPU Intel Core i7 7700HQ allowed to shorten the time of learning in the network more than to 20 times.

5.2 Topology of Applied Convolutional Neural Networks

For each group of associated images, the same topology of convolutional neural network was applied. The multilayer network without feedback contains two repeated areas each of which contains two convolutional layers and one subsampling layer. These layers provide the detection of features of the associated images.

At the end of the network, there is a classifier consisting of one of the fully connected layers of 521 neurons and the output layer from k neurons. The number of neurons of the output layer is changed depending on the number of recognizable classes at the beginning of the program work.

To exclude the possibility of relearning the net, the layers, realizing the method of Dropout ("decimating" the neurons) preventing neuron mutual adaptation in the process of learning, were added.

The number of neurons on the layers depends on the size (in pixels) of minisamlpling according to which the weights are adjusted while forming the feature maps and the size of the output image (in pixels) and can vary.

The Results

The proposed algorithm of fuzzy classification was applied in the work of HR service of the enterprise, which contains about hundred and fifty thousand employees, to analyze the losses of the working time. The considerable number of the staff of the enterprise allows to get the acceptable volume of sample for convolutional networks.

The sources of images were the data from the enterprise electronic clockhouse and the photos of the employees. The data from the clockhouse reflect daily (during the quarter) deviation from the standard time of going through the turnstile at the entrance and exit. Then, these data were transformed into bitmaps. Thus, three types (sources) of images associated with employees were used as follows: a photo and two bitmaps characterizing the time of entrance and exit from the enterprise. Figure 2 shows the deviation of time and bitmaps for 10 employees when they go through the turnstile. The minutes are on the coordinates axes. Ordinal numbers show the number of an employee.

In the learning sample, there were 1000 employees and 100 employees were in the testing one. The classes were characterized by the number of days when the employees were absent at the working place (reasonable and not reasonable excuses were not differentiated). The distribution into classes is given in Table 1.

To justify the appropriateness of application of a few associated images, the learning was performed for three different situations: When all three associated images are available, two types of associated images are available and one type of associated images is available. The results of the classifier work on the testing sample for three different situations are shown in Table 2.

The data in Table 2 demonstrate the general number of employees referred to this or that class and can be applied for statistic estimation of working time losses at the whole enterprise. But they do not give an idea about the accuracy and the completeness of classification that can be estimated with the help of confusion matrix [28].

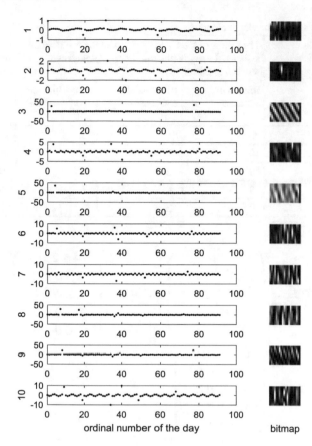

Fig. 2 Deviations of entrance time and their bitmaps

Table 1 The table of class distribution

Number of class	1	2	3	4	5
The number of missed working days in the next quarter	0	1–2	3–5	6–8	More than 8

Confusion matrix is a matrix of $k \times k$ size, the columns of which reflect the factual data and lines are the results of the classifier work.

While filling in the matrix, we incriminate the number at the intersection of a class line which the classifier gave back and the class column to which the object is really referred to. In the case under consideration, we receive three matrixes for three, presented in Table 2, sets of associated images:

Table 2 The results of the classifier work

Number of class	The results of the classifier work at the given set of associated images (the number of employees referred to this class)			Factual number of employees referred to this class
	Bitmap of entrance	Bitmaps of entrance and exit	Photo, bitmaps of entrance and exit	
1	82	80	72	70
2	10	8	12	14
3	8	6	10	8
4	0	4	4	6
5	0	2	2	2

$$CM1 = \begin{pmatrix} 66 & 14 & 0 & 2 & 0 \\ 2 & 6 & 1 & 2 & 0 \\ 0 & 0 & 4 & 2 & 2 \\ 0 & 0 & 0 & 0 & 0 \\ 0 & 0 & 0 & 0 & 0 \end{pmatrix}, \quad CM2 = \begin{pmatrix} 74 & 4 & 2 & 0 & 0 \\ 2 & 5 & 0 & 1 & 0 \\ 0 & 2 & 4 & 0 & 0 \\ 0 & 1 & 0 & 1 & 2 \\ 0 & 0 & 0 & 1 & 1 \end{pmatrix},$$

$$CM3 = \begin{pmatrix} 68 & 4 & 0 & 0 & 0 \\ 1 & 7 & 4 & 0 & 0 \\ 0 & 2 & 8 & 0 & 0 \\ 0 & 0 & 1 & 2 & 1 \\ 0 & 0 & 0 & 1 & 1 \end{pmatrix}.$$

Matrix $CM1$ characterizes the result of classification while learning according to one bitmap of entrance, $CM2$ characterizes according to the bitmaps of exit, and $CM3$ characterizes according to all three associated images. The analysis of the matrix structure shows that with the increase of the number of associated images, using for classification, their form approaches diagonal, that can justify the increase of the classification quality.

Figure 3 shows the results of the classification. For illustrative purpose, the actual number of employees of a relevant class is joined by a dashed line. Asterisks stand for the number referred to the class according to one bitmap of entrance. Crosses stand for the number referred to the class according to the bitmaps of entrance and exit. Circles stand for the number referred to the class according to all these three associated images.

The analysis of Fig. 3 shows that the classification according to three images gives the closest group near the dashed line of the actual classification and points to the reasonability of the increase in associated images groups when classification is made.

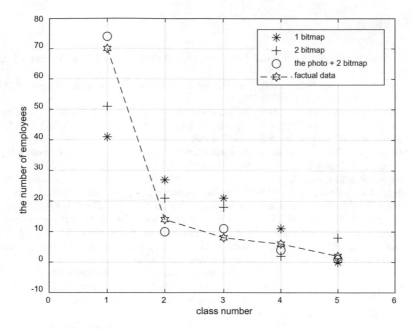

Fig. 3 The results of classification

6 Conclusion

In this paper, we report about the analysis of the modern state of the theory and practice of image recognition which was carried out during the work.

The relevance of application of deep neural networks for solving this problem is also shown.

Scientific novelty of the research is in the proposed architecture of the fuzzy classification system based on the recognition of images associated with an object. The recognition is done by convolutional neural networks.

The approbation of the algorithm, realizing the proposed fuzzy classifier, was performed at the enterprise to estimate the losses of the working time during the quarter. When a number of groups of images associated with an employee increases, we can see that the error of classification has a tendency to decrease. In the example under consideration, it was decreased from 44% (when there was one group of associated images) to 10% (when there were three groups).

So, it is stated that the proposed algorithm of fuzzy classifier based on the convolutional neural networks can find the application in the systems of decision-making support in different spheres, where the data about the object under investigation can be presented not only in numeric form but also in the form of different images as well.

390 A. Puchkov et al.

References

1. Hkalilov, D.: Marketing in Social Networks, 2nd edn. Mann, Ivanov and Febber, Moscow (2014)
2. Zlobina, N., Zavrazhina, K.: Marketing in social networks: modern tendencies and outlook. Sci. Tech. Bull. **6**(233), 166–172 (2015)
3. Kasim, A.A., Wardoyo, R., Harjoko, A.: Batik classification with artificial neural network based on texture-shape feature of main ornament. Int. J. Intell. Syst. Appl. (IJISA) **9**(6), 55–65 (2017). https://doi.org/10.5815/ijisa.2017.06.06
4. Ostrovsky, A.: Fuzzy clastarization of electronic information resources of project repository at automatic designing. Dissertation, State Technical University in Ulyanovsk (2010)
5. Dli, M., Puchkov, A., Malevich, E.: The solving of inverse problems on the base of fuzzy inverse Kalman's algorithm. Software products, systems and algorithms. Electron. J. **2** (2016)
6. Kosko, B.: Fuzzy systems as universal approximators. IEEE Trans. Comput. **43**(11) (2004)
7. Boyarinov, Y., Stoyanov, O., Dli, M.: The application of neuro-fuzzy method of group argument account for building the models of social economic systems. Softw. Prod. Syst. **3**, 7–11 (2006)
8. Puchkov, A., Dli, M.: The improvement of diagnostic system of a boiler unit on the base of neuro-fuzzy algorithms. Int. Sci. Technol. J. **7**, 47–50 (2016)
9. Korchagin, V., Krossovsky, V.: Neuron networks classification in distribution of the repairing fund of units according to the test results. News State Univ. Tula. Tech. Sci. **5**, 127–133 (2013)
10. Zhuravlyov, Y., Ryazanov, V., Senko, O.: Detection. Mathematical Methods, Software System. Practical Application. Phases, Moscow (2005)
11. Fedotova, N.: The Theory of Features of Images Detection on the Base of Stochastic Geometry and Functional Analysis. PHYSMATLIT, Moscow (2010)
12. Kumar, S., Tripathi, B.K.: On the root-power mean aggregation based neuron in quaternionic domain. Int. J. Intell. Syst. Appl. (IJISA) **10**(7), 11–26 (2018). https://doi.org/10.5815/ijisa.2018.07.02
13. Nayak, S.C.: Development and performance evaluation of adaptive hybrid higher order neural networks for exchange rate prediction. Int. J. Intell. Syst. Appl. (IJISA) **9**(8), 71–85 (2017). https://doi.org/10.5815/ijisa.2017.08.08
14. Hinton, G., Deng, L., Yu, D., Dahl, G., Mohamed, A., Jaitly, N., Senior, A., Vanhoucke, V., Nguyen, P., Sainath, T., Kingsbury, B.: Deep neural networks for acoustic modeling in speech recognition. IEEE Signal Process. Mag. **29**(6), 82–97 (2012)
15. LeCun, Y., Kavukcuoglu, K., Farabet, C.: Convolutional networks and applications in vision. In: Proceedings of 2010 IEEE International Symposium on Circuits and Systems (ISCAS), pp. 253–256. IEEE (2010)
16. Lee, H., Grosse, R., Ranganath, R., Ng, A.Y.: Convolutional deep belief networks for scalable unsupervised learning of hierarchical representations. In: Proceedings of the 26th Annual International Conference on Machine Learning, pp. 609–616. ACM (2009)
17. Capsule networks from Hinton, images processing, machine learning. https://habr.com/company/recognitor/blog/343726/. Accessed 2018/05/05
18. Candrot, E., Sanders, J.: CUDA Technology on Examples. Introduction to Programming of Graphic Processes. DMK Press, Moscow (2016)
19. Oppenheim, A.V., Schafer, R.W.: From frequency to quefrency: a history of the cepstrum. IEEE Signal Process. Mag. **21**(5), 95–106 (2004)
20. Ostapets, A.: Decision rules for the sets of networks of probable classifiers while solving the problems of classifications with intersection classes. Mach. Learn. Data Anal. **2**(3), 276–285 (2016)
21. Kruglov, V., Dli, M., Golunov, R.: Fuzzy Logic and Artificial Neuron Networks. Physics and Mathematics Publ, Moscow (2001)
22. Zhang, M.L., Zhou, Z.H.: ML-KNN: a lazy learning approach to multi-label learning. Pattern Recogn. **40**(7), 2038–2048 (2007)

23. Kotomin, A.: Recognition of voice command with the use of convolutional neuron networks. In: Digest of Annual Youth Conference: Knowledge-Intensive Information Technologies, SIT-2012, pp. 17–28, Pereslavl-Zalesky (2012)
24. Smith, S.: Digital Signal Processing: Practical Handbook for Engineers and Scientists. Dodeka-XXI, Moscow (2008)
25. Ha, L.M.: Convolutional neuron network for solving the problem of classification. MFTI Works **8**(3), 91–97, https://mipt.ru/upload/medialibrary/659/91_97.pdf. Accessed 2018/04/30
26. Geron, A.: The Applied Machine Learning with the Help of Scikit-Learn and Conceptions, Tools and Techniques for Intetellectual Systems Creation. Dialektika, Moscow (2018)
27. Izotov, P., Suhkonov, S., Golovashkin, D.: The technology of realization of neural networks algorithm in CUDA environment on the example of handwritten digit recognition. Comput. Opt. **34**(2), 243–251 (2010)
28. Classifier evaluation (accuracy, completeness, F-measure). http://bazhenov.me/blog/2012/07/21/classification-performance-evaluation.html/. Accessed 2018/05/15

Modeling of Intellect with the Use of Complex Conditional Reflexes and Selective Neural Network Technologies

M. Mazurov

Abstract Is considered the mathematical model of the formation of the simplest artificial intelligence on the basis of the creation of difficult conditional reflexes with use of selective neural network technologies. A mathematical model based on complex third-order conditioned reflexes is practically implemented. Is substantiated the possibility of generalization of the mathematical model of artificial intelligence on the basis of the formation of complex conditioned reflexes of any nth order. The mathematical model of the formation of temporary connections at simultaneous inclusion of conditional and unconditional incentives is proved. This justification is based on the hypothesis of the formation of communication channels with each inclusion of conditional and unconditional stimuli and the formation of communication channels between the neural centers of conditional and unconditional stimuli. Are substantiated possible applications in the field of robotics, organization of intellectual speech, translation from foreign languages. Compares some of the theoretical approaches to the modeling of intelligence. A simple model of intelligence based on complex conditioned reflexes, using material devices that are implemented in the brain in the form of certain subsystems, is proposed. These subsystems are: ganglia, cerebellum, hippocampus, neocortex, and other subsystems. The proposed structural scheme of the model of intelligence, including the real material of the device.

Keywords Conditioned reflexes · Unconditioned reflex · Communication channels · Neural center · Dendrites · Mathematical model · Models of intelligence

1 Introduction

Complex conditioned reflexes are a simple "automatic" form of intelligence controlling the behavior of an object—animal, human, robot. Therefore, understanding the mechanism of creating complex conditioned reflexes can be useful in the design of

M. Mazurov (✉)
Russian Economic University G.V. Plekhanova, Moscow, Russia
e-mail: mazurov37@mail.ru

© Springer Nature Switzerland AG 2020   393
Z. Hu et al. (eds.), *Advances in Artificial Systems for Medicine and Education II*,
Advances in Intelligent Systems and Computing 902,
https://doi.org/10.1007/978-3-030-12082-5_36

the intelligence of robots that must possess "automatic" intelligence, while human intelligence is not available to them. Complex conditioned reflexes underlie the intellectual properties of ordinary speech, the basis of intellectual translation from foreign languages, intellectual medical diagnostics, and other cases. Biological modeling of intelligence is different from modeling artificial intelligence. It is assumed that artificial systems are not required to repeat in their structure and functioning the structure and processes occurring in it, inherent in biological systems. Despite a large number of experimental studies of conditioned and unconditioned reflexes, their mathematical theory has not been adequately studied at present [1–4]. This work is devoted to the study of the biological aspects of complex conditioned reflexes, which can be used as a basis for creating a basic artificial intelligence for robot control systems, machine translation, image recognition, sound and other purposes. At the heart of any biological intelligence lie reflexes: unconditional, conditional, and also their combinations, called instincts. Therefore, this work is devoted to modeling complex conditioned reflexes, as the basis for creating a minimal basic biological intelligence and artificial intelligence. Priority in the study of the conditioned reflex belongs to Pavlov [1–4], who established the mechanisms of formation of the conditioned reflex and the ways of its realization. The physiological basis for the onset of conditioned reflexes is the formation of functional temporary connections in the higher sections of the central nervous system. Closure of connections between foci of excitation, caused by conditioned and unconditioned stimuli, occurs both horizontally along the conducting paths of the cortex and along the vertical: along the cortex-subcortex, the cortex. The first model, explaining the closure of the conditioned reflex connection from indifferent and unconditioned stimuli in the cortex, was proposed by Pavlov, modified by Asratyan, P. K. Anokhin [1–4]. There is a significant number of mathematical models to explain the occurrence of conditioned reflexes on the basis of abstract logical devices. These models made it possible to obtain discrete characteristic functions simulating a discrete increase in the probability of the appearance of a conditioned reflex as the number of repetitions of the stimulus increased [5, 6]. However, these models were not relevant to real biological objects and subsequently did not develop.

This work is devoted to the study of the biological aspects of complex conditioned reflexes, which can be used as a basis for creating a basic artificial intelligence for robot control systems, machine translation, image recognition, sound, and other purposes.

Consider the mathematical model of the formation of complex conditioned reflexes of the third order and the generalization of the mathematical model to conditioned reflexes of any nth order. The formation of a third-order reflex—a food reaction to light, sound, touching the body with a touch button is illustrated in Fig. 1.

Consider the mathematical model of complex conditioned reflexes using the example of complex conditioned reflexes of the third order, we begin with its justification.

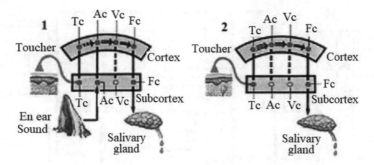

Fig. 1 Formation of a third-order reflex. 1—the process of formation of the conditioned reflex; 2—process of realization of the conditioned reflex. The following notation is used in the figure: Tc—tangent center; Ac—auditory center; Vc—the visual center; Fc—food center

2 Mathematical Model of Conditional Reflexes of the Third Order

The proposed mathematical model is based on experimental studies and the theory of Pavlov's conditioned reflexes, a neurophysiological hypothesis about the closure of temporary connections based on the growth of selective dendritic communication channels in the neural network of the brain [7, 8], due to the predominant growth of the excitatory activity of dendrites involved in the process of excitation and impulse activity. Let us consider the structure of the mathematical model of conditioned reflexes of the third order. To model the mechanisms of complex conditioned reflexes and simple "automatic" intelligence based on them, we will use selective neural networks. In the model, we use selective neurons and selective neural networks, presented in [9, 10] and patents [11–14]. Selective properties can be realized in neural networks, working both on binary input signals, and on impulse, spike input signals. The model includes selective neurons, selective neural networks, selective neural centers (NC), approximated by one neuron, since neural networks are formed by a group of identical neurons. In the formation of an unconditioned salivation reflex to the inputs of neural networks, in the model under consideration, information on the three properties of meat can be given in the first approximation: smell, taste, color. At the output of these neural networks, we obtain three input signals of the food neural center of the PC. In the presence of these three signals, the object is recognized—this is meat and there is a salivation effect.

For a conditioned reflex of the third order, it is necessary to excite three NCs: an acoustic AC, a light VC, and a tangential center TC. We assume that their excitation occurs consistently from three types of different external influences. The order of excitations can vary. For example, each of the centers can be successively excited and the latter excites an unconditioned salivation reflex. This case is illustrated in Fig. 2. Excitations of NC occur consistently from three types of different external influences. The last NC-FC excites an unconditioned reflex salivation. Figure 2 shows how the input signals are fed to a selective neuron or a selective neural pool and below the

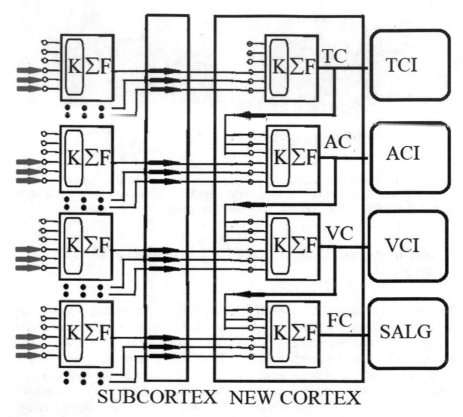

SUBCORTEX NEW CORTEX

Fig. 2 Structure of the mathematical model of third-order reflexes. The model includes selective neurons, selective neural networks, selective neural centers. The following designations are used: K—cluster of neural connections; TCI—indicator of excitation of the tangent center; ACI—excitation indicator of the auditory center; VCI—an indicator of excitation of the visual center; SAL-G—salivary gland. Excitations of neural centers occur consistently from three types of different external influences. The last food center FC excites an unconditioned reflex salivation

image of the neurons are shown two paths through which signals from other sensing devices are fed. It is assumed that the signal from the axon at the output of the selective neuron or the selective neural pool can branch and can be associated with a variety of other neurons. A schematic representation of the axon branching on the axonal terminus is shown in Fig. 3.

In general, a mixed sequential–parallel action is possible. We denote the sequential excitation of an object with $n \ n \ \to$, a number parallel to both $n \ \downarrow$. Then the possible forms of excitation can be represented in the form: (1) $(1 \to 2 \to 3 \to$ UNCR)—consecutive excitation; (2) $(1 \downarrow 2 \downarrow 3 \downarrow$ UNCR)—parallel excitation; (3) $(1 \to 2 \to 3 \downarrow$ UNCR) is one of the mixed excitations. It is not difficult to determine that the total number of excitation options for a conditioned reflex of the third order

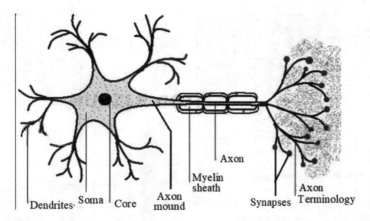

Fig. 3 Schematic representation of the axon branching on the axon terminus. The case is considered when all conditional excitations act in parallel, let us assume that their total action excites the NC of the unconditioned reflex

is equal to the number of permutations of the three elements $N = n!$. For $n = 3$ $N = 6$.

Similarly, as in the cases described above, conditioned reflexes of higher orders can be formed. The formation of temporary connections in a mathematical model is performed with the simultaneous inclusion of a conditioned and unconditioned stimuli. It is assumed that each connection involves the formation of several communication channels—dendrites and axons— between the output of the NC of the conditioned stimulus and the input of the neural center of the unconditioned stimulus. In this case, the electrical potential of the total stimulus may exceed the threshold value. This leads to the fact that stimulation by a conditioned stimulus will lead to the appearance of an unconditioned reflex. Let us examine in more detail the dynamics of the formation of conditioned reflexes of higher orders. With one stimulation, a beam of communication channels appears, which is illustrated by Figure 1, in Fig. 4 on the top right. Conditionally, we assume that the bundle of communication channels is equal to 1. For double and triple simultaneous stimulations, the communication channel beams increase, which is illustrated, respectively, by the numbers 2 and 3. Similarly, n simultaneous stimulations will lead to the appearance of n beams of communication channels. An illustration of the appearance of temporary connections is illustrated in Fig. 4. The figure uses the same notation as in Fig. 2.

In Fig. 4, digits 0, 1, 2, 3 indicate the states of temporary connections. Figure 4 shows: (a) the absence of temporary links; (b) one link is shown in the presence of one combined effect of stimuli—conditional and unconditional; (c) two links are shown in the presence of two joint actions of stimuli—conditional and unconditional; (d) three links are shown in the presence of three joint actions of stimuli—conditional and unconditional. According to preliminary agreement, the appearance of three new connections causes the appearance of an unconditioned reflex—the salivation of the

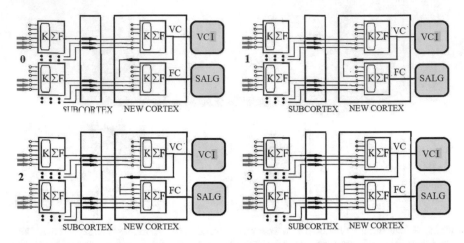

Fig. 4 Structure of the mathematical model of the formation of temporary connections of the conditioned reflex. 0—there is no stimulus; 1—unconditional and conditioned stimuli (sound stimuli) are applied once; 2—unconditional and conditioned stimuli (sound stimuli) are applied two times; 3—unconditional and conditional (sound) stimuli are applied three times

salivary gland, in view of the excess of the threshold level of stimulation equal to three units.

3 Mathematical Theory of Complex Conditional Reflexes in Electronic Neural Networks

Here we will consider the mathematical theory of complex conditioned reflexes in selective neural networks. Selective neurons and selective neural networks are described in [9, 10] and patents [11–14]. Let Z the reaction to the stimulus that caused the unconditioned reflex. We represent it in the form

$$Z = f_1(\mathbf{y}_1)$$
$$\mathbf{y}_1 = f_2(\mathbf{y}_2), \ \mathbf{y}_2 = f_3(\mathbf{y}_3), \dots, \mathbf{y}_{n-1} = f_n(\mathbf{y}_n)$$
$$Z = f_1(f_2(f_3 \dots f_n(\mathbf{y}_n) \dots))$$

To describe the first-order SD, we use the relation

$$Z = f_1(f_2(\mathbf{y}_1)), \quad \mathbf{y}_1 = f_1(\mathbf{y}_2) = F_1\left(\sum_{k \in K} y_{ik} y_{kj} - U_{\Pi i}\right)$$

where $\mathbf{y}_i = (y_{i1}, \dots, y_{in})$—$i$ the response of the system (eye, ear, and others) to input signals or perception vectors, $n \in N$; $U_{\Pi i}$—excitation thresholds; K—selective clusters of internal communication channels—dendrites tuned to certain input signals; F_i—nonlinear threshold functions; Z—a transformation characterizing the formation of an unconditioned reflex (for example, salivation and others). Clusters can be configured, for example, on vertical or horizontal lines of the input object. Note that in the proposed mathematical model of complex conditioned reflexes, conditional reflexes from unconditional reflexes are usually created. That is, a mathematical model is realized.

$$CR_1 \Rightarrow CR_2 \Rightarrow \ldots \Rightarrow CR_n \Rightarrow UNCR$$

Here we use the notation CR_n n-ty—the conditioned reflex, UNCR—the unconditioned reflex. Of conditioned reflexes, conditioned reflexes can again be created. The following relations $CR_1 \Rightarrow CR_2 \Rightarrow \ldots \Rightarrow CR_n \Rightarrow CR_{n+1}$ are realized. Here we use the notation CR_{n+1}—the conditioned reflex $(n + 1)$ of order.

4 Application of the Mathematical Theory of Complex Conditional Reflexes Using Electronic Network Technologies

We note some applications of artificial intelligence, which are based on the use of complex conditioned reflexes and selective neural network technologies. (1) The field of robotics. Controlling robots for orientation—movement in a complex external environment is a challenge. Intelligence is required by robots to manipulate objects, navigate with localization problems (locate, explore nearby areas) and plan traffic [15]. Figure 5 shows the structure of the robot android, in which mathematical models can be used. (2) Intellectual speech, based on the analysis of meaning, rather than spelling rules. Our fast speech is also a habit of talking, we do not pay attention to the quickly pronounced words of our native language, since we are "used" to it. But we should start speaking in a foreign language, which we do not know well, how we will have to think about every word. The habit is explained by the formation of a conditioned reflex or a whole complex of conditioned reflexes. (3) Intelligent translation from foreign languages, also based on the analysis of meaning, not spelling rules. (4) Intellectual recognition in various areas of practice, for example, in medicine. For example, in work [16] recognition of contour medical curves is realized on the example of ECG electrocardiograms.

5 Modeling of More Complex Types of Intellect

Prologue for modeling more complex types of intelligence was the work of academic P. K. Anokhin, devoted to the theory of functional systems of the organism as a whole

Fig. 5 Structure of the robot android, in which mathematical models of the formation of intelligence can be used based on the use of complex conditioned reflexes and selective neural network technologies

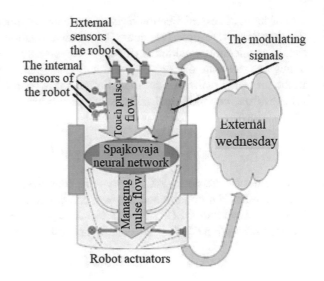

[4, 17]. Anokhin formulated a new approach to understanding the functions of the whole organism. Instead of classical physiology of organs, traditionally the following anatomical principles, the theory of functional systems proclaims the systemic organization of human functions from the molecular up to the social level.

6 Synthesis of a Model of Simple Intellect

The main task of modeling the intellect and thinking of the authors of the models examined was to create a virtual system without a specific material realization in the hardware. For example, K. V. Anokhin manipulates with virtual kogami, commas, cognitomas, arguing in the end that all this is realized in the form of neurons, neural connections, and substances in the brain [18, 19]. Chernavskaya and Chernavsky use intangible objects, such as information (by the way, until now it is not exactly established what this substance represents) [20, 21]. Information is processed in non-material neural layers—plates, realizing various types of connections "black", "gray", "white". Material realization is not discussed. In this paper, we propose a model of intelligence, thinking, using specific devices that are implemented in the brain in the form of specific systems: neural centers, ganglia, cerebellum, hippocampus, neocortex, and other subsystems. The structural scheme of the developed model of intelligence, including real material devices, is shown in Fig. 6.

The proposed model of intelligence and thinking is only the first material approximation. We note the elements by which it differs in the aggregate. These are the following elements: 1. Scanning systems. Scans the state of sleep and wakefulness, the current scan during wakefulness; 2. The biological clock system necessary for the operation of scanning systems; 3. Engrams, in general, having a distributed charac-

Fig. 6 Structural diagram of the developed model of intelligence in the first approximation, including real material devices. The following notation is used in figure: H is hours; CD is comparison device; SS is scanning system; STS is stop-system; PS is playback system; CS is control system; R is receptor; DL is deep learning; ENG is the engram; UNCR is unconditioned reflex; SI is a simple intellect

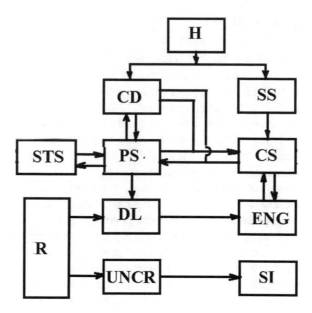

ter; 4. A playback system that implements the commands submitted to the executive bodies and to the deep learning path. Teams submitted to the deep learning pathway implement clear or diffuse images from engrams; 5. Control systems of various types, specific devices and the location of these systems are completely unknown; 6. A system for comparing information coming under the influence of control systems from engrams, as well as a comparison with newly arrived information; 7. System of deep learning. The exact device and location of the systems are completely unknown.

In this case, we do not discuss which real brain structures include individual parts of our model of an intellectual system of a higher level than simple intellect. As an initial example, it can be said that the engram is mainly concentrated in the neocortex; control systems in the cerebellum, hippocampus and others; the localization of the scanning system is not exactly established; the principle of the action of deep training in neural networks connecting the perceiving systems with engrams is not fully established; localization of the comparison systems is not defined. Approximately the localization of the clock has been established recently—this is the suprachiasmatic nucleus.

Localization of structures that implement a simple intellect, we discussed in the first part of the work. The organization of the intellect and its connection with the brain, its location in the brain is also not known. This is one of the fascinating secrets of nature. The most popular works in this field are the works of Lashley [22], Pribram [23], Nobel laureates Wiesel and Hubel [24], K. V. Anokhin [18] and many others. Some applications of artificial intelligence are presented in [25–30].

7   Conclusion

This work is devoted to the study of biological aspects, which can be used as a basis
for the creation of basic artificial intelligence for robot control systems, machine
translation of image recognition, sound and other purposes. We discuss some applications in which the simulation of intelligence based on complex conditioned reflexes
and selective neural network technologies can be used.

Is considered the mathematical model of the formation of the simplest artificial
intelligence on the basis of the creation of difficult conditional reflexes with use
of selective neural network technologies. A mathematical model based on complex
third-order conditioned reflexes is practically implemented. Is substantiated the possibility of generalization of the mathematical model of artificial intelligence on the
basis of the formation of complex conditioned reflexes of any nth order. The mathematical model of the formation of temporary connections at simultaneous inclusion
of conditional and unconditional incentives is proved. This justification is based on
the hypothesis of the formation of communication channels with each inclusion of
conditional and unconditional stimuli and the formation of communication channels
between the neural centers of conditional and unconditional stimuli. Are substantiated possible applications in the field of robotics, organization of intellectual speech,
translation from foreign languages. Compares some of the theoretical approaches
to the modeling of intelligence. A simple model of intelligence based on complex
conditioned reflexes, using material devices that are implemented in the brain in the
form of certain subsystems, is proposed. These subsystems are: ganglia, cerebellum,
hippocampus, neocortex and other subsystems. The proposed structural scheme of
the model of intelligence, including the real material of the device.

References

1. Asratyan, E.A.: Ivan Petrovich Pavlov, p. 456. Science, Moscow (1974)
2. Asratyan, E.A.: Essays on the Physiology of Conditioned Reflexes, p. 360. Nauka, Moscow
   (1970)
3. Sechenov, I.M.: Reflexes of the Brain, p. 352. AST, Moscow (2015)
4. Anokhin, P.K.: Essays on the Physiology of Functional Systems, p. 447. Medicine, Moscow
   (1975)
5. Brynes, S.N., Napalkov, A.V., Svechinsky, V.B.: Neurocybernetics, p. 172. Izd. Med. Lit.,
   Moscow (1962)
6. Gaase-Rappoport, M.G.: Automata and Living Organisms, p. 224. GIFML, Moscow (1961)
7. Gutman, : Dendrites of Nerve Cells Theory of Electrophysiology Function, p. 144. Mokslas,
   Vilnius (1984)
8. Jeffrey, C., Johnston, M., Johnston, D.: Plasticity of dendritic function. Curr. Opin. Neurobiol.
   15(3), 334–342 (2005)
9. Mazurov, M.E.: Selective neural networks for the recognition of complex objects. In: Mathematical Biology and Bioinformatics: Proceedings VI International Conference, pp. 82–83.
   MAX Press, Moscow (2016)

10. Mazurov, M.E.: Neural selectivity in neural network systems, selective neurons and neural networks. In: Mathematical Biology and Bioinformatics: Works VI International Conference, pp. 84–85. MAX Press, Moscow (2016)
11. Mazurov, M.E.: A pulse neuron close to the real. Patent for Invention No. 2598298, 30 Aug 2016
12. Mazurov, M.E.: Neuron modeling the properties of a real neuron. Patent for Invention No. 2597495, 11 July 2014
13. Mazurov, M.E.: A single-layer perceptron based on selective neurons. Patent for Invention No. 2597497, 13 Jan 2015
14. Mazurov, M.E.: A single-layer perceptron modeling the properties of a real perceptron. Patent for Invention No. 2597496, 22 Aug 2016
15. Akkar, H.A.R., Mahdi, F.R.: Adaptive path tracking mobile robot controller based on neural networks and novel grass root optimization algorithm. Int. J. Intell. Syst. Appl. (IJISA) 9(5), 1–9 (2017). https://doi.org/10.5815/ijisa.2017.05.01
16. Mazurov, M.E.: Intelligent recognition of electrocardiograms using selective neuron networks and deep learning. In: International Conference of Artificial Intelligence, Medical Engineering, Education, pp. 182–198. Moscow, Russia (2017)
17. Anokhin, P.K.: Systemic Mechanisms of Higher Nervous Activity, p. 453. Science, Moscow (1979)
18. Anokhin, K.V.: Cognite: a hyper-network model of the brain. In: Proceedings of the 5th International Conference on Cognitive Science, Kaliningrad (2012)
19. Anokhin, K.V.: Neural mechanisms of memory: synaptic and genomic hypotheses. J. High. Nerv. Act. I.P. Pavlova 61(6), 660–674 (2011)
20. Chernavskaya, O.D., Chernavskii, D.S.: Naturally constructivist approach to modeling thinking. Biophysics 61(1), 201–208 (2016)
21. Chernavskaya, O.D., Chernavskii, D.S., Karp, V.P., Nikitin, A.P.: On the Approach to the Modeling of Thinking from the Standpoint of the Dynamic Information Theory. Sat. Ed. Redko, pp. 29–88. URSS, Moscow (2014)
22. Lashley, K.S.: The Role of the Mass of Nervous Tissue in the Functions of the Brain (1932)
23. Pribram, K.: Languages of the Brain. Experimental Paradoxes and Principles of Neuropsychology, p. 64. Ed. "Progress", Moscow (1975)
24. Hubel, D.H., Wiesel, T.H.: Receptive fields, binocular interaction and functional architecture in the cat's visual cortex. J. Physiol. 160, 106–154 (1962)
25. Mansor, M.A., Kasihmuddin, M.S.M., Sathasivam, S.: Enhanced Hopfield network for pattern satisfiability optimization. Int. J. Intell. Syst. Appl. (IJISA) 8(11), 27–33 (2016). https://doi.org/10.5815/ijisa.2016.11.04
26. Mishra, N., Soni, H.K., Sharma, S., Upadhyay, A.K.: Development and analysis of artificial neural network models for rainfall prediction by using time-series data. Int. J. Intell. Syst. Appl. (IJISA) 10(1), 16–23 (2018). https://doi.org/10.5815/ijisa.2018.01.03
27. Karande, A.M., Kalbande, D.R.: Weight assignment algorithms for designing fully connected neural network. Int. J. Intell. Syst. Appl. (IJISA) 10(6), 68–76 (2018). https://doi.org/10.5815/ijisa.2018.06.08
28. Akkar, H.A.R., Jasim, F.B.A.: Intelligent training algorithm for artificial neural network EEG classifications. Int. J. Intell. Syst. Appl. (IJISA) 10(5), 33–41 (2018). https://doi.org/10.5815/ijisa.2018.05.04
29. Lytvyn, V., Vysotska, V., Peleshchak, I., Rishnyak, I., Peleshchak, R.: Time dependence of the output signal morphology for nonlinear oscillator neuron based on Van der Pol model. Int. J. Intell. Syst. Appl. (IJISA) 10(4), 8–17 (2018). https://doi.org/10.5815/ijisa.2018.04.02
30. Awadalla, M.H.A.: Spiking neural network and bull genetic algorithm for active vibration control. Int. J. Intell. Syst. Appl. (IJISA) 10(2), 17–26 (2018). https://doi.org/10.5815/ijisa.2018.02.02

M. E. Mazurov is Professor of Russian Economic University, Doctor of physical and mathematical Sciences. His research interests include neuroinformatics, methods, deep learning, neural networks, nonlinear dynamics, synchronization in nonlinear dynamical systems, identification of mathematical models of nonlinear dynamic systems. The author of the discoveries in the theory of nonlinear oscillations has over 100 scientific publications.

Phase Equilibria in Fractal *Core-Shell* Nanoparticles of the Pb₅(VO₄)₃Cl–Pb₅(PO₄)₃Cl System: The Influence of Size and Shape

Alexander V. Shishulin, Alexander A. Potapov and Victor B. Fedoseev

Abstract The investigation of phase equilibria in nanosystems has become one of the new trends in chemical thermodynamics. For binary alloys with limited mutual solubilities, a reduction of the system's volume can be accompanied by significant changes of solubilities and the appearance of unusual metastable phases. To comprehend the structural characteristics and the thermodynamical stability of nanoalloys and to improve their performance, a knowledge of their phase equilibria, that considers their size and shape dependence, is critically needed. Due to the fact that experimental investigations are extremely challenging to perform at the nanoscale, useful predictions can be provided by nanothermodynamical simulations. In this paper, size- and shape-dependent phase equilibria in *core-shell* nanoparticles of the Pb₅(VO₄)₃Cl–Pb₅(PO₄)₃Cl stratifying solid solution are calculated. Those compounds are subjects of biochemistry and geochemistry, known as natural minerals vanadinite and pyromorphite. In a *core-shell* structure, two heterogeneous states are possible. At the nanoscale, one of them is metastable and the size- and shape-dependent properties of co-existing phases are different in each state. The shape of each phase is modeled by its shape coefficient which is equal to the ratio between surface areas of the phase and the sphere of the same volume. The thermodynamical stability of heterogeneous states depends on shape coefficients of both phases. The geometrical properties of *core-shell*-nanoparticles are also determined by using the fractal geometry.

A. V. Shishulin (✉) · V. B. Fedoseev
Razuvaev Institute of Organometallic Chemistry, Russian Academy of Sciences, Nizhny Novgorod, Russia
e-mail: Chichouline_Alex@live.ru

A. A. Potapov
Kotel'nikov Institute of Radio, Engineering and Electronics, Russian Academy of Sciences, Moscow, Russia

IREE Joint Laboratory of Fractal Method & Signal Processing, Department of Electronic Engineering, College of Information Science and Technology, Jinan University, Guangzhou, China

© Springer Nature Switzerland AG 2020
Z. Hu et al. (eds.), *Advances in Artificial Systems for Medicine and Education II*,
Advances in Intelligent Systems and Computing 902,
https://doi.org/10.1007/978-3-030-12082-5_37

Keywords Nanoparticles · Nanothermodynamics · Phase equilibria · Vanadinite · Pyromorphite

1 Introduction

When considering phase transformations in systems of colloid size (ex. nanoparticles, nanodrops), we should take into account some specific effects. The equilibrium compositions and volumes of co-existing phases in a nanosystem depend on a system's size [1–7], shape [8–13], and thermodynamical properties of its external boundaries [14]. These effects are reflected in phase diagrams as a change of phase transition temperatures and could be explained by a significant increase of surface energy contribution to the system's full energy. A great number of phase diagrams obtained for macro-sized systems could not be used correctly while analyzing the processes in systems of a very small volume.

Size-dependent decreasing melting temperature of nanoparticles is one of those effects which has been observed experimentally [15–17] and simulated by using a variety of theoretical approaches [17–20]. Size-dependent solubilities for some binary systems obtained by methods of molecular dynamics and equilibrium chemical thermodynamics denote the substantial distinctions in equilibrium compositions in nanoscopic and macroscopic systems. Those distinctions occurred in an experiment [7] when nanograined ZnO–Mn system was characterized and a considerable increase of Mn solubility was noticed. The shape influence on the phase equilibria in nanosystems with different geometrical configuration was regarded using the example of spherical nanoparticles and nanotubes with various crystallographic orientation [8] or nanoparticles in the shape of regular polyhedra [9–11]. The thermodynamic conditions leading to the formation of *core-shell* and *janus* geometrical structures were also observed [12]. Nevertheless, at present, it seems impossible to systematize the obtained results due to a small number of investigated systems, differences in implemented models and their incompleteness and, in particular, due to the lack of consideration of possible metastable states occurrence. Add to this, some authors suggested using the thermodynamical approach while simulating phase equilibria in nanoparticles of 5–10 nm radii. The applicability of the thermodynamical approach for such small systems must be discussed separately [13]. Simulation of shape-dependent phase equilibria in binary metal nanoparticles with metastable states taken into account was performed by our group [14]. There, we suggested an approach for describing a wide range of geometrical structures using a continuous parameter generalizing various versions of isochoric transformations of the shape, including smooth deformations. It is also shown that a phase diagram of a nanosized system also depends on the nature of the surrounding environment which sets the value of the surface free energy on its outer boundary [15].

The goal of this research is simulating the size- and shape dependent phase equilibria in nanoparticles of the $Pb_5(VO_4)_3Cl$–$Pb_5(PO_4)_3Cl$ stratifying solid solution with a *core-shell* configuration. In a small number of related works of other authors

[9–12], only shape-dependent equilibria between liquid and solid phases were considered in case of binary alloy nanoparticles of very simple shapes (spheres, cubes, icosahedrons etc.). In this paper, the approach suggested previously [14] for binary metal nanoalloys with phase transitions in the solid state is being highly developed and for the first time applied to predict the phase composition of alloy nanoparticles from the apatite family. The real shapes are usually complicated; the methods of fractal geometry are used for their consideration.

The compounds of the formulae above are subjects of biochemistry and geochemistry, known as natural minerals vanadinite and pyromorphite, respectively [21]. As the experimental data shows, this system can be described using the regular solid solution model. The components are boundedly mutual soluble up to the temperature of ~740 K [22].

2 Thermodynamic Description of Phase Equilibria in Systems of a Very Small Volume

To observe the influence of the system's size on the phase equilibria thermodynamically, let us consider a spherical particle with 50 mol% of $Pb_5(VO_4)_3Cl$ at 600 K. Its composition and temperature belong to the heterogeneous region in the phase diagram [22], so the mixture becomes stratified with the formation of a *core-shell* structure with two co-existing phases. The phase touching the system's external boundary is further denoted by the index s (*shell*) irrespective of its composition, and the phase in the center is denoted by the index c (*core*). Conservation conditions of matter for a closed binary thermodynamic system are given by:

$$n_1 = n_2 = 0.5n, \quad x_{1f} = \frac{n_{1f}}{n_{1f} + n_{2f}}, \quad x_{2f} = 1 - x_{1f} \tag{1}$$

where n and n_i ($i = 1, 2$) are the total quantity of substance and the number of moles of components 1 and 2 in the system, respectively; component 1 is $Pb_5(VO_4)_3Cl$ and component 2 is $Pb_5(PO_4)_3Cl$. n_{if} is the number of moles of component i in phase f ($f = c, s$) and x_{if} is its concentration in phase f.

The geometrical properties of the *core-shell* structure are determined by:

$$V = 0.5n(V_1 + V_2), \quad V_f = n_{1f}V_1 + n_{2f}V_2, \quad R_c = \left(\frac{3}{4\pi}V_c\right)^{1/3},$$

$$R_s = \left(\frac{3}{4\pi}V\right)^{1/3}, \quad A_f = 4\pi R_f^2 \tag{2}$$

where V is the system's total volume, V_1 and V_2 are the molar volumes of components 1 and 2; $V_1 = 202.18 \times 10^{-6}$ m^3/mol, $V_2 = 181.67 \times 10^{-6}$ m^3/mol; V_f, R_f and A_f are the volume, the radius and the surface area of phase f, respectively.

The equilibrium phase composition is obtained by finding the minima of the total Gibbs free energy of the two-phase mixture including the energy contribution of all interphase boundaries:

$$g = (n_{1c} + n_{2c})G(x_{1c}, T) + (n_{1s} + n_{2s})G(x_{1s}, T) + g_{surf} \qquad (3)$$

where $G(x, T)$ is the molar Gibbs function of the $Pb_5(VO_4)_3Cl$–$Pb_5(PO_4)_3Cl$ system obtained using the following solid solution model [22]:

$$G(x, T) = 17,000x(1-x)^2 + 17,000x^2(1-x) + RT(x \ln x + (1-x)\ln(1-x)) \qquad (4)$$

where x is the molar fraction of vanadinite in the melt, T is the temperature, and R is the ideal gas constant. The energy contribution of all interphase boundaries is given by:

$$g_{surf} = \sigma_s A_s + \sigma_{cs} A_c \qquad (5)$$

where σ_s and σ_{cs} are the specific surface energies on the external boundary and on the boundary between *core*- and *shell*-phases, respectively. In the absence of reliable experimental or theoretical data, we use the linear approximation $\sigma(x) = \sigma_1 x + \sigma_2(1-x)$ and for the boundary between *core*- and *shell*-phases the approximation $\sigma_{cs} = 0.5(\sigma(x_c)+\sigma(x_s))$ [23] is used. It is necessary to mention that there are extremely few studies on surface energies for wide ranges of compounds from the apatite family in different surrounding environments and at different temperatures. This gap needs strongly to be filled. An attempt to systematize the obtained results for various apatite crystals was made in [23]. In our calculations, we accept specific surface energies of the pure components to be $\sigma_1 = 0.125$ J/m^2, $\sigma_2 = 0.185$ J/m^2, these values correlate to the data given in [24] and the estimates for vanadinite and pyromorphite made using the Jerakhov's law [25].

The Gibbs function for the bulk system has two minima of equal depth corresponding to the state with a pyromorphite-based solid solution in the *core*-position and, vice versa, to the state, in which the *core*-phase is vanadinite-rich. As the size of the system decreases down to the nanometer values, the minima shift. This fact is a consequence of substantial differences in equilibrium compositions and volumes of co-existing phases between the systems of different radii. The state with a higher surface energy (with a vanadinite-based *core*-phase) becomes metastable with a decrease in the particle's radius. Similarly, the other state with a lower energy is called "stable" below. The mutual solubilities and ratios of equilibrium volumes of *core*- and *shell*-phases differ in those states. For both states, they were obtained by minimizing the Gibbs functions for different sizes of the system (see Eqs. 1–5). Size-dependent vanadinite solubility in pyromorphite and volume fraction of the *core*-phase in the structure for the state with a vanadinite-based solid solution in the *core*-position at 600 K are plotted in Fig. 1.

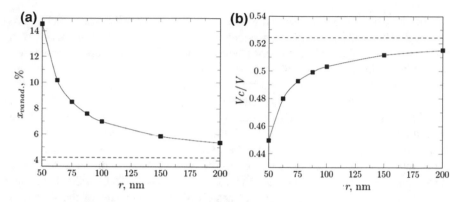

Fig. 1 Size-dependent: **a**—vanadinite solubility in pyromorphite and **b**—the volume fraction *core*-phase in the structure. Both curves are plotted for the state, in which the *core*-phase is formed with a solid solution of $Pb_5(PO_4)_3Cl$ in $Pb_5(VO_4)_3Cl$, at 600 K. Corresponding values for the bulk system are shown with the dashed line

As Fig. 1 shows, for the state with a $Pb_5(VO_4)_3Cl$-based solid solution in the *core*-position, a size decrease from the bulk state down to 50 nm is accompanied by a dramatic increase in vanadinite solubility in pyromorphite (by more than 10 mol%). The volume fraction of the *core*-phase in the structure decreases by 7.5%. Pyromorphite solubility in vanadinite also decreases by 2 mol%. For another state (the solvent in the *core*-phase is $Pb_5(PO_4)_3Cl$), simulation shows a slight decrease in $Pb_5(PO_4)_3Cl$, solubility in $Pb_5(VO_4)_3Cl$ (less than 1 mol%) and, vice versa, an increase in $Pb_5(VO_4)_3Cl$ solubility in $Pb_5(PO_4)_3Cl$ (~1 mol%). The volume fraction of the *core*-phase in the stable state also increases by ~1 mol%. A binary heterogeneous system has the possibility to decrease the Gibbs energy by transferring the component with a maximal molar volume and a minimal specific surface energy ($Pb_5(VO_4)_3Cl$) into the *shell*-phase. This process occurs until the gain from a decrease in the surface energy is equal to the losses from the violation of the bulk equilibrium composition of co-existing phases. It is necessary to underline that the solubilities, the ratios of equilibrium phase volumes, and the whole phase diagrams in nanosized systems are determined by specific surface free energies of pure components and depend nonlinearly either on their values or on the difference between them [15].

3 The Influence of Shape

To observe the shape effect in the $Pb_5(VO_4)_3Cl$–$Pb_5(PO_4)_3Cl$ system, let us consider a particle with non-spherical interphase boundaries. Its volume is equivalent to the volume of a sphere of 75 nm radius of equimolar composition at $T = 600$ K. When the system is at the phase equilibrium, a single inclusion which does not touch the outer boundary appears in the particle (the formation of a *core-shell* structure takes place). The general case is observed where both *core-* and *shell*-phases can be non-spherical. The shape of the *shell*-phase can be determined by the boundaries of a pore, a matrix, or the grain boundaries which emerge during dispersion processes, and can be modified during the deformation. The shape of the *core*-phase can be non-spherical, for example, due to an elastic stress field anisotropy emerging during the growth process of a new phase. At constant volume, the surface areas of interphase boundaries A_f, $f = c, s$ are determined by their shape and are of minimum value when the *core-* and the *shell*-phases are both spherical. The degree of their shape's deviation from the spherical shape can be used as a parameter for the calculation. Let us introduce the shape coefficient k_f in the form of the ratio of the surface area of the considered figure A_f and the sphere of the same volume A_{f0}. On that basis, the introduced shape coefficient generalizes all isochoric changes of the shape. In this case, the change of shape reduces to adding k_f factor to the last of the Eq. (2).

High k_f values can be obtained if the shapes of co-existing phases are fractal. A fractal phase is characterized by its fractal dimension D [26–29] which correlates its volume and surface area:

$$A_f = C V_f^{2/D}, \quad k_f = \frac{C}{4\pi} \frac{V_f^{2/D}}{(3V_f/4\pi)^{2/3}} \tag{6}$$

where C is a numerical coefficient. In calculations below, we accepted C to be 4π, that corresponds to structures with a central symmetry. For fractal structures, $D \leq 3$ and can be non-integer. Formation of fractal structures is generic for many non-equilibrium processes. The most classic examples of fractals are dendrite-like, "amoeba-like," or "porcupine-like" particles. The shape coefficient k_f as a function of a particle's volume (see Eq. 6) is plotted in Fig. 2 for three fractal dimensions. Volumes of fractal particles are determined by their effective radii–radii of the spheres of the same volume. There are also examples of "amoeba-like" and "porcupine-like" particles with various fractal dimensions and volumes in Fig. 2b.

The dependences of vanadinite solubility in pyromorphite and the volume fraction of the *core*-phase in the structure on the shape of the outer boundaries (set by k_s) at varying k_c are plotted in Fig. 3. Both curves correspond to the state with a vanadinite-based solid solution in the *core*-position at 600 K. For this state, the higher the values of k_c and k_s are, the higher vanadinite solubility in pyromorphite and the lower the volume fraction of the *core*-phase are. In addition to this, the higher the value of k_c is, the more sensitive to a change of the shape of the outer boundaries the system is (the higher the value of derivative $\partial x_{\text{vanad.}}/\partial k_s$ is, see Fig. 3). For the state with a vanadinite-based solid solution in the *core*-position, simulation shows less significant dependences of mutual solubilities on the shape coefficients k_c and k_s. However, the transition from a configuration with $k_c = 1.5$, $k_s = 1$ to a configuration with $k_c = 1$, $k_s = 1.5$ leads to a nearly 1% decrease in $Pb_5(VO_4)_3Cl$ solubility in $Pb_5(PO_4)_3Cl$. At constant k_s, the increase of k_c also leads to a slight decrease in the vanadinite solubility in pyromorphite.

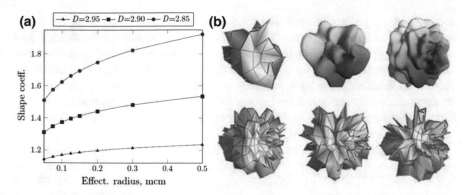

Fig. 2 **a**—Shape coefficient as a function of the nanoparticle's volume and fractal dimension and **b**—examples of "amoeba-like" and "porcupine-like" nanoparticles of $D = 2.95$ (**b**, top row) and $D = 2.90$ (**b**, bottom row)

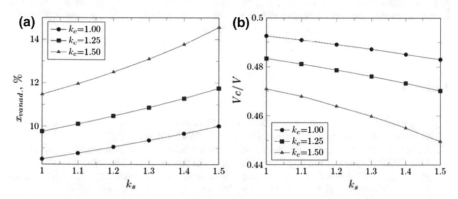

Fig. 3. Shape-dependent: **a**— vanadinite solubility in pyromorphite and **b**—the volume fraction of the *core*-phase in the structure. Both curves are plotted for the state, in which the *core*-phase is formed with a solid solution of $Pb_5(PO_4)_3Cl$ in $Pb_5(VO_4)_3Cl$ at 600 K

4 Conclusions

In case of nanoparticles of stratifying solid solutions, the size and shape are factors which determine the number of co-existing phases, their equilibrium volumes, and compositions. In spherical nanoparticles of the $Pb_5(VO_4)_3Cl–Pb_5(PO_4)_3Cl$ system with a *core-shell* configuration, two heterogeneous states are possible. These possible states are distinct in the component which prevails in the *core*-phase. The shift of phase equilibria with a decreasing particle's size is different in two states. If the *core*-phase is a vanadinite-based solid solution, a significant increase of vanadinite solubility in pyromorphite and a decrease of the pyromorphite solubility in vanadinite with decreasing size are expected. Meanwhile, the increase of $Pb_5(VO_4)_3Cl$ solubility in $Pb_5(PO_4)_3Cl$ at 600 K in the considered external environment reaches the value of 10%. The volume fraction of the *core*-phase decreases with decreasing size. For the state with a pyromorphite-based solid solution in the *core*-position, the trends are opposite to the given above. The dependences of the mutual solubilities, sets of co-existing phases, and their volume fractions on the shape of both *core*- and *shell*-phases were also revealed. In another paper of us [30], it has been found that a substantial isochoric increase in the outer boundary's surface area leads to only one thermodynamically stable heterogeneous state remaining in the system. At the same time, the only homogeneous stable state in the system is a result of a great increase in the *core*-phase's surface area.

The obtained results can be useful for the simulation of phase transitions in nanograined ceramic materials during thermal or thermomechanical treatments or the decomposition of solid solutions inside a pore of a nanocomposite. Predicting the phase equilibria is fundamental to understanding how materials behave at tiny volume scales and to designing materials with improved performance.

Acknowledgements The authors acknowledge the support from the Russian Foundation for Basic Research (RFBR) (project № 18-08-01356). V. B. Fedoseev acknowledges the support from the Russian Science Foundation (project № 15-13-00137-p.).

References

1. Magnin, Y., Zappelli, A., Amara, H., Ducastelle, F., Bichara, C.: Size dependent phase diagrams of nickel-carbon nanoparticles. Phys. Rev. Lett. **115**(20), 205502–205506 (2015)
2. Dahan, Y., Makov, G., Shneck, R.Z.: Nanometric size-dependent phase diagram of Bi-Sn. CALPHAD: Comput. Coupling Phase Diagrams Thermochem. **53**, 136–145 (2016)
3. Park, J., Lee, J.: Phase diagram reassessment of Ag-Au system including size effect. CALPHAD: Comput. Coupling Phase Diagrams Thermochem. **32**(1), 135–141 (2008)
4. Monji, F., Jabbareh, M.A.: Thermodynamic model for prediction of binary alloy nanoparticle phase diagram including size effect. CALPHAD: Comput. Coupling Phase Diagrams Thermochem. **58**, 1–5 (2017)
5. Fedoseev, V.B., Shishulin, A.V., Titaeva, E.K., Fedoseeva, E.N.: On the possibility on the formation of NaCl-KCl solid-solution crystal from an aqueous solution at room temperature in small-volume systems. Phys. Solid State **58**(10), 2095–2100 (2016)
6. Shishulin, A.V., Fedoseev, V.B.: Size effect in the phase separation of Cr-W solid solutions. Inorg. Mater. **54**(6), 546–549 (2018)
7. Straumal, B., Baretzky, B., Mazilkin, A., Protasova, S., Myatiev, A., Straumal, P.: Increase of Mn solubility with decreasing grain size in ZnO. J. Eur. Ceram. Soc. **29**(10), 1963–1970 (2009)
8. Ghasemi, M., Zanolli, Z., Stankovski, M., Johansson, J.: Size- and shape-dependent phase diagram of In-Sb nano-alloys. Nanoscale **7**, 17387–17396 (2015)
9. Cui, M., Lu, H., Jiang, H., Cao, Z., Meng, X.: Phase diagram of continuous binary nanoalloys: size, shape and segregation effects. Sci. Rep. **7**, 1–10 (2017)
10. Guisbiers, G., Mendoza-Cruz, R., Bazán-Díaz, L., Velázquez-Salazar, J.J., Mendoza-Pérez, R., Robledo-Torres, J., et al.: Electrum, the gold–silver alloy, from the bulk scale to the nanoscale: synthesis, properties, and segregation rules. ASC Nano **10**(1), 188–198 (2015)
11. Guisbiers, G., Mendoza-Pérez, R., Bazán-Díaz, L., Mendoza-Cruz, R., Velázquez-Salazar, J.J., Yakamán, M.J.: Size and shape effects on the phase diagrams of nickel-based bimetallic nanoalloys. J. Phys. Chem. **121**(12), 6930–6939 (2017)
12. Guisbiers, G., Khanal, S., Ruis-Zepeda, F., Roque de la Puente, J., Yakamán, M.J.: Cu-Ni nano-alloy: mixed, core-shell or janus nano-particle. Nanoscale **24**(6), 14630–14635 (2014)

13. Tovbin, YuK: Lower size boundary for the applicability of thermodynamics. Russ. J. Phys. Chem. A **86**(9), 1356–1369 (2012)
14. Fedoseev, V.B., Potapov, A.A., Shishulin, A.V., Fedoseeva, E.N.: Size and shape effect on the phase transitions in a small system with fractal interphase boundaries. Eurasian Phys. Tech. J **14**(1), 18–24 (2017)
15. Shishulin, A.V., Fedoseev, V.B., Shishulina, A.V.: Environment-dependent phase equilibria in a small volume system in case of decomposition of Bi-Sb solid solutions. Butlerov Commun. **51**(7), 31–37 (2017). (In Russian)
16. Gusev, A.I., Rempel, A.A.: Nanocrystalline Materials. Fizmatlit, Moscow (2000). (In Russian)
17. Gladkikh, N.T., Dukarov, S.V., Kryshtal, A.P., Larin, V.I., Sukhov, V.N., Bogatyrenko, S.I.: Surface Phenomena and Phase Transitions in Condensed Thin Films. V. N. Karazin Kharkiv National University, Kharkiv (2004). (In Russian)
18. Alymov, M.I., Shorshorov, MKh: The influence of size factors on the melting temperature and surface tension of ultra-disperse particles. Metally **2**, 29–31 (1999). (In Russian)
19. Rusanov, A.I.: Phase Equilibria and Surface Phenomena. Khimiya, Leningrad (1967). (In Russian)
20. Gusarov, V.V.: The thermal effect of melting in polycrystalline systems. Thermochim. Acta **256**(2), 467–472 (1995)
21. Betkhtin, A.G.: The Course of Mineralogy. Gosudarstvennoye nauchno-tekhnicheskoye izdatel'stvo po geologii i okhrane nedr, Moscow (1956). (In Russian)
22. Chernorukov, N.G., Knyazev, A.V., Bulanov, E.N.: Isomorphism and phase diagram of the $Pb_5(VO_4)_3Cl$–$Pb_5(PO_4)_3Cl$ system. Russ. J. Inorg. Chem. **55**(9), 1463–1470 (2010)
23. Hourlier, D., Perrot, P.: Au-Si and Au-Ge phase diagrams for nanosystems. Mater. Sci. Forum **653**, 77–85 (2010)
24. Rakovan, J.: Growth and surface properties of apatite. Rev. Mineral. Geochem. **48**(1), 51–86 (2002)
25. Lazarev, S.Y.: Assessment of substance properties by the criteria of surface energy, hardness and energy density. Metalloobrabotka **2**, 38–42 (2003). (In Russian)
26. Kalinin, S.V., Gorbachev, D.L., Borisevich, A.Y., Tomashevitch, K.V., Vertegel, A.A., Markworth, A.J., et al.: Evolution of fractal particles in systems with conserved order parameter. Phys. Rev. E **61**(2), 1189–1194 (2000)
27. Katunin, A.: Construction of fractals based on catalan solids. Int. J. Math. Sci. Comput. (IJMSC) **3**(4), 1–7 (2017). https://doi.org/10.5815/ijmsc.2017.04.01
28. Nayak, S.R., Mishra, J.: On calculation of fractal dimension of color images. Int. J. Image Graph. Sig. Process. (IJIGSP) **9**(3), 33–40 (2017). https://doi.org/10.5815/ijigsp.2017.03.04
29. Khemis, K., Lazzouni, S.A., Messadi, M., Loudjedi, S., Bessaid, A.: New algorithm for fractal dimension estimation based on texture measurements: application on breast tissue characterization. Int. J. Image Graph. Sig. Process. (IJIGSP) **8**(4), 9–15 (2016). https://doi.org/10.5815/ijigsp.2016.04.02
30. Fedoseev, V.B., Shishulin, A.V.: Shape effect in layering of solid solutions in small volume: bismuth-antimony alloy. Phys. Solid State **60**(7), 1398–1404 (2018)

Application of Neural Networks for Controlling the Vibrational System Based on Electric Dynamic Drive

R. F. Ganiev, S. S. Panin, M. S. Dovbnenko and E. A. Bryzgalov

Abstract One of the main ideas of wave mixing is based on the feasibility to organize the complex quasi-one-directed flotations of liquid medium due to its interaction with solid bodies (the mixture working organs) which are dipped into mixing medium and are making oscillations relatively to it. In other words, it is possible to organize the transfer of the mixing staff throughout the whole volume due to oscillatory impact alone. Whereby this method of flotation organization permits to realize complex differently directed flotations (up to opposite flotations) in one volume with relatively substantial shift of liquid medium, absence of dead zones, and diffusion of transverse waves in the volume. This mode of medium motion permits to secure the intensive mixing in combination with wave impact which in its turn permits to get qualitatively new results related to the conversion of physical and rheological properties of mixing medium. In this paper, the experiment concerning the possibility to implement a control system for the resonant mode of a wave mixer with the application of neural network technology based on electrodynamic excitation is set up. A feasibility study for application of this technology as a control system for operating modes of the mixing unit with the aim to increase the mixing quality is conducted.

Keywords Neural net · Nonlinear wave mechanics

1 Introduction

For many years, the Scientific Center for Nonlinear Wave Mechanics and Technology of the Russian Academy of Sciences has been carrying out fundamental and applied research to find out fundamentally new possibilities for using wave and oscillatory processes to introduce this technology into various industries. In recent years, a technology has been developed on the basis of the Scientific Center of the North-Caucasian Scientific and Technical University of the Russian Academy of Sciences,

R. F. Ganiev · S. S. Panin · M. S. Dovbnenko (✉) · E. A. Bryzgalov
Mechanical Engineering Research Institute of the Russian Academy of Sciences, Moscow, Russia
e-mail: dovbnenko_ms@mail.ru

© Springer Nature Switzerland AG 2020
Z. Hu et al. (eds.), *Advances in Artificial Systems for Medicine and Education II*,
Advances in Intelligent Systems and Computing 902,
https://doi.org/10.1007/978-3-030-12082-5_38

which makes it possible to intensify various processes of mixing viscous, multi-phase, and free-flowing media. This technology allows to optimize technological processes in many industries such as oil and gas production, oil refining, chemistry and petrochemistry, construction, engineering, and food industry. A new class of wave machines and devices has been developed to realize open wave effects. One of such classes of machines is resonant wave systems with mobile working bodies. In this paper, one of the possible configurations of the described machines and apparatus is considered, the general view of which is shown in Fig. 1. The installation is based on an electrodynamic drive, the rotor of which is connected with a spring providing the system with a fixed natural oscillation frequency. Also on the rotor, there is a working area in which the process of mixing takes place. A detailed description of the work of this system, as well as theoretical calculations, can be found in [1]. Here we will dwell only on the basic principles of work. By feeding a sinusoidal alternating voltage to the stator winding, synchronous oscillations are set in the system. It is known from [1] that the control of the amplitude of mechanical oscillations in such systems can be performed by varying the amplitude of the supply voltage. When the exciting and natural frequencies coincide, a resonance regime sets in. This mode is characterized by an increase in the amplitude of the oscillations of the working member, and also by the steady-state phase difference between the actual oscillations and the exciting force equal to $\pi/2$ rad. The mixing efficiency directly depends on the accuracy of resonance mode maintenance [1]. In practice, it turns out that the natural frequency of the system can change during operation, which, in turn, leads to a decrease in the amplitude of the oscillations and a decrease of mixing efficiency. Such phenomena are associated with several factors: during the work, the stirred medium is heated, which in turn reduces the viscosity of the working medium, as a result of which the attached mass changes. Such changes in the future will be called "dynamic," that is, such changes in the parameters of the system that arise directly in the process of work. It is also possible to distinguish static factors: different degree of loading of the working area and dispersion in the characteristics of viscosity of working media [2]. From all that has been said above, we can distinguish two main problems and, correspondingly, two directions in the management: the search for the initial resonant frequency and compensation for dynamic parameter changes during the operation of the plant. To solve the control problem at the Scientific Center for Nonlinear Wave Mechanics and Technology of the Russian Academy of Sciences, under the guidance of Academician Ganiev RF, a prototype frequency converter was developed to study the control of oscillations in installations based on electrodynamic drives. The power part of the frequency converter is based on the classical scheme of a three-phase inverter, the only difference being the replacement of a three-phase IGBT bridge by a single-phase bridge (the so-called H-bridge). The bridge is controlled by pulse-width modulation. The controller simulates a sinusoidal signal of a specified frequency and duty cycle to ensure the required quality and voltage level of the electrodynamic drive. The AVR ATMega 2560 chip is used as the controller of the power system of the frequency converter. It also implements the output of service information on the display and monitoring of the power section parameters, such as bridge temperature and DC bus voltage. If necessary, it is also possible to

Fig. 1 Units general layout

use the built-in amplitude measurement system. It should be noted that all software was developed by employees of the Scientific Center of Nonlinear Wave Mechanics and Technology of the Russian Academy of Sciences.

2 Output to the Resonance Frequency

In any closed control system, there is a link responsible for transmitting information about the state of the control object. In general terms, such a link is called the feedback sensor [3–11]. Based on the nature of the process to be controlled, the accelerometer becomes a natural choice. The parameters of the accelerometer are selected based on the dynamic characteristics of the oscillatory system. The technique is reduced to selecting the parameters of the sensor in such a way that, on the one hand, it provides a sufficient range of operating accelerations, and, on the other hand, it maintains compatibility with the used analog-to-digital converter in the parameter mV/g. At too small values, it becomes necessary to use additional hardware solutions in the form of instrument amplifiers to obtain a correct output signal. In the case under consideration, a sensor with an integrated amplifier AP2028-100-01 was used.

The choice of the mounting location of the sensor also has an important role. It is necessary to provide a seat for the sensor at a strictly fixed distance from the center of the oscillation axis, and the sensor itself must be installed tangentially to this axis. The importance of observing the distance to the axis is justified by adjusting the sensor in the frequency converter and bringing the actual amplitude taken from the sensor in the form of a voltage signal to the units of angle measurement. For correct operation of the system, it is first necessary to search for a resonance frequency. As stated above, the natural frequency of the system is a known value, but only if the work area is empty. In this regard, the initial search is a kind of calibration of the system before launching. The search for the resonance frequency is accomplished by passing through a known frequency range lying within predetermined limits relative to the theoretical (calculated) natural frequency of the system. We will express it as follows:

$$W = \{\omega_t | \omega_t \in \mathfrak{R} \wedge (\omega_t - \varepsilon_l) < \omega_t < (\omega_t + \varepsilon_r)\}$$

where W is the set of all possible frequencies in a neighborhood given by positive constants ε_l and ε_r that are different from zero and calculated with respect to the natural frequency ω_t. At start-up, it is assumed that the vicinity is large enough. This assumption is obvious due to the dependence of the frequency from the parameters of the medium.

Figure 2 shows the graphs of the amplitude–frequency characteristics of the plant, obtained after loading liquids with different viscosities into the working region. An apparent displacement of the resonant frequency and a drop in the amplitude of the oscillations are seen. The presented graphs also reflect the fact that the system is linear. During the control in this case, the approach to the resonance frequency can be realized from two sides. The amplitude of the oscillations is now expressed in terms of the ADC sample units as follows:

$$A = \sqrt{\sum_{n=1}^{N} (a_n)^2}$$

where N is the number of samples per period of one oscillation and a_n is the instantaneous value of the amplitude in units of the ADC. It depends on the accuracy of the approximation of the sinusoidal signal, more precisely, on the number of points specified for one oscillation period. After the resonance frequency is found, the transition to the basic mode of operation is realized.

3 Principles Laid Down in Management

There are standard approaches to building management systems. They are considerably general and applicable in most cases. One such approach is the construction

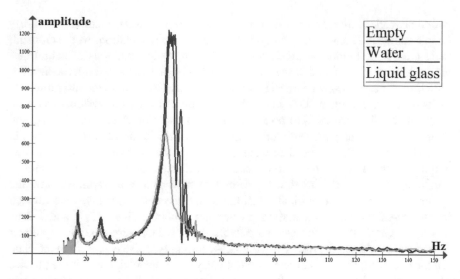

Fig. 2 Frequency response for different types of media

of control systems on various types of controllers, the most popular of which is the PID (proportional–integral–differential) regulator. The general theory of construction is known and accessible. The main idea of regulators is to measure the difference between the set value of the regulated parameter and its actual value. This difference is called an error. Based on this error, the output control is calculated. Such an approach is theoretically applicable in the case of control of the oscillatory system, but in practice situations have been revealed where the system loses stability. It should be noted that the control problem cannot be limited only to maintaining a given amplitude, it also extends to the retention of the resonance regime, since work in it reduces energy costs for operation of the installation. With small-amplitude assignments, the system can support it without being at resonance. As mentioned above, changes in the viscosity and attached mass parameters affect the natural frequency. Such changes are very difficult to predict and take into account in the mathematical model due to their stochastic nature. To solve this problem, an experiment was executed with the introduction of a classifier of operating modes into the control loop of the system. The idea is to preprocess the phase and amplitude signals and further decide whether to simply adjust or search for a resonance frequency. For completeness of the experiment, along with other modes, an idle mode was introduced. In this case, the system is not regulated. The main management cycle consists of several stages. In the first step, the phase difference and the actual amplitude of the oscillations are measured. The phase difference is measured by the zero-crossing method. This is an easy-to-implement and fairly accurate method. Its accuracy depends on the frequency of the signal sampling. In the described case, a self-made device was used. Once the measured values are obtained, the system attempts to classify the mode on which it is now located, and then the control action follows. As the classifier, an artificial neural network of direct propagation was chosen. The choice is justified by a fairly simple architecture and, as a result, quick program implementation. At the moment, there are a huge number of ready-to-use software implementations of neural networks, but due to the fact that the classifier had to be launched on a microcontroller with limited computing resources, it became necessary to develop the program code of its own. An important stage in the construction of intelligent systems based on artificial neural networks is to obtain a relevant sample of values for training and their preliminary analysis. During the experiments with the installation described in the framework of this work, data were obtained on the behavior of the amplitude and the phase difference between the exciting and actual oscillations under various conditions and operating modes. Representing these data in a graphical form, the operating modes were empirically identified, and the data themselves are divided into the so-called samples with the same number of points. Figures 3, 4, and 5 show a graphic representation for each case. Note that the data were obtained by direct measurement without the use of any filters. Accordingly, noise is present in the data. On the one hand, these phenomena do not carry any information about the physics of processes and they can be filtered; on the other hand, no filter, whether soft or hard, can guarantee accurate filtering. And since the phase difference plays a huge role in this task, the use of filters is not entirely justified, since in most cases they are phase shifting. Training classifier is a classic method of back-propagation errors. The method was described in 1974 by A. I. Galushkin and in parallel with him by J. J. Verbes. Over the past time, the method has improved and evolved, but the main ideas embedded in it have remained. The essence of the algorithm is to spread the error from the outputs of the neural network to its inputs and sequentially compute its gradients. A detailed description of the method can be found in [12].

As it was said before, during the operation of the mixer dynamic changes in the parameters of oscillations occur. The reasoning is constructed as follows: there are operating modes that do not require frequency and amplitude corrections (Fig. 3). This is possible under the condition that either the working area is empty and the blades do not experience additional resistance, or the medium parameters vary slowly enough. Conversely, in case of changes much more time will be needed to achieve a positive result of mixing. Obviously, such a regime does not require additional control by the system, but it cannot be excluded from consideration.

To build the classifier, we use a three-layer neural network of direct propagation [12]. Figure 5 shows the distribution of some experimental data. Axis of abscissas represents amplitude value and axis of ordinates—phase value. Different operating modes are presented by color. Blue color represents "normal" mode, red and green—represent the need to regulate frequency and amplitude. As a result of numerical experiments, the following configurations were obtained on the basis of data gathered to identify the minimum number of layers and neurons, on the one hand, and the maximum identifiability rate on the other, as shown in Table 1.

Once the model has been obtained that satisfies the quality of the identifiability, it must be transferred to the controller of the frequency converter. To do this, the software provides the ability to generate the source code of the network, which included both the implementation of the most artificial neural network and its parameters.

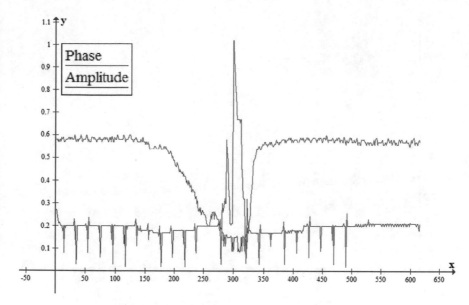

Fig. 3 Decay of amplitude and phase caused by changes in environmental parameters

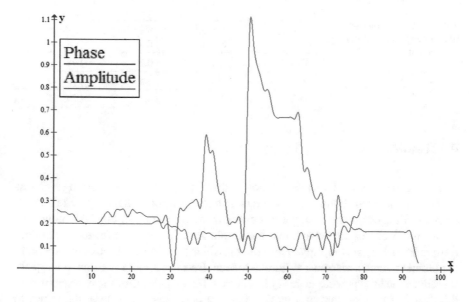

Fig. 4 Need to search for a resonant frequency

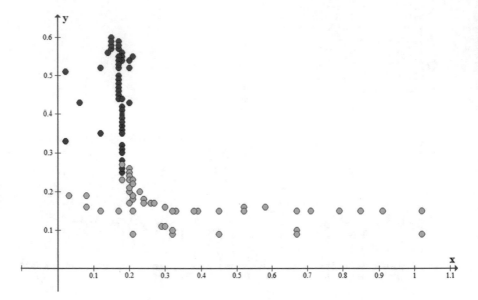

Fig. 5 Distribution of experimental data on the plane

Table 1 Configurations of neural networks used in the experiment

Topology	Number of examples	Identifiability rate	Accuracy (%)
2–2–1	241	0.2	98
2–4–1	241	0.5	96
4–2–1	241	0.3	99

4   Results

To test the obtained system, the situation with sharp loading of the working area was artificially modeled. In this case, there is a sharp decrease in the amplitude and the corresponding departure from the resonance mode of operation. Figure 6 shows two curves representing the behavior of the amplitude in the case of disabled classifier (purple line) and, accordingly, using the classifier (red line). Slumps correspond to the moment when a viscous medium is loaded. Because it is impossible to uniquely reproduce the boot process to verify both cases, the magnitude of the recessions may differ. It can be seen that in a work without a classification system, the return to normal mode takes more time. This can be explained in the following way: after the working area has been loaded, the operating elements of the mixer begin to experience resistance, as a result of which the amplitude falls. The system begins to compensate for the decline, raising the amplitude of the supply voltage, and changing its frequency. Since both processes occur conditionally at the same time, then after the frequency is found, a sharp surge of amplitude inevitably occurs, which must also

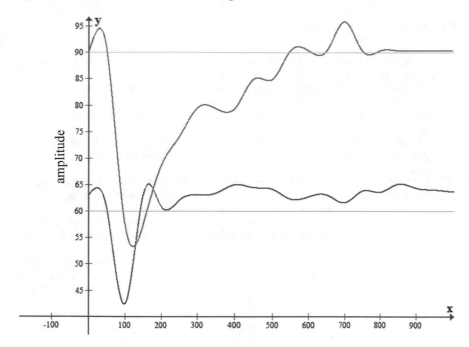

Fig. 6 Results of experiments

be compensated. Such bursts can lead to breakdown of the main components of the installation. To avoid this kind of situations, we choose slow rate of compensation, which, in turn, affects the response of the system. In the case of using a classification system, such situations can be avoided in advance by laying down the action algorithm for different cases and comparing it with some experimental sample. In this case, after detection, the supply voltage is reduced, a new frequency pass is made and then the regulators are turned on.

5 Conclusions

Drawing conclusions, we can say that the system obtained is promising, but it requires improvements and a more detailed formalization of the problem. In the process of experiments in addition to the positive results, the regimes were also obtained during which the stability of mixing was broken. It should also be noted that the application of neural networks requires sufficient amount of computational resources. Hence, for better efficiency, the controllers or the microprocessor systems of bigger bulk storage, higher digit capacity, and bigger clock speed are needed.

References

1. Ganiev, R.F., Ukrainski, L.E.: Regular and chaotic dynamics. In: Nonlinear Wave Mechanics and Technology, 712 p. Scientific and Publishing Center (2008)
2. Panin, S.S.: Development of a wave mixer for mixing high-viscosity liquids. Prob. Mach. Build. Mach. Reliab. (2), 61–70 (2012)
3. Agagu, T., Tran, T.: Context aware recommendation methods. Context-Aware Recomm. Methods 9(10), 1–12 (2018)
4. Bilgaiyan, S., Aditya, K., Mishra, S., Das, M.: A swarm intelligence based chaotic morphological approach for software development cost estimation. Context-Aware Recomm. Methods 9(10), 13–22 (2018)
5. Christian Ameh, A., Olaniyi, O.M., Dogo, E.M., Aliyu, S., Arulogun, O.T.: Nature-inspired optimal tuning of scaling factors of Mamdani fuzzy model for intelligent feed dispensing system. Context-Aware Recomm. Methods 9(10), 57–65 (2018)
6. Panin, S.S., Dovbnenko, M.S.: Computer simulation of control processes of the wave mixing machine witch electrodynamic drive. In: Oscillations and Waves in Mechanical Systems, pp. 39–40 (2017)
7. Ganiev, R.F., Ukrainski, L.E., Panin, S.S., Ganiev, O.R., Ganiev, S.R., Pustovgar, A.P.: Wave technology in the building industry, chap. in monograph. In: Nonlinear Wave Mechanics and Technologies: Wave and Oscillatory Phenomena on the Basis of High Technologies. Begell, New York, Connecticut, pp. 475–481 (2012)
8. Panin, S.S., Kyrmenev, D.V., Jakovenko, N.I., Bryzgalov, E.A.: Investigation of the impact of wave actions on the processes of grinding of solid granular media, computer simulation of control processes of the wave mixing machine witch electrodynamic drive. In: Oscillations and Waves in Mechanical Systems, 34 p. (2012)
9. Ibraheem, I.K., Al-Hussainy, A.A.-H.: A multi QoS genetic-based adaptive routing in wireless mesh networks with Pareto solutions. Int. J. Comput. Netw. Inf. Secur. (IJCNIS) 10(9), 1–9 (2018). https://doi.org/10.5815/ijcnis.2018.09.01
10. Shrivastava, N., Varshney, P.: Implementation of Carlson based fractional differentiators in control of fractional order plants. Int. J. Intell. Syst. Appl. (IJISA) 10(9), 66–74 (2018). https://doi.org/10.5815/ijisa.2018.09.08
11. Thara, S., Athul Krishna, N.S.: Aspect sentiment identification using random Fourier features. Int. J. Intell. Syst. Appl. (IJISA) 10(9), 32–39 (2018). https://doi.org/10.5815/ijisa.2018.09.04
12. Haykin, S.: Neural Networks. A Comprehensive Foundation, 2nd edn, 1104 p. (trans. with English). Publishing house "Williams" (2006)

A Dynamic Power Flow Model Considering the Uncertainty of Primary Frequency Regulation of System

Daojun Chen, Nianguang Zhou, Chenkun Li, Hu Guo and Ting Cui

Abstract In view of the problem that uncertainty of primary frequency regulation (PFR) has not been considered in the existing research on the dynamic power flow, which may cause fault, this paper presents a stochastic dynamic power flow model that takes into account the uncertainty of the primary frequency regulation coefficient and the total adjustment amount of the AGC system, the switching process of allocation and control mode. By analyzing the uncertainties of primary frequency regulation coefficient and the regulation of AGC and AVC, as well as considering the active coordination equation of AGC unit with load frequency advanced control strategy, this model solves the problems of existing dynamic power flow model. The impact of the uncertain variables is addressed by the method of Monte Carlo simulation. Applying the proposed model to the IEEE 39-bus system to simulate the power flow distribution of cutting unit, load disturbance, and so on, the results prove that the model is practical and precise.

Keywords Dynamic power flow · The primary frequency regulation · Automatic power generation control · Uncertainty · Monte Carlo simulation

1 Introduction

In the general power flow model, there is an assumption of an ideal power balanced node, which is regarded to play a role of balancing the system power by supplying deficient power or absorbing excess power to maintain the system frequency stable. In theory, the balanced node represents an infinite system connected to power system. Nevertheless, there is no generator with infinite capacity in the actual power grid. In recent years, a large number of distributed power supplies with intermittent and

D. Chen (✉) · C. Li · H. Guo · T. Cui
State Grid Hunan Electric Power Corporation Research Institute, Changsha 410007, China
e-mail: 1531755324@qq.com

N. Zhou
State Grid Hunan Electric Power Corporation, Changsha 410014, China

© Springer Nature Switzerland AG 2020
Z. Hu et al. (eds.), *Advances in Artificial Systems for Medicine and Education II*,
Advances in Intelligent Systems and Computing 902,
https://doi.org/10.1007/978-3-030-12082-5_39

random characteristics are connected to the power grid, which made the disturbance of system becomes more frequent. When power system suffers large disturbance, it is very likely that the output power of balanced node cannot meet the requirement of power balance or even the output power is negative if general power flow model is used, which is not in accordance with the actual operating state.

In the actual power system, all units and loads share the unbalanced power in the system, not only by the ideal balancing unit for power consumption or compensation. Relevant researches on how to distribute unbalanced power between units and loads have been made by scholars worldwide. Reference [1] proposed a method of setting up a multi-balancing generator to distribute unbalanced power. Although the model is improved on the basis of the conventional power flow model, it still assumes an ideal balanced node essentially and is not in line with the actual power grid. References [2–4] present a dynamic power flow model, which considers that the unbalanced power in the system is proportionally distributed according to the static frequency regulation coefficient of loads and units. Considering the primary frequency regulation of the system, the model can well adapt to the actual operation of the power grid and is widely used in the simulation training of dispatcher [5].

At present, the dynamic power flow model is mainly studied from two aspects. One is combining other power system analysis models with dynamic power flow model, such as probabilistic power flow analysis that accounts for the dynamic power flow characteristics [6, 7] and continuous power flow analysis that considers the dynamic power flow characteristics [8] and so on. Based on the dynamic power flow model, the another research is considering the impact of the secondary frequency regulation (SFR) and the limits of the frequency modulation capacity on the power flow distribution, as shown in Ref. [5]. However, the influence of the coefficient uncertainty on the power flow calculation result is not fully considered as the frequency regulation coefficient is regarded as a constant in this method.

In recent years, scholars have done a great deal of research on the static frequency regulation coefficient. Studies have shown that the types and proportions of the integrated load can directly affect the static frequency regulation coefficient [9–12]. As the loads in the power grid are widely distributed, large in quantity, and fast in change, the loads are highly uncertain, under which case it is not possible to accurately calculate the actual static frequency regulation coefficient of load at any moment. Besides, the setting of parameters also greatly affects the participation and the performance of units involved in PFR [13–15]. In order to maintain stability, the power grid has mandatory instructions for all generators involved in PFR to participate. Meanwhile, a large number of evaluation indexes and practical evaluation methods have been proposed for reward or punishment to encourage generator units to participate in PFR [14, 16]. As the units continually change the output according to the frequency regulation signal, it will produce a lot of wear and tear that is not conducive to its stable operation, and it is uneconomical for units to participate in frequency regulation; many units set unreasonable parameters so as to be artificially removed from PFR. It is found that the main reason why the ability of PFR in system differs greatly from expect is that some generator units do not work as expected [12,

17–22]. In reference [17], it has been pointed out that some units even input frequent regulation signals but actually contribute none in serious frequent fluctuations.

In conclusion, uncertainties exist in the participation of units and actual power regulation of the load. If all the units are considered to be involved in PFR and the frequent regulation coefficients of units and load are set as empirical values for dynamic power flow calculation, the result obtained would be greatly different from the actual situation. Therefore, in order to get the actual state of the system accurately, it is crucial to further study the uncertainty of the PFR. Based on the previous models, this paper considers the SCF of the load and generations as random variables and proposes a stochastic dynamic power flow model which takes the uncertainty of the SCF into account.

In recent years, simulation method [23–25], analog method, and approximate method have been considered as three main methods in analysis of stochastic power flow calculation. The analytic method refers to the method of obtaining the probability distribution of the unknown state through the convolution under the assumption that the input random variables are independent, such as the semi-invariant method. The simulation method refers to the method that obtains the statistical characteristic values and distribution characteristics of variables by largely sampling each random variable according to its probability distribution model and giving each sample a deterministic calculation, such as the Monte Carlo simulation method. The analog method refers to the method of approximating the statistical properties of variables by using the numerical features of input variables, such as the point estimation method.

A stochastic dynamic power flow model is proposed and solved by Monte Carlo simulation method in this paper. General situations such as cutoff and load disturbance have been set up in IEEE 39-bus system and verified the effectiveness and practical significance of the model. The comparison of different power flow states under random parameters is analyzed using different probability distribution models, which has further verified the influence of uncertainty of the static coefficient on power flow.

2 System Static Frequency Regulation Characteristics and the Uncertainty of Its Factor

2.1 System Static Frequency Regulation Characteristics

The system static frequency regulation characteristic, which is also called the characteristic of system primary frequency regulation, refers to the characteristic that the active power changes with the frequency deviation. It can be divided into generator unit characteristics and load characteristics [7].

Fig. 1 Curve of USRC

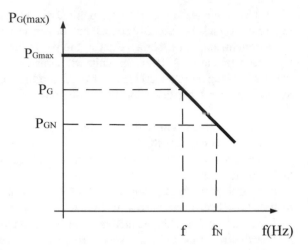

2.1.1 Static Frequency Regulation Characteristics of Generator Unit (USRC)

The USRC can be expressed as:

$$P_G = P_{GN} - k(f - f_N) \tag{1}$$

where f_N and f represent the rated power and actual power of system, respectively. $P_{G}\text{max}$, P_{GN}, and P_G refer to the maximum, rated, and actual active power outputs of generator, respectively. k_G represents the unit static frequency regulation factor of generator, which is also called the unit power regulation factor (URP) of generator, that is, the absolute value of the slope of the broken line in the graph.

The curve of generator static frequency regulation characteristics is shown in Fig. 1.

Notice that the output of the generator is the maximum output when the generator is fully loaded and unaffected by the change of frequency, which is shown as the part that is parallel to the horizontal axis in Fig. 1, and the generator under this condition no longer has the ability of PFR. In addition, the total URP is the sum of URP of all the generators if a node is connected with more than one generator unit.

2.1.2 Load Static Frequency Regulation Characteristics (LSRC)

The LSRC can be expressed as:

$$P_L = P_{LN} + k_L(f - f_N) \tag{2}$$

Fig. 2 Curve of LSRC

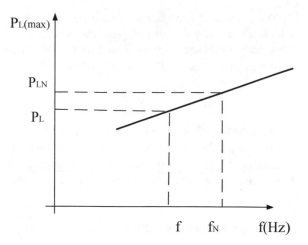

where P_{LN} and P_L represent the rated power and actual power of load, respectively. k_L represents the unit static frequency regulation factor of load, which is also called the unit power regulation factor (URP) of load, that is, the slope of the characteristic curve. Its per-unit value is as follows:

$$k_{L*} = \frac{\Delta P_L/P_{LN}}{\Delta f/f_N} = k_L \frac{f_N}{P_{LN}} \tag{3}$$

The curve of LSRC is shown in Fig. 2.

2.2 Analysis of the Uncertainty of System Static Frequency Regulation Factor (SSRF)

2.2.1 The Uncertainty of LSRF

Different loads have different load–frequency characteristic; the corresponding primary frequency regulation characteristics are not identical. Ignoring the impact of system voltage fluctuations, the relationship between the active power of the integrated load and frequency can be expressed as [10]:

$$P_L = P_{LN}\left(a_0 + a_1 \frac{f}{f_N} + a_2 \left(\frac{f}{f_N}\right)^2 + \cdots\right) \tag{4}$$

where $a_i(i = 1, 2, \ldots)$ represents the percentage that load proportional to the ith system frequency of the rated load and satisfies $a_0 + a_1 + a_2 + \cdots = 1$.

Due to the load characteristic that is proportional to the third or higher frequency which is rare, only the case of quadratic is considered here for simplified analysis. Therefore, combined with the definition of URP of load shown in formula (3), the per-unit value of the URP of the integrated load can be expressed as:

$$k_{L^*} = a_1 + 2a_2 f_* + 3a_2 f_*^2 \tag{5}$$

It can be seen that the composition of the load and the deviation of the frequency are two major factors that affect the URP of the integrated load. The typical value of URP of different power consumers in different seasons is given in reference [23]. Generally, the per-unit value of load's URP is considered to satisfy [9]: $k_{L^*} \in [0, 3]$.

2.2.2 The Uncertainty of the USRF

The inherent speed control system of generator achieves the function of PFR. There are two key factors that affect the performance of its static frequency regulation [13, 17]:

Adjusting dead zone (ADZ): When the generator conducts PFR, it is allowed not to act within a certain range of frequency deviation. The frequency range is the ADZ of generator, which can be set artificial. ADZ is used mainly to avoid the excessive repose to small frequency disturbance and thus to protect the generator. Man-made settings of the ADZ can control whether the generator participates in PFR. For some special operating generators (such as the ones that only need to operate with base load), it ensures that they do not participate in PFR by setting a larger range of ADZ, like nuclear power. The relationship between the range of ADZ and generator's PFR performance is shown in Fig. 3. In the figure, Δf_{max} represents the setting range of ADZ, and Δf and ΔPG represent offset frequency relative to rated value and the change of generator's active power output, respectively.

Fig. 3 Relationship between the range of ADZ and generator's PFR performance

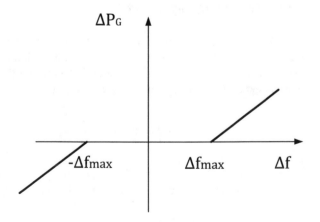

Relative regulation coefficient reflects static frequency–power characteristic of generator, which is the reciprocal of generator's unit adjust power and can be expressed as:

$$\sigma = \frac{\Delta f}{\Delta P} = \frac{\Delta f / f_N}{\Delta P / P_N} \frac{f_N}{P_N} = \sigma_* \frac{f_N}{P_N} \tag{6}$$

In formula (6), σ and σ_* represent regulation coefficient of generator and relative regulation coefficient of generator.

For a system with multiple generators, if frequency offset is given, the smaller the regulation coefficient, the larger the variation of the active power output of the corresponding generator in the system, and the more unstable the generator is. PFR mainly aims at short-period power disturbance with small adjustment amplitude, during which exists no large frequency fluctuation. If σ_* is set too large, the corresponding power output of generator gets too small and PFR is ineffective

Based on the analysis above and the relevant studies in references [12, 17, 18], it can be seen that for a generator unit (in this paper, it refers to generator unit with enough capacity and not only base load is required to carry) in PFR, the following situations may cause the generator fails to meet the requirements:

(1) ADZ is set too large, causing the generator unit does not respond to the large frequency disturbance;
(2) Relative regulation coefficient is set too large, which causes the change of power output too small when large frequency offset occurs in system and basically fails to take effect;
(3) Putting abnormal generator unit to the PFR circuit, which cause the actual power output has not changed during PFR and the function did not play.

It can be seen that there is uncertainty in whether a generator unit truly participates in PFR.

3 Stochastic Dynamic Power Flow Model

3.1 Deterministic Dynamic Power Flow Model

Deterministic dynamic power flow model can be expressed as:

$$S = g(X_0 + \Delta X, Y) \tag{7}$$

In formula (7), g refers to node power equation, Y refers to the network structure of the system, X_0 and ΔX represent the increment of the state variable vector and the post-disturbance variable of the system, respectively, which includes the voltage amplitude, phase angle, and system frequency, and S represents the injected power vector of the node after perturbation, including active and reactive components, which can be expressed as follows:

$$
\begin{cases}
P = P_{Gf} - P_{Lf} \\
\quad = [P_{G0} - k_G(f - f_N)] - [DP + P_{L0} + k_L(f - f_N)] \\
Q = Q_{G0} - Q_{L0}
\end{cases} \tag{8}
$$

In formula (8), P_{G0} and Q_{G0} represent active output and reactive power output under normal condition, respectively; P_{L0} and Q_{L0} represent the active and reactive power absorbed by load under normal condition; ΔP represents the load disturbance occurs in normal condition; P_{Gf} and P_{Lf} represents the actual active power output and actual load absorption when in PFR after disturbance, respectively.

However, the concept of a balanced node does not exist in the dynamic power flow model. Therefore, the active imbalance equation of the original balanced node should be added to the mode compared to conventional power flow equations to get system frequency.

It is noteworthy that the model is defined as follows: (1) The expression "after perturbation" is only for the steady-state problem, which refers to the new equilibrium point achieved after PFR without regard to the process from the normal state to the new equilibrium state. (2) The relation between frequency and active power is considered as strong coupling, while the weak coupling is considered between reactive powers. Therefore, the characteristics of reactive power adjustment are neglected. It is considered that the original balanced node has enough reactive power to ensure the node voltage amplitude is constant.

3.2   Stochastic Dynamic Power Flow Model

Considering the uncertainties of USRF and LSRF, a stochastic dynamic power flow model is proposed in this paper based on the deterministic dynamic power flow model studied, which is expressed as follows:

$$
S = g(X_0 + \Delta X, Y, \xi) \tag{9}
$$

In formula (9), ξ denotes the vector of random variables in the system, which contains the USRF of all generators with PFR capability in the system.

It is assumed that whether a generator participates in PFR subject to 0–1 distribution based on the analysis above. The probability of 1 is p, which indicates that the generator is normally involved in PFR as required, and USRF is k_G; the probability of 0 is $q = (q = 1 - p)$, which indicates that the generator does not participate in PFR as expected. USRF obeys the [0, 3] uniform distribution or normal distribution, that is, $\mu = 1.5$ and $\sigma = 0.5$.

4 Monte Carlo Simulation Method

The model proposed in this paper is solved by the Monte Carlo simulation method, and the steps are as follows:

(1) Set the total number of samples at N.
(2) Establish the probability model of random variables. The random variables in the proposed model include URP of generator units and load.
(3) Take N groups of samples from all the random variables in the system according to their probability distribution.
(4) Conduct a deterministic dynamic power flow calculation on each group of samples to get the corresponding power flow distribution data.
(5) Based on the result of step (4), calculate the cumulative distribution function and statistical eigenvalue of the variable.

5 Case Study

In order to verify the practicability and validity of the proposed model, we take the IEEE 39-bus system as an example to simulate and analyze some cases, such as offline and load disturbances.

On the purpose of carrying on the simulation and analysis efficiently, the following modifications are made to the example:

(1) Set node 2 only operate with base load and has no capability of PFR.
(2) Set each power plant at node 3 to have five generator units. The capacity of generator units of each power plant is equally allocated.

The stochastic model is solved by Monte Carlo simulation method and the sampling time $N = 1000$. Scheme A and Scheme B are defined to compare and analyze the influence of the uncertainty of primary frequency regulation coefficient on the dynamic power flow distribution, where Scheme A corresponds to the deterministic dynamic power flow model (hereinafter called the deterministic model for short), while Scheme B corresponds to the stochastic dynamic power flow model (hereinafter called the stochastic model for short). The relevant parameters of PFR in the two schemes are as follows:

Scheme A: The σ_* of the connected units at node 1 and node 3 are determined as 3 and 4%, respectively. The URP of load is $k_{L*} = 1.5$.

Scheme B: The σ_* of the generator units at node 1 and node 3 are 3 and 4%, respectively. The probability of the generator unit participating in PFR is $p = 0.9$. The URP of load follows the uniform distribution in the interval $[0, 3]$, that is, $k_{L*} \sim U(0, 3)$.

5.1 Offline Simulation

There are five generator units in the system. The simulations when the generator at node 1 is off-grid are as follows. Table 1 lists the system states after disturbance, respectively, solved by the two schemes. Then, the system frequency f and the deviation rate related to nominal frequency (calculated as $\Delta f / f_N \times 100\%$) are shown in Table 2. Notice that the data corresponding to scheme B is the expected value of the variable.

It is shown in Table 1 that only the generator unit at node 2 maintains the power output unchanged due to the lack of the capacity of PFR when units are offline. In comparison with the power flow results of the deterministic model at other nodes, the active output of generator, the active power absorbed by load, and the phase angle of the node obtained by the stochastic model all have some changes. Meanwhile, it can be seen from Table 2 that the system frequency calculated by stochastic model decreases more by 0.0185% than the system frequency calculated by the deterministic model, which results from taking the uncertainty of PFR into account; the expected value of system's URP becomes smaller, and a new balance can only be found through a larger frequency deviation under the same disturbance.

Table 1 System state after offline

Node	θi (°)		P_{Li} (MW)		P_{Gi} (MW)	
	A	B	A	B	A	B
1	0	0	0	0	67.375	67.275
2	10.084	10.092	0	0	163.000	163.000
3	5.4764	5.4854	0	0	89.043	89.061
4	2.2542	2.2508	0	0	–	–
5	3.7126	3.7066	89.7993	89.774	–	–
6	2.4954	2.5039	0	0	–	–
7	1.1065	1.1153	99.7770	99.750	–	–
8	4.2139	4.2219	0	0	–	–
9	−4.1009	−4.0949	124.7212	124.6879		

Table 2 System frequency after offline

	Scheme A	Scheme B
f (Hz)	49.9257	49.9164
$\Delta f / f_N \times 100\%$	−0.1487%	−0.1672%

5.2 Load Disturbance

Simulating the case of load disturbance at node 5 as follow: Table 3 shows the actual active power (expressed as $PL9$) absorbed by node 9 when the load disturbance increments are +5, +10, and +14%, respectively. The total active output of the power plant (expressed as $PG1$) and the frequency f at node 1 are also shown in Table 3.

It can be seen from Table 3 that the difference of calculated results obtained by Scheme A and Scheme B gradually becomes more and more distinct. This indicates that the influence of the coefficient uncertainty on the power flow distribution increases with the disturbance. In particular, it can be found from frequency under + 14% disturbance that the frequency calculated in Scheme A meets the requirement and the expected frequency calculated in Scheme B does not meet the requirement due to lower than the minimum allowable frequency if the allowed frequency fluctuation for the system is ±0.2 Hz. Thus, considering the uncertainty of PFR is of great importance to correctly calculate the power flow and judge the system operating state.

In order to further analyze the effect of the uncertainty of primary frequency regulation coefficient on the power flow distribution, three different disturbances are set in the system. The frequency and the active power output of node 1 obtained in Scheme B are shown in Figs. 4 and 5. As shown in Fig. 4, the greater the disturbance is, the larger the range of frequency fluctuation is. Meanwhile, on the one hand, the probability of the system frequency being lower than the minimum allowable value of 49.8 Hz is quite large when the disturbance is +14%; on the other hand, when the disturbance is +10%, although the expectation of the frequency is 49.84 Hz which meets the requirement of not lower than 49.8 Hz, there is also a certain risk of out of limit. It is shown in Fig. 5 that the fluctuation of unit output increases with the increase of disturbance and the influence of uncertainty on unit output is distinct.

5.3 Effect of Coefficient Probability Model

In order to analyze the relationship between the probabilistic model and the result of stochastic dynamic power flow, Scheme C and Scheme D are set on the basis of Scheme A and Scheme B. Set $p = 0.8$ and $q = 0.2$ in Scheme C, and the load's URP

Table 3 System state under three disturbance conditions

	Load disturbance + 5%		Load disturbance +10%		Load disturbance + 14%	
	A	B	A	B	A	B
P_{L9} (MW)	124.7361	124.7046	124.4711	124.4102	124.2583	124.17145
P_{L9} (MW)	83.5659	83.4609	95.2263	95.0543	104.5913	104.3212
f (Hz)	49.9296	49.9209	49.8590	49.8418	49.8022	49.7778

Fig. 4 Frequency of node 1

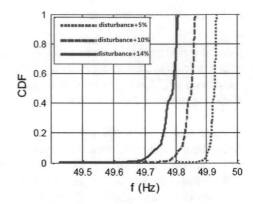

Fig. 5 Active power output of node 1

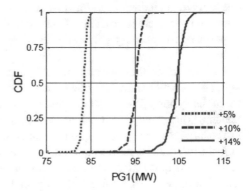

in D follows the normal distribution with the expectation of 1.5 and the mean square deviation of 0.5. The other values in Scheme D are same as Scheme B.

When there is a 30 MW power disturbance at node 5, the active power output of generator units (the generator units at node 2 are not listed due to lack of the capacity of PFR), the actual active power absorbed by loads and the system frequency are listed in Table 4. It can be seen from Table 4 that the results of power flow calculation under different probability models are different. Compared with Scheme B, Scheme C, and Scheme D, it can be found that the probability of the generator unit participating in PFR has an evident impact on the result of power flow.

Because the URP of load is further less than that of generator unit, generator units play a major role in the process of PFR. Therefore, whether the generator unit participates in PFR has a great impact on the system capacity of PFR. The less the probability of the generator unit participating in PFR is, the worse the system capacity of PFR is, the larger the deviation of the frequency is, and the more the changes in power flow distribution under the same disturbance.

Table 4 Different probability distributions of power flow

Variables	Node	Scheme A	Scheme B	Scheme C	Scheme D
PG (MW)	1	94.1137	93.9231	93.6704	93.9372
	3	92.3057	92.3525	92.3525	92.3348
PL (MW)	5	119.6374	119.5945	119.5356	119.5935
	7	99.5971	99.5538	99.4832	99.5501
	9	124.4964	124.4364	124.3632	124.4380
f (Hz)		49.8657	49.8487	49.8272	49.8493

6 Conclusion

A stochastic dynamic power flow model is proposed considering the uncertainty of primary frequency regulation coefficient, and Monte Carlo simulation method is used to solve the model in this paper.

The simulation results on the IEEE 39-bus system show that the uncertainty of primary frequency regulation has a great impact on the calculation of the power flow distribution and judgement of the actual system operation state. The influence of the uncertainty of PFR on power flow distribution is further proved by comparing the simulation data of different probability distribution models.

References

1. Ji-lai, Y., Jiang, W., Zhuo, L.: Improvement on usual load flow algorithms of power system. Proc. Chin. Soc. Electr. Eng. **21**(9): 88–93 (2001)
2. Ramanathan, R., Ramchandani, H., Sackett, S.A.: Dynamic load flow technique for power system simulators. IEEE Trans. Power Syst. **1**(3), 25–30 (1986)
3. Bing, L.: The research of dynamic load flow model for power system dispatcher training simulator. Proc. EPSA **9**(03):33–39 (1997, in Chinese)
4. LANG, B.: Research on dynamic load flow model for power system dispatcher training simulator. J. North. Jiaotong Univ. **26**(04), 112–118 (2002). (in Chinese)
5. Sun, Y., Hang, N.: A power flow model considering the static control characteristic of units in power systems. Proc. CSEE **30**(10): 43–49 (2010, in Chinese)
6. Zhu, X., Liu, W., Zhang, J., et al.: Probabilistic load flow method considering function of frequency modulation. Proc. CSEE **34**(1): 168–178 (2014, in Chinese)
7. Qu, F.: Stochastic Load Flow Calculation for Power system Base on Point Estimate Method. Northeast Electric University (2011, in Chinese)
8. Sun, H., Li, Q., Zhang, M., et al.: Continuation power flow method based on dynamic power flow equation. Proc. CSEE **31**(7): 77–82 (2011, in Chinese)
9. Shan, Y., Jiang, Y., Cai, S.: Primary frequency regulation performance analysis for generating units of different types. East China Electr. Power **3**(06), 1242–1245 (2014). (in Chinese)
10. Zhang, J., Li, H., Xie, H.: Stability of frequency regulating system for hydropower units in grid-connected operation. Power Syst. Technol. **8**(9), 57–62 (2009). (in Chinese)
11. Gao, L., Dai, Y., Wang, J.: An online estimation method of primary frequency regulation parameters of generation units. Proc. CSEE **15**(16), 62–69 (2012, in Chinese)

12. Chen, L., Chen, H.: Analysis on primary frequency regulation of generator units in Guangdong power system. Guangdong Electr. Power **11**(08), 8–12 (2008). (in Chinese)
13. Ma, B., Zhou, W., Wang, T.: An experimental investigations on applicable criterion for AC corona discharge evolvement of rod-plane gap based on UV digital image processing. Autom. Electr. Power Syst. **32**(24), 52–55 (2008, in Chinese)
14. Dai, Y., Zhao, T., Gao, L.: Research on characteristics of power system primary frequency control operating on power plants. Electr. Power **39**(11), 37–41 (2006, in Chinese)
15. Du, L., Liu, J., Lei, X.: The primary frequency regulation dynamic model based on power network. In: IEEE, International Conference on Power System Technology, pp. 1–8 (2006)
16. Yang, G.: Evaluation of primary frequency modulation performance of generator units. Electric Power. **1**(03), 32–35 (2013). (in Chinese)
17. Yu, L.I.U.: Review on algorithms for probabilist load flow in power system. Autom. Power Syst. **12**(7), 127–135 (2014). (in Chinese)
18. Singh, R.K., Choudhury, N.B.D., Goswami, S.K.: Optimal allocation of distributed generation in distribution network with voltage and frequency dependent loads. In: Industrial and Information Systems, 2008. IEEE Region 10 and the Third international Conference on Kharagpur (2008)
19. Zhao, X.: Dynamic Modeling and Probabilistic Rotor Angle Stability Assessment of Asynchronized Generator in a Multimachine System. College of Electrical Engineering of Chongqing University (2009, in Chinese)
20. Khani Maghanaki, P., Tahani, A.: Designing of fuzzy controller to stabilize voltage and frequency amplitude in a wind turbine equipped with induction generator. Int. J. Mod. Educ. Comput. Sci. (IJMECS) **7**(7), 17–27 (2015). https://doi.org/10.5815/ijmecs.2015.07.03
21. Ramji, T., Ramesh Babu, N.: Comparative analysis of pitch angle controller strategies for PMSG based wind energy conversion system. Int. J. Intell. Syst. Appl. (IJISA) **9**(5), 62–73 (2017). https://doi.org/10.5815/ijisa.2017.05.08
22. Abebe, Y.M., Mallikarjuna Rao, P., Gopichand Nak, M.: Load flow analysis of a power system network in the presence of uncertainty using complex affine arithmetic. Int. J. Eng. Manuf. (IJEM) **7**(5), 48–64 (2017). https://doi.org/10.5815/ijem.2017.05.05
23. Mihaela, O.: Numerical simulation methods of electromagnetic field in higher education: didactic application with graphical interface for FDTD method. Int. J. Mod. Educ. Comput. Sci. (IJMECS) **10**(8), 1–10 (2018). https://doi.org/10.5815/ijmecs.2018.08.01
24. Al-Mashaqbeh, I.: Computer simulation instruction: carrying out chemical experiments. Int. J. Mod. Educ. Comput. Sci. (IJMECS) **6**(5), 1–7 (2014). https://doi.org/10.5815/ijmecs.2014.05.01
25. Wang, J., Dou, W., Shi, K.: A scalable simulation method for network attack. Int. J. Wireless Microw. Technol. (IJWMT) **1**(3), 21–28 (2011). https://doi.org/10.5815/ijwmt.2011.03.04

A Data Model of the Internet Social Environment

Andriy Peleshchyshyn and Oleg Mastykash

Abstract In the chapter, there was made a conceptual chart of the Internet social environment platform. It was developed the following models: a data model, a service model, a navigation model, a presentation model, and a visual model. The Internet social environment platform's data was divided into four groups according to such categories: content, frequency of changes, distribution statues, and owner. In the chapter, the process of data transformation from a visual model till a data model was analyzed. It was also highlighted key nodes page of the users platform and made their detailed analysis. The basic principles of formation and preservation of nodes of these platforms are analyzed. A comparative analysis of the presence of key nodes of the most common virtual community's platforms has been made. The Internet social environment, which is most suitable for the analysis, is established.

Keywords Social environment platform · Virtual community data model · Community node · Post · User profile

1 Introduction

Research in the field of the Internet social environments and network analysis has been conducted for many years. Nowadays, more than one models of the organization of the Internet social environment has already been built. Fedushko in his work [1] analyzed the model, which is based on the theory of graphs, the structural equivalence, and the random graphs; he identified perspective ways for improving them and outlined such meaning as: actor, communication, dyads, triads, subgroups, group, social network, structural and composite variables, networks of independence,

A. Peleshchyshyn · O. Mastykash (✉)
Social Communications and Information Activities Department, Lviv Polytechnic National University, Lviv, Ukraine
e-mail: mastykash@itstep.org

A. Peleshchyshyn
e-mail: apele@ridne.net

© Springer Nature Switzerland AG 2020
Z. Hu et al. (eds.), *Advances in Artificial Systems for Medicine and Education II*,
Advances in Intelligent Systems and Computing 902,
https://doi.org/10.1007/978-3-030-12082-5_40

social offer, and social role. The chapter focuses on the analysis of the online social networks. Such authors in their works [2] wrote general description of the scientific social network and gave meaning of the social network, as an online service or a platform. In article [3], there is a literary review of scientific sources, where we can see the modeling aspects of the functioning social network, a description of their characteristics, and a prediction of the future behavior of the social network, which depends on the internal or external factors. In spite of making researches, nowadays, we do not have united unified model ISEP, which allows the developers of software to write the really quality applications for the analysis of the platforms' data. In our opinion, we feel acute deficiency of the systematic methods' presentation and the analysis' algorithms ISEP, which are adapted to the modern applied research in economy, sociology and politology [4–6]. The built model ISEP will be a foundation for such researches.

One of the most important and basic aim in the user's page analysis is modeling a custom's data environment. A construction of the algorithm and the principle of parsing the page are depended on the environment model's architecture. Modeling platforms of the Internet social environment gives us understanding the process of their function and allows predicting the distribution of information. The construction of unified model also has got an information content, which allows to isolate, to classify, and to structure heterogeneous data of the Internet social environment. According to this, we can write the high-quality software of the analysis data of the Internet social environment platform (ISEP).

2   Conceptual Graph of ISEP

On the basis of the analysis [7, 8], the conceptual graph of ISEP was constructed (Fig. 1).

The **data model** is a system of meanings and rules for data's representation [9]. The data model describes the character, structure, and architecture of the ISIP's data and the mechanisms of work with them.

The platform's data allows characterizing it at each level of its presentation. Types of data are presented in Fig. 2. On this picture, we did not take into account the system data of the ISEP administrators, and the data of the ISEP settings, because they do not allow us to characterize the user platform, so they do not make a significant contribution to the subject area of the study.

The primary task of the social environmental platform of the Internet is to save generated/received data. Repositories where the data is saved are follows:

- Database;
- Cookies;
- Sessions;
- Cloud storage, BLOB, storage;
- DOM document tree.

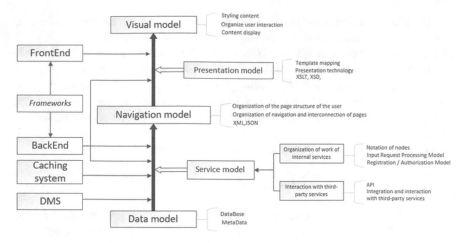

Fig. 1 Conceptual graph of ISEP

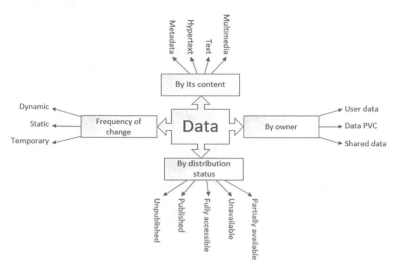

Fig. 2 Types of data

The most part of the platform's data are saved in the database. According to the structure of data organization, the database model is divided into the following types: relational, non-relational, blockade, network, and hierarchical. The database management systems (DMSs) are used for data interaction. According to the DB models' division, the DMS is also divided into relational and non-relational. In most ISIPs, the relational DB is used at the lowest level of the technology stack [10, 11]. For example, Facebook uses the MySQL database. If in real time it is necessary to intensively process very large volumes of information, then relational DMS, in comparison with non-relational, works with greater delays. Therefore, for the adequate operation of the DB, the platform's architects implement a multilayered system for processing incoming requests, which reduces the number of hits to the DB. The load balancer corresponds to a Web server is responded for controlling the load on the platform's datacenters, which redirects client requests to less loaded servers. Above this layer, there is a data caching system that allows faster processing by the server of the incoming request. If the ISIP saves very large volumes of unstructured data, then it will be advisable to use non-relational DB [12]. All saved data have to be of a specified type.

At the formal level, the data model can be represented as a tuple type:

$$DM = \langle \{Data_i, Relation_i\}_{i=0}^{N}, DD, SD \rangle \tag{1}$$

where DM is a data model, $Data$ is a data, $Relation$ is relationship between data, DD is a data dictionary (data typing), and SD is a data section (publication, news).

The **service model** defines the rules for transforming the data from the data model to the navigation model and identifies the main functionality of the platform. Depending on how the service model is organized, such will be the direction of the ISIP and its functionality.

The user of the ISIP platform, interacting with its functionality, generates a large amount of data. User data types are indicated at the service model level. User data can be presented as follows:

$$Data(User_i) = \langle Direct Data(User_i), Inh Data(User_i) \rangle \tag{2}$$

where $Data(User_i)$ is a user data, $Direct Data(User_i)$ is a data specified by the user directly, and $Inh Data(User_i)$ is a derivative data (those which the user did not enter by him/herself, but they appeared due to his/her activity).

$$Direct Data(User_i) = \left\langle \begin{array}{l} \{PersonalData_j(User_i)\}_{j=0}^{N^{(PD)}}, \{PrivateData_j(User_i)\}_{j=0}^{N^{(PRD)}}, \\ \{NodeData_j(User_i)\}_{j=0}^{N^{(ND)}}, \{Comment_j(User_i)\}_{j=0}^{N^{(COMM)}}, \\ \{UsersData_j(User_i)\}_{j=0}^{N^{(UD)}} \end{array} \right\rangle \tag{3}$$

where $PersonalData(User_i)$ is a personal data (data that uniquely characterize the user),), $PrivateData(User_i)$ is a private data (these are data that are only available to the user and his/her partner, such data without), $NodeData(User_i)$ is the data of their own posts on the page, $Comment(User_i)$ is a user comment, and $UsersData(User_i)$ is distributed messages of other users (news, photographs, videos)

$$InhData(User_i) = \left\langle \begin{array}{l} \{AppData_j(User_i)\}_{j=0}^{N^{(AD)}}, \{NodesData_j(User_i)\}_{j=0}^{N^{(ND)}}, \\ \{GroupData_j(User_i)\}_{j=0}^{N^{(GD)}}, \{MediaData_j(User_i)\}_{j=0}^{N^{(MD)}}, \end{array} \right\rangle \tag{4}$$

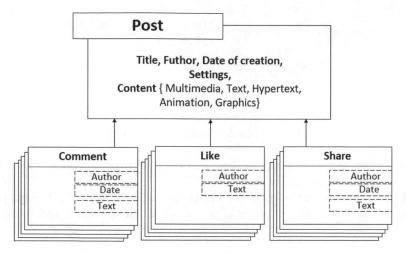

Fig. 3 Structure of the user's post

where $AppData(User_i)$ is a application platform data, $NodesData(User_i)$ is a data about linked nodes: friends, groups, $GroupData(Useri)$ is posting groups where he is a member, and $MediaData(User_i)$ is a multimedia user data.

The service model defines the main and additional units of the ISIP:

$$Unit(ISEP) = \left\langle \begin{matrix} \{UserProfile_j(ISEP)\}_{j=0}^{N^{(UP)}}, \{RelationUsers_j(ISEP)\}_{j=0}^{N^{(RU)}}, \\ \{PageGroup_j(ISEP)\}_{j=0}^{N^{(PG)}}, \{Msg_j(ISEP)\}_{j=0}^{N^{(MSG)}}, \\ \{Publication_j(ISEP)\}_{j=0}^{N^{(PUB)}} \end{matrix} \right\rangle \quad (5)$$

where $Unit(ISEP)$ is a ISEP unit, $UserProfile_j(ISEP)$ is a user profile, $RelationUsers_j(ISEP)$ is a connection between the user and the node next to it, $PageGroup_j(ISEP)$ is a group page ISEP, $Msg_j(ISEP)$ is a user message, and $Post_j(ISEP)$ is a user post.

A post is a record on the ISEP which has a strictly defined structure and type. Types of posts:

$$Post = \{Publication, News, Media, Article, API\} \quad (6)$$

where *Publication* is a user publication, *News* are the user news, *Alert* is a notification, *Media* is a multimedia content, *Article* is a user article, and the *API* is a specification of the additional functionality of the ISEP.

Also, the service model specifies the structure of all ISEP nodes. The structure of the post is depicted in Fig. 3:

A discussion model can be represented as a tuple type:

$$Post = \left\langle \begin{matrix} Title, Author, CreateDate, Content, \\ \{Comment\}, \{Like\}, \{Share\}, \end{matrix} \right. \quad (7)$$

$$CommentText = Content = \left\langle \begin{array}{l} \{Image\}, \{Video\}, \{Reference\}, \\ \{Share\}, \{Text\}, \end{array} \right\rangle \quad (8)$$

$$Comment = \langle Author, Date, CommentText \rangle \quad (9)$$

$$Like = \langle Author, Date \rangle \quad (10)$$

$$Share = \langle Author, Date, AuthorText \rangle \quad (11)$$

Will there be an additional functional on the platform and what exactly it will be—it depends on the service model.

The **navigation model** provides navigation rules between ISEP pages and describes the user's page structure of the platform. The current model forms the rule:

$$Rules(ISEP) = \left\langle \begin{array}{l} \{PageNavigation_j(ISEP)\}_{j=0}^{N^{(PN)}}, \{Search_j(ISEP)\}_{j=0}^{N^{(S)}}, \\ \{PostNavigation_j(ISEP)\}_{j=0}^{N^{(PSN)}}, \{Menu_j(ISEP)\}_{j=0}^{N^{(M)}}, \end{array} \right\rangle$$
$$(12)$$

where *Rules* are the navigation rules of ISEP, *PageNavigation* is navigation rules for pages and nodes of ISEP, *Search* is navigation search rules: search by groups, by users (according to a certain category), *PostNavigation* is navigation rules for posts, and *Menu* is navigation rules of the community menu.

Navigation rules for pages specify through the rules for forming a graph of links between nodes ISEP. Navigation search rules specify the rules for finding information by groups, by users according to a certain category. The list of search category is also given in the navigation model [13]. Navigation rules for posts describe the rules for creating a hierarchical structure of posts (the ability to comment on a post, commentary, etc.) and the transition between them; possibility to go to the author's page; the author of the post to the comment. The menu allows you to navigate the community platform using various menu types (main menu, additional menu, and context menu).

Also, the navigation model specifies the rules for constructing the platform user page and defines the user data model. According to it, the user page can be divided into the following parts:

- The part that can be edited by the user, for example, page settings, personal records, etc;
- The part that can be seen by the user, but cannot be edited (or it is able to edit once when creating a page), for example, login, attachment mail, etc;
- A public part of a page that is readable by everyone or a certain circle of people, for example, disclosed personal information;
- That part, which everyone can see and supplement, modify or comment on, rate or share, for example, user posts, advertisements, etc.

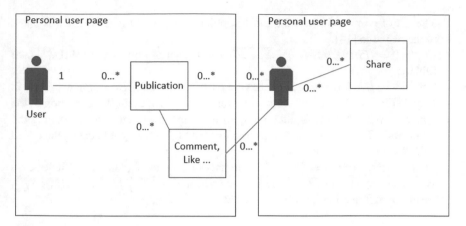

Fig. 4 Sharing of the publication

The presentation model specifies the ISIP data mapping template and chooses the technology for presenting these data:

$$
UserPage = \left\langle \begin{array}{c} \{UserProfile_j\}_{j=0}^{1}, \{RelationUsers_j\}_{j=0}^{N^{(RU)}}, \\ \{RefPageGroup_j\}_{j=0}^{N^{(RPG)}}, \{Msg_j(ISEP)\}_{j=0}^{N^{(MSG)}}, \\ \{Publication_j(ISEP)\}_{j=0}^{N^{(PUB)}} \end{array} \right\rangle \quad (13)
$$

The places where you can post a discussion are shown on this model. This can be: user pages, group page, specially designed part of the page (wall), custom page, etc. The rules for spreading the post are set on a presentation model; for example, one user post can be placed on the page and may also be on other pages shared in publications and messages. It can be shared by the author or by another user who has access to it. As a result, it is possible to count those who shared and whom was shared the article. So, one post of one user can be on many pages (1 ...*). The post cannot have many authors (Fig. 4).

When we keep track of important post, we will get the numbers of messages. Based on the comments, we can make a rating of publications and posts. We can also draw a graph of user activity, getting numbers of his comments for different publications on different pages and other action.

3 A Comparative Analysis of the Availability Platform's Nodes of the Internet Social Environment

Key nodes are the nodes that can help to identify the community user unique.

The list of communities is the most popular in the world in 2017; it was taken from the resource [14]:

Let *ISIP* = *{ISIPi}*, where $i = \overline{1\ldots 10}\rightarrow$ —the set of comparable ISIPs, shown in Table 1.

Let *Node* = *{Node_j}* where $j = \overline{1..8}$ is the list of key nodes *ISEP_i*, shown in Table 2.

Each ISEP node has its own impact factor. Nodes form the full description of the user profile. When the coefficient is larger, its weight is greater too in shaping the user's portrait. The user's personal profile has the highest weight because the nodes contain personal user information.

Let $P = \{Pk\}$ where $k = (1\ldots 10)$ is the list of elements N_j, shown in Table 3.

We introduce the results of the presence of nodes $\{N_j\}$ in the matrix $\{X_{ij}\}$. Elements are defined by the following rule:

$$X_{ij} = \begin{cases} 0, \text{ if } j \text{ node available in the environment } i; \\ 1, \text{ if } j \text{ node is not in the environment } i. \end{cases}$$

Table 1 A list of ISIPs

Symbolic designation	Name	Web address	Monthly audience
ISIP1	Facebook	facebook.com	2,000,000,000
ISIP2	YouTube	youtube.com	1,000,000,000
ISIP3	Instagram	instagram.com	700,000,000
ISIP4	Twitter	twitter.com	313,000,000
ISIP5	Reddit	reddit.com	250,000,000
ISIP6	Vine	vine.co	200,000,000
ISIP7	Ask.fm	ask.fm	160,000,000
ISIP8	Pinterest	pinterest.com	150,000,000
ISIP9	Tumblr	tumblr.com	115,000,000
ISIP10	Google+	plus.google.com	111,000,000

Table 2 Available key ISEP

Symbolic designation	Name	Influence factor
Node1	Personal profile	11
Node2	The user's wall	1
Node3	Posts	1
Node4	Friends	1
Node5	Communities	1
Node6	Photographs	1
Node7	Events	1
Node8	Recommendations	1

Table 3 User profile

Symbolic designation	Name	Influence factor
Element1	Photograph	1
Element2	F. L. M.	1
Element3	Address of residence	1
Element4	Teaching	1
Element5	Work	1
Element6	Status	1
Element7	Looks	1
Element8	Interest	1
Element9	Contact Information	1
Element10	Life events	1

Table 4 Informative ISEP completeness

Virtual community	Node1	Node2	Node3	Node4	Node5	Node6	Node7	Node8	$\sum\limits_{j=1}^{8} X_{ij}$
ISIP1	10	1	1	1	1	1	1	1	17
ISIP2	4	1	0	1	0	1	1	1	9
ISIP3	6	1	1	1	1	1	1	1	13
ISIP4	5	1	1	1	1	1	1	1	12
ISIP5	4	1	1	1	1	1	1	0	10
ISIP6	2	0	0	1	1	1	0	1	6
ISIP7	3	0	0	1	1	1	0	1	7
ISIP8	5	1	1	1	1	1	1	0	11
ISIP9	4	0	0	1	1	1	0	0	7
ISIP10	6	1	1	1	1	1	1	0	12

As we can see in Table 4, the most informative ISEP is Facebook, LinkedIn. Despite the fact that YouTube is on the second place for the user's audience, it is not the best ISIP for analysis.

4 Conclusions

In the chapter, a comparative analysis was made of the presence of key nodes in the most common of the Internet social environment platforms, which forms the completeness of the user profile description. We found out which IEIPs are best suited for analysis. We also made the review and analysis of literary sources which are related to the construction of models of the organization of the Internet social environment. The ISEP conceptual graph was constructed in the following order: the ISEP data model was built at first, which shows the general structure of the data organization; then, a service model is constructed that defines the main functional of the ISEP and the interaction rules with the previous model; a navigational model is the next one, that defines the structure of the user page and organizes the relationship between ISEP pages; then, a presentation model is constructed that specifies a ISEP mapping template, and the last one is a visual model that sets ISEP style styling rules and displays its content.

References

1. Korobiichuk, I., Fedushko, S., Juś, A., Syerov, Y.: Methods of determining information support of web community user personal data verification system. automation 2017. Adv. Intell. Syst. Comput. **550**, 144–150 (2017)
2. Gupta, P., Kamra, A., Thakral, R., Aggarwal, M., Bhatti, S., Jain, V.: A proposed framework to analyze abusive tweets on the social networks. Int. J. Mod. Educ. Comput. Sci. (IJMECS) **10**(1), 46–56 (2018). https://doi.org/10.5815/ijmecs.2018.01.05
3. Nistor, N.: Participation in virtual academic communities of practice under the influence of technology acceptance and community factors. A learning analytics application. Comput. Hum. Behav. 339–344 (2014)
4. Meligy, A.M., Ibrahim, H.M., Torky, M.F.: Identity verification mechanism for detecting fake profiles in online social networks. Int. J. Comput. Netw. Inf. Secur. (IJCNIS) **9**(1), 31–39 (2017). https://doi.org/10.5815/ijcnis.2017.01.04
5. Kajitori, K.: Generating code for simple dynamic web applications via routing configurations. Int. J. Mod. Educ. Comput. Sci. (IJMECS) **9**(11), 1–12 (2017). https://doi.org/10.5815/ijmecs.2017.11.01
6. Hu, Z.: An ensemble of adaptive neuro-fuzzy Kohonen networks for online data stream fuzzy clustering. barXiv preprint arXiv:1610.06490, 12–18 (2016)
7. Zavuschak, I.: Methods of processing context in intelligent systems. Int. J. Mod. Educ. Comput. Sci. (IJMECS) **10**(3), 1–8 (2018). https://doi.org/10.5815/ijmecs.2018.03.01
8. Peleshchyshyn, A., Mastykash, O.: Analysis of the methods of data collection on social networks. In: International Scientific and Technical Conference "Computer Science and Information Technologies", 05–08 Sept 2017
9. Korzh, R., Fedushko, S., Peleschyshyn, A.: Methods for forming an informational image of a higher education institution. Webology **12**(2), Article 140 (2015). Available at: http://www.webology.org/2015/v12n2/a140.pdf
10. Syerov, Y., Fedushko, S., Loboda, Z.: Determination of development scenarios of the educational web forum. In: 2016 XIth International Scientific and Technical Conference Computer Sciences and Information Technologies (CSIT). Lviv, pp. 73–76 (2016)
11. Kim, J., Hastak, M.: Social network analysis. Int. J. Inf. Manage.: J. Inf. Prof. 86–96 (2018)
12. Hu, Z., Gnatyuk, S., Koval, O., Gnatyuk, V., Bondarovets, S.: Anomaly detection system in secure cloud computing environment. Int. J. Comput. Netw. Inf. Secur. (IJCNIS) **9**(4), 10–21 (2017). https://doi.org/10.5815/ijcnis.2017.04.02
13. Borgatti, S.P., Everett, M.G., Johnson, J.C.: Analyzing Social Networks. Sage (2018)
14. Kraut, R., Resnick, P.: Building Successful Online Communities: Evidence-Based Social Design. Massachusetts Institute of Technology. MIT Press, p. 283 (2012)

Anomaly Detection of Distribution Network Synchronous Measurement Data Based on Large Dimensional Random Matrix

Zhongming Chen, Yaoyu Zhang, Chuan Qing, Jierong Liu, Jiaqi Tang and Jingzhi Pang

Abstract In order to identify faults of power system measurement subsystem, a method based on deep belief network is proposed in this paper. Firstly, data from actual measurement system is collected and divided into training and test samples. And then, the data is used to train a deep belief network. Finally, the model's fault diagnosis results and actual samples' labels are combined as a cross-validation set to test the deep belief network. The results show that the method based on deep belief network proposed in this paper can be more stable and reliable identification of electric power measurement equipment fault diagnosis.

Keywords Deep belief network · Measurement equipment · Fault diagnosis · Pattern recognition

1 Introduction

With the continuous expansion of the scale of the flexible DC transmission system, safety and stability of the measurement subsystem have become much more important [1]. Measurement subsystem undertakes the task of electrical detection, which is usually composed of electronic transformers, merging units, digital watt-hour meter, and other components [2]. Any fault of these components will lead to inaccuracy measurement, even the paralysis of the entire high-pressure measurement subsystem [3]. In recent years, many scholars have taken part in a research about measurement subsystem fault diagnosis. They found that some faults of the measurement system are difficult to directly detect [4], of which the more hidden faults existed in the primary side of the current transformer caused by short-circuit fault [5]. Because the load current which flowthrough the primary side of the transformer is changing with the load. It is difficult to determine what causes to reducing electrical energy, low

Z. Chen (✉) · Y. Zhang · C. Qing · J. Liu · J. Tang · J. Pang
Guangdong Power Grid Co., Ltd. Foshan Power Supply Bureau, Foshan 528200, Guangdong, China
e-mail: 102131694@qq.com

© Springer Nature Switzerland AG 2020
Z. Hu et al. (eds.), *Advances in Artificial Systems for Medicine and Education II*,
Advances in Intelligent Systems and Computing 902,
https://doi.org/10.1007/978-3-030-12082-5_41

load or short circuit on the primary side. For this problem, Ref. [6] regards the whole electricity system and the metrology system as a network, establishing a primary side short-circuit fault detection model of the current transformer, and proposes a fault detection method. Reference [7] obtains a good fault discrimination effect under the system of low load. However, when the system load is large, the saturation of the measuring transformer appears, and the detection signal cannot reflect the system impedance correctly. Reference [8] concludes that the current transformer secondary winding terminal voltage and the current through the ammeter winding current ratio of the current meter with ammeter are related to short-circuit fault closely. The simulation results show it correctly. The analysis found that there are many kinds of faults in the power measurement system and some faults cannot be distinguished by a single signal.

Regarding the issues above, a power system measurement equipment fault diagnosis method based on deep belief network is proposed in this paper. A deep belief network is a new kind of a deep learning neural network based on the Bayesian network and the limited Boltzmann machine. Each DBN network has many hidden layers and can automatically abstract the complex features of data from low to high by the layer-by-layer unsupervised greedy learning method so as to realize the complicated data classification problem. DBN has been used in MNIST dataset recognition [9], voice recognition [10], and object recognition [11–13]. Reference [14] has also proved that the network has some data reasoning ability, and it can still maintain some robustness after the observed data is disturbed by noise.

To test the effectiveness of this method proposed in this paper, it is also compared with traditional fault diagnosis methods such as backpropagation (BP) neural network [15–17] or support vector machine (SVM) [18–21]. The result shows that DBN performs better in accuracy and speed than BPNN and SVM, because of its nonlinear fitting of high-dimensional functions, strong expression, and fast convergence.

2   Construction of Deep Belief Network for Measurement System Fault Diagnosis

2.1   Artificial Neural Network

Artificial neural network (ANN) is a research hotspot in the field of artificial intelligence since the 1980s. It abstracts the neural network of human brain from the information processing viewpoint, building a simple model and composing different networks according to different connection modes. Neural network is a computing model, which is composed by a large number of nodes (or neurons) linking with each other. Each node represents a specific output function. The connection between each node represents a weight, which is equivalent to the memory of the artificial neural network. The output of the network depends on how the network is connected, the weights, and the incentive functions. The network itself is usually built by a natural

Fig. 1 Diagram of artificial
neural network

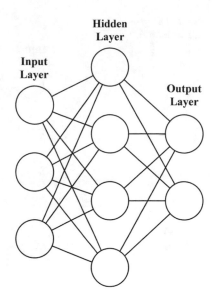

algorithm or function approximation, or the expression of a logical strategy. Diagram
of artificial neural network is shown in Fig. 1.

2.2 Recurrent Neural Network

The traditional neural network can be described as a directed acyclic graph from
input to output, and the input data is outputted to the next layer after the mapping
transformation of each layer. In the complex data processing, it is impossible to
obtain very a good classification effect. For this, scholars proposed recurrent neural
network. Recursive neural network nodes can constitute a directional ring so that
hidden layers can also be linked. Hopfield neural network is a cyclic neural network.
When the signal is inputted, the state of the neurons will continue to change and finally
tend to be stable or show periodic shocks. The schematic diagram of recurrent neural
network is shown in Fig. 2.

2.3 Boltzmann Machine and Restricted Boltzmann Machine

An important application of the recurrent neural network is the Boltzmann machine.
Boltzmann machine is a system state perception machine proposed by Geoffrey
Hinton and Terry Sejnowski in 1985 based on random recurrent neural network and
Markov random field model. Boltzmann machine has no concept of output layer.

Fig. 2 Diagram of recurrent
neural network

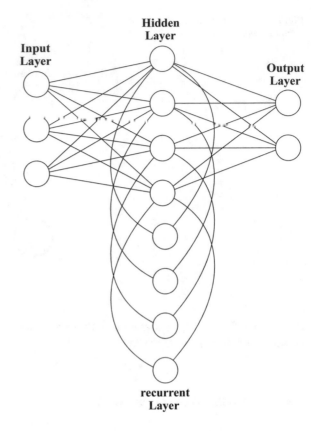

Fig. 3 Diagram of
Boltzmann machine

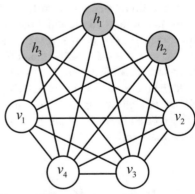

Boltzmann machine is mainly used to learn the "internal representation" of a group
of data. Diagram of Boltzmann machine is shown in Fig. 3.

In order to increase the output node in the Boltzmann machine, scholars proposed
restricted Boltzmann machine (RBM) network model. Instead of a deep model,
the RBM itself has a single layer of latent variables that may be used to learn a

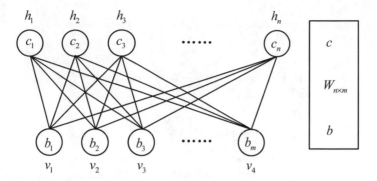

Fig. 4 Diagram of RBM

representation for the input, which is the foundational of many deeper models. Here, its units are organized into large groups called layers, while the connectivity between layers is described by a matrix, and the connectivity is relatively dense.

In Fig. 3, it consists of m visible units $V = (V_1, \ldots, V_m)$, representing observable data and n hidden units $H = (H_1, \ldots, H_n)$, capturing dependencies between observed variables. The random variables (V, H) take values $(v, h) \in \{0, 1\}^{m+n}$, and the joint probability distribution under the model is given by the Gibbs distribution $p(v, h) = \frac{1}{Z} e^{-E(v,h)}$ with the energy function

$$E(v, h) = -\sum_{i=1}^{n}\sum_{j=1}^{n} \omega_{ij} h_i v_i - \sum_{j=1}^{m} b_j v_j = \sum_{i=1}^{n} c_i h_i \tag{1}$$

For all $i \in \{i, \ldots, n\}$ and $j \in \{1, \ldots, m\}$, $\omega_{i,j}$ is a real valued weight associated with the edge between units v_j and h_i while b_j and c_i are real valued bias terms associated with the jth visible and the ith hidden variable, respectively.

The graph of RBM shown in Fig. 4 is as follows:

According to Bayesian network theory:

$$p(h|v) = \prod_{i=1}^{n} p(h_i|v)$$

$$p(v|h) = \prod_{i=1}^{m} p(v_i|h) \tag{2}$$

So the visible variable can be calculated as follows:

$$p(v) = \frac{1}{Z} \sum_{h} p(v, h) = \frac{1}{Z} \sum_{h} e^{-E(v,h)}$$

$$= \frac{1}{Z} \sum_{h_1} \sum_{h_2} \cdots \sum_{h_n} e^{\sum_{j=1}^{m} b_j v_j} \prod_{i=1}^{n} e^{h_i \left(c_i + \sum_{j=1}^{m} \omega_{i,j} v_j \right)}$$

$$= \frac{1}{Z} e^{\sum_{j=1}^{m} b_j v_j} \sum_{h_1} e^{h_1 \left(c_1 + \sum_{j=1}^{m} \omega_{1,j} v_j \right)} \cdots \sum_{h_2} e^{h_n \left(c_n + \sum_{j=1}^{m} \omega_{n,j} v_j \right)}$$

$$= \frac{1}{Z} e^{\sum_{j=1}^{m} b_j v_j} \prod_{i=1}^{n} \sum_{h_1} e^{h_1 \left(c_1 + \sum_{j=1}^{m} \omega_{1,j} v_j \right)}$$

$$= \frac{1}{Z} \prod_{i=1}^{n} \sum_{i=1}^{n} e^{h_i \left(c_i + \sum_{j=1}^{m} \omega_{i,j} v_j \right)}$$

$$= \frac{1}{Z} \prod_{j=1}^{m} e^{b_j v_j} \prod_{i=1}^{n} \left(1 + e^{c_i + \sum_{j=1}^{m} \omega_{i,j} v_j} \right) \tag{3}$$

2.4 Deep Belief Network

A deep belief network (DBN) is that multiple RBM models are connected in series. A deep belief network is a kind of generative models with several layers of latent variables. The latent variables are typically binary, while the visible units may be binary or real. Every unit in each layer is connected to every unit in each neighboring layer, though it is possible to construct more sparsely connected DBNs. The connections between the top two layers are undirected, while the ones between all other layers are directed, with the arrows pointed toward the layer that is closest to the data.

In this paper, the BPN network is composed of three-layer RBM networks and two-layer BP network. The activation function of each BP network is a sigmoid function. The BP network is used for adjusting the weight of deep belief network output and reducing the error by using gradient descent. RBM network is used to extract the characteristics of the signal pattern recognition.

In summary, the structure of DBN network topology for lightning current identification constructed in this paper is shown as the following figure: The orange node represents the input layer, which is directly connected to the input signal. The number of input neurons is the same as the size of the lightning wave signal window. The nodes with concentric circles represent the probabilistic hidden cell. The purple nodes represent the backed input cell, which is the output layer of the deep belief network and the input layer of the BP Network. The yellow color indicates the output layer of the DBN network (Fig. 5).

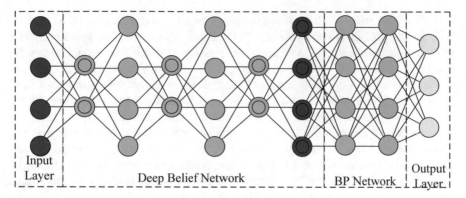

Fig. 5 Deep belief network for lighting identification

2.5 DBN Training Process for Fault Diagnosis

DBN learning process is divided into two steps: unsupervised training process and supervised training process. During unsupervised training process, DBN initializes all parameters of the entire network and applies greedy algorithm to training RBM layer by layer. RBM training for each hidden layer is the input layer of the next RBM to be trained, and the initial parameters of the whole DBN network are obtained step by step.

Then, the error is propagated from top to bottom to each RBM by BP classifier, and the initial parameters of RBM are adjusted to realize the supervised training of the whole network.

RBM training is the core of the DBN learning process. The initialization of network parameters is realized by layer-by-layer learning of RBM. The initialized network parameters are not optimal parameters; however, they are usually close to the optimal parameters. This avoids the drawback that the BP algorithm is prone to fall into the local optimum and the training time is too long due to the random initialization of the network parameters when training the DBN.

3 Sample Data Processing

Voltage, current, and other data are connected in measurement subsystem, with different dimensions and orders of magnitude. When the level of each index varies greatly, the direct analysis of the original index will highlight the role of higher index in the comprehensive analysis and weaken the effect of the lower index. Therefore, in order to ensure the reliability of the results, the original indicator data needs to be standardized.

Before data analysis, data normalization is usually needed and uses the standard data for analysis. Data standardization is the index of statistics. The standardization of data mainly includes two aspects: data processing and non-dimensional treatment. Data with chemotaxis mainly solves the problem of a different nature of data; the direct addition of a different nature of the indicators cannot reflect the combined results of different forces correctly. It must be first considered that the nature of the inverse index data is changed so that all the indicators on the evaluation program have the role of chemotaxis. Data dimensionless processing mainly solves the data comparability. After standardization process, the original data is converted to non-dimensional index measurement value. It means that the index values are in the same amount of level, which can be a comprehensive evaluation analysis. In this paper, min–max standardization method is adopted.

The min–max standardization method, also known as dispersion normalization, is a linear transformation of the raw data, leaving the result in the [0, 1] interval with the following conversion functions:

$$x^* = \frac{x - x_{min}}{x_{max} - x_{min}} \tag{4}$$

where x_{max} is the maximum sample data and x_{min} is the minimum sample data. After each type of data is standardized, it can be transformed into a unified dimension.

4 Fault Diagnosis Method Performance Test

In this paper, 1908 samples are selected from the actual monitoring data, including 1300 training samples and 608 test samples, and the training and test samples are normalized, respectively.

In this paper, the number of visible layer units of Boltzmann at each layer is set as 8 and the unit of hidden layer is set as 30. The number of layers of the model is set as 1, 3, and 5 layers. The model has five output nodes, representing the operation of five conditions. Sigmoid is defined as activation function, the learning rate alpha = 1, the maximum number of traversal = 5, batch size = 5, and momentum = 0.

After confirming the depth of input, output, and hidden layer nodes of the network, which are, respectively, trained by deep belief network, BPNN, and SVM. Training results are used to characterize the model recognition accuracy. Table 1 shows three kinds of structures of DBN and its test results. Table 2 shows the comparison in test error between DBN, SVM, and BPNN.

It can be concluded from Table 1 that the different structures of DBN model lead to a significant difference between training and testing accuracy. The fault diagnosis error of a single-layer constrained Boltzmann machine (which can be regarded as a special DBN) is only 0.996%, which performs much better than that of the three-layer and five-layer model in recognition errors. Because the actual sample data records are only eight dimensions, the number of features learned by each hidden layer unit is

Table 1 Deep confidence network troubleshooting results

Model structure	Training error (%)	Test error (%)
One layer	0.1033	0.0996
Three layers	53.3840	53.3845
Five layers	46.6654	46.6154

Table 2 Comparison of different learning algorithm fault diagnosis

Model	The number of error samples	Error (%)
DBN	1	0.0996
SVM	13	2.1382
BPNN	17	2.7961

too limited. When the number of Boltzmann machine is only one layer, it is enough to complete the task of learning and extracting features of the fault samples. In this paper, the fast disparity algorithm is used in DBN. When the number of Boltzmann planes is 3, 5, or more, the error will propagate to the next layer with the characteristics. The error accumulation leads to a great deal of output classification result impact, which cannot meet accuracy requirements of fault diagnosis of measurement subsystem.

In the meantime, this paper also builds SVM and BPNN. The same training and test samples are applied for multi-type fault identification and diagnosis. As mentioned earlier, because of the limited and imbalanced sample data, BPNN recognition error is higher than the DBN. The simulation results show that there is a wide gap between SVM and DBN in multi-class classification accuracy.

5   Conclusion

Power system measurement equipment fault diagnosis method based on deep belief network is studied in this paper, comparing with traditional shadow networks such as BP neural networks and SVM. The data from actual measurement subsystem is collected to verify the paper's viewpoint with a deep learning algorithm. The test results show that the DBN model with one layer performs better than three layers and five layers in fault diagnosis accuracy and test speed.

References

1. Hurwitz, J.: The importance of monitoring all your smart meters. Electron. Eng. Prod. World **13**(1), 23–24 (2016)
2. Xiao, Y., Jiang, B., Zhao, W., et al.: Research on IEC61850 based verification technologies of digital watt-hour meter. Electr. Metering Instrum. **7**(4), 121–128 (2014)
3. Xu, W., Zhao, W., et.al.: A method for electrical energy measurement in consideration of inter-harmonics. Power Syst. Technol. (2016)

4. Solanics, P., Kozminski, K., Bajpai, M., et al.: The impact of large steel mill loads on power generating units. IEEE Trans. Power Deliv. **15**(1), 24–30 (2000)
5. Lin, G., Zhou, S., Sun, W., et al.: Calibration technology of nontraditional electric energy measuring equipment in digitalized substation. Proc. CSU-EPSA **23**(3), 145–149 (2011)
6. Femine, A.D., Gallo, D., Landi, C., et al.: Advanced instrument for field calibration of electrical energy meters. IEEE Trans. Instrum. Meas. **58**(3), 618–625 (2009)
7. Wang, L., Lei, M., Zhang, S.: Research on electric energy algorithm by IEC 61850-9-1 protocol. Electr. Meas. Instrum. **49**(2), 13–18 (2012)
8. Chen, R., Wang, Z., Kong, Z., et al.: Study of a new algorithm of digital wattmeter calibration. Electr. Meas. Instrum. **49**(9), 18–23 (2012)
9. Salakhutdinov, R., Hinton, G.E.: Deep Boltzmann machines. In: AISTATS, vol. 1, p. 3 (2009)
10. Mohamed, A.R., Dahl, G.E., Hinton, G.: Acoustic modeling using deep belief networks. IEEE Trans. Audio Speech Lang. Proc. **20**(1), 14–22 (2012)
11. Nair, V., Hinton, G.E.: 3D object recognition with deep belief nets. In: International Conference on Neural Information Processing Systems, Curran Associates Inc., pp. 1339–1347 (2009)
12. Zhou, S., Chen, Q., Wang, X.: Active deep learning method for semi-supervised sentiment classification. Neurocomputing **120**(10), 536–546 (2013)
13. Zhou, S., Chen, Q., Wang, X.: Fuzzy deep belief networks for semi-supervised sentiment classification. Neurocomputing **131**(9), 312–322 (2014)
14. Zaremba, W., Sutskever, I.: Learning to Execute. Eprint Arxiv (2015)
15. Yakkali, R.T., Raghava, N.S.: Neural network synchronous binary counter using hybrid algorithm training. Int. J. Image Graph. Signal Process. (IJIGSP) **9**(10), 38–49 (2017). https://doi.org/10.5815/ijigsp.2017.10.05
16. Al-Maqaleh, B.M., Al-Mansoub, A.A., Al-Badani, F.N.: Forecasting using artificial neural network and statistics models. Int. J. Educ. Manag. Eng. (IJEME) **6**(3), 20–32 (2016). https://doi.org/10.5815/ijeme.2016.03.03
17. Praynlin, E., Latha, P.: Performance analysis of software effort estimation models using neural networks. Int. J. Inf. Technol. Comput. Sci. (IJITCS) **5**(9), 101–107 (2013). https://doi.org/10.5815/ijitcs.2013.09.11
18. Thu, T.N.T., Xuan, V.D.: Supervised support vector machine in predicting foreign exchange trading. Int. J. Intell. Syst. Appl. (IJISA) **10**(9), 48–56 (2018). https://doi.org/10.5815/ijisa.2018.09.06
19. Gopalan, N.P., Bellamkonda, S.: Pattern averaging technique for facial expression recognition using support vector machines. Int. J. Image Graph. Signal Process. (IJIGSP) **10**(9), 27–33 (2018). https://doi.org/10.5815/ijigsp.2018.09.04
20. Desai, P., Kulkarni, G.R.: Use of API's for comparison of different product information under one roof: analysis using SVM. Int. J. Inf. Technol. Comput. Sci. (IJITCS) **10**(6), 11–22 (2018). https://doi.org/10.5815/ijitcs.2018.06.02
21. Ahmad, M., Aftab, S.: Analyzing the performance of SVM for polarity detection with different datasets. Int. J. Mod. Educ. Comput. Sci. (IJMECS) **9**(10), 29–36 (2017). https://doi.org/10.5815/ijmecs.2017.10.04

The Hurst Exponent Application in the Fractal Analysis of the Russian Stock Market

Alexander Laktyunkin and Alexander A. Potapov

Abstract The main purpose of this report is to investigate dynamics and behaviour of the financial time series for the Russian market using the fractality conception which was initially introduced by Mandelbrot [1]. The fractals have already been proved themselves as a model which describes experimental data better than previously used conventional theories in such fields of science like radiolocation, natural resources investigations, distant sounding, navigation, meteorology, information processing from unmanned aerial vehicles (UAV) and synthetic aperture radars (SAR), medicine and biology [2, 3]. At the same time, there is no fractal unified theory of the financial data behaviour. There are just few separate works devoted this topic; however, some worthy efforts were already done in this area for the last years [4–6]. It was shown that price changes rather obeyed to the Levi flight rules than to the Gaussian distribution and also we could watch the evolution from the effective market hypothesis to the fractal market hypothesis which can better explain the market crashes, especially during crises. Here, we applied the fractal approach to the Russian young market (about 20 years of history, 2 significant crises) and calculated Hurst exponents for some stocks and indexes to prove the fractality.

Keywords Fractality · Self-similarity · Financial markets · Price changes' distributions · Prices correlation

A. Laktyunkin (✉) · A. A. Potapov
V.A. Kotelnikov Institute of Radio Engineering and Electronics, Russian Academy of Sciences, Moscow, Russia
e-mail: laktyun@gmail.com

A. A. Potapov
e-mail: potapov@cplire.ru

A. A. Potapov
Joint-Laboratory of JNU-IREE RAS, Jinan University, Guangzhou, China

© Springer Nature Switzerland AG 2020 459
Z. Hu et al. (eds.), *Advances in Artificial Systems for Medicine and Education II*,
Advances in Intelligent Systems and Computing 902,
https://doi.org/10.1007/978-3-030-12082-5_42

1 Introduction

Scientific analysis of financial markets has rather evident practical application besides pure academic significance: understanding of financial markets' behaviour let investors get big returns from their investments during stable periods and, what is more important, to avoid market collapses and crises.

Classical theories do not explain stock markets' collapses or its capabilities to further reviving. Nevertheless, they promise a possibility of predictions and control of free market oscillation. Contrary to this, the chaos theory and fractal geometry considers the market as a complex system which is capable adapting to changes in the environment during its evolution of time. Such processes are remarkable for its long-term stability. However, the condition of this stability is uncertainty in a short term.

In this work, we posit that the Russian stock market possesses fractal features and so it obeys the rules of the fractal market hypothesis (FMH). The work objective is confirmation of nonlinearity of the Russian stock market and detection of fractal features.

The fractal market hypothesis was worked out in the beginning of nineties of XX. It was to be an alternative for the effective market hypothesis (EMH). FMH imparts a particular significance to information influence and investment horizons in investors' behaviour.

The main five assumptions for FMH were proposed by Peters [7]:

(1) The market is made by a great number of individuals with many different investment horizons.
(2) Information has different effects on different investment horizons.
(3) A basic factor which has an effect on market stability is liquidity (it balances demand and supply). The liquidity is reached when market consists of many investors with many different investment horizons.
(4) Prices reflect a combination of the short-term technical analysis and the long-term fundamental estimation.
(5) There will be no long-term trends if risk is not related to an economical cycle. Trade, liquidity and information for the short-term investment horizon will dominate.

2 The Hurst Exponent Analysis

For estimation of fractality, we will apply the R/S method.

An approach for investigation of fractal time series was proposed by Mandelbrot and is based on researches performed by the English scientist Hurst. It is based on analysis of parameter's range (difference between maximum and minimum on the segment under consideration) and a standard deviation. Hurst proposed a new parameter—the Hurst exponent (H) [8].

The Hurst approach let us find out if the financial time series are random walks. First of all, we need to define the range:

$$X_{t,N} = \sum_{i=1}^{t} (e_i - M_N)$$

where $X_{t,N}$ is the accumulated deviation for N periods, e_i is an increment in year i, M_N is average e_i for N periods.

Then, the range is difference between the maximum and the minimum levels of accumulated deviation:

$$R = \max(X_{t,N}) - \min(X_{t,N})$$

where R is the range of deviation $X_{t,N}$, $\max(X_{t,N})$ is the maximum value for $X_{t,N}$, $\min(X_{t,N})$ is the minimum value for $X_{t,N}$.

To compare different types of time series, Hurst divided this range on the standard deviation of initial observations. This 'standard range' must increase with time. Hurst introduced the following relation:

$$R/S = (a \cdot N)^H$$

where R/S is the standard range, N is number of observations, a is a constant, and H is the Hurst exponent.

In accordance with statistical mechanics, the Hurst exponent must be equal 0.5, if the series represents a random walk. In other words, the range of accumulated deviations must increase proportionally to square root of time N. When H differs from 0.5, it means that observations are not independent. Each observation has memory about every preceding event. It is not the short-term memory which is often called 'Markovian'. It is the long-term memory, and it should remain permanently. Recent events have a stronger effect than old events, but the residual effect of the latter is always considerable. In the long-term scale, a system which produces the Hurst statistics is the result of a long flow of interconnected events. What happens today has an effect on the future.

There are three different cases for the Hurst exponent:

(1) $H = 0.5$,

It implies random series. Events are random and not correlated. The present has no effect on the future. The probability density function can be a normal curve, but it is not a mandatory condition. The R/S-analysis can classify arbitrary series irrelative of a particular type of corresponding distribution.

(2) $0 \leq H < 0.5$,

This interval corresponds to anti-persistent or ergodic series. Such a system type is often called as 'return to the average'. If a system demonstrates growth in the preceding period, then probably it shows decay in the next period. And vice versa, if it has been decreasing then a close raise is imminent. Stability of such an anti-persistent behaviour depends on how close H is to zero. Such series is more unsteady or volatile than random series, since it consists of frequent changes from decay to raise. In spite of wide propagation of conception of return to the average in economical and financial literature, there are few anti-persistent series found for the moment.

(3) $0.5 < H < 1$.

Here, we have persistent or trend-stable series. If series ascend (descend) in the preceding period, then probably it will keep this behaviour during some time in the future. Trend stability or the persistent force increases when H approaches 1. The closer H to 0.5 the noisier series and the less noticeable its trend. Persistent series are the generalized Brownian motion or the shifted random walk. The shift value depends on how much H is bigger than 0.5.

Persistent time series are more interesting classes since it has been turned out that they appear not only in nature in abundance but they are inherent to financial markets as well.

Let us find the logarithm of $R/S = (a \cdot N)^H$:

$$\log\left(\frac{R}{S}\right) = H \cdot (\log N + \log a)$$

If we find the slope R/S in the double logarithmic coordinates as a function of N, then we get the H estimation. This estimation is not related to any assumptions about an underlying distribution.

For very big amount of observations N, one can expect convergence of series to $H = 0.5$ since the memory effect decreases to the level when it is noticeable no more. In other words, in case of long series of observations one can expect that its properties become indistinguishable from properties of a common Brownian motion or common random walk because the memory effect disappear.

Hurst also proposed formula for estimation of the H value basing on the R/S value.

$$H = \log\left(\frac{R}{S}\right) / \log\left(\frac{n}{2}\right)$$

where n is the observations amount.

In this formula, it is supposed that a from relation $\log\left(\frac{R}{S}\right) = H \cdot (\log N + \log a)$ is equal to 0.5.

Feder showed [9] that this empiric law has a tendency to overestimate H when it is higher than 0.7 and on the contrary to underestimate if it is equal or lower than 0.4. However, for short series, where regression is not possible, this empiric law can be used as a reasonable approach.

Fractal dimension of time series or accumulated changes for a random walk is 1.5. Fractal dimension of a curve line is 1, and fractal dimension of a geometric plane is 2. Thus, fractal dimension of a random walk is between a curve line and a plane.

The Hurst exponent can be transformed to the fractal dimension using the following formula [10]:

$$D = 2 - H$$

Thus if $H = 0.5$, then $D = 1.5$. Both values describe an independent random system. Value $0.5 < H \leq 1$ will correspond to a fractal dimension which is closer to a curve line. This is a persistent time series which produces a smoother and less jagged line as compared to a random walk. Anti-persistent value H ($0 < H < 0.5$) gives us a higher fractal dimension and more broken line, respectively, as compared to a random walk. And therefore, it describes the system which is more susceptible to changes.

Even if an abnormal value of H is found, there is a valid question if its estimation is substantiated. One can doubt, if there were enough data. To resolve this issue, there is a simple test. Virtually, the estimation of H which differs from 0.5 significantly has two possible explanations:

(1) The time series under investigation have the short-term memory. Each observation correlates with further observations to a certain extent.
(2) Analysis of such a kind is inconsistent, and the abnormal value of H does not mean that the long-memory effect takes place.

One can verify validity of results by means of a random mixing of the data and as a result the order of observations will be totally different as compared to the initial series. Since the observations remain the same, their frequency distribution will remain the same as well. Then, we need to calculate the Hurst exponent for the mixed data. If the series are really independent, then the Hurst exponent will not change since there was no long-term memory effect or a correlation between observations. In this case, mixing the data does not affect the qualitative characteristics of the data.

If there was the long-term memory effect, then the data order is rather important. The mixed data destroy the system structure. At that, the estimation of H will be much lower and will approach to 0.5 even if the frequency distribution of observations does not change.

3   Choosing the Data

It is incorrect to evaluate the Hurst exponent for the entire data range since the series has a finite memory and begins following random walks. Theoretically, the process with the long-term memory is supposed to originate from the infinitely remote past. But in the chaos theory, it is stated that in every nonlinear system always there is a point where the memory about initial conditions is lost. 'The loss point' is similar to the end of the natural period of the system. Based on this, it is supposed that processes with the long-term memory in the majority of systems are not infinite. They have a limit. This memory length depends on the structure of a nonlinear dynamic system which generates fractal time series.

We should determine the quantity of data which is sufficient for analysis. Feder says that modelling over points' number less than 2500 can be challenged. Peters considers the sufficient quantity of data when the system natural period can be easily distinguished. In addition to this, the chaos theory states that data of ten cycles are enough [11]. If we can estimate the cycle length, then we should use the number of points which is ten times higher than the cycle length as a valid data set. So, we take the daily prices for the period 2006–2018 what corresponds to the total amount of 3000 points.

4 Results

Diagrams for the Hurst exponent dependence on the calculation range have been plotted. These diagrams have been plotted for eight stocks of the blue-chip companies and for two indexes: Moscow Exchange (MOEX) and Russian Trade System (RTS). These two indexes describe the state of the Russian Market on the whole and are calculated on the basis of 30 and 50 leading emitters, respectively. Also, the first index is based on the rubles estimation while the second one is based on the USD estimation. Also the comparison with a randomly distributed value has been plotted (Fig. 1).

For each emitter/index, the diagrams have been plotted for three different scales. We have analysed daily, hourly and minutely closing prices (Figs. 2, 3, 4, 5, 6, 7, 8, 9, 10 and 11).

Fig. 1 Top diagram corresponds to a normally distributed random value (value changes), and bottom diagram is a typical SNGS daily diagram for half a year (2013–2014)

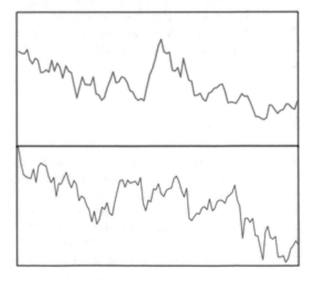

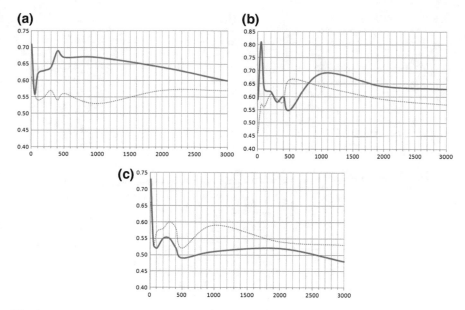

Fig. 2 **a** AFLT daily closing prices. **b** AFLT hourly closing prices. **c** AFLT minutely closing prices

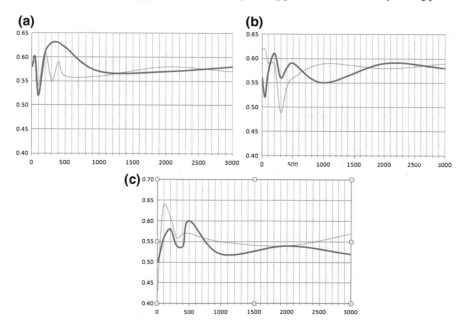

Fig. 3 **a** CHMF daily closing prices. **b** CHMF hourly closing prices. **c** CHMF minutely closing prices

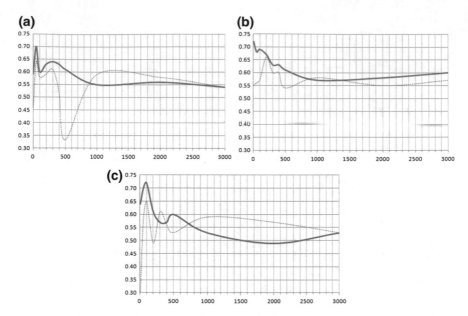

Fig. 4 **a** GAZP daily closing prices. **b** GAZP hourly closing prices. **c** GAZP minutely closing prices

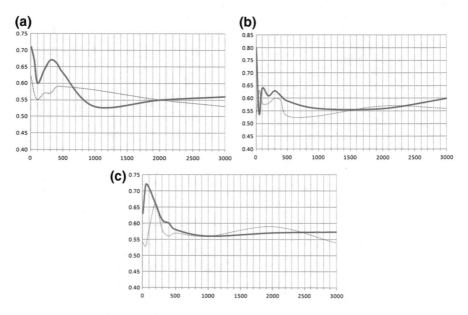

Fig. 5 **a** MOEX daily closing prices. **b** MOEX hourly closing prices. **c** MOEX minutely closing prices

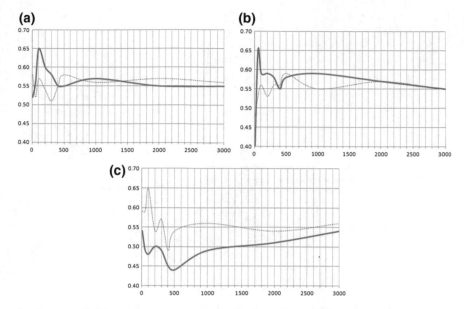

Fig. 6 **a** MTSS daily closing prices. **b** MTSS hourly closing prices. **c** MTSS minutely closing prices

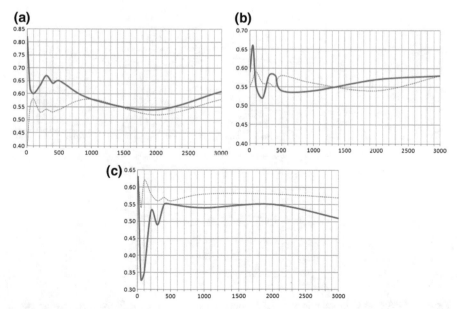

Fig. 7 **a** RTKMP daily closing prices. **b** RTKMP hourly closing prices. **c** RTKMP minutely closing prices

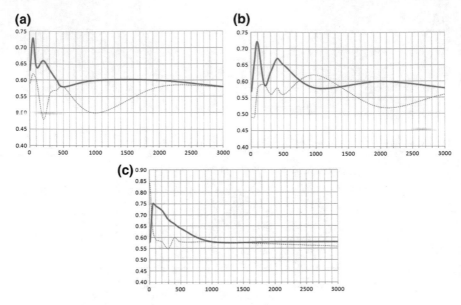

Fig. 8 **a** RTSI daily closing prices. **b** RTSI hourly closing prices. **c** RTSI minutely closing prices

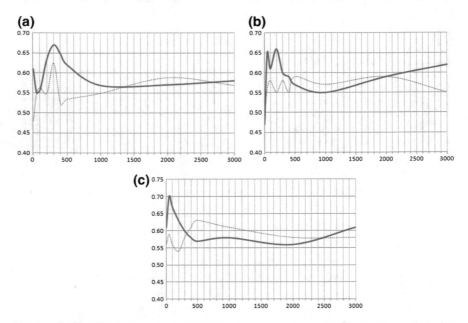

Fig. 9 **a** SBER daily closing prices. **b** SBER hourly closing prices. **c** SBER minutely closing prices

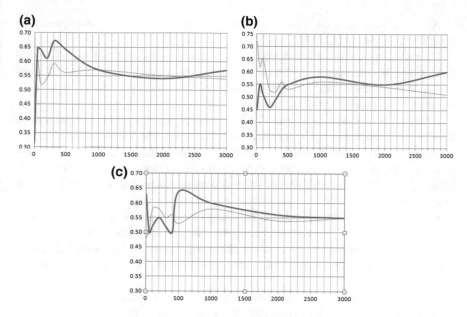

Fig. 10 **a** SIBN daily closing prices. **b** SIBN hourly closing prices. **c** SIBN minutely closing prices

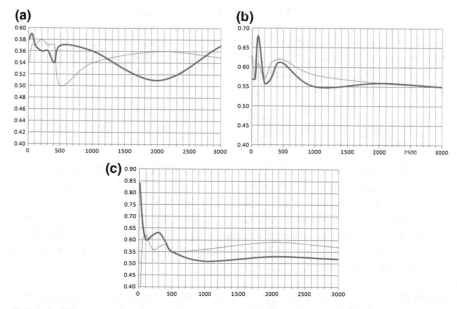

Fig. 11 **a** SNGS daily closing prices. **b** SNGS hourly closing prices. **c** SNGS minutely closing prices

As for the RTS index daily closing prices, the form of the plotted data agrees with some previous works [12, 13].

On each diagram, there is a dotted line calculated on the basis of mixed data. It is shown to demonstrate the difference between the random walk and the memory effect. In most cases, we see disappearance of the memory effect or at least its significant decrease after mixing the data.

All the daily diagrams look similarly. They have typical scatter of H values around 0.6–0.7 in the range under 400 points, and the following smooth decrease. There are some cases such as SNGS which do not look like the other patterns. H has almost nothing different from the random walk pattern for this case. It can be understood better by looking at the diagrams of daily closing prices for this emitter (Fig. 1). It looks like a random walk.

If we look at the hourly and minutely diagrams, we can see that the memory period is reduced as compared to daily ones. The minutely diagrams 'remember' approximately 50 previous values only. However, a general view remains similar with daily diagrams, that is, we observe the self-similarity or a similar behaviour at different scales.

The period of the memory effect has been determined. It varies from 200 to 400 points for daily prices and decreases for hours and minutes. For all scales, the Hurst exponent approaches 0.5 for large ranges of points. However, we found a typical feature for most of daily diagrams. For the range of 3000 points, we observe a small but noticeable increase of the Hurst exponent again. Most likely, it is due to the anomaly of 2008—a financial crisis happened at that time. In spite of high amount of points, we can clearly see deviation from a random behaviour when increasing the daily price range under investigation.

5   Conclusions

Thus, we have made sure about existence of fractal properties of the Russian Stock Market using the Hurst exponent estimation. It has been found significantly higher than 0.5 for most of cases. It can be explained by the effect of the long-term memory.

It should be also noted that all the obtained results are valid within a limited time range in the future. The market situation can change in a year or two (400 points), and we will observe other values of the Hurst exponent. Although, qualitative conclusions will be probably the same.

Acknowledgements The authors prepared this article while working on the project 'Leading Talents of Guangdong Province', №. 00201502 (2016–2020) in the JiNan University (China, Guangzhou).

References

1. Mandelbrot, B.B., Hudson, R.L.: The (Mis)Behavior of Markets: A Fractal View of Risk, Ruin, and Reward. Basic books (2004)
2. Tyagi, T., Dubey, H.M., Pandit, M.: Multi-objective optimal dispatch solution of solar-wind-thermal system using improved stochastic fractal search algorithm. Int. J. Inf. Technol. Comput. Sci. (IJITCS) **8**(11), 61–73 (2016). https://doi.org/10.5815/ijitcs.2016.11.08
3. Wang, Z.: Cloud theory and fractal application in virtual plants. Int. J. Intell. Syst. Appl. (IJISA) **3**(2), 17–23 (2011). https://doi.org/10.5815/ijisa.2011.02.03
4. Rayevnyeva, O., Stryzhychenko, K.: Investigation of the wave nature of the Ukrainian stock market. Int. J. Intell. Syst. Appl. (IJISA) **4**(1), 1–10 (2012). https://doi.org/10.5815/ijisa.2012.01.01
5. Voit, J.: The Statistical Mechanics of Financial Markets. Springer, Berlin (2005)
6. Blackledge, J.M.: Application of the fractal market hypothesis for macroeconomic time series analysis. ISAST Trans. Electron. Signal Process. **1**(2), 1–22 (2008)
7. Peters, E.: Fractal Market Analysis. Applying Chaos Theory to Investment & Economics. Wiley, New York (1994)
8. Hurst, H.E.: Long-term storage of reservoirs. Trans. Am. Soc. Civil Eng. **116** (1951)
9. Feder, E., Fractals. M.: Mir (1991)
10. Mandelbrot, Benoit B.: Self-affinity and fractal dimension. Phys. Scr. **32**, 257–260 (1985)
11. Williams, B.: New trading dimensions. M.: IK Analitika (2000)
12. Strygin, A.Y., Shvedov, A.S.: Analysis of Fractal Properties of Economic and Financial Process in the Russian Economics. Graduation Paper. Higher School of Economics, Saint-Petersburg (2004)
13. Kalyagina, L.V., Razumov, P.E.: Fractality of the Russian market. Soc. Econ. Humanit. J. Krasnoyarsk State Agrarian Univ. **1**, 15–23 (2015)

Development of Models and Methods of Virtual Community Life Cycle Organization

Olha Trach and Andriy Peleshchyshyn

Abstract This article presents the models and methods for organization of virtual community life cycle. Organization of virtual community life cycle is necessary for the successful and effective creation of the virtual community. The virtual life cycle model of the virtual community based on analysis of the life cycles of related branches of knowledge, information content and users by including in the model a number of special stages and directions of virtual community life cycle organization is developed. This model helps structure the project execution challenges faced by developers of virtual communities. A model for organizing the life cycle of a virtual community based on Petri network is developed. Model to map the links between elements and improve the organization of parallel life cycle processes is used for improving the effectiveness of community management. The method of formation of planned indicators is developed, and formal criteria for the identification of critical and important indicators are constructed to prevent an imbalance during the formation of a planned indicator of the tasks of the virtual community life cycle organization. The developed models and methods reduce the number of executors, financial and time costs, and improve the overall process of creating a virtual community.

Keywords Virtual community · Life cycle · Organization · Social networks · Indicators

1 Introduction

Today, virtual communities are one of the elements of creating an information society. The virtual community is a developed type of society, which functions on the Internet

O. Trach (✉) · A. Peleshchyshyn
Social Communications and Information Activities Department, Lviv Polytechnic National University, Lviv, Ukraine
e-mail: olya@trach.com.ua

A. Peleshchyshyn
e-mail: apele@ridne.net

© Springer Nature Switzerland AG 2020   473
Z. Hu et al. (eds.), *Advances in Artificial Systems for Medicine and Education II*,
Advances in Intelligent Systems and Computing 902,
https://doi.org/10.1007/978-3-030-12082-5_43

to satisfy the interests of group members, to communicate among members of the group, to help with the execution of tasks, etc.

The constant growth of the number of virtual communities and their intensive development increase in the volume of data and users in the global information space, which increases the need to organize projects and events management, to meet business needs, taking into account advertising and commercial benefits, and requires predictable management of the community development process. Virtual communities form the unique content and are an effective medium for information exchange. In addition, the virtual communities are the basis of the participants; they create information content and interact with each other.

Frequently, virtual communities emerge spontaneously, unpredictably, sometimes without clear goals and objectives.

Their creators or end customers do not think about during their first existence efficiency, ensuring quality management. This necessitates the development of models and methods for virtual community life cycle organization.

Formalization of the life cycle provides a qualitative approach to planning processes, analysis, development and management of the virtual community.

Actual, for the qualitative creation of virtual communities is research in the following areas: management of virtual communities [1]; socio-demographic characteristics of users of the virtual community [2–4]; information security in virtual communities [5–7]; create and manage the content of the virtual community [8–10]; site metrics and methods that take into account the ratio of site metrics [11].

However, studies on the virtual community life cycle organization are incomplete and episodic. The researchers represented only the conceptual models of the virtual life cycle of the virtual community, projecting them based on already known models of life cycles in other fields of knowledge. Mostly, researchers focus on analyzing information content and users. The stages and directions of the virtual community life cycle organization are not explored. As a result, the transition between the stages of the virtual community life cycle organization is unmanaged. Virtual communities have certain features that are taken into account during the virtual community life cycle organization.

Therefore, the purpose of the given article is to develop methods and tools for organizing the life cycle of the virtual community, based on the existing stages and support areas of the virtual community life cycle organization. Developed methods and facilities will be able to predict, structure, create and manage community and to significantly reduce the time and cost of creating virtual communities.

2 Virtual Community Life Cycle Organization

Creating a virtual community involves virtual community life cycle organization. The life cycle of a virtual community is a period from the planning of creating a virtual community to its complete elimination. Taking into account the features of the virtual community, they organize the life cycle through stages and auxiliary directions: stage—step process of organization life cycle of the virtual community; the direction—the focus of the implementation phase of the virtual community life cycle organization [12].

On the basis of any process of creation and management of information systems, certain basic standards and processes are laid. The virtual community is no exception. This set of standards and processes is reflected in the life cycle.

Taking as a basis the life cycles of production of the product, investment project, Web sites and software, the following stages of the virtual community life cycle organization are highlighted: planning, analysis, design, development, testing, implementation, operation, complex verification, conservation of the project and liquidation.

For qualitative decision-making on the transition between the stages of the virtual community life cycle organization, there are four verifications: success of the test phase of a comprehensive verification (occurring after the verification stage), urgent reengineering, decision to preserve the project and making a decision on the expediency of restoring the virtual community.

For the effective functioning and development of the virtual community, the stages and characteristics of the virtual community are identified, which are termed directions of the life cycle of the virtual community: user direction, information direction, resource direction and reputation direction.

The process of virtual community life cycle organization is distributed, the individual components of which are executed by the virtual community life cycle organization, which should be divided into levels. Managers of all stages are a virtual community manager. The virtual community manager completes and manages the team, which includes the executives of the stages and directions—the specialists in the required qualifications and industry—and the analyst—a specialist who is responsible for analyzing the domain of the virtual community and reference communities.

2.1 A Formal Model of Virtual Community Life Cycle Organization

The formal model of virtual community life cycle organization, formed on the basis of analysis of life cycles of related fields of knowledge, information content and users of the virtual community, includes a series of special stages and directions for virtual community life cycle organization.

A formal model of virtual community life cycle organization (VCLCO):

$$OrgLifeCycle(Com) = \langle Stage(Com), Dr(Com), Cell(Com) \rangle \qquad (1)$$

The components of VCLCO model are *Stage(Com)* stages of virtual community life cycle organization, *Dr(Com)* directions of virtual community life cycle organization and *Cell(Com)* structure of the cell of intersection of stages and directions of virtual community life cycle organization.

$$Stage(Com) = \{Stage_i(Com)\}_{i=1...N^{(Stage)}} \tag{2}$$

where $Stage_i$ is j-th direction of the VCLCO, and $N^{(Stage)}$ is the number of stages in the VCLCO.

$$Stage_{Com} = \left\langle \begin{matrix} Plan_{com}, \ Analys_{Com}, \ Design_{Com}, \ Devel_{Com}, \ Test_{Com}, \\ Implement_{Com}, \ Expl_{Com}, \ ComVer_{Com}, \ Exprec_{Com}, \ Liq_{Com} \end{matrix} \right\rangle \tag{3}$$

where $Plan_{Com}$ is planning stage; $Analys_{Com}$ is stage of analysis; $Design_{Com}$ is design stage; $Devel_{Com}$ is stage of development; $Test_{Com}$ is testing stage; $Implement_{Com}$ is implementation stage; $Expl_{Com}$ is stage of operation; $Comver_{Com}$ is comprehensive verification stage; $Exprec_{Com}$ is project conservation stage; and Liq_{Com} is liquidation stage.

$$Dr(Com) = \left\{ Dr_j(Com) \right\}_{j=1...M^{(Dr)}} \tag{4}$$

where Dr_j is i-th the stage of the VCLCO, and $M^{(Dr)}$ is the number of stages in the VCLCO.

$$Dr = \langle Us_i, \ Inf_i, \ Rs_i, \ Rp_i \rangle \tag{5}$$

where Us is user direction; Inf is information direction; Rp is reputation direction; and Rs is resource direction.

$$Cell(Com) = \left\{ Cell_{ij}(Com) \right\}_{ij=1...L^{(Cell)}} \tag{6}$$

where $Cell_{ij}$ is ij-th cell of VCLCO, and $L^{(Cell)}$ is the number of cells in the VCLCO.

$$C(Cell_k) = \left\langle \begin{matrix} Task(Cell_k), \ Performer(Cell_k), \ Time(Cell_k), \\ Document(Cell_k), \ Finances(Cell_k) \end{matrix} \right\rangle \tag{7}$$

where $Task(Cell_k)$ is task of directions; $Performer(Cell_k)$ is performers of tasks directions; $Time(Cell_k)$ is time to perform directions tasks; $Document(Cell_k)$ is documentation; and $Finances(Cell_k)$ is financial support for the tasks of the directions.

2.2 Networking Model for the Virtual Community Life Cycle Organization

Petri networks are used to display parallel processes in modeling complex and large projects and are the main tool for simulation in project work. The task of virtual community life cycle organization must be performed at the stage of all directions in parallel processes [13–15]. Therefore, it is advisable to present the virtual community

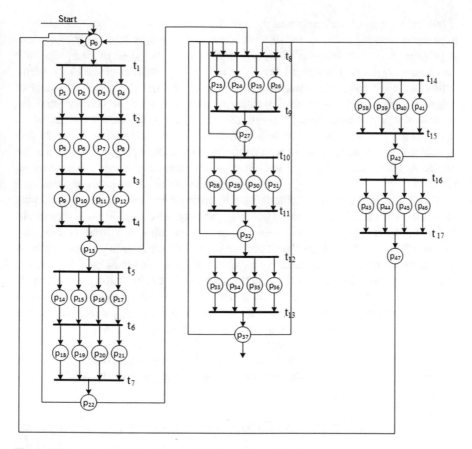

Fig. 1 Petri network of virtual community life cycle organization

life cycle organization through the Petri network (Fig. 1). The network shown refers to a case in which tasks are performed in all directions; however, it can be modified in accordance with the created community.

The virtual community life cycle organization of Petri networks has this form:

$$N = (P, T, I, O) \tag{8}$$

where $P = \{p_0 \ldots p_{47}\}$ is set of positions, $T = \{t_1 \ldots t_{17}\}$ is set of transitions, I is input function, and O is output function.

The set of transitions in the VCLCO includes the following actions: t_1—transition from planning stage to analysis stage; t_2—transition to the implementation stage of the design; t_3—transition to the implementation stage of the development; t_4—the transition from the development stage to the process of verifying the success of the creation of the virtual community; t_5—transition to the stage of testing; t_6—transition to the implementation stage; t_7—the transition from the implementation stage to verifying the success of the implementation of the virtual community; t_8—transition to execution stage exploitation; t_9—the transition from the operation stage to the testing of the success of the operation; t_{10}—transition to the stage of comprehensive verification; t_{11}—the transition from the integrated verification phase to the verification of the success of the comprehensive verification; t_{12}—transition to a stage of urgent reengineering; t_{13}—transition from urgent reengineering to verification of the success of urgent reengineering, t_{14}—transition to the implementation phase of the conservation project; t_{15}—the transition from the conservation phase of the project to verifying the expediency of restoring the virtual community; t_{16}—transition to the implementation phase of the liquidation; t_{17}—transition to creating a new community.

Transitions in the virtual community life cycle organization are: p_0—planning for creating a virtual community (defining the purpose and objectives of the virtual community); p_{13}—checking the success of creating a virtual community; p_{22}—verification of the implementation of the virtual community; p_{27}—verifying the success of the virtual community exploitation; p_{32}—verification of the success of complex verification; p_{37}—checking the success of reengineering; p_{42}—verifying the expediency of restoring the virtual community; p_{47}—shutting down the virtual community project; p_1, p_2, p_3, p_4, p_5, p_6, p_7, p_8, p_9, p_{10}, p_{11}, p_{12}, p_{14}, p_{15}, p_{16}, p_{17}, p_{19}, p_{20}, p_{21}, p_{23}, p_{24}, p_{25}, p_{26}, p_{28}, p_{29}, p_{30}, p_{31}, p_{33}, p_{34}, p_{35}, p_{36}, p_{38}, p_{39}, p_{40}, p_{41}, p_{43}, p_{44}, p_{45}, p_{46}—perform tasks of user, information, resource and reputation directions at the stages of the virtual community life cycle organization.

Formally, the position of the network of the organization of the VCLCO is as follows:

$$P_i = \langle IndIn_i, Task_i, IndPlan_i, IndOut_i \rangle \tag{9}$$

where $Task_i = \{Task_{ik}\}_{k=1}^{N_i}$ is a set of tasks directly to VCLCO at the site, and N_i is number of tasks of the i-th node;

$IndPlan = \{IndPlan_{ik}\}_{k=1}^{N_i}$ is a set of scheduled task indicators for directing the VCLCO at the site, and N_i is number of tasks of the i-th node;

$IndOut_j = \{IndOut_{jk}\}_{k=1}^{N_j}$ is a set of output indicators to perform tasks directly in the VCLCO, and N_k is number of output indicators of k-th task;

$IndIn_j = \{IndIn_{jk}\}_{k=1}^{N_j}$ is a set of input indicators for the task of directing the VCLCO, and N_k is number of input indicators of k-th task.

2.3 The Method of Forming a Plan of Indicators for Task Directions of the Virtual Community Life Cycle Organization

The purpose of the method of forming the planned indicator of the objectives of the virtual community life cycle organization is to correctly formulate the planned

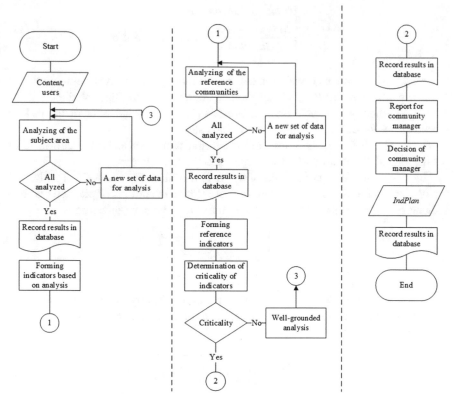

Fig. 2 Method of formation of the planned indicator of tasks directions VCLCO

indicator of the virtual community life cycle organization. Taking into account the reports of the analyzed of the subject area and reference communities, the results of the records in the database of the VCLCO (Fig. 2).

The basis for the formation of the target is the determination of the criticality of the indicators. When defining criticality, the indicators of the tasks of directions of virtual community life cycle organization are divided into three categories: critical, uncritical and important.

Critical indicators of the tasks of the virtual community life cycle organization. Execution of tasks of directions is carried out; the level of relative deviation of the input indicator is admissible, that is:

$$\Delta k_i^{Dr} \leq \Delta^* k_i^{Dr} \tag{10}$$

$$\Delta k_i^{Dr} = \begin{cases} 0, \textit{якщоIndPlan} < IndIn \\ \dfrac{IndPlan - IndIn}{IndPlan}, IndPlan > IndIn \end{cases} \qquad (11)$$

where $\Delta^* k_i^{Dr}$ is tolerance which is set by the manager of virtual community life cycle organization; *IndPlan* is the planned indicator of the task of directing the virtual community life cycle organization; *IndIn* is a real indicator of the task of directing the VCLCO, formed in the previous stage.

$Dr \in \{Us, Inf, Rp, Rs\}$ is user, informational, reputation and resource directions for the virtual community life cycle organization, respectively.

Uncritical indicators of the tasks of the virtual community life cycle organization are defined as follows. Let ρ is scattering of relative deviations of planned and real indicators of the tasks of the virtual community life cycle organization:

$$\rho = \left[\sum_{i=1}^{N} \frac{1}{N} (\Delta k_i^{Dr})^2 \right]^{1/2} \qquad (12)$$

Important indicators of the tasks of the virtual community life cycle organization. Determined if its individual value exceeds the total scattering: $Ind_i^{Dr} > \rho$, moreover Ind_i^{Dr} is calculated by the formula:

$$Ind_i^{Dr} = \sqrt{\sum (Ind_i^{Dr})^2 * w_i^{Dr}} \qquad (13)$$

where $Ind_i^{Dr} = \begin{cases} 0, \textit{якщоIndPlan} < IndIn \\ (IndPlan - IndIn), IndPlan > IndIn \end{cases}$; w_i^{Dr} is weight factor, which is determined by the manager, $0 \geq w_i \leq 1$, $w_i \in w$, $\sum_{w_i \in w} w_i = 1$; *IndPlan* is a planned indicator of task direction, and *IndIn* is a real indicator of task direction.

2.4 Implementation of Models and Methods for Virtual Community Life Cycle Organization

The official community of the department of social communication and information activities (SCIA), Lviv Polytechnic National University, in the social network Facebook worked without the use of methods and models. The introduction of models and methods for virtual community life cycle organization took place at the stage of managing the virtual community. When applying the methods of virtual community life cycle organization, the number of implementers of the virtual community life cycle organization of the department has decreased SCIA in the social network Facebook (Fig. 3).

Proper distribution of tasks between performers improved community management results and established a relationship between the implementers of the life cycle

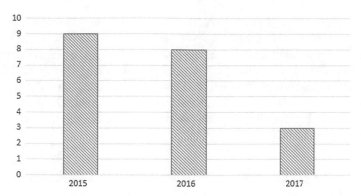

Fig. 3 Indicators of performance of the virtual community life cycle organization

of the virtual community. Constantly preserving the intermediate results of the virtual community provides an opportunity to compare and analyze work on community management and promotion.

The effectiveness of the virtual community life cycle organization is a significant reduction in the time and cost of creating virtual communities. This is confirmed by the actual research data and the results obtained from the implementation of models and methods for virtual community life cycle organization.

The indicator of the effectiveness of the virtual community life cycle organization is determined as follows:

$$Efficiency(Com) = \frac{N^{(Task)}}{N^{(Arm)}} \tag{14}$$

where $N^{(Task)}$ is number of tasks for the virtual community life cycle organization, and $N^{(Arm)}$ is number of implementers of the virtual community life cycle organization.

Reducing the number of virtual community life cycle implementers, with a large number of parallel and predictable tasks, reduces the financial cost of creating a community. The results obtained can improve the overall process of creating a virtual community of 30–40% depending on the specifics of the virtual community.

3   Conclusions

Developing a virtual community life cycle model allows to structure the execution of project tasks, which are faced by the developers of the virtual community. A model for organizing the life cycle of a virtual community based on Petri network, which is used to map the links between elements and improve the organization of parallel life cycle processes, allows improving community management efficiency. The method of formation of planned indicators is developed, which allows preventing imbalance when creating a planned indicator of the tasks of the virtual community life cycle organization. The effectiveness of a virtual community life cycle organization there is a significant reduction of time and financial cost of creating virtual communities. This is supported by research evidence and results from the introduction of methods and tools of the virtual community life cycle organization in the virtual community of the department of SCIA.

References

1. Azizifard, N.: Social network clustering. Int. J. Inf. Technol. Comput. Sci. (IJITCS) **6**(1), 76–81 (2014). https://doi.org/10.5815/ijitcs.2014.01.09
2. Syerov, Y., Fedushko, S., Loboda, Z.: Determination of development scenarios of the educational web forum. In: Proceedings of the XIth International Scientific and Technical Conference (CSIT 2016), Lviv, pp. 73–76 (2016)
3. Korobiichuk, I., Fedushko, S., Juś, A., Syerov, Y.: Methods of determining information support of web community user personal data verification system. In: Automation (2017)
4. ICA: Advances in Intelligent Systems and Computing, vol. 550, pp. 144–150. Springer (2017)
5. Mottaghi, S., Farhadi, F., Dokohaki, S., Farhadi, M.M.: Artificial intelligence design waterfronts and particular places management to improve relationships between people. Int. J. Intell. Syst. Appl. (IJISA) **5**(9), 58–66 (2013). https://doi.org/10.5815/ijisa.2013.09.07
6. Hu, Z., Gnatyuk, S., Koval, O., Gnatyuk, V., Bondarovets, S.: Anomaly detection system in secure cloud computing environment. Int. J. Comput. Netw. Inf. Secur. (IJCNIS) **9**(4), 10–21 (2017). https://doi.org/10.5815/ijcnis.2017.04.02
7. Hu, Z., Khokhlachova, Y., Sydorenko, V., Opirskyy, I.: Method for optimization of information security systems behavior under conditions of influences. Int. J. Intell. Syst. Appl. (IJISA) **9**(12), 46–58 (2017). https://doi.org/10.5815/ijisa.2017.12.05
8. Hu, Z., Gnatyuk, V., Sydorenko, V., Odarchenko, R., Gnatyuk, S.: Method for cyber incidents network-centric monitoring in critical information infrastructure. Int. J. Comput. Netw. Inf. Secur. (IJCNIS) **9**(6), 30–43 (2017). https://doi.org/10.5815/ijcnis.2017.06.04
9. Howard, R.: How to: manage a sustainable online community (electronic resource). Mode of access: http://mashable.com/2010/07/30/sustainable-online-community/
10. Sihare, S.R.: Roles of E–content for E–business: analysis. Int. J. Inf. Eng. Electron. Bus. (IJIEEB) **10**(1), 24–30 (2018). https://doi.org/10.5815/ijieeb.2018.01.04
11. Kraut, R., Resnick, P.: Building Successful Online Communities: Evidence-Based Social Design Massachusetts Institute of Technology. MIT Press, p. 283 (2012)
12. Peleshchyshyn, A., Mastykash, O.: Analysis of the methods of data collection on social networks. In: Proceedings of the XIth International Scientific and Technical Conference "Computer Sciences and Information Technologies" (CSIT-2017), Lviv, 05–08 Sept 2017, pp. 89–91
13. Trach, O., Peleshchyshyn, A.: Development of directions tasks indicators of virtual community life cycle organization. In: Proceedings of the XIth International Scientific and Technical Conference "Computer Sciences and Information Technologies" (CSIT-2017), pp. 127–130 (2017)
14. Trach, O., Peleshchyshyn, A.: Functional-network model of tasks performance of virtual communication life cycle directions. In: Proceedings of the XIth International Scientific and Technical Conference 'Computer Science and Information Technology' (CSIT 2016), pp. 108–110 (2016)

15. Narayanan, M., Cherukuri, A.K.: Verification of cloud based information integration architecture using colored Petri nets. Int. J. Comput. Netw. Inf. Secur. (IJCNIS) **10**(2), 1–11 (2018). https://doi.org/10.5815/ijcnis.2018.02.01

A Formal Approach to Modeling the Characteristics of Users of Social Networks Regarding Information Security Issues

Andriy Peleshchyshyn, Volodymyr Vus, Solomiia Albota
and Oleksandr Markovets

Abstract The article focuses on formalization and identification of characteristics of users of social networks, which are treated as of great importance concerning national security. In particular, a group of individual identification data, its network identification, the state security characteristics groups are considered. Two components for describing network characteristics in terms of state security: the user activity log and social portrait have been included into the model. Characteristics that describe the level of user interaction with other users and with communities are regarded. The atomic components of the model have been reduced to the measuring values and data, which are suitable for direct storing in the relational database. A typical data model in the entity–relationship diagram format has been provided.

Keywords Social networks · User activity · Socially significant content · Roles of users

1 Introduction

Social networking today remains a keystone of state information space, still the most vulnerable to different types of social and communicative threats.

Improving social networking security policy requires a variety of appropriate measures, information and technological as well: computer programs, information systems, and services, high-tech human–machine systems together with the elements of artificial intelligence.

Implementation of such systems is impossible without subject area formalization as the basis for their algorithmic efficiency and corresponding data modeling.

A. Peleshchyshyn · V. Vus · S. Albota · O. Markovets (✉)
Social Communications and Information Activity Department, Lviv Polytechnic
National University, Lviv, Ukraine
e-mail: oleksandr.v.markovets@lpnu.ua

A. Peleshchyshyn · V. Vus · S. Albota · O. Markovets
Applied Linguistics Department, Lviv Polytechnic National University, Lviv, Ukraine

© Springer Nature Switzerland AG 2020
Z. Hu et al. (eds.), *Advances in Artificial Systems for Medicine and Education II*,
Advances in Intelligent Systems and Computing 902,
https://doi.org/10.1007/978-3-030-12082-5_44

Such formalization involves the allocation of a sufficiently complete set of characteristics, which have been grouped by semantic closeness. Description of characteristics should not be limited to the verbal one, but include a system of notations and categorization in the form suitable for further implementation within the structure of the relational database. Verification of such a model can be obtained by providing a typical structure of the relational database, which has been represented in the diagrams, which aim at adopting approaches to security enhancement of state information space and are provided thoroughly below.

2 Relation Works

There are various current approaches to classify users of social networks, namely in academic writings [1–6], but the classifications mentioned are particularly focused on the issue of effectiveness in virtual communities. Several specific categories of community members concerning their behavioral characteristics have been singled out in the research papers [7]. The methods for assessing users of social networks, based on the nature of their interaction with state executive authorities, have been selected in the academic papers [8]. Such approaches are quite close to the issue of national information space security, taking into account the general principles of the classifications—behavior in virtual community, the nature of an influence on state institutions. Yet, the studies mentioned above do not cover certain, important, related to social and information security aspects: the assessment of the level of influence of individual users on the social or virtual community, meaningful activity characteristics, the willingness to cooperate in terms of fundamental security tasks, and the strategic management of hostile structures.

3 A Specific Security Model for the Social Networking User

A range of specific models of social networking users, aimed at information activity problem solving, has been recently developed. Such models are given in the academic papers [9–11]. However, it is not allowed to use models directly concerning security issues because of their specialization. Furthermore, some security-relevant aspects have not been taken into account concerning the models (particularly, social connections of Internet users), since they are focused more on traditional Web forums than on social networks.

The components of user model are grouped into separate content-sharing features. Such groups are the following: identifier and personal data; activity; content readers; content providers; user communities; controlled resources.

Thus, the following tuple marks a user:

$$User_i = \langle UI_i, UA_i, UF_i, US_i, UC_i, UR_i \rangle \tag{1}$$

where the elements of the tuple are the corresponding components of the model. These components are considered in detail below (Table 1).

Table 1 Indicators of user identification (*UI* group)

Indicator	Symbol	Data type	Comment
Unique identifier	*UIId*	Character string	Identifier for the internal DB of SN users
UIS indicators of network identification (repeatedly)			
Theme	*UITh*	Key words	Subject matter covering specific activity
SN address with a profile located	*UISN*	URI	The SN network address, main URI platform
Identifier of user profile	*UISNId*	Character string	Identifier in terms of SN
Address of user profile	*UISNA*	URI	User profile address
Network name	*UISName*	Character string	User name (nickname)
UIR indicators of physical identification			
Name of person	*UIRName*	Full name	
A means of communication	*UIRTele*	Feature group	Mobile, e-mail, messenger, etc.
Network and technical data	*UIRNet*	Feature group	IP addresses, etc.
Demographic data	*UIRDem*	Feature group	Age, sex, education, language, etc.
Juridical data	*UIRJur*	Feature group	Addresses, documents, ITN, etc.
UIV indicators of virtual identification			
The level of virtualization	*UIVirt*	[0,1]	Ability of an individual to bind and influence user behavior
The level of intellectualization	*UIVIntel*	[0,1]	Software intelligence implementing a bot
Typical tasks	*UIVBT*	List of types	Commenting, transmission, responding, etc.

3.1   Identifier and Personal User Data

A crucial thing is that a set of the network identification indicators for each user is not unique. The part of features is a set of tuples with elements of the type given. Hence, several records with the profile address, identifier, nickname, etc., can be defined for a single person:

$$UI_i = \langle UIID_i, UIS_i, UIR_i, UIV_i \rangle \tag{2}$$

where $UIS_i = \{\langle UISN_{ij}, UISNId_{ij}, UISA_{ij}, UISName_{ij} \rangle\}_{j=1}^{N_i^{(UIS)}}$, $N_i^{(UIS)}$ is a number of the network identifiers of the ith person.

Taking into account that information on the actual network, rather than physical identification, is considered as a primary one in a multitude of tasks concerning the Internet social structure analysis, the following set is of importance:

$$UIS = \bigcup_{i=1}^{N^{(UI)}} \{\langle UISN_{ij}, UISNId_{ij}, UISA_{ij}, UISName_{ij} \rangle\}_{j=1}^{N_i^{(UIS)}} \tag{3}$$

where $N^{(UI)}$ is a number of physical users.

The *UIS* set represents a set of all network identities (roughly, a set of SN virtual individuals).

3.2   The State Security Characteristics

When forming a database of users within the tasks of information security, it is significant to consider its specific features related to the state security.

The group features are listed in Table 2.

The group indicators are difficult to define. But they will be possible to be automatically processed, as the artificial intelligence technologies in terms of the natural language processing to assess judgments and identify feelings are being developed.

Table 2 National security indicators (*US*)

Indicator	Symbol	Data type	Comment
Attitude to the state	*USG*	[−1,1]	[Anti-government … patriotic]
Independence of judgments	*USIn*	[0,1]	[Fully managed … independent]
Position stability	*USSt*	[0,1]	[Constantly changing position … no changes]
Willingness to engage in dialogue	*USDl*	[0,1]	[Not ready … open]

In general, current technological capacity and scientific discoveries [12] allow to automate some time-consuming tasks.

The integrated position indicator of user flexibility can be determined the following way based on the indicators of Table 2:

$$UserFlex(User_i) = USIn_i * USSt_i * USDl_i. \qquad (4)$$

This indicator reflects the user ability to perceive the thoughts of opponents in terms of discussions, to change their views in the course of argumentative discussion. The indicator is further used to define the methods of interaction with opinion leaders of different perspectives.

3.3 The Formal Description of User Activity

The specific forms of user activity of the social networking can be currently manifested in various forms. They are summarized on the basis of the typical functions of social networks. The following forms of activity for the formation of content are (Table 3):

- Publishing new content;
- Posting a comment;
- Content retransmission;
- Content evaluation;
- Influence action.

Table 3 Indicators of user activity (*UA* group)

Indicator	Symbol	Data type	Commentary
Placement of author content	UAUC	A set of records	Unique author content
Socially relevant author content	UAIC	A set of records	Efficient content for mass consumer
Comments	UACom	A set of records	
Retransmitted content	UART	A set of records	Links or reposts
Content rating system	UAOM	A set of records	"Likes", short messages, limited by rating
Influence action	UAIA	A set of records	
The average frequency of content placement	UACF	Natural number	Number of messages per checking period (week or month)

The author content is formalized as:

$$UA_i = \langle UAUC_i, UAIC_i, UACom_i, UART_i, UART_i, UAOM_i, UAIA_i \rangle \quad (5)$$

where the components of the tuple are the relations described below.

Content is created by a user, and it is placed in the SN:

$$UAUC_i = \left\{ \left\langle ID_{ij}^{(UAUC)}, URI_{ij}^{(UAUC)}, Home_{ij}^{(UAUC)}, Content_{ij}^{(UAUC)}, Date_{ij}^{(UAUC)} \right\rangle \right\}_{i=1}^{N_i^{(UAUC)}}$$
$$(6)$$

where $ID_{ij}^{(UAUC)}$ is unique content identifier; $URI_{ij}^{(UAUC)}$ is network content address; $Home_{ij}^{(UAUC)}$ is the base (main) address of the resource where the resource is located; $Content_{ij}^{(UAUC)}$ is information content (text, images); $Date_{ij}^{(UAUC)}$ is date of publication; $N_i^{(UAUC)}$ is a number of author content records of the ith user.

The indicator displays content deliberately created for spreading awareness through mass by a user.

From the formalization viewpoint, socially significant content has a similar structure of content elements and is a subset of the whole content author:

$$UAIC_i \subseteq UAUC_i \quad (7)$$

Comments include the outcomes of user reactive actions in terms of text concerning the new content of users, which are being tracked. Accordingly, their relation is described in the following way:

$$UACom_i = \left\{ \left\langle ID_{ij}^{(UACom)}, URI_{ij}^{(UACom)}, MainID_{ij}^{(UACom)}, TargetID_{ij}^{(UACom)}, \right. \right.$$
$$\left. \left. , Content_{ij}^{(UACom)}, E_{ij}^{(UACom)} \right\rangle \right\}_{i=1}^{N_i^{(UACom)}}$$
$$(8)$$

where $ID_{ij}^{(UAUC)}$ is unique content identifier; $URI_{ij}^{(UACom)}$ is the network address of the comment (if it can be determined); $MainUri_{ij}^{(UACom)}$ is identifier of the main post to which the comment is written; $TargetID_{ij}^{(UACom)}$ is identifier of the target content to which the comment is written; $Content_{ij}^{(UACom)}$ is information content (text, images); $Date_{ij}^{(UACom)}$ is date of publication; $N_i^{(UACom)}$ is a number of content author records of the ith user.

Content retransmission corresponds to the comments according to the structure of its components. The differences regarding the technical aspect of the implementation—retransmitted content for readers is being provided autonomously and may consider to be the one of authors. Its relation is described as follows:

$$UART_i = \left\{ \left\langle ID_{ij}^{(UART)}, URI_{ij}^{(UART)}, MainID_{ij}^{(UART)}, TargetID_{ij}^{(UART)}, \right. \right.$$
$$\left. \left. Content_{ij}^{(UART)}, Date_{ij}^{(UART)} \right\rangle \right\}_{i=1}^{N_i^{(UART)}} \tag{9}$$

where the description of the tuple components is similar to the comment components.

Content ratings describe user responses to content that do not contain a content component but emotional one. The relation is as follows:

$$UAOM_i = \left\{ \left\langle ID_{ij}^{(UAOM)}, MainID_{ij}^{(UAOM)}, TargetID_{ij}^{(UAOM)}, \right. \right.$$
$$\left. \left. Opinion_{ij}^{(UAOM)}, Date_{ij}^{(UAOM)} \right\rangle \right\}_{i=1}^{N_i^{(UAOM)}} \tag{10}$$

where the description of the tuple components is similar to the comment components. On the contrary, $Opinion_{ij}^{(UAOM)} \in [-1, 1]$ is a formalized description of the user response to the content. The value "-1" is completely negative and "1"—completely positive. The indicator is induced due to the lack of a single mechanism of emotional transfer.

Influence action relates to the administrative responses of users with additional rights to the user content placed in SN. The source of influence is either the content owners or users with specific rights (administrators and community moderators). The following relation is given:

$$UAIA_i = \left\{ \left\langle ID_{ij}^{(UAIA)}, MainID_{ij}^{(UAIA)}, TargetID_{ij}^{(UAIA)}, \right. \right.$$
$$\left. \left. Action_{ij}^{(UAIA)}, Date_{ij}^{(UAIA)} \right\rangle \right\}_{i=1}^{N_i^{(UAIA)}} \tag{11}$$

where the description of the tuple components is similar to the comment components. On the contrary, $Action_{ij}^{(UAIA)} \in [0, 1]$ describes the change within the content visibility as a result of moderator actions. The value "0" corresponds to the complete elimination of the content visibility, and "1"—maximization of the content visibility to 100% of the wide audience, consuming information from the $Home_{ij}^{(UAUC)}$ main page, where $MainID_{ij}^{(UAIA)} = ID^{(UAUC)}$. The induced indicator does not reflect the full spectrum of possible support actions owing to their diversity and a lack of digital footprint. However, most actions of moderator can be restrained to the scheme given.

In fact, *TargetID* is a set that describes a single address field for all content types in social networking

$$TargetID = ID^{(UAUC)} \cup ID^{(UACom)} \cup ID^{(UART)}$$
$$\cup ID^{(UACom)} \cup ID^{(UAOM)} \cup ID^{(UAIA)}. \tag{12}$$

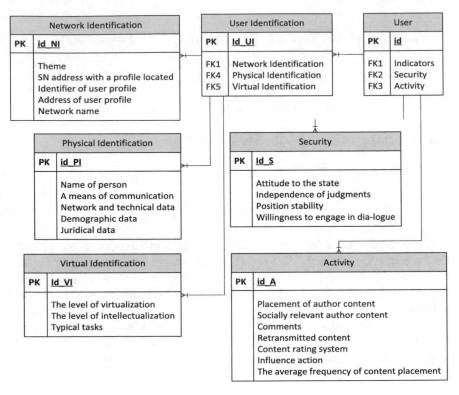

Fig. 1 Database structure

In this case, for each of the expressions mentioned above, there is a restriction:

$$TargetID_{ij} \in TargetID. \tag{13}$$

In fact, *TargetID* is a set that describes a single address field for all content types in social networking.

Based on models, a typical database structure in the entity–relationship diagram format, which is presented in Fig. 1, is provided.

The database structure allows to verify the model by implementing it as a component of a comprehensive information system for protecting the informational state space.

3.4 The Formal Description of the User Social Portrait

As it was previously mentioned, every user of the social networking is characterized not only by the content that is being generated, but also by the social networking system with other users. They are formalized by combining the SN user into a social portrait account in the following way.

Readers (consumers) of content—SN users who have signed up for the subscription to automatically obtain new content, which was created by a user. The subscription technology prevails at the present time via tracking systems and friends in social networks. What is more, traditional e-mail subscriptions or electronic messengers are actively used for Web forums and blogs. Consequently, the rapidly growing popularity of Telegram messenger for the content subscriptions is worth noting. In addition, RSS feeds and Web browser notifications are used.

Concerning current technological features, the effective monitoring of third parties is subject to subscriptions through social networks. The information regarding e-mail and messenger users is also available for owners of the platform. Users who make automated access through RSS and browser notifications are not generally subject to accounting.

The content consumers will be formally described as a subset of all sets of SN virtual individuals:

$$UF_i \subset UIS. \tag{14}$$

Respectively, content providers are also a subset of that set:

$$US_i \subset UIS. \tag{15}$$

Content providers will be considered as those who deliver content to the ith user concerning technologies described above.

Another hybrid form of social ties is community participation. Communities provide communication between users under another scheme with intermediate distribution centers of resources. Each community user makes use of the content community to deliver and consume content. All community members are consumers of content created within the community.

The sets for each user are defined in the following way:

Communities with the right to review—a set of community network identifiers $UComR_i = \{CIN_j, \text{ where } i\text{th user is a reader}\}_{j=1}^{N^{(Com)}}$, $N^{(Com)}$ is a number of communities.

Communities with the right to publish—a set of community network identifiers $UComP_i = \{CIN_j, \text{ where } i\text{th user is a publisher}\}_{j=1}^{N^{(Com)}}$, $N^{(Com)}$ is a number of communities.

Controlled resources—a set of community network identifiers $UComC_i = \{CIN_j, \text{ where } i\text{th user moderates content}\}_{j=1}^{N^{(Com)}}$, $N^{(Com)}$ is a number of communities.

4 Conclusion

The article deals with the formalization of the main subjects of information activity and confrontation in social networks: users and communities. Formalization was represented by the theoretical multiple descriptions defining a system of relations and attributes.

The formal description is a basis for designing a database of information activities in the processes of protecting the social media from harmful influences on both national and global levels. Numerical attributes should be treated as of great importance when processing specific algorithms for automated detection of aggression, counteraction to such aggressive actions and classification of information threats in social networks. The structure of the database allows to verify the model by implementing it as a component of a comprehensive information system for protecting the informational state space.

References

1. Mumtaz, D., Ahuja, B.: A lexical approach for opinion mining in twitter. Int. J Educ. Manage. Eng. (IJEME) 6(4), 20–29 (2016). https://doi.org/10.5815/ijeme.2016.04.03
2. Limsaiprom, P., Praneetpolgrang, P., Subsermsri, P.: Visualization of influencing nodes in online social networks. Int. J. Comput. Netw. Inf. Secur. (IJCNIS) 6(5), 9–20 (2014). https://doi.org/10.5815/ijcnis.2014.05.02
3. El Marrakchi, M., Bensaid, H., Bellafkih, M.: E-reputation prediction model in online social networks. Int. J. Intell. Syst. Appl. (IJISA) 9(11), 17–25 (2017). https://doi.org/10.5815/ijisa.2017.11.03
4. Meligy, A.M., Ibrahim, H.M., Torky, M.F.: Identity verification mechanism for detecting fake profiles in online social networks. Int. J. Comput. Netw. Inf. Secur. (IJCNIS) 9(1), 31–39 (2017). https://doi.org/10.5815/ijcnis.2017.01.04
5. Khatoon, M., Aisha Banu, W.: A survey on community detection methods in social networks. IJEME 5(1), 8–18 (2015). https://doi.org/10.5815/ijeme.2015.01.02
6. Gururaj, H.L., Swathi, B.H., Ramesh, B.: Threats, consequences and issues of various attacks on online social networks. Int. J. Educ. Manage. Eng. (IJEME) 8(4), 50–60 (2018). https://doi.org/10.5815/ijeme.2018.04.05
7. Syerov, Y., Fedushko, S., Loboda, Z.: Determination of development scenarios of the educational web forum. In: 2016 XIth International Scientific and Technical Conference Computer Sciences and Information Technologies (CSIT), pp. 73–76. Lviv (2016)
8. Markovets, O., Korzh, R., Yarka, U.: Research of means used in communication of Internet users with local authorities. East.-Eur. J. Enterp. Technol. 3(9) (63), 38–41 (2013)
9. Korzh, R., Fedushko, S., Peleschyshyn, A.: Methods for forming an informational image of a higher education institution. Webology 12(2), Article 140 (2015). Available at http://www.webology.org/2015/v12n2/a140.pdf
10. Markovets, O., Peleschyshyn, A.: Modeling of citizen claims processing by means of queuing system. Int. J. Comput. Sci. Bus. Inform. (UCSBI). India: IJCSBI.ORG. 15(1), 36–46 (2015)
11. Peleshchyshyn, A., Korzh, R.: Basic features and a model of university units: university as a subject of information activity. East.-Eur. J. Enterp. Technol 2(2), 27–34 (2015)
12. Peleshchyshyn, A., Korzh, R., Tymovchak-Maksymets, O.: Advanced search query for identifying web-forum threads relevant to given subject area. In: Modern Problems of Radio Engineering, Telecommunications and Computer Science—Proceedings of the 11th International Conference, TCSET'2012, Article 229 (2012)

Advanced Morphological Approach for Knowledge-Based Engineering (KBE) in Aerospace

A. Bardenhagen, M. Pecheykina and D. Rakov

Abstract The need of innovative solutions for new engineering system structures has brought interest of investigators and developers for systematic innovation. The potential of knowledge-based engineering using the advanced morphological approach (AMA) is discussed in this paper. This approach is evaluated in a case study of new engineering solutions in aerospace. Based on this case study, the proposed approach shows a significant potential in comparison with competing methods. The structural synthesis and parametric optimization programs resulting from this approach can be implemented into Model-Based Systems Engineering (MBSE)

Keywords Knowledge engineering · System theory · Computer-Aided Innovations (CAI) · Conceptual design support · Solutions space · Morphological matrix · Model-Based Systems Engineering (MBSE)

1 Introduction

One of the key stages in the aircraft design process is the conceptual design phase. Conceptual design is considered to be the most difficult engineering design phase, with its success depending to a great extent on the expertise of the designer. Automation and "intellectualization" of some aspects of this phase would be of immense practical benefit [1]. During the conceptual phase, the designer must devise an initial design which (a) incorporates "working principles" or physical solutions for all required "essential" features of the problem and which (b) has been evaluated to be

A. Bardenhagen
Institute of Aeronautic and Astronautic, Berlin Technical University, Berlin, Germany

M. Pecheykina
Moscow Power Engineering Institute, National Research University, Moscow, Russia

D. Rakov (✉)
Institute of Machines Science named after A. A. Blagonravov of the Russian
Academy of Sciences, Moscow, Russia
e-mail: rdl@mail.ru

© Springer Nature Switzerland AG 2020
Z. Hu et al. (eds.), *Advances in Artificial Systems for Medicine and Education II*,
Advances in Intelligent Systems and Computing 902,
https://doi.org/10.1007/978-3-030-12082-5_45

495

Fig. 1 Change in project cost, cost influence, and uncertainty of information during project execution

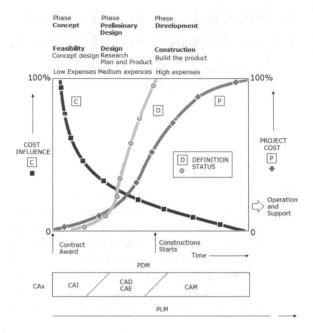

acceptable and feasible [2]. This is the phase of the design process "that makes the greatest demands on the designer, and where there is the most scope for striking improvements and where the most important decisions are taken" [3].

Therefore, conceptual design is the fundamental and indispensable forerunner of the more detailed design phases. It is well known that the right design concept is the key factor influencing the majority of product life-cycle cost and defines the level of product innovation. But an excellent detailed design based upon a poor and inappropriate design concept can never compensate the shortcomings of that concept. Conceptual design is the first and early phase of the design process, involving the generation of solutions, of engineering concepts and of design principles to satisfy the functional requirements for a given design problem. As more than only one solution of a problem exists, improved designs can be identified within the defined design space if the set of potential Engineering Solution (ES) can be enlarged compared to present possibilities [4]. As shown in Fig. 1, the largest information uncertainty exists during the concept phase and then decreases toward the development phase. The accumulated project costs are minimal at the concept stage, but the impact of Engineering Solutions decided during this phase is maximal. Typically, the conceptual design phase absorbs only around 5% but determines around 70% of the total project cost (Fig. 2). Therefore, the conceptual design is the basic phase of design process. Computer-aided innovation (CAI), which can be considered as part of knowledge-based engineering [5], supports identification and evaluation of ES during conceptual design [6].

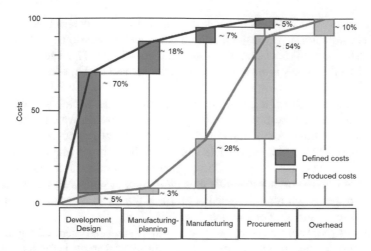

Fig. 2 Dependence of cost, information, and solution corrections at various project stages. *Source* Airbus, design to cost, Hamburg 1990

The more variants of ES are analyzed, the higher are the quality of the study and the confidence to achieve the project requirements and objectives. For this reason, the choice and the consideration of alternative variants is the main task of the design process.

The great uncertainty of information during conceptual design leads to a consideration of "crude" models and multi-variant design solutions, i.e., parallel processing of a number of alternative variants. Such a detailed system and mathematical study need to reproduce the interaction of external and internal factors during the design process. The automation of the process itself creates patterns in the solution space influenced by the defining process characteristics.

2 Structural and Parametric Synthesis

The design of a device (system, process) is a set of two main tasks: the definition of (a) the structure (structural synthesis) and (b) of parameter range for the synthesized structure (parametric synthesis or parametric optimization). The solution strategies for these two tasks are different: Parametric synthesis tasks are usually reduced to the determination of solutions satisfying the metric criteria, making them formally resolved while the task of structural synthesis is absolutely different. The latter cannot be generally allocated to the class of formally solvable problems. The result of structural synthesis is the choice of the rational structure of the object. Structural synthesis requires to work with uncertain structural connections, non-metrical attributes of the structural elements, and quality criteria. The objective function of a structural synthesis does not correspond to the main requirements of usual optimization methods because (1) it is discontinuous or cannot always be determined; (2) it exists in operator notation; (3) it is not based on analytical expressions; and (4) it is not differentiable, not unimodal, not separable, and not additive [7]. The solution of the structural synthesis task is the main and exclusive subject of the researcher's creative activity.

The specifics of structural synthesis tasks consist of the discreteness of variables and the presence of conditionally logical limitations. To this, we will add the need to work with multiple conflicting criteria. The essence of project research consists in the purposeful alteration of characteristic values for variants improving the initial ones. The very notion of "the best" in project tasks is undefined and vague since a number of criteria are not quantifiable and/or conflict with each other. The main difficulty in the search for a designed of an ES is the uncertainty of the results due to incomplete information on evaluation criteria [7, 8].

At present, there are many methods to search and synthesize engineering solutions, including processing Big data in Internet [9] and structural analysis for the realization of scientific and technical ideas [10].

The most common method among the discursive techniques is morphological analysis [11, 12]. By frequency of use, morphological methods are the first among ranks of discursive approaches. Thus, according to statistics compiled in 2009 by German scientists, the total number of companies using the morphology is more than 40%, while regular use is done by more than 20% [13]. Morphological synthesis is regarded as a methodology to streamline the problem to be solved. Morphological analysis is a method developed by Zwicky for exploring all the possible solutions to a multi-dimensional, non-quantified problem complex [14]. Zwicky applied this method to such diverse tasks as the classification of astrophysical objects and the development of jet and rocket propulsion systems. More recently, morphological analysis has been extended and applied by a number of researchers in the USA and Europe in the field of future studies, engineering system analysis, and strategy modeling. The morphological approach serves as a standard when new systems are being designed [15, 16].

3   Advanced Morphological Approach in Aerospace

The disadvantage of morphological methods is the impossibility to analyze all potential variants—as the theoretical number of all possible variants in a morphological array can be enormous (up to hundred thousand or even millions variants). To reduce the order of magnitude (Dimensions) of a morphological array, the advanced morphological approach (AMA) on the cluster analysis basis was developed [17–20].

As an example, the number of missions for unmanned aircraft systems (UAS) systems is legion, especially in the civil field. The mission requirements, as defined by the customer, place demands on the system determining principally shape, size, performance, and cost of the air vehicle itself as well as of the overall UAS operating system. For the air vehicle, some of the most important parameters involved are briefly discussed below [21].

The studied UAS shell has the performance to fulfill the following mission:

- UAS with civil mission (e.g., observation, atmospheric research, and communication node.)
- Flight altitude: 12–20 km (above Jetstream)
- Long flight in the stratosphere (as long as possible on station; ideal: >1 Week)
- No range requirement—defined position behold within area of 4 km^2
- Max. 10 m/s wind during climb—no jetstream—max. 10 m/s wind on position for 4 h/day flight time
- Payload 1 kg, constant 50 W electric energy consumption.

The morphological matrix and the criterion table were created (Table 1). The morphological matrix contains 248,832 potential UAS variants. First, 12,000 variants were generated. 256 variants (about 2% of 12,000) were selected for analysis and grouped into 16 clusters (Figs. 3 and 4). In the solution space, 15 reference variants (Fig. 5) [21] are included.

In the cluster analysis, following four areas were identified to be of interest for further investigations:

1. Aerodynamic configurations with energy storage on board as well as external power supply and with aerodynamic or thrust vector flight control.
2. Investigations of hybrid (lift) UAS.
3. Aerodynamic configurations or helicopter configurations with power supply by cable.

The program "Lane" was used for parametric calculations of different UAS configurations (Fig. 6) [22].

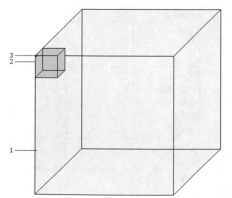

1 - Total number of variants - 248832

2 - Number of variants for generating - 12000

3 - Number of variants for clustering - 256

Fig. 3 Generated and choice variants

Table 1 Morphological matrix

Category	P_X	Attribute (descriptors)	Option P_X^1	Option P_X^2	Option P_X^3	Option P_X^4
Lift	1	Lift	Aerodynamic	Thrust	Aerostatic	
Thrust	2	Thrust	Coupled to lift generation	Independent from lift generation		
Energy storage	3	Internal energy storage	Non	Chemical, reversible (e.g., LiPo battery)	Chemical, irreversible (e.g., fuel tank)	Mechanic (e.g., fly-wheel)
Energy supply	4	External energy supply	Non	Continuous (e.g., solar, microwave)	Interrupted, discontinuous (e.g., tank)	
Power generation	5	Engines	Electric	Internal combustion (e.g., diesel engine)	Gas turbine	Reaction engine (e.g., rocket motor)
	6	Engines	Single engine	Twin engines	More than 2 engines	
Flight control	7	Flight height control	Aerodynamic (e.g., elevators)	Changing of thrust	Aerostatic	
	8	Flight directional control	Aerodynamic (e.g., rudder)	Thrust imbalance (e.g. two engine)		
Fuselage	9	Fuselage	No	One fuselage	Twin-boom	
Geometric characteristics wing	10	Increasing the wing area	No	Yes		
	11	Wing area control	No	Yes (e.g. to maximize solar radiation usage)		
Flight guidance	12	Trajectory	Constant height	Changing height		
	13	Guidance	Remote controlled	Autonomous		

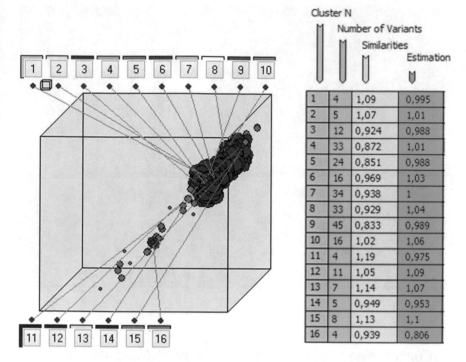

	Cluster N	Number of Variants	Similarities	Estimation
1		4	1,09	0,995
2		5	1,07	1,01
3		12	0,924	0,988
4		33	0,872	1,01
5		24	0,851	0,988
6		16	0,969	1,03
7		34	0,938	1
8		33	0,929	1,04
9		45	0,833	0,989
10		16	1,02	1,06
11		4	1,19	0,975
12		11	1,05	1,09
13		7	1,14	1,07
14		5	0,949	0,953
15		8	1,13	1,1
16		4	0,939	0,806

Fig. 4 Solutions space with 16 clusters

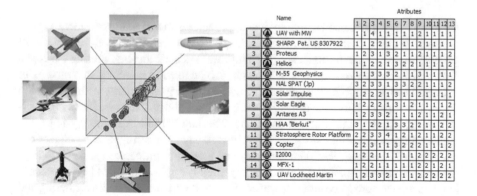

		Name	Atributes												
			1	2	3	4	5	6	7	8	9	10	11	12	13
1		UAV with MW	1	1	4	1	1	1	1	1	2	1	1	1	1
2		SHARP Pat. US 8307922	1	1	2	2	1	1	1	1	2	1	1	1	1
3		Proteus	1	2	3	1	3	2	1	1	2	1	1	1	2
4		Helios	1	1	2	2	1	3	2	2	1	1	1	1	2
5		M-55 Geophysics	1	1	3	3	3	2	1	1	3	1	1	1	1
6		NAL SPAT (Jp)	3	2	3	3	1	3	3	2	1	1	1	1	2
7		Solar Impulse	1	2	2	2	1	3	1	1	2	1	1	1	1
8		Solar Eagle	1	2	2	2	1	3	1	2	1	1	1	1	2
9		Antares A3	1	2	3	3	2	1	1	1	2	1	1	2	1
10		HAA "Berkut"	3	1	2	2	1	3	3	2	2	1	1	2	2
11		Stratosphere Rotor Platform	2	2	3	3	4	1	2	1	2	1	1	2	2
12		Copter	2	2	3	1	1	3	2	2	2	1	1	1	2
13		I2000	1	2	2	1	1	1	1	2	2	2	2	2	2
14		MFX-1	1	2	2	1	1	1	1	1	2	2	1	2	1
15		UAV Lockheed Martin	1	2	3	3	2	1	1	1	2	2	2	2	2

Fig. 5 Reference variants in the solutions space

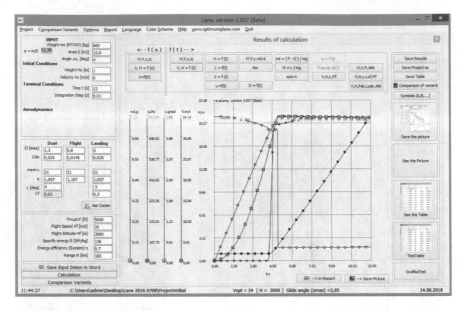

Fig. 6 Screenshot of Lane program

4 Concept Generation Using the Advanced Morphological Approach

Up to now, the advanced morphological approach was used mainly to identify new potential system variants (e.g., flight vehicles) based on high-level variables (lift generation, thrust generation, energy storage, etc.) describing engineering concepts. Taking these optimized variants, AMA can also be used in a second step on a more detailed level to combine the primarily engineering driven variants with production capability and capacity assessments during concept phase. Thus, it allows cost estimation and strategic production decisions. The key behind this strategy is to link engineering design principles (e.g., for stringer design and stringer-skin connection) with materials and generic manufacturing and assembly processes (milling, rolling, riveting, and bonding). Automotive industry already applied this idea [23]. Broadening the scope to aerospace industry and combining it with the AMA as a clustering and efficient selection process will lead to a novel approach to assess production aspects in aerospace during conceptual design within digital CAI and MBSE environment.

Taking the load-carrying structure of a big UAS or an aircraft as an example, the process for e.g. a typical aircraft fuselage shell (a major component assembly) starts with its functional decomposition in two main components: (1) The panel taking longitudinal and internal pressure loads and (2) the circumferential structures, i.e. frames attached to the panel, stiffening the fuselage (global buckling), introducing loads (floor) and taking radial loads (partially internal pressure). This functional decomposition in "panel" and "frame" represents a first-order design principle. In the second-order design principle, the panel can be described e.g., as (a) a monocoque panel, (b) skin with stringer, and (c) a sandwich design.

After this decomposition down to single-part level and their corresponding assembly principles the methodology according to [24] can be used. First, characteristic ranges of properties (length, width, thickness, etc.) are selected for each single part of the load-carrying structure. For a specific single part, based on its properties, combinations of suitable materials and manufacturing processes are identified. To all feasible combinations, the suitable joining and assembly methods are linked. In the second step, these single parts are combined to sub-assemblies or modules taking their compatibility regarding, e.g., materials and joining processes into consideration. This results in known solution patterns that can be manufactured and assembled. In both steps, formalized dependencies help to exclude unrealizable solution patterns. In the third step, these are assembled to sub-assemblies or modules forming component assemblies. Those are combined in the fourth step to major component assemblies. The data taken from them fill morphological boxes for each sub-step. Their data then become sub-assemblies/modules, component assemblies and major component assemblies along the assembly process. Thus, the first step equals the usage of a design catalogue while the following assembly steps are leading to morphological boxes.

The prerequisite for all these combinations is the availability of in databases stored formalized knowledge in the area of typical single parts, manufacturing processes, materials, assembly/joining processes—a typical but huge Knowledge-Based Engineering (KBE) effort. Also the respective compatibilities are stored in the database based on verified criteria. In addition, the database shall also link cost to material manufacturing and assembly variants. As the number of potential variants created as combinations within the four sub-steps described above is tremendous, the AMA methodology is used to identify solution clusters and select the variants fulfilling best given objective functions (e.g., weight and cost).

5   Conclusion

A major objectives of advanced morphological approach are:

- Expansion of the number of potential variants,
- Their clustering,
- Efficient selection for the solution space synthesis in order to increase innovative solutions in Engineering Design.

The technique demonstrates in the case study the power of the approach for design concept generation.

In addition, the proposed approach clarifies and organizes the structuring of the deciding task. The validity of decision making is increased while a multitude of variants, among which the selection is carried out, can be handled. This improves the process quality to select/identify the optimum engineering systems to be developed.

The programs (structural synthesis and parametrical optimization) can be further implemented into Model-Based Systems Engineering (MBSE) [24, 25].

Acknowlegdments This research project was funded by the Deutsche Forschungsgemeinschaft (DFG, German Research Foundation)—407995419.

References

1. Potter, S., Culley, S.J., Darlington, M.J., Chawdhry, P.K.: Automatic conceptual design using experience-derived heuristics. In: Research in Engineering Design, vol. 14, no. 3, pp. 131–144. Springer, London (2003). https://doi.org/10.1007/s00163-003-0034-4
2. Pahl, G., Beitz, W.: Engineering Design—A Systematic Approach, 2nd edn. Springer, New York (1996)
3. French, M.: Conceptual Design for Engineers, 2nd edn. Springer, New York (1985)
4. Chakrabarti, A., Bligh, T.P.: An approach to functional synthesis of solutions in mechanical conceptual design. In: Part I: Introduction and Knowledge Representation Engineering Design Centre. UK Research in Engineering Design. Department of Engineering, University of Cambridge, pp. 127–141 (1994)
5. Kendal, S.L., Creen, M.: An Introduction to Knowledge Engineering, p. 290. Springer, London (2007). ISBN 978-1-84628-475-5. https://doi.org/10.1007/978-1-84628-667-4
6. Werner, D., Weidlich, C., Guenther, R., Blaurock, B., Joerg, E.: Engineers' CAx education—it's not only CAD. Comput. Aided Des. **36**(14), 1439–1450 (2004). https://doi.org/10.1016/j.cad. 2004.02.011
7. Mishin, V.P., Osin, M.I.: Introduction to Aircrafts Design. Mashinostroenie, Moscow (1978). (in Russian)
8. Rakov, D., Sinyev, A.: The structural analysis of new technical systems based on a morphological approach under uncertainty conditions. J. Mach. Manuf. Reliab. **44**(7), 74–81 (2015). https://doi.org/10.3103/S1052618815070110
9. Hashimova, K.K.: The role of big data in internet advertising problem solution. Int. J. Educ. Manage. Eng. **6**(4), 10–19 (2016). https://doi.org/10.5815/ijeme.2016.04.02
10. Aliyev, A.G., Shahverdiyeva, R.O.: Structural analysis of the transformation processes of scientific and technical ideas and knowledge into innovations in technoparks. Int. J. Eng. Manuf. **7**(2), 1–10 (2017). https://doi.org/10.5815/ijem.2017.02.01
11. Zwicky, F.: Discovery, Invention, Research—Through the Morphological Approach. The Macmillan Company, Toronto (1969)
12. Levin, M.: Modular System Design and Evaluation. Springer, New York (2015). https://doi. org/10.1007/978-3-319-09876-0
13. Smerlinski, M., Stephan, M., Gundlach, C.: Innovationsmanagement in hessischen Unternehmen. Eine empirische Untersuchung zur Praxis in klein- und mittelständischen Unternehmen. Discussion Paper on Strategy and Innovation, Marburg, Juni 2009. ISSN 1864-2039
14. Zwicky, F.: Discovery, Invention, Research Through the Morphological Approach. McMillan, New York (1969)
15. VDI 2222: Methodic development of solution principles. Verein Deutscher Ingenieure (1997)
16. VDI 2221: Systematic approach to the development and design of technical systems and products. Verein Deutscher Ingenieure (1993)
17. Hu, Z., Bodyanskiy, Y.V., Tyshchenko, O.K., Samitova, V.O.: Possibilistic fuzzy clustering for categorical data arrays based on frequency prototypes and dissimilarity measures. Int. J. Intell. Syst. Appl. **9**(5), 55–61 (2017). https://doi.org/10.5815/ijisa.2017.05.07
18. Karabutov, N.: Frameworks in problems of structural identification systems. Int. J. Intell. Syst. Appl. **9**(1), 1–19 (2017). https://doi.org/10.5815/ijisa.2017.01.01

19. Fahim, A.: A clustering algorithm based on local density of points. Int. J. Mod. Educ. Comput. Sci. **9**(12), 9–16 (2017). https://doi.org/10.5815/ijmecs.2017.12.02
20. Klimenko, B., Rakov, D.: Analysis and synthesis of innovative engineering solutions and technologies based on advanced morphological approach. In: Hu, Z., Petoukhov, S., He, M. (eds.) Advances in Artificial Systems for Medicine and Education. AIMEE 2017. Advances in Intelligent Systems and Computing, vol. 658, pp. 274–283. Springer, New York (2018). https://doi.org/10.1007/978-3-319-67349-3_26
21. Bardenhagen, A., Gavrilina, L.V., Klimenko, B.M., Pecheykina, M.A., Rakov, D.L., Statnikov, I.N.: A comprehensive approach to the structural synthesis and evaluation of engineering solutions in the design of transportation and technological systems. J. Mach. Manuf. Reliab. **46**, 453–462 (2017). https://doi.org/10.3103/S105261881705003X
22. Rakov, D.: The program LANE for modeling future aircrafts. In: The conference "From Innovation to the technique of the future", 11-th Specialized Exhibition Products and Dual-use Technology, Moscow, Russia, 23–26 Nov 2010 (in Russian)
23. Hasenpusch, J.: Methodik zur Beurteilung eigenschaftsoptimierter Karosseriekonzepte in Mischbauweise. Springer, New York (2018). https://doi.org/10.1007/978-3-658-22227-7
24. Estefan, J.: Survey of Candidate Model-Based Systems Engineering (MBSE) Methodologies, rev. B. Seattle, WA, USA: International Council on Systems Engineering (INCOSE) (2008). INCOSE-TD-2007-003-02. Available at: http://www.omgsysml.org/MBSE_Methodology_Survey_RevB.pdf
25. Eigner, M., Roubanov, D., Zafirov, R.: Modellbasierte virtuelle Produktentwicklung. Springer, Berlin (2014). https://doi.org/10.1007/978-3-662-43816-9

Problems of Intelligent Automation of Unmanned Underground Coal Mining

Andrey M. Valuev and Ludmila P. Volkova

Abstract The article deals with the field of artificial intelligence in two aspects: (1) in the broad sense, the interaction between man and machine, the automation of processes, intellectual automation are considered; (2) specifically it treats neural networks applications, in particular for the recognition of echolocation signals. The possibilities for underground coal mining without the permanent presence of workers in areas of working faces (the so-called technology of unmanned coal mining) are studied, in this respect advantages of plow cutter technology for thin seams being shown as well as needs of its perfection. Main objectives for automation of techno-logical and logistical operations during underground coal excavation are analyzed; in particular, to achieve greater maneuverability of extraction units by controlling their positions on the bottoms of coal seams. The paper also notes that the main of con-trolling positions of machines that result from non-parallelism of the drifts, assume the use of technique of neural networks being proposed as the means of adjusting these positions. It is shown that the dynamism of the operation mode of the electric drive when mining thin seams demands application of tensioning devices for plow chain to reduce dynamic loads in the electromechanical system in question. As for the operation system of unmanned coal mining as a whole, intelligent control with the possibility of creating self-organizing systems is discussed and proposed.

Keywords Intelligent mechatronics and robotics · Intelligent automation · Underground coal mining · Unmanned extraction technology · Plow cutter · Operation unit positioning and stabilization · Neural networks

A. M. Valuev (✉) · L. P. Volkova
Mechanical Engineering Research Institute of the Russian Academy of Sciences,
4, Malyi Kharitonievsky Pereulok, 101990 Moscow, Russia
e-mail: valuev.online@gmail.com

L. P. Volkova
e-mail: volkova_lp@mail.ru

A. M. Valuev · L. P. Volkova
The National University of Science and Technology MISiS, 6, Leninsky Prospekt,
119991 Moscow, Russia

© Springer Nature Switzerland AG 2020
Z. Hu et al. (eds.), *Advances in Artificial Systems for Medicine and Education II*,
Advances in Intelligent Systems and Computing 902,
https://doi.org/10.1007/978-3-030-12082-5_46

507

1 Introduction

In many areas of science and industry, advanced technologies are now widely used, with the dominant role of information technologies, but the coal industry experiences a serious crisis. Nevertheless, the forecasts related to the energy resources of mankind assign a significant role to the intensification of coal extraction in the future by creating not only automated, but also robotic complexes for the underground coal mining. This is especially true for thin seams located at big depths where working conditions for people are particularly severe and the cost of coal mining is still very high. The ideas of the "unmanned excavation," which inspired the creation of promising plans for automation of the coal industry in previous years, were never fully implemented and their implementation is still a challenge for scientists and engineers. The development of information technologies, however, esp., computer-aided design, flexible automated and robotic production facilities and intelligent information technologies gives a chance for automation of technological processes in question.

When creating robotic complexes for coal mining that are sufficiently adapted to the difficult conditions of coal mining, it is necessary to simultaneously use the capabilities of software and hardware. In this case, choosing control options in accordance with the set of information flows provided by the equipment for feedback to the object in real time can be carried out at the expense of the software. In addition, the control algorithms can be modified by implementing various strategies for finding the optimal control mode. The final choice of possible control options, if necessary, in many cases can be implemented in hardware. In this case, the decisive role is played by the ability of the entire system to assess the current situation and to adjust the weights of individual indicators used in the corresponding evaluation function that is implemented in a particular case.

When speaking about the ways of excavating coal without the presence of workers in the active face, the term "unmanned extraction" is usually used, which means a technology with a high level of technology and organization, eliminating time-consuming manual labor in the face, high productivity, and safety of work [1]. And it is customary to distinguish between two groups of methods of unmanned extraction, the first with fastening and the second without fastening of the bottom-hole space. In the first case, the presence of people in the face is allowed only when extraction of coal stops, i.e., during preventive and repair work, installation and dismantling of equipment. In the second case, a person cannot be in the face in terms of technology of work and safety.

The specific interest of the papers is related to conditions of working out steep coal seams, the experience for which is now reflected in the works that are being conducted in China, Ukraine, and Russia. For example, at the mines of China, as reported in [2], there are more than 2000 cleaning faces, and the level of mechanization of the excavation is 41.8% (one- and two-way narrow-cutting coal combines and coal plows). And in China, 40% of the reserves are concentrated in reservoirs with a capacity of less than 2 m.

The main problem discussed here is the stabilization of the operating unit in the workspace by creating complex feedback indicating the properties of a heterogeneous mining environment at the nearest points.

As to measurement techniques and equipment serving the essential part for operation control systems, it is given a great deal of attention, with various solutions [3]. The efficiency of the entire monitoring system proposed in [4], assessed in the aspect of the method of designing adaptive observers and the criteria for structural identification of the system serves for us a recommendation to implement a similar solution. Due to the complexity of the structural identification of such a dynamic object as a plow unit, and the complexity of its design, it is advisable to use this method with modifications caused by the present conditions. In this case, it is important to coordinate the control of related subsystems at a semantic level. Therefore, the approach described in [5] is of interest, as applied to the information channels of the two control subsystems.

Some features of the design and operation modes can be adjusted by analyzing the results of studies of some analogues in other subject areas. For example, it is interesting that the purpose of one study was to develop an algorithm that provides a potential for measuring the quality of soil treatment in real time using image processing [6]. The photos were taken at three levels of chamber height for nine different sizes of soil aggregates. During operation, this allows you to adjust instrument parameters in real time. The development of this method makes it possible to obtain the desired plow design with the lowest possible operating costs. These ideas can be used to control the unit in terms of hypsometry and seam thickness.

Due to above said, two main objectives are considered: the choice of proper blocks of the necessary control system based on existing technical possibilities of obtaining the proper data and to integrate them in a sole system. For the latter, solutions based on neuro-computation are proposed. The most proper means of measurement of the necessary data for the problem in question are discussed too.

2 Opportunities for Creating Systems with "Unmanned Extraction" with the Use of Plow Cutters

Unmanned excavation without fastening of the bottom-hole area is made by drilling and auger installations, combines, vibration equipment, as well as by drilling and blasting method. This type of extraction process may employ plow cutters as well.

Plow cutting is treated, as a rule, as the method of narrow-cut coal extraction, in which the separation of minerals from the massif by working bodies of plow cutters is carried out by thin shavings (0.15–0.3 m thick, 0.4–0.5 m with active cutters).

It is believed that the most effective area of application of the plow extraction, in comparison with that by a combine, is thin (up to 1.5 m) and very thin seams.

Plow extraction may be fulfilled in combination with mechanized supports or as a part of an aggregate unit. Characteristic feature of plowing is the placement of drives of the plow cutter and conveyor either at the ends of the working face or in adjacent mine workings. Conveyor moves after the extraction of coal by the working body, performing the functions of the delivery machine and the base for directional movement of the working body along the face. The existing technology of plowing does not require the presence of a person in the bottom-hole space, as when cutting coal by a combine. Therefore, for mines with thin layers where large coal reserves are concentrated and coal is extracted at great depth, face equipment is often located in a confined space where it is difficult and dangerous for people to find themselves, most of all in the conditions of increased danger of methane explosion, the expediency of plowing is obvious even now. In addition, at present, it remains an urgent problem of coal extraction in an automated mode without the constant presence of a person, which in turn is related to the need to improve the treatment facilities, and to automatize the technological operations in mine faces.

Plow cutters operate in reverse cyclic mode with frequent starts and load surges. The presence of extended elastic bonds (i.e., branches of the plow chain) in the electromechanical system of the plowing units creates the conditions for the occurrence of oscillations, which substantially increase the dynamic loads of the elements of the system and accelerate their wear and tear, and also reduce the durability and reliability of the entire installation as a whole. The amplitudes of the vibration with deformations of the elastic elements can reach dangerous values, which is accompanied by cutting off the safety pins or the rupture of the plow chain leading to forced downtime of the extracting equipment to replace them [7]. To study the dynamics of the functioning of such a complex technical object, mathematical models can be used that reduce the cost of experimental verification of the adopted design decisions.

3 Technical and Informational Problems of Design of Production Systems with "Unmanned Extraction"

The useful time for excavating coal by plowing units that can be used in thin layers is usually no more than 40–60 min per shift. The rest of the time is spent on repair and preventive works, preparatory operations for tensioning chains of plows and conveyors, setting safety elements. Certain prospects in these conditions were associated with the introduction of a regulated automated drive of plowing machines to reduce dynamic loads in the electromechanical system [7]. Plow cutter installation has a fairly simple design of the executive body. The problems of creating automated design systems for robotic complexes for the extraction of coal are associated with the use of modern energy-saving and labor-saving technologies.

The problems of parallelization of search operations that are common for many modern computer systems may be especially acute in the context of the problem in question due to the complexity of the operating conditions of coal extracting machines. Reducing the cost of computing for interprocessor exchange is achieved, as in many cases, by differentiating the search levels by implementing parallel algorithms for selecting the optimal control mode. Control algorithms can be developed at the information level [8].

When the plow moves on the end sections of the longwall where the stiffness coefficient of the working branch of the plow chain cannot be considered as constant, since it acquires a clearly displayed nonlinear character. In this case, dynamic loads in the electromechanical system of plowing units are significantly increased. As noted in [9], the elaborated complex structure of the electric drive control system in this case ceases to be effective without additional measures, such as the use of automatic tensioning devices, which in itself is problematic. An important role in the control of automated excavating systems should be played by tracking the formation hypsometry. In the synthesis of the optimal control structure, it is also necessary to take into account the random nature of the change in the load on the cutters of the plow.

We must emphasize the great advantages of coal mining technology with the use of plow units compared with coal mining by the combine for conditions of thin seams. At the same time, for the automation of technological operations for such conditions it is necessary to improve the treatment complexes with the intelligent control. In this case, the greatest maneuverability of the unit in the plane of the formation is achieved when controlling the position of the base of the unit. However, problems arise in controlling the unit when it moves in the formation plane, which are related to the non-parallelism of the drifts. To adjust the position of the base of the unit in the face space in accordance with the non-parallelism of the drifts, it is proposed to use neural networks.

The theory of neural networks involves questions of control theory and parallel computing algorithms. The practice of using neural networks shows that they are most effective in those areas where the formalization of the computational process is impossible. The process of solving the problem is represented as the functioning in time of a certain dynamic system that accepts a vector of input data, which is formed as a result of summing the input signal of the neurocontroller and corrective feedback on the output signal. Corrective feedback, in turn, is formed as two signals: one signal directly from the neuro-controller output, as well as an additional feedback signal, which affects on the input of the neuro-controller across the algorithm of backpropagation. Schematically, it can be presented in Fig. 1.

To construct such a system, it is necessary to define objects that act as inputs (the input data element, the initial value of the determined quantities, etc.) and outputs (the solution itself or its characteristics) of the neural network signals. In addition, it is necessary to determine the desired output signal, and then the network structure: the number of layers, the connections between the layers, and the weights on the inputs of the neurons of the input and hidden layers.

Further actions include the definition of the error function and the target function which depends on the error. As it is known, one of the advantages of neural networks is their ability to generate a nonlinear model of the process based on the learning outcomes of the network. Thus, the neural network is an adaptive system whose life cycle consists of two independent phases—the phase of learning the network and the phase of the network. Training is considered complete when the network

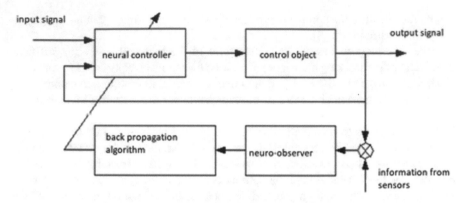

Fig. 1 General scheme of control with the aid of neural controllers

correctly performs the transformation on test cases and further training does not cause a significant change in the customizable weighting factors. However, in this case it is crucial that the system is highly dynamic. In addition, the control must be carried out in real time. So the issues of functional stability come to the forefront. In this connection, methods that are used to provide the functional stability of a system operating in real time, given in [10], are of interest.

The problem exists to scrutinize operating modes of mining machines operating in difficult mining and geological conditions, as well as in the field of self-adjusting control systems for plow aggregates and installations. This task becomes even more urgent for the development of a complex aggregate of systems subject to changing external influences. Further research should be directed to solving the actual problem of automation of the underground coal mining process, especially from mid-latitude and thin seams with a large depth of occurrence.

Investigations of operating modes of mining machines operating in difficult mining and geological conditions, as well as analysis of developments in the field of self-adjusting control systems, allow us to conclude that it is necessary to search for new ways to implement optimal control regimes for plow aggregates and plants. This task becomes even more active when considering the prospects for the development of complexes for unmanned coal mining, the management principles of which can be constructed using constantly changing external factors affecting the control system, and also taking into account the internal dynamics of the system that is reflection of external influences.

The most serious problems in the operation of the plow unit are related to the need for efficient operation in conditions of non-parallel drifts. In this regard, the actual task is to visualize the process of passing the unit along the drifts as the slaughter line moves to control the "bend" of the unit in order to track the non-parallelism of the drifts. As for the possibility of controlling the unit, the problem is connected with achieving the greatest maneuverability of the unit in the formation plane. This is achieved by controlling the deflection of the base of the unit [11]. In the light of the creation of conditions for the extraction of coal without the constant presence of people in the slaughter area, the intelligent management of the purification unit becomes reasonable.

When implementing the unmanned extraction of coal, the problem arises to automate the movement control of the plow complex or aggregate for continuous coal extraction (in the working interval). The experimental sample of the frontal aggregate was developed with the participation of specialists from the Moscow State Mining University (chair of mining machines and equipment) and was tested in industrial conditions at the Yubileynaya mine.

During the research, the possibilities of using informational modeling methods arise, which make it possible to replace expensive field experiments with computational ones. Modern information technologies also open significant opportunities in the development of applied science, where the use of artificial intelligence methods is especially promising [12].

When testing the unit, it was moved with a mechanized control of the unit's base in order to increase its maneuverability when moving to another slaughter. But when coal was cut out, it was necessary to stop the unit to adjust the length of the base of the unit and adjust it to the length of the lava, which changed due to the non-parallelism of the drifts. The problem of non-parallelism of drifts is associated with different rock strengths when drifting drifts, limited by the capabilities of tunneling equipment [13].

To solve the problems of unmanned coal mining in the project, the problems of controlling the unit can be considered, in which two subsystems can be distinguished: For the first of them, the electric actuator of the plowing executive body acts as the control object, and for the second it is the hydraulic drive of the support. The questions of intellectualization of management, considered in this work, are connected with the management of the unit base by the effect on the hydraulic drive of the support in accordance with the scheme (Fig. 2).

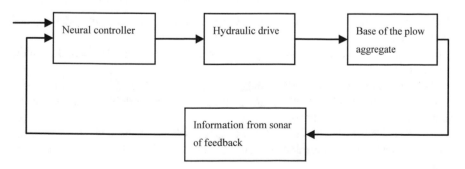

Fig. 2 Scheme of controlling the position of the base of aggregate

4 Design of the Control System for the Movement of the Unit

In view of the previously described operating conditions of the plow aggregate, as well as the above control scheme, it has been proposed as a variant of the solution to design a system for analyzing video images from sensors that are located on the extreme sections of the base of aggregate. Based on the analysis of information from sensors, a decision is made to automatically adjust the position of the unit relative to the drift. And as sonar sensors for monitoring the location of the drifts, sonars are chosen. In this case, the effect of echolocation based on the analysis of the frequency of the signal reflected from the monitored surface is used. Sonar based on the received echo signal forms a video image, which, in the future, will be available for the software.

To organize the control process as a whole and to adjust the position of the device base, it is necessary to formalize the description of the business processes that occur during the operation of the plow unit, when the drift is not parallel. The process of designing the block motion control system was carried out using the IDEF0 representation.

The following blocks are designed to decompose the process at high level:

A1—definition of the new position of the base of the unit—with the help of sonars and planned software, the current position of the base of the unit relative to the drifts is assessed;
A2—alignment of the position of the unit relative to the drifts—the software, using the system of directed movement of the unit in the plane of the formation, changes the position of the base of the unit to correct the angle of the assembly and the drift;
A3—continued production—the plow aggregate continues its work, without the necessary long stops.

In turn, the process block A1 is represented with the following actions:

A11—signal feeding—the software starts the sonar, which in turn generates a signal toward the rock;
A12—receiving the response echo-signal—sonar receives a signal reflected from the surface of the rock;
A13—the formation of the video image—sonar on the basis of the received echo signal forms a video image, which, in the future, will be available for the software, as well as for an outside observer;
A14—interpretation of the received video image—the software with the help of the received video signal determines the current position of the drift in relation to the base of the unit.

Sonars in this model perform functions: generation/reception of audio signals, as well as the formation of a video image, which will be further processed by the software to obtain an estimate of the current position of the drifts relative to the base of the unit. Incorrect position estimation of the unit—occurs when a software error occurs during the determination of a new location of the unit base. This may be due to the structure and parameters of the hydraulic drive control circuits and the drive of the plow aggregate.

The dynamism of the operation modes of the plow aggregate leads to a decrease in the reliability of the equipment. To increase the strength, durability, and reliability of structural elements by protecting them from dynamic overloads, attempts are currently being made, which are mainly reduced to the improvement of mechanical structures [14]. In addition, the frequency of failures in the mining process is analyzed depending on the type of mining machine used for mining operations, both plow installations and plow units. At the same time, measures are proposed to increase the efficiency of coal mining. To reduce the costs of failures, maintenance teams are prescribed to regularly monitor the use and operation of machines in a rational and efficient manner. It is this kind of activity that will reduce downtime and, as a consequence, increase the efficiency of the mining enterprise [15].

5 Conclusions

The paper presents some solutions in the field of creating the info-technical system of unmanned underground extraction of coal. They demonstrate more advantages from the perspective of robotization, intellectualization, and the organization in the future coal mining without the presence of people for the plow aggregates. At present, however, achievement in the research of dynamic regimes of plow aggregates and installations stay insufficient. We hope that the progress in this area coupled with advances of CAD and CAM will give sufficient progress in underground robotization relatively soon.

References

1. Mining encyclopedia. http://www.mining-enc.ru/b/bezlyudnaya-vyemka/. (In Russian)
2. Remezov, A.V., Anufriev, A.V.: Foreign experience of application of technological schemes for mining sloping and steeply inclined coal seams in mines [Electronic resource]. http://science.kuzstu.ru/wpcontent/Events/Conference/Other/2015/gd/gd2015/pages/Articles/3/2.pdf. (In Russian)
3. Hu, Q., Zhang, D., Liu, W.: Precise positioning of moving objects in coal face: challenges and solutions. Int. J. Digit. Content Technol. Appl. 7(1), 213 (2013)
4. Karabutov, N.: Structural-parametrical design method of adaptive observers for nonlinear systems. Int. J. Intell. Syst. Appl. (IJISA) 10(2), 1–16 (2018). https://doi.org/10.5815/ijisa.2018.02.01
5. Bildstein, A., Feng, J.: A channel theory based 2-step approach to semantic alignment in a complex environment. Int. J. Mod. Educ. Comput. Sci. (IJMECS) 9(9), 1–12 (2017). https://doi.org/10.5815/ijmecs.2017.09.01
6. Ajdadi, F.R., Gilandeh, Y.A., Mollazade, K., Hasanzadeh, R.P.: Application of machine vision for classification of soil aggregate size. Soil Tillage Res. 162(1), 8–17 (2016)

7. Volkova, L.P.: Investigation of the dynamics of plowing units on the model under conditions of variable rigidity of the working branch of the chain. Min. Inform. Anal. Bull. (Sci. Tech. J.). (OB6), 552–566 (2011). (In Russian)
8. Haken, H.: Information and Self-organization. A Macroscopic Approach to Complex Systems, Second Enlarged edn. Springer-Verlag, Berlin (2000)
9. Volkova, L.P.: On application of intellectual technologies in control systems of the electric drive of a plow. Sci. J. "Inform. Math." (1), 12–19 (2005). (In Russian)
10. Tolubko, V., Vyshnivskyi, V., Mukhin, V., Haidur, H., Dovzhenko, N., Ilin, O., Vasylenko, V.: Method for determination of cyber threats based on machine learning for real-time information system. Int. J. Intell. Syst. Appl. (IJISA) **10**, 8 (2018). https://doi.org/10.5815/ijisa.2018.08.02
11. Pastoev, I.L.: Structure and functions of the movement system of the purification unit along the mineral strata. Proc. High Sch. Min. J. **11**, 23–28 (1985). (In Russian)
12. Volkova, L.P., Pankrushin, P.Yu.: Features of the control of the plow aggregate in conditions of non-parallelism of drifts. Min. Inform. Anal. Bull. (Sci. Tech. J.). (6) (2013). (In Russian)
13. Kantovich, L.I., Pastoev, I.L.: The problem of controllability of automated aggregates and complexes when working on gently sloping layers without the presence of people in the face. Min. Inform. Anal. Bull. (Sci. Tech. J.). (OV1), 410–420, (2010). (In Russian)
14. Plow system sets new low-seam coal production record (2014). http://www.coalage.com/news/product-news/3797-plow-system-sets-new-low-seam-coal-production-record.html#. VkiijF70xdg
15. Biały, W.: Application of quality management tools for evaluating the failure frequency of cutter-loader and plough mining systems. Arch. Min. Sci. **62**(2), 243–252 (2017)

Structural and Parametric Control of a Signalized Intersection with Real-Time "Education" of Drivers

Anatoliy A. Solovyev and Andrey M. Valuev

Abstract The article is devoted to the problem of intelligent traffic control at inter-sections of road networks of large cities. The problem of optimization of a signalized intersection is set that is based on joint choice of the phase separation of passage directions and duration of traffic light cycle phases. The detailed representation of a crossroads is introduced, and formulas for evaluating structural safety for an arbitrary scheme of its passage are introduced. On their basis, the combinatorial problem of structural optimization of a traffic light cycle is set that enables the establishment of most safe phase-separation schemes. The similar optimization problem of phases' duration choice according to established traffic demand is set as well. An approach to the solution of the latter on the basis of detailed simulation of traffic through an intersection is offered. The problem of "education" of drivers that provides the efficiency of the proposed control system is discussed.

Keywords Traffic flow · Crossroads · Signalized intersection · Phase separation · Safety · Intensity · Intelligent control system · Computational experiments · Driver's education

1 Introduction

The paper is aimed at evolving the concept an intelligent traffic management system and focuses on regulated intersections. This particular interest is due to the crucial role

A. A. Solovyev · A. M. Valuev (✉)
Mechanical Engineering Research Institute of the Russian Academy of Sciences, 4, Malyi Kharitonievsky Pereulok, 101990 Moscow, Russia
e-mail: valuev.online@gmail.com

A. A. Solovyev
e-mail: aa.solovjev@yandex.ru

A. A. Solovyev
Moscow Institute of Physics and Technology (State University), 9, Institutskiy per, 141700 Dolgoprudny, Moscow, Russia

© Springer Nature Switzerland AG 2020   517
Z. Hu et al. (eds.), *Advances in Artificial Systems for Medicine and Education II*,
Advances in Intelligent Systems and Computing 902,
https://doi.org/10.1007/978-3-030-12082-5_47

they play in both traffic safety and the throughput of urban road networks (URNs). Unlike the existing approaches to solving this problem, the proposed approach is based on the simultaneous choice of the optimal intersection structure from the set of permissible configurations and its parameters, with joint consideration of the intensity and traffic safety at the intersection. The configuration of a crossroads is treated as the set of possible directions of motion on it; for signalized intersections, these directions are not in action simultaneously but shared between phases of the traffic light cycle.

Usually, the security problem is treated separately not taking it into account in the traffic control optimization [1–4]. At present, the choice of duration of the traffic light cycle phases due to distribution of traffic demand between possible directions is most traditional, although new methods have been proposed for their solution, the Artificial Bee Colony method [5] among them. With [5] technique, the developed model optimally schedules green light timing in accordance with traffic condition in order to minimize the average waiting time at the cross intersection, however, not taking in account traffic safety. The proposed structural control consists in altering schemes of passage at some phases (and, perhaps, or exclusion of existing ones) within the given set of motion directions, giving preference, if possible, to more safe schemes. In addition, the paper discusses the need to teach drivers the proper behavior in the system of intelligent traffic management with such features.

Section 2 presents a formal description of the regulated intersection. Section 3 introduces the structural safety index of a regulated intersection. A combinatorial optimization problem of structural safety on a set of admissible intersection configurations on a control loop is formulated. A practical example of a crossroads is considered as well. Section 4 is devoted to the joint optimization of safety and traffic intensity. An algorithm for solving the problem is given. Section 5 discusses the importance of driver's behavior training adapted to reception of real-time traffic information.

2 Formal Description of a Signalized Intersection and Traffic Organization on It Affecting Its Throughput and Safety

An intersection (a crossroads), both signalized and non-signalized, must be treated as an area of the road surface divided into a set of short directed lanes. Practically, each crossroads gives drivers the possibility to change the direction of their motion. Separation of directions on a signalized crossroads by phases of the traffic light phases excludes simultaneous motion along intersecting lanes. On the other hand, to provide the possibility of changing directions, some of lanes must either merge or branch; places of lanes' merging or branching are treated as borders of corresponding lanes, whereas in the consideration of a signalized crossroads' places of lanes' intersection are ignored.

Legalized motion of a vehicle through an intersection takes place within lanes, and change of a lane is admissible only at its end as it was determined above. Motion of a single vehicle may be represented with the trajectory of its characteristic point, namely the middle of its front bumper. Further, we consider the vehicle motion, namely as the motion of this point and treat it as the motion along axes of subsequently passed lanes. So, a signalized intersection is structurally and geometrically characterized with a set of these axes that begin or end either on the intersection border or in its *singular points* (SPs) of "merging"- or "separation"-type [6]. Entering lanes at a signalized intersection begin from stop-line forming a part of the intersection border; for exiting lanes, their ends have conditional definition, due to which, however, all existing SP must be located within the intersection border. The introduced formal description of the signalized crossroads is illustrated with Fig. 1.

Traffic organization at a regulated intersection, i.e., "a complex of legal and techno-organizational measures and administrative actions on traffic management" on it [7] consists in the use of technical means such as road markings and traffic lights.

With road markings, the set of lanes in the intersection area and their borders are determined and shown. In this paper, the system of lanes and their markings are assumed to be given. The first problem in question is the choice of the structure of the traffic light cycle (TLC), i.e., a set of its phases and corresponding schemes of phase separation [7] (the latter determining the sets of movement directions at the intersection for all phases, which together cover the entire set of permitted directions,

Fig. 1 Entire set of possible directions of motion at the intersection of Profsoyuznaya and Obrucheva streets. ∇ denotes SP of separation and △—SP of merging

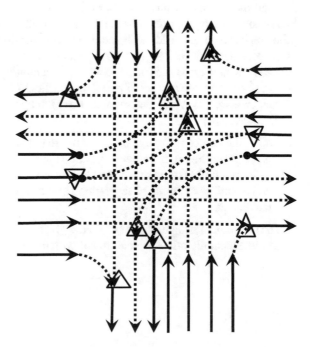

detailed to the road lanes). Parameters of traffic light regulation are the durations of
TLC phases. Effective and safe use of a controlled intersection assumes optimization
of the structure and parameters of traffic light regulation under traffic safety restrictions. First, we assess structural safety of possible phase separation of directions
that gives possibility to recommend or not to recommend certain schemes of phase
separation.

Traffic organization restricts the freedom of movement of vehicles, since any
vehicle must move along the axis of the current road lane. The potential danger to
the vehicle motion through a singular point depends on the type of the SP. The point
of separation requires attention from the driver for the timely start of maneuver. The
potential danger consists in getting a blow from a preceding or succeeding car, as
when driving along the road lane. The passage of a point of confluence results the
danger of a sliding lateral collision or a collision between the front and rear bumpers
of neighboring cars. The competition of vehicles for the passage of a "crossing"-
type SP is threatened by direct collision, accompanied by the most severe emergency
consequences. The connection of the accident with the passage of an SP of a different
type can be traced on specific examples in [2–4].

3 Structural Optimization of the Traffic Light Cycle

The purpose of the controlled intersection is to exclude SPs of the "intersection"-type
from all the scheme of phase separation of paths, and to minimize the number of
SPs of the other types, primarily the merging points. For a formal statement of the
problem, it is required to define the safety index of the flows passing through a section
of the road network with a given travel pattern. Based on the qualitative difference
between the three types of SPs, the following quantitative indicators can be proposed.
For crossroads treated as a set of possible directions of traffic and a specific scheme
of their passage (as it was determined above), the following structural parameters
are introduced [6]:

(1) n_m is the number of merge points of the road lanes for the scheme, and $n_{m\,max}$
is for the entire set of possible directions of motion;
(2) n_s is the number of points of road lanes separation for the scheme, and $n_{s\,max}$
is the number of points of separation for the entire set of possible directions of
movement.

The indicator of structural safety S for the given controlled crossroads with definite
set of movement directions is determined with the relationship

$$S = \alpha_1 \times \alpha_2 \tag{1}$$

where

$$\alpha_1 = \sqrt[4]{1 - n_m/(n_{m}\,max + 1)}, \tag{2}$$

$$\alpha_2 = \sqrt{1 - n_s/(n_{s\,max} + 1)}. \tag{3}$$

The introduced formulas, based on the ranking of SP in terms of their degree of danger, allow us to compare the structural safety of various schemes of phase separation. To compare the possible structures of the traffic light cycle, the sum of the indices S divided by the number of phases is proposed as an estimate.

The set Φ of possible directions of motion at the intersection is a finite set, $\Phi = \{\Phi_j\}_{j=\overline{1,n}}$. Numbering TLC structures (TLCSs) in question by l, denoting by K_l, the number of phases in the lth TLCS, and by K_{max}, the maximum allowable number of phases of a TLC, and using notation $\Phi_i^l \subseteq \Phi$ for the set of directions admitted by the phase-separation scheme for the i-th phase of the l-th structure, S_i^l for the index (1) for it, we represent the problem of structural optimization of the traffic light cycle in the form

$$\frac{\sum_{i=1}^{K_l} S_i^l}{K_l} \to \max, \tag{4}$$

$$K_l \le K_{max}, \tag{5}$$

$$n^*(\Phi_i^l) = 0, \tag{6}$$

$$\bigcup_{i=1}^{K_l} \Phi_i^l = \Phi. \tag{7}$$

Here the expression in (6) denotes the number of SPs of intersection for the i-th phase. This problem is a combinatorial optimization problem. With a reasonable value of $K_{max} \le 4$ corresponding to the usual practice of traffic signaling at typical intersection, the number of options in it, in view of condition (7), does not actually depend on the number of lanes on one road. Indeed, in this case, in order to avoid exceeding the number K_{max}, all the lanes of the same road intersecting with the other road should be included in one scheme of phase separation, as it is in practice. After receiving the solution of the problem (4)–(7), a set of suboptimal phase-separation schemes must be found. The letter set contains schemes that may serve as the most adequate when the optimal scheme.

Figure 2 shows possible sets of directions for the first phase on the crossroads shown in Fig. 1. In this and next figures below admitted directions are plotted with solid lines and prohibited ones with dotted lines. The values of the indicator S for them are 1 and 0.882, respectively. Two other phases are shown in Fig. 3.

It may be shown that the scheme of phase separation determined with sets of directions shown in Figs. 2a, 3a, and 3b is structurally optimal for the case when $K_{max} = 3$. The value of the indicator (4) is 0.936, and for the scheme shown in Figs. 2b, 3a, and 3b, it is equal to 0.896; so this modification may be treated as a suboptimal scheme.

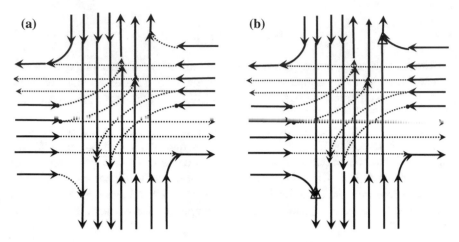

Fig. 2 Two variants of the set of directions for the first phase

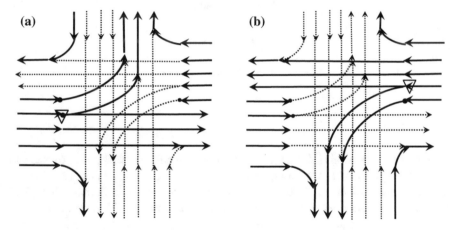

Fig. 3 Sets of directions for the second and the third phases

To obtain the optimal scheme with the uttermost safety indicator value 1, we must forbid direct motion in both directions by "horizontal" lanes that are second from left. It may be done without change of road markings by installation the separate traffic lights for the first and the second lanes in both horizontal directions. In this case, forbidding direct motion on any of these lanes that are second from left may be non-permanent and act only when the desired intensity of left turns is relatively great. To implement this organizational solution, it is necessary, however, to inform drivers about temporary prohibition of direct motion on this (these) lanes. This requirement is only a particular case of more general one: to change the structure of the traffic light cycle on a certain crossroads for the sake of better satisfaction of the traffic demand, it is necessary in many cases inform about this change in advance all drivers potentially concerned with it.

4 Traffic Light Optimization in Relation with Traffic Intensity

More precisely, the optimization of the TLC structure must be carried out taking into account the intensity of the flows through the intersection. For the lth TLCS, the security of the intersection for each vehicle provided that the direction of its passage through the intersection takes place at the ith phase, in view of the definitions introduced, is estimated by the quantity S_i^l. Consequently, for a certain intensity of the passage of the aggregate flow Q_i^l on the ith phase, the total safety index is the sum of $S_i^l \cdot Q_i^l$ (in fact, this formula is not universal but it is correct when traffic flows for all directions are known definite values). However, for the comparability of the values of the total safety indicators for different TLCSs, it is necessary to relate indicators $S_i^l \cdot Q_i^l$, which differ in their number and values, with the traffic intensity through the intersection, independent of the configuration.

More accurately, we must take into account traffic demand for all directions admitted by the crossroads. It may be established only empirically, but with modern technical means of traffic monitoring it is not a hard task. In the case when each admissible direction (to go straight, to turn left or right) corresponds to the separate lane(s) on each entrance to the intersection, the number of drivers choosing a definite direction for a traffic cycle duration may be approximately established in such a way. If the length of queue on the corresponding lane stays constant from one cycle to the next one, then this number is an average number of vehicles passing the direction for a cycle. In the other case, it is the sum of the average number of passed vehicles and the average increment of the number of vehicles in the queue on the corresponding lane(s).

For known traffic demand differentiated for all possible directions, we principally have the possibility for optimization of TLC concerning both its structure and parameters. Let the definite scheme of phase separation is chosen. Let Q_j be the total traffic demand for the direction Φ_j, $\Phi_j \in \Phi_i^l$. Let T be the duration of TLC for the intersection in question; we assume that this value is determined independent of TLCS. Then for the time T, the queue of $L_j = Q_j T$ vehicles for the jth direction forms for one cycle at the entry of the intersection. For the chosen lth TLCS, we may evaluate the time T_i^l for which these queues vanish for all directions Φ_j, $\Phi_j \in \Phi_i^l$. This evaluation may be performed on the basis of traffic modeling, either by numerical method, or, more likely for many cases, by computer simulation.

There are many approaches to traffic modeling [8]. For crossroads, either mesoscopic [9] or microscopic models [10–12] may be applied. For our aim, microscopic "leader-following" models [12–14] seem to be the most adequate.

The desired simulation, as it is demonstrated in [13] for a non-controlled crossroads, may be efficiently fulfilled. It must be emphasized that the above-proposed minimization of conflicts and therefore minimization of SPs number leads to splitting of the entire traffic flow on each phase into a set of simple independent traffic flows concerning, mostly, one or two directions. For example, for the Phase 1 shown in Fig. 1, we have in both variants eight independent flows, six of which embrace the only direction, and the rest two ones either one or two directions. For the second and the third phases, we have eight independent flows as well, six of which embrace the only direction. So, modeling of such flows is much simpler and less time-consuming than for typical non-controlled crossroads and may be performed in real time even by simulation. Moreover, since the needed evaluation has only one or two parameters L_j, the computation for known crossroads may be fulfilled in advance and tabulated, and then valuations for the actual data may be obtained by interpolation.

When valuations of T_i^l are obtained for current situation, the conclusion may be made whether the lth TLCS may guarantee that the existing traffic demand may be satisfied with it. The condition for that is the following:

$$\sum_{i=1}^{K_l} T_i^l \leq T. \tag{8}$$

When there are more than one suboptimal TLSC, then one of them is chosen that maximizes

$$J_l(Q) = \sum_{i=1}^{K_l} S_i^l \sum_{\Phi_j \in \Phi_i^l} Q_j. \tag{9}$$

If condition (8) may be satisfied with no TLCS, then the TLCS must be chosen that minimizes its violation.

Our calculations show cases in which one of the two variants shown in Fig. 2 is preferable.

5 The Necessity for Real-Time "Education" of Drivers to Provide the Possibility of the Intelligent Traffic Light Control

There is the principal obstacle for the implementation of the proposed control system. It consists in insufficiency and low quality of information available for drivers. Both printed city maps for drivers and the Internet resources, such as interactive online Google maps and Yandex maps give only the principal organization of passage of intersections, namely whether it is possible to turn right or left or to change the direction to the opposite on the road axis near the certain intersection. To realize the movement in the desired direction on the intersection, the driver must reach in time the lane admitting this direction. If the use of some entry lanes may change in time as it was proposed switching from the passage to left only, simultaneously strait and to left or strait only as it was discussed for Fig. 3, makes the task of the lane choice for a driver more difficult.

The solution of the problem is the implementation of Internet-based online navigational tools that show the present scheme of crossroads passage. These means are treated by us as the way of real-time "education" of drivers. But in advance, drivers must be principally educated about the necessity of use of these Internet services. Much more important these means must be for automatic vehicles; for them, the choice of lanes must be based on data transfer from these services. For multi-lane roads, recommendations may be transferred for registered users that provide more uniform distribution of vehicles between lanes that provide better use of URN resources.

Besides, they should be supplemented by information and communication technologies (ICT) that propose the recommendations as to the choice of route plans first proposed for public transport travels [15]. The paper [16] demonstrates the possibility to avoid gridlocks heavy for cargo traffic nearby great ports. To get more reliable results, it must develop simulation techniques altogether with technologies of detection [17].

6 Conclusions

The combinatorial optimization problem of the structural safety of a regulated signalized intersection is formulated on the basis of the introduced safety parameters for the passage of singular points. An example of a crossroads of two Moscow avenues is considered. The formalization of the problem of optimizing the parameters of the traffic signal regulation taking into account the traffic flow intensities on the permissible directions of motion is given as well.

The practical solution of the problem involves real-time monitoring of the traffic situation at the intersection in order to obtain actual information on the length of the queues.

At the same time, effective intelligent traffic management is impossible without solving the problem of driver training in behavioral skills and decision making in conditions of increased information requirements. Providing the education of drivers in this sense, the traffic control system will be able to use operational changes in the organization of traffic, including partial changes of intersection passage schemes, thereby increasing the efficiency of using the resources of the road network, but not causing conflicts.

References

1. Tarko, A.P.: Use of crash surrogates and exceedance statistics to estimate road safety. Accid. Anal. Prev. **45**(1), 230–240 (2012)
2. Rifaat, S.M., Tay, R., De Barros, A.: Effect of street pattern on the severity of crashes involving vulnerable road users. Accid. Anal. Prev. **43**(1), 276–283 (2011)

3. Laureshyn, A., Svensson, Å., Hydén, C.: Evaluation of traffic safety, based on micro-level behavioural data: theoretical framework and first implementation. Accid. Anal. Prev. **42**(6), 1637–1646 (2010)
4. Shabalin, I.V., Ertman, S.A., Ertman, Y.A.: Assessment of the transport danger of a crossroads as a basis for analyzing the effectiveness of the scheme for organizing road traffic, in Problems of the functioning of transport systems. In: Proceedings of the All-Russian Scientific and Practical Conference of Students, Postgraduates and Young Scientists (with international participation), Tyumen Industrial University, Tyumen, pp. 360–373 (2014) (in Russian)
5. Adebiyi, R.F.O., Abubilal, K.A., Tekanyi, A.M.S., Adebiyi, B.H.: Management of vehicular traffic system using artificial bee colony algorithm. Int. J. Image Graph. Signal Process. **9**(11), 18–28 (2017). https://doi.org/10.5815/ijigsp.2017.11.03
6. Solovyev, A.A., Valuev, A.M.: On the structural complexity and estimation of the crossroad throughput, in control of the development of large-scale systems (MLSD'2016). In: Vassiliev, S.N., Tsvirkun, A.D. (eds.) Materials of the Ninth International Conference, vol. 2, pp. 98–101. ICS RAS, Moscow (2016) (in Russian)
7. Methodical recommendations on the design of traffic lights on highways, p. 5. Federal Road Agency (Rosavtodor), Moscow (2013) (in Russian)
8. Treiber, M., Kesting, A.: Traffic Flow Dynamics: Data, Models and Simulation. Springer, Berlin-Heidelberg (2013)
9. Di Gangi, M., Cantarella, G.E., Pace, R.D., Memoli, S.: Network traffic control based on a mesoscopic dynamic flow model. Transp. Res. Part C Emerg. Technol. **66**, 3–26 (2016)
10. Derai, S., Ghoul, R.H.: Control isolated intersections with hybrid petri nets and hybrid automaton. EEA—Electroteh. Electron. Automat. **65**(3), 112–116 (2017)
11. Babicheva, T.S.: The use of queuing theory at research and optimization of traffic on the signal-controlled road intersections. Procedia Comput. Sci. **55**, 469–478 (2015)
12. Glukharev, K.K., Ulyukov, N.M., Valuev, A.M., Kalinin, I.N.: On traffic flow on the arterial network model. In: Kozlov, V.V., et al. (eds.) Traffic and Granular Flow'11, pp. 399–412. Springer, Berlin-Heidelberg (2013)
13. Kalinin, I.N., Glukharev, K.K.: The study of integral characteristics of crossroads using microscopic models of traffic flows. Komp'yuternye issledovaniya i modelirovanie [Computer Research and Simulation], vol. 6, no. 4, pp. 523–534 (2014) (in Russian)
14. Valuev, A.M., Solovyev, A.A.: Modeling of dependencies characterizing the dynamics of traffic flows. Informatizatsiya i svyaz' [Informatization and Communication], no. 2, pp. 106–113 (2018) (in Russian)
15. Zhou, C., Weng, Z., Chen, X., Zhizhe, S.: Integrated traffic information service system for public travel based on smart phones applications: a case in China. Int. J. Intell. Syst. Appl. **5**(12), 72–80 (2013). https://doi.org/10.5815/ijisa.2013.12.06
16. Efiong, J.E.: Mobile device-based cargo gridlocks management framework for urban areas in Nigeria. Int. J. Educ. Manage. Eng. **7**(6), 14–23 (2017). https://doi.org/10.5815/ijeme.2017.06.02
17. Goyal, K., Kaur, D.: A novel vehicle classification model for urban traffic surveillance using the deep neural network model. Int. J. Educ. Manage. Eng. **6**(1), 18–31 (2016). https://doi.org/10.5815/ijeme.2016.01.03

Some Aspects of the Method for Tourist Route Creation

Nataliya Shakhovska⊙, Khrystyna Shakhovska and Solomia Fedushko⊙

Abstract In the article, the conceptual model of tourists' guide system is created. Proposed model is based on clustering of tourists' object by user request, finding the fastest way to the chosen objects taking into account switching of traffic lights. The MongoDB is chosen for database creation. Information about interested tourists place is taken using Goggle API. The modified grid clustering method is used for location of suitable district. After that, the shooters path to chosen object is proposed. For this purpose, the Dijkstra's algorithm is modified using addition weight of nodes in the graph. The obtained results of the Dijkstra's algorithm and the developed method show that the proposed method allows to work with a two-weighted graph and look for routes in the city, taking into account the duration of switching traffic lights. Although the route in two algorithms turned out to be the same, the marks are quite different.

Keywords Information system · Information technology · Hierarchical clustering · Modified Dijkstra's algorithm · NoSQL database

1 Introduction

Nowadays, people have fast-moving lifestyle. They are traveling, moving out, having business trip, etc. That is why, it is important to save our time. Therefore, it becomes an idea to deal with problem of lacking time. The main problem in big city is the huge count of tourist object that cannot be displayed on the map.

N. Shakhovska (✉) · K. Shakhovska · S. Fedushko
Lviv Polytechnic National University, Lviv 79013, Ukraine
e-mail: nataliya.b.shakhovska@lpnu.ua

K. Shakhovska
e-mail: khrystyna.shakhovska.kn.2016@lpnu.ua

S. Fedushko
e-mail: felomia@gmail.com

© Springer Nature Switzerland AG 2020
Z. Hu et al. (eds.), *Advances in Artificial Systems for Medicine and Education II*,
Advances in Intelligent Systems and Computing 902,
https://doi.org/10.1007/978-3-030-12082-5_48

In this paper, we come up with idea to create a guide system for the city. This guidebook can help to find exactly that place which you need and to get there in the fastest way. We will choose a place by using cluster analysis. Modified Dijkstra's algorithm helps to find the optimal way.

We will cluster city/region using parameters, such as restaurants, cafes, gas stations, ATMs, pharmacies, groceries. After checking parameters, which you need, system finds more suitable place. Next, our guidebook finds the fastest (time) path including possible stops on traffic light.

The main problems of routing building are huge amount of data and special requests from users. Moreover, the routing building in real time in modern application is possible only for local region. That is why the purpose of paper is to create original method for geodata clustering taking to account parameters such as traffic lights, famous places to be founded.

2   Related Works

Despite the serious achievements of researchers in the field of cluster analysis, the developed algorithms are too static for the dynamic formation of a geodatabase, which, in fact, is the base of knowledge. In the knowledge base, the links between the knowledge components are multidisciplinary, multivalued, so the responsible removal or addition of the knowledge component requires verification of all possible connections in which the knowledge that is being deleted or acquired can potentially participate. Such verification is of a global nature and, as a rule, leads to a reorganization of the knowledge structure.

The universal multiparameter clustering methods proposed in the literature for geolocating and knowledge locating mainly deal with non-spatial characteristics and have very limited power in the recognition of spatial patterns. Spatial measurements (e.g., expressed as latitude and longitude or x and y coordinates) cannot be simply processed as two additional parameters in a multiparameter data space. Spatial measurements that do not depend on each other have real significance. Their unique and complex relationships cause difficulties for clustering, which does not always recognize these specific relationships [1].

Clustering methods defined for spatial data have been used as an important tool in geographic analysis. Various approaches to spatial clustering have been developed, including statistical approaches [2], Delaunay triangulation [3, 4], variable density method [5], grid-based separation [6], random walk methods [7, 8]. Existing spatial methods of clustering, however, work well only with a small data space (usually two or three-dimensional space, plus an aggregated geo-linked feature). Spatial clustering methods often use measures that are meaningful for practice, for example, road distance or transit time, and can address some complex situations, for example, geographic obstacles [9]. Such unique methods of clustering are difficult to unify within the multiparameter clustering methods.

Geospatial data sets, which are often encountered at the present time, have a high parametric dimension (two- or three-dimensional space plus a sufficiently large number of features). Such data sets are often collected from multiple data sources that are thematically different and could be used for various purposes [10]. The purpose of the clustering subspace (or projective clustering) is to identify subsets (subsets of the dimensions of the original multidimensional data space) that contain meaningful clusters, and then look for clusters in each subspace [11]. Although several methods of clustering subspaces have been proposed lately [11–14], none of them is practical enough to successfully analyze real geospatial data sets. To achieve efficiency in the study of large and multidimensional geospatial data sets, it is necessary to develop a highly interactive analysis environment that combines the best human and machine capabilities [15]. With the help of computational methods, it is possible to process large amounts of data for a certain type of images very quickly with mechanical accuracy and consistency, but they have a limited ability to interpret complex images and when adapting to different data sets. On the contrary, people can visually select complex images very quickly, comprehend them (evaluate and interpret images), and generate hypotheses for further analysis [15]. A knowledge discovery system for processing the flow of geospatial data sets must, therefore, automate the selection of elements.

CLIQUE [11], ORCLUS [12], ENCLUS [13], and DOC [14] can serve as examples of existing methods for clustering sub-spaces. However, the identification of subspaces that contain clusters remains a difficult research problem for two reasons. First, existing methodologies of clustering subspaces try to find clusters and associated subspaces simultaneously. Thus, the identification of the relevant subspaces depends to a large extent on the particular clustering algorithm. This identification may also depend on several subjective input parameters of the clustering algorithm. For example, CLIQUE needs an interval number ξ and a threshold of density τ; for ORCLUS, it is necessary to specify the number of clusters k and the dimensionality of subspaces p; and DOC needs a characteristic length w, a threshold of density α, and an equilibrium factor β. All these parameters are critical for algorithms, but problematic if you need to fix them in advance; in fact, they correspond to strong hypotheses about how clusters will be declared, or what types of clusters are of interest. Secondly, existing methods of subspace clustering cannot perform a hierarchical clustering in each subspace and cannot adapt well to a variety of application data sets and models of various scales. The user must execute the algorithm many times with different settings of one or more parameters to get a complete view of the data set.

The method proposed in ICEAGE [16, 17] is effective for hierarchical spatial clustering (for two-dimensional spatial points); it fully provides an interactive study of hierarchical clusters, achieving complexity $O(n \log n)$ and without using any index structure. It has advantages both in front of AMOEBA and in front of OPTICS. The method can generate optimal spatial ordering of clusters to store hierarchical clusters and encode spatial proximity information as much as possible. It is based on the Delaunay triangulation and the minimal binding tree method and overcomes the effect of a single reference by selecting boundary points for special processing.

In [18], association rules mining method is proposed for data analysis. Papers [1, 19, 21] used clustering methods based on fuzzy logic, paper [20] proposed uncertainty process.

The goal of paper is to develop the method of tourist route using improving grid clustering and Dijkstra algorithm.

3   Main Idea

The main idea is to find by the tourist query the group of tourist objects, such as restaurant, theaters, museum, café, monument, to localize peace with the biggest count of tourist objects and after that to find the shortest path to chosen object taking into account roads and traffic lights.

3.1   Database Creation

Firstly, we have to get information about touristic objects, roads, lights on the roads. The Google Maps API allows you to use most of the features available in Google Maps site to create applications with the ability, among other things, to recover the latitude and longitude of an address, add a marker on a map, calculating the distance between two points.

Geodata mostly is synonym of Big data, because it has complexity structure, huge amount, and specific algorithms for processing and analysis. That is why we choice MongoDB as the type NoSQL database. The code is open and can be used on different platforms and language customers. Furthermore, MongoDB uses JSON to describe and store documents.

The touristic objects are saved as a vector of type, coordinates, description in GeoJSON format.

```
"type": "FeatureCollection",
"features": [
  {
    "type": "Feature",
    "properties": {},
    "geometry": {
      "type": "Point",
      "coordinates": [ 24.02622, 49.84403]
    }
  },
  "properties": {
    "name": "Lviv Opera House",
    "obj_type": "theatre" "  }
]
```

Script for adding new elements in database is follows as:

```
> db.test_collection3.insert({_id : 100, pos: {lng :
126.9, lat : 35.2 } , type : "restaurant"})
> db.test_collection3.insert({_id : 200, pos: {lng :
127.5, lat : 36.1 } , type : "restaurant"})
> db.test_collection3.insert({ id : 300, pos: {lng :
128.0, lat : 36.7 } , type : "national park"})
```

Script for finding restaurants not far 1000 m:

```
db.test_collection3.createIndex({pos :
"geoHaystack", type : 1 } ,{bucketSize : 1 } )
db.runCommand({geoSearch : "test_collection3" ,
search : {type: "restaurant" } ,
near : [24, 49] ,
maxDistance : 1000 } )
```

3.2   Cluster Building

In our case, we have assumptions about the number of clusters, that is why non-hierarchical algorithms are recommended.

Non-hierarchical methods exhibit higher resistance to noise and emissions, incorrect metric selection, and the introduction of insignificant variables into the set involved in clustering. The price that you have to pay for these benefits of the method is the word "a priori." The analyst must pre-determine the number of clusters, the number of iterations, or the stop rule, as well as some other clustering parameters.

Our parameters and places we will take from Google Maps.

The base of proposed algorithm is the grid clustering method, in which the map is divided into squares of a certain size (varying for each zoom level), and in each such square, the markers are grouped. For a specific marker, a cluster is created, to which markers are added within the boundaries of the cluster square. The process is repeated until all markers are included in the nearest grid marker clusters, taking into account the zoom level of the map.

If the marker is within the boundaries of several existing clusters, the method determines the distance from it to each cluster and adds a marker to the nearest cluster using fuzzy C-mean approach:

(1) Initialization is carried out by accidental filling of the matrix of belonging F with the preservation of the conditions of normalization $\sum_{i=1}^{c} \mu_{ki} = 1$ and go to step 1 or random filling of cluster centroids V_i.

(2) For each iteration, we calculate:

$$V_i = \frac{\sum_{k=1}^{M} \mu_{ki}^m * X_k}{\sum_{k=1}^{M} \mu_{ki}^m}, \quad i = \overline{1, c},$$

$$D_{ki} = \sqrt{\|X_k - V_i\|^2}, \quad k = \overline{1, M}, \quad i = \overline{1, c},$$

$$\mu_{ki}^m = \frac{1}{\sum_{j=1}^{c} \left(\frac{D_{ki}}{D_{kj}}\right)^{2/m - 1}}, \quad k = \overline{1, M}, \quad i = \overline{1, c}.$$

At the end of each iteration, the condition of reaching accuracy is checked $\max_{k=\overline{1,M}, i-\overline{1,c}} (|\mu_{ki} - \mu_{ki}^*|) < \varepsilon$ where μ_{ki}^* value calculated at the previous iteration.

The result of clustering is the list of tourist object with geodata, road to these objects, and traffic lights on them.

3.3   Modified Dijkstra's Algorithm

In order to be able to work with information on the duration of traffic light and the frequency of its switching as one of the elements of the map, we introduce the concept of a two-weighted graph.

Definition: Two-weighted graph we call directed graph $V(E)$, which have weight of nodes and arrows:

$$V \to Z, E \to Z \tag{1}$$

where V is set of nodes, E is set of arrows, and Z is set of real number.

Path $V_1 V_2$ in two-weighted graph we call a sequence of arrows, which connect the adjacent nodes, started from V_1 and ended V_2.

Weight of node of two-weighted graph in our case consists of five elements:

$$\text{Weight}\langle V_i \rangle = \langle D_i, R_i, G_i, RF_i, GF_i \rangle \tag{2}$$

Dijkstra's marks D: $D_i = \langle W_i; V_{i-1} \rangle$, duration of red light $R \to Z$, duration of green light $G \to Z$, duration of red light from start point of way $RF \to Z$, duration of green light from start point of way $GF \to Z$.

Dijkstra's marks is weight of node V_i, is found by modified Dijkstra's algorithm. Denotes cost of way W_i to node V_i and node V_{i-1}, as started point to V_i:

$$W_i = E(V_{i-1}, V_i) + W_{i-1}. \tag{3}$$

Example of two-weighted graph is shown on the Fig. 1.

Modified Dijkstra's algorithm for two-weighted graph consists of sequence of next steps:

Step 0. Put Dijkstra's mark in the next way: [0;0]—for start node, [∞; ∞]—for else nodes. All nodes of graph regard as unreviewed.

Step 1. Until set of unreviewed nodes is not empty:

    1.1. From unreviewed nodes, choose node V_{i-1} with smallest value of W_{i-1} Dijkstra's marks.

    1.2. For all nodes, which are adjacent with V_{i-1}, count Dijkstra's marks:

$$W_i = \min(W_i; W_{i-1} + E(V_{i-1}, V_i) + S), \tag{4}$$

$$S = \begin{cases} R - RF - W_{i-1} \bmod (R + G) \bmod R, \, GF = 0 \\ R - (W_{i-1} \bmod (R + G) \bmod G - (G - GF)), \, RF = 0 \end{cases}$$

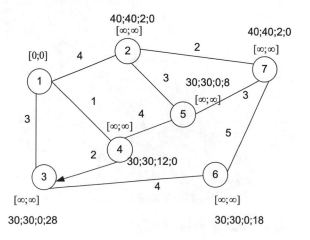

Fig. 1 Two-weighted graph

1.3. Node V_{i-1} regard as viewed.
1.4. Go to step 1.1.

Step 2. For last node, show cost of path.

4 Results

Now we will analyze the results of clustering. The marked base graph is shown on
the Fig. 2.

After that the area with the biggest count of objects is chosen.

Let us compare the results of the developed algorithm with the results of the
Dijkstra's algorithm. The comparison will be carried out on the same array of data.

Let us show each stage of the algorithms. We will search the path from the node 1
to the node 7. The weight of the arrow indicates the length of the path in kilometers
between the traffic lights.

After that, we modify nodes with the smallest value of marks (Fig. 3).

At the beginning, we count how many seconds will take the road. Then, the result
in seconds is added to that indicator of the current burning time, which is different
from zero, formula (3). For example, to the top 3, the duration of the movement was
15 s. From the beginning, the green light burned for 28 s, and the total burning time
of green was 30 s. So, while the traveler gets to the traffic light, the light will turn
red and will burn $15 - 2 = 13$ s. Burning red lasts for 30 s. So, for another $30 - 13$
$= 17$, the traveler will wait for the traffic light. Therefore, the weight of the node 3
is $15 + 17$ waiting time $= 32$ s (Fig. 4).

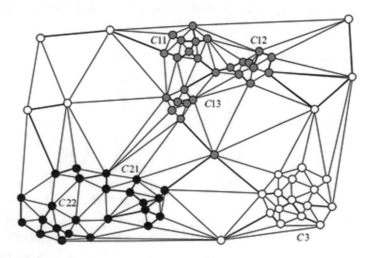

Fig. 2 The grouped tourists' objects

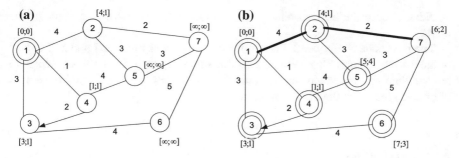

Fig. 3 The weights on the first step (**a**) and on the last step (**b**)

Fig. 4 The second step of modified Dijkstra's algorithm

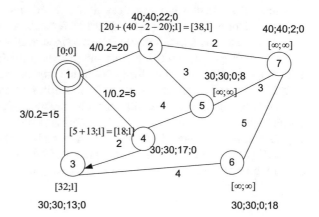

Fig. 5 The final path, and finally, we reached the same point 7 in 48 s. Path: 1–2–7.

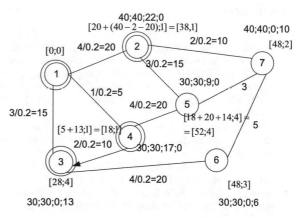

Next, choose node with the lowest value of the Dijkstra's marks, that is, the node number 4. From it, we can get in 3 in $18 + 10 = 28$ s. The traffic light during 17 s will burn red light and the next 13 s—green. At the time of arrival on the traffic light number 3 will be a green light and the traveler can move on. Consequently, the new value of the node is less than the previous and is 28 s.

Next, choose the next current node—number 3 with the smallest value of the mark.

The next node is number 2. From it is access to nodes number 7 and 5. The cost of the road to 5 equals $38 + 15 = 53$ s. At that moment, the traffic light will switch to red and then again in green and will burn $53 - (30 - 8) - 30 = 1$ s. But the length of time is greater, so the mark is not rewritten (Fig. 5).

5 Conclusions

The modified grid clustering method allows us to build clusters with city tourists' objects and locate the most appropriated area for user. After the object chosen system builds the fastest path to this object. The obtained results of the Dijkstra's algorithm and the developed method show that the proposed method allows to work with a two-weighted graph and look for routes in the city, taking into account the duration of switching traffic lights. Although the route in two algorithms turned out to be the same, the marks are quite different. The practice of applying geodatabase clustering methods shows that successful clustering is possible only with an interactive approach using visual feedback, which provides adjustment of parameters during the research process; therefore, the development of effective imaging techniques is no less important than the improvement of clustering methods.

References

1. Hu, Z., Bodyanskiy, Y.V., Tyshchenko, O.K., Samitova, V.O.: Possibilistic fuzzy clustering for categorical data arrays based on frequency prototypes and dissimilarity measures. Int. J. Intell. Syst. Appl. (IJISA) **9**(5), 55–61 (2017). https://doi.org/10.5815/ijisa.2017.05.07
2. Kumar, A., Sharma, R.: A genetic algorithm based fractional fuzzy PID controller for integer and fractional order systems. Int. J. Intell. Syst. Appl. (IJISA) **10**(5), 23–32 (2018). https://doi.org/10.5815/ijisa.2018.05.03
3. Estivill-Castro, V., Lee, I.: Amoeba Hierarchical clustering based on spatial proximity using Delaunay diagram. In: 9th International Symposium on Spatial Data Handling, Beijing, China, 2000, pp. 26–41
4. Zhang, C., Xiao, W., Tang, D., Tang, J.: P2P-based multidimensional indexing methods: a survey. J. Syst. Softw. **84**(12), 2348–2362 (2011)
5. Bohm, C., Railing, K., Kriegel, H. P., Kroger, P.: Density connected clustering with local subspace preferences. In: Fourth IEEE International Conference on Data Mining, 2004. ICDM'04, 2004, pp. 27–34
6. Wang, Y., Wu, X.: Heterogeneous spatial data mining based on grid. In: Lecture Notes in Computer Science. vol. 4683, pp. 503–510. Springer, Heidelberg (2007)
7. Berkhin, P.: A survey of clustering data mining techniques. In: Grouping Multidimensional Data, pp. 25–71. Springer, Berlin, Heidelberg (2006)
8. Oyana, T.J., Dai, D.: Automatic cluster identification for environmental applications using the self-organizing maps and a new genetic algorithm. Geocarto Int. **25**(1), 53–69 (2010)
9. Bithi, A.A., Ferdaus, A.A.: Mining sequential patterns from mFUSP—tree. Int. J. Inf. Technol. Comput. Sci. (IJITCS) **7**(7), 77–89 (2015). https://doi.org/10.5815/ijitcs.2015.07.09

10. Veres, O., Shakhovska, N.: Elements of the formal model big date. In: 2015 XI International Conference on Perspective Technologies and Methods in MEMS Design (MEMSTECH), 2015, pp. 81–83
11. Agrawal, R., Gehrke, J., Gunopulos, D., Raghavan, P.: Automatic sub-space clustering of high dimensional data. Data Min. Knowl. Discovery **11**(1), 5–33 (2005)
12. Aggarwal C., Yu, P.: Finding generalized projected clusters in high dimensional spaces. In: ACM SIGMOD International Conference on Management of Data, 2000, pp. 70–81
13. Liu, G., Li, J., Sim, K., Wong, L.: Distance based subspace clustering with flexible dimension partitioning. In: IEEE 23rd International Conference on Data Engineering, ICDE 2007, 2007, pp. 1250–1254
14. Procopiuc, C.M., Jones, M., Agarwal, P.K., Murali, T.M.: A Monte Carlo algorithm for fast projective clustering In: ACM SIGMOD International Conference on Management of Data. Madison, Wisconsin, USA, 2002, pp. 418–427
15. Shakhovska, N., Vysotska, V., Chyrun, L.: Features of E-learning realization using virtual research laboratory. In: Scientific and Technical Conference "Computer Sciences and Information Technologies (CSIT), 2016 XIth International, 2016, pp. 143–148
16. Ivanov, Y., et al.: Adaptive moving object segmentation algorithms in cluttered environments. In: The Experience of Designing and Application of CAD Systems in Microelectronics, Lviv, 2015, 2015, pp. 97–99. https://doi.org/10.1109/cadsm.2015.7230806
17. Guo, D., Peuquet, D.J., Gahegan, M.: ICEAGE: Interactive clustering and exploration of large and high-dimensional geodata. Geoinformatica **3**(7), 229–253 (2003)
18. Shakhovska, N., Kaminskyy, R., Zasoba, E., Tsiutsiura, M.: Association rules mining in big data. Int. J. Comput. **17**(1), 25–32 (2018)
19. Madaan, J., Kashyap, I.: A novel handoff necessity estimation approach based on travelling distance. Int. J. Intell. Syst. Appl. (IJISA) **10**(1), 46–57 (2018). https://doi.org/10.5815/ijisa.2018.01.06
20. Shakhovska, N., Medykovsky, M., Stakhiv, P.: Application of algorithms of classification for uncertainty reduction. Przeglad Elektrotechniczny **89**(4), 284–286 (2013)
21. Hu, Z., Bodyanskiy, Y.V., Tyshchenko, O.K., Tkachov, V.M.: Fuzzy clustering data arrays with omitted observations. Int. J. Intell. Syst. Appl. (IJISA) **9**(6), 24–32 (2017). https://doi.org/10.5815/ijisa.2017.06.03

Experimental Substantiation of Soft Cutting Modes Method

P. A. Eremeykin, A. D. Zhargalova and S. S. Gavriushin

Abstract This article concerns problems related with experimental substantiation of a new promising method of thin-walled workpieces processing. The method is based on applying a software system that allows predicting technological deformations during processing. Using the system a technologist is able to pick rational cutting modes taking into consideration deformation predictions. The task of experimental verification consists in the development of the experimental scheme and the measurement sequence. This article describes an experimental facility, stages and results interpretation. Emphasis is placed on a deformed workpiece profile analysis. The article highlights the issues arising during the experiment organization. Authors suggest solutions using available general-purpose equipment.

Keywords Thin-walled part · Numerical modeling · Turning · Software development · Finite element analysis (FEA)

1 Introduction

Many fields of modern industry require extremely accurate, lightweight and yet inexpensive components. This kind of parts is frequently applied in aerospace, power machine building and other areas. The specified properties are as desirable as controversial. Regularly when a technologist faces a problem of providing these characteristics, he sacrifices one to meet another two. Some applications are not tolerant of providing only two characteristics, so each of such task claims some intellectual approaches.

P. A. Eremeykin
Mechanical Engineering Research Institute of the Russian Academy of Sciences, Moscow
101990, Russia
e-mail: eremeykin@gmail.com

A. D. Zhargalova (✉) · S. S. Gavriushin
Bauman Moscow State Technical University, Moscow 105005, Russia
e-mail: azhargalova@bmstu.ru

© Springer Nature Switzerland AG 2020
Z. Hu et al. (eds.), *Advances in Artificial Systems for Medicine and Education II*,
Advances in Intelligent Systems and Computing 902,
https://doi.org/10.1007/978-3-030-12082-5_49

The solving this problem in general formulation is too hard, so we will limit types of considered parts with thin-walled shells that are body of revolution. Under this restriction, it is reasonable to suppose that the workpieces are treated with turning.

The main problem that occurs during development a technology for thin-walled parts is plastic deformation. When the processed workpiece is thin-walled, it is deformed under the action of cutting forces. The processing of deformed workpiece leads to nonuniform material removal. So the final part has geometry flaws.

There are some traditional approaches that allow processing low-rigid workpiece: clamping with expanding mandrel or soft (unhardened) jaws, filling the hollow part with fusible technological aggregate. A modern additive technology is also considerable solution for accurate thin-walled parts production [1]. The main disadvantages of the listed approaches are increased requirements for technological equipment and greater machine-setting time.

Recently a new promising method of turning thin-walled workpieces has been suggested [2]. The method is called "soft" cutting modes. The idea the method is based on is providing required characteristics by rational cutting modes assignment. As authors of the method believe, the cutting modes assignment backed with numerical modeling of the treatment process makes it possible to achieve good accuracy for thin-walled workpiece, and at the same time, the costs growth is quite small. The key thing of the new approach the authors emphasize is the developed integrated software system that allows quickly estimate technological deformations that correspond to processing at given cutting and clamping conditions. Using the system a technologist is able to pick cutting modes specific to given situation and dedicated for the workpiece he deals with. The system input is workpiece dimensions and material properties, cutting and clamping forces, and the output is deformed state. Generally speaking, the open-source software libraries nowadays allow constructing sophisticated systems at reasonable efforts and time costs [3, 4]. The article [5] describes functionality of the system and the principal issues that arose during the development.

Every new method must be empirically approved before it can be applied in practice. For the present, there is no empirical substantiation for the soft cutting modes method. In this article, we will consider an experiment that is intended to confirm efficiency of the new approach.

International research teams work intensively on the problem of thin-walled parts processing and offer their own solutions to problems. Although the turning is not prevailing investigation subject, there are some interesting ideas and experiments descriptions in the papers. For example, the article [6] estimates different cutting strategies that influence the result product accuracy. Thus, it turns out that the cutting and clamping forces is not the only factor affecting the technological deformation [7]. The idea of cutting parameters optimization according to some mathematical model was also exploded by researches. The authors of the paper [8] suggest a statistical approach to the problem.

Also, this paper has detailed experiment setting description. The main idea is to estimate errors during the experimental treatment and then reproduce the same conditions with software simulation. The difference between experimental and simulation results will show how good the new method is. Some researchers have made great progress in modeling the mechanics of the cutting process and even tried to model the chip formation during the handling [9–13]. In practice, the modeling of chip formation has arguable value, but surely it is very important to strengthen the theoretical foundation of cutting processes modeling for better understanding the underneath physical phenomenon. So, there is no enough information to corroborate the success of numerical modeling for predict technological deformation.

This investigation is focused on developing the facility, deformations measures scheme and experimental sequencing to approve the applicability of the soft cutting modes method. The paper structure is the following: In the second section, we specify the experimental setup and conditions, the third section is devoted to results description and presentation, and finally, the last section concludes the paper and suggests future research directions and enhancements.

2 Experiment Description

As it was mentioned in the introduction, processing the non-rigid workpieces causes difficulties because of shape deflection during the material removal. In the case of turning a thin-walled workpiece, there are two main factors that induce the deflection: the cutting and clamping forces. These factors are taken into account in the numerical model that is used for implementing the soft cutting modes method.

Our aim is to compare the results of numerical simulation and of the real thin-walled workpiece processing experiment. To properly reproduce the real-cutting conditions with the model, we must control a number of parameters. The experimental facility imposes restrictions on cutting process and helps to carry out the measurements.

To simulate the processing one needs to know workpiece geometrical parameters, material properties, clamping conditions and cutting forces [14]. We chose the simplest low-rigid workpiece shape that is a body of revolution: a hollow thin-walled cylinder. Workpiece dimensions are shown in Fig. 1, and the workpiece material is alloyed steel with characteristics specified in Table 1.

The modeling result is a displacement field like the one shown in Fig. 1. To obtain this result, the finite element analysis (FEA) software system Abaqus was applied. FEA allows calculating complicated deformation states due to meshing the elastic body and considering it like an ensemble of finite elements with known stiffness matrix. Abaqus calculates at each node value of displacement and partitions the range

Table 1 The workpiece material properties

	Value
Poisson ratio	0.33
Elastics modulus (MPa)	2.1e11

Fig. 1 The workpiece
dimensions

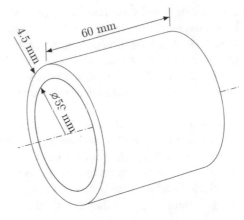

Fig. 2 Sample of simulation
result

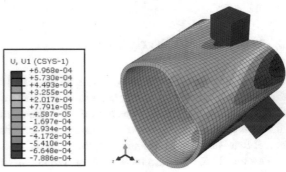

of obtained values such that each value corresponds to one of 12 colors. This model
takes into consideration cutting and clamping force but no temperature processes.
To better explanation of the model see paper [4].

As we cannot obtain the whole displacement field within the experiment and
compare it with the simulation result, we defined particular measurements points
and compare only the displacement values at these points. To position these points,
we defined speculative reference plane that is perpendicular to the workpiece axis.
In Fig. 2, the reference plane is shown with red color. The plane cross-section of
the outer workpiece surface is a circle. In its turn, each such circle defines a set of
measuring points which are separated from each other by angular step of 30°.

Measured displacement at specified points can be compared to corresponded values of simulation result. To carry out the measurements, we developed an experimental facility that is shown in Fig. 4. This facility is based on a regular lathe (1).
The workpiece (2) is holding by a lathe chuck (3) which can be rotated integrally
with the spindle (5). The spindle has an angular dial (4) so one can control position
of the workpiece. Indicator (6) is used to measure the displacement during various
stages of processing. The indicator is held by a stand (7) that is able to move along

Fig. 3 Positioning the
measuring points

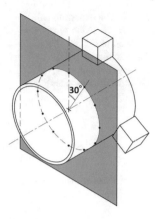

the lathe axis. Moving the stand one can reproduce different positions of controls plane, described above. To control the clamping force, we use a torque wrench.

The experiment is divided into the following four stages:

(a) Before processing; the workpiece is barely clamped, it is in unreformed state
(b) Before processing; the workpiece is strongly fixed and deformed
(c) After processing; the workpiece is strongly fixed
(d) After processing; the workpiece is barely clamped.

At each stage of the experiment, we measure the displacements at four positions of the indicator probe [11], which are shown in Fig. 4 with red dashed lines. Manual rotation of the machine spindle provides different measurement points within the given reference plane like it is shown in Fig. 3.

3 Results

The experimental results can be represented as a number of curves sets. These curves, formed by 12 points, describe dependence of the deflection on the angular position. Each of curve sets corresponds to one of four experiment stages (a–d), specified above, and to one of four reference plane positions: 0, 7, 14, 21 mm (see Fig. 4).

The most representative curves were obtained for the section that is 21 mm far from the end face of the workpiece. This section corresponds to stationary process while sections that are closer to the workpiece end face are affected by the transition process phenomenon. Thus, all article charts are related to 21 mm reference plane positions. Figure 5 shows the dependence of the displacement on the angular position of the measurement probe relative to the workpiece in unclamped state (stages a, b) and Fig. 6 shows the same in clamped state.

As one would expect the initial workpiece profile is not an ideal circle. In Fig. 5, the initial profile of the workpiece is drawn with a green continuous line. When

544   P. A. Eremeykin et al.

Fig. 4 The experimental facility: 1—machine base; 2—workpiece; 3—chuck; 4—machine spindle; 5—angular dial; 6—indicator; 7—stand

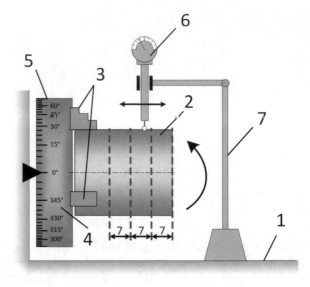

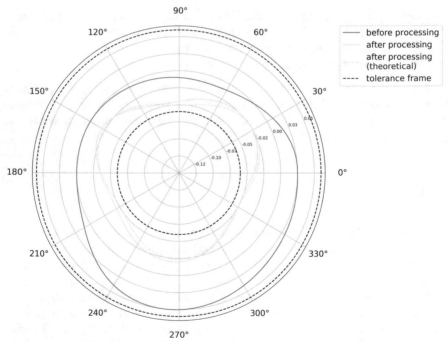

Fig. 5 Measured displacements at unclamped state

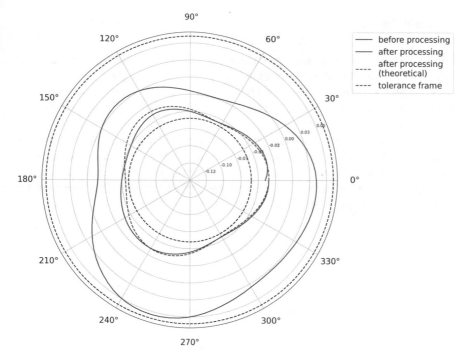

Fig. 6 Measured displacements at clamped state

the workpiece is clamped with a three jaws chuck, the profile acquires minimums at jaws positions (red line in Fig. 6). The clumping force effect on the processing result becomes obvious looking at the blue curve in Fig. 6. This curve corresponds to clamped state of the workpiece right after processing. But the real trouble is the geometrical flaws that are inherent to the result in unchucked state. Continuous yellow line in Fig. 5 corresponds to unchecked workpiece after processing.

One can see that at processed state angular positions of the profile minimums in clamped and unclamped states are mutually shifted by approximately 60°. In other words, the clamped maximum corresponds to unclamped minimum and vice versa. This observation confirms the following physical intuition: The more material protrudes before processing, the more it is cut off.

Theoretical curves are shown by dashed lines of the same color. These curves were derived from the FEA model calculated with Abaqus CAE. Abaqus has a functionality that allows obtaining values of any available field along the specified path. But before applying this functionality one should set a cylindrical coordinate system so that the node movements in the model correspond to the movement of the probe.

The experimental curves are in good agreement with the theoretical ones. Generally speaking, they have the same sequence of minimums and maximums and lie close. The relative error is no more than 24%.

4 Conclusion

In this article, we dealt with a new promising method of processing thin-walled workpieces, called soft cutting modes method. To substantiate the practical workability of the method, we planned and carried out the experiment. The experiment supposes lathing a workpiece and measuring technological deformations. Special equipment was developed for these purposes.

The experiment revealed that empirical and theoretical deformation matches well. Thus, the soft cutting modes method may be implemented on base of a software system that is based on a numerical FEA model. As a possible research course authors consider working on different workpiece configurations, more detailed mathematical justification and rigorous empirical analysis.

References

1. Isaev, A., Grechishnikov, V., Pivkin, P., Kozochkin, M., et al.: Machining of thin-walled parts produced by additive manufacturing technologies. In: 48th CIRP Conference on Manufacturing Systems—CIRP CMS, vol. 41, pp. 1023–1026 (2016). https://doi.org/10.1016/j.procir.2015.08.088
2. Gavriushin, S.S., Zhargalova, A.D., Lazarenko, G.P., Semisalov, V.I.: The method for determining the conditions for thin-walled parts machining. Izvestiya vysshih uchebnyh zavedenij (News High. Educ. Inst.) 11, 53–60 (2015). (In Russian)
3. Nega, A., Kumlachew, A.: Data mining based hybrid intelligent system for medical application. Int. J. Inf. Eng. Electron. Bus. (IJIEEB) 9(4), 38–46 (2017). https://doi.org/10.5815/ijieeb.2017.04.06
4. Oluwole, O.: Design and implementation of an open-source computer-based testing system with end user impact analysis in Africa. Int. J. Mod. Educ. Comput. Sci. (IJMECS) 7(8), 17–24 (2015). https://doi.org/10.5815/ijmecs.2015.08.03
5. Eremeykin, P., Zhargalova, A., Gavriushin, S.S.: A software system for thin-walled parts deformation analysis. In: AIMEE 2017: Advances in Artificial Systems for Medicine and Education, vol. 658, pp. 259–265 (2017). https://doi.org/10.1007/978-3-319-67349-3_24
6. Shamsuddin, K.A.: A comparison of milling cutting path strategies for thin-walled aluminium alloys fabrication. Int. J. Eng. Sci. 2(3), 1–8 (2013)
7. Ratchev, S., Liu, S., Huang, W., Becker, A.: Milling error prediction and compensation in machining of low-rigidity parts. Int. J. Mach. Tools Manuf. 44(15), 1629–1641 (2004). https://doi.org/10.1016/j.ijmachtools.2004.06.001
8. Sridhar, G., Venkateswarlu, G.: Multi objective optimisation of turning process parameters on EN 8 steel using grey relational analysis. Int. J. Eng. Manuf. (IJEM) 4(4), 14–25 (2014). https://doi.org/10.5815/ijem.2014.04.02
9. Scippa, A., Grossi, N., Campatelli, G.: FEM based cutting velocity selection for thin walled part machining. In: 6th CIRP International Conference on High Performance Cutting, HPC2014, vol. 14, pp. 287–292 (2014). https://doi.org/10.1016/j.procir.2014.03.023
10. Huang, Y.A., Zhang, X., Xiong, Y.: Finite element analysis of machining thin-wall parts: error prediction and stability analysis. In: Finite Element Analysis—Applications in Mechanical Engineering, vol. 10 (2012). https://doi.org/10.5772/50374
11. Loehe, J., Zaeh, M.F., Roesch, O.: In-process deformation measurement of thin-walled workpieces. Procedia CIRP 1, 546–551 (2012). https://doi.org/10.1016/j.procir.2012.04.097
12. Mamalis, A., Kundrak, J., Markopoulos, A.: Simulation of high speed hard turning using the finite element method. J. Mach. Form. Technol. 1 (2011)

13. Villumsen, M.F., Fauerholdt, T.G.: Prediction of Cutting Forces in Metal Cutting, Using the Finite Element Method, a Lagrangian Approach. Konferenzbeitrag, LS-DYNA Anwenderforum, Bamberg (2008)
14. Eremeykin, P., Zhargalova, A., Gavriushin, S.S.: Empirical evaluation of technological deformations for "Soft" cutting modes during thin-walled parts turning. Obrabotka metallov (Met. Work. Mater. Sci.) **20**(1), 22–32 (2018). https://doi.org/10.17212/1994-6309-2018-20.1-22-32. (In Russian)

Design of Oxidative Pyrolysis Control Algorithm Based on Fuzzy Safety Area and Center Definition

G. N. Sanayeva, I. E. Kirillov, A. E. Prorokov, V. N. Bogatikov and D. P. Vent

Abstract This article examines an approach to controlling complex dynamic industrial technology using fuzzy logic, which is one of the artificial intelligence "AI" trends helping improve quality of decision making and thereafter the quality of industrial technology management systems. Industrial technologies function under the risk of various accidental violations when process state estimates are complicated and therefore raise a number of control problems, and the solution is aimed at compensating the random prime causes in abnormal situations. Therefore, the solution of dynamic process safety management problems in semi-structured and poorly formalized environments is a relevant and urgent task. An acetylene production control system by oxidative pyrolysis of natural gas is considered with the process peculiarities being taken into account. A sequence of interconnected heaters for initial components and an oxidative pyrolysis reactor is used as control object. The control system has to provide the required pyrolysis gas composition to the reactor outlet in condition that ensures safety of the process. A two-level structure is suggested for the control system process including a local parameter adjustment at the lower level and a local circuit task estimation with a fuzzy controller adjustment at the upper level. Depending on the parameter values determining the process safety, some control actions, i.e., natural gas discharge rate, ratio of natural gas, oxygen and water flow for pyrolysis gas "hardening," are estimated. A control system with a fuzzy controller shows the best performance.

Keywords Control systems · Oxidative pyrolysis · Acetylene · Technological safety · Fuzzy controller · Mathematical modeling

G. N. Sanayeva (✉) · A. E. Prorokov · D. P. Vent
Dmitri Mendeleev University of Chemical Technology of Russia, Novomoskovsk
Institute (D. Mendeleev Branch), Novomoskovsk, Russia
e-mail: gsanaeva@nirhtu.ru

I. E. Kirillov
Apatity Branch of Murmansk State Technical University, Apatity, Russia

V. N. Bogatikov
Tver State Technical University, Tver, Russia

© Springer Nature Switzerland AG 2020
Z. Hu et al. (eds.), *Advances in Artificial Systems for Medicine and Education II*,
Advances in Intelligent Systems and Computing 902,
https://doi.org/10.1007/978-3-030-12082-5_50

1 Introduction

The production of acetylene by oxidative pyrolysis of natural gas is characterized by high explosion and fire risks [1], ensuring technological safety of the process is therefore of particular relevance.

In contrast to other production of acetylene from natural gas in industry (e.g., electric cracking, high-temperature cracking of methane, catalytic cracking of methane), oxidative pyrolysis remains the most popular at the moment due to the use of raw materials (oxygen and natural gas, used both as a starting product for the reaction and as fuel) that is effective in comparison with other methods.

The peculiarities of oxidative pyrolysis are the speed of the main chemical reactions, release of a large amount of heat in a small reactionary volume with partial combustion of raw materials, the dependence of the composition of the resulting product from the fluctuation of composition raw material (natural gas), pile soot elements of equipment, etc. [1].

These factors have an unpredictable effect on the process, and therefore, the traditional control system process (local regulation of the controlled parameters) does not provide the required level of technological safety.

This leads to the need to search for new ways to improve technological safety, in particular, the use of fuzzy logic to ensure the maximum possible amount of acetylene produced as a desired product by stabilizing the main technological parameters in the area corresponding to the technological safety center, which, among other things, will ensure rational use of all the initial products of the oxidative pyrolysis process.

The state of the system at each instance in time t from this interval is characterized by a set of parameters of this system, on which constraints are imposed, depending on the sets of parameters $\{T_i, K_j, U_{pr}\}$ (technological—$\{T_i, i = 1, ..., I\}$, constructive—$\{K_j, j = 1, ..., J\}$, controls—$\{U_{pr}, l = 1, ..., L\}$).

Evaluation of system properties is formed with the help of cognitive mechanisms of expert representations. This view includes the area of possible states (Fig. 1), the laws of system behavior, risk assessment (Fig. 2).

Notations and formulas for determining the state of the system (Fig. 2):

S is the set of all possible states of the system; S_p is the of state of the current moment time of system; $\varphi_i, i = 1, ..., n$ is the set of boundaries of parameter space

Fig. 1 Domain of system state existence

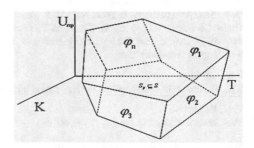

Fig. 2 Safety center
interpretation

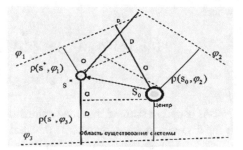

of the technology (it cuts out the space S_p on S); S_o is the point the of the center of safety of the process; s^*—current point;

$l_i = \min (s^*, \varphi_i)$ is the distance from the point s^* to the boundaries φ_i of the working space;

$\Delta_i^* = \min \rho (s^*, s_0)_{\varphi i}$ is the distance from the point s^* until the center of S_0 area relative to the boundaries working space;

$\delta_i = \min \rho(s_0, \varphi_i)$ is the distance from the center S_o to the boundaries of the area of possible states $\varphi_i, i = 1, ..., n$;

$O_i = \delta_{i-i} - l_i$ is degree of danger s^* state for the technological process relative to boundaries $\varphi_i, i = 1, ..., n$;

$D_i = \delta_i - O_i$ is degree of safety s^* state for the technological process relative to boundaries $\varphi_i, i = 1, ..., n$.

Since the functioning of a real process occurs under the constant influence of random disturbances, constant monitoring of the displacement of the operating point of the process and determination of the values of control actions taking into account its current position are required.

As an estimate of the bias, the process safety index is used, which characterizes the degree of fuzzy equality of the current situation and the situation corresponding to the security center. As a safety center, a point may be considered that corresponds to the best regulatory conditions, or, for example, a point whose value is established on the basis of an expert survey of engineers and skilled workers serving the process of oxidative pyrolysis [2, 3].

Formal Definition of "fuzzy" Situation Let be $X = \{x_1, x_2, \ldots, x_n\}$ a set of symbols. Each feature x_i described by a corresponding linguistic variable $< \beta_i, E_i, F_i >$. β_i—name of linguistic variable; $E_i = \left\{ E_i^1, E_i^2, \ldots, E_i^{M_i} \right\}$—term-set of linguistic variable β_i; F_i—base set of linguistic variable β_i.

Fuzzy equality or equivalence. In accordance with the following criteria: the degree of fuzzy inclusion and the degree of fuzzy equality.

The degree equality of the situations is determined by the expression:

$$v\left(\tilde{S}_{Xi}, \tilde{S}_{Xj}\right) = \underset{X}{\&} v\left(\mu_{S_{Xi}}(X), \mu_{S_{Xj}}(X)\right)$$

where

$$v\big(\mu_{S_{Xi}}(X), \mu_{S_{Xj}}(X)\big) = \underset{E_k}{\&}\Big(\mu_{\mu_{S_{Xi}}(X)}(E_k) \rightarrow \mu_{\mu_{S_{Xj}}(X)}(E_k)\Big)$$

$$\mu_{\mu_{S_{Xi}}(X)}(E_k) \rightarrow \mu_{\mu_{S_{Xj}}(X)}(E_k) = \max\Big\{1 - \mu_{\mu_{S_{Xi}}(X)}(E_k), \mu_{\mu_{S_{Xj}}(X)}(E_k)\Big\}$$

It is usually assumed that fuzzy situation the S_i is included in fuzzy situation S_j, if the degree is not less than some threshold of inclusion $tinc \in [0.6; 1]$, determined by the control conditions, that is, $v\big(\tilde{S}_i, \tilde{S}_j\big) \geq t_{\text{inc}}$.

The Degree of Fuzzy Equality If the set of current situations $\tilde{S}_X$ contains such situations $\tilde{S}_{Xi}$ and $\tilde{S}_{Xj}$ that $\tilde{S}_{Xi}$ is of included in $\tilde{S}_{Xj}$ and is of included in $\tilde{S}_{Xi}$, then situations $\tilde{S}_{Xi}$ and $\tilde{S}_{Xj}$ should be treated as one situation. This means that for a given inclusion threshold t_{inc}, situations $\tilde{S}_{Xi}$ and $\tilde{S}_{Xj}$ are approximately the same. Such a similarity of situations is called fuzzy equality, with the degree of fuzzy equality equal to:

$$\mu\big(\tilde{S}_{Xi}, \tilde{S}_{Xj}\big) = v\big(\tilde{S}_{Xi}, \tilde{S}_{Xj}\big)\&v\big(\tilde{S}_{Xj}, \tilde{S}_{Xi}\big)$$

In contrast to the set $^TS_s = \{^TS_1, {}^TS_2, \ldots, {}^TS_n\}$ of current situations, the set $S_s = \big\{\tilde{S}_1, \tilde{S}_2, \ldots, \tilde{S}_n\big\}(n \leq N)$ of reference situations does not contain indistinct equalities for a given threshold of equality of situations. It is assumed that the set S_z is complete. Thus, the situation $\tilde{S}_i$ exists for any input situation S_o. The decision table for this reference situation determines the control solution. This approach is based on the method of situational management [3].

2 Oxidative Pyrolysis Description

The process of oxidative pyrolysis in the acetylene production can be represented by a sequence of the following stages [1, 2]:

- Heating initial components (a heater)
- Mixing (a reactor)
- Oxidative pyrolysis (a reactor)
- "Hardening" of oxidizing pyrolysis products (reactor).

If any of the indicated parameters exceeds the bounds of the appropriate permissible range, the process transfers to "flare piping," i.e., oxidative pyrolysis products burn in a flare and the whole system is blown with nitrogen, regardless of the acetylene content in pyrolysis gas in order to prevent an emergency situation.

The process of oxidative pyrolysis should be carried out with the following restrictions to ensure its technological safety:

- The maximum permissible temperature of the methane–oxygen mixture at the reaction inlet of the natural gas oxidative pyrolysis reactor to prevent the mixture premature ignition or reverse effect from a reaction chamber to a mixer (710 °C)
- The maximum permissible concentrations of oxygen and methane at the reactor outlet to avoid explosive concentrations of oxidative pyrolysis products (0.8 and 9% (vol.))
- The minimum water discharge rate for "hardening" of reaction products—acetylene at reaction temperature is an unstable compound and can decompose to release carbon (soot) and hydrogen, (10 m^3/s)
- The minimum and maximum permissible temperatures of pyrolysis gas at the oxidative pyrolysis reactor outlet (50 and 100 °C).

3 Control System Development

The main task of an oxidative pyrolysis control system is to maintain the required composition of the produced pyrolysis gas under conditions ensuring the process technological safety based on the definition of the safety area and center by stabilizing the of input and output flow parameters within the scopes defined by the appropriate restrictions [3].

To ensure the maximum level of process safety, it is advisable to maintain it at a point maximally distant from the bounds specified by the process restrictions—the security center. Geometrically, the security center is a point of intersection of normals to the lines corresponding to technological restrictions. However, the security center point may not be in compliance with the regulation conditions providing the goal pyrolysis gas composition, namely the content of acetylene as the required end product. In this regard, it is appropriate to introduce the notion of a security area as an area where the required pyrolysis gas composition is guaranteed at some reasonable decrease of security index. Boundaries of the security area are determined by the distances from the values of technological restriction parameters to the current values of process parameters providing the specified level of technological safety in dynamics. Thus, the oxidative pyrolysis control problem consists of determining some control vector transferring the technological process into a security area.

Figure 3 shows the process of oxidative pyrolysis as a control object where G_{NG} and G_{O2} are discharge rates of natural gas and oxygen per reaction, m^3/s; G_b is a discharge rate of natural gas per heater burner, m^3/s; G_{WH}—water flow rate for pyrolysis gas "hardening," m^3/s; T_{NG} and T_{O2} are temperatures of natural gas and oxygen after a heater, K; C_{C2H2}, C_{O2}, C_{CH4} are contents of acetylene, oxygen, and methane in pyrolysis gas at the reactor outlet, % (vol.); T_{PG} is pyrolysis gas temperature at the reactor outlet, K. It should be noted that the distributor grid temperature T_{dg}, as one of technological restrictions, is determined by natural gas and oxygen temperatures at the heater outlet.

To control technological safety of the oxidative pyrolysis process, a periodic analysis of the pyrolysis gas composition is performed with a view to determining

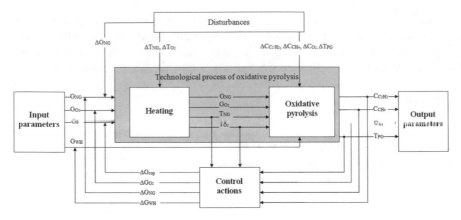

Fig. 3 The process of oxidative pyrolysis as a control object

the potentially explosive concentrations of methane, oxygen, acetylene in pyrolysis gas as well as the temperature of oxygen and natural gas at the heater outlet and that of pyrolysis gas at the reactor outlet are being constantly monitored. Values of these parameters enter the controller input where the security index is determined as a point in the multidimensional space formed by technological constraints. Notably, the maximum safety index values for methane and oxygen correspond to their minimum values and those for the natural gas and oxygen temperatures at the reactor inlet and the reactor outlet pyrolysis gas temperatures match admissible midrange values.

After comparing the actual values of technological parameters with those calculated using the mathematical model [3], the control actions are computed to change the natural gas and oxygen discharge (or their ratio) for providing the oxidative pyrolysis reaction, the natural gas flow to the heater burner, water flow for "hardening" the pyrolysis gas with a view to improving a safety index [4].

Since the real operation is under constant exposure of random disturbances, it is required to constantly monitor the operating point shift and to define values of control actions consistent with its current position. The technological process safety index is used as a shift estimate. It describes the degree of fuzzy equality of a current state and a safety center point. A point that corresponds to the best operation conditions, or, for example, a point which value is based on an expert survey of engineers and skilled workers providing the run of oxidative pyrolysis can be referred as a safety center.

To determine the appropriate way of controlling the oxidative pyrolysis process, a control system is generated (Fig. 4), with the required output flow characteristics being achieved by means of stabilization of internal control parameters depending on the values of output parameters (content of acetylene, methane, and oxygen in pyrolysis gas; pyrolysis gas temperature at the reactor outlet) based on the prediction of fuzzy inference and the vectoring of control actions, including:

- Natural gas discharge per reaction

- Oxygen discharge per reaction
- Natural gas discharge per burner
- Water flow for "hardening" the pyrolysis gas.

At the lower level of the proposed control system, controllers C_1 and C_2 regulate the inlet flow of natural gas and oxygen to reactor R; controllers C_3 and C_4 regulate the supply of natural gas to the heater burner to control the temperature of natural gas and oxygen at the outlet of heater H; controller C_5 regulates the water supply for "hardening" the pyrolysis gas to control the pyrolysis gas temperature at the reactor outlet.

The upper level is a basic one as regards the ensuring of process safety: It provides the calculation of tasks for the first level controllers with due regard to safety center.

Due to technological peculiarities of the oxidative pyrolysis process, the upper level controller is expedient to be designed by applying the fuzzy logic programming [5, 6], since such controllers, in some cases, are able to provide higher quality parameters of transient processes in comparison with standard ones [7, 8]. In this context, the following linguistic variables are designated: "natural gas discharge per burner," "natural gas discharge per reaction," "oxygen discharge per reaction," "temperature deviation of distribution grid," "deviation of methane content in pyrolysis gas," "acetylene content in pyrolysis gas," "deviation of oxygen content in pyrolysis gas," "pyrolysis gas temperature deviation," "water flow for hardening the pyrolysis gas."

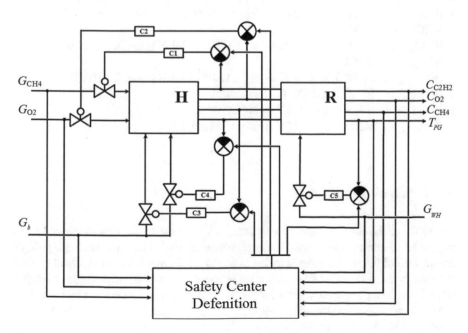

Fig. 4 Two-level control system based on safety center definition

Terms "LOW" (small), "MIDDLE" (medium), "HIGH" (many) are chosen for the description of linguistic variables, with the method of defuzzification being centroid [7–9].

Linguistic rules for fuzzy controller formulated by IF … AND … THEN … are as follows:

- IF (δT_{dg} is "HIGH") THEN (G_b is "LOW")
- IF (δT_{dg} is "MIDDLE") THEN (G_b is "MIDDLE")
- IF (δT_{dg} is "LOW") THEN (G_b is "HIGH")
- IF (δCO_2 is "HIGH") AND (CC_2H_2 is "LOW") AND (δCCH_4 is "LOW") THEN (GO_2 is "HIGH")
- IF (δCO_2 is "HIGH") AND (CC_2H_2 is "HIGH") AND (δCCH_4 is "LOW") THEN (GCH_4 is "HIGH")
- IF (δCO_2 is "LOW") AND (δCCH_4 is "MIDDLE") AND (CC_2H_2 is "LOW") THEN (G_{NG} is "MIDDLE") AND (GO_2 is "MIDDLE")
- IF (δCO_2 is "MIDDLE") AND (δCCH_4 is "LOW") AND (CC_2H_2 is "LOW") THEN (G_{NG} is "MIDDLE") AND (GO_2 is "MIDDLE")
- IF (δCO_2 is "LOW") AND (δCCH_4 is "MIDDLE") AND (CC_2H_2 is "HIGH") THEN (GO_2 is "HIGH")
- IF (δCCH_4 is "HIGH") AND (δCC_2H_2 is "HIGH") THEN (G_{NG} is "MIDDLE") AND (GO_2 is "MIDDLE")
- IF (δCCH_4 is "HIGH") THEN (GO_2 is "MIDDLE");
- IF (δCCH_4 is "HIGH") AND (CC_2H_2 is "LOW") THEN (GO_2 is "MIDDLE")
- IF (CC_2H_2 is "LOW") AND (δCCH_4 is "HIGH") THEN (G_{NG} is "MIDDLE") AND (GO_2 is "MIDDLE")
- IF (CC_2H_2 is "LOW") AND (δCO_2 is "HIGH") THEN (G_{NG} is "MIDDLE") AND (GO_2 is "MIDDLE")
- IF (δT_{PG} is "LOW") AND (CC_2H_2 is "LOW") THEN (G_{WH} is "HIGH")
- IF (δT_{PG} is "HIGH") THEN (G_{WH} is "HIGH");
- IF (δT_{PG} is "HIGH") THEN (G_{WH} is "LOW");
- IF (δT_{PG} is "MIDDLE") THEN (G_{WH} is "MIDDLE").

4   Research Findings

The study of the control system constructed with the use of both classical regulators used in local control loops and a fuzzy regulator is carried out. The main purpose of application of fuzzy controller is to determine tasks for regulators of the local control systems in the realization of steady-state regimes. As it was mentioned earlier, the rate of transient processes reaches 10^{-6} s^{-1}. Technical implementation management of such a system presents significant difficulties. Therefore, it is proposed to use a fuzzy hybrid system based on fuzzy and classical regulation.

As a result of simulating the proposed version of the control system for the Simulink package in MATLAB system, transient graphs for the parameters were

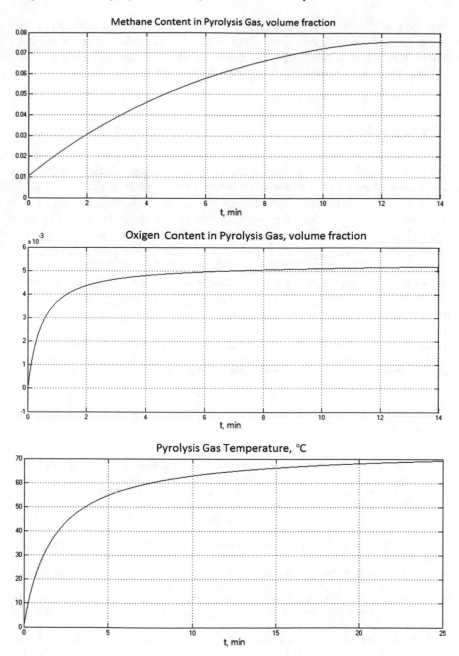

Fig. 5 Transient curves

obtained (Fig. 5). None of the considered cases of exceeding the permissible values of the variables that determine the technological safety of the oxidative pyrolysis process was observed. It is established that the model is sufficiently stable when the process of oxidative pyrolysis is brought to a steady state.

5 Conclusions

The paper considers the approach to the complex dynamic industrial technology management based on the application of fuzzy logic model, which is one of the artificial intelligence techniques. Its implementation can improve the quality of decision making and accordingly the quality of industrial technology control systems. The realization of industrial technologies occurs in the context of various accidental disturbances. However, this causes problems in assessing the process states. They generate a lot of management tasks, the solution of which is aimed at compensation of random prime causes of abnormal situations. In this regard, the solution of problems in safety management of dynamic processes in semi-structured and poor formalized environments is an urgent task.

Acknowledgements The study is performed under financial support of the Russian Federal Property Fund grant *The Research of Risks in Management of Dynamic Processes in Semi-structured and Poor Formalizable Environments*, project No. 17-07-01368.

References

1. Sanayeva, G.N., Prorokov, A.E., Bogatikov, V.N.: Investigation of dynamic behavior of acetylene production by oxidative pyrolysis of natural gas. In: Hu, Z., Petoukhov, S., He, M. (eds.) Advances in Artificial Systems for Medicine and Education. AIMEE, 2017. Advances in Intelligent Systems and Computing, vol 658. Springer, Cham (2018)
2. Melikhov, A.N., Bernshtein, L.S., Korovin, S.Y.: Situational Advisory Systems with Fuzzy Logic, 272 p. Nauka, Moscow (1990)
3. Pospelov, D.A.: Situational Management: Theory and Practice, 288 p. Nauka.-Ch. Ed. fiz.-mat. Lit., Moscow (1986)
4. Mukherjee, S., Bhattacharyya, R., Kar, S.: A novel defuzzification method based on fuzzy distance. Int. J. Fuzzy Syst. Rough Syst. (IJFSRS) **3**(1) (2010)
5. Van Dinh, N., Thao, N.X.: Some measures of picture fuzzy sets and their application in multi-attribute decision making. Int. J. Math. Sci. Comput. (IJMSC) **4**(3), 23–41 (2018). https://doi.org/10.5815/ijmsc.2018.03.03
6. Nasser, A.A., Al-Khulaidi, A.A., Aljober, M.N.: Measuring the information security maturity of enterprises under uncertainty using fuzzy AHP. Int. J. Inf. Technol. Comput. Sci. (IJITCS) **10**(4), 10–25 (2018). https://doi.org/10.5815/ijitcs.2018.04.02
7. Amini, A., Nikraz, N.: Proposing two defuzzification methods based on output fuzzy set weights. Int. J. Intell. Syst. Appl. (IJISA) **8**(2), 1–12 (2016). https://doi.org/10.5815/ijisa.2016.02.01
8. Mekidiche, M., Belmokaddem, M.: Application of weighted additive fuzzy goal programming approach to quality control system design. Int. J. Intell. Syst. Appl. (IJISA) **4**(11), 14–23 (2012). https://doi.org/10.5815/ijisa.2012.11.02

9. Kadhim, M.A., Alam, M.A., Kaur, H.: A multi-intelligent agent system for automatic construction of rule-based expert system. Int. J. Intell. Syst. Appl. (IJISA) **8**(9), 62–68 (2016). https://doi.org/10.5815/ijisa.2016.09.08

Secure Hash Function Constructing for Future Communication Systems and Networks

Sergiy Gnatyuk, Vasyl Kinzeryavyy, Karina Kyrychenko, Khalicha Yubuzova, Marek Aleksander and Roman Odarchenko

Abstract The application of Web technologies and forms of electronic document circulation in the process of information exchange between users though simplifies this process. However, generates a number of new threats for the confidentiality, integrity, and availability of information and the appearance of previously unknown vulnerabilities. One of the most common methods of protection is using the digital certificates that ensure the confidential exchange of data between a client and a server by encrypting and authenticating a digital certificate. A digital certificate is a public key, certified by the EDS of the certification center. Nevertheless, a digital certificate is not just a public key with information, but a so-called signature of a server or Web resource that is implemented using the hash functions. Information technologies development and the emergence of new attack types lead to increasing the amount of disadvantages of existing hash functions. Thus, in the paper a new hashing function was proposed, which was based on the SHA-2 hash function. Improvements involved a number of changes: increased the size of words and an increase in the message digest; at the preprocessing stage, the incoming message is supplemented by a pseudorandom sequence; the numbers of nonlinear functions are increased. The proposed changes allow to reduce the number of rounds in the compression function, which will guarantee at least similar stability indicators with simultaneous increase in data processing speed.

S. Gnatyuk (✉) · V. Kinzeryavyy · K. Kyrychenko · R. Odarchenko
National Aviation University, Kiev, Ukraine
e-mail: s.gnatyuk@nau.edu.ua

V. Kinzeryavyy
e-mail: 0werl0rd@ukr.net

K. Kyrychenko
e-mail: bezverkhayakarina@gmail.com

K. Yubuzova
Satbayev University, Almaty, Republic of Kazakhstan

M. Aleksander
State Higher Vocational School in Nowy Sącz, Nowy Sącz, Poland

© Springer Nature Switzerland AG 2020
Z. Hu et al. (eds.), *Advances in Artificial Systems for Medicine and Education II*,
Advances in Intelligent Systems and Computing 902,
https://doi.org/10.1007/978-3-030-12082-5_51

561

Keywords Information security · Cryptography · Hash function · Digital certificates · SHA-2

1 Introduction

Every year, the level of information security is increasing in organizations of different forms' ownership. First of all, this is due to an increase in the flow of information that is provided in real time with the help of Internet resources. Web-portals of the organization show results of its activities, provide online services and financial services, conduct financial transactions and others. The lion's share of the information circulating in the above-mentioned processes needs to be adequately protected. According to this, another important task is to ensure the proper protection of information during the exchanging data at the expense of Internet resources. One of the most common methods of such protection is the using of cryptographic certificates–digital certificates that ensure the confidential exchange of data between the client and using public key encryption. However, the certificate is not only an open key with information, but also a digital signature of a server or a Web portal that is implemented using hash function. But, the number of cyber attacks, especially on Web portals, has increased in geometric loopholes: blocking access, stealing confidential data, monitoring traffic, etc. At the intelligence stage, hackers monitor the network to identify weaknesses, where you can get the access to users' working machines and ultimately penetrate into the network. Improvement of efficiency of digital certificates, as one of the most common methods of protecting information in the process of exchange and connection, is relevant and needs improvement. Such cyber attacks as DROWN (a vulnerability that allows decrypting ciphertext without a private key) [1], FREAK (vulnerability that allows you to penetrate into the installed encrypted connection and analyze a traffic) [2], LOGJAM (vulnerability that allows reading and modification of data transmitted over a secure communications channel [3]) caused large losses to many Web resource owners, including such giants as Google, Mozilla, and Yahoo and put under the question reliability of digital certificates. Therefore, increasing the reliability of digital certificates, as the most common methods for protecting the exchange of data through communication channels, is relevant and needs to be improved. *The purpose of this work is* the development of secure method for constructing a hash function for using the information security systems of modern communication systems and networks.

2 The Analysis of Modern Approaches to Hash Functions Constructing

One of the most common cryptographic algorithms is hash function. They are necessary for "compressing" information into message digest that represent a bit combinations of fixed length. Hash functions of SHA-2 are very popular in applications related to the systematization, search and protection of information. Digital protocols use public key encryption to authenticate the client and server. At the confirmation stage, the hash function plays the role of the identification mark and is used to ensure the integrity of the data during transmission. That is, the hash function has to make it impossible to fake the certificate, while leaving the same signature of the verification center. Till recently, the SHA-1 hash function was used in digital certificates. In connection with the detection of numerous collisions in SHA-1 [4, 5] and in the most digital certificates [5–7], Microsoft, Google, and others initiated a decision to replace the hash function [8]. Starting in 2016, the SHA-2 hash function is used in the SSL certificate. However, technologies continue to improve the power of technology is increasing, and today many works are devoted to the investigation of cryptographic strength of SHA-2, in particular, the following shortcomings were identified in works [9–14]: collisions for truncated variants SHA-512; finding the first and second prototypes; a birthday attack. The paper proposes a new method for constructing a hash function; it is the prototype of SHA-512. In our opinion, this hash function can allow to improve the efficiency of cryptographic protection of digital certificates when it is applied.

3 Exposition of a New Method for Secure Hash Function Constructing

Preprocessing step. At the preprocessing step, an incoming message M ($M \in V_N$, $V_N \in \{0, 1\}^N$, N is message length M in bits, $N \in Z$, $N < 2^{128}$) are complemented by additional sequence D_l (message length M) and pseudorandom sequence $salt$ (is determined on the basis of M), so that the resulting message length is a multiple of the length data blocks L ($L = 1024 \cdot t'$ bits, $t' \in N$):

$$M_{rez} = (M, D_l, salt) \tag{1}$$

where $M_{rez} \in V_{NN}$, $V_{NN} = N + 128 + N_{salt}$, $D_l = H_{Dl}(M)$, $D_l \in V_{128}$, $salt = H_{Gen}(M)$, $salt \in V_{Nsalt}$, $N_{salt} = 2L - ((N + 128) \bmod L)$, as a function H_{Gen} could be any function of generating a pseudorandom sequence which is based on M, H_{Dl} is function of length M. Based on the completed message M_{rez} will determine the hash value of the message M.

Message M_{rez}, $M_{rez} \in V_{NN}$, broken into kL—bit blocks: $M_{rez} = (m_1, m_2 \ldots m_k)$ where $m_i \in V_L, i = (\overline{1,k})$, $k = (NN)/L$.

Determination step of a hash. The digest of the message is iteratively calculated, processing each one m_i block messages $M_{rez}, m_i \in V_L, i = (\overline{1,k})$ compression function F_g (2), to get the resulting hash (3):

$$h_i = F_g(h_{i-1}, m_i), \quad i = (\overline{1,k}) \tag{2}$$

$$H(IV, M_{rez}) = h_k \tag{3}$$

where $h_0 = IV$, IV is initialization vector, $IV \in V_{L/2}$, h_i is intermediate values of the messages digest $h_i \in V_{L/2}, i = (\overline{1,k})$; H is resulting hash, $H \in V_{L/2}$; F_g is the compression function uses in the hash function.

The compression function F_g is performed in three stages: splitting blocks into words (1), initialization of variables (2), compression (3).

Step 1. Each m_i data block $M_{rez}, m_i \in V_L, i = (\overline{1,k})$, decomposes into 16 words (4):

$$m_i = \left(W_0^i, \ldots, W_{15}^i\right) \tag{4}$$

where $W_j^i \in V_{L/16}, j = \overline{0, 15}$.

On the basis of words which are received $W_j^i, j = \overline{0, 15}$, words are calculated W_u^i (5) $W_u^i \in V_{L/16}, u = \overline{16, 63}$:

$$W_u = W_{u-16} + Delta0(W_{u-15}) + W_{u-7} + Delta1(W_{u-2}), \tag{5}$$

where $Delta0(W_u) = Rotr(W_u, 1) \oplus Rotr(W_u, 8) \oplus SHR(W_u, 7)$, $Delta1(W_u) = Rotr(W_u, 19) \oplus Rotr(W_u, 61) \oplus SHR(W_u, 6)$, $Rotr(x, l)$ is right bitwise cyclic shift of argument x for l-bits; $SHR(x, l)$ is left shift argument x for l-bits.

Step 2. Reinitialization of internal state vectors is performed T (6), $T = (T_1, \ldots, T_8)$, $T_z \in V_{L/16}, z = \overline{1, 8}$:

$$T_z = h_{i-1}^z \tag{6}$$

where $h_{i-1} = \left(h_{i-1}^1, \ldots, h_{i-1}^8\right)$, h_{i-1} is the previous value of the digest, which is fed to the input of the function F_g, $h_{i-1}^z \in V_{L/16}, z = \overline{1, 8}$.

Step 3. At this step, there is a direct compression of the data block $m_i \in V_L, i = \overline{1, k}$, $k = NN/L$, the value of the vectors of the internal state will change each 64 rounds $T = (T_1, \ldots, T_8)$, $T_z \in V_{L/16}, z = \overline{1, 8}$, through their mixing with vectors W_j and constants K_j $j = \overline{0, 63}$.

For each j round, the mathematical actions given in the formulas will be executed (7)–(11), $j = \overline{0, 63}$:

$$F_{g_1} = T_8 \oplus Sigma1(T_5) \oplus Ch(T_5, T_6, T_7) + W_j + K_j \qquad (7)$$

$$F_{g_2} = Sigma0(T_1) \oplus Maj(T_1, T_2, T_3) \qquad (8)$$

$$F_{g3} = JQ(T_3, T_6) \oplus Maj(T_2, T_3) \qquad (9)$$

$$F_{g_4} = SH(T_8, T_7) \oplus Sigma(T_8) \qquad (10)$$

$$T_8 = T_7 + F_{g_4}; \; T_7 = T_6; \; T_6 = T_5; \; T_5 = T_4 + F_{g_1}; \; T_4 = T_3;$$
$$T_3 = T_2 + F_{g_3}; \; T_2 = T_1; \; T_1 = F_{g_1} + F_{g_2} \qquad (11)$$

where T_z are vectors of the internal state, $T_z \in V_{L/16}, z = \overline{1, 8}$; W_j are words which are broken from m_i block; K_j are predetermined constants (if necessary, may change), $K_j \in V_{L/16}$; $Ch(x, y, z)$, $Maj(x, y, z)$, $Sigma0(x)$, $Sigma1(x)$, $Delta0(x)$, $Delta1(x)$, $JQ(x, y)$ and $SH(x, y)$ are nonlinear functions that are described in (12)–(19):

$$Sigma0(x) = Rotr(x, 28) \oplus Rotr(x, 34) \oplus Rotr(x, 39) \qquad (12)$$

$$Sigma1(x) = Rotr(x, 14) \oplus Rotr(x, 18) \oplus Rotr(x, 41) \qquad (13)$$

$$Ch(x, y, z) = (x + y) \oplus (\overline{x} + z) \qquad (14)$$

$$Maj(x, y, z) = (x + y) \oplus (x + z) \oplus (y + z) \qquad (15)$$

$$Delta0(x) = Rotr(x, 1) \oplus Rotr(x, 8) \oplus SHR(x, 7) \qquad (16)$$

$$Delta1(x) = Rotr(x, 19) \oplus Rotr(x, 61) \oplus SHR(x, 6) \qquad (17)$$

$$JQ(x, y) = (\overline{x} + y) \oplus Rotr(x, 13) \oplus SHR(\overline{y}, 17) \qquad (18)$$

$$SH(x, y) = SHR(x, 7) \oplus Rotr(y, 8) \oplus Rotr(\overline{x}, y) \qquad (19)$$

where F_g is intermediate compression function value, $F_{g_o} \in V_{L/16}, o = \overline{1, 4}$.

$Ch(x, y, z)$, $Maj(x, y, z)$, $Sigma0(x)$, $Sigma1(x)$, $Delta0(x)$, $Delta1(x)$, nonlinear functions that were used in the original SHA-2. $JQ(x, y)$ and $SH(x, y)$ are new nonlinear functions that were proposed in this hash function.

After completing the last round, the values of the vectors of the internal state $T = (T_1, \ldots, T_8)$, $T_z \in V_{L/16}, z = \overline{1, 8}$, completely changed as follows:

$$T_z = T_z \oplus h_{i-1}^z, \tag{20}$$

where h_{i-1} is the previous value of the digest, which is fed to the input of the function $F_g \, h_{i-1} = (h_{i-1}^1, \ldots, h_{i-1}^8)$, $h_{i-1}^z \in V_{L/16}$, $z = \overline{1,8}$. The output of the function will be given to the final values of the internal state vectors.

In our opinion, the method for constructing a hash function is developed by adding a pseudorandom sequence *suh* to an incoming message at the preprocessing and non-linear operations $JQ(x, y)$ and $SH(x, y)$ at the step of determining the hash values. It is possible to reduce the total number of rounds in the compression function with similar or better performance and security indicators data in the aspect of resistance to various attacks and neutralization of known vulnerabilities compared with the SHA-2. Theoretical and experimental researches will be conducted for the verification of this statement and cryptanalysis performing in the further works.

4   Experimental Study and Discussion

For the experimental study on the basis of the proposed method, three hash functions with such parameters were constructed: $t' = 1$, $L = 1024 \cdot t' = 1024$, $H \in V_{L/2} = V_{512}$ for BK_1; $t' = 2$, $L = 1024 \cdot t' = 2048$, $H \in V_{L/2} = V_{1024}$ for BK_2; $t' = 3$, $L = 1024 \cdot t' = 3072$, $H \in V_{L/2} = V_{1536}$ for BK_3. As a function H_{Gen} for BK_i, $i = \overline{1,3}$ cryptographic algorithm SNOW 2.0 was selected. The software implementation of proposed hash functions was carried out as console tool using the programming language C++. Development environment was Microsoft Visual Studio 2013 (Release Version).

Therefore, to study the statistical characteristics of hash functions, these were investigated in the statistical test NIST STS [15]. Also proposed hash functions were compared with the results of the benchmark generator of pseudorandom sequences BBS and some block symmetric ciphers (Kalyna [16], Luna [17], Neptun [17]), which worked in counter mode. Note that for this research, based on the developed hash functions and the SHA-512 function, stream ciphers were constructed to generate required length files for NIST STS statistical tests.

In Figs. 1, 2, and 3, the statistical portraits of the passage of statistical tests are given, and in Table 1 the results of the study were showed. It showed that the proposed functions of healing passed a comprehensive control in accordance to NIST STS.

Also, the study of the rate characteristics for developed hash functions BK_i, $i = \overline{1,3}$ was carried out. To do this, several files of different sizes were randomly selected and hash code was scanned for each selected file while measuring the hash code time (see Table 2).

All experiments were performed using the system with following characteristics: Intel (R) Core (TM) i3-6100 processor, 3.7 GHz processor, and a 4-GB RAM based on the 64-bit Windows 7 Service Pack 1.

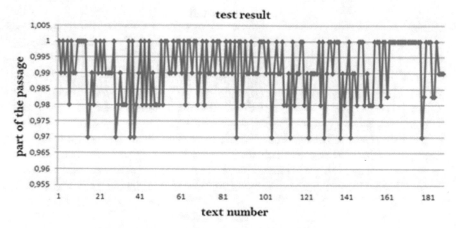

Fig. 1 Statistical portrait of the stream cipher on the basis of BK_1

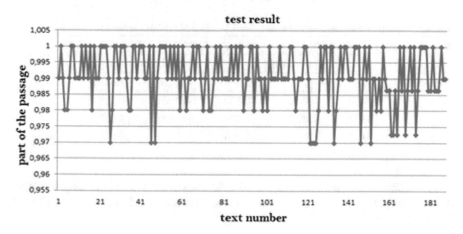

Fig. 2 Statistical portrait of the streaming cipher on the basis of BK_2

Table 1 Sequence testing results

Generator	Number of tests in which the testing was completed	
	99% sequence	96% sequence
BBS	133 (70.70%)	188 (100%)
Kalyna	136 (72.30%)	188 (100%)
Luna	141 (75.00%)	188 (100%)
Neptun	140 (74.46%)	188 (100%)
SHA-512	137 (72.87%)	188 (100%)
BK_1	141 (75.00%)	188 (100%)
BK_2	140 (74.46%)	188 (100%)
BK_3	142 (75.53%)	188 (100%)

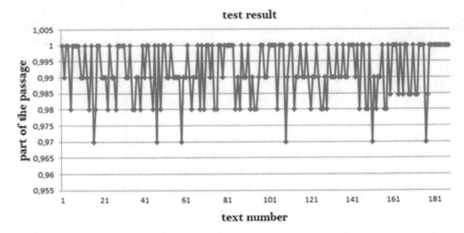

Fig. 3 Statistical portrait of the streaming cipher on the basis of BK_3

Table 2 Results of the study for the speed characteristics of the hash functions

Hash function	File 1, 1 MB		File 2, 10 MB		File 2, 100 MB	
	t (s)	v (MB/s)	t (s)	v (MB/s)	t (s)	v (MB/s)
SHA-512	0.015	68.26	0.145	70.62	1.38	74.20
BK_1	0.012	85.33	0.101	101.38	0.926	110.58
BK_2	0.011	93.09	0.098	104.48	0.902	113.52
BK_3	0.011	93.09	0.094	108.93	0.879	116.49

According to the obtained results of the developed hash functions' speed charac-teristics BK_i, $i = \overline{1, 3}$, are better than well-known and widely used SHA-512 hash function.

5   Conclusions

The paper proposes a new method for secure hash function constructing, which can be used to improve the effectiveness of cryptographic protection of digital certificates in the future. It will provide a more reliable exchange of confidential information on the network. The method requires further research to test the performance on different platforms, the resistance to common methods of cryptanalysis. In the following works, it is planned to conduct the above-mentioned studies and compare them with the parameters of the hash functions of the SHA-2 series.

References

1. Aviram, N., Schinzel, S., Somorovsky, J.: DROWN: breaking TLS using SSLv2. In: Proceedings of the 25th USENIX Security Symposium, p. 18 (2016). [Online]. Available: https://drownattack.com/drown-attack-paper.pdf |Date accesses: April 2018|
2. Green, M.: Attack of the week: FREAK (or 'factoring the NSA for fun and profit'). [Online]. Available: https://blog.cryptographyengineering.com/2015/03/03/attack-of-week-freak-or-factoring-nsa/ |Date accesses: April 2018|
3. Duncan, B.: Weak Diffie-Hellman and the Logjam attack. [Online]. Available: https://weakdh.org/ |Date accesses: April 2018|
4. Karpman, P., Peyrin, T., Stevens, M.: Practical free-start collision attacks on 76-step SHA-1. [Online]. Available: https://eprint.iacr.org/2015/530 |Date accesses: April 2018|
5. Sanadhya, S., Sarkar, P.: 22-step collisions for SHA-2. [Online]. Available: http://arxiv.org/abs/0803.1220 |Date accesses: April 2018|
6. Kohlar, F., Schage, S.: On the security of TLS-DH and TLS-RSA in the standard model1, p. 50 (2013). [Online]. Available: http://eprint.iacr.org/2013/367.pdf |Date accesses: April 2018|
7. Meyer, C., Schwenk, J.: Chair for network and data security Ruhr-University Bochum. Lessons learned from previous SSL/TLS attacks. A brief chronology of attacks and weaknesses, p. 15. [Online]. Available: http://eprint.iacr.org/2013/049.pdf |Date accesses: April 2018|
8. Castelluccia, C., Mykletun, E.: Improving secure server performance by re-balancing SSL/TLS handshakes, p. 11 (Published in Proceeding ASIACCS '06 Proceedings of the 2006 ACM Symposium on Information, Computer and Communications Security, pp 26–34)
9. Mendel, F.: Improving local collisions: new attacks on reduced SHA-256, p. 17. [Online]. Available: https://eprint.iacr.org/2015/350.pdf |Date accesses: May 2017|
10. Dobraunig, C., Eichlseder, M.: Analysis of SHA-512/224 and SHA-512/256, p. 30. [Online]. Available: https://eprint.iacr.org/2016/374.pdf |Date accesses: May 2017|
11. Gnatyuk, S., Kovtun, M., Kovtun, V., Okhrimenko, A.: Search method development of birationally equivalent binary Edwards curves for binary Weierstrass curves from DSTU 4145-2002. In: Proceedings of 2nd International Scientific-Practical Conference on the Problems of Infocommunications. Science and Technology (PIC S&T 2015), pp. 5–8, Kharkiv, Ukraine, 13–15 Oct 2015
12. Hu, Z., Gnatyuk, S., Koval, O., Gnatyuk, V., Bondarovets, S.: Anomaly detection system in secure cloud computing environment. Int. J. Comput. Netw. Inf. Secur. (IJCNIS) 9(4), 10–21 (2017). https://doi.org/10.5815/ijcnis.2017.04.02
13. Hu, Z., Gnatyuk, V., Sydorenko, V., Odarchenko, R., Gnatyuk, S.: Method for cyberincidents network-centric monitoring in critical information infrastructure. Int. J. Comput. Netw. Inf. Secur. (IJCNIS) 9(6), 30–43 (2017). https://doi.org/10.5815/ijcnis.2017.06.04
14. Gnatyuk, S., Okhrimenko, A., Kovtun, M., Gancarczyk, T., Karpinskyi, V.: Method of algorithm building for modular reducing by irreducible polynomial. In: Proceedings of the 16th International Conference on Control, Automation and Systems, pp. 1476–1479, Gyeongju, Korea, 16–19 Oct 2016
15. NIST Special Publication 800-22: A statistical test suite for random and pseudorandom number generators for cryptographic applications. [Online]. Available: https://nvlpubs.nist.gov/nistpubs/legacy/sp/nistspecialpublication800-22r1a.pdf
16. Oliynykov, R., Gorbenko, I., Kazymyrov, O., Ruzhentsev, V., Kuznetsov, O., Gorbenko, Y., Dyrda, O., Dolgov, V., Pushkaryov, A., Mordvinov, R., Kaidalov, D.: DSTU 7624:2014. National Standard of Ukraine. Information Technologies. Cryptographic Data Security. Symmetric Block Transformation Algorithm. Ministry of Economical Development and Trade of Ukraine (2015). (In Ukrainian)
17. Gnatyuk, S., Kinzeryavyy, V., Iavich, M., Prysiazhnyi, D., Yubuzova, Kh.: High-performance reliable block encryption algorithms secured against linear and differential cryptanalytic attacks. In: Proceedings of the 14th International Conference on ICT in Education, Research and Industrial Applications. Integration, Harmonization and Knowledge Transfer, vol. II: Workshops, pp. 657–668, Kyiv, Ukraine, 14–17 May 2018

Code Obfuscation Technique for Enhancing Software Protection Against Reverse Engineering

Sergiy Gnatyuk, Vasyl Kinzeryavyy, Iryna Stepanenko, Yana Gorbatyuk, Andrii Gizun and Vitalii Kotelianets

Abstract Software reliability is an actual problem caused with absence of software protection techniques, especially absence of software code protection from reverse engineering. Computer piracy and illegal software usage makes big damage for state economy. The development of new approaches and modification of existed obfuscation technologies is actual task directed to growing efficiency of secure coding and protection against reverse engineering. In this paper, authors present the obfuscation method for software protection, which ensures protection from reverse engineering. The method is based on a new sequence of obfuscation transformations. Also software tool StiK was developed, and based on the submitted sequence of operations, pseudocode for protection method was created. Experimental study was conducted according to the presented technique. Experimental results show the efficiency and generality of the proposed method (StiK obfuscator is 10% faster as well as 1.37 times more protected than analogues). Consequently, the developed technique can be used to prevent or at least hamper interpretation, decoding, analysis, or reverse engineering of software.

Keywords Information security · Software code protection · Obfuscation technique · Obfuscation · Obfuscation transformations · Secure coding · Obfuscation algorithm

S. Gnatyuk (✉) · V. Kinzeryavyy · I. Stepanenko · Y. Gorbatyuk · A. Gizun · V. Kotelianets
National Aviation University, Kiev, Ukraine
e-mail: s.gnatyuk@nau.edu.ua

V. Kinzeryavyy
e-mail: v.kinzeryavyy@gmail.com

I. Stepanenko
e-mail: stepanenko.iryna.v@gmail.com

Y. Gorbatyuk
e-mail: pravo@nau.edu.ua

A. Gizun
e-mail: caesar07@meta.ua

© Springer Nature Switzerland AG 2020
Z. Hu et al. (eds.), *Advances in Artificial Systems for Medicine and Education II*,
Advances in Intelligent Systems and Computing 902,
https://doi.org/10.1007/978-3-030-12082-5_52

1   Introduction

Software security against computer pirates and unauthorized users is actual problem during few decades. It makes serious damage to software engineering industry. This problem is amplified by speed multimedia and Internet technologies' development because the quantity of ways to get non-license content grows every day. Modern world is characterized by break of software every year, month, and day. It costs billions of US dollars. Attacks with code study and its hidden vulnerabilities detection are the most serious.

In the core of every software tools, there is intellectual property of its developers. Software protection against unauthorized usage, modifying, and copying is the most important issue in modern information and communication systems. Computer piracy and illegal software usage causes big damage for state economy particularly in hi-tech sector.

Development of effective security methods for codes is very important to prevent non license products expansion. It is one of the main tasks for companies' developers as well as for state policy in IT sector.

Today, there are many approaches to this problem solving. These are encryption, watermarking, etc., but no one gives guaranteed result. Also there are many obfuscation transformations in scientific and technical literature [1], but detailed obfuscation security methods description are absent. But this technology is one of prospect methods to make hard illegal code study and modification. From this viewpoint, the development of new approaches and modification of existed obfuscation technologies are actual tasks directed to growing efficiency of secure coding and protection against reverse engineering.

Modern software development companies should provide high-quality and secure information product to their customers. An important aspect of software development is to guarantee its reliability and integrity. Illegal intruders try to get source code and bypass the licensing stage. Therefore, it is necessary to protect not only the software overall but also the source code. Obfuscation protection technique could be used to confuse the program code and complicate the analysis and comprehension of operation algorithms furthermore preserving the functional program features. One of the main problems of software reliability is absence of the developed and implemented software protection methods, especially absence of software code protection from reverse engineering [1–3].

2   Analysis of Modern Obfuscation Techniques

The analysis of modern approaches to obfuscation was carried out [1, 3–6], and it has showed a lot of attention from researchers to the basic obfuscation process requirements, described categories of obfuscation distribution transformations, and presented obfuscation protection methods.

Modern obfuscation methods for secure coding were analyzed in review papers [7, 8], and papers [9, 10] contain effective software security techniques with some elements and procedures of code obfuscation.

However, a complex mechanism for source code protection that uses most of the known obfuscation transformations is absent.

Consequently, to decrease the probability of the reverse engineering process implementation, it is important to develop a reliable technique of obfuscation protection of executable software files.

The *main purpose* of research is to create a reliable obfuscation technique for software security that provides program code protection against the reverse engineering process.

To achieve purpose, the following tasks were completed:

- New obfuscation technique of software protection was developed;
- Software tool for source code protection was created;
- Experimental study of created software tool was carried out.

Next parts of paper contain theoretical background and experimental study of proposed obfuscation technique.

3 Theoretical Background of Proposed Technique

The obfuscation technique called StiK was developed based on a new sequence of obfuscation transformations to solve mentioned problems [7, 8].

The input data for this technique is:

- *Source Code A*;
- *Obfuscation Code Structure Transformation $S = (S_1, \ldots, S_6)$, S_i* is one of the transformations: the restructuring of the arrays; the clone method; modification of loop conditions; the dead code method; use of mark <<goto>>; the parallel code method, $i = \overline{1, 6}$;
- *Obfuscation Transformation of Variables $V = (V_1, \ldots, V_4)$, V_j* is one of the transformations: inheritance relations modification, splitting or merging of variables, huge variables, converting static data to procedure, $j = \overline{1, 4}$;
- *Obfuscation Punctuation Transformation $P = (P_1, P_2)$, P_k* is one of the transformations: inversion of code elements, token removing $k = \overline{1, 2}$.
- In this technique, was determined the obfuscation sequence which must be accomplished to provide effective software protection from the reverse engineering process, to complicate the analysis process and comprehension of the program code algorithm.

Code Obfuscation Technique Description

Stage 1. First, the source code A is loaded into a memory fragment. This code is divided into logical structures $A = (A_1, \ldots, A_n)$, (A_x—the logical structure of the code A, $n \in N$, $x = \overline{1, n}$). The amount of the logical structures is determined by the program code size. Then logic units A_x, $x = \overline{1,n}$ are being transformed using some obfuscation structure transformations S_i, $i = \overline{1,6}$. The number of transformations for each logical structure is determined accidentally. The obfuscation transformation number for each A_x should be at least three. The final part is to combine all the elements into the code and to test its availability.

Stage 2.   Next, the resulting code B was divided into logical structures $B = (B_1, \ldots, B_m)$ (B_y—the logical structure of the code B, $m \in N$, $y = \overline{1,m}$). Logical structures B_y, $y = \overline{1,m}$ are being processed by obfuscation variable transformation V_j, $j = \overline{1,4}$ (the obfuscation transformation number for each B_y should be at least two). Obtained transformed structures were combined into program code C, and its availability was tested.

Stage 3.   The resulting code C is being divided into logical structures $C = (C_1, \ldots, C_g)$ (C_z—the logical structure of the code C, $g \in N$, $z = \overline{1,g}$). Next, logical structures C_z, $z = \overline{1,g}$ are being transformed by obfuscation punctuation transformations P_k, $k = \overline{1,2}$. As a result of these stages, the software code D is created and its availability is tested.

To provide more reliable software protection mechanism using more marks <> after each obfuscation stage of the StiK method was proposed. This item increases the stability index of the software code and the average difference index between the transformation code and the source code. To implement it the violation of program logic and unreadable tangled code were used. The output software code is completely modified; however, the program preserves a whole functionality of the source code, presented in Fig. 1.

A pseudocode was developed for the StiK protection method. The basic procedures according to the scheme are presented in Fig. 2.

A procedure *OpenFile* () is used to download the program code. Apply to its input the file name that will be transformed.

A *FunRand* () is applied to generate a pseudorandom sequence whereby the corresponding indexes were obtained that will be used for the code transformation.

A procedure *DivFunction* () is used to separate the program code into logical structure.

CodeStructure (), *VariableFun* (), *PunctuationFun* () procedures are applied for each stage of obfuscation transformation. Data about logical structures and generated obfuscation transformation indices in appropriate procedures were entered.

A *FunGoTo* () is an unconditional branch instruction, which is applied several times in each transformation.

A procedure *AssociationF* () is used to combine separate logical structures into a single code.

A procedure *Cheking* () is applied for sanity check of the generated code.

A procedure *WriteFile* () saves the transformed code as a new file.

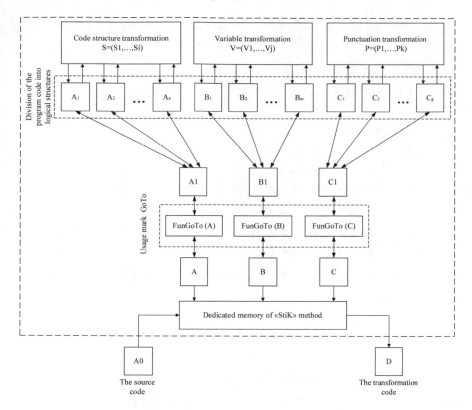

Fig. 1 Scheme of the obfuscation protection method

4 Experimental Study and Discussion

The software tool was developed based on the submitted sequence of operations and created pseudocode for protection method. The StiK software tool was created using C++ programming language, in the Visual Studio 2013 programming environment. Obfuscation transformations were defined from each category: the dead code method, usage of mark <<goto>>, huge variables, token removing.

Experimental study was conducted to determine the speed characteristics of the obfuscation process and the stability of the software code to the reverse engineering process. The experimental two-part methodology was developed. The input data for these experiments was ten files with source code; they were transformed using the developed software tool.

Using this console tool and two-part experimental methodology, the efficiency of StiK was studied in following manner:

1. Obfuscation rate was assessed (Experiment 1).
2. Security of code against reverse engineering was assessed (Experiment 2).

Input: *NameFile* is the file name with the source code, transformations S, P, V.
Output: *NameFileNew* is the file name with the transformation code.

1. $A = OpenFile(NameFile);$
2. $\{A_x\} = DivFunction(A)$, $A = (A_1,...,A_n)$, A_x is the logical structure A, $n \in N$, $x = \overline{1,n}$;
3. $for(x = 1; x \le n; x++)$
    3.1. $\{A_x\} = FunGoTo(\{A_x\})$
    3.2. $for(x_1 = 1; x_1 \le 3; x_1++)$
            3.2.1. $\{i\} = FunRand(S)$;
            3.2.2. $A_x = CodeStructure(A_x, S, i)$, $i \in \overline{1,6}$;
    3.3. $\{A_x\} = FunGoTo(\{A_x\})$
    3.4. $B = AssociationF(\{A_x\})$;
4. $\{B_y\} = DivFunction(B)$, $B = (B_1,...,B_m)$, B_y is the logical structure B, $m \in N$ $y = \overline{1,m}$;
5. $for(y = 1; y \le m; y++)$
    5.1. $\{B_y\} = FunGoTo(\{B_y\})$
    5.2. $for(y_1 = 1; y_1 \le 2; y_1++)$
            5.2.1. $\{j\} = FunRand(V)$;
            5.2.2. $B_y = VariableFun(B_y, V, j)$, $j \in \overline{1,4}$;
    5.3. $\{B_y\} = FunGoTo(\{B_y\})$
    5.4. $C = AssociationF(\{B_y\})$;
6. $\{C_z\} = DivFunction(C)$, $C = (C_1,...,C_g)$, C_g is the logical structure C, $g \in N$ $y = \overline{1,g}$;
7. $for(z = 1; z \le g; z++)$
    7.1. $\{C_g\} = FunGoTo(\{C_g\})$
    7.2. $for(z_1 = 1; z_1 \le 1; z_1++)$
            7.2.1. $C_z = PunctionFun(C_z, P, z)$;
    7.3. $\{C_g\} = FunGoTo(\{C_g\})$
    7.4. $D = AssociationF(\{C_z\})$;
8. $Cheking(D)$
9. $WriteFile(D, NameFileNew)$

Fig. 2 Pseudocode for the obfuscation protection method

Experimental studies were fulfilled using computer system with processor Intel (R) Core (TM) i3-6100, 3.7 GHz, and 4 GB RAM based on 64-bit operation system Windows 7 Service Pack 1. Also all results were compared with similar results of well-known and effective SmartAssembly obfuscator (see Table 2).

Experiment 1. To realize this experiment input / transformed files size as well as time for obfuscation were fixed. Obfuscation rate was assessed using expression $v_i = O_i/t_i, i = \overline{1, 10}$, where v_i is obfuscation rate for ith file; O_i is transformed ith file size, t_i is time for ith file obfuscation. For all assessed obfuscators, the parameter of middle rate was calculated using expression $v_{\text{cep}} = \sum_{i=1}^{n} v_i/n$.

In Table 1, results of Experiment 1 are presented.

In accordance to Table 1, StiK obfuscator is **10% faster** than SmartAssembly.

Experiment 2. During this study, ten executed files were analyzed and these files were protected using SmartAssembly and StiK obfuscators. Coefficient of code growing was calculated using formula $k_i = CT_i/CP_i, i = \overline{1, 10}$, where CT_i is number of processes in transformed code, CP_i is number of processes in input code.

Table 2 contains results of Experiment 2.

In accordance to Table 2, StiK obfuscator is **1.37 times more protected** than SmartAssembly (middle coefficient for StiK is equal 5.95 and for SmartAssembly this coefficient is 4.33).

5 Conclusions

In this paper, new obfuscation method for software protection was proposed and it was based on a new sequence of obfuscation transformations. It allows to provide software protection from the reverse engineering. Also the software tool has been developed according to StiK method, and the experimental studies have been carried out.

As a result, authors found that the average difference index between the transformation code and the source code was 36.21% and the average speed characteristics for obfuscation process were 140.6 Kb per second. In accordance to experimental results, StiK obfuscator is 10% faster as well as 1.37 times more protected than analogues.

However, in the future, implementation of more obfuscation transformations is planning and also comparative analysis with existing obfuscation programs will be carried out.

Table 1 Results of obfuscation rate assessing (Experiment 1)

File	1	2	3	4	5	6	7	8	9	10	Middle rate
File size (kB)	4.09	0.90	5.18	2.68	3.08	1.34	1.51	1.18	0.72	3.55	–
t (s) SmartAssembly	0.1	0.008	0.14	0.012	0.029	0.042	0.028	0.016	0.038	0.027	
t (s) StiK	0.103	0.038	0.12	0.058	0.061	0.033	0.041	0.022	0.017	0.058	
File size (kB) SmartAssembly	4.95	1.86	10.0	3.08	3.42	3.86	3.01	3.56	1.31	3.68	
File size (kB) StiK	8.23	3.75	14.7	6.97	5.86	4.71	4.96	5.27	3.95	5.33	
Rate (kB/s) SmartAssembly	49.5	202.5	75.5	256.6	118.9	91.7	107.3	205.6	34.1	136.3	127.8
Rate (kB/s) StiK	82.9	99.7	132.6	138.1	96.1	162.7	125.9	239.5	232.4	95.9	**140.6**

Table 2 Results of security of code against reverse engineering assessing (Experiment 2)

Categories	1	2	3	4	5	6	7	8	9	10	Middle
Number of processes in input code	14	4	39	22	16	17	31	8	5	28	18.4
Number of processes in StiK	55	19	383	58	38	123	157	81	49	106	106.9
Number of processes in SmartAssembly	47	26	187	39	67	84	85	46	32	79	69.2
Coefficient for StiK	3.93	4.75	9.82	2.64	2.38	7.24	5.06	10.13	9.80	3.79	**5.95**
Coefficient for SmartAssembly	3.36	6.50	4.79	1.77	4.19	4.94	2.74	5.75	6.40	2.82	**4.33**

References

1. Yadegari, B., Johannesmeyer, B., Whitely, B., Debray, S.: A generic approach to automatic deobfuscation of executable code. In: IEEE Symposium Security and Privacy (S&P), 18 p. (2014)
2. Buzatu, F.: Methods for obfuscating Java programs. J. Mobile, Embed. Distrib. Syst. **4**, 25–30 (2012)
3. Hu, Z., S, Gnatyuk, Koval, O., Gnatyuk, V., Bondarovets, S.. Anomaly detection system in secure cloud computing environment. Int. J. Comput. Netw. Inform. Secur. (IJCNIS) **9**(4), 10–21 (2017). https://doi.org/10.5815/ijcnis.2017.04.02
4. Hu, Z., Gnatyuk, V., Sydorenko, V., Odarchenko, R., Gnatyuk, S.: Method for cyberincidents network-centric monitoring in critical information infrastructure. Int. J. Comput. Netw. Inform. Secur. (IJCNIS) **9**(6), 30–43 (2017). https://doi.org/10.5815/ijcnis.2017.06.04
5. Danik, Y., Hryschuk, R., Gnatyuk, S.: Synergistic effects of information and cybernetic interaction in civil aviation. Aviation **20**(3), 137–144 (2016)
6. Garg, V., Srivastava, A., Mishra, A.: Obscuring mobile agents by source code obfuscation. Int. J. Comput. Appl. **61**(9), 46–50 (2013)
7. Stepanenko, I., Kinzeryavyy, V., Nagi, A., Lozinskyi, I.: Modern obfuscation methods for secure coding. Ukrainian Sci. J. Inform. Secur. **22**(1), 32–37 (2016). https://doi.org/10.18372/2225-5036.22.10451
8. Kinzeryavyy, V., Stepanenko, I., Lozinskyi, I.: Obfuscation method for software code protection. Ukraine Eng. Acad. J. **2**, 81–85 (2016)
9. Jeet, K., Dhir, R.: Software module clustering using hybrid socio-evolutionary algorithms. Int. J. Inform. Eng. Electron. Bus. (IJIEEB) **8**(4), 43–53 (2016). https://doi.org/10.5815/ijieeb.2016.04.06
10. Kaur, J., Tomar, P.: Clustering based architecture for software component selection. Int. J. Mod. Educ. Comput. Sci. (IJMECS) **10**(8), 33–40 (2018). https://doi.org/10.5815/ijmecs.2018.08.04

Modern Method and Software Tool for Guaranteed Data Deletion in Advanced Big Data Systems

Sergiy Gnatyuk, Vasyl Kinzeryavyy, Tetyana Sapozhnik, Iryna Sopilko, Nurgul Seilova and Anatoliy Hrytsak

Abstract In today's digitally empowered world, confidential data protection is a bit difficult especially when the user lacks knowledge about the importance of data sanitization before discarding, selling, or donating devices with memory (storage). To protect data, generally, users encrypt files and folders, hard drives, password protect flash drives and encrypt SD card on their smartphone to avoid unauthorized access to their confidential data. But almost all of these devices are discarded, sold, or donated without proper data sanitization at the end of their life cycle. Deleted data can be easily recovered from any storage media whether it's a USB thumb drive, HDD, SSD, SD card, etc. Interestingly, data from a formatted, encrypted, and broken or crushed storage media can also be recovered using data recovery software and services. Therefore, to protect and safeguard your confidential data from an unauthorized access, you must erase the files/folders and wipe the empty spaces (unused space) on your hard drive volume as the empty spaces might contain previously deleted sensitive and private data that can be easily recovered. This paper investigates international standards and methods of guaranteed data deletion that exist in different countries, a comparative characteristic of data deletion software. Also in this study, the method of guaranteed data deletion that has been improved due to the developed pseudorandom sequence generator develops algorithms and software for this method and conducts experimental researches of developed solutions to verify their effectiveness. The developed software module STM Shredder refers to

S. Gnatyuk (✉) · V. Kinzeryavyy · T. Sapozhnik · I. Sopilko · A. Hrytsak
National Aviation University, Kyiv, Ukraine
e-mail: s.gnatyuk@nau.edu.ua

V. Kinzeryavyy
e-mail: v.kinzeryavyy@gmail.com

T. Sapozhnik
e-mail: tanya.sapojnik@gmail.com

I. Sopilko
e-mail: pravo@nau.edu.ua

N. Seilova
Satbayev University, Almaty, Republic of Kazakhstan
e-mail: seilova_na@mail.ru

© Springer Nature Switzerland AG 2020 581
Z. Hu et al. (eds.), *Advances in Artificial Systems for Medicine and Education II*,
Advances in Intelligent Systems and Computing 902,
https://doi.org/10.1007/978-3-030-12082-5_53

information security particularly in Big Data concept. The results of experiments show its ability to expand the capabilities of information processing and storage systems for large files (faster by 30%).

Keywords Information security · Data protection · Guaranteed data deletion · Data deletion software · Data recovery · Big Data

1 Introduction

Today, the issue of security is very important. Computer technologies are used everywhere and became part of the document flow. The focus of information security takes when it is introduced, processed, transferred, and stored, when the issue of protection of deleted information remains open.

As the data transmission through public and private networks increases, it becomes increasingly important to protect the privacy of information stored and exchanged between computers. Information concealment may be needed in different situations. The confidentiality of the user's work at the computer must be ensured by the concealment and secrecy of the work. It should provide cleanup of different parts of the computer. Mentioned issues are critical for big amount of data like Big Data concept (Fig. 1).

There are often situations in which information stored on media or hard disks needs to be deleted without the possibility of their recovery. Many people think that it is enough to format the hard drive, clean the trash, or just delete private files or directories. Such operations allow only complicate access, but the information remains not deleted. The file on the computer is located in different parts of the disk in the form of fragments. All information about which sectors of the drive match this file are stored in a special database is FAT. To restore such a deleted file, it is enough to find a previous copy of FAT, returning the deleted record to the place [1].

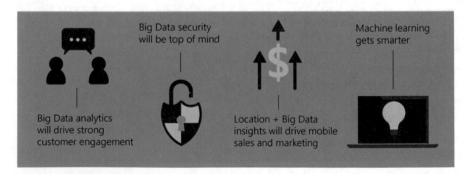

Fig. 1 Big data trends

2   Related Works and Existed Approaches on Guaranteed Data Deletion

Today, there are many methods of guaranteed data deletion, which are standardized in almost all advanced countries, issued national standards, norms and rules governing the use of software tools for the deletion of information and describing the mechanisms for its implementation.

Methods of deletion of information are described as sequence of operations intended to carry out software and/or hardware irreversible data deletion, including residual information [2].

Theoretically, the easiest way of deletion an output file is its full rewrite by a binary mask of eight binary units, zeros, or any other arbitrary numbers, thereby making its restoration impossible for software tools available to the user [3].

To ensure that there is no possibility of data recovery, there are international standards for data deletion, which are described below. Table 1 gives a brief description of each analyzed method.

Russian Standard GOST P50739-95, the algorithm conducts one cycle of rewrite with pseudorandom numbers.

Table 1   Comparative analysis of modern methods for data deletion

Method	Number of rounds	Advantages	Disadvantages
Russian standard GOST R50739-95	1	The fastest method	Outdated. Rewrite data can be used only 1 time
German Standard VSITR	7	More reliable than DoD	Does not guarantee data deletion
Bruce Schneier's method	7	The method is more reliable, compared with German, with the same number of cycles	The reliability of information deletion is low
Peter Gutman's method	35	Absolutely reliable, it isn't possible to restore any byte	Takes long period of time
American standard DoD 5220.22-M	3	Balanced reliability and speed	Not reliable
Canadian RCMP TSSIT OPS-II	7	Performs 7 cycles, which ensures greater reliability of data deletion	Not completely tested, does not exist as an independent software
American standard NAVSO P-5239-26	3	Used for MFM-encoded devices	Small number of cycles, not always effective, some files can be restored

National Standard of the US Department of Defense DoD 5220.22M. Option E, in accordance with this standard, involves two cycles of recording; in the ECE version, data overwriting is performed seven times.

American National Standard NAVSO P-5239-26. This standard provides three cycles of data rewrite: The first one is #01, in the second is #7FFFFFF, in the third is a sequence of pseudorandom numbers.

German National Standard VSITR, from the first to the sixth consecutively written bytes #00 and #FF, in the seventh #AA [3, 4].

Canadian RCMP TSSIT OPS-II contains seven rewrite passes.

Bruce Schneier's algorithm: #FF is recorded in the first cycle, #00 in the second cycle, and pseudorandom numbers in the five other cycles. It is considered one of the most effective algorithms.

Peter Gutmann's method, developed by Peter Gutmann, is considered to be the most reliable due to the presence of 35 cycles [4–7].

After standards analyzing, we can say that the efficiency and duration of data deletion depend on the length of the rounds used in the methods. Among the reviewed methods, the method of Peter Gutmann proved to be most effective, but it takes a lot of time. The fastest is the Russian standard, but the data deleted by it can be restored. From viewpoint of related works analysis, modern approaches to guaranteed data deletion are presented in [8–13], where mathematical simulations as well as commercial software tools were investigated.

Paper [8] proposes ErasuCrypto, a lightweight secure deletion framework with low block erasure and data migration overhead, it integrates both erasure-based and encryption-based data deletion methods and flexibly selects the more cost-effective one to securely delete invalid data.

In [9], the authors propose a page-level deletion operation called scrubbing, and main idea is to directly overwrite an invalid data page to turn all the bits in the page to 0, thus deleting the invalid data.

The study [10] provides a new scheme that aims to improve the security of FADE by using the Trusted Platform Module that stores safely keys, passwords, and digital certificates for Big Data and Clouds.

In [11], a novel SDN-based Big Data management approach was proposed with respect to the optimized network resource consumption such as network bandwidth and data storage units.

Also, the work [12] focuses on secure data deletion by proposing an assured data deletion scheme which fulfills verifiable data deletion as well as flexible access control over sensitive data. The proposed in mentioned paper protocol takes advantage of attribute-based encryption, whose security can be proved under standard model and also the theoretical analysis and the implementation results demonstrate the feasibility of authors' proposal.

In [13] using a variety of crypto-ransomware samples, authors employ reverse engineering and dynamic analysis to evaluate the underlying attack structures and data deletion techniques employed by the ransomware.

Today, there is a large amount of software that can be used to restore practically all data. Any software contains information deletion algorithms that represent the sequence of operations intended to delete data. However, these methods are outdated for modern tools and technologies. Therefore, the development of new and improved existing methods of guaranteed data removal is actual, which will allow to quickly and effectively delete all information without the possibility of its restoration.

The *purpose of this work* is to increase the effectiveness of guaranteed data deletion with the help of a new software module based on the improved method of guaranteed data deletion for advanced Big Data systems.

3   Tools for Guaranteed Data Deletion

Those described international standards and methods of guaranteed data deletion are used in various software tools, which are intended to completely delete all the files and folders of the software tool, as well as records in the computer's registry.

These software tools also contain their own data deletion algorithms. Each of the described utilities has its own interface and functions, as well as depending on whether the paid version is free, or on its capabilities.

For the experimental study, the most famous utilities operating on Windows as well as on Linux were used.

One such software is *Secure-Delete,* fast and easy to use, but there are problems with removing on SSDs and flash drives.

Hardwipe is another data deletion software which contains several rewriting options, but some algorithms that are embedded in the tool are not always effective.

Freeraser Setup is a tool, which includes three data deletion modes: fast, reliable, and uncompromising, the latter is available only in the paid version; the other two are not effective.

Files Terminator Free uses various removal methods, but the process itself takes a lot of time. The tool can delete both entire folders and individual files.

File Shredder is convenient and easy to use data deletion software, it contains five removal algorithms, but after performing the operation leaves temporary files, somewhat slow in comparison with other tools.

The Privacy Eraser Free tool not only deletes data using three known algorithms, but also provides detailed information for each workflow.

Shred is a simple and structured way of deleting data with the ability to set the number of required cycles, but not all types of files can be removed from the first time, does not delete directories.

After analyzing software for guaranteed data deletion, we can conclude that some utilities have quickly and efficiently deleted files of any size. Some of them work too long and not always effective. In addition, algorithms with a small number of cycles are not always useful, and algorithms with more cycles take a lot of time and do not always have the desired result.

4 Proposed Method of Guaranteed Data Deletion

After the analysis of standards and tools for the removal and recovery, a new method of guaranteed data deletion is proposed named *STM Schredder*. It allows effectively deletion data and is based on an improved algorithm of guaranteed information deletion. The basis of the guaranteed information deletion method is posed as several options for overwriting: replacing data with zeros, replacing data with pseudorandom numbers, and replacing those numbers whose values are derived from special tables. The algorithm (interconnected steps) of the method is presented in Fig. 2.

The structure of the method can be described as follows:

– *Step 0*. Start—entering the input data.

Inputs can be files of different formats, such as *jpeg, pdf, docx, txt, xls, mp3, zip, dng, psb, png, avi, mpeg,* and many others with different sizes, for example, from 1 KB to 1 GB. At that, they can be located both on the main computer and on any media.

– *Step 1*. Rewrite information by 1 and 0 in the sectors of the hard drive or SSD disk (Fig. 3 [4–7]).
– *Step 2*. Rewrite by ones, 1 is written at each byte disk sector.
– *Step 3*. Rewrite inverted information data. In byte sector where 1 is located, this byte will be rewritten as 0 and vice versa.
– *Step 4*. Rewriting information is executed as bit sequence that has a length of 32 bits, which was generated using security—developed pseudorandom number generator.
– *Step 5*. Overwrite byte sectors by 0.
– *Step 6*. Shutdown. After passing five stages of rewrite, the information is completely cached in the sectors of the disk and deleted.

By using the developed method, recovery will be impossible, even if using special software to recover deleted information. Thus, the developed method consists of five

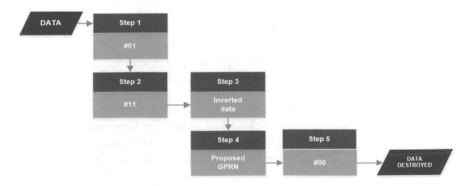

Fig. 2 Method of guaranteed data deletion: steps of realization

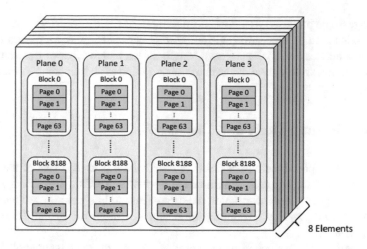

Fig. 3 The four-level hierarchy of 64GiB SSD with 8 elements (each element contains 4 planes, each comprising 2048 blocks of 64 4KiB pages)

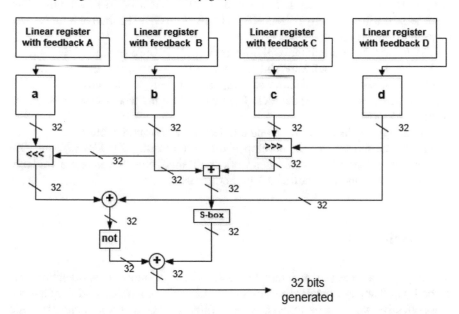

Fig. 4 Proposed registers scheme

steps that can increase the removal speed, which differs from international guaranteed data deletion methods like Bruce Schneier method, German or Canadaian standards, that use seven stages.

Using the developed pseudorandom number generator, which is used in the fourth step, allowed to improve efficiency of data deletion without possibility of recovery as opposed to the American Standard DoD, NAVSO, and Russian, which are using only three stages. The proposed pseudorandom number generator security, which is used in Step 4, is based on four shift registers with linear feedback A, B, C, and D, each of which generates 32 bits, has its own length, order, and number of transactions.

Figure 4 shows a structural diagram of the operation of all four generator registers.

5   Experimental Study and Discussion

Based on the proposed method of deleting data, which is used in one of the stages, a pseudorandom number generator, a console software module *Shredder STM* was developed for guaranteed information deletion. This module was realized using C++ programming language in developing environment Microsoft Visual Studio 2013 (release version).

An experimental research was conducted to confirm the achievement of the goal. The speed and efficiency of the guaranteed removal of information from the *Shredder STM* software were compared. To conduct an experiment on five PCs, three files were deleted with a capacity of 36.5 KB, 9.1 MB, 7.96 GB. The time of removal was measured, and the speed was calculated.

To compare the results of the experiment, three files were used which were deleted by the *Shredder STM* software and another software tools, and then the average value of the speed of data deletion was compared. To verify the effectiveness, special software was used to recover the files. They were checking whether the deleted files could be restored.

The testing results of the speed and efficiency are shown in Table 2.

The results showed that the developed software module *STM Shredder* on 10% slower than the famous software when works with small files. Nevertheless, for large files the software module showed a 30% better result.

6   Conclusions

In this paper, an analysis of international standards devoted to data deletion was carried out. They are standardized in almost all major countries, and they are the most effective way to remove data. By analyzing the standards, it is clear that the efficiency and time of deleting data depend on the length of rounds used in the methods.

Also, in order to develop our own software module, the analysis was conducted which shown the comparative characteristics of software for guaranteed data removal that allowed to describe the main advantages and disadvantages of each method.

Based on the analysis, the method of data deletion, which uses five stages of data rewrite, has been improved. One of them is using the proposed pseudorandom sequence generator. An experimental study was conducted, which showed that the developed software module STM Shredder would allow to not only increase the

Table 2 The testing results of the speed and efficiency

Tools' name	File size	File type	Time for deleting	Speed, KB/s	Recovery (foremost, final data)
Secure-Delete	36.5 KB	.pdf	3 s	12.1	Not recovered
	9.1 MB	.mp3	57 s	159.6	Not recovered
	7.96 GB	.mkv	1 h 23 min	1586	Not recovered
Hardwipe	36.5 KB	.pdf	1 s	35.6	Not recovered
	9.1 MB	.mp3	3 s	3033	Not recovered
	7.96 GB	mkv	26 min 7 s	5041	Recovered with mistakes
Freeraser Setup	36.5 KB	.pdf	2 s	17.8	Not recovered
	9.1 MB	.mp3	54 s	168	Not recovered
	7.96 GB	.mkv	1 h 15 min	1755	Not recovered
File Shredder	36.5 KB	.pdf	2 s	17.8	Not recovered
	9.1 MB	.mp3	3 s	3033	Not recovered
	7.96 GB	.mkv	1 h 48 min	1219	Not recovered
Privacy Eraser Free	36.5 KB	.pdf	1 s	35.6	Recovered
	9.1 MB	.mp3	4 s	2275	Recovered
	7.96 GB	.mkv	22 min 33 s	5838	Recovered
Shred	36.5 KB	.pdf	1 s	35.6	Not recovered
	9.1 MB	.mp3	6 s	1516	Not recovered
	7.96 GB	.mkv	3 min 17 s	41.578	Recovered with mistakes
Eraser	36.5 KB	.pdf	3 s	11.8	Not recovered
	9.1 MB	.mp3	5 s	1820	Not recovered
	7.96 GB	.mkv	51 min	2581	Recovered
Shredder STM	36.5 KB	.pdf	3 s	19.5	Not recovered
	9.1 MB	.mp3	7 s	1353	Not recovered
	7.96 GB	.mkv	1 h 9 min	2077	Not recovered

effectiveness of using pseudorandom number generator, but also to reduce the time of overwriting operation and will be useful for advanced Big Data systems (faster by at least 30% for large files).

References

1. Gapak, O.: Definition of the length of the period of generators of pseudorandom sequences based on shift registers with feedback and transfer. Model. Inf. Technol. **73**, 92–97 (2014)

2. Bem, M., Gorodsky, I., Sutton, G., Rodionenko, O.: Protection of Personal Data Legal Regulation and Practical Aspects. Kyiv, 220 p. (2015)
3. Kozhenevsky, S.: Methods of guaranteed data deletion on hard magnetic disks. Access mode: http://www.epos.ua/
4. Hu, Z., Gnatyuk, S., Koval, O., Gnatyuk, V., Bondarovets, S.: Anomaly detection system in secure cloud computing environment. Int. J. Comput. Netw. Inf. Secur. (IJCNIS) 9(4), 10–21 (2017). https://doi.org/10.5815/ijcnis.2017.04.02
5. Shredding, or how to remove information without restoration? Archive of articles, 2011. Access mode. http://arhiv-statey.pp.ua/index.php?newsid=5934
6. Hu, Z., Gnatyuk, V., Sydorenko, V., Odarchenko, R., Gnatyuk, S.: Method for cyberincidents network-centric monitoring in critical information infrastructure. Int. J. Comput. Netw. Inf. Secur. (IJCNIS) 9(6), 30–43 (2017). https://doi.org/10.5815/ijcnis.2017.06.04
7. Tikhomirov, A., Kinash, N., Gnatyuk, S., Trufanov, A., Berestneva, O., et al.: Network society: aggregate topological models. In: Communications in Computer and Information Science. Springer International Publishing, vol. 487, pp. 415–421 (2014)
8. Liu, C., Khouzani, H.A., Yang, C.: ErasuCrypto: a light-weight secure data deletion scheme for solid state drives. In: Proceedings on Privacy Enhancing Technologies, pp. 132–148 (2017)
9. Wei, M., Grupp, L.M., Spada, F.E., Swanson, S.: Reliably erasing data from flash-based solid state drives. In: 9th USENIX Conference on File and Storage Technologies, pp. 105–117 (2011)
10. Igarramen, Z., Hedabou, M.: FADETPM: novel approach of file assured deletion based on trusted platform module. In: 2017 3rd International Conference of Cloud Computing Technologies and Applications, #17577517 (2017). https://doi.org/10.1109/cloudtech.2017.8284727
11. Chaudhary, R., Aujla, G., Kumar, N., Rodrigues, J.: Optimized big data management across multi-cloud data centers: software-defined-network-based analysis. IEEE Commun. Mag. 56(2), 118–126 (2018). https://doi.org/10.1109/MCOM.2018.1700211
12. Yu, Y., Xue, L., Li, Y., Du, X., Guizani, M., Yang, B.: Assured data deletion with fine-grained access control for fog-based industrial applications. In: IEEE Transactions on Industrial Informatics, pp. 1–11 (2018). https://doi.org/10.1109/tii.2018.2841047
13. Zimba, A., Wang, Z., Simukonda, L.: Towards data resilience: the analytical case of crypto ransomware data recovery techniques. Int. J. Inf. Technol. Comput. Sci. (IJITCS) 10(1), 40–51 (2018). https://doi.org/10.5815/ijitcs.2018.01.05

Computer Implementation of the Fuzzy Model for Evaluating the Educational Activities of the University

N. Yu. Mutovkina ⓘ

Abstract The article describes the computer implementation of an objective assessment of the university's educational activities. It is expertly established that the effectiveness of educational activity is an integral indicator, consisting of three integrated characteristics: The effectiveness of educational activities in terms of methodological support, the quality of the educational process, and the effectiveness of educational activities in the part of the level of vocational training of graduates. Each of these characteristics consists of individual indicators, expressed in different units of measurement. Different perceptions by individuals of the importance of each of these indicators, as well as a wide variation in understanding how effective educational activity should look, make the development of such a valuation model relevant. The model of estimation is based on the ideas of the theory of fuzzy logic and fuzzy sets. The model is developed in the Fuzzy Logic Toolbox package of the MATLAB software environment. The model is open for modification. Any registered user can change the model at his discretion by adding or removing evaluation parameters, linguistic rules. The work is of interest to a wide range of readers, since the evaluation of the educational activity of universities is a question that many people, somehow connected with the sphere of higher education, are asking.

Keywords Computer modeling · Expert judgments · Fuzzy logic · Educational activity · Membership functions

1 Introduction

In the conditions of dynamic changes in the labor market, stricter requirements for the training of modern specialists, to their professional competencies, competition between educational institutions are increasing. This is especially acute during the admission campaign. According to the Federal Law "On Education in the Russian

N. Yu. Mutovkina (✉)
Tver State Technical University, Tver, Russia
e-mail: letter-boxNM@yandex.ru

© Springer Nature Switzerland AG 2020
Z. Hu et al. (eds.), *Advances in Artificial Systems for Medicine and Education II*,
Advances in Intelligent Systems and Computing 902,
https://doi.org/10.1007/978-3-030-12082-5_54

Federation" [1] and the Rules for Admission to Higher Educational Institutions, the applicant can apply to five universities for three training courses in each. So in modern conditions, not a university chooses a student, but a student chooses a university. This is especially true for students who scored a large number of points in the Unified State Exam. The parameters of the choice of the university, as a rule, are the following: university status, career prospects, the number of budget places, the cost of training, the level of interest of the university in the employment of graduates, the conditions for providing a hostel, the presence or absence of internal examinations, university Olympiads, the presence or absence of a military department, and the location of university. Obviously, universities with a good (above average) reputation will gain a greater number of the best students. The reputation of the university is formed depending on how successfully educational activities are carried out in it. This is a complex and multifaceted notion, which includes both teaching and methodological support, the organization of the learning process, as well as mastery, the qualification of the teaching staff, the training of students in the ability to perceive broadcast information, and so on. The most indicative criteria for the effectiveness of educational activities are the level of employment of graduates of the university, as well as the proportion of those graduates who have continued their studies in magistracy and graduate school.

In the face of increasing competition between universities, the actual issue is determining the effectiveness of the educational activities of universities. Consumers of this information are both external and internal users. External users include: entrants and their immediate relatives; state authorities regulating higher education in the country; and representatives of business structures. Internal users are: students of the evaluated university, its faculty. All listed users have the right to receive reliable information about the university, its activities, its successes, and failures.

The originality of the article is the formation of a system of indicators that most fully characterizes the educational activity of any university. In addition, the proposed system of linguistic rules possesses uniqueness. This system was created by the author of the article on the basis of the opinions and conclusions of representatives of all interested groups listed above.

The significance of the work is that, guided by the model proposed by the author, anyone can evaluate the educational activity of universities and draw conclusions for themselves. In addition, the user can, at his own discretion, change the model, adding to it those estimated parameters that he deems important.

The aim of the work is to demonstrate the computer method of implementation of the theory of fuzzy sets and the theory of fuzzy logic for objective evaluation of the effectiveness of educational activities of the university through the development and implementation of the author's evaluation model.

Evaluation of the success and effectiveness of educational activities is a complex and responsible procedure that can be carried out both with the help of classical statistical methods and through a system of fuzzy logical inference. Further, the author of the article discusses the advantages of the model of fuzzy estimation of the effectiveness of educational activity describes the model and the results of its work.

2 Theoretical Provisions for the Formation of the Evaluation Procedure

To assess the effectiveness of the activities of educational institutions, systems for assessing the quality of education at different managerial levels are designing. These systems allow to receive to the various subjects of the educational process reliable information on the quality of the services provided, to correlate the expectations of consumers and their satisfaction with the quality of education.

The system of assessing the quality of education, as a rule, is a set of procedures for internal and external evaluation of the results of educational activities of its participants and the factors that affect these results. These factors include the features of educational services and the conditions of the educational process. External evaluation mechanisms include procedures for independent examination, and internal assessment of the quality of education includes monitoring the dynamics of development of subjects of the educational process according to the relevant parameters of the quality assessment system [2].

In assessing the effectiveness of educational activities of higher education institutions, a statistical approach is widely used, which is as follows:

(1) Experts determine a set of parameters for assessing educational activities, for which there is a certain statistical information. This set usually does not include qualitative parameters, since their measurement is difficult;
(2) For each parameter, experts are assigning a weighting factor that indicates the importance of the parameter relative to the importance of other parameters in the set. Weight factors can be changed depending on the tasks of monitoring;
(3) Experts set benchmarks there are the values of the parameters below which the indicators are considered unacceptable. The reference indicators are determined for each parameter of the information-evaluation model of the effectiveness evaluation. This can be: the average value of the parameter for all universities; minimum (or maximum) possible value of the parameter; parameter value determined by regulatory documents;
(4) Calculation of deviations of actual values of parameters from their reference values:

$$\Delta_i^{abs.} = x_i - x_i^* \to 0 \tag{1}$$

where x_i is the value of the i-th parameter of the effectiveness of the educational activity of the university; x_i^* is its reference value;
(5) A comparison of the values of the indicators obtained for a particular university with the indicators on a sample of higher educational institutions. For each parameter, the average value is calculated for all units of the aggregate that are in the sample. In addition, the mean square deviation of each parameter is calculated:

$$\Delta_i^{\text{relative}} = \left(\frac{x_i - \tilde{x}_i}{s_i} \right) \cdot k_i \tag{2}$$

where x_i is the value of the i-th parameter of the effectiveness of the educational activity of the university; $\tilde{x}_i$ is sample mean on the parameter; s_i is standard deviation of the i-th parameter; k_i is weight coefficient of the i-th parameter;

(6) The calculation of the final effectiveness (FE) of the educational activities (FA) of the university by the formula:

$$\text{FE}_{\text{EA}} = \sum_{i=1}^{n} \Delta_i^{\text{relative}} \tag{3}$$

where n is the number of parameters defined in the first stage.

The foregoing approach, undoubtedly, proved itself well in the conduct of valuation procedures, but it has any shortcomings. For example, all parameters must be quantitative; it is necessary to spend a lot of effort and money to conduct a statistical study; all statistical data must be reliable; expert assessments are used in the approach, which means that verification of their consistency is necessary [3]. If expert estimates are not agreed, then they cannot be used; otherwise, the results of the assessment will be distorted.

In solving these problems, the evaluation of the effectiveness of educational activity on the basis of the linguistic rules of fuzzy logic can help. The methodology of fuzzy logic and fuzzy sets increasingly finds its application in valuation activities [4, 5].

3   Creating an Evaluation Model

In developing a fuzzy model for evaluating the educational activities of the university, the author of the article was guided by the theoretical positions of fuzzy sets and fuzzy logic, set forth in [6–8] and other sources, and therefore, these points are not considered in detail here.

It is expertly established that the effectiveness of the educational activity of the university (Y) is most fully described by three parameters: the result of educational activity in terms of methodological support (X_1); the quality of the teaching and educational (communicative) process (X_2); the effectiveness of educational activities in terms of the level of vocational training of graduates (X_3) [9]. In this case, the parameter X_1 can be considered as an intermediate product of educational activity, and the parameter X_3—as its final product (Fig. 1).

In turn, each of these parameters includes estimates of other particular parameters. Thus, each of these parameters becomes an output variable, and for it, there are input variables, namely: For the parameter, X_1 there are three input variables (x_{11}, x_{12}, and x_{13}); for the parameter X_2, there are five input variables (x_{21}, x_{22}, x_{23}, x_{24}, and

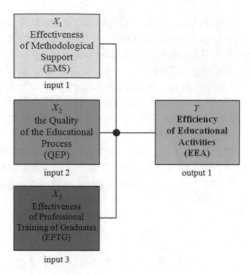

Fig. 1 General scheme for evaluating the educational activities of the university

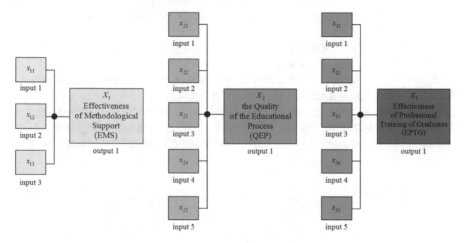

Fig. 2 Internal nesting of parameters X_1, X_2 and X_3

x_{25}); for the parameter X_3, there are five input variables (x_{31}, x_{32}, x_{33}, x_{34}, and x_{35}) (Fig. 2).

The output variable X_1 has such input variables as follows: x_{11} is the coefficient of the students' provision with educational and methodical literature; x_{12} is the average percentage of the provision of electronic educational and methodological complexes of disciplines (on the university); x_{13} is degree of conformity of educational and methodical complexes to modern requirements of directions and training profiles, %.

The output variable X_2 has input parameters as follows: x_{21} is the share of full-time teachers in the total number of teachers for the reporting period, %; x_{22} is the share of teachers with academic degrees during the reporting period, %; x_{23} is average age of full-time university teachers, years; x_{24} is the condition and equipment of classrooms (the percentage of their willingness to conduct classes); x_{25} is the state of computer resources and the availability of Internet access, %. The last two parameters are determined by the results of the questionnaire survey of teachers and students.

The input parameters for the output variable X_3 are as follows: x_{31} is percentage of graduates, who have received grades "good" and "excellent" on the state examination; x_{32} is the percentage of graduates, who received marks "good" and "excellent" on the protection of the final qualifying work (diploma); x_{33} is percentage of diplomas with distinction; x_{34} is percentage of graduates, who found work during the year after graduation of university; x_{35} is percentage of bachelor's graduates, who have continued their studies in magistracy and graduate school.

Each of the input parameters for X_1, X_2, and X_3 shown in Fig. 2 is expressed as a fuzzy set with three levels: low, medium, and high. Each of the parameters X_1, X_2, and X_3, being the input to Y, is a fuzzy set with five levels: low, below average, medium, above average, and high. For example, fuzzy sets of parameter X_2 are given in Table 1.

Scales similar to those shown in Table 1 are compiled for each parameter and in accordance with the levels of their nesting. Fuzzy sets of the output parameter are indicated in Table 2.

Further, in accordance with the theory of fuzzy logic [6], a rule base is created for the parameters X_1, X_2, X_3, and Y. This procedure begins in the reverse order, i.e., with variables of internal nesting. For each block of rules, all possible options are sorted out. Examples of rules are presented in Table 3.

Parameters x_{34} and x_{35} can be considered as alternatives. The average age of full-time university teachers (x_{23}) for less than forty years is recognized as low with confidence 1.0, and the age of more than seventy years is too large also with

Table 1 Fuzzy sets of parameter X_2, %

Level	0–20	21–40	41–60	61–80	81–100
Low	1.0	0.8	0.6	0.4	0.2
Below average	0.5	1.0	0.7	0.55	0.25
Medium	0.2	0.7	1.0	0.75	0.3
Above average	0.2	0.45	0.75	1.0	0.65
High	0.1	0.3	0.55	0.8	1.0

Table 2 Fuzzy sets of parameter Y, %

Level	0–19	20–39	40–59	60–79	80–100
Low	1.0	0.8	0.6	0.4	0.2
Below average	0.5	1.0	0.7	0.55	0.25
Medium	0.2	0.7	1.0	0.75	0.3
Above average	0.2	0.45	0.75	1.0	0.65
High	0.1	0.3	0.55	0.8	1.0

Table 3 Examples of linguistic rules

Rule block	Rule number	The formulation
X_1	1	IF x_{11}—low, AND x_{12}—low, AND x_{13}—low, THEN X_1—low;
	11	IF x_{11}—high, AND x_{12}—medium, AND x_{13}—medium, THEN X_1—above average;
	32	IF x_{11}—high, AND x_{12}—high, AND x_{13}—high, THEN X_1—high;
X_2	1	IF x_{21}—low, AND x_{22}—low, AND x_{23}—low, AND x_{24}—low, AND x_{25}—low, THEN X_2—low;
	19	IF x_{21}—medium, AND x_{22}—medium, AND x_{23}—medium, AND x_{24}—low, AND x_{25}—medium, THEN X_2—medium;
X_3	20	IF x_{31}—medium, AND x_{32}—high, AND x_{33}—low, AND x_{34}—medium, AND x_{35}—medium, THEN X_3—above average;
	37	IF x_{31}—high, AND x_{32}—high, AND x_{33}—medium, AND x_{35}—high, THEN X_3—high;

confidence 1.0. After all parameters are investigated, their definition areas and terms are given, the rule bases are created, and the fuzzy evaluation model is implemented in the Fuzzy Logic Toolbox.

4 Implementation of Fuzzy Estimation Model in the MATLAB Package

There are created four files: EMS.fis, QEP.fis, EPTG.fis, and EEA.fis (according to the abbreviations in Fig. 1). these files are entered the necessary number of input variables, their names, the intervals of their action (the domain of definition), and the corresponding terms. The **FIS editor** makes it possible to describe the basic properties of the fuzzy inference system. In this case, the following characteristics were chosen: the Mamdani algorithm (because of its features, indicated in [10]), method "prod" is method of algebraic product of the degree of truth of connected fuzzy statements, method "probor" is algebraic sum of degrees of truth of connected fuzzy statements, method of inferring the conclusion is prod, method "probor" is used to aggregate the values of the membership function of each of the output linguistic variables of the conclusions of fuzzy rules, method "mom" is the method of the average maximum is used to perform the defuzzification. In the editor of rules of fuzzy output (**Rule Editor**), all rules are set according to the principle of full enumeration. Fig. 3 shows a window for displaying rules for a variable Y.

The fuzzy visualization window for each set of listed rules, and accordingly for each input variable, appears when the **Rule Viewer** fuzzy rule viewer is launched. Thus, Fig. 4 shows the graphical interface of the rule system for the variable Y.

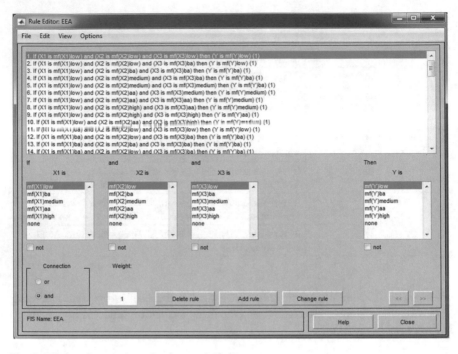

Fig. 3 Window for entering rules for a variable Y

With the help of the model realized in this way, it is possible to objectively evaluate the effectiveness of the university's educational activities.

5 Conclusions

The educational activity of the university is a key indicator of his effectiveness, status. Since higher education is one of the key conditions for the success of a person in the future, his professional growth, people quite seriously decide on the choice of a university. The universities themselves in a highly competitive environment are also interested in assessing his own educational activities. Such an assessment allows identifying the strengths and weaknesses, threats and possible risks for the educational institution. Therefore, the model proposed in this paper is the very tool by which this problem can be solved.

In the basis of the model for evaluating the educational activity of the university, both classical statistical procedures and methods of fuzzy mathematics can be used. The paper presents a computer evaluation model developed in the Fuzzy Logic Toolbox package of the MATLAB software environment. The model is based on a system of estimated parameters (Figs. 1 and 2) and on a system of linguistic rules, examples of which are presented in Table 3.

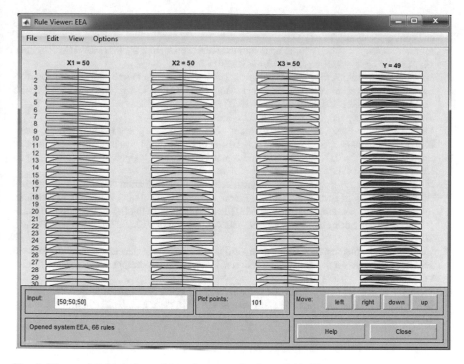

Fig. 4 The graphical interface of the rule system for the variable Y

For comparison and analysis of the results obtained with the help of classical statistical methods and fuzzy logic methods, computational experiments in MATLAB have been implemented, confirming the identity of the efficiency estimates based on a clear and fuzzy model, which indicates the adequacy of the latter. The proposed model is open for modification and can be used in evaluating other areas of activity the university: research, international, financial and economic, administrative and control activity.

Acknowledgements The reported study was funded by RFBR according to the research project No. 17-01-00817A.

References

1. Federal Law No. 273-FZ of December 29, 2012 "On Education in the Russian Federation" (as amended and supplemented, which effecting from 08.07.2018). http://www.consultant.ru/document/cons_doc_LAW_140174/ (date of access: 22.07.2018)
2. Pavlov, E.N.: Monitoring of the effectiveness of participants in an educational project using fuzzy logic. In: Pavlov, E.N., Shuykova, I.A. (eds.) Vestnik of the VSU. Series: System Analysis and Information Technologies, vol. 2, pp. 69–74 (2014)

3. Saati, T.: Decision-making. Method of the analysis of hierarchies. In: Saati, T (ed.) M.: Radio and communication, 278 p (1993)
4. Mitra, M., Das, A.: A fuzzy logic approach to assess web learner's joint skills. IJMECS **7**(9), 14–21 (2015). https://doi.org/10.5815/ijmecs.2015.09.02
5. Sunish Kumar, O.S.: A fuzzy based comprehensive study of factors affecting teacher's performance in higher technical education. Int. J. Mod. Educ. Comput. Sci. (IJMECS) **5**(3), 26–32 (2013). https://doi.org/10.5815/ijmecs.2013.03.04
6. Zadeh, L.A.: Fuzzy sets. Inf. Control **8**, 338–353 (1965)
7. Mutovkina, N.Y.: Fuzzy complex assessment of activities of the agent in multi-agent system. In: Hu, Z., Petoukhov, S., He, M. (eds.) Advances in Artificial Systems for Medicine and Education. AIMEE 2017. Advances in Intelligent Systems and Computing, vol. 658, Springer, Cham. pp. 303–314. https://doi.org/10.1007/978-3-319-67349-3_29
8. Mutovkina, N.Yu., Kuznetsov, V.N., Klyushin, A.Yu.: The formation of the optimal composition of multi-agent system. In: Hu, Z., Petoukhov, S., He, M. (eds.) Advances in Artificial Systems for Medicine and Education. AIMEE 2017. Advances in Intelligent Systems and Computing, vol. 658. Springer, Cham. pp. 293–302; https://doi.org/10.1007/978-3-319-67349-3_28
9. Kushel, E.S.: Implementation of the competitive strategy for the development of educational activities of the university. In: News of Higher Educational Institutions. Geology and Exploration. vol. 2, pp. 74–77 (2013)
10. Mamoria, P., Raj, D.: Comparison of mamdani fuzzy inference system for multiple membership functions. Int. J. Image, Graph. Signal Process. (IJIGSP) **8**(9), 26–30 (2016). https://doi.org/10.5815/ijigsp.2016.09.04

Temperature Field Simulation of Gyro Unit-Platform Assembly Accounting for Thermal Expansion and Roughness of Contact Surfaces

Mikhail V. Murashov

Abstract This chapter demonstrates the applicability of the previously developed finite element model of contact between rough bodies to macroproblems using the example of a model problem of the temperature estimation of gyro unit-platform assembly. Thermal contact conductance of the joint was preliminarily obtained for the range of operating contact pressures using the calculation by the column model for the micron-size roughness area. The problem was solved for bodies of unequal elastic-plastic hardening materials. Then, the macroproblem of warming up the gyro unit-platform assembly was solved accounting for the uneven distribution of pressure and correspondingly uneven thermal contact conductance over the real contact area. The significant influence of thermal expansion and tightening of the bolts fixing the gyro unit to the platform was determined.

Keywords Thermal contact conductance · Gyro unit · Finite element method · Elastic-Plastic deformation · ANSYS

1 Introduction

Artificial intelligence approach and precision technologies bring constantly increasing accuracy of assembled structures, and therefore, determination of their temperature state is an actual problem in engineering. For example, in machinery engineering, errors caused by thermal expansion are one of the most frequently overlooked and most frequent in causing difficulties in understanding [1]. The temperature field of assembled structures made of high thermal conductivity materials depends significantly on the thermal conduction of the contacts.

It should be emphasized that for constructions with a large number of contact surfaces, such as aircraft engines and internal combustion engines, it is sufficient to take the same value of thermal contact conductance for large groups of surfaces.

M. V. Murashov (✉)
Bauman Moscow State Technical University, Moscow 105005, Russia
e-mail: murashov@bmstu.ru

© Springer Nature Switzerland AG 2020
Z. Hu et al. (eds.), *Advances in Artificial Systems for Medicine and Education II*,
Advances in Intelligent Systems and Computing 902,
https://doi.org/10.1007/978-3-030-12082-5_55

601

This will not result in a significant error, since the heat has alternative transmission paths across many neighboring surfaces. However, individual values for the thermal contact conductance of a particular contact are usually required for systems with contacts critical for the transfer of heat. These are generally contacts of guidance systems devices and objects of precision instrument engineering, the electronic, and nuclear industry.

The problem of determining the thermal and deformed state of the construction of platform inertial navigation systems for launch vehicles is demonstrative [2, 3]. For such constructions, the requirements for deflection of the device axes due to temperature deformations may be less than $1''$, and the demand for the compliance of gyro units with their temperature regime may be to an accuracy of less than 1 °C. Additionally, while the possibility of heat removal in flight conditions is limited, the construction should not overheat. The decisive influence on the temperature state of such a construction and on providing it with the required accuracy is the thermal resistance in the contacts.

The study of thermal contact conductance has been practiced for more than 100 years, beginning with early work [4–6]. Semi-empirical, statistical, and deterministic models of heat conduction through contact have been developed, for example, [7–11]. A lot of highly cited articles [12–17] have been published on this topic. However, the key task of developing a reliable method for predicting the parameters of contact heat transfer has not been solved to date.

It has been experimentally confirmed [18, 19] that under the influence of contact pressure plastic deformation occurs and, consequently, hardening of the material of the asperities takes place. And beginning with [20], models of surface deformation appear that take into account the plastic response of the material. Some models provided for a simplified ideally plastic response of the material, for example, [21]. In fact, though, the material can be hardened both in the process of contact deformation and in cold work during preliminary surface treatment. The need to take hardening into account in models of deformation of rough surfaces was indicated in studies [22–26]. Attempts were made to take into account hardening of the surface in statistical models [27–29]. In addition to taking into account the initially hardened surface layer, it is essential to take into account the indentation size effect, which is manifested for a roughness of several microns and less [30].

Finite element modeling of rough surfaces has a number of significant advantages over statistical methods—it allows for accounting for the influence of the shape of asperities and of the change in the properties of materials during deformation. The absence of a stress–strain curve for asperities on micron scales is critical for calculating their deformation, since the available reference data obtained by means of macro experiment are not suitable. To overcome this problem, one can refer to indentation, drawing an analogy between the penetration of a roughness peak and the indenter penetration. Based on the results of indentation, the properties of the surface in which the penetration is performed can be obtained. Since the properties of an individual asperity are required, indentation needs to be made to a separate vertex, and not, as is usually done, to a polished smooth surface. Recovery of the stress–strain curve is possible only from the subsequent numerical simulation. Such experiment and simulation of penetration into the element of roughness were carried out for the first time in [31] where a comparison between the experiment on penetration of the Berkovich indenter into a separate roughness peak and the finite element modeling of this process suggests a method for determining the plastic properties of the asperities in a form in which they can be used for modeling the deformation of rough surfaces for problems of thermal contact conductance. The obtained properties, in addition to hardening, describe the existing cold work hardening and indentation size effect.

On the basis of [31], a column micromodel of heat transfer through a rough contact was developed, covering all these phenomena. A description of this model and the results of studying it are presented in [11]. However, to date it has not been shown that this model makes possible the inclusion of the results of calculating the thermal contact conductance for rough surfaces into the calculations of macroscale constructions. This possibility is demonstrated in this current work using the example of calculating the distribution of temperatures for the gyro unit-platform assembly as a typical module of the platform inertial navigation system that significantly affects its temperature state. In addition, the effect of thermal expansion of the construction and the uneven distribution of pressure from the clamp force of the bolts on the actual contact area are considered. Axial and bending external loads can modify the contact area [32], but here, their absence is assumed.

To solve this problem, the dependence of the thermal contact conductance on the pressure is calculated using a column model describing the contact of rough surfaces, herein called the micromodel. Then, the indicated dependence is transferred to the model of the gyro unit-platform assembly module, herein called the macromodel, in view of the absence of the description of parts less than a millimeter in size.

In actual constructions, the thermal contact conductance is not the same at different points of contact and is also transient. The solution for macromodels will open the opportunity to evaluate and analyze the transient behavior of the value of the thermal contact conductance. Based on the results of preliminary calculations, the thermal contact conductance varies with the change in the temperature field of the constructions, which also follows from the physical meaning of the phenomenon. This, apparently, can explain the noticeable change in the measured thermal contact conductance as a function of the deformation rate in the hot forging of metal, described in [33]. Faster loading forms a different temperature field of the die and the billet and, consequently, changes the thermal contact conductance. Another factor leading ultimately to transient thermal contact conductance is thermal expansion. These phenomena can have a significant impact specifically on macroscale constructions.

2 Macromodel of Construction

In the right-handed Cartesian reference system, let us consider the hypothetical construction of a gyro unit attached to a platform (Fig. 1). The model of the gyro unit is a

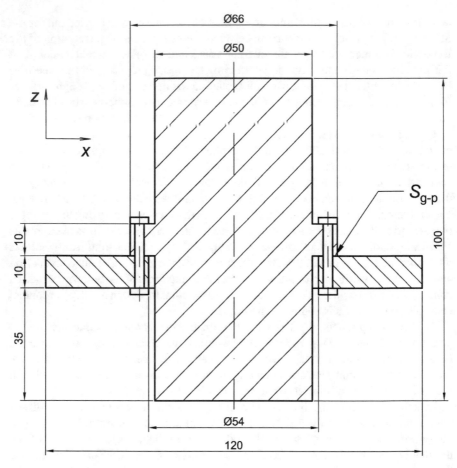

Fig. 1 Geometry of macromodel (cross section shown, dimensions in mm). S_{g-p}—contact surface of gyro unit-platform

solid cylinder with a flange, made of AISI 1020 steel. Inside the gyro unit, we assume the presence of a thermostatic system maintaining on a constant level the temperature of some internal part of the device. At the same time, the temperature of the device body is not constant and is determined, among other things, by interaction with the platform. The model does not consider the internal parts and the thermostatic system as separate bodies, but it takes into account that this system generates heat with a power of 20 W, corresponding to a specific heat generation rate $q = 94{,}832$ W/m^3 for the given volume of the gyro unit. Heat transfer by radiation is neglected. The gyro unit is fixed from the flange onto the square platform of 1050 UNS A91050 Aluminum with four M3 bolts made of AISI 5140 steel. Half-length of the platform side is $L1 = 0.06$ m. Distance from the bottom surface of the gyro unit to the bottom surface of the platform is $L2 = 0.035$ m. The behavior of the bolt material is considered linearly elastic. The clamp force of each bolt F is 2000 N. The joints of the parts are in a vacuum. Isotropic dry friction according to the Amontons–Coulomb law with a friction coefficient $\mu = 0.5$, constant in the process of deformation, is taken into account in the joints. Initially, the contact surfaces of the macromodel are ideally flat. The nominal contact area of the gyro unit-platform is 1.103×10^{-3} m^2. The model material properties are presented in Table 1.

The gyro unit-platform assembly model is constrained at the four corners of the lower platform plane as follows. One corner is constrained from displacement along all three axes x, y, z. The corner adjacent to it is constrained along axis z and along the axis orthogonal to the connection line of the first and second corners. The third and fourth corners are constrained only along axis z. Thus, these constraints do not oppose thermal expansion, i.e., do not cause additional thermal stresses.

On the right wall of the platform, heat is removed with an intensity of $h = 500$ W/(m^2 K) and coolant temperature $T_{coolant} = 20$ °C. Thermal expansion was taken into account when calculating the assembly model. Reference temperature is $T_0 = 20$ °C. The coupled problem of heat conductance and deformation in a stationary setting was solved. All loads were applied simultaneously in a single load step.

Let us move on to the index designations of axes of the Cartesian reference system $x_i, i = 1, 2, 3$. The following mathematical model corresponds to the problem, including equilibrium equations (1), the generalized Hooke's law (2), the flow rule (3), thermal strain relation (4), strain-displacement relations (5), von Mises yield condition (6), ratio for calculation of contact pressure of augmented Lagrangian method (7), Amontons–Coulomb friction law (8) on the contact surface, structural

Table 1 Macromodel material properties

Part	Material grade	Modulus of elasticity E, GPa	Poisson ratio ν	Yield strength, MPa	Linear coefficient of thermal expansion α, 1/K	Thermal conductivity k, W/(m K)
Gyro unit	AISI 1020 steel	212	0.36	280	11.1×10^{-6}	86
Bolts	AISI 5140 steel	212	0.3	–	12×10^{-6}	41
Platform	1050 UNS A91050 aluminum	71	0.32	55.7	23.5×10^{-6}	210

boundary conditions (9)–(12), heat conduction equation (13), and thermal boundary condition (14)

$$\sigma_{ji,j} = 0; \tag{1}$$

$$\varepsilon_{ij} = \frac{1+\nu}{E}\sigma_{ij} - \frac{\nu}{F}\delta_{ij}\sigma_{kk}; \tag{2}$$

$$d\varepsilon_{ij}^{P} = s_{ij}d\lambda; \tag{3}$$

$$d\varepsilon_{ij}^{T} = \alpha(T - T_0); \tag{4}$$

$$\varepsilon_{ij} = \frac{1}{2}\left(u_{i,j} + u_{j,i}\right); \tag{5}$$

$$(\sigma_1 - \sigma_2)^2 + (\sigma_2 - \sigma_3)^2 + (\sigma_3 - \sigma_1)^2 = 2\Phi\left(\varepsilon^{P}\right)^2; \tag{6}$$

$$p(x_i) = K\delta + \lambda_c, \quad x_i \in S_c; \tag{7}$$

$$\tau_c \le \mu p(x_i), \quad x_i \in S_c; \tag{8}$$

$$\sigma_{33}^{n}(x_3) = \frac{F}{A_n}, \quad x_3 \in S_n^{b}, n = 1 \ldots 4 \tag{9}$$

$$u_i|_{x_i=(L_1,-L_1,L_2)} = 0; \tag{10}$$

$$u_1|_{x_i=(-L_1,-L_1,L_2)} = 0, \quad u_2|_{x_i=(-L_1,-L_1,L_2)} = 0; \tag{11}$$

$$u_3|_{x_i=(-L_1,L_1,L_2)} = 0, \quad u_3|_{x_i=(L_1,L_1,L_2)} = 0; \tag{12}$$

$$k\Delta T + q = 0; \tag{13}$$

$$-k\frac{\partial T}{\partial x_1} = h(T - T_{\text{coolant}}), \quad x_i \in S_{\text{conv}} \tag{14}$$

where σ_{ij} and ε_{ij} are the Cartesian components of tensors of stress and strain, u_i are the components of displacements vector, δ_{ij} is the Kronecker delta, s_{ij} are the components of stress deviator tensor, T is the temperature, σ_1, σ_2, σ_3 are the principal stresses, λ, λ_c are Lagrange multipliers, $\Phi\left(\varepsilon^{P}\right)$ is the function of the material's stress–strain curve, p is the contact pressure, K is the contact stiffness, δ is the contact gap size, τ_c is the tangential stress on the contact surface, S_c is the surface of contacts, n is a bolt index, S_n^{b} is a cross section of bolt n, A_n is an area of S_n^{b}, Δ. the Laplace operator, and S_{conv} is the cooled platform surface.

The problem is solved using ANSYS finite element software. For the simulation of bolts, pretension type of analysis and PRETS179 finite elements are used. A geometric model is created in ANSYS or can be imported from the computer-aided

Fig. 2 Finite element mesh
of the assembly model

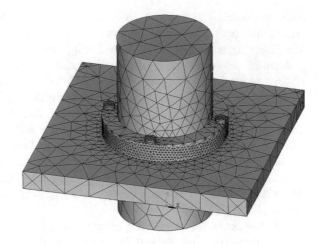

design (CAD) system. The finite element mesh of the assembly model is shown in
Fig. 2.

To determine the contact pressure and gaps for all contact areas, the Standard
contact type is selected in the ANSYS options. In this case, the constraints of the
construction should not allow it to "fall to pieces." In addition, the rotational degrees
of freedom of the model should be constrained, since if, for example, concentrated
forces are applied, their direction can "go off-course" in the case of unforeseen
rotations in the process of quasi-static loading. Augmented Lagrange method is used
as a contact algorithm with Gauss point contact detection method.

3 Micromodel of Contact Surfaces Interaction

On the contact finite elements of the macromodel, thermal contact conductance is
specified as the TCC parameter of the ANSYS program.

In the assembly, it is possible to define the contact of the gyro unit-platform, which
is the most influential in the heat transfer, and the contacts of the gyro unit—bolts and
the platform—bolts. As calculations have shown, the latter do not make a significant
contribution to the heat transfer, although this issue requires additional complicated
research. Therefore, in this model problem for these contacts the TCC parameter is
estimated to be 100,000 W/(m^2 K) and the ANSYS standard contact type is used.
Further, the main focus will be on the contact of the gyro unit-platform.

Since the bolts form a field of contact pressures markedly different from uniform, the thermal contact conductance is specified tabularly as a function of pressure
TCC(p) where p is the contact pressure on a specific finite element. This dependence
is obtained from the calculation of the column micromodel of contact rough bodies
with dimensions $90 \times 90 \times 1125$ μm described in [11], taking into account the
indentation size effect (ISE). In contrast to [11], the model of contact of bodies from

different materials 1050 UNS A91050 aluminum and AISI 1020 steel is considered in this paper. The contact surfaces are formed using the modified Weierstrass-Mandelbrot fractal function and have a roughness Ra equal to 3 μm. The coupled problem of deformation and heat conductance is solved. In calculating the deformation, taking into account the elastoplastic behavior of materials with isotropic hardening, a quasi-static approach with gradually applied loads is used. The hardening curve $\Phi(\varepsilon^P)$ for 1050 UNS A91050 aluminum is given multilinearly from the data obtained in the experiment for the analogous AD1 aluminum [31]. In view of model nature of the problem being solved, to account for the ISE, the hardening curve for the cold-worked AISI 1020 was not obtained from a physical experiment, but was calculated by increasing the above-mentioned hardening curve of the cold-worked AD1 aluminum in proportion to the exceeding of yield strength of one material over another, i.e., 5.03 times.

With this contact, the steel body, being more hard, deforms slightly and, practically without changing shape, penetrates into the aluminum body. In this case, the near-surface finite elements of the soft material deform to a greater extent, and smaller loading steps are required to avoid the appearance of distorted finite elements. The single contact spot persists up to a pressure of 5 MPa, and the real contact area turns out to be less than that of aluminum–aluminum contact.

Figure 3 shows the temperature distribution at an external pressure of 20 MPa. It can be seen that high thermal conductivity of aluminum makes the steel protrusion

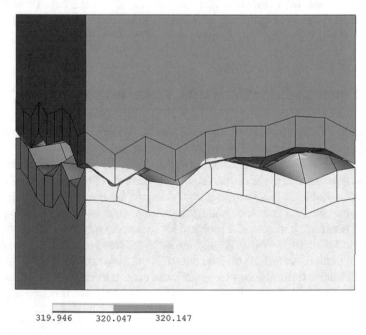

319.946   320.047   320.147

Fig. 3 Distribution of temperatures in the micromodel. The warming up of the asperity of the upper body of steel can be seen

in the foreground warm up entirely to the depth of its penetration into the aluminum. Here, we see a picture similar to that previously discussed in the problem of the contact of asperities with a gas flow around them [34]. Similar to the situation when the presence of lateral gas heating of the asperities increased the heat transfer, in this case lateral contact with the aluminum should result in greater heat conductance of the protrusions compared with end contact of the same real contact area. This is characteristic to some extent for contacts of materials of significantly different hardness.

The calculation results are shown in Figs. 4 and 5. The values of the thermal contact conductance were calculated up to an external pressure of 20 MPa in increments of 2.5 MPa. For higher pressures, the thermal contact conductance was calculated by extrapolation of the data from the 0–20 MPa section. This is a model problem, and therefore, a refined calculation with the second-level roughness was not carried out.

Based on the obtained values of thermal contact conductance (Fig. 5), the values of the table specifying the TCC(p) parameter of the gyro unit-platform assembly model were formed.

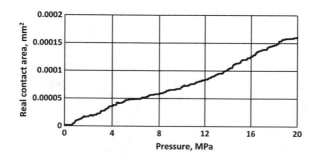

Fig. 4 Real contact area depending on the pressure for contact AISI 1020 steel—1050 UNS A91050 aluminum

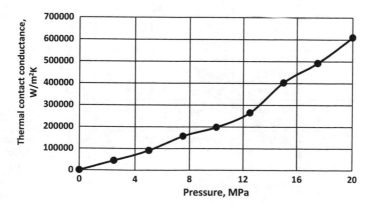

Fig. 5 Thermal contact conductance depending on the pressure for contact AISI 1020 steel—1050 UNS A91050 aluminum

4 Computational Experiments on the Macromodel and Results

A number of computational experiments were carried out, the conditions and results of which are summarized in Table 2. The temperature field of the macromodel was determined, and then, to compare experiment results, the thermal contact conductance of the gyro unit platform contact region averaged over the nominal area was calculated by the relation

$$\overline{\alpha_c} = \frac{\bar{q}}{\overline{T_2} - \overline{T_1}}$$

where $\bar{q}$ is the average heat flux density over the area of finite elements of the lower nominal surface, $\overline{T_1}$, $\overline{T_2}$ are the average temperatures over the area of finite elements of the nominal lower and upper contact surfaces, respectively. Experiments were

Table 2 Results of computational experiments for the contact of gyro unit-platform in the macro-model

Experiment No.	TCC, W/(m² K)	Contact type of gyro unit-platform connection	Averaged over the nominal area thermal contact conductance of gyro unit-platform connection $\overline{\alpha_c}$, W/(m² K)	Maximum temperature of gyro unit, °C	Note
1	157,000	No separation	137,770	60.69	–
2	Tabularly specified	No separation	32,368	61.23	–
3	157,000	Standard	26,848	61.35	–
4	Tabularly specified	Standard	20,765	61.53	
5	157,000	Standard	31,798	61.22	Friction coefficient $\mu = 0.3$
6	Tabularly specified	Standard	3389	65.6	Clamp force $F = 100$ N

conducted to evaluate the influence of various factors, while the most appropriate model should be considered as the experiment No. 4 model.

The TCC parameter was set as a constant value equal to 157,000 W/(m² K) or as the above-tabulated dependence on pressure TCC(p). The value 157,000 W/(m² K) is obtained from the graph in Fig. 5 for the average pressure from the bolt clamp force of 7.3 MPa, calculated by dividing the sum of the clamp forces of each bolt (2000 N) by the nominal contact area.

To evaluate the effect of thermal expansion, experiments were carried out for two types of contact behavior, Standard and No separation. For the Standard contact type, the contact heat transfer occurred strictly in the real contact area (Fig. 6), for which the TCC parameter was set. Thus, this type of contact reflects the influence of change in shape from thermal expansion. For the No separation contact type, movement of the contact surfaces along the contact plane is allowed, but separation of the surfaces is not permitted and the real contact area is equal to the nominal one. Thus, in the case of No separation contact, the change in shape of the surfaces from thermal expansion is not reflected in temperature results since it does not affect the thermal contact conductance. In this case, the heat transfer occurs over the entire nominal contact area. The wide use of this type of contact in actual practice is due to the significantly lower computational complexity and, accordingly, solution time. The No separation contact type was set on the gyro unit-platform connection in experiments Nos. 1 and 2.

The calculations were carried out with the assumption of small displacements, since the accounting for large displacements for experiment No. 6 resulted in a

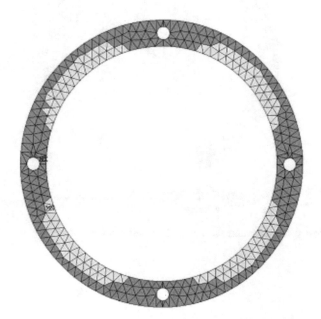

Fig. 6 Finite element mesh for nominal contact area (real contact area is darkened)

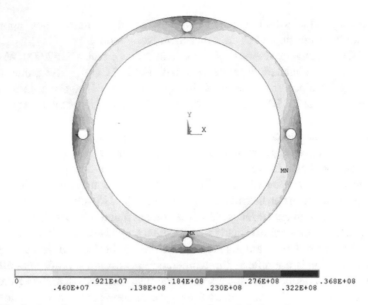

Fig. 7 Distribution of contact pressure on contact surface of platform, Pa

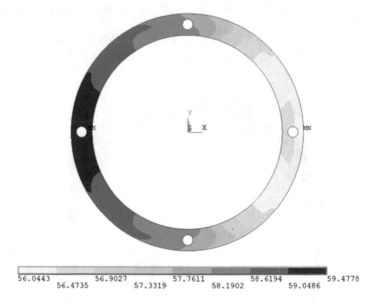

Fig. 8 Distribution of temperature on contact surface of platform, °C

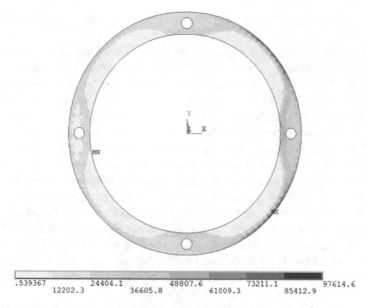

Fig. 9 Distribution of heat flux density on contact surface of platform, W/m²

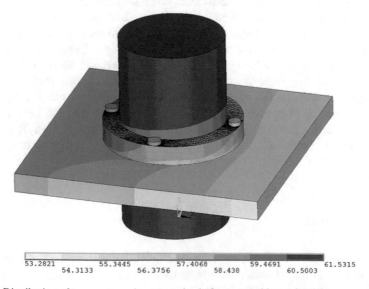

Fig. 10 Distribution of temperatures in gyro unit-platform assembly model, °C

change in the averaged thermal contact conductance of 0.1%, which is considered insignificant.

The first and second experiments set the heat transfer throughout the whole of the nominal contact area. In this case, setting the dependence of the thermal contact conductance TCC on the contact pressure p obtained in the micromodel led to a decrease of 4.3 times in the averaged thermal contact conductance of the macromodel.

The models used in the computational experiments Nos. 3 and 4 take into account the effect of thermal expansion on the real contact area. Because of the change in shape of the cylindrical body of the gyro unit, tangency takes place in the form of a narrow ring along the outer edge of the nominal contact area. Also, areas near the bolts are in direct contact. The real contact area was 56% of the nominal area (Fig. 6). As is clear from a comparison of experiments Nos. 3 and 1, the averaged thermal contact conductance decreased by more than 5 times just due to accounting for the real contact area at a constant TCC of 157,000 W/(m^2 K). Under the same conditions and using the TCC(p) dependence (experiments Nos. 2 and 4), the averaged thermal contact conductance decreased noticeably less, by 56%, which can be considered a result of thermal expansion without the direct influence of contact pressure. Repetition of the result of 56% is a random coincidence in this case.

Experiment No. 5 showed that the use of a friction coefficient 0.3 instead of 0.5 led to a slight increase in the thermal contact conductance (by 18%). Thus, the friction coefficient has a noticeable effect on the conductance of the actual contact.

The thermal contact conductance is significantly affected by clamp force of the bolts. Experiment No. 6 showed that using the conditions of experiment No. 4 and decreasing the clamp force from 2000 to 100 N, the averaged thermal contact conductance decreased by more than 6 times. A similar effect in real structures can arise in the case of more complicated connections, for example, with clasps [35].

The distribution of contact pressures, surface temperatures, and heat fluxes for the contact platform surface in experiment No. 4 is shown in Figs. 7, 8, and 9. Figure 10 shows the distribution of temperatures throughout the entire model of the gyro unit-platform assembly under the same conditions.

5 Conclusions

The model problem of the contact of the gyro unit-platform demonstrates the applicability of the developed simulation method for the contact of rough surfaces at the micro-level [11] to macroscale objects.

It is determined that essential factors affecting the real contact area and, consequently, the temperature distribution in the macroscale (i.e., curving the surfaces) are both thermal expansion and uneven pressure due to the tightening of the bolts. They lead to the change of up to 4 times in averaged thermal contact conductance and should be included in thermal contact models developed. The degree of this influence depends significantly on the clamp force of the bolts.

With the allowable tolerance in the temperature of the gyro unit often not more than 1 °C, even such a simplified model of the device gives a temperature change of more than 1 °C with different models of thermal contact (see Table 2). In most cases of contacts in constructions due to uneven pressure distribution, form deviations, and thermal expansion, the application of thermal contact conductance as a constant for the surface does not correspond to the physics of the process and can lead to significant errors.

It could be noted that the thermal contact conductance is not used as a separately calculated value in the applied approach to the calculation of the temperatures of macro construction parts, which contributes to an increase in the accuracy and clarity of the simulation.

References

1. Slocum, A.H.: Precision Machine Design. Society of Manufacturing Engineers, Dearborn (1992)
2. Chen, Y., Huang, X., Kang, R.: Generation of accelerated stability experiment profile of inertial platform based on finite element. Chin. J. Aeronaut. **25**(4), 584–592 (2012). https://doi.org/10.1016/S1000-9361(11)60422-6
3. Das, S.K., Pal, D., Kumar, V., Nandy, S., Banerjee, K., Mazumdar, C.: Stochastic characterization of a MEMs based inertial navigation sensor using interval methods. Int. J. Image, Graph. Signal Process. (IJIGSP) **7**(7), 24–32 (2015). https://doi.org/10.5815/ijigsp.2015.07.04
4. Ott, L.: Untersuchungen zur frage der erwärmung elektrischer maschinen. In: Mitteilungen über forschungsarbeiten auf dem gebiete des ingenieurwesens, insbedere aus den laboratorien der technischen hochschulen, vols. 35–36, pp. 53–107. Springer, Berlin (1906)
5. Barratt, T.: Thermal and electrical conductivities of some of the rarer metals and alloys. Proc. Phys. Soc. London **26**(5), 347–371 (1913). https://doi.org/10.1088/1478-7814/26/1/335
6. Barratt, T.: The magnitude of the thermal resistance introduced at the slightly conical junction of two solids, and its variation with the nature of the surfaces in contact. Proc. Phys. Soc. London **28**, 14–20 (1915). https://doi.org/10.1088/1478-7814/28/1/302
7. Ciavarella, M., Delfine, V., Demelio, G.: A "re-vitalized" Greenwood and Williamson model of elastic contact between fractal surfaces. J. Mech. Phys. Solids **54**, 2569–2591 (2006). https://doi.org/10.1016/j.jmps.2006.05.006
8. Bahrami, M., Yovanovich, M.M., Culham, J.R.: Thermal contact resistance at low contact pressure: effect of elastic deformation. Int. J. Heat Mass Transf. **48**(16), 3284–3293 (2005). https://doi.org/10.1016/j.ijheatmasstransfer.2005.02.033
9. Thompson, M.K.: A multi-scale iterative approach for finite element modelling of thermal contact resistance. Ph.D. thesis, Massachusetts Institute of Technology (2007)
10. Murashov, M.V., Panin, S.D.: Modeling of thermal contact conductance. In: Proceedings of the International Heat Transfer Conference IHTC14, Washington, DC, USA, vol. 6, pp. 387–392 (2010). https://doi.org/10.1115/ihtc14-22616
11. Murashov, M.V., Panin, S.D.: Numerical modelling of contact heat transfer problem with work hardened rough surfaces. Int. J. Heat Mass Transf. **90**, 72–80 (2015). https://doi.org/10.1016/j.ijheatmasstransfer.2015.06.024
12. Archard, J.F.: Elastic deformation and the laws of friction. Proc. R. Soc. Lond. A Math. Phys. Sci. **243**(1233), 190–205 (1957). https://doi.org/10.1098/rspa.1957.0214
13. Greenwood, J.A., Williamson, J.B.P.: Contact of nominally flat surfaces. Proc. R. Soc. Lond. A Math. Phys. Sci. **295**, 300–319 (1966). https://doi.org/10.1098/rspa.1966.0242

14. Persson, B.N.J.: Theory of rubber friction and contact mechanics. J. Chem. Phys. **115**, 3840–3861 (2001). https://doi.org/10.1063/1.1388626
15. Cooper, M.G., Mikic, B.B., Yovanovich, M.M.: Thermal contact conductance. Int. J. Heat Mass Transf. **12**(3), 279–300 (1969). https://doi.org/10.1016/0017-9310(69)90011-8
16. Mo, Y., Turner, K.T., Szlufarska, I.: Friction laws at the nanoscale. Nature **457**, 1116–1119 (2009). https://doi.org/10.1038/nature07748
17. Jackson, R.L., Green, I.: A finite element study of elasto-plastic hemispherical contact against a rigid flat. J. Tribol. **127**(2), 343–354 (2005). https://doi.org/10.1115/1.1866160
18. Dieterich, J.H., Kilgore, B.D.: Imaging surface contacts: power law contact distributions and contact stresses in quartz, calcite, glass and acrylic plastic. Tectonophysics **256**(1–4), 219–239 (1996). https://doi.org/10.1016/0040-1951(95)00165-4
19. Raji Reddy, D., Laxminarayana, P., Reddy, G.C.M., Reddy, G.M.S.: Process parameters influence on impact toughness and microstructure of pre-heat treated friction welded 15CDV6 alloy steel. Int. J. Eng. Manuf. (IJEM) **6**(5), 38–47 (2016). https://doi.org/10.5815/ijem.2016.05.05
20. Abbot, E.J., Firestone, F.A.: Specifying surface quality: a method based on accurate measurement and comparison. Mech. Eng. **55**, 569–572 (1933)
21. Yan, W., Komvopoulos, K.: Contact analysis of elastic-plastic fractal surfaces. J. Appl. Phy. **84**, 3617–3624 (1998). https://doi.org/10.1063/1.368536
22. Yovanovich, M.M., Rohsenow, W.M.: Influence of surface roughness and waviness upon thermal contact resistance. Technical Report No.6361-48, Massachusetts Institute of Technology, Cambridge, Massachusetts (1967)
23. Majumdar, A., Bhushan, B.: Fractal model of elastic-plastic contact between rough surfaces. J. Tribol. **113**(1), 1–11 (1991). https://doi.org/10.1115/1.2920588
24. Yovanovich, M.M., Hegazy, A.: An accurate universal contact conductance correlation for conforming rough surfaces with different micro-hardness profiles. AIAA Paper 83-1434 (1983)
25. Hegazy, A.A.-H.: Thermal joint conductance of conforming rough surfaces: effect of surface micro-hardness variation. Ph.D. thesis, University of Waterloo (1985)
26. Tabor, D.: The Hardness of Materials. Clarendon Press, Oxford (1951)
27. Zavarise, G., Borri-Brunetto, M., Paggi, M.: On the reliability of microscopical contact models. Wear **257**, 229–245 (2004). https://doi.org/10.1016/j.wear.2003.12.010
28. Jackson, R.L.: The effect of scale dependent hardness on elasto-plastic asperity contact between rough surfaces. Tribol. Trans. **49**(2), 135–150 (2006). https://doi.org/10.1080/05698190500544254
29. Polonskiy, I.A., Keer, L.M.: Scale effects of elastic-plastic behavior of microscopic asperity contacts. J. Tribol. **118**(2), 335–340 (1996). https://doi.org/10.1115/1.2831305
30. Manika, I., Maniks, J.: Size effects in micro- and nanoscale indentation. Acta Mater. **54**(8), 2049–2056 (2006). https://doi.org/10.1016/j.actamat.2005.12.031
31. Murashov, M.V., Kornev, YuV: Elastoplastic deformation of a roughness element. Tech. Phys. **84**(3), 75–81 (2014). https://doi.org/10.1134/S1063784214030189
32. Coria, I., Martín, I., Bouzid, A.-H., Heras, I., Abasolo, M.: Efficient assembly of bolted joints under external loads using numerical FEM. Int. J. Mech. Sci. **142–143**, 575–582 (2018). https://doi.org/10.1016/j.ijmecsci.2018.05.022
33. Lu, B.S., Wanga, L.G., Huang, Y.: Effect of deformation rate on interfacial heat transfer coefficient in the superalloy GH4169 hot forging process. Appl. Therm. Eng. **108**, 516–524 (2016). https://doi.org/10.1016/j.applthermaleng.2016.07.167
34. Murashov, M.V., Panin, S.D.: Permanence of basic assumption in models of heat transfer in contacts. Eng. J. Sci. Innov. **3**(15), (2013). (in Russian) https://doi.org/10.18698/2308-6033-2013-3-732
35. Fang, Z., Du, D., He, K., Shu, W., He, Q., Xiang, B., Xiao, H.: Study on contact fatigue life prediction for clasp joint structure of mooring buoy. Int. J. Eng. Manuf. (IJEM) **2**(2), 29–35 (2012). https://doi.org/10.5815/ijem.2012.02.05

The Primary Geo-electromagnetic Data Preprocessing Received from a Modified Geophysical Automatic Station

Roman Kaminskyj, Nataliya Shakhovska⬤ and Lidia Savkiv

Abstract The results of preliminary processing of the primary data obtained from the geophysical automatic station are presented. Descriptive statistics and simulation of time series were used as processing methods. The presentation of the results by multidimensional graphs allowed to reveal the phenomenon of coincidence in the first indicators of descriptive statistics, and in the second, the coincidence of the model's coefficients for the daily measurements of the natural electric field. This phenomenon represents that the days with practically identical values of indicators and coefficients of the model are manifested. However, the essence of this phenomenon needs new additional research.

Keywords Time series · Prediction · Research of constant natural electric field · Preliminary processing of primary geo-electromagnetic data · Method of natural electric field

1 Introduction

Various geophysical methods and techniques are used to study and research the physical processes and phenomena occurring in the surface layers and depths of the Earth, as well as on its surface and in the near-Earth space. Such studies are organized in a variety of ways: It may be systematic scientific research in the form of continuous monitoring or geophysical observations, or periodic studies of individual regions or

R. Kaminskyj · N. Shakhovska (✉)
Lviv Polytechnic National University, 79013 Lviv, Ukraine
e-mail: nataliya.b.shakhovska@lpnu.ua

R. Kaminskyj
e-mail: kaminsky.roman@gmail.com

L. Savkiv
Carpathian Branch of the Institute of Geophysics named after S. I. Subbotin,
National Academy of Sciences of Ukraine, Lviv, Ukraine

© Springer Nature Switzerland AG 2020
Z. Hu et al. (eds.), *Advances in Artificial Systems for Medicine and Education II*,
Advances in Intelligent Systems and Computing 902,
https://doi.org/10.1007/978-3-030-12082-5_56

areas. Each of these areas provides important primary information for further study of these processes.

Continuous geophysical observations are conducted continuously in stationary and temporary observation points. For today, the collection and registration of physical fields occurs mainly automatically, through the creation and replenishment of various databases. According to regular and long-term observations, it is possible to evaluate and analyze the field parameters and their characteristics, study the dynamics of changes over a long period of time, and control the critical indicators. Usually, such studies are provided by international networks, world and national centers, consortia, geological, ecological, and other services, associations, and agencies. The mentioned organizations, in addition to other types of data, work with geomagnetic, seismic, magnetotelluric observations, form the corresponding archives, present the visual representation of such data online, and, in addition, provide special or direct access to them. However, along with such a large-scale study of the world level of geophysical information, local studies of individual regions or individual parts of the Earth's surface are also of considerable interest in the field of geophysics. Such local studies are useful in that they allow deeper, more precise, and detailed study of a particular region, to explore local changes or features specific to this area. In addition, such information can serve as an important supporting, additional information for the processing and analysis of global geophysical data.

2 Analysis of Recent Research and Publications

An overview of literary sources suggests that many sites and organizations of world-class and global scale have long worked, analyzed, and presented the results of geophysical observations in real time: the international network of geomagnetic observatories INTERMAGNET [1, 2], the National Environmental Information Centers (NCEI, Asheville, North Carolina, USA) [3, 4], International Consortium of Seismological Institutes (IRIS, USA) [5, 6], US Geological Survey (USGS) [7]. Much attention is given to seismological data both in Ukraine and abroad [5, 6, 8–13].

However, local electromagnetic observations and issues of processing this type of data of a regional nature are currently very poorly covered. Interesting on this is the work [14], which addresses issues of organization of geo-electric monitoring at the Boulder Magnetic Observatory of the US Geological Survey. An example of statistical analysis and processing of the data of the network of atmospheric measurements is work [15]. The similar nature of the submission of daily data and its processing is work [14], which is devoted to the analysis of the dynamics of sales of oil, using the average price, the scope of price values, and sales. In addition, the analysis and simulation of the trend in the daily time series is also carried out and the visualization of the regularities is presented graphically [16–18].

3 The Purpose and Objectives of the Research

Preferably, automatic stationary observation points for geophysical data are structures formed by a set of special sensors, recording equipment, and means of transmitting information to destinations. The resulting data is encoded appropriately and transmitted in real-time mode to the relevant organization for processing and storage. The peculiarity of such data, in our case, is that the collection of such data is already a long time, and the measurement of the values of geophysical parameters is carried out continuously and at rather short intervals of time, for example, in this case, every 5 min for an hour, day, month, year and, yes, from year to year. Obviously, such volumes of data require special mathematical methods and computer software. For further transmission of the received data to the central state authorities, they carry out their preliminary processing in the units of service points of observation. The data obtained directly from the measurement of the parameters of the geophysical situation, in particular, the electrical component, may already be processed, processed, and interpreted by staff in certain ways and submitted to the higher authority as a result of preliminary processing, obviously along with the original values of the observed process.

World-class systems and networks, their arrangement, structure, and principle of work are very good; however, the issue of organizing permanent local studies is mainly described very superficially. Unfortunately, the description of local geo-electric stations and posts is practically absent, but most importantly there are no recommended processing procedures and express interpretation of the data received. Nowadays, the development of information technology for the primary processing and interpretation of the received geophysical data from stationary observation points, which automatically arrives at the appropriate centers of their processing, is an important and relevant scientific and practical task. Therefore, the purpose of this research is to focus on the choice of simple methods of primary data processing and the development of appropriate mathematical models of the dynamics of their development in terms of creating a special Web site for relevant services and researchers.

4 The Proposed Method

Typical features of permanent local electromagnetic studies can be considered on the example of the CSG "Nyzhnie Selyshche" (Transcarpathian region, Khust District, geographical coordinates: 48.197472 °N, 23.456028 °E). The station is located on the territory of Transcarpathian seismic zone; therefore, in addition to other geophysical measurements, continuous electromagnetic observations are conducted at the station.

For research, the method of a natural electric field (PEP) is used, the essence of which is to determine the parameters of the PEP in two mutually perpendicular directions: north–south (N–S) and west–east (W–E). To do this, an appropriate measuring

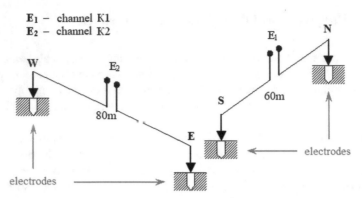

Fig. 1 Scheme of the measuring device for continuous research of PEP at the CSG "Nyzhnie Selyshche"

device is deployed on the territory of the regime geophysical station. It consists of two pairs of electrodes, which are located strictly in the direction of N–S and W–E. The electrodes of each pair are fixed at a certain distance, namely: The electrodes of the direction N–S are at a distance of 60 m from each other, W–E 80 m. The scheme of installation is depicted in Fig. 1. Two pairs of electrodes continuously measure the analog values of the channel voltages $K1$ and $K2$ corresponding to the potential difference $E1$ and $E2$ on the electrodes of the direction of N–S ($E1$) and W–E ($E2$). In the future, these signals through the analog-to-digital converters will be digitized.

During the day, the measuring geophysical equipment automatically records the value of channel voltages from both directions in mV for 5 min, and hourly—the temperature at °C. At the end of each day, all registered signals are written in the form of several arrays of sequential values, and for further processing and archiving, they are automatically sent to the local server.

Although this form is understandable for professionals, it is quite uncomfortable, even for preprocessing, since it requires additional conversion to standard formats. In other words, it is necessary to exclude non-informative indicators of the hour, date, and temperature and to create an appropriate presentation for them, for example, in the form of individual columns, certain labels, etc. In addition, the values themselves should be brought to the desired look. Obviously, as a result of such data structuring, their processing allows the use of various mathematical specialized packages.

4.1 Data Preparing for Interpretation

It has been shown above that the data received is extremely large, which creates considerable difficulties for their processing. Indeed, the first moment in processing and interpreting the data is as follows. Only within a day, the sample contains 288 options that are sufficient for a representative statistical survey. If we consider data for a month (30 days), the sample size will be 8640 values, and for the year (365 days)—103,680 values. Obviously, the analysis of such data should be done using shorter time intervals of observations and a general conclusion to be made based on the results obtained, for example, within a day, a month. Then, comparing daily data, we get the situation during the month, and comparing monthly data, we get the situation during the year, etc.

The second point is that the data of these observations, in fact, is a continuous–discrete random process, or, in other words, they can be classified as non-stationary equivalent stationary time series. In this case, the processing of such a time series requires the construction of a mathematical model, which should characterize its general tendency and the corresponding parameters of its structure. The statistical rapid processing of the received electromagnetic data is proposed to be divided into several stages.

Stage 1. Submission of daily dynamics of indicators of descriptive statistics in the form of a general schedule of lunar duration.
Stage 2. Simulation of the daily dynamics of the process development with the justification of the choice of a mathematical model.
Stage 3. Determination of the role of the dynamics of the coefficients of the model, approximating the daily trend, by identifying the coincidence of models at different points of the investigated process.

4.2 Determination of Indicators of Descriptive Statistics

The geophysical data represent the values of the dynamics of the parameters of a constant electric field, that is, the sample $X = \{x : x_1, x_2, \ldots, x_n$ and $n = 288\}$ characterizes its value during the day. The most common methods of data processing in this case are methods for determining the descriptive statistics. These indicators describe and provide the basic statistical properties of the data. They allow us to summarize the primary results and are used to solve the following two problems:

- Show the *overall* nature of the data set.
- Show *what is* and *how much* data varies among themselves.

Among the indicators that are included in it, we used the following: For solving the first problem, these are indicators of the central trend—the arithmetic mean, the median, and, for the second, the variation indicators, which include the mean-square deviation and scale (interval). The dispersion rate was not used, since it is the square of the mean-square deviation. In addition, instead of the variance index, the coefficient of variation is used. Its function in this study is the definition of the variability of the researched indicator—the value of the electric field as a ratio of the mean-square deviation to the mean arithmetic.

Given a significant amount of data within one day of the rice, there are several daily graphs of data in the order of their receipt in time. In practical, the form of the tendencies of changing the values of the electric field is very similar between them, and it shows the influence of the unknown factor at the moment, which within the limits of the day influences in the same way on the magnitude of the electric field. It follows from the graph that in the first such interval the values of the electric field increase monotonically (Fig. 2). In the following two such intervals, these values decrease monotonically, and in the latter, they grow again, with the same monotony.

In this study, conducted by the methods of descriptive statistics, indicators of descriptive statistics, including the coefficient of variation for the month of July 2017, are obtained.

The results characterize the central tendency and variation of data within one month. The meaning of other indicators of descriptive statistics, such as standard error, asymmetry, excess, minimum, maximum, amount, and number in this study is not used. The fact is that the standard error here is determined by the relation $\sigma/\sqrt{n}$ where σ, n—mean-square deviation and amount of data, respectively. Since for all daily data the quantity is the same, in this case it is quite sufficient as an indicator of the variation of the mean-square deviation.

Figure 3 shows daily charts. On the charts, a substantially nonlinear change in the indicators of the central tendency is very clearly expressed. For it, there is a pivot point, which means the presence of the fashion, which confirms the data of descriptive statistics.

An important visual point is that the nature of the variation is very similar. According to the results of daily indicators of descriptive statistics, a multidimensional graphical representation of them is constructed. Taking into account the significant difference between the values of the mean, mode, median from the mean-square deviation, magnitude, and variation coefficient, all values were given to the unit interval

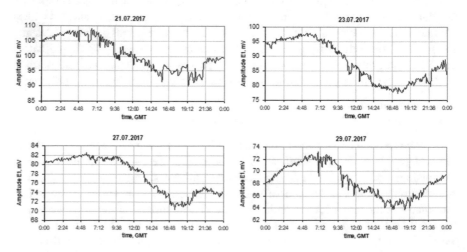

Fig. 2 Daily dynamics of indicators of natural electric field

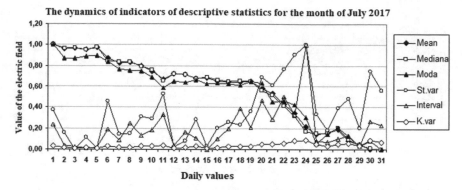

Fig. 3 Dynamics of indicators of descriptive statistics within one month

in their samples by the formula: $\tilde{x} = \dfrac{x_i - x\text{min}}{x\text{max} - x\text{min}}$ where $\tilde{x} \in [0, 1]$ reduced to a single interval value x_i.

The results of such a valuation are presented in the form of a multidimensional graph in Fig. 3. Submission of results of the descriptive statistics in the way that is done in Fig. 3 is not traditional, but its use has revealed some coincidence of descriptive statistics for some days. The behavior of the levels (absolute values of the electric field measurements) during the month can be considered quasi-synchronic, since the trend pattern for each day is actually repeated, but the exception may include only one–three days. The existence of such coincidences may be a purely accidental event, but it may also be the result of the absence of any influence, that is, the state of the environment can be considered normal, and all others as a result of the influence of various factors. There may be a lot of options for their interpretation, and therefore, research in this direction has a certain scientific meaning to obtain a scientific result.

4.3 Analysis of Trends in Daily Time Series

The obtained data can be represented in the form of a graph of the equidistant time series, depicted in Fig. 4.

A graphical representation gives a general view of its structure, indicating the nature and type of stationarity and oscillatory component. This series can be formally submitted as a plural

$$Y(t) = \{y_j : y_j = y(t_j), \quad j = 1, 2, \ldots, 288, \quad t_j = 5\,\text{min}, \quad t_j \in T\},$$

orderly random variables and characterize (define describe) as linear non-stationary process, with daily periodicity, duration $\Delta t = 288 \cdot 5\,\text{min} = 24\,\text{h}$ where T is set of 5 min intervals of time. Since the form of the levels of the time series is repeated, the differences in its behavior can be estimated using the values of the coefficients

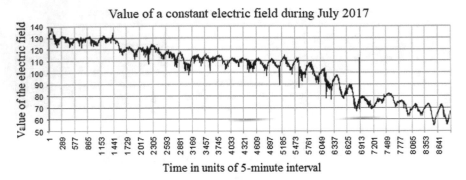

Fig. 4 Value of a constant electric field during July 2017

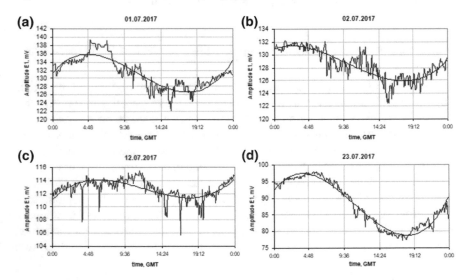

Fig. 5 Trends in daily data approximated by polynomials of the third degree

of the model, which approximates the daily levels. In this case, visually, the trend of daily values has an inflection point, which is predominantly in the middle of the day, as shown in Fig. 5.

At first sight, such approximating function $y(t)$ as a trend model, it is advisable to try the polynomial of the third degree: $y(t) = a_3 \cdot t^3 + a_2 \cdot t^2 + a_1 \cdot t + a_0$. It is the property of a third-order polynomial, namely, to reproduce an addiction that has an inflection point, making it the simplest function for the approximation of these data. The main disadvantage of polynomials is that it is virtually impossible to provide them with the ratios of semantic loading or physical interpretation. However, an attempt is made to make such an approximation. In Fig. 6, graphs of approximated trends in daily data are presented.

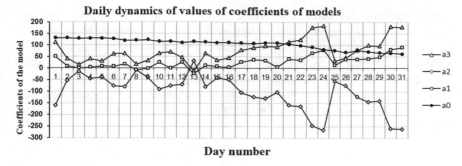

Fig. 6 Change of the coefficients of the model of the day

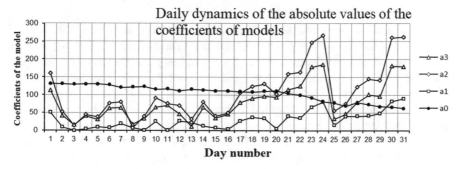

Fig. 7 Dynamics of absolute values of coefficients of models

As in the case of indicators of descriptive statistics, the approach was used—the representation of the values of the coefficients of the polynomial on one graph, depicted in Fig. 6. The peculiarity of the results is that there are also specific points in which the absolute values of the coefficients a_j where $j = 0,1,2,3$, have very close values.

Since the second coefficient a_2 is negative, because of this there is symmetry of the graphs for the coefficients a_2 and a_3. If to submit the given schedule not real values, and absolute, that is to present the values of coefficients a_j, as absolute values $|a_j|$, the graph will look like as shown in Fig. 7.

Already a visual analysis of graphs in Fig. 7 point to special points in which the values of the coefficients of the model coincide. Such coincidences are irregular and not absolute; that is, a complete coincidence does not always take place. For days such as 3, 8, 13, and 25, coincidences are quite close. The nature of these coincidences is difficult to explain using only this data. Obviously, such coincidences can be considered as random first (as a coincidence of random circumstances).

Coefficient a_0 is very close to the arithmetic mean for this time. Its behavior, as can be seen from the graph in Fig. 7, is quite monotonous and relatively smooth. Compared to other months, there is a certain variation of this indicator with a period of several months. This phenomenon also requires relevant research.

Presentation of the behavior of coefficients a_i models on one graph indicates that a_2 and a_3 vary proportionally, and at some stages, they repeat the indicator a_1. Consequently, for some days, there is a similarity in the structures of approximating models. Because the quality of the approximation is based on the determination coefficient $R^2 > 0.89$, we can assume that the correspondence of the model with the daily value is sufficient to use this approach—the establishment of specific points of the dynamics of the natural electric field for a given local mode geophysical automatic station "Nyzhnie Selyshche."

5 Conclusion

Construction of the daily time series from the data obtained showed that the data in the interval of the era are very often repeated in the form of a trend. The polynomial third degree were approximated for the analysis of this phenomenon in a daily equal time series. A multidimensional time series, formed from the coefficients of this polynomial model, also showed the coincidence of their values for several days. These coincidences though do not correspond to such descriptive statistics, but neglect them without knowing the reasons we do not think, because they carry some information and maybe very important. Thus, it can be argued that the use of descriptive statistics and time series methods is appropriate for the processing of primary data [19].

References

1. INTERMAGNET. International Real-time Magnetic Observatory Network (2017). http://www.intermagnet.org http://www.intermagnet.org/index-eng.php. Accessed 05 Aug 2017
2. INTERMAGNET. International Real-time Magnetic Observatory Network (2017). http://www.intermagnet.org/activitymap/activitymap-eng.php. Accessed 05 Sep 2017
3. The National Geophysical Data Center (NGDC) and its sister data centers merged into the National Centers for Environmental Information (NCEI) (2017). http://www.ngdc.noaa.gov http://www.ngdc.noaa.gov/ngdcinfo/aboutngdc.html. Accessed 12 June 2018
4. The National Geophysical Data Center (NGDC) and its sister data centers merged into the National Centers for Environmental Information (NCEI) (2017). ftp://ftp.ngdc.noaa.gov/wdc/geomagnetism/data/observatories/definitive. Accessed 12 Aug 2018
5. Incorporated Research Institutions for Seismology (IRIS) (2017). http://www.iris.edu/hq/ http://ds.iris.edu/seismon/ http://www.iris.edu/hq/programs/gsn. Accessed 05 June 2018
6. Incorporated Research Institutions for Seismology (IRIS) (2017). http://www.usarray.org/researchers/obs/magnetotelluric http://ds.iris.edu/gmap/_US-MT. Accessed 12 June 2018
7. USGS Geomagnetism Program (2018). http://geomag.usgs.gov/ http://geomag.usgs.gov/plots/. Accessed 07 Aug 2018
8. Shakhovska, N., Nych, L., Kaminskyj, R.: The identification of the operator's systems images using the method of the phase portrait. In: Advances in Intelligent Systems and Computing, pp. 241–253 (2017)

9. Logoida, M., Havrysh, B., Anastasiya, D.: Determination of reproduction accuracy of dot area of irregular structure. Int. J. Inf. Eng. Electron. Bus. **10**(1), 9–15 (2018). https://doi.org/10.5815/ijieeb.2018.01.02

10. Ganiev, O.Z., Petrenko, K.V., Sheremet, Y.E., Vakulovich, D.V., Krasny, V.A.: Organization of a seismological point of observation on the island of Zmeiny. Geophysical **33**(2), 122–128 (2011) (in Ukrainian)

11. International Seismological Centre (ISC) (2017). http://www.isc.ac.uk/index.php. Accessed 05 June 2018

12. European-Mediterranean Seismological Centre (EMSC) (2017). https://www.emsc-csem.org. Accessed 08 May 2018

13. Blum, C.C., White, T.C., Sauter, E.A., Stewart, D.C., Bedrosian, P.A., Love, J.J.: Geoelectric monitoring at the Boulder magnetic observatory. Geosci. Instrum. Method. Data Syst. **6**, 447–452 (2017). https://doi.org/10.5194/gi-6-447-2017

14. Yekini, N.A., Oloyede, A.O., Agnes, A.K., Okikiola, F.M.: Microcontroller-based automobile tracking system with audio surveillance using GPS and GSM module. Int. J. Inf. Eng. Electron. Bus. **8**(3), 41–46 (2016). https://doi.org/10.5815/ijieeb.2016.03.05

15. Fuertes, D., Toledano, C., Gonzalez, R., Berjon, A., Torres, B., Cachorro, V.E., Frutos, A.M.D.: CAELIS: software for assimilation, management and processing data of an atmospheric measurement network. Geosci. Instrum. Method. Data Syst. **7**, 67–81 (2018). https://doi.org/10.5194/gi-7-67-2018

16. Shakhovska, N., Kaminskyy, R., Zasoba, E., Tsiutsiura, M.: Association rules mining in big data. Int. J. Comput. **17**(1), 25–32 (2018)

17. Shakhovska, N., Syerov, Y.: Web-community ontological representation using intelligent dataspace analyzing agent. In: Xth International Conference on the Experience of Designing and Application of CAD Systems in Microeletronics (CADSM-2009), pp. 479–480 (2009)

18. Shakhovska, N., Vysotska, V., Chyrun, L.: Features of e-learning realization using virtual research laboratory. In: XIth International Scientific and Technical Conference on the Computer Sciences and Information Technologies (CSIT), 2016, pp. 143–148 (2016)

19. Hu, Z., Mashtalir, S.V., Tyshchenko, O.K., Stolbovyi, M.I.: Clustering matrix sequences based on the iterative dynamic time deformation procedure. Int. J. Intell. Syst. Appl. **10**(7), 66–73 (2018). https://doi.org/10.5815/ijisa.2018.07.07

Statistical Analysis of Probability Characteristics of Precipitation in Different Geographical Regions

Maria Vasilieva, Andrey Gorshenin and Victor Korolev

Abstract Some results of the statistical analysis of the observed regularities in some characteristics of the precipitation process are presented in the paper. The importance of this problem is emphasized by that the information concerning the regularities in duration of wet and dry periods plays a significant role in predicting floods and preventing them. It is demonstrated that the fluctuations of durations of wet or dry periods can be successfully modeled by the negative binomial probability distribution. The parameters of this distribution as well as moment characteristics such as expectation, variance, and variation coefficient are calculated for numerous stations in Europe, Russia, and neighboring countries. The values of these parameters of statistical regularities inherent in the data under study can significantly improve the efficiency and accuracy of intellectual data analysis by introducing additional features at the learning stage without any actual increase of the amount of available data.

Keywords Statistical analysis · Precipitation · Wet periods · Dry periods · Negative binomial distribution

M. Vasilieva · A. Gorshenin (✉) · V. Korolev
Faculty of Computational Mathematics and Cybernetics, Lomonosov Moscow State University, Moscow, Russia
e-mail: agorshenin@frccsc.ru

M. Vasilieva
e-mail: vasmaral95@gmail.com

V. Korolev
e-mail: vkorolev@cs.msu.ru

A. Gorshenin · V. Korolev
Federal Research Center "Computer Science and Control" of the Russian Academy of Sciences, Moscow, Russia

V. Korolev
Hangzhou Dianzi University, Hangzhou, China

© Springer Nature Switzerland AG 2020   629
Z. Hu et al. (eds.), *Advances in Artificial Systems for Medicine and Education II*,
Advances in Intelligent Systems and Computing 902,
https://doi.org/10.1007/978-3-030-12082-5_57

1   Introduction

Statistical analysis of meteorological data often uses inadequate mathematical models of statistical regularities of observed phenomena. For example, it is widely believed that the length of rainy intervals measured in days has the geometric probability distribution (e.g., see [1]).

However, this model can hardly be considered fair in view of its disagreement with real data. The basis for this model is a prejudice related to the generally accepted interpretation of the geometric distribution within the framework of classical scheme of Bernoulli trials: the distribution of the number of rainy days taken for "success" to the first day with zero rainfall, which is equivalent to "failure." Nevertheless, in this scheme, the independence of the tests is assumed, while as a result of the statistical analysis of meteorological data it was obtained that the sequence of rainy intervals does not possess this property. In connection with this, the classical Bernoulli scheme cannot be applied to mathematical modeling of precipitation phenomena.

It turned out that the duration of rainy intervals is well described by the negative binomial distribution with the shape parameter, as a rule, less than one. It was demonstrated for Potsdam and Elista (see, e.g., papers [2, 3]) that the duration of rainy periods in these points has the negative binomial distribution with the shape parameter $r \approx 0.847$ and $r \approx 0.876$, respectively. Also in those papers, it was suggested to explain this phenomenon by the known property of the negative binomial distribution, which is both mixed geometric distribution and mixed Poisson distribution.

As regards the mixed Poisson model, it is known (see, e.g., [4]) that discrete chaotic stochastic processes are best described by the Poisson distribution. In this case, the mixing gamma distribution of the negative binomial model characterizes the statistical patterns of random changes of external factors.

Moreover, as regards the mixed geometric representation of the negative binomial model, it was shown in [5] that any negative binomial distribution with the shape parameter r less than one is a mixed geometric distribution. This representation of the negative binomial distribution can be interpreted in terms of Bernoulli trials with a random probability of success. Initially, the probability of success is determined, and then a random variable is considered as the number of successes until the first failure in the sequence of Bernoulli trials with a given value of the random probability of success. With this interpretation, we can consider the sequence of rainy intervals conditionally independent for a fixed value of random variable that determines the value of the probability of success. This probability of success varies, say, depending on the season, from period to period and is determined by the configuration of various factors external to the local system under investigation. Within this reasoning, the observed sample of durations of wet periods can be treated as non-homogeneous containing geometrically distributed observations with different probabilities of success.

The negative binomial model of the duration of rainy intervals makes it possible to propose asymptotic approximations to some precipitation characteristics, namely to the distribution of the total amount of precipitation that fell during one rainy period (see [5]) and the distribution of the extreme daily rainfall. It should be noted that the values of the parameters of statistical regularities inherent in the data under study can significantly improve the efficiency and accuracy of intellectual data analysis by introducing additional features at the learning stage without any actual increase of the amount of available data.

2   Initial Data

The data were obtained from NOAA's National Centers for Environmental Information (NCEI). NCEI (https://www.ncdc.noaa.gov) hosts and provides access to one of the most significant archives in the world, with comprehensive oceanic, atmospheric, and geophysical data. An example of precipitation of data in Prague is demonstrated in Fig. 1.

GHCN-Daily (Global Historical Climatology Network) is the database that addresses the critical need for historical daily temperature, precipitation, and snow records over global land areas. GHCN-Daily is a composite of climate records from numerous sources that were merged and then subjected to a suite of quality assurance reviews. The archive includes over 40 meteorological elements including tempera-

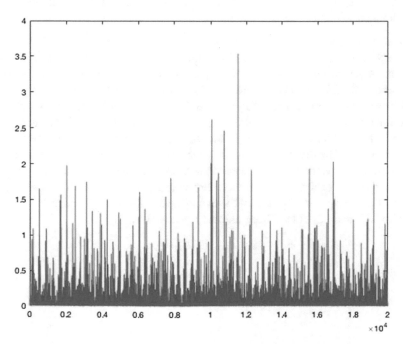

Fig. 1 Daily measurements of precipitation over a period of more than fifty years in Prague

ture daily maximum/minimum, temperature at observation time, precipitation, snowfall, snow depth, evaporation, wind movement, wind maximums, soil temperature, cloudiness, and more.

Each record represents all selected observations (values) available for a given station day. The initial section of each record is ordered as follows with the following definitions: station, station name, geographic location, and date. Station (17 characters) is the station identification code, Station name (max 50 characters) is the name of the station (usually city/airport name). Geographic location (31 characters) is the latitude (decimated degrees w/Northern hemisphere values >0 and Southern hemisphere values <0), longitude (decimated degrees w/Western hemisphere values <0 and Eastern hemisphere values >0) and elevation above mean sea level (tenths of meters). Date is the year of the record (4 digits) followed by month (2 digits) and day (2 digits).

The five key values are as follows:

- PRCP, Precipitation (mm or inches as per user preference, inches to hundredths on Daily Form).
- SNOW, Snowfall (mm or inches as per user preference, inches to tenths on Daily Form).
- SNWD, Snow depth (mm or inches as per user preference, inches on Daily Form).
- TMAX, Maximum temperature (Fahrenheit or Celsius as per user preference, Fahrenheit to tenths on Daily Form).
- TMIN, Minimum temperature (Fahrenheit or Celsius as per user preference, Fahrenheit to tenths on Daily Form).

Data on meteorological stations from six hundred geographical points in Russia and neighboring countries (Ukraine, Kazakhstan, Belarus, Lithuania, Latvia, Estonia, Tajikistan, Moldova, Georgia, Kyrgyzstan, Armenia, Azerbaijan, Uzbekistan, and Turkmenistan) were taken from the All-Russia Research Institute of Hydrometeorological Information. The data array was the daily sum of precipitation, measured in millimeters with an accuracy of 0.1 mm. The time interval could vary depending on the location of the meteorological station, but on average, the observation period was fifty years. The quality of the data leaves much to be desired, since in a sufficiently large number of days the measurements were not performed or were rejected, such incorrect indicators were excluded from consideration. In view of these circumstances, there are gaps in the data of the experiments, the study of which is a separate complex task and has not yet been uniquely resolved.

3 Statistical Analysis of Data

3.1 Negative Binomial Distribution

A random variable N_{rp} is said to have the negative binomial distribution with parameters $r > 0$ ("shape") and $p \in (0, 1)$ ("success probability"), if

$$\mathbb{P}(N_{r,p} = k) = \frac{\Gamma(r+k)}{k!\,\Gamma(r)} \cdot p^r (1-p)^k, \quad k = 0, 1, 2, \ldots$$

where $\Gamma(r)$ is the Euler gamma function

$$\Gamma(r) = \int_0^\infty x^{r-1} e^{-x} \mathrm{d}x.$$

Fitted negative binomial distribution for Prague data is represented in Fig. 2.

3.2 Parameters of Negative Binomial Distribution in Different Geographical Regions

By the definition of the negative binomial distribution, two parameters were approximated for 330 European cities and 800 towns in Russia and neighboring countries. Two types of data were considered: wet and dry periods (with and without precip-

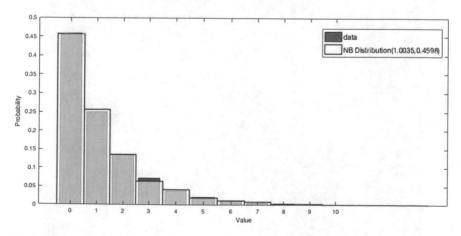

Fig. 2 Histogram of wet periods and fitted negative binomial distribution (Prague)

itation, respectively). Parameter approximations have been obtained by MATLAB function *fitdist*.

Parameter r takes values in the range from 0.3 to 1.9 for wet periods and from 0.2 to 1.1 for dry periods. Estimated parameter p takes values from 0.01 to 0.6 for wet periods and from 0.02 to 0.4 for dry periods. An example for several stations in Austria is shown in Table 1.

For visualization of maps and obtained values, *basemap*, extension of Python's library *matplotlib*, was used. Estimations of parameters for European countries are demonstrated in Figs. 3, 4, 5, and 6.

Table 1 Parameters r and p for wet periods in Austria

Station	r	p
Feuergokel	0.474	0.112
Graz	0.837	0.404
Innsbruck	0.94	0.353
Klagenfurt	0.426	0.21
Kremsmuenste	0.927	0.346
Salzburg	0.885	0.280
Sonnblick	0.532	0.105
St. Poelten	0.435	0.181
Venna	0.906	0.51
Villacheralpe	0.473	0.102
Wien	0.8565	0.402

Wet periods, parameter R: Europe

Fig. 3 Wet periods, parameter r (Europe)

Wet periods, parameter P: Europe

Fig. 4 Wet periods, parameter *p* (Europe)

Dry periods, parameter R: Europe

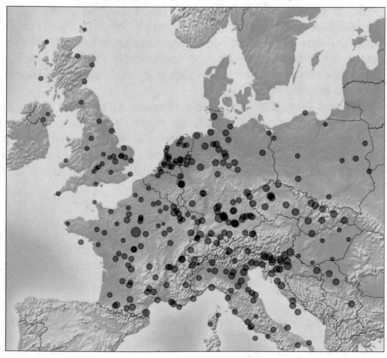

Fig. 5 Dry periods, parameter *r* (Europe)

Dry periods, parameter P: Europe

Fig. 6 Dry periods, parameter p (Europe)

According to the obtained results, we can make some conclusions and retrieve patterns in distribution of data. An interesting conclusion can be made concerning the value of the estimated parameters. For geographic points located in lowland, the shape parameter r of the negative distribution of wet periods takes a value less than one. On the other hand, parameter r in terrain near mountains or large water areas can take a value greater than one.

This pattern can be explained by special characteristics of climate like humidity, pressure, and wind strength that determine the mechanism of cloud formation. To make further investigation on this subject, specific features can be used, probably, something like characteristics mentioned earlier. Another way to discover new patterns is to process data from uncovered geographical territories: Africa, North America, South America, Antarctic, and Australia.

3.3 Expectations, Variances, and Coefficients of Variation of Negative Binomial Distribution in Different Geographical Regions

Expectation, variance, and coefficient of variation of the negative binomial distribution can be obtained using following formulas (where $X \sim \mathbf{NB}(r, p)$):

$$\mathbb{E}X = \frac{r(1-p)}{p}, \quad \mathbb{D}X = \frac{r(1-p)}{p^2}, \quad \mathbb{CV} = \frac{\sqrt{\mathbb{D}X}}{\mathbb{E}X} = \frac{1}{\sqrt{r(1-p)}}.$$

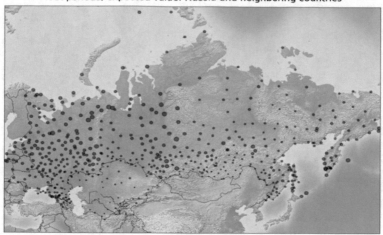

Fig. 7 Wet periods, expectation (Russia and neighboring countries)

Fig. 8 Wet periods, variance (Russia and neighboring countries)

The values of these quantities for Russia and neighboring countries are demonstrated in Figs. 7, 8, 9 and 10.

4 Conclusion

It was demonstrated that the fluctuations of durations of wet or dry periods could be successfully modeled by the negative binomial distributions in different geographical

Wet periods, coefficient of variation: Russia and neighboring countries

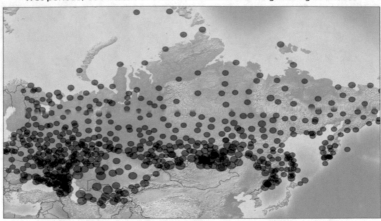

Fig. 9 Wet periods, coefficient of variation (Russia and neighboring countries)

Dry periods, coefficient of variation: Russia and neighboring countries

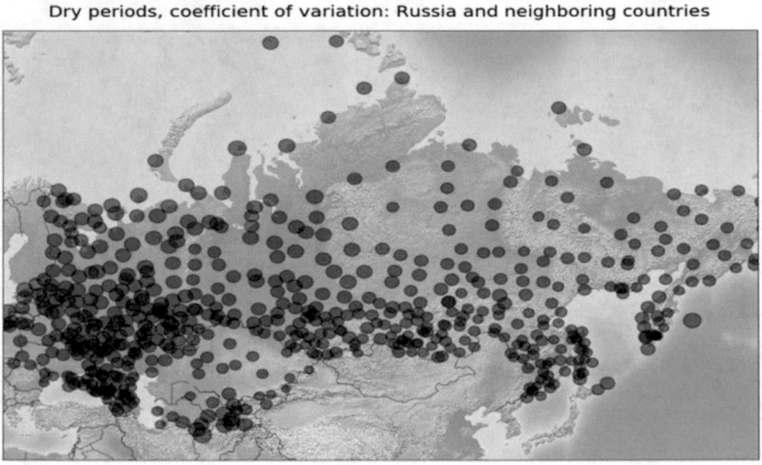

Fig. 10 Dry periods, coefficient of variation (Russia and neighboring countries)

regions. The parameters of the distributions as well as moment characteristics such as expectation, variance, and variation coefficient are calculated for numerous stations in Europe, Russia, and neighboring countries. The corresponding software solution has been implemented using Python programming language.

The regularities are stable in the plains, whereas there are significant stochastic fluctuations in the foothills and mountains. The obtained results can be used for the analysis of distribution of precipitation. In the further researches, more data from different geographic areas (see, e.g., [6, 7]) will be analyzed, and new software tools will be implemented and compared with existing analogues (e.g., [8–10]). The more sophisticated models will be developed and applied for various real tasks including problems of risks estimation [11, 12].

Acknowledgements The research was supported by the Russian Science Foundation, project 18-11-00155.

References

1. Zolina, O., Simmer, C., Belyaev, K., Gulev, S., Koltermann, P.: Changes in the duration of European wet and dry spells during the last 60 years. J. Clim. **26**, 2022–2047 (2013). https://doi.org/10.1175/JCLI-D-11-00498.1
2. Gorshenin, A.K.: On some mathematical and programming methods for construction of structural models of information flows. Informatika i ee Primeneniya **11**(1), 58–68 (2017). https://doi.org/10.14357/19922264170105
3. Korolev, V.Yu., Gorshenin, A.K.: The probability distribution of extreme precipitation. Dokl. Earth Sci. **477**(2), 1461–1466 (2017). https://doi.org/10.1134/S1028334X17120145
4. Kingman, J.F.C.: Poisson Processes. Clarendon Press, Oxford (1993)
5. Korolev, V.Yu.: Analogs of Gleser's theorem for negative binomial and generalized gamma distributions and some of their applications. Informatika i ee Primeneniya **11**(3), 2–17 (2017). https://doi.org/10.14357/19922264170301
6. Lazri, M., Ameur, S., Brucker, J.M.: Analysis of the time trends of precipitation over mediterranean region. Int. J. Inf. Eng. Electron. Bus. (IJIEEB) **6**(4), 38–44 (2014). https://doi.org/10.5815/ijieeb.2014.04.06
7. Tang, G., Long, D., Hong, Y., Gao, J., Wan, W.: Documentation of multifactorial relationships between precipitation and topography of the Tibetan Plateau using spaceborne precipitation radars. Remote Sens. Environ. **208**, 82–96 (2018). https://doi.org/10.1016/j.rse.2018.02.007
8. Olaiya, F., Adeyemo, A.B.: Application of data mining techniques in weather prediction and climate change studies. Int. J. Inf. Eng. Electron. Bus. (IJIEEB) **4**(1), 51–59 (2012). https://doi.org/10.5815/ijieeb.2012.01.07
9. Diez-Sierra, J., del Jesus, M.: A rainfall analysis and forecasting tool. Environ. Model Softw. **97**, 243–258 (2017). https://doi.org/10.1016/j.envsoft.2017.08.011
10. Mishra, N., Soni, H.K., Sharma, S., Upadhyay, A.K.: Development and analysis of artificial neural network models for rainfall prediction by using time-series data. Int. J. Intell. Syst. Appl. (IJISA) **10**(1), 16–23 (2018). https://doi.org/10.5815/ijisa.2018.01.03
11. Costa, V., Fernandes, W.: Bayesian estimation of extreme flood quantiles using a rainfall-runoff model and a stochastic daily rainfall generator. J. Hydrol. **554**, 137–154 (2017). https://doi.org/10.1016/j.jhydrol.2017.09.003
12. Garcia-Barron, L., Morales, J., Sousa, A.: A new methodology for estimating rainfall aggressiveness risk based on daily rainfall records for multi-decennial periods. Sci. Total Environ. **615**, 564–571 (2018). https://doi.org/10.1016/j.scitotenv.2017.09.305

The Knowledge Management Approach to Digitalization of Smart Education

Natalia V. Dneprovskaya, Nina V. Komleva and Arkadiy I. Urintsov

Abstract The paper deals with knowledge management as an approach to the organization of the development of the content of smart education. The purpose of the study is to develop an approach to improving the information support for the content development of higher educational programs and training materials. The system of higher education faces the challenge of supporting the development of the digital economy through training and research outcomes. Given the high pace of development of digital technologies and methods of their use, the creation of the appropriate content of education is a time-consuming task that requires a quick solution. Currently, in the studies of smart education focus on methodical aspects of the conduct of training practices, but the process of content development goes to the background, like something what is in competence of only author or lecturer. However, it is the issues of the content of education that become the most relevant. Technologies and methods of knowledge management able to provide the necessary level of speed and quality of development of educational materials.

Keywords Smart education · Digital economy · Knowledge management · Open educational resources · Content assessment

1 Introduction

The educational systems are facing new challenges caused by digital transformations in society and economy. Much attention is paid to the search for new technological and organizational solutions dictated by the requirements of the digital environment.

N. V. Dneprovskaya (✉) · N. V. Komleva · A. I. Urintsov
Plekhanov Russian University of Economics, Moscow, Russia
e-mail: Dneprovskaya.NV@rea.ru

N. V. Komleva
e-mail: Komleva.NV@rea.ru

A. I. Urintsov
e-mail: Urintsov.AI@rea.ru

© Springer Nature Switzerland AG 2020
Z. Hu et al. (eds.), *Advances in Artificial Systems for Medicine and Education II*,
Advances in Intelligent Systems and Computing 902,
https://doi.org/10.1007/978-3-030-12082-5_58

At the same time, the greatest changes occur in the content of educational programs and research. The pace of these changes is constantly increasing. Accelerating digitalization requires new approaches to content development that will support the development of smart education:

- To provide participation of students in the e-learning process at the expense of a currently central content of the curricula;
- To bring the content of education programs in compliance with the needs of digital economics development through the integration with professional associations;
- To hold off the increasing load on professors and students through algorithmization of routine operations in formatting the content to meet varying requirements of the LMS.

The purpose of the study is to develop an approach to improving the information support for the content development of higher educational programs and training materials.

The present-day trends in education give evidence of transition to a new paradigm of instruction where university education becomes a mass phenomenon. Thanks to the IT (Information Technology), a professor can teach hundreds of thousands of online students all over the world simultaneously, while the educational resources are open and available to everyone [1, 2]. Intensive use of the IT in the sphere of education leads to a replacement or transfer of some elements of educational process into the electronic medium which does not comply yet with the needs of digital economy [3, 4]. On one side, high requirements are placed on the actuality of education content and on the other side—high requirements for the results of training, namely, digital competence of students [5, 6]. Without doubt, the education system for digital economy is a major prerequisite for its development, namely, the training of students who will create and apply the algorithms for processing the digital data in various branches of economy, produce high-tech goods and services, and introduce innovations [7, 8].

The training of students, who will meet the requirements of digital economy development, is feasible only through the use of the achievements of e-learning development and transition to smart education [9]. A great number of new aspects arise in education which calls for reconsideration. While the problems of e-learning tools are covered adequately and new learning environments and approaches to organization of training process in these environments are created, the matters pertaining to working out the information content of a training course is a competence of professors who are not in view of e-learning specialists. A search for tools for timely development and actualization of the education content becomes urgent with the advance of digital economy.

Appropriate effect in working out the information content and in the area of training can be attained with the help of a knowledge management approach (KM) [10, 11]. The use of the KM allows a professor to save time to get the common view on knowledge resources of the specific subject and select appropriate educational resources for his students.

2 Smart Education Requirements to Educational Materials

The smart education implies a transition to new technology and methodology stage of learning by which is meant the organized and realized (through the use of engineering innovations and Internet) interaction between the subject matter, student, professor, and other participants of the process aimed at formation of a systematic multidimensional vision of the subject matter of science, including its different aspects (economic, legal, social, technological, etc.). In Korea Republic, the concept of smart education has been approved as a standard [12].

The smart education is a concept that proposes an integrated modification of all educational processes as well as methods and techniques utilized in these processes [13]. The concept "smart" adopted for education brought into existence such technologies as a "smart blackboard", "smart course", "smart campus", etc. [14]. Each of these technologies allows us to construct anew the process of development of educational materials, its delivery and actualization [15, 16]. The training becomes feasible not only in a lecture hall, but also at any other place. With the increase in the variability of means, forms, and the place of training, tendencies toward the individualization of instruction in the development of the educational environment, resources, and the educational process are clearly manifested, Fig. 1. But the role of a major element in the educational process will be played by the active academic content stored in the KM of an institution of higher education and available at any time and in any place for use and upgrade.

The concept of smart education is also based on the idea of individualization of training that can be realized only through creation by a professor of educational materials intended for a particular student [17]. There are the major requirements for the content of smart education:

1. Use of relevant information in the education program for solution of training tasks. The rate and quantity of information traffic in education and in any professional activity are increasing rapidly. The existing teaching materials shall be supplemented with current information to prepare the students for solution of practical tasks, for work under real conditions rather than practice with drill exercises and models.
2. Organization of self-guided cognitive, research, and project activities for students. This requirement is a key one in training the specialists who will be prepared for a creative search of the solution of professional tasks and for independent information and research activities.
3. Realization of training process in the distributed learning environment. At present time, the learning environment is not limited by the university territory or by the LMS boundaries. The training process shall be continuous, including the instruction in professional environment with the use of applicable means.
4. Interaction of students with a professional association. The professional environment is not only a customer interested in training of specialists, but he also becomes an active participant of the training process.

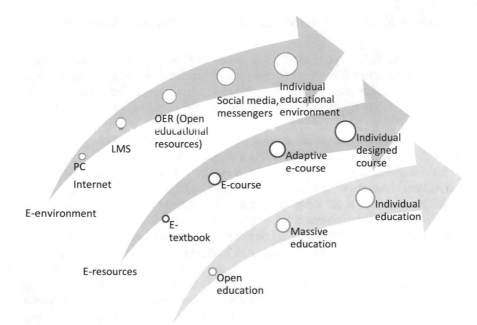

Fig. 1 Trends in the individualization of the educational environment, content, and learning

5. Flexible educational paths, individualization of training. The sphere of education is extended considerably due to attraction into the system of education the working citizens, frequent change of the kind of professional activity, and intensive development of know-how. The students attending the e-learning course at the university are well aware of the need to study. The university's mission is to provide adequate educational services in accordance with the requirements and possibilities of students.

6. Great variety of educational activities require that wide possibilities be provided for students in studying the education programs and training courses with application of adequate tools in the training process and in accordance with their health, material, and social conditions.

Realization of the smart education requirements on outdated approaches for creation of instructional and methodic materials will not bring a desired effect. Instructional resources of new type are required that will constitute an integrated teaching material created and updated based on technological innovations with a systematic description of knowledge gained in the topical area [18]. At present, it is not sufficient just to know the subject, but it is necessary to update the knowledge at all times, because the rate of appearance of new information is colossal and it keeps growing. As for the smart learning content, it is necessary to furnish the educational programs with current information aimed at formation of competences needed for digital economy. The online relations between the practice and the content of education programs can be arranged with the help of the KM.

3 Knowledge Management Approach

The highlighted requirements for content of smart education can be achieved at the expense of the KM approach to design knowledge repository (KR) and collaborative tools for either lecturers or students. The KR supposed to have every object of knowledge is identified and described. The array of identical objects allows them to be combined, so a unique content can be created to satisfy the needs of every student. Such approach is convenient both for creating the information content and for its actualization.

Also, the KM approach offers to the student's new possibilities such as work in the communities of practice where they can see how the professionals settle the real problems. The creation of the KR makes it possible to integrate KM tools with communities of practice and co-authorship instruments. As a result, the KR can be supplied with currently central knowledge and the process of content development can be improved.

By the KM is meant a combination of technologies, methods and sources of knowledge that provides the conditions for a free creation, accumulation, spreading, and use of the knowledge by the fellow workers of an organization [19]. In university, the LMS as learning environment and KR are of great importance for educational and research activities of this institution. The integration of these systems will contribute to attaining new synergetic effects. The e-learning is considered in the knowledge management theory as one of the methods for spreading knowledge in the organization [20], while the KM is used at the university for creating an efficient e-learning environment [21].

The elaboration of the KM is aimed at formation of a KR to support the actions of the university professors and fellow workers in creation, accumulation, storage, search, and use of the education content and its components [22]. So, the KM key element is a KR containing the education materials, each being described with the help of metadata system. It is the common system of metadata, describing the professional activities and education materials, that allows us to convert the practical tasks into competences, the content of professional activity into ontology, and business-cases into the content of education program.

The prototype of content development system basis on KM was designed and tested at the Moscow State University of Economics, Statistics and Informatics, after reorganization it is Plekhanov Russian University of Economics. The pilot version of the KR was integrated with university's LMS Moodle, which allowed it to be tested in the training process. The results of the pilot implementation made it possible to reveal the interest of either lecturers or students in KR, to receive approval and recommendations from the faculty for further development.

The KR users are divided into three categories: professors and research workers, management of the institution, technical support personnel, and students. The professors and research workers replenish the knowledge warehouse and update the education materials. The main advantage offered by the KR to a professor is that the required content for a curriculum can be drawn up using the materials stored in the knowledge warehouse. Elaboration of the content takes less time and information can be selected from a large set of data.

4 Content Evaluation in Knowledge Management Systems

LMS and KR form a new expanded educational environment, in which many professionals and students participate, a lot of materials are published and stored [23]. Thus, the question arises about the selection of educational materials, from evaluation. Methods of forming a meta-description for each knowledgeable object are very laborious and do not keep up with the rapidly increasing information flow.

Ever more information resources are posted on the Internet for free use and can be connected through KR with learning environment. A major role, including for corporate training, belongs to open educational resources [24]. Processing of large volume of materials at a KM system is impossible without application of intellectual technologies to support the process and evaluation of labor costs of relevant tasks of ordering and rating of heterogeneous information.

All information coming to a KM system is stored at the repository of objects as files in various formats or web-links. Information contained in the repository is analyzed by the community members. They publish, discuss, edit, evaluate, and vote for materials.

One peculiarity of information content is that it is hard to quantitatively evaluate its characteristics as there is no universal definition of quality content indicators. This is due to individual peculiarities of each information environment and heterogeneity of materials accumulated in it, absence of accurate quantitatively measurable characteristics of the content. Obviously, the ultimate solution is constructed based on the set of criteria, with some of them depending on multiple characteristics.

Fuzzy logic methods along with neural network concept are recognized among the most efficient and promising instruments for content processing [25]. There exists vast experience of their applications in various subject areas. Using fuzzy sets, it is possible to draw conclusions based on content parameters of multivalent and inaccurate nature:

- Hard to formalize evaluation criteria that are related to poorly structured complex models and imply a need to operate unclear input and output data and their confidence level;
- Parameters that are hard to formally define. For example, "relevant material", "helpful material";

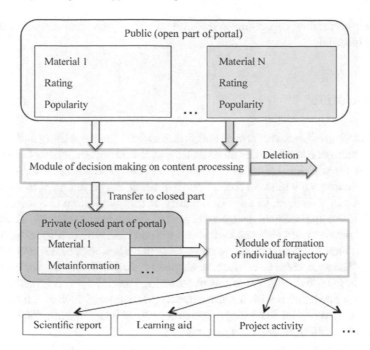

Fig. 2 Major stages of content processing

- Evaluation results with intermediate values: "Quality of the material is below average", "popularity of the content is rather low".

Based on the fuzzy logic apparatus, it is possible to optimally implement the system of making decisions on content processing with incomplete and unclear information and on the basis of arbitrary number of parameters. The mechanism of content processing in an information environment can be divided into the stages, Fig. 2.

1. Establishment of values of characteristics of a material posted to the public part of the information environment (total number of views, number of views within a week, user rating, moderators' recommendations, results of a forum vote).
2. Conditioned on whether values of characteristics reach a given level (vote threshold, number of views), the rating of the material in question is calculated using a formula that takes into account ratings of users who evaluate it, votes of moderators and vote threshold needed for rating evaluation. The content can be assigned a fixed rating following a moderator's recommendation.
3. Based on the rating, moderator's recommendations and a few other characteristics, content processing is implemented using fuzzy inference algorithm.

Ideally, a recalculation of a user rating and an update of dynamic rating tables is implemented with each new evaluation, although it can also be done according to a schedule or on a due request [26]. Also, the threshold number of evaluations to participate in the rating and the threshold for exclusion from the rating are fixed but subjects to corrections depending on the total number of users and median evaluation value of materials.

5 Conclusion

The elaboration of the KM approach comprises a knowledge repository and a module for elaboration of education content will allow us to update the content of curriculum in accordance with the dynamics of changes and development of digital economy. The KM elaboration will make it possible to decrease the period for elaboration and adaptation of curriculum, to raise the quality of this curriculum due to the use of current information and help the students master the education material. The unified integrated platform will allow us to draw up the curricula with real-time updating to overcome discrepancies between the needs of digital economy and the content of the curricula. The algorithms of content development provide the means for adaptation of curricula to individual peculiarities of students. The KM approach functioning is based on a knowledge repository, the important feature of which is a metadata of education materials. The knowledge repository can be integrated into the e-learning environments widely employed at present.

Introduction of the KM into the sphere of action of the institutions of higher education has a strategic goal of upgrading the higher education in Russia: raising of the competitive edge of the Russian information and communications technologies and educational services, and formation of a solid basis for development of education and science, extension of the knowledge network.

The achievements and advantages of e-learning will grow upon integration with the KM. In realization of the KM, of great importance are both the technical issues and the aspects of preparation of professors for work in the e-learning environment.

Acknowledgements The research has been funded by the Russian President's grant for state support of leading scientific schools No. NSh-5449.2018.6, "Study of digital transformation of the economy".

References

1. Parry, M.: Online bigger classes may be better classes. The chronicle of higher education, http://chronicle.com/article/Open-Teaching-When-the/124170. Accessed 15 Sept 2018 (2013)
2. Fetaji, B., Fetaji, M., Ebibi, M., Kera, S.: Analyses of impacting factors of ICT in education management: case study. Int. J. Mod. Educ. Comput. Sci. (IJMECS) **10**(2), 26–34 (2018). https://doi.org/10.5815/ijmecs.2018.02.03
3. Townsend, A.: Smart Cities: Big Data, Civic Hackers, and the Quest for New Utopia, p. 384. W.W. Norton & Company, New York City (2013)
4. Kumar, M., Singh, A.J., Handa, D.: Literature survey on educational dropout prediction. Int. J. Educ. Manag. Eng. (IJEME) **7**(2), 8–19 (2017). https://doi.org/10.5815/ijeme.2017.02.02

5. Kotevski, Z., Tasevska, I.: Evaluating the potentials of educational systems to advance implementing multimedia technologies. Int. J. Mod. Educ. Comput. Sci. (IJMECS) **9**(1), 26–35 (2017). https://doi.org/10.5815/ijmecs.2017.01.03
6. Tikhomirova, N., Tikhomirov, V., Maksimova, V., Telnov, Y.: The competence approach to the creation and updating of academic knowledge in the smart economy. In: 8th International Conference on Intellectual Capital, Knowledge Management and Organizational Learning (ICI-CKM), pp. 563–570 (2011)
7. Hammad, R., Ludlow, D.: Towards a smart learning environment for smart city governance. In: 2016 IEEE/ACM 9th International Conference on Utility and Cloud Computing (UCC), Shanghai, pp. 185–190 (2016)
8. Sriram, B.: Learner's satisfactions on ICT innovations: Omani learners viewpoints. Int. J. Mod. Educ. Comput. Sci. (IJMECS) **7**(6), 24–29 (2015). https://doi.org/10.5815/ijmecs.2015.06.04
9. Elhoseny, H., Elhoseny, M., Abdelrazek, S., Riad, A.M., Hassanien, A.E.: Ubiquitous smart learning system for smart cities. In: 2017 Eighth International Conference on Intelligent Computing and Information Systems (ICICIS), Cairo, pp. 329–334 (2017)
10. Zhao, R., Zhang, C.: An ontology-based knowledge management approach for e-learning system. In: 2009 International Conference on Management and Service Science, Wuhan, pp. 1–4 (2009)
11. Shehabat, I., Berrish, M.: E-learning content enhanced by active knowledge management techniques. In: 2013 IEEE 63rd Annual Conference International Council for Education Media (ICEM), Singapore, pp. 1–8 (2013)
12. Noh, K.S., Ju, S.H., Jung, J.T.: An exploratory study on concept and realization conditions of smart learning. Korea Soc. Digit. Policy Manag. **9**(2), 79–88 (2011)
13. Merzon, E.E., Ibatullin, R.R.: Architecture of smart learning courses in higher education. In: 2016 IEEE 10th International Conference on Application of Information and Communication Technologies (AICT), Baku, pp. 1–5 (2016)
14. Balonin, N.A., Petoukhov, S.V., Sergeev, M.B.: Matrices in improvement of systems of artificial intelligence and education of specialists. In: Hu, Z., Petoukhov, S., He, M. (eds.) Advances in Artificial Systems for Medicine and Education. AIMEE 2017. Advances in Intelligent Systems and Computing, vol. 658. Springer, Cham (2018)
15. Im, D.U., Lee, J.O.: Mission-type education programs with smart device facilitating. Int. J. Multimedia Ubiquit. Eng. **8**(2), 81–88 (2013)
16. Uthayakumar, C., Sarukesi, K.: An adaptive e-learning system using knowledge management. In: 3rd International Conference on Trendz in Information Sciences & Computing (TISC2011), Chennai, pp. 70–74 (2011)
17. Brusilovsky, P., Somyürek, S., Guerra, J., Hosseini, R., Zadorozhny, V., Durlach, P.J.: Open social student modeling for personalized learning. IEEE Trans. Emerg. Top. Comput. **4**(3), 450–461 (2016)
18. Huang, N.: Analysis and design of University teaching evaluation system based on JSP platform. Int. J. Educ. Manag. Eng. (IJEME) **7**(3), 43–50 (2017). https://doi.org/10.5815/ijeme.2017.03.05
19. Keishing, V., Renukadevi, S.: A review of knowledge management based career exploration system in engineering education. Int. J. Mod. Educ. Comput. Sci. (IJMECS) **8**(1), 8–15 (2016). https://doi.org/10.5815/ijmecs.2016.01.02
20. Okamoto, T., Nagata, N., Anma, F.: The knowledge circulated-organisational management for accomplishing e-learning. Knowl. Manag. E-Learn. Int. J. **1**(1), 6–17 (2009)
21. Karna, N., Supriana, I., Maulidevi, N.: Implementation of e-learning based on knowledge management system for Indonesian academic institution. In: 2016 1st International Conference on Information Technology, Information Systems and Electrical Engineering (ICITISEE), Yogyakarta, pp. 43–48 (2016)
22. Dneprovskaya, N., Shevtsova, I., Bayaskalanova, T., Lutoev, I.: Knowledge management methods in online course development. In: Novotná, J., Jancarik, A. (eds.) Academic Conferences and Publishing International Limited, pp. 159–165 (2016)

23. Vilariño, A.B.L., García, I.P.: An e-learning platform for integrated management of documents based on automatic digitization. IEEE Revista Iberoamericana de Tecnologias del Aprendizaje **8**(2), 48–55 (2013)
24. Amindoust, A., Ahmed, S., Saghafinia, A., Bahreininejad, A.: Sustainable supplier selection: a ranking model based on fuzzy inference system. Appl. Soft Comput. J. **12**(6), 1668–1677 (2012)
25. Hu, Z., Bodyanskiy, Y.V., Tyshchenko, O.K., Tkachov, V.M.: Fuzzy clustering data arrays with omitted observations. Int. J. Intell. Syst. Appl. (IJISA) **9**(6), 24–32 (2017). https://doi.org/10.5015/ijisa.2017.06.03
26. Komleva, N., Khlopkova, O.: Obrabotka kontenta v informacionnyh sredah na osnove nejro-nechetkoj modeli prinyatiya reshenij. Ekonomika, statistika i informatika. Vestn. UMO Sci. Pract. J. **5**, 188–192 (2013). (In Russian)

Algorithms for Agreement and Harmonization the Creative Solutions of Agents in an Intelligent Active System

N. Yu. Mutovkina⬤ and V. N. Kuznetsov⬤

Abstract The article deals with the methods of interaction of agents in an abstract intellectual active system when they generate creative solutions. Often collective decision making in active systems is accompanied by conflict situations. To eliminate conflicts, minimize them, algorithms for agreement and harmonization the creative solutions of intellectual agents are proposed. These algorithms depend on the strength of the contradictions revealed between the agents and the psycho-behavioral type of each of them. The description of algorithms is based on the possibility of changing the behavior of the same agents under different conditions. Agents are understood as intellectual entities (physical or programmatic), whose behavior is similar to human behavior. Therefore, in the course of reasoning, the already proven postulates from sociology, psychology and other social sciences are also used.

Keywords Intelligent active system · Agent · Algorithm · Psycho-behavioral type · Agreement · Harmonization

1 Introduction

Activation of creative activity depends on many factors, not only objective, but also subjective. Social and economic transformations, accumulation of new knowledge in various branches of science, strengthening of the role of information in society—all this creates new requirements for a person. Each individual has to adapt and evolve in a changing reality that is changing at an unimaginable pace. Objective factors here are: the place and time of birth of a person, his family, the environment, i.e., all that he cannot choose. Subjective factors include: the individual's psychological readiness for learning, his adaptability to the new collective, the specificity of communicating with new people and so on. Unlike objective factors, an individual can control subjective factors, strengthening or, conversely, limiting the influence of the latter

N. Yu. Mutovkina (✉) · V. N. Kuznetsov
Tver State Technical University, Tver, Russia
e-mail: letter-boxNM@yandex.ru

© Springer Nature Switzerland AG 2020
Z. Hu et al. (eds.), *Advances in Artificial Systems for Medicine and Education II*,
Advances in Intelligent Systems and Computing 902,
https://doi.org/10.1007/978-3-030-12082-5_59

on his life. Many people manage to cope with the negative influence of subjective factors due to their own willpower, as well as a favorable coincidence. However, this statement is true for people. And what about intelligent agents are artificial physical or software entities with their own goals, opinions, interests and acting in one system for the sake of achieving a user-defined common goal? It's no secret that in recent times, the attention of human society to the problem of creating artificial intelligence and managing it is growing rapidly. At the same time, a person compares artificial intelligence with his own intellect, empowers artificial intelligence with the ability to think, express his will, have his own opinion, work in a team, generate new ideas, offer solutions to non-standard problems, etc. One of the most important issues in this case is to ensure the productive interaction of agents in a system whose dominance is the avoidance of conflict situations between agents. It is especially difficult to ensure a conflict-free interaction of agents when solving complex, non-standard tasks and proposing creative solutions for them.

Disagreements and discrepancies between agents' actions in the intelligent active system (IAS) generate conflicts between agents. Conflicts due to their destructive features adversely affect the functioning of the system. Thus, a negative exponential effect is obtained: Mismatch generates conflicts; conflicts are the causes of increased disagreement, which generates new conflicts.

In this paper, the authors present their understanding of how to agree and harmonize the creative decisions of agents in IAS. The object of the research is intellectual agents who jointly develop possible solutions to the creative problem, submit them for discussion and evaluate them, choose the best solution, analyzing all possible gains and risks. The subject of the study is the methods of fuzzy logic and soft computing, the latest achievements of sociology, psychology, semantics and other related sciences.

The purpose of this article is to develop algorithms for agreement and harmonizing actions and decisions of intellectual agents. The application of these algorithms will avoid conflict situations in IAS.

2   Theoretical Aspects of a Research

In article [1], the authors proposed methods for avoiding conflicts in the IAS. These methods are based on the formation of the optimal composition of the system and can be successfully applied in the event that the supply of candidates for participation in the system exceeds the demand for them. Thus, it is possible to choose from a variety of candidates who have applied for participation, the bests, and the most suitable for the joint solution of tasks. These agents have all the necessary competencies and personal characteristics to perform the tasks, assigned to them, in conditions of constructive, conflict-free interaction at the same time. It's good to have a choice. What if there's no choice? What if there are enough agents available to solve the task, but lack of mutual understanding makes it impossible for them to work together?

The proposed algorithms for agreement and harmonization are a tool for eradicating such a problem.

Agreement is a procedure that allows finding a solution suiting all participants of the system to one degree or another. However, some of the agents who agreed with this decision may experience negative feelings: discomfort, anxiety, own insignificance, etc. The agreement of the decision does not guarantee that all the agents with it actually agree. They can only pretend that they are satisfied with the adopted option, but at the same time have internal contradictions. Nevertheless, agreement is the most common way out of the crisis situation.

Harmonization can be defined as an "enhanced" agreement, as a result of which all agents experience only positive emotions. The adopted alternative suits all agents almost equally.

In what follows we use the following notation.

Let each agent in the system be described by a set of goals, links and parameters:

$$a_i = \langle Z, G_i, Y_{P_i}, Y_{r_i} \rangle, \tag{1}$$

where $A = \{a_1, a_2, \ldots, a_i, \ldots, a_n\}$ is a lot of agents in the IAS;

$Z = \{z_1, z_2, \ldots, z_k, \ldots, z_w\}$ is a lot of tasks (jobs, functions) for the implementation of which IAS is created;

$G_i = \{g_1, g_2, \ldots, g_l\}$ is a lot of local target functions of agent;

$Y_{P_i} = \{y_1^{p_i}, y_2^{p_i}, \ldots, y_m^{p_i}\}$ is a lot of professional characteristics (competencies) of the agent i;

$Y_{r_i} = \{y_1^{r_i}, y_2^{r_i}, \ldots, y_m^{r_i}\}$ is a lot of personal qualities of agent i, defining his psycho-behavioral type r_i. Such qualities are: benevolence, truthfulness, self-confidence, rationality and others.

The agent's condition depends on his perception of the goals and objectives of the system, his own goals and interests, his own intellect (knowledge and ability to apply them), as well as emotions and feelings that the agent experiences while in the system. The agent's perception of the environment, as well as the effectiveness of his work in the system, directly depends on the agent's interaction with other agents. Interaction is characterized by the establishment of multilateral dynamic relations between agents. Due to the interaction with each other, agents can change their condition over time, from the worst to the best, and vice versa. The main characteristics of the interaction of agents are the direction of their actions, selectivity, intensity and dynamism of contacts with other agents [2, 3].

At the heart of the construction of algorithms for agreement and harmonization is the concept of the psycho-behavioral type of the agent, the relationship of each agent to other participants in the system and to himself, as well as the management influence that changes the agent's type and attitude [4]. The authors attempted to formalize these concepts, as shown in Fig. 1.

This figure shows the color scales symbolizing the status (type) of the intelligent agent, and the corresponding fuzzy numerical interval corresponding to it. The zone of shades of blue in the first scale is typical for the evading agent, the green shades

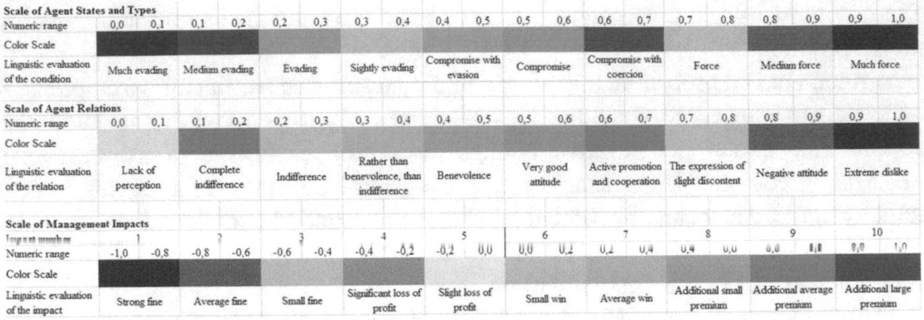

Scale of Agent States and Types

Numeric range	0,0	0,1	0,1	0,2	0,2	0,3	0,3	0,4	0,4	0,5	0,5	0,6	0,6	0,7	0,7	0,8	0,8	0,9	0,9	1,0
Color Scale																				
Linguistic evaluation of the condition	Much evading		Medium evading		Evading		Slightly evading		Compromise with evasion		Compromise		Compromise with coercion		Force		Medium force		Much force	

Scale of Agent Relations

Numeric range	0,0	0,1	0,1	0,2	0,2	0,3	0,3	0,4	0,4	0,5	0,5	0,6	0,6	0,7	0,7	0,8	0,8	0,9	0,9	1,0
Color Scale																				
Linguistic evaluation of the relation	Lack of perception		Complete indifference		Indifference		Rather than benevolence, than indifference		Benevolence		Very good attitude		Active promotion and cooperation		The expression of slight discontent		Negative attitude		Extreme dislike	

Scale of Management Impacts

	1		2		3		4		5		6		7		8		9		10	
Numeric range	-1,0	-0,8	-0,8	-0,6	-0,6	-0,4	-0,4	-0,2	-0,2	0,0	0,0	0,2	0,2	0,4	0,4	0,6	0,6	0,8	0,8	1,0
Color Scale																				
Linguistic evaluation of the impact	Strong fine		Average fine		Small fine		Significant loss of profit		Slight loss of profit		Small win		Average win		Additional small premium		Additional average premium		Additional large premium	

Fig. 1 Fuzzy scales of psycho-behavioral types, relations and impacts

belong to the agent of a compromise type, and the orange and red colors represent the agent of the force type. Agents can relate to themselves differently, to other agents, decisions made in the system, and also to react differently to control actions. The second scale is a scale of relations, starting with complete indifference and ending with extreme hostility. The third scale shows possible control effects transferred from the world of people (organizational systems) to the world of intellectual agents. This world also has its own fines, losses, wins and bonuses. In the works [5–7], various management actions involving a system of fines and premiums were considered. However, since issues of agreement and harmonization are considered here, the authors consider it unacceptable to influence agents in this way. Agreement in the IAS is achieved by the introduction of incentives, which motivating of agent. These incentives are the agent's belief that if he does not agree to act in accordance with the adopted option (plan), he will not receive any income and will regret the lost profit. Therefore, it is logical that in the developed algorithms, only the agent's beliefs in obtaining (or not receiving) a certain gain or in the probability of lost profit act as the coordinating influences.

The objects of agreement in the IAS can be: plans for further actions, the option of solution to the problem, option of interaction of agents, distribution of tasks and functions among agents, etc. The objects of harmonization are the intelligent agents themselves.

3 Features of Algorithms of Agreement and Harmonization

Since the procedures for agreement and harmonization strongly depend on the specifics of the problem, task, being solved in the IAS, there are no clear, unambiguously established rules for carrying out these procedures. Algorithms for agreement and harmonization have their own peculiarities for each situation. Due to its ability to the transforming and take into account the features of the development of events, algorithms for agreement and harmonization creative solutions have an alternative

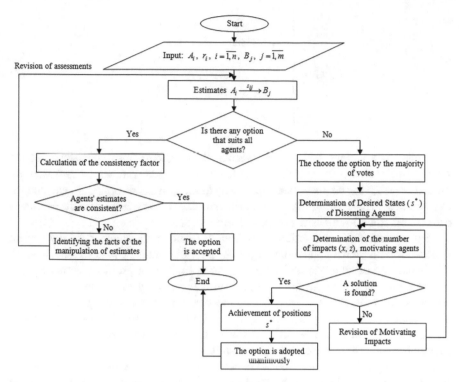

Fig. 2 Scheme of the algorithm for agreement a variant of further actions in the IAS

name—algorithmic regulations [8]. If the executor of the algorithm, as a rule, is an automaton, the intellectual essence always acts as the executor of the algorithmic regulation.

If the object of agreement is a variant of further actions in the IAS, then the agreement algorithm can be represented, as in Fig. 2.

The initial data are: the number of agents, the psycho-behavioral type of each of them (determined from the observation of the agent's behavior over a period of time) and possible variants for the development of the system (there may be one or more variants). Each agent gives his assessment showing his attitude to each of the variants B_j, $j = \overline{1, m}$, according to the relationship scale (Fig. 1). So the matrix of relations is formed:

$$
C = \begin{bmatrix}
 & B_1 & B_2 & \dots & B_j & \dots & B_m \\
A_1 & s_{11} & s_{12} & \dots & s_{1j} & \dots & s_{1m} \\
A_2 & s_{21} & s_{22} & \dots & s_{2j} & \dots & s_{2m} \\
\dots & \dots & \dots & \dots & \dots & \dots & \dots \\
A_i & s_{i1} & s_{i2} & \dots & s_{ij} & \dots & s_{im} \\
\dots & \dots & \dots & \dots & \dots & \dots & \dots \\
A_n & s_{n1} & s_{n2} & & s_{nj} & & s_{nm}
\end{bmatrix}.
\tag{2}
$$

If there is a variant, that suits all agents, then the degree of consistency of the relationship matrix is calculated and a conclusion is made whether the estimates are agreed or not. If the estimates are sufficiently agreed, then this option is accepted for execution. If facts of manipulation of agent estimates are revealed, then the estimates are revised and all steps of the algorithm are repeated.

If there is not of such variant, which would suit to all agents, then the variant is chosen, with which agreed the largest number of agents. The desirable states s^* of dissenting agents are establishing and the optimization problem is solving, the required data in which are the type and number agreeing of actions, the scale of which is shown in Fig. 1. Depending on the situation, the task of optimization can be a linear or non-linear programming task with fuzzy constraints [9]. In the event that the agent's beliefs have been reasoned, the agents will change their original judgments about this option and agree with it.

In IAS, as well as in any human collective, there can be agents that like, do not like, cause extreme dislike or do not cause any emotions until a certain time. And all these perceptions can change over time, because in IAS, as in the world, there is nothing permanent. The reconciliation of the mutual perception of intellectual agents is much more difficult than the coordination of their opinions on any issue, whether it is a variant of the development of the system, the choice of the priority task, the method for solving it and so on. Here, the essence of the problem lies in the deep structures of the subconscious of agents [10], although at this stage of the development of artificial intelligence, the conflict of intellectual agents can be successfully solved by redistributing resources, tasks and self-organization of agents in terms of their independent formation of teams (working groups). Of course, the condition must be fulfilled, that all agents can solve all the tasks. The algorithm of agreement for the described situation is shown in Fig. 3.

The determining factor of agreement in the algorithm presented in Fig. 3 is the matrix of benevolence of agents in relation to each other, which has the form:

$$
F = \begin{bmatrix}
 & A_1 & A_2 & \dots & A_i & \dots & A_n \\
A_1 & f_{11} & f_{12} & \dots & f_{1i} & \dots & f_{1n} \\
A_2 & f_{21} & f_{22} & \dots & f_{2i} & \dots & f_{2n} \\
\dots & \dots & \dots & \dots & \dots & \dots & \dots \\
A_i & f_{i1} & f_{i2} & \dots & f_{ii} & \dots & f_{in} \\
\dots & \dots & \dots & \dots & \dots & \dots & \dots \\
A_n & f_{n1} & f_{n2} & \dots & f_{ni} & \dots & f_{nn}
\end{bmatrix}.
\tag{3}
$$

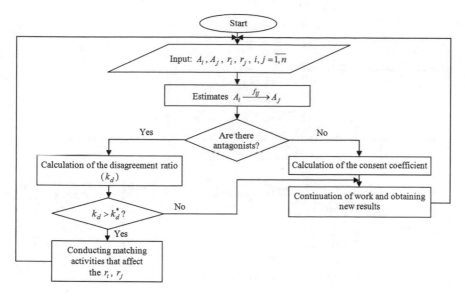

Fig. 3 Scheme of the algorithm of agreement of mutual perception of agents

Elements of matrix (3) $f_{ij} \in [0; 1]$ show the ratio of agent i to agent j. The scale of relations, shown in Fig. 1, applies here. Diagonal elements of the matrix (3) also take values based on the recorded rules and show the attitude of the agent himself to his activity in the IAS. These elements can be interpreted as self-evaluations of agents. If the matrix (3) is dominated by negative estimates (from gray, blue and red ranges), then it is obvious that there are antagonistic agents in the system. If this is the case, the disagreement ratio k_d is calculated and compared with the critical value k_d^*. When the condition $k_d > k_d^*$ is met, the matching activities are carried out, which positively affect the values r_i, r_j and force the agent i to change his attitude to individual agents of the system for the better. Agreement and harmonization of agents' interaction are carried out through the application of certain management actions $v_i(t) \in V$, where V is a set of possible management actions. These impacts can be divided into two types: external and internal actions. External actions are carried out by agents, surrounding of agent i. Internal influence is exerted on itself by the agent i, guided by its own self-consciousness.

As noted above, the harmonization algorithm is much more difficult to develop and execute than the algorithms of agreement. Harmonization cannot be observed in formal groups. In addition to the common goal, the necessary condition for harmonization is the existence of common interests and mutual sympathy, which is a sign of an informal group. From performing work in an informal group, agents should receive only positive emotions. Therefore, harmonization is considered to be achieved only if there is a complete absence of negative emotions in the IAS.

Fig. 4 shows an approximate harmonization algorithm. It is assumed that all agents have all the necessary competencies to solve any of the assigned tasks.

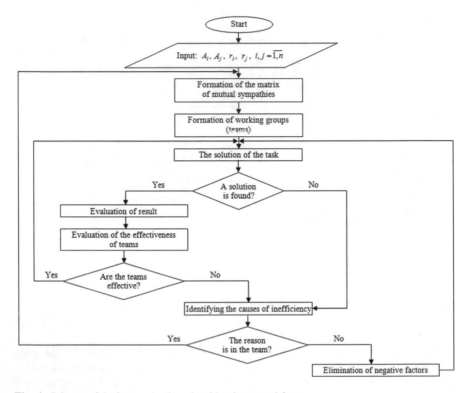

Fig. 4 Scheme of the harmonization algorithm in general form

The algorithm, presented in Fig. 4, is based on the ability of the IAS to self-organize. Agents, who have certain sympathy for each other, are teamed up to perform assigned tasks. At the same time, the results of their work can be shared by teams in absentia, on a specially designated portal. If the performance of any of the teams is not good enough, then the reasons for this are clarified. If the reason is the non-optimal composition of the team, then it is checked for agreement on the mutual perception of the agents. It is impossible to exclude the possibility of pseudo-sympathy, when agent i at the stage of formation of the matrix of mutual likes was sure that he would get along with agent j. However, in reality, this turns out not the case at all.

If it is the influence of other negative factors, for example, the absence of funds, then this influence is eliminated by the usual methods adopted in society. And then the solution of problems in IAS continues.

4 Conclusions

The developed algorithms are a tool for management the behavior of agents in IAS so that their joint work brings the most useful result. The central part of the algorithms is the application of such management actions that can redirect the energy of the conflicting agents to a peaceful course—on the way to achieve the stated goal of the IAS.

At the same time, the installation is implemented in the agents' minds, ensuring that the winnings of the system as a whole will increase the winnings of each of them, and vice versa, if the task is not resolved in time, the system will not receive any winning, this means that the system participants too will not receive the win.

Agents, based on their own interests, can inform each other of inaccurate information, ignore or distort important information. To increase the effectiveness of IAS activities, it is necessary to exclude such manifestations of activity, in time to identify false reports and participants transmitting these messages. Thus, agents should be convinced that the message of the information which is not corresponding to true, first of all is not favorable to them; the agent's win depends on the winning of the IAS; IAS has a Nash equilibrium: To deviate one from one's positive behavior, strategies are not beneficial to any of the agents, because in this case they will worsen their positions in the system and will receive a smaller gain.

When agents interact, one of the main problems may be the mismatch of their goals. Each decision taken in the IAS is aimed at achieving one or more goals. It is important to agree on goals so that they are all achievable; to exclude those goals that cannot be achieved simultaneously with the goals recognized as the most important. To do this, the method of constructing an agreed target tree is often used, which is a directed graph with expertly defined weights of importance.

An important moment in the agreement is not only the settlement of the relationships of participants in the system, but also ensuring the internal coordination of the agent, eliminating internal contradictions in it. This is successfully handled by the harmonization algorithm, which involves communication the agent only with those agents that cause positive emotions in it.

Presence and application of management mechanisms, algorithms of agreement and harmonization at each significant stage of IAS operation allow to obtain the best result and to solve difficult, non-standard, and creative tasks with coordinated interaction of agents.

Acknowledgements The reported study was funded by RFBR according to the research project No. 17-01-00817A.

References

1. Mutovkina, N.Y., Kuznetsov, V.N., Klyushin A.Y.: The formation of the optimal composition of multi-agent system. In: Hu, Z., Petoukhov, S., He, M. (eds.) Advances in Artificial Systems

　for Medicine and Education. AIMEE 2017. Advances in Intelligent Systems and Computing, vol. 658, pp. 293–302. Springer, Cham. https://doi.org/10.1007/978-3-319-67349-3_28

2. Pujari, S., Mukhopadhyay, S.: Petri net: a tool for modeling and analyze multi-agent oriented systems. Int. J. Intell. Syst. Appl. (IJISA) **4**(10), 103–112 (2012). https://doi.org/10.5815/ijisa.2012.10.11

3. Zaharija, G., Mladenović, S., Dunić, S.: Cognitive agents and learning problems. Int. J. Intell. Syst. Appl. (IJISA) **9**(3), 1–7 (2017). https://doi.org/10.5815/ijisa.2017.03.01

4. Mutovkina, N.Y., Semenov, N.A.: A model of modification of intelligent agents types in the any logic system dynamics methodology. Softw. Syst. **31**(1), 145–151 (2018) https://doi.org/10.15827/0236-235X.031.1.145-151

5. Burkov, V.N., Irikov V.A.: Models and Methods of Management of Organizational Systems. In: Kulba, V.V. (ed.). Nauka, Moscow, 270 p. (1994)

6. Novikov, D.A.: Stimulation in Socio-Economic Systems (Basic Mathematical Models). Institute of Control Sciences of the Russian Academy of Sciences, Moscow, 216 p. (1998)

7. Novikov, D.A.: Theory of Management of Organizational Systems. The Moscow Psychological and Social Institute, Moscow, 584 p. (2005)

8. Landa, L.N. Algorithmization in teaching. In: Landa, L.N. (ed. and co enters) Under the Society. Gnedenko, B.V., Biryukov, B.V. M.: "Enlightenment" (article), 524 p. (1966)

9. Erfanian, H.R., Abdi, M.J., Kahrizi, S.: Solving a linear programming with fuzzy constraint and objective coefficients. Int. J. Intell. Syst. Appl. (IJISA) **8**(7), 65–72 (2016). https://doi.org/10.5815/ijisa.2016.07.07

10. Bandler, R.: Guide to Personal Change. Eksmo, Moscow, 208 p. (2010)

Designing an Information Retrieval System in the Archival Subdivision of Higher Educational Institutions

Tetiana Bilushchak, Andriy Peleshchyshyn, Ulyana Yarka and Zhanna Myna

Abstract The article deals with the design of the information retrieval system (IRS) in the archival subdivisions of higher educational institutions (HEIs). A functional model has been constructed in order to formalize the process of automation of the archival search for documents of the HEI graduates, employees. The functional model of the IRS has been designed with the help of structural modeling using the DFD tools. The expediency of the introduction and use of the information retrieval system of the archive in archival subdivisions of higher educational institutions has been analyzed, in particular, on the example of the Lviv Polytechnic National University. Due to the developed information retrieval system in the form of a database, taking into account the International Standard Archival Description (ISAD (G)), the search for archival information improves, and the performance of the department staff is facilitated to serve the population in solving issues of social importance.

Keywords Information retrieval archival system · Template · Automation · Archival subdivision · Informatization of higher educational institution · ISAD (G) · Verification · Design · Structural modeling

T. Bilushchak (✉) · A. Peleshchyshyn · U. Yarka · Z. Myna
Social Communications and Information Activities Department,
Lviv Polytechnic National University, Lviv, Ukraine
e-mail: t.bilushchak@gmail.com

A. Peleshchyshyn
e-mail: apele@ridne.net

U. Yarka
e-mail: ulyarka74@gmail.com

Z. Myna
e-mail: zhanna.shijaniuk@gmail.com

© Springer Nature Switzerland AG 2020
Z. Hu et al. (eds.), *Advances in Artificial Systems for Medicine and Education II*,
Advances in Intelligent Systems and Computing 902,
https://doi.org/10.1007/978-3-030-12082-5_60

1 Introduction

Formulation of the problem. In today's conditions of global development of the information society, social institutions and educational institutions need to use automation systems based on information technology. The use of up-to-date computer technologies will make it possible to facilitate and improve the efficiency of work of archival employees of HEIs. Automation can not only shorten working hours, but also create a fundamentally new tool in servicing population in solving tasks of social importance—verification of personal data in the archival subdivisions of universities. The rapid growth of volumes and accumulation of information leads, to a certain extent, to an increase in time the employee spends on searching for archival information in documents and searching for the very documents.

Thus, the challenge of development of the information retrieval archival system, which automates the processes of collecting and extracting information, arises.

Formulation of the objectives. On the basis of the foregoing, one can form the task of the study, which consists in developing the advantages and expediency of using the information retrieval archival system by the staff of the university's archive. In order to construct IRAS efficiently, it has been suggested to consider it through DFD, which is a tool of modeling the functional requirements to the system being designed. The result of development of the functional model of the system is a hierarchy of data flow diagrams that describe the process of transformation of information from its introduction into the system to delivery to the user.
The purpose of the article is to create a project on structuring of the information retrieval system for archival subdivisions of higher education institutions, taking into account the International Standard Archival Description (ISAD (G)), which will allow the application of international rules and technologies that will be understandable (accepted) in the European information space.

Analysis of research and publications shows that the study of the issue and prospects of the automation of search, checking large volumes of information inquiries of the institutions, is relevant [1–6]. However, studies of the problem of finding large volumes of documents and information arrays in the archival subdivision of higher educational institutions are incomplete. The application of international archival standards in the development of the information retrieval system has not been fully studied.

2 Comprehensible Use of the Information Retrieval System in the Archival Subdivision of the Lviv Polytechnic National University for Making a Quick Multidisciplinary Inquiry

Currently, in Ukraine and abroad, there are still systems with the help of which one can confirm or refute studying of a person in a higher educational institution. However, these systems do not provide such multidimensional archival information for verifying and obtaining data about a graduate or an employee of the university as a direct inquiry to the university's archival subdivision. For example, Sweden has a Ladok database that contains student data from the majority of the 50 higher education institutions in Sweden. This system allows the employer, with the help of a certain assigned registration number, verify the authenticity of obtaining educational documents. In addition, the experience of verifying the data such as the Norwegian Registry of Diplomas Vitnemålsportalen will be of interest for us. However, it should be noted that the work of this system was not the initiative of the Institute, but of the Ministry of Education and Science of Norway. As to the functioning of the verification of educational documents on the territory of Ukraine, this is the Information and Production System (IPS) "Osvita." The IPS "Osvita" is fully integrated with the unified state electronic database in the USEDE system. The software tools of the IPS "Osvita" ensure the verification of the authenticity of educational documents. However, if a graduate does not have information about the series and number of the diploma and other details of the lost diploma, it is necessary to address the archive of the educational institution.

Further consideration of access to the verification of educational papers of graduates will be analyzed on the example of the USA. Despite the fact that higher education in the USA is decentralized, there is a centralized approach for keeping records of students in each institution. Given the number of third-party services suggested to institutions, there are plenty of options of electronic systems for student data and records, including specially developed systems that are specific to universities, commercially developed general systems that allow institutions to access the records of each other. Regardless of the type of the system, there is one absolute information: The data belongs to the students, and the student must allow access. This is specifically outlined in the Family Educational Rights and Privacy Act (FERPA). FERPA determines the confidentiality of records. All access to student accounts is governed by FERPA, regardless of the method or means of access. According to FERPA, the student must allow access to data [7].

Therefore, with the growth of large volumes of archival funds in universities and the need to ensure their quick use in order to obtain complete and reliable archival information, the expediency and necessity of using the information retrieval system arise. The information retrieval system (IRS) is a set of methods and tools for storing and searching for documents, information about them, or certain facts. The space for experimental research has been the archive of the Lviv Polytechnic National University. Based on the analysis, the problems of searching for large volumes of documented and information arrays have been discovered. One of the problems of the archival functioning of the structural subdivision of the Lviv Polytechnic National University is a large number of information inquiries (confirmation of graduation from the educational institution, confirmation of taking courses of disciplines with allocated hours and grades, information on the length of service, on salaries, etc.). It has been established that the archive of the Lviv Polytechnic National University processed more than 623 inquiries in 2017. As information search is done through paper media, which greatly complicates and slows down the work of the staff, it is suggested to change the technology of manual search for electronic means using the information retrieval archival system.

In each university, the archival structural subdivision, among other major tasks, makes searches in response to inquiries for verification of personal data. The structure of the inquiry stages in the archival subdivision of the Lviv Polytechnic National University is shown in Fig. 1.

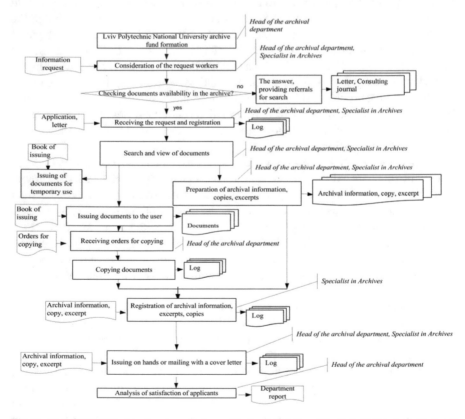

Fig. 1 Algorithm of the inquiry stages in the archival subdivision of the Lviv Polytechnic National University

In order to establish common requirements for determination of the quality of services rendered by enterprises, institutions, local governments, archival information, extracts, and copies to fulfill the biographical and factual inquiries by the archival subdivisions of the higher educational institutions are implementing a quality management system. The regulatory framework for the development of such systems is the international standard ISO 9001-2008 "Quality Management System" [8, 9]. The process of providing archival references, excerpts, copies in the archive of the Lviv Polytechnic National University includes receiving a letter of the applicant, studying the inquiry, search and review of documents, registration of archival information, excerpts, copies, and delivery of archival information, excerpts, and copies.

3 Proposals on Technical Implementation and Main Functional Features of the Information Retrieval Archival System

The domain of the information retrieval archival system is that it enriches the prospects for the user in the operation of an array of archival information as a means of using the relevant databases in the structural subdivisions of the HEI, which allows only limited access to documents. Thus, the information retrieval archival system will allow: search by a person (full name, year of birth, or period of work and study) regardless of his/her place of stay in the past; search by a person in all documents; search by the object; organize the geographical and chronological filter of documents; automatic filling out of the document template for retrieval from the archive.

The information retrieval archival system in the form of a database is created to: improve the conditions of citizens' servicing; improve the efficiency of work—the speed of servicing, the convenience of searching for information, eliminating obstacles in obtaining services due to a human factor (corruption, bureaucracy, physical overload, etc.); and store archival documents.

It is suggested to create the information retrieval system in the form of a database. The development of the information retrieval system facilitates the laborious and routine work of the search problem and will be implemented as a means of reliable data storage, which makes it convenient and quick to find the necessary information [10].

Let us consider the diagram of data flows for a better understanding of the process of automation of archival search in the structural subdivisions of the Lviv Polytechnic National University (Fig. 2).

Data flow diagrams are the main means of modeling functional requirements of the system being designed. With their help, these requirements are divided into functional components (processes) and are presented as a network linked by data flows. The main purpose of such tools is to demonstrate how each process transforms its input data into output data, as well as identify the relationship between these processes. An important role in the model belongs to a special type of DFD—a context diagram that simulates the system in the most common way. The context diagram represents the interface of the system with the surroundings, namely the information flows between the system and the external entities, which it should be linked with. It identifies these external entities, and, as a rule, one process that reflects the main purpose or nature of the system as close as possible.

Figure 2 depicts a context diagram with one main process—"Automation of archival search for documents of graduates, employees of HEIs"—which is connected with three external entities through data flows: University graduate, employee is a person who applies for educational papers or their renewal. The employee of the archival subdivision is a person who considers the issue of renewal of educational papers, controls and verifies data transmitted to the automated system, and receives a statistical report on the system performance. Archivist of the informational profile is a person who is an expert in the field of archives, and is engaged in structuring, modernization, filling out of the database, and in the event of mistakes during testing of the DB deletes it upon receiving the remarks.

Let us consider DFD level 1 of decomposition of the process "Automation of archival search for documents of graduates, employees of HEIs" (Fig. 3).

Figure 3 depicts the decomposition of the context diagram. It shows five processes: "Receiving an inquiry for a reference," "Data correction," "Development of information retrieval help system," "Generation of the reference," and "Generation of statistical reporting."

A further step in the construction of DFD is the division of the allocated processes into subprocesses, that is, the decomposition of level 2 is done. The second-level decomposition diagram reflects the division of "Development of information retrieval help system" process into five subprocesses: "Receiving input informa-

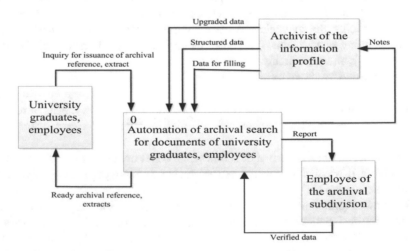

Fig. 2 Context diagram "Automation of archival search for documents of graduates of higher educational institutions"

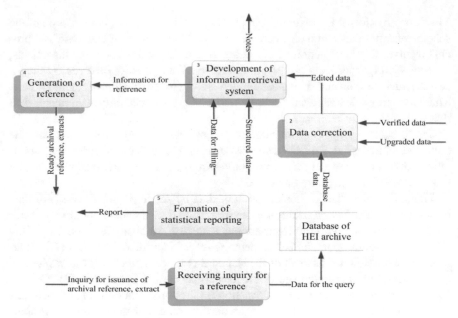

Fig. 3 DFD level 1 of decomposition of the process "Automation of archival search for documents of graduates, employees of HEIs"

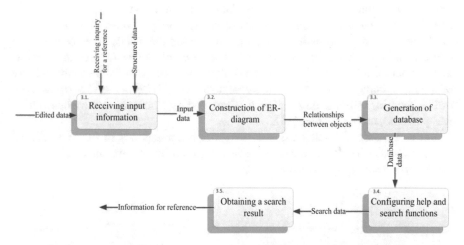

Fig. 4 DFD level 2 decomposition of the "Development of information retrieval help system" process

tion," "Construction of ER diagram," "Generation of Database," "Configuring Help and Search Functions," and "Obtaining a Search Result" (Fig. 4).

The "Receiving input information" subprocess involves receiving edited, structured data and data about contents, resulting in a set of input data of the database.

The "Construction of ER diagram" subprocess involves the very construction of the scheme of input data and their interrelation. The "Generation of Database" subprocess involves the construction of a database with existing links between the objects.

The "Configuring Help and Search Functions" subprocess involves creating a search structure from an existing database, resulting in the creation of reference data. The "Obtaining a Search Result" subprocess implies the availability of aggregate data that will be used for reference.

The "Receiving input documentation" subprocess (Fig. 4) involves not only the collection of necessary data, but also that these data have a certain order and presentation. That is the view (Fig. 5). One should carry out the systematization of data by sections; create the structure of each level.

Figure 6 depicts the "Systematization of data by sections" subprocess. For this, the proposed information retrieval system presented as database should be developed in accordance with the International Standard Archival Descriptions (ISAD (G)), which should contain 26 elements (Fig. 7) identified in ISAD (G), organized in seven volume sections of the standard (Fig. 6), namely (1) *"Identification block"*—contains information needed to identify the unit of description; (2) *"Content block and structure"*—contains information on the topic and ordering of materials in the unit of description, a brief abstract on the content of the document, systematic indices and classification codes, subject headings; (3) *"Block of access conditions and use"*—contains information on the legal status of the document, states the conditions that limit or affect the availability of archival materials; (4) *"Context block"*—contains information about the origin and storage of the unit of description; (5) *"Block of notes"*—contains important information that needs to be recorded and that cannot be placed elsewhere; (6) *"Block additional materials"*—contains information about the directly related archival materials to the unit of storage: availability of originals, copies, or reissues of archival materials; (7) *"Block of descriptive control"*—contains information about how, when, and by whom the archive description was prepared, archivist notes [11].

Let us consider the design of the IRS, which corresponds to the international standard ISAD (G), the elements of the description of the document should contain the following sections, which are shown in Fig. 7.

Fig. 5 DFD level 3 decomposition of the "Receiving input documentation" process

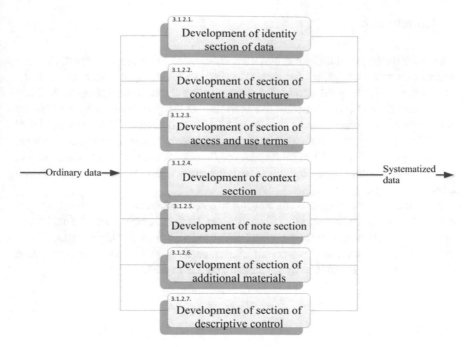

Fig. 6 DFD level 4 of decomposition of "Systematization of data by sections" subprocess

Elements of describing						
IDENTIFICATION BLOCK	**CONTENT BLOCK AND STRUCTURE**	**BLOCK OF ACCESS CONDITIONS AND USE**	**CONTEXT BLOCK**	**BLOCK OF NOTES**	**BLOCK OF ADDITIONAL MATERIALS**	**BLOCK OF DESCRIPTIVE CONTROL**
*Code(s)	*The scope and content	*Conditions that regulate access	*Creator's name	*Notes	*Availability and location of originals	*Notes of archivist
*Title(s)	*Information on examination, utilization and storage	*Conditions that regulate reproduction	*History of device / creator biography		*Availability and location of copies	*Terms or conditions
*Date(s)	period [units of description]	*Language of documents / graphics system	*Archive history		*Units of description, related to the main	*Date(s) of description
*Level of description	*Additional income	*Physical characteristics and technical requirements	*The immediate primary source or source of transmission		*Notice of publication	
*Content and media of unit of description (quantity, volume or size)	*Organization system	*Search engines				

Fig. 7 Schematic representation of the elements of the description according to the international standard ISAD (G) and their sections

4   Conclusion

Thus, with the help of DFD structural modeling tools, the process of developing the information retrieval system for archival subdivisions of higher educational institutions has been designed taking into account the International Standard Archival Description (ISAD (G)). A detailed description of the process is presented by the four levels of decomposition of the context DFD. The result of the construction of the DFD hierarchy has been to make the requirements to the information retrieval archival system of the HEI units understandable at each level of detalization, and to divide these requirements into parts with precisely defined relationships between them.

Thus, the proposed project on the development of the information retrieval archival system that is able to accumulate information and output it via the multidimensional inquiries will allow the staff of the archival subdivision of the higher educational institution to increase the efficiency of implementation of inquiries for verification of personal data of the university graduates and employees.

References

1. Zavuschak, I.: Methods of processing context in intelligent systems. Int. J. Mod. Educ. Comput. Sci. (IJMECS) **10**(3), 1–8 (2018). https://doi.org/10.5815/ijmecs.2018.03.01
2. Kadda, B., Ahmed, L.: Semantic annotation of pedagogic documents. Int. J. Mod. Educ. Comput. Sci. (IJMECS) **8**(6), 13–19 (2016). https://doi.org/10.5815/ijmecs.2016.06.02
3. Chen, Y., Cao, J.: TakeXIR: A type-ahead keyword search XML information retrieval system. Int. J. Educ. Manag. Eng. (IJEME), **2**(8), 1–5 (2012). https://doi.org/10.5815/ijeme.2012.08.01
4. Hatano, K., Kinutani, H., Yoshikawa, M., Uemura, S.: Information retrieval system for XML documents. Database and expert systems applications. DEXA **2453**, 758–767 (2002). https://doi.org/10.1007/3-540-46146-9_75
5. Syerov, Y., Fedushko, S., Loboda, Z.: Determination of development scenarios of the educational web forum. In: XIth International Scientific and Technical Conference Computer Sciences and Information Technologies (CSIT). Lviv, pp. 73–76 (2016)
6. Korobiichuk, I., Fedushko, S., Juś, A., Syerov, Y.: Methods of determining information support of web community user personal data verification system. In: Automation 2017. Advances in Intelligent Systems and Computing, vol. 550, pp. 144–150. Springer, Berlin (2017)
7. Identifying the challenges of differences in models of online databases. Available at: http://www.aacrao.org/docs/default-source/PDF-Files/gdn-final-symposium-white-paper—online-databases-eicher-funaki-johansson-koenig.pdf?sfvrsn=0
8. Myna, Z., Yarka, U., Peleschyshyn, O., Bilushchak, T.: Using international standards of quality management system in higher educational institutions. In: XIIIth International Conference (TCSET 2016), Lviv, pp. 834–837 (2016)
9. ISAD (G): 1999 General international standard archival description. Available at: http://www.icacds.org.uk/eng/ISAD(G).pdf

10. Bilushchak, T., Myna, Z., Yarka, U., Peleshchyshyn, O.: Integration processes in the archival section of Lviv Polytechnic National University. In: XIth International Scientific and Technical Conference Computer Sciences and Information Technologies (CSIT), Lviv, pp. 200–203 (2017)
11. ISO 9001-2008. Quality management systems. Available at: http://www.plitka.kharkov.ua/certs/433_iso9001.pdf

The Manner of Spacecraft Protection from Potential Impact of Space Debris as the Problem of Selection with Fuzzy Logic

B. V. Paliukh⊙, V. K. Kemaykin⊙, Yu. G. Kozlova⊙ and I. V. Kozhukhin⊙

Abstract Space debris (SD) is a real danger for spacecraft (S) in orbit operation. Taking this danger into account is a S flight safety requirement. Plenty of measures are provided in the course of constructional design and engineering of S. They differ in scope and results of protection in maintenance of security for S. This paper proposes the method to estimate a protection manner with regard to the importance of the parameters characterizing one. Solving this problem carries out using fuzzy analytic hierarchy process (ranking) (FAHP) and fuzzy logical conclusion (FLC).

Keywords Space debris · Manner of spacecraft protection · Fuzzy logical conclusion · Knowledge base · Fuzzy analytic hierarchy process · Paired comparison matrix

1 Introduction

Constructive and technological decisions, as well as organizational and controlling one, for protection against space debris (SD) impact increase autonomy of S operation when these decisions are implemented on a spacecraft (S) board.

The S protection problem is resolved in two stages:

1. A package of measures is formed and implemented in S engineering and production stage [7].

B. V. Paliukh · V. K. Kemaykin · Yu. G. Kozlova (✉) · I. V. Kozhukhin (✉)
Tver State Technical University, 22, emb. of A. Nikitin, Tver 170026, Russian Federation
e-mail: jul_kozl@mail.ru

I. V. Kozhukhin
e-mail: kozhukhin@mail.ru

B. V. Paliukh
e-mail: pboris@tvstu.ru

V. K. Kemaykin
e-mail: vk-kem@mail.ru

© Springer Nature Switzerland AG 2020
Z. Hu et al. (eds.), *Advances in Artificial Systems for Medicine and Education II*,
Advances in Intelligent Systems and Computing 902,
https://doi.org/10.1007/978-3-030-12082-5_61

2. A specific set (manner) for protection among all implemented measures is selected and used when a threat of SD impact occurs in orbit operation of S.

The goal of this paper is to improve selection of a measure package (a manner of spacecraft protection) from potential impact of space debris with complex integrated control system (CICS). The CICS on a S board is designed to solve a wide range of problems such as location and tracking of SD objects using S electro-optical means, selection of SD objects which are hazardous to S, forecasting the results of the impact caused by SD objects on a S, selection and use a S protection manner among all implemented measures based on overall assessment of the situation. Implementing the mentioned functional in automatic real-time mode of CICS is a burning question.

2 The Modern Approach

There are three types of S protection from potential impact of SD such as passive, active, and operational (the IADC standard) [8].

The passive protection includes strengthening of an instrument container and elements by applying protection covers (i.e., thermal blanket) or mounting bumper shields to armor S body and its elements. It knows the protection manner allowing us to form a bumper shield, to separate one from a protected S, and to steer one to potentially hazardous objects [9]. To select the passive protection measures, it is necessary to consider sizes, materials, and velocities of SD objects, which can strike S, as well as a period of bumper shield deploy. At present, the problem of S protection from small SD can be resolved only bumper shield use.

The active protection is based on simultaneous use of sensing and warning equipment on a possible collision and emergency programs. The emergency programs include a maneuver of S withdrawal from a hazardous zone to another orbit, avoidance maneuver from a collision with a hazardous SD particle and returning to an orbit. It is possible to change orbits of others body to avoid collision. The requirements to active protection implementation are available fuel reserve (energy store), desired time for program execution and further S operation, available time before possible collision with SD and time, required to computation and execution of a program.

The operational protection provides execution of control command (program) set, connected with changing an onboard system operation mode, configuration layout diagram and S orientation from center of mass to minimize S destruction damage from SD incident particles. The requirements of the operational protection are time before possible collision, limiting control command (program) execution, available inboard fuel reserve as well as sizes, materials, and velocities of SD objects, which can have an effect on S.

A package of the measures, formed in engineering and production stage, enables to provide and implement measures beforehand. These measures are sufficient to protect different S [1–5]. This supposes selection of materials and S configuration layout diagram, definition of characteristics for onboard basic systems and placement of these systems in an instrument container, emergency programs development to change over the standby mode. The solutions in this stage concern with search for compromise between the permitted levels of S damage, caused by a collision with SD, the added mass of protection means and acceptable reduction of the workload mass. The methodology of this operation is group ordering of S protection manners (measures) from potential impact of SD. They are Pareto-noncomparable among themselves.

$$X = \{\{x_3, x_5, x_9\}, \{x_2, x_4, x_8\}, \{x_1, x_6, x_7\}\}$$

The package of measures, including the first subset manners, which are noncomparable (optimum) for separate criteria, is selected to implement in engineering and production stage.

The problem of selecting and using the manner (the package of specified measures), which is sufficient one for protection and adequate one to the reality situation, must be resolved in S orbit operation. This stage specifies an efficiency estimate of manners with regard to the importance of the parameters such as available time before a possible collision with SD, an available fuel reserve and a requirement to further S use and existence. The efficiency estimate of manners is multicriterion analysis problem. It has arisen in engineering and production stage and must be specified during S operation, when a real threat of collision exists.

The traditional methods of multiparameter analysis [10] provide transformation of the vector of manner evaluation parameters to the scalar overall index with regard to weight coefficients. They have been weighed early and must reflect the importance of each parameter and its contribution to the integral index. The essential restriction of this approach is bad suitability to expert judgements and consideration of real settings in space, when a real treat emerges.

The suggested method contains two stages.

The first stage has done earlier on Earth. It requires neither quantitative estimation of manner parameters nor the procedure of their scalarization. In this stage, it is used expert information in terms of pair comparisons as follows:

- The responsiveness: the protection manner 1 is much the same as the protection manner 3.
- The productivity: the protection manner 2 is better than the protection manner 1.
- The economy: the protection manner 3 is much better than the protection manner 2 and so on.

The second stage is realized in real time on S board. The knowledge base, introducing in CICS and containing expert views about the importance of the parameters in different conditions of the real-life environment, is used in this stage.

Source data are: $S = \{s_1, s_2, ..., s_v\}$ is a set of alternative manners (a package of measures) to protect S from SD; $C = \{c_1, c_2, c_3\}$ is a set of manner evaluation parameters; $F = (f_1, f_2, f_3)$ is the importance of the indicators of manner evaluation parameters, determined from the real-life environment. The task is to order manners from set S by parameters from set C with regard to their importance F in real time, that is, in the onboard control system cycle after detecting SD. By $\mu^l(s_v)$ let us denote a number with the range [0,1]. This number characters s_v with respect to c^l. The more $\mu^l(s_v)$, the higher estimate of a manner with respect to c^l. Then, the parameter c^l can be expressed as a fuzzy set is designated on ground set as follow:

$$c^l = \left\{ \frac{\mu^l(s_1)}{s_1}, \frac{\mu^l(s_2)}{s_2}, \frac{\mu^l(s_3)}{s_3} \right\} \tag{1}$$

where $\mu^l(s_v)$ is grade of membership s_v to a fuzzy set c^l.

3 The First Stage

To define the grade of membership from (1), we early form paired comparison matrixes of each parameter for each manner from a set S. Total number of matrixes is defined a number of parameters and equals l.

For parameter c^l, the matrix of parameter is written as:

$$A^l = \begin{array}{c} \\ s_1 \\ s_2 \\ s_n \end{array} \begin{array}{ccc} s_1 & s_2 & s_n \\ a^l_{11} & a^l_{12} & a^l_{1n} \\ a^l_{21} & a^l_{22} & a^l_{2n} \\ a^l_{n1} & a^l_{n2} & a^l_{nn} \end{array} \tag{2}$$

where an element a_{ij} is estimated by an expert on nine-point Saaty's scale [11, 15].

This matrix allows us to range and order of a set S by each evaluation parameter c^l. To compute ranks, according to the method [6, 13], it is necessary to find an eigenvector of the matrix (2). One can use the procedure reviewed in [12, 14] for first approximation of rank desired characteristics. This procedure supposes that the matrix (2) has following characteristics:

- It is diagonal one, that is, $a_{ii} = 1$, $i = 1, \ldots v$. The elements a_{ii} are symmetric with respect to the main diagonal and are as related $a_{ii} = 1/a_{ii}$;
- It is transitive, that is, $a_{ik}a_{kj} = a_{ii}$.

These properties allow us to define all elements of the matrix (2), when we know elements of one row. If we know a row k, that is, the elements a_{kj}, then an arbitrary element a_{ij} is defined as:

$$a^l_{ij} = \frac{a^l_{kj}}{a^l_{ki}}, i, j, k = \overline{1 \ldots v} \tag{3}$$

When all elements of the matrix (2) are defined, the grades of membership, required for fuzzy set (1) are calculated according to the formula [12]:

$$\mu^l(s_v) = \frac{1}{a_{v1}^l + a_{v2}^l + \cdots + a_{vk}^l} \tag{4}$$

According to Bellman and Zadeh's principle [12], the best manner is one with guaranteed estimate for c_1, c_2, c_3 parameters.

Thus, a fuzzy set, required for estimate, is defined as intersection (as the effectiveness parameter of the manner).

$$K_{sv} = c_1 \cap c_2 \cap c_3$$

Since the intersection operation ($\cap$) corresponds with the minimum operation, we obtain:

$$K_s = \left\{ \frac{\min\limits_{l=1...3} \mu^l(s_1)}{s_1}, \frac{\min\limits_{l=1...3} \mu^l(s_2)}{s_2}, \frac{\min\limits_{l=1...3} \mu^l(s_v)}{s_v} \right\} \tag{5}$$

The best manner is one with the largest grades of membership (the numerator).

4 The Second Stage

The estimation of parameters by their importance must be conducted in real time during situation analysis. This estimation assumes the greater importance in autonomy orbital operation. By f_1, f_2, f_3, let us denote the importance of the coefficients (or ranks) for manner evaluation parameters. Ranking with the paired comparison method is unacceptable here, since experts do not include in the CICS operation cycle. In real time, the coefficients f_l can be obtained by solving fuzzy logical equation systems. These equations are based on the knowledge matrix or the logical statement system which is isomorphic to the knowledge matrix. These equations allow us to obtain the values membership function for fixed values of the input variables. The input values are linguistic variable ones such as available time before possible collision with SD and desired time for executing the relevant protection manner, available fuel reserve on S board and desired fuel burn for executing the relevant protection manner, predictive S damage probability and the requirements to protection and further S existence. They are represented as membership functions. The output values are the importance of the coefficients for manner evaluation parameters.

The knowledge base is formed by the rules [3]:

1. The dimension of this matrix equals $(n + 1)N$, where $(n + 1)$ is number of columns while $N = k_1 + k_2 + \ldots k_m$, m is number of rows;
2. The first n columns of the matrix correspond to input variables x_i, $i = 1 \ldots n$, and $(n + 1)$th column correspond to the values of f_j ($j = 1 \ldots m$);

3. Each row of the matrix represents a combination of input variable values, this combination classified by an expert as corresponding to one of the possible values of the output variable. The first k_1 rows correspond to output variable value d_1, the second k_2 rows correspond to value d_2, and the last k_n rows correspond to value d_m;

4. The element a_i^{jp}, standing at the intersection of the column i and row j, corresponds to linguistic estimation of the parameter x_i in j_p row of the fuzzy knowledge base. The linguistic estimate a_i^{jp} is chosen from the term-set of the variable x_i, that is, $a_i^{jp} \in A_i, i = 1 \dots n, j = 1 \dots m, p = 1 \dots k_j$.

Fuzzy logic equations are derived by replacing linguistic terms a_i^{jp} and f_j with corresponding membership functions. The operations ∩ and ∪ are derived by replacing with min and max operations accordingly.

The system of logical equations can be written in the following compact way:

$$\mu^{f_l}(x_1, x_2, \ldots, xn) = \overset{k_j}{\underset{p=1}{\vee}}\left[\overset{n}{\underset{i=1}{\wedge}} \mu^{a_i^{jp}}(xi)\right], \quad l = \overline{1, m}$$

When calculating, one should replace the logical operations AND (∨) and OR (∧) over the membership functions with min and max operations accordingly:

$$\mu^{f_l}(x_1, x_2, \ldots, x_n) = \underset{p=\overline{1,k_j}}{\max}\left\{\underset{i=\overline{1,n}}{\min}\left(\mu^{a_i^{jp}}(xi)\right)\right\}, \quad j = \overline{1, m}$$

In the presence of the importance of the coefficients $f_l, l = 1 \dots 3$, the formula (5) is written as:

$$K_s = \left\{\frac{\underset{l=\overline{1\dots3}}{\min}\ \mu^l(s_1)^{f_l}}{s_1}, \frac{\underset{l=\overline{1\dots3}}{\min}\ \mu^l(s_2)^{f_l}}{s_2}, \frac{\underset{l=\overline{1\dots3}}{\min}\ \mu^l(s_v)^{f_l}}{s_v}\right\} \qquad (6)$$

where the degree f_l shows fuzzy set concentration in compliance with the measure of the importance of the parameter c_l.

The algorithm of problem solving involves:

1. Considering the evaluation parameters for alternative manners as fuzzy sets. They are designated on the ground set of fault handling manners using membership functions.

2. Defining membership functions for fuzzy sets in terms of expert information. This obtains by paired comparison of fault handling manners using nine-point Saaty's scale.

3. Evaluating effectiveness of fault handling manners. It is based on fuzzy set interaction (Bellman and Zadeh's principle [12]).

4. Evaluating importance of the parameter manner using fuzzy logic inference. The obtained weights are regarded as concentration degree of appropriate membership functions.

5. Evaluating effectiveness for the selected manner with regard to real situation.

5 The Discussion of Results

To verify the method, we examined a bounded set of alternative manners. They are presented in Table 1.

We define evaluation following parameters for manners: c_1 is responsiveness of the manner; c_2 is economy of the manner; c_3 is productivity of the manner;

The selected set of parameters is not closed and can be added depending on the requirements for S protection.

Statements by evaluation parameters for manners are obtained by prior review results:

- Responsiveness of the manner: decisive superiority S_3 and S_2 over S_1; weak superiority S_3 over S_2; significant superiority S_1 over S_2.
- Economy of the manner: decisive superiority S_1 over S_3; significant superiority S_2 over S_1.
- Productivity of the manner: significant superiority S_2 over S_1; almost decisive superiority S_1 over S_3; almost weak superiority S_2 over S_3.

The aforecited experts statements correspond with paired comparison matrixes, derived with using Saaty's scale [6]:

$$A(c_1) = \begin{array}{c} \\ s_1 \\ s_2 \\ s_3 \end{array} \begin{array}{ccc} s_1 & s_2 & s_3 \\ 1 & 7 & 7 \\ 1/7 & 1 & 3 \\ 1/7 & 1/3 & 1 \end{array}$$

$$A(c_2) = \begin{array}{c} \\ s_1 \\ s_2 \\ s_3 \end{array} \begin{array}{ccc} s_1 & s_2 & s_3 \\ 1 & 1/5 & 1/7 \\ 5 & 1 & 1/5 \\ 7 & 5 & 1 \end{array}$$

$$A(c_3) = \begin{array}{c} \\ s_1 \\ s_2 \\ s_3 \end{array} \begin{array}{ccc} s_1 & s_2 & s_3 \\ 1 & 5 & 1/6 \\ 1/5 & 1 & 1/2 \\ 6 & 2 & 1 \end{array}$$

Table 1 Bounded set of alternative manners

S_1	Performing avoidance maneuver (the emergency program for elevation of perigee on 20 km)
S_2	Expansion a bumper shield toward incident SD particles
S_3	Changing over to the standby mode

With the help of paired comparison matrixes and the formula (4), we obtain:

$$c_1 = \left\{ \frac{0.125^{0.68}}{s_1}, \frac{0.24^{0.68}}{s_2}, \frac{0.68^{0.68}}{s_3} \right\} = \left\{ \frac{0.24}{s_1}, \frac{0.37}{s_2}, \frac{0.77}{s_3} \right\},$$
$$c_2 = \left\{ \frac{0.75}{s_1}, \frac{0.16}{s_2}, \frac{0,08}{s_3} \right\}, \quad c_3 = \left\{ \frac{0.16}{s_1}, \frac{0.59}{s_2}, \frac{0.11}{s_3} \right\}$$

With the help of fuzzy sets for c_1–c_3 and the model (5), we obtain:

$$K_S = \left\{ \frac{0.125}{s_1}, \frac{0.16}{s_2}, \frac{0.08}{s_3} \right\}$$

This indicates about superiority S_2 over S_1 and S_3.

When we detected a SD object, which is potentially dangerous because of approach with S, we must assume the protection measures (or use a manner) from the onboard protection measures. Thus, we must specify the estimate of efficiency for manners. When SD object has detected, CICS computes the required source data in real time. Timing data of a possible collision: available time before possible collision; desired time for execution of each manner (each package of measures), realized on S board. The material characteristics of a possible collision: available fuel reserve; desired fuel burn for executing of each manner (each package of measures), realized on S board. The probabilistic characteristics of a possible collision: the probability of S collision with SD; the damage probability (risk) of S elements or the whole S; the probability of saving S functionality, which is determined the requirements for protection and further S operation.

The indicated above parameters are presented by linguistic variables. The fuzzy knowledge matrixes, combined to the fuzzy knowledge base of CICS, work out for these variables.

When S is functioning, the problem is solved for onboard control system in an operation cycle with CICS. It is used the fuzzy logic inference with a knowledge base of the importance of manner parameters for solving the problem.

To perform calculations by each parameter of a manner, it is used such rules as:

IF Available time before possible collision = short AND Time for execution of the manner (the package of measures) = middle, THEN The responsiveness = obvious (clear) superiority;

IF Available fuel reserve = large AND Desired fuel burn for a manner (a package of measures) = average THEN The economy = slight (weak) superiority;

IF The probability of S collision = strong AND The damage probability (risk) = strong AND The probability of saving S functionality = strong THEN The productivity = significant superiority;

Total number of production rules in a knowledge base is 28.

The following propositions, concerning the importance of manner parameters, have obtained on basis of required rules after computing:

- obvious (clear) superiority c_1 over c_2 and c_3;
- slight (weak) superiority c_3 over c_2;

The paired comparison matrixes are made on basis of Saaty's scale allow us to obtain the following calculated values: $f_1 = 0.68; f_2 = 0.2; f_3 = 0.12$. The estimate of parameters with regard to ranks:

$$c_1 = \left\{ \frac{0.125^{0.68}}{s_1}, \frac{0.24^{0.68}}{s_2}, \frac{0.68^{0.68}}{s_3} \right\} = \left\{ \frac{0.24}{s_1}, \frac{0.37}{s_2}, \frac{0.77}{s_3} \right\}$$

$$c_2 = \left\{ \frac{0.75^{0.2}}{s_1}, \frac{0.16^{0.2}}{s_2}, \frac{0.08^{0.2}}{s_3} \right\} = \left\{ \frac{0.94}{s_1}, \frac{0.69}{s_2}, \frac{0.6}{s_3} \right\}$$

$$c_3 = \left\{ \frac{0.16^{0.12}}{s_1}, \frac{0.59^{0.12}}{s_2}, \frac{0.11^{0.12}}{s_3} \right\} = \left\{ \frac{0.08}{s_1}, \frac{0.94}{s_2}, \frac{0.77}{s_3} \right\}$$

The effectiveness of the manner is derived as:

$$K_S = \left\{ \frac{0.24}{s_1}, \frac{0.37}{s_2}, \frac{0.6}{s_3} \right\}$$

This characterizes the effectiveness of the manners with a glance of the situation. The situation is defined by available and desired fuel reserve, time, as well as the requirements to further operation.

There is preference to execute the manner s_3 when the requirement to responsiveness of protection from possible impact of SD dominates.

6 Conclusions

Engineering and production of S provides for realization of the requirements for protection from possible impact of SD. The list of measures for execution of the specified requirement is included in onboard technical decisions. Some measures of passive protection have actuated since a S creation moment. They provide for protection from objects less 1 cm in size. The measures of active and operation protection provide for more reliable protection, but they require selection depending on situation for using. The selected package of measures is a manner of protection. It must be valid and correspond to the possible impact level and the damage level caused by this impact. This paper proposes a variant of S independent operations when a protection manner is formed by CICS in automatic mode on basis of real-time accounting of situation if a hazard exists. The introduced method is based on fuzzy logic theory as well as the principle of objectives and restrictions merging. The obtained estimates of efficiency for a manner are implemented the principle of guaranteed results with balanced criteria, using for estimate. In real time, it is specified the estimate of efficiency for each detected SD object with a glance a rule set for required and existed resource restrictions to reliability, time, and fuel burn on implementing potential protect manners. The example of estimating efficiency is presented. The example shows that the determined requirement to responsiveness causes by changing selection of the implemented manner.

Acknowledgements The research was done within the government task of the Ministry of Education and Science of the Russian Federation. The number for the publication is 2.1777.2017/4.6.

References

1. Kobylkin, I.F., Selivanov, V.V.: Materials and Structures of Light Armor Protection. BMSTU, Moscow, Russia (2014). (in Russian)
2. Gerasimov, A.V., Pashkov, S.V., Khristenko, Yu. F.: Protection of spacecraft from anthropogenic and natural fragments. Experiment and numerical simulation [Bull. Tomsk State Univ. Math. Mech.] **4**(16), 70–78 (2011). (in Russian)
3. Ryan, S., Christiansen, E.L.: NASA/TM-2009-000000. Honeycomb vs. Foam: Evaluating a Potential Upgrade to ISS Module Shielding for Micrometeoroids and Orbital Debris. NASA, 2009
4. Destefanis, R., Amerio, E., Briccarello, M., Belluco, M., Faraud, M., Tracino, E., Lobascio, C.: Space environment characterisation of Kevlar: good for bullets, debris and radiation too. Univ. J. Aeronaut. Aerosp. Sci. **2**, 80–113 (2014)
5. Destefank, D., Lambert, M., Schäfer, F., Drolshagen, G., Francesconi, D.: Debris shielding development for the ATV integrated cargo carrier. In: Proceedings of the 4th European Conference on Space Debris. ESA/ESOC: Darmstadt, Germany, 2005, pp. 453–458. Available at: http://adsabs.harvard.edu/full/2005ESASP.587..453D.html. Accessed 27 July 2018
6. Muhametzianov, I.Z.: Fuzzy logic inference and fuzzy method of fuzzy method of hierarchy analysis in decision support systems: an application to the evaluation of the reliability of technical systems. Cybern. Program. **2**, 59–77 (2017). https://doi.org/10.7256/2306-4196.2017.2. 21794
7. OST 134-1031-2003. Space technique products. General requirements to aerospace systems security against mechanical impacts caused by particles of natural and anthropogenic origin. (in Russian)
8. Zelentsov, V.V.: A Spacecraft Protection From Impact of Small Space Debris Fragments. BMSTU, Moscow, Russia (2015). (in Russian). https://doi.org/10.7463/0615.0778339
9. The Military Academy of Strategic Rocket Troops after Peter the Great: The Method of protection of spacecraft. 2294866, (2007). Available at: http://www.findpatent.ru/patent/229/2294866.html. Accessed 27 July 2018
10. Lotov, A.V., Pospelova, I.I.: Multicriterion Decision-Making Problems. MSU, Moscow, Russia (2008). (in Russian)
11. Borisov, A.N., Krumberg, O.A., Fedorov, I.P.: Decision Making Based on Fuzzy Models: Examples of Use. Riga, USSR, Znanie (1990). (in Russian)
12. Rotshteyn, A.P.: Intellectual Technologies of Identification: Fuzzy Logic, Genetic Algorithms, Neural Networks. Universum: Vinnitsa, Ukraine, 1999
13. Hu, Z., Bodyanskiy, Y.V., Tyshchenko, O.K., Tkachov, V.M.: Fuzzy clustering data arrays with omitted observations. Int. J. Intell. Syst. Appl. (IJISA) **9**(6), 24–32 (2017). https://doi.org/10. 5815/ijisa.2017.06.03
14. Bodyanskiy, Y.V., Tyshchenko, O.K., Kopaliani, D.S.: An extended neo-fuzzy neuron and its adaptive learning algorithm. Int. J. Intell. Syst. Appl. (IJISA) **7**(2), 21–26 (2015). https://doi. org/10.5815/ijisa.2015.02.03
15. Sharma, S., Kumar, S., Singh, B.: Hybrid intelligent routing in wireless mesh networks: soft computing based approaches. Int. J. Intell. Syst. Appl. (IJISA) **6**(1), 45–57 (2014). https://doi. org/10.5815/ijisa.2014.01.06

Structural Synthesis of Spatial *l*-Coordinate Mechanisms with Additional Links for Technological Robots

V. A. Glazunov, G. V. Rashoyan, A. K. Aleshin, K. A. Shalyukhin
and S. A. Skvortsov

Abstract The article is devoted to the structural synthesis of mechanisms of parallel structure, belonging to the class *l*-coordinate mechanisms. Studies of the mechanisms of the parallel structure in connection with their wide application are conducted in many countries of the world. The proposed synthesized mechanism structures open up prospects for their wider application, in particular, in technological installations, devices for measurement and testing, additive technologies and manipulators of medical purpose. The introduction of additional intermediate links in the synthesized structural circuits of mechanisms, as well as the removal of drives outside the working space, can significantly expand their functional capabilities. This circumstance serves to develop the theory of synthesis and analysis of mechanisms of parallel structure.

Keywords Parallel structure mechanisms · *l*-coordinates · Intermediate links · Structural synthesis

1 Introduction

This article presents mechanisms of parallel structure that extend the classification of *l*-coordinate mechanisms proposed by Arzumanyan and Koliskor [1] and differ in

V. A. Glazunov · G. V. Rashoyan · A. K. Aleshin · K. A. Shalyukhin (✉) · S. A. Skvortsov
Mechanical Engineering, Research Institute of the Russian Academy of Sciences,
4, Malyi Kharitonievsky pereulok, 101990 Moscow, Russian Federation
e-mail: constmeister@gmail.com

V. A. Glazunov
e-mail: vaglznv@mail.ru

G. V. Rashoyan
e-mail: gagik_r@bk.ru

A. K. Aleshin
e-mail: aleshin_ak@mail.ru

S. A. Skvortsov
e-mail: 1691skvorcov@mail.ru

© Springer Nature Switzerland AG 2020
Z. Hu et al. (eds.), *Advances in Artificial Systems for Medicine and Education II*,
Advances in Intelligent Systems and Computing 902,
https://doi.org/10.1007/978-3-030-12082-5_62

that the output link and the base are connected not only by means of six kinematic chains 6-*SPS* (where *S* is spherical and *P*—translational kinematic pairs) but also by additional links. In the synthesized structural circuits of mechanisms, the drives can be located outside the working area. This circumstance is due to the change in the boundaries of the working zone and its configuration. It is very important to expand the functionality of these mechanisms. This corresponds to the basic principles of the theory of synthesis and analysis of parallel structure mechanisms [2–14].

It is known that in general, *l*-coordinate mechanisms constitute a small part of parallel structure mechanisms. In conditions where the scope of the parallel structure mechanisms is extensive and covers such fields of application as engineering, additive technologies, measuring and testing equipment and technologies, as well as medical manipulators for surgical purposes and rehabilitation—any classification extension of parallel mechanisms is a step towards parallel structure mechanisms broad application. From this point of view, the declared topic of the article is relevant.

2 Example of the *l*-Coordinate Mechanism

Consider an *l*-coordinate mechanism in which six translational drives, corresponding to six *l*-coordinates, are located between three mounting points on the base and three on the output link (Fig. 1). The mechanism has six degrees of freedom, in accordance with the Somov–Malyshev formula:

$$W = 6n - 5P_5 - 4P_4 - 3P_3 - 2P_2 - P_1 \tag{1}$$

where n is the number of mechanism movable links; P_1, P_2, P_3, P_4, P_5—the number of five, four, three, two, one movable kinematic pairs.

When calculating W, to eliminate local mobility, six spherical kinematic pairs are replaced by two moving pairs:

Fig. 1 *l*-coordinate mechanism with six translational drives

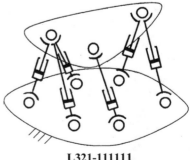

L321-111111

$$n = 13, \quad P_5 = 6, \quad P_4 = 6, \quad P_3 = 6$$
$$W = 6 \times 13 - 5 \times 6 - 4 \times 6 - 3 \times 6 = 6.$$

In the article, in calculating W, to eliminate local mobility, six spherical kinematic pairs are replaced by two mobile pairs.

3 *l*-Coordinate Mechanism with Additional Links

Consider the *l*-coordinate mechanism (Fig. 2), with *l*-coordinates attached to the three intermediate links, which does not change the mechanism DOF number.

The drives are located outside the working area, and these additional links pass through the base or the output link (Fig. 2). The connecting rod is coupled with that link, through which it passes, the spherical joint with the translational pair located inside.

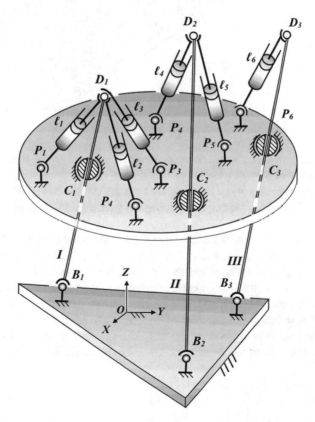

Fig. 2 *l*-coordinate mechanism with three intermediate links

Fig. 3 *l*-coordinate
mechanism with two
intermediate links

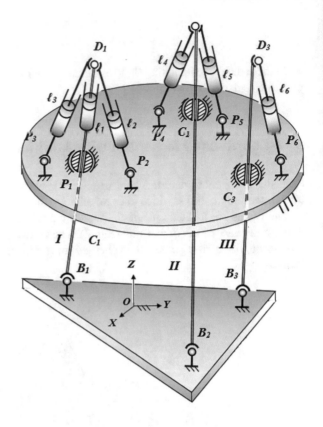

In accordance with the above-mentioned information, points B_1, B_2 and B_3 (Fig. 2) are located on the base with which the OXYZ fixed coordinate system is connected. Three connecting input rods, I, II and III, are associated with the output link by spherical joints C_1, C_2 *and* C_3, having an opening—a female element of the translational kinematic pair.

The *l*-coordinates l_1–l_6 express the displacements in the drives, which are located between the points P_1, P_2, P_3, P_4, P_5, P_6 and points D_1, D_2, D_3, respectively.

Such an arrangement of the drives shows that they are connected to the output link and the input rods, and the base is connected only to the said rods. This structure is in demand for mobile robots working in extreme areas; for example, in space, there is a necessity to withdraw engines out of an aggressive environment.

One can see that the motion of the output link corresponds to six degrees of freedom; for this, we use the Somov–Malyshev formula (1), in which:

$$n = 16, \quad P_5 = 6, \quad P_3 = 15, \quad P_2 = 3$$
$$W = 6 \times 16 - 5 \times 6 - 3 \times 15 - 2 \times 3 = 96 - 81 = 15,$$
$$W = W_{\text{main}} + W_{\text{loc}} = 6 + 9, \quad \text{where } W_{\text{main}} = 6, W_{\text{loc}} = 9.$$

Fig. 4 *l*-coordinate mechanism with two intermediate links (variant 2)

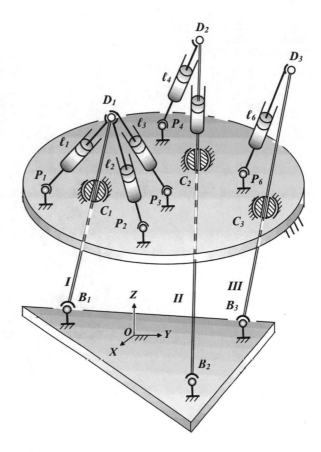

In this case, the local mobility is related to the rotation of the kinematic circuits $P_1D_1, P_2D_1, P_3D_1, P_4D_2, P_5D_2, P_6D_3$ and additional links *I, II, III* around their own axes. These mobilities do not affect the kinematics of the mechanism. They can be eliminated by replacing the spherical hinges at the points $P_1, P_2, P_3, P_4, P_5, P_6$ of the output link and at the B_1, B_2, B_3 points of the base on the double-hinged joints (e.g. a spherical hinge with a finger). In this case, here and below, we obtain the following result:

$$n = 16, \quad P_5 = 6, \quad P_4 = 9, \quad P_3 = 6, \quad P_2 = 3$$
$$W = 6 \times 16 - 5 \times 6 - 4 \times 9 - 3 \times 6 - 23 = 96 - 90 = 6.$$

Developing the proposed approach to the formation of new *l*-coordinate mechanisms, we point out that one of the drives may be rigidly coupled to the rod input, so that their axes coincide (Fig. 3). Here all the notations correspond to the above. One can see that the number of DOF is also equal to six:

Fig. 5 *l*-coordinate
mechanism with two
intermediate links

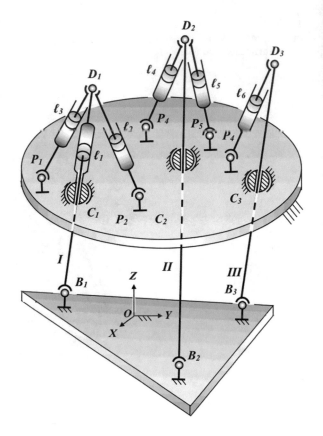

$$n = 15, \quad P_5 = 6, \quad P_4 = 8, \quad P_3 = 6, \quad P_2 = 2$$
$$W = 6 \times 15 - 5 \times 6 - 4 \times 8 - 3 \times 6 - 2 \times 2 = 90 - 84 = 6.$$

In the same way as in the previous case, another actuator rod associated with another *l*-coordinate may be connected to the input shaft (Fig. 4). Here again all the notations correspond to the previous ones, and the number of DOF is again equal to six:

$$n = 15, \quad P_5 = 6, \quad P_4 = 8, \quad P_3 = 6, \quad P_2 = 2$$
$$W = 6 \times 15 - 5 \times 6 - 4 \times 8 - 3 \times 6 - 2 \times 2 = 6.$$

Arguing further in this way, it is possible to indicate the case when the additional rod-input link is connected to the actuator stem conjugated with two more *l*-coordinates (Fig. 5). The designation, as in the previous cases, is analogous; the number of DOF is again six:

$$n = 15, \quad P_5 = 6, \quad P_4 = 8, \quad P_3 = 6, \quad P_2 = 2$$
$$W = 6 \times 15 - 5 \times 6 - 4 \times 8 - 3 \times 6 - 2 \times 2 = 90 - 84 = 6.$$

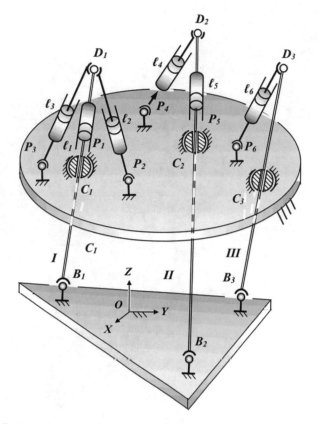

Fig. 6 *l*-coordinate mechanism with one intermediate link

You can continue to consider the synthesized mechanisms in such a way that two drives (Fig. 6) and three drives (Fig. 7) will coincide with the rods. The number of DOF in all such cases is six.

$$n = 14, \quad P_5 = 6, \quad P_4 = 7, \quad P_3 = 6, \quad P_2 = 1$$
$$W = 6 \times 14 - 5 \times 6 - 4 \times 7 - 3 \times 6 - 2 \times 1 = 84 - 78 = 6.$$

4 Conclusions

In conclusion, let us dwell on the areas of application of synthesized mechanisms. As noted, their distinguishing feature is the location of engines outside the working space. This means that these devices can be used to manipulate objects of electronic equipment in a vacuum, as well as in space and under water.

Fig. 7 l-coordinate
mechanism without
intermediate links

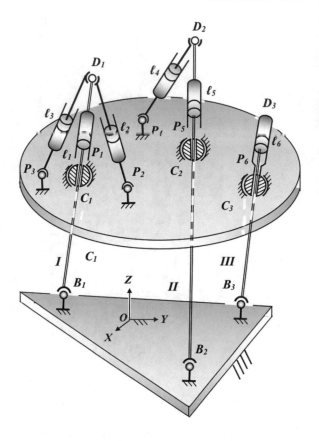

In addition, these devices can work together, for example, in the docking platforms of space and underwater vehicles.

It should be noted that changing the position of the fixing point leads to a change in the parameters of the working area. In addition, changing these parameters is important for the design of these devices. Solving these problems, along with the consideration of kinematics and dynamics, is the subject of further research.

Thus, in this work, a structural synthesis of new l-coordinate mechanisms with drives located between the base (or the output link) and intermediate links of neighbouring kinematic chains is carried out.

Acknowledgements This paper is prepared with financial support of the Russian Basic Research Foundation (the RFFI), Project No. 16-29-04273.

References

1. Arzumanyan, K.S., Koliskor, A.S.: Synthesis of structures of *l*-coordinate systems for research and diagnosis of industrial robots. In: Testing, Monitoring and Diagnosis of Flexible Production Systems. Nauka, Moscow, pp. 70–81 (1988)
2. Gough, V.E.: Contribution to discussion of papers on research in automobile stability, control and tyre performance. In: Proc. Auto Div. Inst. Mech. Eng., pp. 392–394 (1956–1957)
3. Stewart, D.A.: Platform with six degrees of freedom. In: Proceeding of Institute of Mechanical Engineering, pp. 371–386 (1965)
4. Hunt, K.H.: Geometry of robotic devices. Inst. Eng. Austral Mech. Eng. Trans. **7**(4), 213–220 (1982)
5. Merlet, J.-P.: Parallel Robots. Kluwer Academic Publishers, 372 p. (2000)
6. Glazunov, V.A., Koliskor, A.S., Krainev, A.F., Model, B.I.: Principles of classification and methods of analysis of spatial mechanisms with parallel structure. Probl. Mech. Eng. Mach. Reliab. **1**, 41–49 (1990)
7. Glazunov, V.A., Koliskor, A.S., Krainev, A.F.: Spatial Mechanisms of Parallel Structure, p. 95. Nauka, Moscow (1991)
8. Rashoyan, G.V., Shalyukhin, K.A., Gaponenko, E.V.: Development of structural schemes of parallel structure manipulators using screw calculus. IOP Conf. Ser. Mater. Sci. Eng. **327**(4), 042090 (2018)
9. Antonov, A.V., Glazunov, V.A., Aleshin, A.K.: J. Mach. Manuf. Reliab. Том: 47 Выпуск: 2 Стр.,121–127
10. Glazunov, V.A., Brio, S., Arakelyan, V.: A new class of manipulation mechanisms of parallel-cross-over structure. Classification and research. Directory. Eng. J. **4**, 35–40 (2008)
11. Salehi, A., Piltan, F., Mirshekaran, M., Kazeminasab, M., Esmaeili, Z.: Comparative study between two important nonlinear methodologies for continuum robot manipulator control. Int. J. Inf. Technol. Comput. Sci. (IJITCS) **6**(4), 66–80 (2014). https://doi.org/10.5815/ijitcs.2014.04.08
12. Joumah, A.A., Albitar, C.: Design optimization of 6-RUS parallel manipulator using hybrid algorithm. Int. J. Inf. Technol. Comput. Sci. (IJITCS) **10**(2), 83–95 (2018). https://doi.org/10.5815/ijitcs.2018.02.08
13. Kumar, V., Sen, S., Roy, S.S., Das, S.K., Shome, S.N.: Inverse kinematics of redundant manipulator using interval Newton method. Int. J. Eng. Manuf. (IJEM) **5**(2), 19–29 (2015). https://doi.org/10.5815/ijem.2015.02.03
14. Polishchuk, M., Opashnianskyi, M., Suyazov, N.: Walking mobile robot of arbitrary orientation. Int. J. Eng. Manuf. (IJEM) **8**(3), 1–11 (2018). https://doi.org/10.5815/ijem.2018.03.01

Characteristics Analysis for Corporate Wi-Fi Network Using the Buzen's Algorithm

Elena V. Kokoreva and Ksenia I. Shurygina

Abstract The paper presents the results of the wireless broadband access systems analytical modeling. The purpose of this study is to evaluate the quality of service (QoS) parameters in a channel of a corporate network based on the IEEE 802.11 standard. Mathematical apparatus of queuing networks was chosen as a simulation tool, as it had proved its effectiveness in calculating the characteristics of infocommunication systems of various purposes and any dimensionality. The authors examined various research methods, developed the conceptual, algorithmic and programming network architecture models in the form of a closed homogeneous queuing network and calculated its characteristics. The Buzen's algorithm was adapted to obtain the quality of service parameters for a network with Wi-Fi technology. Time-probability characteristics of the service process in the wireless broadband access systems were obtained during the simulation, such as transmission delay and network throughput for the different numbers of active subscribers. The results are presented as graphs and can be used both for designing new wireless networks and for efficient traffic management in existing ones.

Keywords Broadband access system · Wi-Fi · IEEE 802.11 · Queuing network · Queuing system · Multiplicative form · Buzen's algorithm

1 Introduction

In modern broadband access systems, effective traffic management is required to ensure the guaranteed quality of service (QoS). Concerning this, an important problem, solved with the application of the mathematical modeling methods, becomes the

E. V. Kokoreva (✉) · K. I. Shurygina
Siberian State University of Telecommunications and Information Sciences, 86, Kirova st., 630009 Novosibirsk, Russian Federation
e-mail: elen.vik@gmail.com

K. I. Shurygina
e-mail: miraclele@yandex.ru

© Springer Nature Switzerland AG 2020   693
Z. Hu et al. (eds.), *Advances in Artificial Systems for Medicine and Education II*,
Advances in Intelligent Systems and Computing 902,
https://doi.org/10.1007/978-3-030-12082-5_63

evaluation of the quality of service indicators of the system (throughput, utilization, delay, etc.).

The works of Russian and foreign scientists are devoted to the modeling of modern wireless technologies. The most common method is simulation with the use of network simulators (e.g., OPNET Modeler, OMNet ++, NS-2, NS-3), which allow the authors to obtain particular solutions for specific network architectures with determined configurations [1–4].

At the same time, the analysis methods presented in |5| based on the results of queuing theory are too general without being bound to a certain technology and structure.

Modern wireless broadband access systems based on the IEEE 802.11ac WLAN standard are characterized by high-speed data transmission, noise immunity and information security, due to the use of technologies such as OFDM [6], MIMO [7] and new encryption algorithms [8].

Queuing networks are a powerful tool used for the time-probability characteristics analytical modeling of various technical systems, including infocommunications of any type.

The application of analytical modeling based on the queuing networks methodology makes it possible to calculate the characteristics of most modern infocommunication systems of any dimensionality, topology and purpose.

The purpose of this study is to build a corporate wireless LAN model for calculating the quality of service parameters that can be used by network administrators and new system developers.

2 Queuing Network Parameters

Let us consider an infocommunication system model in a form of a closed homogeneous queuing network, which is described by the vector of parameters: $\Gamma = (N, K, \mathbf{M}, \Theta, \mathbf{m}, \mathbf{D}, \boldsymbol{\mu})$ where N is the number of queuing network nodes that are single-channel and multi-channel queuing systems; K is the fixed number of jobs in the closed queuing network; $\mathbf{M}$ is the vector of the service time exponential distributions in the network nodes; $\Theta = \left\| \theta_{ij} \right\|$, $i, j = \overline{1, N}$ is the routing matrix (θ_{ij} is the probability of job transfer from ith node to jth node); $\mathbf{m}$ is the vector that determines the number of servers in the ith queuing system; $\boldsymbol{\mu}$ is the vector of service intensities at the network nodes; $\mathbf{D}$ is the service disciplines vector at the queuing network nodes, where the disciplines come from this set:

- First Come, First Served (FCFS)—the service in order of arrival;
- Last Come, First Served Preemptive Resume (LCFS-PR)—the service in reverse order of arrival;
- Processor Sharing (PS)—uniform distribution of capacity between jobs;
- Infinite Server (IS)—an infinite number of servers;
- Finite Server (FS)—a finite number of servers.

3 Queuing Network Characteristics

For the queuing network analysis, it is necessary, first of all, to define the concept of its state: $s = (k_1, k_2, \ldots, k_N)$ where k_i, $i = \overline{1, N}$ is the number of jobs in the ith queuing system [9–11]. For the closed queuing network, the equality $\sum_{i=1}^{N} k_i = K$ is fulfilled, the state space $S(K, N)$ is a finite value, since there are no external sources and receivers of jobs, which differs it from the open queuing network.

The stationary distribution of the state probabilities is denoted $\pi(k_1, k_2, \ldots, k_N)$ taking into account the normalizing condition:

$$\sum_{s \in S(K,N)} \pi(k_1, k_2, \ldots, k_N) = 1. \tag{1}$$

The input intensities of the jobs entering the ith node are determined by the equilibrium equations of the flows in the closed queuing network:

$$\lambda_i = \sum_{j=1}^{N} \lambda_j \theta_{ji}, \quad i = \overline{1, N}. \tag{2}$$

An important characteristic is the transfer rate: e_i, $i = \overline{1, N}$ is the average number of job transfers to node i:

$$e_i = \frac{\lambda_i}{\lambda}, \quad i = \overline{1, N}, \tag{3}$$

where λ is the network throughput. Transfer rates are derived from (2) to (3) and can be obtained as follows:

$$e_i = \sum_{j=1}^{N} e_j \theta_{ji}, \quad i = \overline{1, N}. \tag{4}$$

Since from N of expressions (4) only $N - 1$ are independent and there is an infinite set of solutions, $e_1 = 1$ is usually used to find an unambiguous solution.

4 Multiplicative Form of Queuing Network State Distribution

The most important role in the queuing networks theory is played by queuing networks, the stationary distribution of which has a multiplicative form, since for them the time-probability characteristics can be obtained in a simple way. This class of networks includes the Gordon–Newell networks, which are the homogeneous closed queuing networks, and their stationary state distribution can be obtained in the following form [9–11]:

$$\pi(k_1, k_2, \ldots, k_N) = \frac{1}{G(K, N)} \prod_{i=1}^{N} \frac{x_i^{k_i}}{\beta_i(k_i)}, \tag{5}$$

where $G(K, N)$ is a normalizing constant:

$$G(K, N) = \sum_{k \in S(K,N)} \prod_{i=1}^{N} \frac{x_i^{k_i}}{\beta_i(k_i)}; \tag{6}$$

$x_i = \frac{e_i}{\mu_i}$, $i = \overline{1, N}$, and the function: $\beta_i(k_i) = \begin{cases} k_i!, & k_i \leq m_i \\ m_i! \, m_i^{k_i - m_i}, & k_i \geq m_i \end{cases}$, $i = \overline{1, N}$ depends on the number of jobs at the network nodes. In this case, the marginal state probabilities of the queuing systems forming the network are defined as:

$$\pi_i(s) = \sum_{k_i = s} \pi(k_1, k_2, \ldots, k_N). \tag{7}$$

5   Buzen's Algorithm

For the networks of large dimensionality with a complex topology and a great number of jobs, calculating the normalizing constant (6) and the stationary state probability distribution of the queuing network (5) requires a considerable amount of computing resources and time. Therefore, in practice, the special methods of calculation are used, one of which is the recurrent Buzen's algorithm [10–12].

The calculation of the normalizing constant $G(K, N)$ (6) is performed as a sequence of iterations to calculate the function values:

$$G_n(s) = \sum_{\sum_{i=1}^{n} k_i = s} \prod_{i=1}^{n} X_i(k_i), \quad n = \overline{1, N}, \quad s = \overline{0, K}, \tag{8}$$

where $X_i(k_i) = \frac{x_i^{k_i}}{\beta_i(k_i)}$.

The normalizing constant then will be: $G(K, N) = G_N(K)$.

The value of $G_N(K)$ can be obtained recurrently with the following equation:

$$G_n(s) = \sum_{j=0}^{s} X_n(j) G_{n-1}(s - j), \quad n = \overline{2, N}, \quad s = \overline{1, K} \tag{9}$$

with the starting conditions: $G_n(0) = 1$, $n = \overline{1, N}$ and $G_1(s) = X_1(s)$, $s = \overline{1, K}$.

The marginal probabilities can be recurrently obtained from (5) and (7):

$$\pi_i(s) = \frac{X_i(s)}{G(K, N)} \cdot G_N^{(i)}(K - s), \quad i = \overline{1, N}, \quad s = \overline{0, K}, \tag{10}$$

where $G_N^{(i)}(s)$ is an auxiliary variable denoting the normalizing constant for a network of N nodes, excluding the node i:

$$G_N^{(i)}(s) = G_N(s) - \sum_{j=1}^{s} X_i(j) \cdot G_N^{(i)}(s - j), \quad i = \overline{1, N} \tag{11}$$

with the starting conditions: $G_N^{(i)}(0) = 1$, $i = \overline{1, N}$.

At the same time:

$$G_{N-1}(s) = G_{N-1}^{(N)}(s) = G_N^{(N)}(s), \quad s = \overline{0, K}. \tag{12}$$

The job input intensities at ith nodes can be determined:

$$\lambda_i(K) = e_i \frac{G_N(K - 1)}{G_N(K)}, \quad i = \overline{1, N}. \tag{13}$$

Further, the utilization of the nodes and the other characteristics are determined according to the known queuing theory formulas [13].

Utilization of the ith queuing system:

$$\rho_i = \frac{\lambda_i}{m_i \mu_i}, \quad i = \overline{1, N}. \tag{14}$$

The average number of jobs at the ith node:

$$\overline{K_i} = \sum_{s=1}^{K} s \cdot \pi_i(s), \quad i = \overline{1, N}. \tag{15}$$

The average reaction time is determined by applying Little's formula:

$$\overline{T_i} = \frac{\overline{K_i}}{\lambda_i}, \quad i = \overline{1, N}. \tag{16}$$

6 Modeling a Corporate Wireless Local Network

Let us consider a wireless broadband access system based on Wi-Fi technology of IEEE 802.11ac standard, the structure of which is shown in Fig. 1.

The model based on the infrastructure described above contains the following components:

- Local Ethernet network with Internet access;
- Several wireless access points (AP);
- Web server, multimedia server and DBMS server;
- Network interfaces with specified throughputs;
- Several mobile subscribers: laptops, smartphones and PCs equipped with wireless network interface controllers (WNIC).

The model of the wireless LAN described above is a closed queuing network shown in Fig. 2.

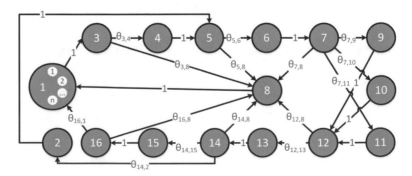

Fig. 1 An example of the corporative Wi-Fi network architecture

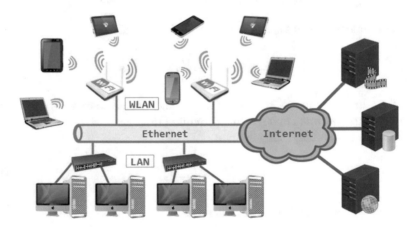

Fig. 2 A model of the Wi-Fi network

The closed queuing network nodes are the time delays at the nodes and interfaces of the simulated system [14, 15]. Transfers of the job from node to node are determined by the routing matrix Θ. Node 1 denotes the Wi-Fi network subscribers and the traffic generated by them; node 2 takes into account the background traffic generated by subscribers of the local Ethernet network; nodes 3, 5, 7, 12, 14, 16 simulate the communication lines of various types: wired and wireless; node 8 determines the losses in data transmission channels due to the interferences and in the buffer queues of the network switches as a result of their overflow. The remaining nodes imitate the various physical components of the network, including servers, which are the queuing systems with an infinite number of serving devices.

7 Modeling Results

As a result of the analytical modeling, the graphs showing the relationship between the quality of service (QoS) parameters and the network load were obtained for different subscribers quantities (5, 10 and 20), as shown in Figs. 3, 4, 5 and 6. The calculations were made with the system of mathematical and engineering calculations Mathcad.

It can be seen in the figures that as the load increases, the time-probability characteristics of the network go up, but remain within the limits of acceptable values for modern mobile broadband access systems.

Figure 3 illustrates the changes in the E2E transmission delay. It can be seen that an increase in the number of subscribers leads to some deterioration of this parameter, but overall the value does not exceed the one specified by the IEEE 802.11ac standard.

Figure 4 shows the relationship between the network load and the loss probability, which undergoes minor changes both with the increase in the number of jobs in the

Fig. 3 Relationship between the delay and the network load

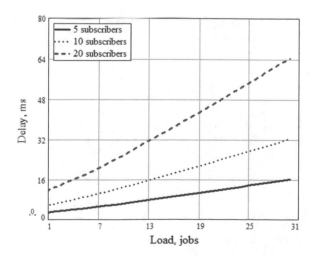

Fig. 4 Relationship between the loss probability and the network load

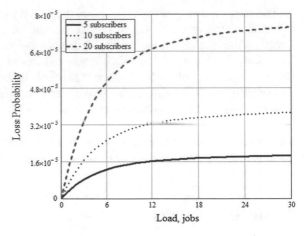

Fig. 5 Relationship between the throughput and the network load

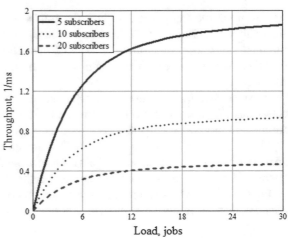

Fig. 6 Relationship between the utilization and the network load

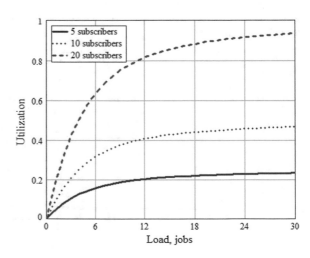

network and with the increase in the number of active subscribers, and does not exceed the values of the order of 10^{-5}.

As can be seen in Fig. 5, the system performance, measured by the number of jobs serviced per unit of time, goes up with the increase in the input load and decreases with the increase in the number of active subscribers. This can be explained by the fact that a larger number of subscribers generate more network traffic, which leads to an increase in delay and losses in the buffer memory of network switches and due to interference in the radio channel. However, even the lowest performance value corresponding to the service of twenty subscribers, with an average Wi-Fi packet length of 131,072 octets, gives us a minimum transmission rate of 524 Mbps, which is acceptable for this technology.

The utilization shown in Fig. 6 behaves predictably with an increase in the input load and the number of subscribers, and its value in the worst case does not exceed one, which means that the stationarity of the system is not disturbed.

Figures 7, 8, 9 and 10 illustrate the system behavior when transmitting various types of traffic depending on the number of active network subscribers.

As shown in Figs. 7 and 8, the prioritized voice traffic has the least delay: The packets will be lost with the least probability, and the throughput curve for the voice traffic in Fig. 9 passes above the others, respectively.

Figures 9 and 10 show that the channel is mainly loaded with multimedia traffic, which has the largest volume; therefore, its losses occur more often than the losses of other traffic types.

The obtained results show that for a network with a given configuration, the quality of service parameters suits the limitations defined by the IEEE 802.11ac standard, and therefore, this network meets QoS requirements and can be implemented in practice.

Fig. 7 Relationship between the delay and the number of subscribers

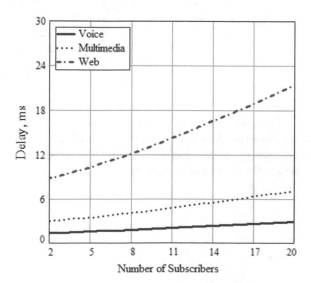

Fig. 8 Relationship between the loss probability and the number of subscribers

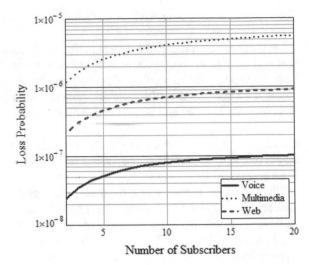

Fig. 9 Relationship between the throughput and the network load

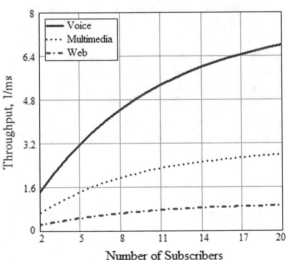

The developed model is universal and allows to obtain the time-probability characteristics for the systems of any configuration by specifying the corresponding input parameters: routing matrix, intensities and service disciplines at the queuing network nodes, the number of servers, etc.

Fig. 10 Relationship
between the utilization and
the network load

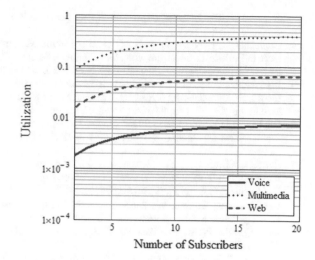

8   Conclusions

The authors built the conceptual, algorithmic and software models of the corporate
wireless LAN based on the IEEE 802.11ac standard.

The queuing networks mathematical apparatus is applied as a simulation tool.

Buzen's algorithm (or convolution algorithm), adapted to the real operating conditions of wireless communication networks, was used to calculate the time-probability
characteristics.

The resulting characteristics allow one to evaluate the quality of Wi-Fi network
service parameters, such as delay (ms), throughput (ms^{-1}), network load factor and
loss probability.

These results can be applied by network administrators or operators of existing
communication networks to manage traffic in order to improve the level of services
provided, as well as by designers—to create efficient network architectures.

References

1. Iskounen, S., Nguyen, T.-M.-T., Monnet, S.: WiFi-direct simulation for INET in OMNeT++.
   In: OMNet++ Community Summit. University Pierre and Marie Curie (UPMC). Paris, 2016.
   https://arxiv.org/ftp/arxiv/papers/1609/1609.04604.pdf (date of circulation: 05/08/2018)
2. Swanlund, E., Loodu, P., Chowdhury, S.: Analysis and performance evaluation of a Wi-Fi
   network using ns-2. In: ENSC 427 Communication Networks. School of Engineering Science:
   Simon Fraser University, 2013. http://www2.ensc.sfu.ca/~ljilja/ENSC427/Spring13/Projects/
   team11/Analysis_and_Performance_Evaluation_of_a_Wi-Fi_Network_presentation.pdf (date
   of circulation: 05/08/2018)
3. Zhang, C., Chau, R., Sun, W.: Wi-Fi network simulation OPNET. In: ENSC 427 Communication Networks. School of Engineering Science: Simon Fraser University, 2009. http://www2.

ensc.sfu.ca/~ljilja/ENSC427/Spring09/Projects/team8/427_Project_Report_Final.pdf (date of circulation: 05/08/2018)

4. Giupponia, L., Hendersonb, T., Bojovica, B., Miozzoa, M.: Simulating LTE and Wi-Fi Coexistence in Unlicensed Spectrum with ns-3. Computer Science. Cornell University, 2016. https://arxiv.org/ftp/arxiv/papers/1604/1604.06826.pdf (date of circulation: 05/08/2018)

5. Kozyrev, D.: Analysis of a repairable redundant system with PH distribution of restoration times of its elements. In: XXXI International Seminar on Stability Problems for Stochastic Models, pp. 43–44. Institute of Informatics Problems, RAS, Moscow, 23–27 Apr 2013

6. Parekha, C.D., Patel, J.M.: OFDM synchronization techniques for 802.11ac WLAN. Int. J. Wirel. Microwave Technol. (IJWMT) 8(4), 1–13 (2018). https://doi.org/10.5815/ijwmt.2018.04.01

7. Sur, S.N., Ghosh, D.: Channel capacity and BER performance analysis of MIMO system with linear receiver in nakagami channel. Int. J. Wirel. Microwave Technol. (IJWMT) 3(1), 26–36 (2013). https://doi.org/10.5815/ijwmt.2013.01.03

8. Liu, H., Zhang, H., Xu, W., Yang, Y., Xu, M.: A new secure strategy in small-scale IEEE 802.11 wireless local area networks with web authentication and virtual local area network. Int. J. Comput. Netw. Inf. Secur. (IJCNIS) 3(2), 19–25 (2011). https://doi.org/10.5815/ijcnis.2011.02.03

9. Kokoreva, E.V.: Overviewing the analyzing queuing networks methods for the infocommunication systems modeling. In: The Modern Telecommunications Problems": Proceedings of Russian Scientific and Technical Conference, pp. 721–730. SibSUTIS, Novosibirk, 2016

10. Kokoreva, E.V.: Reviewing the closed homogeneous queuing networks characteristics calculation methods. Mod. Sci. Res. Eng. 3(3), 50–59 (2016)

11. Bolch, G., Greiner, S., de Meer, H., Trivedi, K.S.: Queueing Networks and Markov Chains: Modeling and Performance Evaluation with Computer Science Applications, 2nd edn, 896 p. Wiley (2006)

12. Buzen, J.: Convolution algorithms for closed queueing networks with exponential servers. Commun. ACM 16(9), 527–531 (1973)

13. Giambene, G.: Queuing Theory and Telecommunications: Networks and Applications. Springer Science + Business Media, New York (2014). https://doi.org/10.1007/978-1-4614-4084-0

14. Kokoreva, E.V.: Analyzing the queuing networks for the broadband access systems characteristics modeling. In: Applied Information Systems: III All-Russian Scientific-Practical Conference: Conference Proceedings, pp. 57–64. UlTSU, Ulyanovsk, 2016

15. Kokoreva, E.V.: The analysis of quality of service parameters in the fourth generation mobile networks. Softw. Syst. Comput. Methods 3, 35–44 (2018). http://e-notabene.ru/ppsvm/article_26920.html (date of circulation: 05/08/2018)

The Application of Elements of Information Theory to the Problem of Rational Choice of Measuring Instruments

I. A. Meshchikhin and S. S. Gavriushin

Abstract While operating complex technical structures and structures, there is a need for an operative evaluation of the current characteristics of the monitoring object for making an informed decision on the possibility of further operation, maintaining loading statistics. The task of rapid assessment of the state parameters allows to solve the monitoring system. When developing a system for monitoring complex structures and structures, it is necessary to organize a measurement of a set of state parameters: for example, combinations of acting forces, moments, and pressure. When developing a monitoring system, an effective solution of the measurement problem is necessary: by indirect measurements, the most accurately measured parameters and the stress of its stressed state. The method proposed in this work allows solving the inverse measurement problem: to determine the rational composition of the measuring instruments and the requirements to them based on the design model of the operation of the structures.

Keywords Information theory · Load monitoring · Indirect measurements

1 Introduction

Load monitoring is an important component of the system for structural health monitoring [1]. The ultimate goal of load monitoring is to record the current state and maintain load statistics for the entire period of operation for efficient management of the operation object, estimate the residual resource, and identify the dangerous combination of loads.

The result of measurement is usually connected with the criterion of the state of the monitoring object indirectly: the metrological chain includes both measurement errors and state restoration. Increasing the accuracy of determining the state of the system as a whole is possible due to the development of data processing

I. A. Meshchikhin · S. S. Gavriushin (✉)
Bauman University, ul. Baumanskaya 2-Ya, 5, Moscow 105005, Russia
e-mail: gss@bmstu.ru

© Springer Nature Switzerland AG 2020   705
Z. Hu et al. (eds.), *Advances in Artificial Systems for Medicine and Education II*,
Advances in Intelligent Systems and Computing 902,
https://doi.org/10.1007/978-3-030-12082-5_64

methods, increasing the accuracy of measuring instruments, and by optimizing the composition of measuring instruments. The development of methods for the calculation of the choice of measuring instruments will improve the quality of the restored state of structures and structures with a minimum composition and requirements for measuring instruments and algorithms for data processing, and the ability to give a calculation estimate to the potential of sensing will allow optimizing the design by the criteria of observability of its deformed state.

In general, the required loading is defined as the superposition of base loads, the number and type of which is determined from the analysis of operating conditions. Each basic loading in turn is complex, difficult to measure directly [2]: the effect of distributed wave load on berthing facilities, loads from the action of forces of payload weight, water level drop for shut-off structures is effectively measured indirectly, initially recording deformations in a set of design points. Thus, the problem of loading monitoring is largely reduced to the problem of indirect measurement, within which the desired combination of loads is determined by means of a model and measurement results—identification parameters [3]. The result of the development of a monitoring system is a sensor network [4] with a calculated rationale for the composition and placement of measuring instruments. The initial data for the proposed version of the calculation is the model of errors in the measurement of measuring instruments and the mathematical model of the operation of the monitoring object [5]. We represent the model of an object in the form of a linear operator and define the identification parameters as: $\delta = Af + Be$, where A is matrix of the parameters considered, B is matrix of parameters not taken into account.

The following requirements are imposed on the parameters used to identify the object:

1. The values of the identification parameters for a given set of active loads should be maximum. Identification parameters are measured with a certain accuracy, depending on the method of measurement. A set of loads is understood as a set of various combinations of simultaneously acting on the object of impacts, determined by the conditions of the object's operation. At small values ($\delta_i < \Delta$) of the measurement result, it is impossible to distinguish between a useful signal and noise. Numerically, this criterion can be represented as $abs(\delta) \to$ max;

2. The identification parameters should, if possible, provide a linearly independent response of the structure for a given set of loads. With a small value $\delta_i(1 - \langle \delta_i, \delta_j \rangle / abs(\delta_i \delta_j)) < \Delta$, it is not possible to distinguish the measurement results under loads i and j. Numerically, this criterion can be represented as minimum scalar product $\langle \delta_i, \delta_j \rangle \to$ min, $i \neq j$;

3. The closest proximity of the range of change of identification parameters. Numerically, this criterion can be represented as the number of conditionality of the matrix A: $S_{\min}(A)/S_{\max}(A) \to$ max, where S is the singular numbers of the matrix A. When estimating nonsymmetric ones, including rectangular matrices A and B, it is more convenient to use singular numbers when working in the field of real numbers;

4. The minimum impact of factors that are not taken into account with respect to those considered. Numerically, this criterion can be represented as $S_{\max}(BA^{-1}) \to \max$;

5. Minimal sensitivity of the criterion to positioning errors. Information value of the monitoring object. The requirement of maximum signal-to-noise ratio and linear independence of responses to a linearly independent complex of loads can be formulated as the requirement of the maximum amount of information in the recorded signal. For each design, there is such a set of loads that the result of the measurement on the action of its element will be less than the measurement error. We denote the minimum loading scale by $F_{\min}$. There is also such a region in the loading space that the probability of realizing the load outside it is lower than the predetermined threshold value. We denote the maximum loading scale by $F_{\max}$. The amount of information is defined as the logarithm of the number of states, the registration of which is possible, taking into account the measurement error. In other words $I = \mathrm{Ln}(F_{\max}/F_{\min})$ [bit], where is the maximum loading, the realization of which is possible (from the analysis of loading conditions or bearing capacity) $F_{\min} = \Delta/S$, Δ where is the measurement error, is the model coefficient. In the event that a combination of loads is acting on the structure (each combination is equally probable), the information content of the monitoring system is defined as the total information obtained from the measurement at n points. The root-mean-square signal level for each channel with equiprobable realization of either of two combinations of loads is:

$$S_1^{\Sigma} = \sqrt{\frac{\int_0^{S_2}\left(1-\frac{F_2^2}{S_2^2}\right)S_1^2 \mathrm{d}F_2}{S_2}} = S_1\sqrt{\frac{2}{3}},$$ where S_i is the singular numbers of the operator connecting the result of measurements with the restored load. The coefficient $\sqrt{\frac{2}{3}}$ is also valid for an arbitrary number of loads. Thus, the rms informativity of the model is numerically equal to $I_{\Sigma} = \sum_{i=1}^{n} I_i^{\Sigma} = \sqrt{\frac{2}{3}}\sum_{i=1}^{n}\ln\left(\frac{F_{\max} S_i}{\Delta}\right) = \sqrt{\frac{2}{3}}\ln\left(\prod_{i=1}^{n}\left(\frac{F_{\max} S_i}{\Delta}\right)\right) = \sqrt{\frac{2}{3}}\ln\left(\frac{F_{\max}\|A\|}{\Delta^n}\right).$

The response matrix A is formed by columns. Each column of the response matrix is represented as a vector in the parameterization parameter space. The component of the response matrix A_{ij} is numerically equal to the value of the generalized displacement (identification parameter) caused by the action of this load. The numerical value of the determinant is interpreted as an oriented volume of a parallelepiped whose edges are given by the column vectors of the response matrix [6]. The geometric interpretation for the 2×2 response matrix is shown in Fig. 1. The matrix operator A produces a linear mapping from the loading space to the measurement space. In particular, it maps a unit circle into an ellipse, the lengths of the principal axes of which are equal to the moduli of the eigenvalues (singular numbers) of the matrix A, and their orientation to its eigenvectors.

It is assumed that for the measured values, a certain noise level is determined by the error of the measuring instruments used. The signal is suitable for identifying the load if all the components of the measurement vector exceed in absolute value

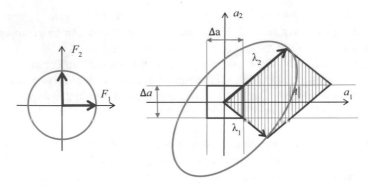

Fig. 1 Geometric interpretation of the procedure for selecting a set of informative parameters for $n = 2$

of noise rms. The quality of the choice of the operator A is conveniently estimated from the ratio of the modulus of its minimum eigenvalue to the noise level. In the future, to evaluate the error in identifying the parameters of loading, we will use the condition:

$$\sqrt{\frac{2}{3}} \sum_{i=1}^{n} \ln\left(\frac{F_{max} S_i}{\Delta}\right) > 1$$

Let us consider the approach presented here using the example of a cantilevered beam as the most transparent for demonstration and understanding.

We present the following requirements to the monitoring object:

- Maximum information:

$$I_{\Sigma} = \sqrt{\frac{2}{3}} \sum_{i=1}^{n} \ln\left(\frac{F_{max} S_i}{\Delta}\right) \rightarrow max$$

- Equal accuracy of recovery of various components of the load vector $\begin{Bmatrix} P_1 \\ P_2 \end{Bmatrix} = P$:

$$S_2/S_1 = \frac{min(S)}{max(S)} \rightarrow max,$$

where S is singular numbers of the matrix A.

The calculation scheme is shown in Fig. 2.
With numerical estimates, $L_1 = L_2 = 0.5$ m.
Flection from the action of force $P1$:

$$\delta = \frac{P_1}{3EJ}\left(L_1^3 + 3x L_1^2\right)$$

Fig. 2 Calculation scheme

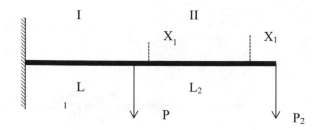

Flection from the action of force *P2*:

$$\delta = \frac{P_1}{EJ}\left(\frac{lx^2}{2} - \frac{x^3}{6}\right)$$

Response matrix:

$$A = \frac{1}{EJ}\begin{bmatrix} \frac{l_1^3}{3} + x_1 l_1^2 & \frac{x_1^3}{6} - \frac{l_2 x_1^2}{2} \\ \frac{l_1^3}{3} + x_2 l_1^2 & \frac{x_2^3}{6} - \frac{l_2 x_2^2}{2} \end{bmatrix}$$

The choice of identification parameters is largely due to the type of loading model. The optimal measurement result should be similar to the unit matrix.

In this case, the construction of independent functions of a pair of forces of equal range—the model of the experimental plan—is an ideal reduced model of constructions that must have eigenvalues that are close in value.

The Pareto front is shown in Fig. 3.

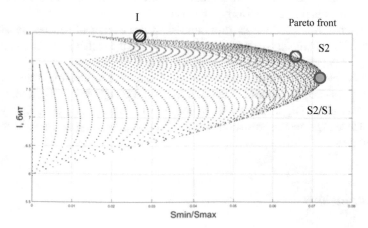

Fig. 3 Pareto front and particle solutions

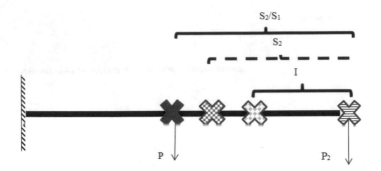

Fig. 4 Values of the identification parameters

The solution vector is called Pareto optimal if there is no other solution each criterion of which is greater than the one found. The set of Pareto optimal solutions is also called the Pareto front.

Figure 3 shows the locations of the various locations of the tilt points on the structure:

- maximum response—maximum of the informative criterion, bit;
- uniform response—maximum of the criterion S_2/S_1;
- compromise variant (dotted line).

To select the identification parameters that are optimal for the two criteria, simultaneously, we introduce a resolving rule of the form $S_2 = \frac{S_2 S_1}{S_1}$.

Thus, the scalarized criterion has the form $S_2 = \|S_{\min}\| \rightarrow \max$.

It can be seen from Fig. 4 that the maximum value of S_2 is an acceptable compromise between the choice of identification parameters that satisfy the criterion of maximum information and uniform response.

This approach is particularly pronounced in the design of sensors sensitive to force, which mechanically decouples the useful signal through the channels. A typical geometry of sensors of this type is shown on Fig. 5.

In the development of this product, the problem of optimization of the objective functions presented in the article was solved. As a result, the configuration of the product was obtained, which makes it possible to restore the active combination of three forces and moments with minimum weight according to the indications of strain gauges.

2   Conclusions

In the linearized form, the model of the monitoring object is a matrix, and the criteria for the quality of the model's choice are expressed as requirements for its singular values. The analysis of the problem made it possible to isolate the image of an ideal

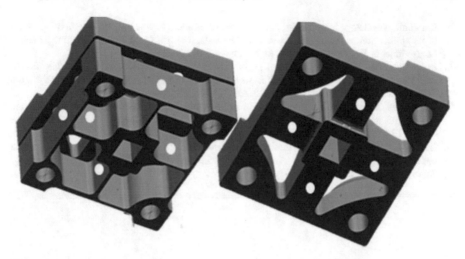

Fig. 5 General view of the forces of the moment sensor. Marker shows the location of the strain gages

model—a single matrix with a maximum multiplier. When the problem is generalized to a nonlinear class [7, 8], the result can be represented in geometric images: when the unit sphere is mapped into the identification parameter space, the result of the mapping should be as close as possible to the unit sphere.

The design of a monitoring system with the proper parameters is to determine the mutual placement of strain gauges. The forecast of states whose experimental observation is difficult (structural destruction) should be based on preliminary modeling, the compaction of the result of which is proposed to be used to estimate the current state of the design object.

The proposed approach allows designing sensor networks that are rational for recording the loaded state of a product with a given geometry, and optimizing the geometry of the product, and integrating the sensor functions into the product design.

References

1. Vengrinovich, V.L.: Monitoring of technical condition. The analysis of risks in technical systems. In: Nondestructive Testing and Diagnostics, p. 3. 2 (2014)
2. Klyuchev, A.O., et al.: Hardware and Software of Built-In Systems Textbook. SPb.: SPbSU ITMO (2010)
3. Meshchikhin, I.A., Gavriushin, S.S., Zaitsev, E.A.: Monitoring of technical designs on the basis of reduced finite element models. Izvestiya VUZov. Mech. Eng. **9** (2015)
4. Mezhin, V.S., Obukhov, V.V.: Practice of application of modal tests for the purposes of verification of finite-element models of the design of rocket and space equipment. Space Technol. Technol. **1**, 86–91 (2014)

5. Kanunnikova, E.A., Meshchikhin, I.A.: Parametric model of structural loading and algorithm of its application in the estimation of maximum stresses. In: Questions of electromechanics. Proceedings of VNIIEM, Moscow, No. 137, pp. 15–22 (2013)
6. Rouhollah Pour, H., Asgari Marnani, J., Tabatabei, A.A.: A novel method for crack detection in steel cantilever beam using wavelet analysis by combination mode shapes. Int. J. Image Graph. Signal Process. (IJIGSP) 10(4), 1–12 (2018). https://doi.org/10.5815/ijigsp.2018.04.01
7. Bhavyarani, M.P., Mahadeva Swamy, U.B., Shrynik Jain, M.B.: Inter integrated WSN for crude oil pipeline monitoring. Int. J. Comput. Netw. Inf. Secur. (IJCNIS) 10(3), 37–51 (2018). https://doi.org/10.5815/ijcnis.2018.03.03
8. Gao, F., Hou, A., Yang, X.: Numerical analysis of dynamic mechanical properties for rock sample under strong impact loading. Int. J. Inf. Eng. Electron. Bus. (IJIEEB) 2(2), 10–16 (2010). https://doi.org/10.5815/ijieeb.2010.02.02

Power System Transient Voltage Stability Assessment Based on Kernel Principal Component Analysis and DBN

Zhang Guoli, Gong Qingwu, Qian Wenxiao, Lu Jianqiang, Zheng Bowen, Gao He, Zheng Tingting, Wu Liuchuang, Chen Wenhui, Liu Xu, Wang Bo and Qiao Hui

Abstract Aiming at the transient voltage stability assessment of power system, this paper proposes a method based on kernel principal component analysis (KPCA) and deep belief network (DBN) for power system transient voltage stability assessment. Firstly, a set of 45-dimensional eigenvectors that can reflect the transient voltage stability of the power system is constructed. The feature vector set is reduced based on KPCA, the feature vector dimension and the filtering redundancy feature are reduced, then the dimensionality-reduced feature vector is input into the DBN network. The training process consists of pre-training and fine-tuning is performed to optimize the DBN grid structure parameters. The simulation results of the 10-machine 39-node in New England show that the method can reduce the dimension of input data, remove redundant features, reduce the error rate, and test time of transient stability assessment. It can accurately and quickly judge the steady state voltage state of the power system.

Keywords KPCA · DBN · Power system transient voltage stability assessment · Deep learning · Error rate

1 Introduction

With the scale of the power system growing constantly and the complexity of the power grid structure increasing frequently, the stable operation of the power system faces a more severe condition. Accurate and rapid transient stability assessment is of great significance for the safe and stable operation of the system and the planning of the power grid construction [1–5].

Z. Guoli · Q. Wenxiao · L. Jianqiang · Z. Bowen · G. He · Z. Tingting
State Grid Inner Mongolia Eastern Power Company Limited Electric Power
Research Institute, Hohhot 010000, China

G. Qingwu · W. Liuchuang (✉) · C. Wenhui · L. Xu · W. Bo · Q. Hui
School of Electrical Engineering, Wuhan University, Wuhan 430072, China
e-mail: 289004577@qq.com

© Springer Nature Switzerland AG 2020
Z. Hu et al. (eds.), *Advances in Artificial Systems for Medicine and Education II*,
Advances in Intelligent Systems and Computing 902,
https://doi.org/10.1007/978-3-030-12082-5_65

Traditional transient voltage stability analysis of power systems mainly contains time domain simulation method [6] and transient energy function method [7]. The former is computationally intensive and slow; the latter is difficult for a complex system to get the energy function fulfilling requirements [8]. Neither of these can achieve the accuracy and rapidity of transient voltage stability assessment.

Under the circumstance that there are many unknown factors in the principle of affecting transient voltage stability [9], transient voltage stability assessment based on machine learning can be independent of specific grid physical models, avoid complex instability mechanisms during the evaluation process. It can update the classification parameters through a large number of offline training to deal with nonlinear problems, which has the advantages of high accuracy, short time consumption, and strong online real-time application. However, the machine learning algorithms are limited to shallow learning methods such as artificial neural networks [10] and support vector machines [11]. When those algorithms solve complex classification problems, the generalization ability is limited. For high-dimensional raw electrical quantities, it takes too long to meet the requirements of transient voltage stability evaluation speed.

So this paper proposes a transient voltage stability assessment method based on kernel principal component analysis (KPCA) and deep belief network (DBN). From the measured electrical physical quantity that reflects the grid structure, operating state and transient voltage stability of the power system. We get 45-dimensional eigenvectors as feature vector sets, then the feature vector based on KPCA is dimension-reduced, and the dimension-reduced feature vector is input into the DBN network. The training process consists of pre-training and fine-tuning is used to optimize the DBN grid structure parameters. The simulation results of the New England 10-machine 39-node shows that the proposed method can effectively reduce the data dimension, the error rate and time of transient voltage stability assessment. It can be used as a method for power system transient voltage stability assessment.

2   Kernel Principal Component Analysis

Principal component analysis (PCA) is a method that utilizes a linear dimension reduction [12]. However, the power system is a typical nonlinear system. After the fault occurs, the electrical physical quantity has a strong nonlinear relationship. It is inefficient to describe it with a linear relationship. Therefore, nonlinear dimension reduction of the electrical quantity should be considered. KPCA uses the idea of kernel function to introduce a nonlinear mapping function to map the space of the original sample to a higher dimensional space, in which space PCA is performed.

Given a matrix $X = [x_1, x_2, \cdots, x_n]$, vector $x_i \in R^m$, nonlinear feature mapping $\Phi : R^m \rightarrow H$, where H is a Hilbert space called feature space, we get $\Phi(x) = [\Phi(x_1), \Phi(x_2), \cdots, \Phi(x_n)]$ on the feature space H. When the mean value of $\Phi(x_i)$ is zero, the sample of covariance matrix in the feature space H is showed as [13]:

$$C = \frac{1}{n} \sum_{i=1}^{n} \Phi(x_i)\Phi(x_i)^{\mathrm{T}} = \frac{1}{n}\Phi(x)\Phi(x)^{\mathrm{T}} \tag{1}$$

And then calculate the linear PCA in $\Phi(x)$ [12], we can get

$$C\omega = \frac{1}{n}\Phi(x)\Phi(x)^{\mathrm{T}}\omega = \lambda\omega \tag{2}$$

And define $K = \Phi(x)^{\mathrm{T}}\Phi(x)$, $K = (k_{ij})_{n \times n}$, $k_{ij} = (\Phi(x_i) \cdot \Phi(x_j))$, where K is a positive semidefinite kernel matrix, kernel function can calculate the inner product of on the feature space H and does not need to get the mapping function $\Phi(x)$, we get:

$$Ku = \Phi(x)^{\mathrm{T}}\Phi(x)u = \tilde{\lambda}u \tag{3}$$

Multiply both sides by $\Phi(x)$: $(\Phi(x)\Phi(x)^{\mathrm{T}})(\Phi(x)u) = \tilde{\lambda}(\Phi(x)u)$, we find that

$$nC(\Phi(x)u) = \tilde{\lambda}(\Phi(x)u) \tag{4}$$

By comparing and analyzing formula (2) and formula (4), we can find $\tilde{\lambda} = n\lambda$, $\omega = \Phi(x)u$. So the principal components for vector x_i can be calculated by

$$Z = \omega^{\mathrm{T}}\Phi(x) = u^{\mathrm{T}}\Phi(x)^{\mathrm{T}}\Phi(x) = u^{\mathrm{T}}K \tag{5}$$

Gaussian radial basis kernel function, neural network kernel function and polynomial kernel function are main kernel functions. The calculation of Gaussian radial basis kernel function is simple and the effect of classification is good. So chose Gaussian radial basis as the kernel function, it is represented as:

$$K_{ij} = \exp\left(-\frac{\|x_j - x_i\|}{2\delta^2}\right) \tag{6}$$

3 Deep Belief Network

DBN was proposed by Geoffrey Hinton in 2006, and the structure is shown in Fig. 1. It is a probabilistic generation model consisting of a number of restricted Boltzmann machines (RBM) and a supervised backpropagation network, which is used to establish a joint distribution between observation data and labels [14, 15]. The training process consists of pre-training and fine-tuning. Pre-training: Unsupervised RBM training is performed layer by layer from bottom to top, and the output of the underlying RBM hidden layer is used as the input of the upper layer RBM visible layer. Fine-tuning: Training on BP network under supervision and the actual classi-

Fig. 1 DBN model structure

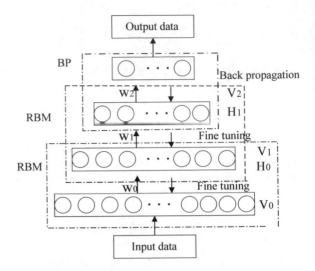

fication state is compared with the expected one, then the error will be obtained and propagated downward to fine-tune the DBN parameters.

3.1 Pre-training

The RBM consists of a visible layer and a hidden layer [16]. The neurons between the visible layer and the hidden layer are bidirectionally connected. As shown in Fig. 2, the visible layer and the hidden layer are, respectively, represented, W is the connection weight between the adjacent two layers, the visible layer and the hidden layer are fully connected, and the neurons of the same level are not connected. And the elements are binary variables whose state takes 0 or 1.

For a specific set of (v, h), the RBM energy function is:

Fig. 2 RBM model structure

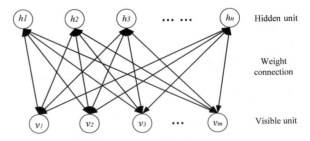

$$E(v, h|\alpha) = -\sum_{i=1}^{m} a_i v_i - \sum_{j=1}^{n} b_j h_j - \sum_{i=1}^{m}\sum_{j=1}^{n} w_{ij} v_i h_j \qquad (7)$$

In the formula, $\alpha = (w, a, b)$ is the parameter of the RBM model, a_i and b_j represent, respectively, the offset values of the visible and hidden layer units. w_{ij} represents the connection weight of visible unit v_i and hidden layer unit h_j. The state distribution of the RBM is subject to a regular distribution, so the joint probability distribution of any particular group of (v, h) is:

$$P(v, h|\alpha) = \frac{1}{Z(\alpha)} \exp(-E(v, h|\alpha)) \qquad (8)$$

In the formula, $Z(\alpha) = \sum_v \sum_h \exp(-E(v, h|\alpha))$ is the normalization term.

Since the neurons in the same RBM layer are not connected, the activation states of the neurons in each hidden layer are independent of each other, in which case the activation probability of the first hidden layer neurons in the hidden layer is:

$$P(h_j = 1|v, \alpha) = \sigma(b_j + \sum_{i=1}^{m} v_i w_{ij}) \qquad (9)$$

In the formula, $\sigma(x) = 1/(1 + \exp(-x))$ is a sigmoid function, similarly, the activation probability of the i-th visible unit v_i is:

$$P(v_i = 1|h, \alpha) = \sigma(a_i + \sum_{j=1}^{n} h_j w_{ij}) \qquad (10)$$

RBM fits the given training data set by training and learning the value of parameter $\alpha = (w, a, b)$. The parameter α can be obtained by the maximum log-likelihood function on the training set (assuming the number of samples is M), so:

$$\alpha^* = \arg\max_{\alpha} L(\alpha) = \arg\max \sum_{i=1}^{M} \ln P(v^i|h, \alpha) \qquad (11)$$

We can get the formula by using the contrast divergence algorithm proposed by Hinton as follows:

$$\Delta w_{ij} = \eta(< v_i h_j >_{\text{data}} - < v_i h_j >_{\text{model}}) \qquad (12)$$

$$\Delta a_i = \eta(< v_i >_{\text{data}} - < v_i >_{\text{model}}) \qquad (13)$$

$$\Delta b_j = \eta(< h_j >_{\text{data}} - < h_j >_{\text{model}}) \qquad (14)$$

Among them, η is the learning rate of forward training, $< v_i h_j >_{\text{data}}$ is the expectation of training data distribution, and $< v_i h_j >_{\text{model}}$ is the expectation on the distribution defined for the RBM model. This paper uses the Gibbs one-time sampling data distribution as the distribution definition of the model definition, utilize the above RBM's consulting algorithm to train all RBM structures unsupervised layer by layer.

3.2 Fine-Tuning

BP network is the last layer of DBN, the training is divided into two parts: In the first step, the output feature vector of the upper layer RBM is accepted, and the predicted classification state is obtained; in the second step, the actual classification state is compared with the expectation to obtain an error, and the error is propagated downward to finely adjust the DBN parameters.

4 Power System Transient Voltage Stability Assessment Based on Kernel Principal Component Analysis and DBN

The overall flow of power system transient voltage stability assessment based on KPCA and DBN is shown in Fig. 3. Including: build an input feature set, DBN-based transient voltage stability evaluation and evaluation index calculation. Through the massive data samples obtained by offline simulation, a complete transient voltage stability evaluation model is trained offline, and then, the test set is used to test and evaluate the performance of the model.

4.1 Build an Input Feature Set

An important step in the evaluation of transient voltage stability based on machine learning is to construct scientific and reasonable feature quantities. Analysis of the classical voltage indicators in the references [17–22]: transient voltage reactive power sensitivity, transient energy function, voltage collapse proximity indicator, and other indicators, find that the voltage stability problem involves the following electrical response: reactive power Q_i, active power P_i and current I_i of the branch, the active power P_b and reactive power Q_b of the node, the reactive power Q_g of the generator, the voltage amplitude V_m and phase angle φ of the node. In the actual power grid, the number of branches and the number of nodes are extremely large, if the above-mentioned electrical quantity of the whole network is directly used as an input feature, a dimensional disaster will be caused. Therefore, this paper selects the statistics of the above eight electrical quantities as input. At the same time, the dynamic

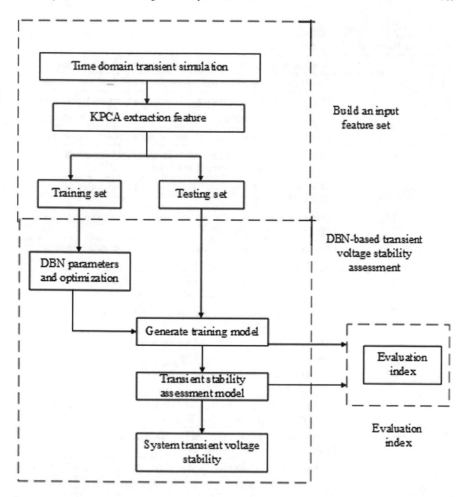

Fig. 3 Power system transient voltage stability assessment process based on kernel principal component analysis and DBN

development process of transient voltage stability is comprehensively considered, and the input feature quantity of DBN is constructed at multiple times that consist of the moment before the occurrence of system failure, the moment of fault occurrence and the moments of fault removal. The 45-dimensional feature is constructed. As shown in Table 1: Feature 1–11 is the characteristic of the power system steady state before the fault occurs that react the influence of the operation mode of the power system on the transient voltage stability. Feature 12–23 is the characteristic quantity of the moment when the fault occurs, and the instantaneous power balance is instantaneously broken when the reaction fault occurs, which affects the transient voltage. Feature 24–36 shows the effect of a moment when fault on transient voltage

Table 1 45-Dimensional feature

Input feature number	Feature description
Feature 1	The sum of the reactive power of the generator before the fault
Feature 2–3	The sum of active and reactive power of the node before the fault
Feature 4–5	The sum of active and reactive power of each branch before the fault
Feature 6–8	Maximum, minimum, and mean values of node voltage before failure
Feature 9–11	Maximum, minimum, and variance of branch current before failure
Feature 12–14	The maximum value, mean, and variance of reactive power acceleration of each generator at t0 fault removal time
Feature 15–17	The maximum, minimum, and variance of the node voltage at t0 fault removal time
Feature 18–20	The maximum, minimum, and variance of the node voltage change rate at the time of t0 failure
Feature 21–23	The maximum, minimum, and variance of the branch current change rate at the time of t0 fault occurrence
Feature 24–26	The maximum value, mean, and variance of reactive power acceleration of each generator at t1 fault removal time
Feature 27–29	The maximum, minimum, and variance of the node voltage at the time of t1 fault removal
Feature 31–33	The maximum, minimum, and variance of the node voltage change rate at the time of t1 fault removal
Feature 34–36	The maximum, minimum, and variance of the branch current change rate at the time of t1 failure
Feature 37–39	The maximum value, mean, and variance of the additional reactive power of each generator from t0 to t1
Feature 40–42	The maximum value, mean, and variance of the phase change of the voltage of each node from t0 to t1
Feature 43–44	The sum of the active and reactive changes of the branch from t0 to t1
Feature 44–45	The sum of the active and reactive changes of the node from t0 to t1

is removed. Feature 37–45 indicates the impact of power imbalance on transient voltage during the duration of the fault removal.

The 45-dimensional feature constructed by the KPCA is reduced in dimension, the feature dimension of the DBN input is reduced, and the redundant features are eliminated to improve the efficiency of the algorithm.

4.2 DBN-Based Transient Voltage Stability Assessment

This part is divided into three sections: (1) Pre-training: With the 36-dimensional feature vector of the whole training sample data as input, the RBM is trained from bottom to top layer by layer; (2) Fine-tuning: With the BP network trained under the supervision, the actual classification state is compared with the expectation to get the error, and the error is propagated downward to finely adjust the DBN parameters to generate the training model; and (3) Test the generated training model by the test sample and use the evaluation error rate and test time as evaluation indicators.

4.3 Evaluation Index

In order to meet the requirements of accuracy and rapidity of power system transient stability assessment, this paper adopts the error rate and test time of transient voltage stability assessment as evaluation indicators.

5 Simulation Analysis

The simulation example in this paper is a New England power system with 10 generators, 39 busbars and 46 AC lines. The reference power is 100 MVA and the reference voltage is 345 kV. The load has nine modes of operation: 80, 85, 90, 95, 100, 105, 110, 115, 120% of the standard load, at the same time, the output of the generator will be correspondingly changed. The power flow calculation is performed in nine operating modes, and the fault is set for transient calculation in the case of power flow convergence. Three-phase short-circuit or single-phase short-circuit faults are set at 0, 10, 20, 30, 40, 50, 60, 70, 80, 90% from the head of the line. Simulation time is set: The fault occurrence time is 0.2 s, and the fault removal time is 0.4 s.

Power system transient voltage stability assessment is a two-class problem, encoding data labels based on transient voltage stability criteria: The bus voltage is continuously below 0.75 pu for no more than 1s [23].

From the above simulation results, 4000 effective samples were selected, among which there were 2251 samples of transient voltage stability and 1749 samples of transient voltage instability.

5.1 KPCA Feature Dimension Reduction

The KPCA algorithm is used to reduce the dimension of the 4000 samples obtained by simulation [24–29]. The eigenvalues of the covariance matrix are calculated to determine the principal component contribution rate, and the corresponding eigenvectors are calculated. The contribution rate is shown in Table 2. It can be seen from Fig. 4 that the variance contribution rate after the 12th principal component is particularly small.

Considering the dimension reduction effect and the retained information ratio, we find that the 12 principal components with the largest contribution rate of the

Table 2 Main component information

Ingredient	Eigenvalues	Variance contribution rate (%)	Cumulative variance contribution rate (%)
1	4.907696	27.9908	27.9908
2	2.72689	15.5527	43.5435
3	2.172659	12.3916	55.9351
4	1.647293	9.3933	65.3304
5	1.469822	8.3831	73.7134
6	0.945379	5.3919	79.1054
7	0.748274	4.2677	83.3731
8	0.689828	3.9344	87.3075
9	0.50175	2.8617	90.1692
10	0.392448	2.2383	92.4075
11	0.306102	1.7458	94.1533
12	0.262741	1.4985	95.6519

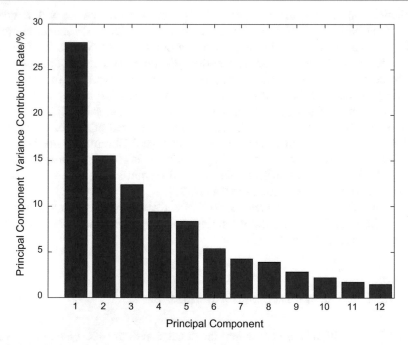

Fig. 4 Contribution rate of variance of each principal component

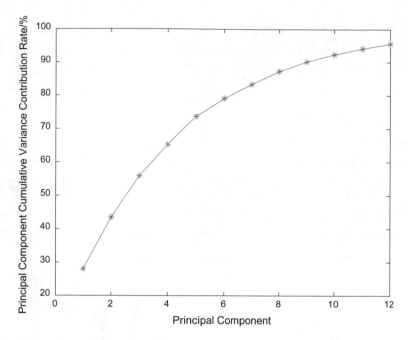

Fig. 5 Cumulative variance contribution rate of each principal component

principal component variance can represent more than 95% of the original data. The eigenvalue, variance contribution rate, and cumulative variance contribution rate of each principal component are shown in Table 2. Figures 4 and 5. The 12 principal components can represent more than 95% of the information and reduce the dimension from 45 to 12 dimensions, which greatly improve the efficiency of the algorithm.

5.2 Determine the DBN Grid Structure

The 12-dimensional feature quantity sample set extracted by KPCA is trained in MATLAB, including 3000 training samples and 1000 test samples. The DBN hidden layer structure is 25–20–8, the learning rate is 1, the momentum is 0.5, the number of supervised training is 100, the training step is 50, then the appropriate DBN grid parameters are searched. DBN different unsupervised times simulation results are shown in Table 3 and Fig. 6.

It can be seen from Table 3 and Fig. 6 that the number of unsupervised iterations is increased, and the error rate of transient stability assessment has a tendency to decrease first and then rise. When the number of unsupervised iterations is small, the feature extraction of the original data is not fully obvious. As the number of times

Table 3 DBN different unsupervised times simulation results

Unsupervised training times	Error rate (%)
1	7.3
2	7.3
3	6.8
4	6.6
5	6.2
6	4.7
7	8.3
8	8.7
9	9.1
10	9.4

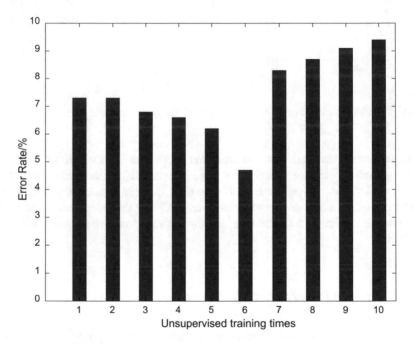

Fig. 6 Error rate of different unsupervised training times of DBN

increases, the feature extraction is better, and the error rate declines. However, after the number of iterations reaches six and reaches the minimum value of 4.7%, the error rate increases as the number of iterations increases. The possible reason is that excessive extraction leads to the loss of some important feature information.

It can be seen that when the number of unsupervised training times is six, the error rate is the lowest, and the number of supervised trainings and the number of batches are set under this DBN structure. The simulation results are shown in Table 4.

Table 4 DBN different supervised training times and batch number simulation results

Supervised training times and batches	Error rate (%)
100–50	4.7
200–100	6
200–50	2.9

When the number of batches remains the same, the more training times, the number of weight adjustments is higher; When the number of trainings is constant, the smaller the number of batches, the number of weight adjustments is higher, and the error rate is lower.

5.3　DBN Training Results Analysis

Therefore, in the simulation of this paper, DBN selects the structure of 4.2: The hidden layer has three layers, the number of neurons is 25–20–8, the learning rate is 1, the momentum is 0.5, the number of unsupervised training is 6, the number of supervised training is 200, and the training step is 50. The parameters of stacked autoencoder (SAE): The number of supervised training is 200, and the training step is 50. The activation function is sign. Least squares support vector machine (LSSVM) selects radial basis function RBF, regularization parameter, kernel parameter (bandwidth in the case of the RBF_kernel) is 0.001, whose training sample is 3000 and test sample is 1000. The transient voltage stability evaluation results under different models are shown in Table 5 and Fig. 7.

From Table 5 and Fig. 7 we can find: (1) After KPCA dimension reduction, the evaluation error rate of each model is reduced at least by 0.4%, indicating that KPCA dimension reduction can eliminate redundant components of data and improve the accuracy of classification; (2) After KPCA dimension reduction, the test time of each model is reduced, which indicates that the KPCA dimension reduction reduces the dimension of the input feature vector, which can reduce the test time of the model and is more conducive to the requirements of the transient stability evaluation of the transient system; (3) DBN has lower error rate than SAE and LSSVM, and the

Table 5 Results of transient voltage stability evaluation under different models

Training model	Error rate (%)	Test time (s)
DBN	3.3	0.002
SAE	6.5	0.006
LSSVM	3.9	0.135
KPCA + DBN	2.9	0.001
KPCA + SAE	3.2	0.004
KPCA + LSSVM	3.4	0.124

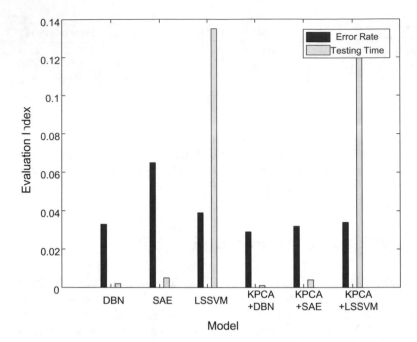

Fig. 7 Transient voltage stability evaluation results under different models

test time is much smaller than SAE and LSSVM. Its error rate is only 2.9% and test time is only 0.001 s, which can meet the accuracy and rapidity of power system transient stability assessment. Therefore, the transient voltage stability evaluation based on KPCA and DBN can judge whether the system can maintain transient stable operation within 0.1 s after fault removal. It can be used as an auxiliary means of stabilization and complements other types of stability methods.

6 Conclusion

This paper proposes a method based on KPCA and DBN for power system transient voltage stability assessment. And in the New England 10-machine 39-node standard power grid, the simulation study was carried out, and the following conclusions were obtained:

(1) The dimensionality reduction method based on KPCA algorithm can retain a large amount of effective information through nonlinear mapping, at the same time, it can also reduce the dimension of input data, remove redundant features, reduce the error rate of transient stability assessment, and test time.

(2) DBN-based power system transient stability assessment has the advantage of low error rate and test time. Its error rate is only 2.9% and test time is only

0.001 s. Meanwhile, the error rate is lower than SAE and LSSVM and the test time is much smaller than SAE and LSSVM, with higher accuracy and lower test time. So it can be used as a transient voltage stability evaluation method for power systems, and it can judge the system voltage stability state within 0.1 s after fault removal and can be used as a voltage stability auxiliary stabilization measure.

The proposed method based on KPCA and DBN for power system transient voltage stability evaluation has low error rate and short test time, which can meet the requirements of power system transient voltage stability accuracy and rapidity. But how to better integrate the actual grid model with deep learning requires more research.

References

1. Slootweg, J.G., Kling, W.L.: The impact of large scale wind power generation on power system oscillations. Electr. Power Syst. Res. **67**(1), 9–20 (2003)
2. Flueck, A.J., Chiang, H.D., Shah, K.S.: Investigating the installed real power transfer capability of a large scale power system under a proposed multiarea interchange schedule using CPFLOW. IEEE Trans. Power Syst. **11**(2), 883–889 (1996)
3. Liu, W., Lund, H., Mathiesen, B.V.: Large-scale integration of wind power into the existing Chinese energy system. Energy **36**(8), 4753–4760 (2011)
4. Aigner, T., et al.: The effect of large-scale wind power on system balancing in Northern Europe. IEEE Trans. Sustain. Energy **3**(4), 751–759 (2012)
5. Caralis, G., Papantonis, D., Zervos, A.: The role of pumped storage systems towards the large scale wind integration in the Greek power supply system. Renew. Sustain. Energy Rev. **16**(5), 2558–2565 (2012)
6. Yang, D., Ajjarapu, V.: A decoupled time-domain simulation method via Invariant subspace partition for power system analysis. IEEE Trans. Power Syst. **21**(1), 11–18 (2006)
7. Xiang, L., Wang, X., Wang, X.: Characteristics of power system transient stability assessment. Power Syst. Prot. Control **36**(6), 26–31 (2008)
8. Chan, K.W., Zhou, Q., Chung, T.S.: Transient stability margin assessment for large power system using time domain simulation based hybrid extended equal area criterion method. In: International Conference on Advances in Power System Control, Operation and Management IET, vol. 2, pp. 405–409 (2000)
9. Yue, Z., et al.: Transient stability assessment of power system based on deep learning technology. Electr. Power Constr. (2018)
10. Dai, R., Zhang, B.: Transient stability analysis based on ANN. Autom. Electr. Power Syst. (2000)
11. Ye, S., et al.: Dual-stage feature selection for transient stability assessment based on support vector machine. Proc. CSEE **30**(31), 28–34 (2010)
12. Moon, H., Phillips, P.J.: Computational and performance aspects of PCA-based face-recognition algorithms. Perception **30**(3), 303–321 (2001)
13. Liu, C.: Gabor-Based Kernel PCA with Fractional Power Polynomial Models for Face Recognition. IEEE Computer Society (2004)
14. Chen, Y., Zhao, X., Jia, X.: Spectral–spatial classification of hyperspectral data based on deep belief network. IEEE J. Sel. Top. Appl. Earth Obs. Remote Sens. **8**(6), 2381–2392 (2015)
15. O'Connor, P., et al.: Real-time classification and sensor fusion with a spiking deep belief network. Front. Neurosci. **7**(7), 178 (2013)

16. Kuremoto, T., et al.: Time series forecasting using a deep belief network with restricted Boltzmann machines. Neurocomputing **137**(15), 47–56 (2014)
17. Sun, Q., et al.: A dynamic reactive power reserve optimization method to enhance transient voltage security. Proc. CSEE **35**(11), 1523–1528 (2015)
18. Fouad, A.A., Vittal, V.: The transient energy function method. Int. J. Electr. Power Energy Syst. **10**(4), 233–246 (1988)
19. Narasimhamurthi, N., Musavi, M.: A generalized energy function for transient stability analysis of power systems. IEEE Trans. Circuits Syst. **31**(7), 637–645 (1984)
20. Chebbo, A.M., Irving, M.R., Sterling, M.J.H.: Voltage collapse proximity indicator. behaviour and implications. IEE Proc. C Gener. Transm. Distrib. **139**(3), 241–252 (1992)
21. Deng, G., et al.: Local voltage-stability margin based on short-circuit capacity. In: Power and Energy Engineering Conference, pp. 1–4. IEEE (2010)
22. Gao, B., Morison, G.K., Kundur, P.: Voltage stability evaluation using modal analysis. IEEE Trans. Power Syst. **7**(4), 1529–1542 (1992)
23. China Southern Power Grid Corporation: Q/CSG 11004-2009 Southern Power Grid Safety and Stability Calculation Analysis Guide (2009)
24. Dong, G., Ma, H.: Voltage quality evaluation of distribution network based on probabilistic load flow. Int. J. Mod. Educ. Comput. Sci. (IJMECS) **10**(8), 55–62 (2018). https://doi.org/10.5815/ijmecs.2018.08.06
25. Adebiyi, R.F., Abubilal, K.A., Mu'azu, M.B., Adebiyi, B.H.: Development and simulation of adaptive traffic light controller using artificial bee colony algorithm. Int. J. Intell. Syst. Appl. (IJISA) **10**(8), 68–74 (2018). https://doi.org/10.5815/ijisa.2018.08.06
26. Živanić, J.M., Marković, N.A., Bjelić, S.N.: Simulation of the operation of induction machines at frequencies other than 50 Hz. Int. J. Inf. Technol. Comput. Sci. (IJITCS) **10**(7), 13–21 (2018). https://doi.org/10.5815/ijitcs.2018.07.02
27. Abisoye, B.O., Abisoye, O.A.: Simulation of electric power plant performance using Excel®-VBA. Int. J. Inf. Eng. Electron. Bus. (IJIEEB) **10**(3), 8–14 (2018). https://doi.org/10.5815/ijieeb.2018.03.02
28. Tiwari, J., Singh, A.K., Yadav, A., Jha, R.K.: Modelling and simulation of hydro power plant using MATLAB & WatPro 3.0. Int. J. Intell. Syst. Appl. (IJISA) **7**(8), 1–8 (2015). https://doi.org/10.5815/ijisa.2015.08.01
29. Noubissi, J.-H., Kamgang, J.C., Ramat, E., Asongu, J., Cambier, C.: Meta-population modelling and simulation of the dynamic of malaria transmission with influence of climatic factors. Int. J. Inf. Technol. Comput. Sci. (IJITCS) **9**(7), 1–16 (2017). https://doi.org/10.5815/ijitcs.2017.07.01

Simulation and Analysis of Operating Overvoltage of AC System at ±800 kV Converter Station Based on EMTS/EMTPE

Zheng Ren, Jiaqi Fan, Xiaolu Chen, He Gao, Jianqiang Lu
and Bowen Zheng

Abstract Operating overvoltage in power grid is one of the important reasons of high-voltage equipment damage. It is accompanied by the operation of the power grid equipment, and power grid overvoltage calculation must be applied to the new project commissioning, especially the UHV project in order to understand the overvoltage conditions, take protective measures to prevent damage to equipment. Based on power electronics and electromagnetic transient calculation program, in the paper author established an electromagnetic transient model of the ±800 kV converter station ac system. Simulation analysis is carried out for the possible operating overvoltage during the start-up and debugging of the ac system. It is helpful to provide technical basis for debugging scheme, at the same time provide data for the operation and maintenance of the converter station in the future.

Keywords EMTS · Electromagnetic transient · Operating overvoltage

1 Instruction

At present, energy base in our country is mainly distributed in the Western areas, such as remote areas and large load centers concentrated in the eastern and coastal developed areas. ±800 kV HVDC technology not only has the obvious advantage of long-distance, large-capacity transmission, but also conducive to realize the regional power grid interconnection [1–4]. It is easy to control and adjust the characteristics of effective decrease transmission corridor. Thus, it becomes one of the important development trends of the power grid construction in our country.

Z. Ren (✉) · X. Chen · H. Gao · J. Lu · B. Zheng
Electric Power Research Institute, State Grid East Inner Mongolia Electric Power Company Limited, Hohhot 010000, China
e-mail: 626980096@qq.com

J. Fan
Electric Power Economic Technology Institute, State Grid East Inner Mongolia Electric Power Company Limited, Hohhot 010000, China

© Springer Nature Switzerland AG 2020
Z. Hu et al. (eds.), *Advances in Artificial Systems for Medicine and Education II*,
Advances in Intelligent Systems and Computing 902,
https://doi.org/10.1007/978-3-030-12082-5_66

In this paper, author established an electromagnetic transient model of the ±800 kV converter station ac system. In order to ensure the successful operation of the project, a series of system debugging tests are required to put into operation. In order to ensure the debugging scheme is feasible, and the equipment damage caused by operating overvoltage does not occur in the process, electromagnetic transient simulation should be carried out for the ±800 kV converter station ac system. Then we calculate the operating overvoltage level and provide technical basis for debugging scheme.

2 Operating Overvoltage

Overvoltage is divided into external and internal overvoltage in power system. Internal overvoltage refers to changes in system parameters caused by circuit breaker operation, failure, or other reasons and the voltage increase caused by the conversion or transfer of electromagnetic capacity in the power grid [5]. The internal overvoltage can be divided into operating overvoltage and temporary overvoltage. When in failure or operation, the transient overvoltage produced by the transient process is called the operating overvoltage, and its duration is generally within a few tens of milliseconds. After the end of the transient process, the continuous overvoltage with duration of more than 0.1 s to seconds or even hours is called transient overvoltage, which includes power frequency overvoltage and resonant overvoltage. The external overvoltage refers to the lightning overvoltage. The lightning overvoltage and the internal overvoltage are very different in generating mechanism and limiting method. The main research in this paper is the operating overvoltage of 500 kV system [6].

Capacitance and inductor in power system are energy storage components. When the system operation or fault changes its working state, the transition process of electromagnetic energy oscillation will occur. In this process, the magnetic quantity stored in the inductor will be converted into electric field energy stored in the capacitor element in an instant. It will generate several times transition voltage of the power supply, which is the operating overvoltage. It is high-frequency oscillation, strong damping, and disappear after a few milliseconds to more than ten milliseconds. Its peak value is also characterized by pulse, which is called operation shock wave [7].

2.1 Operating Overvoltage of Transmission Line Closing

Before closing, there will be no ground fault after checking the circuit. The starting voltage of all points of the circuit is zero. After closing, the voltage will rise to the steady state voltage of power frequency within a short period of time. Transmission lines have the characteristics of capacitance, inductance, and other distributed parameters. The closing process is a high-frequency oscillation transition process [8, 9].

In order to simplify the analysis, the equivalent circuit with concentrated parameters is used. As shown in Fig. 1, it is a T-type equivalent simplification circuit with concentrated parameters.

R is the sum of the equivalent resistance of power supply and one-half line resistance, L is the sum of the equivalent inductance of power source and half line inductance, and C is line capacitance.

The power electromotive force is:

$$e(t) = E_m \sin(\omega t + \varphi_0) \tag{1}$$

when B closes, the circuit equation is:

$$u_c + RC \frac{du_c}{dt} + LC \frac{d^2 u_c}{dt^2} = e(t) \tag{2}$$

Solve the differential equation and simplify it:

$$u_c(t) = U_{cm}(\cos \omega t - \cos \omega_0 t) \tag{3}$$

$$U_{cm} = \frac{\frac{1}{\omega c}}{\frac{1}{\omega c} - \omega l} E_m = \frac{1}{1 - \frac{\omega^2}{\omega_0^2}} E_m \tag{4}$$

$$\omega_0 = \frac{1}{\sqrt{LC}} \tag{5}$$

when B is closed, LC series resonant circuit is connected with $e(t)$. According to the above calculation and analysis, the closure process of B is $e(t)$ continuous charging process from L to C, which is constantly oscillating. The overvoltage waveform when the line is closed is shown in Fig. 2.

Fig. 1 T-type equivalent simplification circuit with concentrated parameters

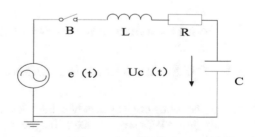

Fig. 2 Typical voltage
waveform of no-load line
closing

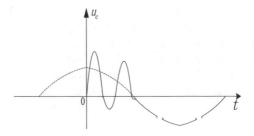

2.2 Operating Overvoltage of Transmission Line Single-Phase Reclosing

Three-phase disconnected line is on one side of the circuit breaker, when the other side is single-phase disconnected and automatic reclosing. The line sound phase is into a no-load running at this time, if the residual charge no leakage out of the line, and a phase of the sound power electric potential maximum polarity, and in contrast to the reference voltage, then the phase will appear high amplitude of overvoltage. The voltage consists of a steady state component and a free component. The free component is composed of multiple harmonic components [10–15].

The variation rule of line terminal voltage with time during the transition process is:

$$u_2(t) = A \cos \omega t + \sum_{k=1}^{\infty} A_k \cos \omega_k t \tag{6}$$

where $A \cos \omega t$ is the steady state component of working frequency, and $A_k \cos \omega_k t$ represents the k-th harmonic component.

3 Model Construction and Simulation Analysis

3.1 Test System and Main Equipment Parameters

The project scale of ac system of a ±800 kV converter station is as follows:

Two 360 MVA sfp-360,000/500 type transformer, YNd11 connection, three 500 kV transmission lines, 500 kV ac filter including four groups of HP24/36 ac filters, four groups of BP11/BP13 ac filters, two groups of HP3 ac filters, ten groups of SC shunt capacitors. Each set of capacity of ten sets of filters is 245 Mvar, and each set of capacity of ten sets of SC shunt capacitors is 365 Mvar. Three sets of 90 Mvar shunt reactors and one set of 90 Mvar shunt capacitors are installed on the low-voltage side of each step-down transformer.

According to the requirements of the electromagnetic transient calculation program, the system equivalent simplification of 500 kV transmission and transformation network of ac field of a ±800 kV converter station is carried out. The simplified system connection is shown in Fig. 3.

Table 1 shows the relevant data of overhead lines of the ±800 kV converter station.

Table 2 shows the approximate sequence parameter of overhead lines of the ±800 kV converter station.

Fig. 3 Simplified wiring diagram of ac field transmission and transformation engineering system of a ±800 kV converter station

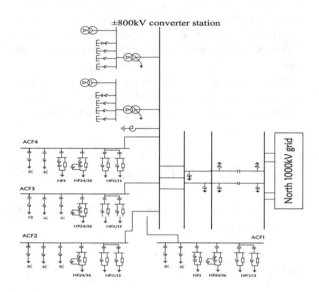

Table 1 Basic information of transmission lines

Name	Length/km	Conductor type
Line I	26.25	4 × JL/G1A-630/45
Line II	26.36	4 × JL/G1A-630/45
Line II	26.10	4 × JL/G1A-630/45

Table 2 Approximate sequence parameter of transmission lines

Name	Sequence parameter	Resistance (Ω/km)	Inductance (Ω/km)	Capacitance (μF/km)
Line I, II, III	Zero sequence	0.041779	0.662876	0.007536
	Positive sequence	0.011931	0.238384	0.014891

3.2 Simulation of No-Load Line Closing Operating Overvoltage

When closing the no-load line, the calculation conditions are as follows:

The statistic operating overvoltage of 500 kV bus and line (operating overvoltage with a probability not exceeding 2%) should not exceed 2 p.u. The dispersion of three-phase circuit breaker is considered as a normal distribution. The circuit breakers on both sides of the circuit of 500 kV transmission lines of converter station have no closing resistance installed. The closing time difference of three-phase circuit breaker is no more than 5 ms, and the statistical times are 100 times.

The results of statistic operating overvoltage of 500 kV lines are as shown in Table 3.

From the results, in the operation overvoltage test of closing no-load line, the maximum overvoltage phase value of line I is 1.69 p.u., which meets the regulation that the phase overvoltage shall not exceed 2.0 p.u.

3.3 Simulation of Transmission Line Single-Phase Reclosing

The results of statistic operating overvoltage of 500 kV lines single-phase reclosing are as shown in Table 4.

Table 3 Statistic operating overvoltage

Operation mode	The line head end		The end of the line	
	Umean	$U2\%$	Umean	$U2\%$
Charging from one side of line I	1.18	1.33	1.34	1.69
Line I is closed	1.02	1.12	1.22	1.55
Line I and line II are closed	0.93	1.06	1.23	1.55

Table 4 Statistic operating overvoltage of 500 kV lines single-phase reclosing

Operation mode	The line head end		The end of the line	
	Umean	$U2\%$	Umean	$U2\%$
Charging from one side of line I	1.23	1.45	1.37	1.62
Line I is closed	1.33	1.53	1.38	1.63
Line I and line II are closed	1.34	1.55	1.38	1.63

From the results, in the operation overvoltage test of 500 kV lines single phase reclosing, the maximum overvoltage phase value of line I is 1.63 p.u., which meets the regulation that the phase overvoltage shall not exceed 2.0 p.u.

3.4 Simulation of Switching AC Filters and Shunt Capacitors

The results of statistic operating overvoltage of switching ac filters and shunt capacitors are as shown in Table 5.

From the results, in the operation overvoltage test of switching ac filters and shunt capacitors, the maximum overvoltage phase value is 1.46 p.u., which meets the regulation that the phase overvoltage shall not exceed 2.0 p.u.

3.5 Simulation of Switching Bus High Resistance

The results of statistic operating overvoltage of switching bus high resistance are as shown in Table 6.

From the results, in the operation overvoltage test of switching bus high resistance, the maximum overvoltage phase value is 1.34 p.u., which meets the regulation that the phase overvoltage shall not exceed 2.0 p.u.

3.6 Simulation of Switching Capacitors and Reactors

The results of statistic operating overvoltage of switching capacitors and reactors are as shown in Table 7.

Table 5 Statistic operating overvoltage of switching ac filters and shunt capacitors

Switch equipment	Closing overvoltage		Cutting overvoltage	
	Umean	$U2\%$	Umean	$U2\%$
H24/36	1.37	1.46	0.98	0.98
BP11/13	1.09	1.26	0.96	0.97
HP3	1.00	1.04	1.02	1.09
SC	1.35	1.43	0.99	1.00

Table 6 Statistic operating overvoltage of switching bus high resistance

Switch equipment	Closing overvoltage		Cutting overvoltage	
	Umean	$U2\%$	Umean	$U2\%$
Bus high resistance	0.94	0.95	1.30	1.34

Table 7 Statistic operating overvoltage of switching capacitors and reactors

Operation	Operating overvoltage	
	500 kV side	66 kV side
Closing capacitors	1.00	1.13
Cutting capacitors	0.92	1.41
Closing reactors	0.92	1.01
Cutting reactors	0.92	1.43

From the results, in the operation overvoltage test of capacitors and reactors of 500 kV side the maximum overvoltage phase value is 1.00 p.u., which meets the regulation that the phase overvoltage shall not exceed 2.0 p.u. In the operation, overvoltage test of capacitors and reactors of 66 kV side the maximum overvoltage phase value is 1.43 p.u., which meets the regulation that the phase overvoltage shall not exceed 4.0 p.u.

4 Summary and Conclusion

Based on EMTS/EMTPE, in the paper author established an electromagnetic transient model of a ±800 kV converter station ac system. Simulation analysis is carried out for the possible operating overvoltage during the start-up and debugging of the ac system. The results show that the operating overvoltage of the ac system of the ±800 kV converter station is within the allowable range of the regulation. It is helpful to provide technical basis for debugging scheme and ensure the ±800 kV converter station commission successfully, at the same time provide data for the operation and maintenance of converter station in the future.

References

1. Bo, L., Xianting, M., Yiling, G.: Theory analysis for distribution transformer to occur ferro-resonant overvoltage. Electr. Saucer Power **30**(4), 288–290 (2008, in Chinese)
2. Wei, W., Shengchang, J., Tao, C., et al.: Theoretical analysis and experimental verification of fundamental frequency ferroresonance. Power Syst. Technol. **33**(17), 226–230 (2009, in Chinese)
3. Fan, L., Caixin, S., Wenxia, S., et al.: The theoretical analysis of chaotic ferroresonance overvoltage in power systems. Trans. China Electrotech. Soc. **21**(2), 103–107 (2006, in Chinese)
4. Yunge, L., Wei, S.: The study of fundamental ferroresonance on neutral-grounded systems by using analytical method, the solution to the power frequency excitation characteristic of nonlinear inductors. Proc. CSEE **23**(10), 94–98 (2003, in Chinese)
5. Kieny, C.: Application of the bifurcation theory in studying and understanding the global behavior of a ferroresonant electric power circuit. IEEE Trans. Power Deliv. **6**(2), 866–871 (1991)

6. Bohmann, L.J., McDaniel, J., Stanek, E.K.: Lightning arrester failure and ferroresonance on a distribution system. IEEE Trans. Power Deliv. **9**(6):1189–1195 (1993)
7. Shott, H.S., Peterson, H A.: Criteria for neutral stability of wye-grounded primary broken-delta-secondary transformer circuit. Trans. AIEE **24**(4), 997–1002 (1941)
8. Fan, L., Wenxia, S., Caixin, S., et al.; Resonance overvoltage based on constant pulse method. Power Syst. Technol. **30**(3), 57–61 (2006, in Chinese)
9. Hua, L., Bin, W., Yanliang, Y., et al.: Influence of excitation characteristics of PT on ferroresonance and detection research. Electr. Measur. Instrum. **538**, 10–12 (2010, in Chinese)
10. Yi, H., Yanxia, Z.: Detection of ferroresonance in neutral non-grounding system based on decomposition characteristic of wavelet packet. Power Syst. Technol. **30**(23), 72–76 (2006, in Chinese)
11. Abebe, Y.M., Mallikarjuna Rao, P., Gopichand Nak, M.: Load flow analysis of a power system network in the presence of uncertainty using complex affine arithmetic. Int. J. Eng. Manuf. (IJEM) **7**(5), 48–64 (2017). https://doi.org/10.5815/ijem.2017.05.05
12. Zhang, M., Xia, C.: A loose wavelet nonlinear regression neural network load forecasting model and error analysis based on SPSS. Int. J. Inf. Technol. Comput. Sci. (IJITCS) **9**(4), 24–30 (2017). https://doi.org/10.5815/ijitcs.2017.04.04
13. Ibrahim, N.M.A., Attia, H.E. M., Talaat, H.E.A., Kasem Alaboudy, A.H.: Modified particle swarm optimization based proportional-derivative power system stabilizer. Int. J. Intell. Syst. Appl. (IJISA) **7**(3), 62–76 (2015). https://doi.org/10.5815/ijisa.2015.03.08
14. Singhal, P., Agarwal, S.K., Kumar, N.: Advanced adaptive particle swarm optimization based SVC controller for power system stability. Int. J. Intell. Syst. Appl. (IJISA) **7**(1), 101–110 (2015). https://doi.org/10.5815/ijisa.2015.01.10
15. Noubissi, J.-H., Kamgang, J.C., Ramat, E., Asongu, J., Cambier, C.: Meta-population modelling and simulation of the dynamic of malaria transmission with influence of climatic factors. Int. J. Inf. Technol. Comput. Sci. (IJITCS) **9**(7), 1–16 (2017). https://doi.org/10.5815/ijitcs.2017.07.01

Simulation Study and Experimental Analysis of Current Closure Overvoltage Caused by High Reactance on the AC Bus at ±800 kV Converter Station

Pengwe Yang, Rui Wang, Cai Xu, Weimng Liu and Yingkun Han

Abstract When high reactance on AC bus in ±800 kV converter station is switching, it will block the current and generate serious overvoltage, which may burn down the equipment. When AC bus in ±800 kV converter station is put into operation with high reactance, simulation calculation must be carried out to understand the situation of overvoltage caused by current closure, so as to arrange the restrictive measures and protect the equipment. In this paper, taking the AC bus high reactance of a ±800 kV converter station as an example, when the high reactance on the bus is disconnected, the overvoltage caused by the current closure is theoretically analyzed in detail, and the EMTS model of the system is established to simulate and calculate the overvoltage of the bus caused by high reactance. The results show that with the increase of the cutoff current value of the circuit breaker, the cutoff overvoltage generated by the shunt will be higher, even exceeding the required value of the regulations. A set of lightning arresters connected in parallel to the high-reactance side can effectively limit the interception overvoltage. During system debugging, field measurement data are collected, which is consistent with simulation data. The results provide valuable reference for the construction and operation of ±800 kV converter station.

Keywords ±800 kV Converter station · EMTS · Intercept overvoltage · Simulation research · High resistance

P. Yang (✉) · C. Xu · W. Liu · Y. Han
State Grid East Inner Mongolia Electric Power Research Institute, Hohhot, Inner Mongolia 100010, China
e-mail: 1641238505@qq.com

R. Wang
State Grid East Inner Mongolia Electric Power Co. LTD., Construction Branch, Hohhot, Inner Mongolia 100010, China

Y. Han
State Grid Shandong Electric Power Research Institute, Jinan 250000, Shandong, China

© Springer Nature Switzerland AG 2020   739
Z. Hu et al. (eds.), *Advances in Artificial Systems for Medicine and Education II*,
Advances in Intelligent Systems and Computing 902,
https://doi.org/10.1007/978-3-030-12082-5_67

1   Instruction

In order to meet the requirements of the system operation mode, the AC bus of the ±800 kV converter station is mostly installed with high bus reactance, and the bus reactance of the converter station is cut with the change of reactive load [1]. A certain amount of energy is stored during the normal operation of the bus with high reactance. Once removed, the energy will be released through the equivalent entrance capacitance of the reactor. Because the capacitance value is small, a high overvoltage will be generated, namely the current closure overvoltage [2, 3]. High overvoltage occurs when high reactance is removed at home and abroad. During the test of shunt reactor for the second circuit resection of Dafang 500 kV, the highest voltage measured in the field is 1.90 p.u. If there is no arrester to limit the overvoltage, the overvoltage level may exceed 2.0 p.u. [4]. A high-reactance side of an Iranian substation is equipped with a lightning arrester, but it is an old silicon carbide lightning arrester. When the high reactance of the 400 kV bus is cut, 3.6 p.u. overvoltage is generated, and the reactor and circuit breaker are burned down. Therefore, when the AC field of the ±800 kV converter station is put into operation, it is necessary to carry out the simulation study of the current closure overvoltage of the high reactance of the AC bus, so as to arrange the corresponding restriction measures as soon as possible, prevent the device from burning due to the high current closure overvoltage, and affect the operation safety of the system.

This article was built with the EMTS program on a ±800 kV converter station and the grid system, a simulation model of calculating the bus reactor on the removal of current closure overvoltage, and probes into the actual operation of current closure overvoltage records, the simulation and measured data are compared, the result shows that installing surge arrester in high head end intercepting overvoltage can be reduce to instruction value.

2   Theoretical Analysis and Limiting Measures of Bus High Impedance Interceptor Overvoltage

2.1   Theoretical Analysis

The reactor is a large inductor, which absorbs a certain amount of reactive power during normal operation and stores it in the inductor. It is generally believed that the load current of the shunt reactor is a small inductive current. When the circuit breaker opens a small inductive current, the interaction between the break and the surrounding circuit usually generates a high-amplitude high-frequency transient voltage, and the resulting high-frequency transient recovery voltage can lead to a harmful effect on the reactance and/or switching equipment [5–9]. The equivalent circuit for removing the high reactance of bus is shown in Fig. 1.

Fig. 1 Equivalent circuit of high reactance of bus

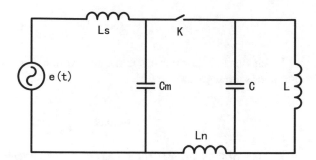

L_s is the equivalent inductance of the system; C_m is equal capacitance of the system; L_n is the equivalent inductance of the connecting line; K is the outlet circuit breaker; C is the equivalent capacitance at the reactor side; L is the inductance of the shunt reactor.

When the current I_L of the reactor A_L naturally exceeds zero, it is cut off, the magnetic energy on the reactor is zero, and then $A_L = \frac{1}{2}L \cdot I_L^2 = 0$. The energy stored on the capacitor is $A_C = \frac{1}{2}C_L \cdot U_L^2$, the voltage U_L on the reactor. The electric energy A_C on the capacitor C is oscillated by high frequency, and its oscillation frequency is $f = 1/\left(2\pi\sqrt{LC}\right)$.

When the reactor current I_L is cut off at nonzero, set the cutoff time of the circuit breaker $e(t) = E_m \cos kt$ and the corresponding phase angle $T = kt_0$. When it is truncated: $e(t) = E_m \cos T$; $U_c = E_m \cos T$; $I_L = I_m \sin T$ after intercepting the current,

$$\frac{1}{2}CU_C^2 + \frac{1}{2}LI_C^2 = \frac{1}{2}CU_m^2 \tag{1}$$

$$U_m = U_C^2 + \frac{L}{C}I_L^2 = \sqrt{E_m^2 \cos^2 T + I_m^2 \frac{L}{C}\sin^2 T} \tag{2}$$

The overvoltage multiples generated after the closure can be obtained:

$$K_n = \frac{U_m}{E_m} = \sqrt{\cos^2 T + \left(\frac{I_m}{E_m}\right)^2 \frac{L}{C}\sin^2 T} \tag{3}$$

$I_m = \frac{E_m}{kL}$, in Eq. (3),

$$K_n = \sqrt{\cos^2 T + \frac{1}{k^2 LC}\sin^2 T} \tag{4}$$

Considering various losses, whether magnetic energy can be converted into electric energy 100% requires the introduction of conversion coefficient Z (about 0.3–0.45), and then Eq. (4) can be rewritten as:

$$K_n = \sqrt{\cos^2 T + \frac{Z}{k^2 LC} \sin^2 T} \tag{5}$$

When the inductor current is the maximum, it is cut off and the overvoltage multiple is the highest $T = 90°$.

$$K_{n\text{max}} - \frac{1}{k}\sqrt{\frac{Z}{LC}} \tag{6}$$

The bus of the converter station has a high resistance capacity of 210 Mvar and rated voltage of 550 kV, and the neutral point is directly grounded. The inductance of single-phase reactor is: $L = \frac{Z_m}{k} = 7.26$ (H). Equivalent capacitance on reactor side C is about $10,000_p F$, take $Z = 0.3$, in Eq. (6), $K_n\text{max} = 6.47$.

It can be seen that the overvoltage of the high reactance of the bus is related to its rated voltage, capacity, and structure, as well as to the stray capacitance of the external connection line and electrical equipment. Since the equivalent inlet capacitance of high reactance is small, the calculated overvoltage multiple is high, and its maximum overvoltage multiple will exceed the insulation level of electrical equipment. Therefore, the corresponding overvoltage limiting measures must be taken to remove the high reactance.

2.2 Restrictions

2.2.1 Zinc Oxide Arrester

The compatibility margin between the protection level of non-gap ZnO arrester and the insulation level of the reactor is 25%. When the cutoff current of the circuit breaker is large, if the generated cutoff wave overvoltage exceeds the volt–ampere characteristic of the arrester, the arrester can absorb some inductance storage energy of the reactor, which consumes little energy. The arrangement of the arrester should be as close as possible to the reactor to reduce the re-ignition overvoltage. Zinc oxide arrester is often used in practical engineering to limit the interception overvoltage [10].

2.2.2 Thyristor

The cutoff resistance of the circuit breaker can limit the high amplitude of the cutoff overvoltage. The switching resistance transfers the main contact to cut off the current, which reduces the amplitude of the truncation current and becomes a "soft truncation," which reduces the cutoff overvoltage. It is not economically or technically advisable to use the thyristor [11] to limit the amount of thyristor needed to intercept overvoltage. In other words, it is not practical to greatly limit the overvoltage of a high-resistance circuit breaker with a shunt resistor.

2.2.3 Phase Separation Device

The selective phase separation device of the circuit breaker can eliminate the reburn-ing of the circuit breaker. The phase-selective tripping device accurately synchronizes the voltage phase of the power supply so as to minimize the arc time and cut off the current when the current first passes zero after the contact separation. However, the phase separation device requires the power side to be equipped with CVT, and the influence of temperature change and aging of operating mechanism on mechanical operation time is taken into account. Therefore, the phase selection closing device is generally not installed in the actual engineering for the high reactance of bus.

3 Simulation and Analysis of Bus High Impedance Interceptor High Voltage

In order to better understand the transient process of removing high reactance, the high reactance removal test of the converter station was simulated and analyzed. The generator, transformer [12–14], lightning arrester, inductance, and capacitance were simulated in detail with EMTS software. The simulation model is shown in Fig. 2. In the simulation experiment, the overvoltage protection without overvoltage and the limitation of overvoltage with arrester are simulated respectively. In the absence of overvoltage protection measures, the relationship between the intercepting overvolt-age on the reactor and the intercepting current is shown in Table 1. The simulation waveform of high overvoltage of the bus is shown in Fig. 3. According to relevant standards, the overvoltage of 500 kV bus and line should not exceed (2.0 p.u.).

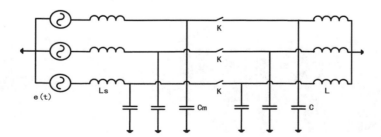

Fig. 2 Simulation model of bus high-resistance resection at converter station

Table 1 Relationship between intercepting overvoltage and intercepting current under unrestricted conditions

I_{ch} (A)	1	5	10	15	20	25	30	35
Maximum overvoltage (p.u.)	1.02	1.23	1.55	1.93	2.42	2.67	2.89	3.2

Table 2 Relationship between intercepting overvoltage and intercepting current under the condition of installing arrester

I_{ch} (A)	1	5	10	15	20	25	30	35
Maximum overvoltage (p.u.)	1.02	1.19	1.42	1.55	1.60	1.68	1.70	1.71

The simulation results show that the interceptor overvoltage of the reactor increases with the increase of the cutoff current. Under the condition of unlimited measures, when the cutoff current exceeds 15 A, the cutoff overvoltage on the reactor will exceed the required value of the regulation. Based on the analysis of intercepting overvoltage limitation measures, we can use the method of installing a lightning arrester on the high-resistance side to limit overvoltage.

The MOA used in this paper is used to limit the overvoltage of high anti-break resistance. The rated voltage is 444 kV, and the residual pressure of the operating shock current shall be less than 900 kV in accordance with gb11032-89, which means that the metal oxide arrester can limit the operating overvoltage to less than 3 times. In the case of the above MOA protection, the relationship between the measured interceptor overvoltage and the interceptor current of the high impedance of the open bus is shown in Table 2.

After installing the arrester to limit the overvoltage, with the increase of the truncation current, the intercepting overvoltage on the reactor is kept at about 1.70 p.u., which is less than the allowable value of the reactor overvoltage. It can be seen that the high-resistance side equipped with lightning arrester is to limit the intercepting overvoltage, an important measure.

Although the current flowing through the arrester under operating overvoltage is generally smaller than that of lightning current, it lasts a long time and requires strict requirements on the through-flow capacity of the arrester. High reactance in normal operation, inductor current is large, storage energy is large, forced truncation of the

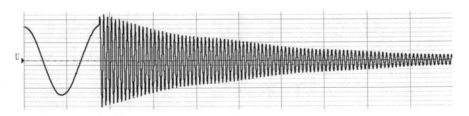

Fig. 3 Simulation waveform of high-voltage overvoltage of the bus

Table 3 Experimental data of high impedance of excision bus

Operation way	Bus overvoltage (p.u.)			High reactance interceptor overvoltage (p.u.)			Truncation current (A)		
	U_a	U_b	U_c	U_a	U_b	U_c	U_a	U_b	U_c
Break-brake 1	1.03	1.000	0.992	1.131	1.033	1.634	2.71	1.22	22.1
Break-brake 2	1.12	1.010	1.014	0.869	1.350	1.091	0.21	15.1	2.43
Break-brake 3	1.02	1.003	1.004	0.869	0.996	0.882	0.72	1.42	0.92
Break-brake 4	1.10	1.000	1.024	1.432	1.225	1.062	17.2	4.41	2.53

arrester flow capacity more stringent requirements. The following is an estimate of the capacity of the arrester to see if it meets the requirement of high resection resistance.

Similarly, the above high reactance is taken as an example. First, the maximum energy stored in high reactance is calculated. High inductance: $L = 7.26$ H High resistance rated current: $I_R = 220$ A. Then, the maximum energy stored by the high-resistance winding during normal operation is: $W = \frac{1}{2}L \cdot I_r^2 = 175$ kJ.

In this paper, the 2 ms square wave flux capacity of 500 kV zinc oxide arrester can reach 1500 A (20 times), and the voltage of MOA under 1500 A is about 850 kV, so the allowed energy is: $W_m = UIt = 2550$ (kJ). It can be seen that the energy that the arrester allows to pass through is much greater than the maximum energy stored by the high-resistance winding. Therefore, it is not a problem to use metal oxide arrester to limit the capacity of high voltage overvoltage.

4 The Actual Measurement and Analysis of Bus High Impedance Interceptor Overvoltage

During the commissioning of the AC station system of the converter station, the bus high-reactance test was carried out, and the data in the operation process were recorded by the same control waveform recorder (TK4024). The recorded data require to be complete and appropriate, and the transient and steady voltage could be read. Field voltage data were obtained from the secondary side of CVT on the high-resistance side, with a variable ratio of 5000. Charged by a single line, under the condition of having a lightning arrester, the high impedance of the bus was tested for 4 times, and the AC bus voltage, high-resistance side voltage, and current waveform were recorded every time. The test data were shown in Table 3.

The high reactance of bus was cut 4 times, and the high-reactance side three-phase arrester acted 2 times. It can be seen from Tables 2 and 3 that the test and simulation overvoltage data of the bus during high-impedance resection meet the requirements of the regulation, and the data trend is consistent. The overvoltage value increases with the increase of the truncation current.

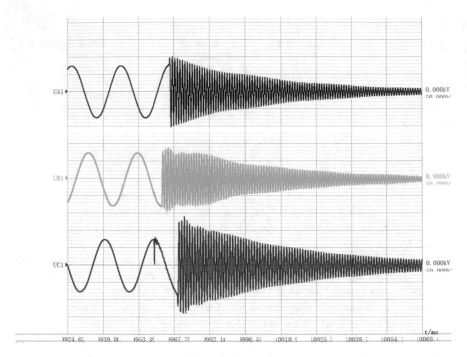

Fig. 4 High-voltage waveform of bus-line

Typical voltage waveforms in the high-resistance test of the bus are shown in Fig. 3, which are basically consistent with the simulation waveforms in the laboratory. They are all high-frequency oscillation attenuation waves with a duration of tens of milliseconds, and the oscillation frequency is determined by the inductance and capacitance of the circuit. During the reburning of phase C during the resection, the circuit breaker fails to open and break, and the high-frequency current in the system begins to oscillate, resulting in the intercepting current, which is 15.8 A. Due to the limitation of the arrester, the intercepting overvoltage of the reactor is 1.634 times, less than the allowable value of high anti-overvoltage, and it will not damage the equipment (Fig. 4).

5 The Excision Method of Bus bar High Resistance

All the above theories and experiments show that the removal of high reactance will result in high overvoltage, which will endanger the insulation of related electrical equipment. Although the zinc oxide arrester has an obvious inhibitory effect on overvoltage, it can make the arrester move frequently and affect the service life of the arrester due to high anti-on–off overvoltage, high energy, and long duration.

On the other hand, up to three times of overvoltage can be generated after arrester limit pressure, which has an impact on other related electrical equipment. In order to further reduce overvoltage and safely and smoothly open high reactance, a method of long-line open high reactance can be adopted. Actually, the circuit breaker is used to cut out the empty line. A lot of practice shows that this operation mode produces low overvoltage multiples and has no effect on the insulation of electrical equipment and transmission lines. It has been proved that the high resistance of bus in the converter station can be removed by the circuit breaker of the higher substation.

6 Conclusions

In this paper, the interceptor overvoltage is analyzed theoretically when the excision of high resistance to bus bar. The simulation results show that installing ZnO arrester on high-resistance side is an effective method to limit the overvoltage of high resistance. In order to further reduce the overvoltage of high anti-break voltage and ensure the safe operation of relevant electrical equipment, the method of removing high resistance with a long line can be adopted to remove the high voltage with a safe and smooth removal of high-voltage reactor.

References

1. Li, Z.-q., Zhou, P.-h., Xiu, M.-h., et al.: Research of AC filter bus connecting shunt-reactor resonance and overvoltage. High Voltage Apparatus 2(46), 27–34 (2010)
2. Xie, G.-r.: Overvoltage of Power System. Water Conservancy and Hydropower Press, Beijing (1983)
3. Weibo, Z., Jinliang, H., Yuming, G.: Overvoltage Protection and Insulation Coordination. Tsinghua University Press, Beijing (2002)
4. General Headquarters of Jinjing 500 kV Transmission and Transformation Engineering: Commissioning of the second transmission and transformation engineering of 500 kV large rooms, Baoding (1987)
5. Xu, G.-z., Zhang, J.-r., Qian, J.-l., et al.: Principle and Application of High Voltage Circuit-Breakers. Tsinghua University Press, Beijing (2000)
6. Ma, Z., Blis, C.A., Penfold, A.R., et al.: An investigation of transient overvoltage generation when switch high voltage shunt reactor by SF6 CB. IEEE Transients Power Deliv. 13(2), 472–479 (1998)
7. Peelo, D.F., Avent, B.L., Drakes, J.E., et al.: Shunt reactor switching tests in BC hydro's 500 kV system. IEEE Proc. C-Gener. Transm. Distrib. 135(5), 420–434 (1988)
8. Yanxin, S.H.I.: Research on vacuum circuit breaker switching shunt capacitor overvoltage. North China Electr. Power 6, 5–8 (2005)
9. Li, B.: Shunt reactors and its switching process. High Voltage Apparatus Technol. (3), 32–35 (1992)
10. Bjelić, S., Bogićević, Z.: Calculation of overvoltage and estimation of power transformer's behavior when activating the reactors. Int. J. Inf. Technol. Comput. Sci. (IJITCS) 6(12), 67–73 (2014). https://doi.org/10.5815/ijitcs.2014.12.09

11. Abdelaziz, A.Y., Ibrahim, A.M.: Protection of thyristor controlled series compensated transmission lines using support vector machine. Int. J. Intell. Syst. Appl. (IJISA) **5**(5), 11–18 (2013). https://doi.org/10.5815/ijisa.2013.05.02
12. Abi, M., Mirzaie, M.: Modelling and detection of turn-to-turn faults in transformer windings using finite element analysis and instantaneous exciting currents space phasor approach. Int. J. Eng. Manuf. (IJEM) **4**(5), 12–23 (2014). https://doi.org/10.5815/ijem.2014.05.02
13. Bjelić, S., Bogićević, Z.: Computer simulation of theoretical model of electromagnetic transient processes in power transformers. Int. J. Inf. Technol. Comput. Sci. (IJITCS) **6**(1), 1–12 (2014). https://doi.org/10.5815/ijitcs.2014.01.01
14. Malviya, R., Saxena, R.K.: Modified approach for harmonic reduction in transmission system using 48-pulse UPFC employing series zig-zag primary and Y-Y secondary transformer. Int. J. Intell. Syst. Appl. (IJISA) **5**(11), 70–79 (2013). https://doi.org/10.5815/ijisa.2013.11.08

Study of Dynamic Reactive Power Support of Synchronous Condenser on UHV AC/DC Hybrid Power System

Bowen Zheng, Jianqiang Lu and He Gao

Abstract As the Inner Mongolia Autonomous Region, UHV AC and DC power transmission and transformation projects have been put into operation successively, the Inner Mongolia area power grid has become UHV AC/DC hybrid power system. However, the development of AC and DC projects is not balanced. The characteristics of "strong DC and weak AC" are obvious. Especially, when the sending end auxiliary thermal power unit is connected to the high voltage AC and DC hybrid power grid, it will be bound to affect the safe and stable operation of the UHV power grid, and the stability of the transmission end system is difficult to be guaranteed. The high capacity synchronous condenser can improve the short circuit ratio of DC power transmission system in AC/DC power grid, enhance the dynamic voltage support capacity of the AC power grid and improve the stability of the power grid. This paper takes the regional power grid of the transmission end of the xilinhaote to Taizhou of jiang su ultra-high voltage dc transmission project (hereinafter referred to as the "Xitai DC") as the research object, using Power System Analysis Software Package simulation software, through simulation to carry out the comparison analysis. From the angle of AC and DC transmission capacity and single end trip of line, it comprehensively evaluates influence of synchronous condenser on the stability level of the power grid. The results show that the synchronous condenser has good dynamic voltage support ability and strong short-time overload ability and can effectively improve the stability level of UHV AC and DC hybrid power grid.

Keywords Synchronous condenser · UHV DC transmission · Voltage stability · Dynamic reactive power

B. Zheng (✉) · J. Lu · H. Gao
State Grid East Inner Mongolia Electric Power Research Institute,
Hohhot, China
e-mail: 121679286@qq.com

© Springer Nature Switzerland AG 2020 749
Z. Hu et al. (eds.), *Advances in Artificial Systems for Medicine and Education II*,
Advances in Intelligent Systems and Computing 902,
https://doi.org/10.1007/978-3-030-12082-5_68

1 Introduction

Since 1967, some developed countries need to transmit long distance and large capacity to some parts of the country, therefore, the research on UHV transmission technology has been started. EHV transmission refers to the transmission technology of AC 1000 kV, DC positive and negative 800 kV and above voltage level, whose biggest feature is that it can transmit electricity with long distance, large capacity and low loss. According to the data provided by the State Grid Corporation, the UHV DC power grid can send 6 million kW electric power, which is equivalent to 5–6 times as much as the existing 500 kV DC power grid. Moreover, the transmission distance is 2–3 times that of the latter, so the efficiency is greatly improved. Especially in the field of long distance and large capacity transmission, UHV DC transmission is usually a better choice, which is superior to AC transmission and UHV AC transmission in terms of economic investment, energy consumption and engineering scale [1].

Although the research on UHV transmission technology started late, China has developed it very rapidly. It's less than 10 years since it started, China has the highest level in the world, and has created a number of world records, and most of the technology advanced UHV AC and HVDC transmission projects have been put into operation smoothly. The interconnection of UHV AC and DC systems and large-scale cross-regional power transmission has become the main characteristics of Chinese power grid. However, in the case of the rapid development of UHV AC and DC technology, especially the step type lifting of the DC transmission, the operating characteristics of the power grid have undergone profound changes, and the contradiction of the "strong DC, weak AC" is prominent [2, 3]. The security of the power grid is facing new challenges.

As a reactive power generation device, a large capacity synchronous condenser has unique advantages to improve the short circuit ratio of the AC power grid receiving end, improve the ultimate power of the DC transmission and enhance the strength and flexibility of the power grid. When there is a failure at the close end of the AC power grid or voltage drops at the converter valve, the large capacity condenser can strongly support excitation supporting voltage and keep system's stability, which wins valuable time for the fault removal [4–6]. In the case that when the fault occurs in the DC system or the power load is suddenly dumped, it can enter into the leading phase operation state and absorb a large amount of the excess reactive power which can inhibit the increase of the system voltage. When the dc system needs to adjust the voltage for steady and normal operation, it can provide continuously adjustable dynamic reactive power support for the AC power grid in late phase or leading phase operation state. Therefore, with the continuous planning and commissioning of UHVDC transmission projects in recent years, the new type of large capacity condenser has been attached great importance by the power grid enterprises again.

2 Overview of Xitai DC Area Grid Structure

2.1 Overview of the Current Situation of AC and DC Projects Engineering

With the UHV AC and DC transmission and transformation projects in the Inner Mongolia Autonomous Region put into operation one after another, Inner Mongolia power grid has become an UHV AC and DC hybrid system. The Xitai DC transmission project starts at Xilinhot converter station in the Inner Mongolia Autonomous Region and ends at Taizhou converter station in Jiangsu Province, whose total route length is 1627.9 km, and the rated transmission power can reach 10 million kW. And through the Shengli substation, it is interconnected with UHV AC transmission engineering. Starting from Xilin Gol League, the Inner Mongolia Autonomous Region, after Ximeng substation, Langfang substation, the project is located in Jinan, Shandong Province, with a capacity of 9 million kW and a transmission distance of 730 km. The schematic diagram of the regional project is shown in Fig. 1.

2.2 Basic Characteristics of Synchronous Condenser

In order to enhance the stability of UHV AC/DC system and meet the requirements of DC large-scale power transmission, large capacity dynamic reactive power compensation equipment must be configured. At present, the main methods of dynamic reactive power compensation are static synchronous compensator, static var compensator and synchronous condenser [7]. The main advantages of the condenser are as follows: The reactive power output is less influenced by the system voltage, strong ability of short-term overload, the operation stability is good, the harmonic gener-

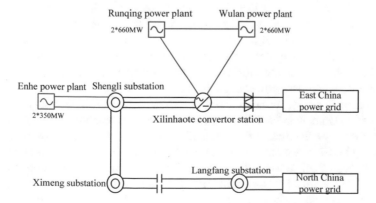

Fig. 1 Schematic diagram of the regional project

ation is almost not produced. The service life is long and generally 30 years. The XiMeng converter station plans to deploy two large capacity condensers, with a rated capacity of 300 Mvar and a self-shunt excitation system, with the maximum excitation capacity of −150 Mvar. The condenser mainly consists of a condenser body, an excitation system, a step-up transformer, a starting system and a cooling system. Compared with the traditional synchronous generator, the main difference of the condenser is that: In structure, the condenser has no original motive and speed regulation system and the ability to regulate the active power. Because the condenser does not carry the load, its axis of rotation is thin, so the length of the air gap and the number of excitation windings decrease, the reactance increases, the standard value of reactance reaches up to about 2; In electrical characteristics, the electromagnetic torque of the condenser is driven torque; in normal operation, the condenser needs to absorb certain active power from the power grid to compensate for its own loss. The average active power ratio is 1.5–5% of its rated capacity [8]. The reactive power regulation method of synchronous condenser is similar to synchronous generator. Under the condition of over excitation, the condenser outputs inductive reactive power; under the condition of under excitation, the synchronous motor absorbs inductive reactive power.

3   Calculation and Analysis of Power Flow Stability

3.1   Analysis of AC Power Supply Capacity

Taking into account the high peak load period in summer and the circumstances of DC line maintenance, it is necessary to consider that the Xitai DC system suspends operation and the matching power supply put into operation to restrict the AC transmission capability. According to the rated power of the generator, we divide the generators of Enhe power plant into one group (hereinafter referred to as A group) and divide the generators of Runqing power plant and Wulan power plant into one group (hereinafter referred to as B group). For different types of generating scheme, by scanning the short circuit faults of UHV AC lines and transformers, we simulate the transient process of system. We find that when the B group starts up to 3 units or more, when the three-phase short circuit fault occurs in the AC system, the sending end generators will have power angle instability problem.

Therefore, in order to ensure the stability of the system, it is necessary to limit the power of the generators [9–11]. In order to achieve as much power as possible for the generators, we first reduce the output of the generators in the A group and use the dichotomy to determine the power limit of the generators, the result can be accurate to 1 MW. We calculate using the same method in the way that two condensers are connected to the grid. The power limit of the generator in different operating modes under the two modes is shown in Table 1. The generator output is not limited in other ways. From the data in the table, we can see that: After two sets of condenser are

Table 1 Maximum output limits of generators under different generating scheme

Generators in group A	Generators in group B	Maximum output limits (MW)	Maximum output limits with two condensers (MW)
1	3	2310	No limits
0	4	2160	No limits
2	3	2470	No limits
1	4	2660	No limits
2	4	2860	No limits

configured in Ximeng converter station, under all operating models, the generator output is not limited and the power rating can be reached.

3.2 Analysis of DC Power Supply Capability

If Xitai DC system runs normally, it will greatly reduce the pressure of the AC system. However, through simulation, in different generating scheme, we find that when DC transmission power exceeds a certain amount, the continuous commutation failure of the inverter side will cause the sending end system to lose stability, as shown in Fig. 2.

When the commutation fails, sending end DC power cannot be sent out, in the short term, there is a large surplus of power, which will impact the sending end AC power grid. Moreover, in the process of commutation failure and recovery, DC

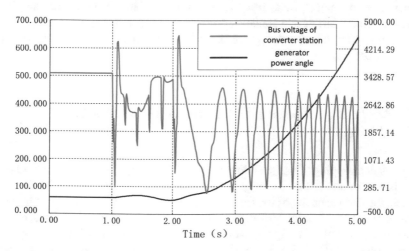

Fig. 2 Voltage fluctuation curve of converter station and power angle curve of generators

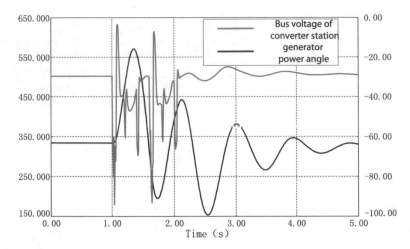

Fig. 3 Voltage fluctuation curve of converter station and power angle curve of generators after DC power is limited

system will absorb a great deal of reactive power from the system, which will also have a great impact on the receiving end power grid. Therefore, in order to ensure the stable operation of the sending end system and do not restrict the power supply of the sending end power, we take measures to restrict DC transmission power to weaken the impact of DC system commutation failure on sending end power grid. When the DC power is limited to a certain amount, the inverter undergoes continuous commutation failures, and the sending end system can remain stable, as shown in Fig. 3. We calculate using the same method in the way that two condensers are connected to the grid. The maximum power limit of the DC transmission in different operating modes under the two modes is shown in Table 2. From the data in the table can see that: After two condensers are configured in Ximeng converter station, the maximum DC transmission power limit is improved at about 30–70 Mvar under different operating modes.

3.3 Analysis of Voltage Fluctuation in Single End Trip of Line

Single-ended tripping of line, also known as line single end unwinding or load rejection, is a common failure in large power grid operation. Especially in UHV AC system, the earth capacity of the transmission line is very large. Once the single end of the line is tripped, it is equivalent to that the UHV long line is no load running, which is equivalent to a large capacity reactive power supply for the system. The system voltage level will be increased greatly, adding the charging power of the line itself, the voltage at the end of the line will be very high, which will seriously affect

Table 2 Maximum limit of DC transmission power under different generating scheme

Generators in group A	Generators in group B	The maximum limit of DC transmission power (MW)	The maximum limit of DC transmission power with two condensers (MW)
0	1	2500	3100
1	0	2100	2800
0	2	2800	3500
1	1	3100	3700
2	0	2200	2900
0	3	4000	4300
1	2	3400	4100
2	1	3300	4000
0	4	4400	4700
1	3	4300	4500
2	2	3600	4300
1	4	4600	4900
2	3	4500	4800
2	4	4900	5300

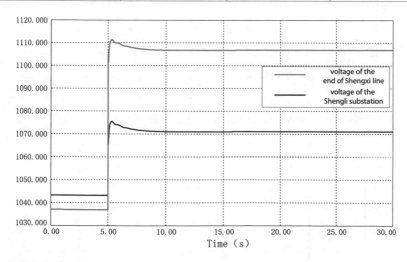

Fig. 4 Voltage fluctuation curve after the trip

the safe operation of the system. We select the line between Shengli substation and Ximeng substation at the end of UHV communication system (hereinafter referred to as the Shengxi line) as the research object, simulate the trip fault of Ximeng substation side in Shengxi line and observe the voltage fluctuation of the head and terminal end of the line, as shown in Fig. 4. From the data, we can see that: After the failure, the voltage of the Shengli substation is increased about 28 kV, the line capacity of the Shengxi line is increased about 36 kV and the terminal end voltage of the line is 1107 kV, which is beyond the high reactance withstand voltage level of the line. The voltage of shengli substation needs to be limited so that the system can run normally. We get the same fault voltage fluctuation when two condensers are running, as shown in Fig. 5. After the failure, the reactive power output of the condenser is shown in

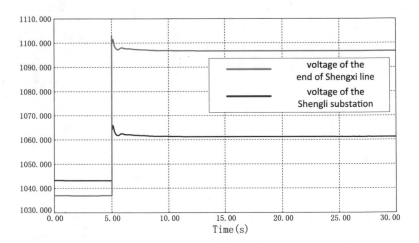

Fig. 5 Voltage fluctuation curve with two condensers after the trip

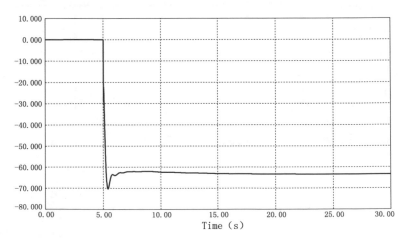

Fig. 6 Reactive power curve of the condenser

Fig. 6, each of which absorbs the reactive power of 63Mvar and plays a great role in supporting reactive power. We simulated the single-ended tripping fault of other lines. The conclusion is almost the same as that of Shengxi line.

4 Summary and Conclusion

This paper taking the regional power grid near Xitai DC as the research object calculates the power flow distribution, voltage change and transient state process of six generators in the sending end auxiliary power plant. The paper focuses on the analysis the influence of large capacity condenser on the stability level of the transmission grid in two aspects of the AC and DC transmission capacity and the line single end tripping after the two 300 Mvar synchronous condensers are configured in Ximeng converter station. The simulation results show that the synchronous condenser has good dynamic voltage support ability and can effectively improve the stability level of the AC and DC hybrid power grid. It provides the basis for the configuration and related research of the dynamic reactive power compensation mode of UHV DC transmission project. It provides an effective reference for the safe operation of large capacity condenser sets and the full play to its function in the AC and DC hybrid power grid.

Acknowledgements The authors thank North China Grid Company Limited for providing research support. Furthermore, we thank the editor as well as reviewers for suggestions and comments that helped to improve the manuscript.

References

1. Luo, L., Li, J., Sun, H.: HVDC wall bushing for solving nonlinear electric field. High Voltage Eng. **28**(5), 3–5 (2002)
2. Teleke, S., Abdulahovic, T., Thiringer, T., Svensson, J.: Dynamic performance comparison of synchronous condenser and SVC. In: IEEE Transactions on Power Delivery (2008)
3. Paris, O., Lewiner, J.: A Finite element method for the determination of space charge distributions in complex geometry. IEEE Trans. Dielectr. Electr. Insul. **7**(4), 556–560 (2000)
4. Katsuya, Y., Mitani, Y., Tsuji, K.: Power system stabilization by synchronous condenser with fast excitation control. In: Proceedings International Conference on Power System Technology (2000)
5. King, P.: The choice of fuel in the eighteenth century iron industry: the Coalbrookdale accounts reconsidered. Econ. Hist. Rev. **64**(1), 132–156 (2011)
6. Duan, J., Sun, Y., Yin, X.: Voltage stability' online prediction using WAMS. High Voltage Eng. **35**(7), 1748–1752 (2009)
7. Pilgrim, J.A., Lewin, P.L., Vaughan, A.S.: Quantifying the operational benefits of new HV cable systems in terms of dielectric design parameters. In: Proceedings of IEEE International Symposium on Electrical Insulation, pp. 261–265 (2012)
8. Adelfio, G., Chiodi, M., D'Alessandro, A., Luzio, D., D'Anna, G., Mangano, G.: Simultaneous seismic wave clustering and registration. Comput. Geosci. **44**, 60–69 (2012)

9. AL-khatib, M.A.S., Lone, A.H.: Acoustic lightweight pseudo random number generator based on cryptographically secure LFSR. Int. J. Comput. Netw. Inf. Secur. (IJCNIS), **10**(2), 38–45 (2018). https://doi.org/10.5815/ijcnis.2018.02.05

10. Agrawal, P., Sinha, S.R.P., Misra, N.K., Wairya, S.: Design of quantum dot cellular automata based parity generator and checker with minimum clocks and latency. Int. J. Mod. Educ. Comput. Sci. (IJMECS) **8**(8), 11–20 (2016). https://doi.org/10.5815/ijmecs.2016.08.02

11. Khani Maghanaki, P., Tahani, A.: Designing of fuzzy controller to stabilize voltage and frequency amplitude in a wind turbine equipped with induction generator. Int. J. Mod. Educ. Comput. Sci. (IJMECS) 7(7), 17–27 (2015). https://doi.org/10.5815/ijmecs.2015.07.03

Wide-Area Feedback Signal and Control Location Selection for Low-Frequency Oscillation

Xiaolu Chen, Xiangxin Li, Zheng Ren and Tingting Zheng

Abstract WAMS provides a high degree-of-freedom for wide-area damping control in the selection of feedback signals and control points. By selecting a feedback signal with strong controllability for the interval oscillation mode and a highly controllable generator set in the entire grid. The wide-area damping controller can more effectively suppress the low-frequency oscillation in the interval. For the main low-frequency oscillation mode of the system in the completed dynamic stability analysis, based on the identification technology, the main analog-to-digital ratio method is used to select the feedback signal, the residual number method is used to select the control point, the correlation gain matrix method is used to select the feedback signal, and the combination of control locations is evaluated overall to form a comprehensive alternative.

Keywords Low-frequency oscillation · Wide-area measurement system · Wide-area control location

1 Instruction

While the grid interconnection brings significant economic and social benefits, its large-scale and complex operational characteristics also pose new challenges to the power operation department. The low-frequency oscillation problem in the weakly damped interval is one of the challenges faced by large-scale grid interconnection, threatening the dynamic security of the grid [1].

The traditional method of suppressing low-frequency oscillations is to apply a power stability stabilizer (PSS) on the excitation side of the generator. However,

X. Chen (✉) · Z. Ren · T. Zheng
Electric Power Research Institute of State Grid East Inner Mongolia Electric Power Company Limited, Hohhot 010000, China
e-mail: 562782819@qq.com

X. Li
North China Dispatching Center of State Grid Corporation, Beijing 100053, China

© Springer Nature Switzerland AG 2020
Z. Hu et al. (eds.), *Advances in Artificial Systems for Medicine and Education II*,
Advances in Intelligent Systems and Computing 902,
https://doi.org/10.1007/978-3-030-12082-5_69

since the PSS uses local signals for feedback, the choice of signals lacks freedom, and coordination between multiple local controllers is also difficult. Therefore, in a system with a large number of PSS installations, the problem of interval low-frequency oscillation is still not effectively solved [1]. The development and popularization of wide-area measurement system (WAMS) technology has brought new opportunities for the study of damping control of low-frequency oscillations in power systems. It has changed the situation that traditional damping control can only select feedback signals locally, so that wide-area signals are used as damping. The controlled input signal is possible [2]. A large number of scholars have shown that wide-area control has obvious advantages over local control for suppressing low-frequency oscillations in weakly damped intervals.

The commonly used low-frequency oscillation detection algorithm can only perform low-frequency oscillation mode detection based on the oscillation data [3]. However, the power system is in steady-state operation for most of the time, without obvious disturbance. The main stimulus in the system is small load fluctuation. In this case, there is no obvious small fluctuation in the PMU measurement data of the system, but the data various oscillation modes similar to noise are superimposed, and these oscillation modes are very helpful for system stability analysis [4]. The development of a WAMS application with fast oscillating mode detection using ambience data (i.e., noise data) facilitates the timely detection of dangerous oscillating modes and damping of the system in accordance with slight disturbances in daily steady-state operation, and at the same time, in order to achieve the danger of discovery for the suppression and elimination of low-frequency oscillations, the method of modal analysis using atmospheric data should also be studied, that is, the participation mode and participation degree of each generator for oscillation.

This paper mainly introduces the wide-area feedback signal and control location selection in wide-area damping control [5]. According to the oscillation mode of the system, based on the identification technology, the feedback signal is selected by the method of the main-mode ratio, the control point is selected by the method of the residual number, and the combination of the alternative feedback signal and the control location is evaluated by the method of the correlation gain matrix. Form a comprehensive alternative.

2 Low-Frequency Oscillation Problems and Research Status

2.1 Low-Frequency Oscillation Problem

The weakly damped low-frequency oscillation problem of interconnected power grids belongs to the category of dynamic stability analysis. In actual operation, different requirements for low-frequency oscillation damping under different size disturbances are proposed. Systematic analysis of low-frequency oscillations is generally based on a power system linearization model. A lot of research and practice have shown that the low-frequency oscillation characteristics of the system disturbance are consistent with the eigenvalue analysis results under the linearization system [6]. This linearization analysis method has good applicability. In recent years, with the development of WAMS technology, the low-frequency oscillation characteristics of the system can be further enriched by performing online detection or off-line analysis such as Prony analysis, random disturbance data analysis, and Hilbert–Huang transformation on phasor measurement unit (PMU) measured data. Analytical means of low-frequency oscillations.

The frequency of the low-frequency oscillation of the power grid is generally in the range of 0.1–2.0 Hz. According to the oscillation participating in the unit and the oscillation frequency, it can be divided into two types: local oscillation and interval oscillation. Local oscillation is the relative sway between a unit or a power-plant unit and other units in the system. The typical oscillation frequency is between 0.7 and 2.0 Hz. With the relative oscillation of the generator, the typical oscillation frequency is 0.1–0.7 Hz. In contrast, the interval oscillation mode problem is more complicated and harmful. Interval oscillation mode will spread throughout the network once it occurs, such as improper control measures will cause serious grid accidents.

2.2 Research Status of Low-Frequency Oscillation Damping Control

According to the different control objects, the damping control can be divided into: (1) generator excitation additional damping control; (2) generator speed regulation additional damping control; (3) high-voltage DC additional damping control; (4) flexible AC transmission system (FACTS) device. Additional damping control, including additional damping control of devices such as Thyristor controlled series compensator (TCSC), static Var compensator (SVC), and static synchronous compensator (STATCOM). Among them, generator excitation additional control is the most common damping control method; the current speed control damping control is still in the research stage, there is no engineering example; the most successful application example of FACTS device additional damping control is the engineering application of TCSC additional damping control in Brazil. The additional damping control of the HVDC transmission system is a unique damping resource in the AC–DC hybrid power grid. The most successful example of DC additional damping control is the US Pacific Linkline project, which increases the transmission power of the AC tie line by 400 MW by suppressing low-frequency oscillations.

Wide-area adaptive damping control is a combination of wide-area damping control and adaptive control. It involves key technologies such as wide-area measurement, wide-area time-delay system damping control, and adaptive control. The following is a brief overview of WAMS and wide-area adaptive damping control with the development status of key technologies.

3 Research Status of Wide-Area Damping Control

3.1 WAMS Development and Application

The wide-area measurement system (WAMS) is a real-time monitoring system based on synchronous phasor measurement technology that targets the dynamic process monitoring, analysis, and control of power systems. WAMS has the characteristics of high-precision synchronous phasor measurement, high-speed communication, and fast response. It is very suitable for real-time monitoring of dynamic processes in long-span interconnected power grids. WAMS has excellent performance in different places, high speed, and synchronization. It has been applied in many fields of power system, including: load identification, state estimation, voltage stability and power angle stability online warning, reactive power integrated control, power flow adjustment and frequency modulation control. It also shows multiple promising directions such as fault location detection, wide area control and protection.

WAMS provides a high degree-of-freedom for wide-area damping control in the selection of feedback signals and control points. By selecting a feedback signal with strong controllability for the interval oscillation mode and a highly controllable generator set in the entire network. The wide-area damping controller can more effectively suppress the low-frequency oscillation in the interval.

3.2 Wide-Area Damping Control System Structure

At present, the damping control structure in the power system is mainly divided into two categories: the first type is local control, and all the traditional PSS are local control. Local control is distributed throughout the system, mainly using local machine end signals such as rotor angle, machine end power, and bus frequency as feedback signals to the local PSS to achieve control of the local subsystem. The design of the controller is based on local models and measurement signals. This distributed local control is effective for suppressing the oscillation of the local subsystem, but often cannot meet the requirements of the high-frequency interconnected system for the low-frequency oscillation damping. The other type is wide-area damping control, and wide-area damping control collects wide-area information from the entire network as a control input signal. Compared with the local machine signal, the wide-area signal can fully reflect the overall dynamic characteristics of the system, and the observability of the interval oscillation mode is strong. Therefore, the wide-area damping control can effectively exert the suppression effect on the interval low-frequency oscillation. These two types of damping control coexist in interconnected systems, where wide-area damping control is usually added to the local damping control, and together with the local PSS constitutes a layered control structure.

Wide-area damping control can be divided into two parts: data acquisition pre-processing module and control module. The data acquisition preprocessing module mainly collects and preprocesses the PMU measurement data distributed in the whole network and transmits it to the control module, which is divided into centralized and distributed. The control module designs the controller and calculates the control output based on the received wide-area signal. The control module is also divided into centralized and distributed. The structure of the wide-area damping control system can be divided into the following three categories according to the distribution of the acquisition preprocessing module and the control module: wide-area centralized control of data collection, wide-area distribution control of data distribution acquisition, and wide-area distribution control of data collection.

4 Wide-Area Feedback Signal and Control Location Selection

The overall process of selecting the feedback signal and the control location is: for the main low-frequency oscillation mode of the system in the completed dynamic stability analysis, based on the identification technology, the main analog ratio method is used to select the feedback signal, and the residual number method is used to select the control point. The method of the correlation gain matrix evaluates the combination of the alternative feedback signal and the control location as a whole to form a comprehensive alternative.

4.1 Identification and Wide-Area Feedback Signal Selection Method

The set of alternative feedback signals is mainly determined by the nodes in the current system where the PMU is installed. All line power signals, generator power angle difference signals, and bus voltage frequency signals that can be measured by the PMU can be used as an alternative.

In the selection of feedback signals, a commonly used power system oscillation identification method, Prony method, is used. The basic principle is briefly described as follows. The Prony method is a method of describing the original sampled signal using an exponential function linear combination model. The Prony algorithm itself has high-frequency resolution and can obtain quantitative information such as frequency, attenuation factor, amplitude, and phase of the signal. It can be used for the analysis of simulation results as well as for real-time measurement of data.

This method was proposed by the French mathematician Prony in 1795, for the form:

$$f(x) = C_1 e^{a_1 x} + C_2 e^{a_2 x} + \cdots + C_n e^{a_n x} \triangleq C_1 \mu_1^x + C_2 \mu_2^x + \cdots + C_n \mu_n^x \quad (1)$$

The model, $f(x)$ consists of N points of equal time interval, and N algebraic equations can be obtained.

$\mu_1, \ldots, \mu_n$ is the root of the following algebraic equation:

$$\mu^n + a_1\mu^{n-1} + \cdots + a_{n-1}\mu + a_n = 0 \tag{2}$$

According to the difference of the number of N, using a solution to the algebraic equations or the least squares method to calculate a linear prediction problem, the coefficient $a_1, \ldots, a_n$ can be obtained, substituted into the above formula, and $\mu_1, \ldots, \mu_n$ can be obtained, and then substituted into the former formula, then the coefficients $C_1, \ldots, C_n$ are obtained. This process is also called the Vandermonde problem, and finally the linear combination model of the original exponential function is obtained.

Use the main-mode ratio indicator to select the alternative feedback signal. The dominant-mode ratio indicator is a quantitative description of the relative intensity of the specified oscillation mode in the signal. It is the observed value of the quantized signal to the specified mode, reflecting the observability of the signal to a particular oscillation mode. The calculation of the main-mode ratio is based on the Prony analysis result of the response signal generated by the excitation in the simulation. For example, for a candidate signal y_i, Prony analysis is performed to obtain $y_i = \sum_{j=1}^{m} A_{ij} e^{\alpha_{ij} t} \cos(2\pi f_j t + \theta_j)$, then the main-mode ratio of the pattern y_i of the signal k is defined as:

$$\text{DMR}_{ik} = 20\lg \frac{\int_0^T |A_{ik} e^{\alpha_{ik} t} \cos(2\pi f_k t + \theta_k)| \, dt}{\sum_{j=1}^{m} \int_0^T |A_{ij} e^{\alpha_{ij} t} \cos(2\pi f_j t + \theta_j)| \, dt} \tag{3}$$

For discrete-time signals, $Y_i = \sum_{j=1}^{m} A_{ij} e^{\alpha_{ij}(n\Delta t)} \cos(2\pi f_j(n\Delta t) + \theta_j)$

$$\text{DMR}_{ik} = 20\lg \frac{\sum_{n=1}^{N} |A_{ik} e^{\alpha_{ik}(n\Delta t)} \cos(2\pi f_k(n\Delta t) + \theta_k)| \Delta t}{\sum_{j=1}^{m} \sum_{n=1}^{N} |A_{ij} e^{\alpha_{ij}(n\Delta t)} \cos(2\pi f_j(n\Delta t) + \theta_j)| \Delta t} \tag{4}$$

If a very accurate calculation is not required, the calculation definition of the main-mode ratio can be simplified. First, the Prony analysis result of the candidate signal is filtered to:

$$y_i \approx y_i' = \sum_{j=1}^{m} A_{ij} e^{\alpha_{ij} t} \cos(2\pi f_j t + \theta_j), \quad (\alpha_{ij} \geq \text{const}, \ j = 1, 2, \ldots, m) \tag{5}$$

$$\text{DMR}_{ik}' = 20\lg \frac{A_{ik}}{\sum_{j=1}^{m} A_{jk}} \tag{6}$$

For discrete-time signals, the simplified calculation defines the same as above.

In the actual calculation, the candidate feedback signal is first determined, then different faults or excitations are set in the system, time-domain simulation calculation is performed, and the response of all candidate feedback signals is recorded, and the response of all candidate signals under different disturbances or excitations is recorded. The curve is Prony analysis. According to the Prony analysis results, the main-mode ratios of the main low-frequency oscillation modes in the system are calculated, respectively. Then, the feedback signals are averaged for the main-mode ratios of the oscillation modes under different faults. The average value selects the partial feedback signal that is best for the appreciability of the mode in different oscillation modes.

4.2 Wide-Area Control Location Selection Method

The range of alternative control points is mainly determined by the structural characteristics of the power grid, and the distribution of the main oscillation modes in the dynamic stability analysis needs to be considered. In the case of ensuring that the distribution of the alternative control points covers the major geographic locations of the system, a generator with a large capacity is preferentially selected as an alternative control point.

Use the residual indicator to select alternative control points. The equation of state of the system is:

$$\begin{cases} \dot{x} = Ax + Bu \\ y = cx \end{cases} \tag{7}$$

The corresponding transfer function is:

$$G(s) = \frac{y(s)}{u(s)} = C(sI - A)^{-1}B \tag{8}$$

Decoupling state variables

$$G(s) = \frac{y(s)}{u(s)} = C\Phi(sI - \Lambda)^{-1}\Psi^T B = \sum_{i=1}^{n} \frac{R_i}{s - \lambda_i} \tag{9}$$

In $R_i = c\phi_i\psi_i^T b$, c reflects the output y on the observability of the pattern i, b reflects the controllability of the input u on the pattern i.

Identifying the transfer function of each candidate control will lead to the same feedback signal under different system faults or specific excitations. Since the feedback signal has been determined in the first step, it is said that the signal is observable to an oscillation mode. It has been determined that the transfer function of all the candidate points to the same feedback signal is recognized, and the residuals of the corresponding oscillation modes are calculated, and the size of the residuals directly reflects the controllability of the control modes for the oscillation modes, so the comparison calculation is obtained. The size of the remaining number can select a control point with better controllability for each oscillation mode.

4.3 Multi-controller Interaction Evaluation Method

After the screening of the candidate feedback signal and the candidate control point is completed, some feedback signals with the best approximation to the mode in each oscillation mode and some control points with the best controllability of the mode are obtained. These feedback signals and control points are combined to form a feedback control loop.

After the control scheme is determined, the controller's parameter design will be carried out by the mature single-input single-output system controller design method. In order to reduce the mutual influence between different control channels, the single-input single-output controller design achieves the desired effect. It is necessary to select the control-channel matching scheme with the least influence on each other when matching the feedback control loop. This effect can be characterized by the associated gain matrix.

Assuming that the transfer function matrix of the system is G, the element λ_{ij} in the definition gain matrix RGA is $\lambda_{ij} = G_{ij} G_{ji}^{-1}$, which is the definition and calculation method of the correlation gain matrix. The correlation gain matrix reflects the relationship between each input signal and each output signal. The value of each element in the RGA matrix calculated at a specific frequency is closer to 1 at the corresponding position of an input/output channel. The channel's signal to this frequency is affected by the interaction of other channels.

Since the feedback channel is configured according to λ_{ij}, the selection criterion is the proximity of λ_{ij} to 1, which is inconvenient to use. Therefore, based on the RGA matrix, an NRGA matrix is proposed for normalizing the elements, where the elements are defined as:

$$\lambda'_{ij} = \begin{cases} 0, & \left(\lambda_{ij} \leq 0\right) \\ \lambda_{ij}, & \left(0 < \lambda_{ij} \leq 1\right) \\ f\left(\lambda_{ij}\right), & \left(1 < \lambda_{ij}\right) \end{cases} \tag{10}$$

where $f(1) = 1$, $\lim_{\lambda \to +\infty} f(\lambda) = 0$, e.g., $f(\lambda) = e^{(1-\lambda)/4}$. All elements in the NRGA matrix are between each other, and the magnitude of the value reflects the closeness to 1. Therefore, selecting the control loop corresponding to the most significant part of the NRGA matrix can best guarantee the control scheme. The interaction between the loops is minimal.

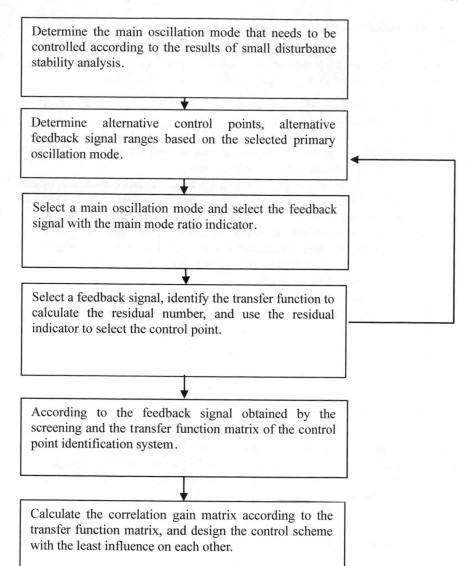

Fig. 1 Control scheme design flow chart

4.4 Overall Selection and Optimization Process

In summary, the overall selection process of the feedback signal and control location of the wide-area control system can be briefly described as follows:

(1) Determine the main low-frequency oscillation mode in the system according to the small disturbance stability analysis result;
(2) According to the configuration of the PMU in the system, the capacity and geographical distribution of each generator set determine the range of alternative feedback signals and alternative control points;
(3) Using the main-mode ratio indicator to filter the candidate feedback signals, and obtaining a relatively large feedback signal of the main-mode ratio in each oscillation mode;
(4) Using the residual indicator to filter the candidate control points, and identify the transfer function from the alternate control point to each feedback signal (the feedback signal is determined by step 3), and obtain the corresponding residual number in each oscillation mode;
(5) Using the correlation gain matrix, the comprehensive screening control effect is good and the control signal pairing scheme with the least influence between each feedback control loop is the least.

The flow chart of this process is shown in Fig. 1.

5 Conclusion

This paper mainly introduces the wide-area feedback signal and control location selection in wide-area damping control. According to the oscillation mode of the system, based on the identification technology, the feedback signal is selected by the method of the main-mode ratio, the control point is selected by the method of the residual number, and the combination of the alternative feedback signal, and the control location is evaluated by the method of the correlation gain matrix. Form a comprehensive alternative.

References

1. Kundur, P.: Power System Stability and Control (2006)
2. Li, W., Gardner, R.M., Dong, J., Wang, L., Xia, T., Zhang, Y., et al.: Wide area synchronized measurements and inter-area oscillation study. In: Power Systems Conference and Exposition, 2009. PSCE '09. IEEE/PES, pp. 1–8. IEEE (2009)
3. Kao, W.S.: The effect of load models on unstable low-frequency oscillation damping in Taipower system experience w/wo power system stabilizers. In: Power Engineering Society Summer Meeting, vols. 3, 16, p. 1585. IEEE (2001)

4. Han, Z.X., Zhu, Z.L., Tian, X.S., Li, F.: Analysis and simulation research on power system low frequency oscillation. In: International Conference on Computer Modeling and Simulation, vol. 2, pp. 223–228. IEEE (2010)
5. Milanovic, J.V.: Damping of the low-frequency oscillations of the generator: dynamic interactions and the effectiveness of the controllers. IEE Proc. Gener. Transm. Distrib. **149**(6), 753–760 (2003)
6. Breulmann, H., Grebe, E., Lösing, M., Winter, G.W., Witzmann, R., Dupuis, P., et al.: Analysis and Damping of Inter-area Oscillations in the UCTE/CENTREL Power System. Researchgate Net (2000)
7. Cai, J., Shen, L.: Adaptive compensation of unknown actuator failures for strict-feedback systems. Int. J. Mod. Educ. Comput. Sci. (IJMECS) **3**(5), 10–17 (2011)
8. Lian, X., Liu, Z., Wang, Z.: A modified T-S model fuzzy adaptive control system based on genetic algorithm. Int. J. Inf. Technol. Comput. Sci. (IJITCS) 3(3), 8–14 (2011)

Voltage Quality Evaluation of Distribution Network and Countermeasure

Wei Deng, Yuan Ji Huang, Liang Zhu, Miaomiao Yu and Rongtao Liao

Abstract With the increasing penetration rate of wind power generation in distribution network, the voltage quality problem caused by wind turbine grid connection cannot be ignored. Firstly, we analyze the influence factors of distributed power supply access to distribution network on voltage impulse. Then, based on the PSCAD, the characteristics of the turbine's grid-connected operation and the voltage quality of the distribution network were simulated and analyzed. Finally, in view of the adverse effects of connection of wind turbine on the distribution network, measures to improve the voltage quality of the distribution network are given, including reasonable selection of the capacity of the wind power plant, suitable location and connection mode, load switching, and using of proper reactive compensation equipment.

Keywords Wind power generation · Distribution network · Voltage quality · PSCAD

W. Deng (✉)
State Gird Hunan Electric Power Company Limited Research Institute, Changsha 410007, Hunan, China
e-mail: 35048542@qq.com

Y. J. Huang
State Grid Hunan Electric Power Company Limited Changsha Power Supply Company, Changsha 410015, Hunan, China

L. Zhu
State Grid Hunan Electric Power Company Limited, Changsha 410004, Hunan, China

M. Yu
Wuhan University, Wuhan, China

R. Liao
Information and Communication Branch of State Grid, Hubei Electric Power CO. LTD., Changsha, China

© Springer Nature Switzerland AG 2020
Z. Hu et al. (eds.), *Advances in Artificial Systems for Medicine and Education II*,
Advances in Intelligent Systems and Computing 902,
https://doi.org/10.1007/978-3-030-12082-5_70

1 Introduction

With the increasingly tight energy supply situation, clean energy has received extensive attention. Among them, wind power has become the fastest growing renewable energy source in recent years with the characters of green and environmental. Traditional wind power utilization is usually centralized power generation, with wind farm connected to power grid, which faces the problem of considerable waste of wind and grid-connected transmission [1–3]. Therefore, the distributed access of wind power has become the focus of scholars at home and abroad. It can be foreseen that with the development of wind power technology, wind power generation, as a distributed generation (DG), will penetrate higher and higher in the future distribution network. However, due to the random variation of wind energy, the power output of the wind turbine is highly volatile, while the traditional medium- and low-voltage distribution network tends to have limit capacity and the grid is weak. Therefore, the change of wind power output will seriously affect the voltage quality of the distribution bus of the distribution network [4–6]. So it is of great significance to study the voltage quality problems caused by wind power access to the distribution network.

In order to improve the grid-connected capacity of wind power, most research on wind power generation is focused on its impact on the system, including the impact of the integration of large-capacity wind farms into the secondary transmission network on the operation and stability of the power system. Positive conclusions and countermeasures have been obtained, which greatly improves the application level of wind power generation. However, there are few researches on distributed wind power generation to connect to low-voltage distribution networks, and it is generally limited to static voltage stability or relay protection for distribution networks [7–9], but less consideration on the quality of power supply. The literature [10] used the short-circuit ratio and stiffness ratio to evaluate the impact of DG access on the voltage quality of the distribution network, but did not consider the specific operational characteristics of the wind turbine and lacked countermeasures.

This paper first analyzes the influencing factors of the distributed power supply distribution network to the voltage impulse and then combines the 13-node example of the distribution network; based on the PSCAD software, the grid-connected operation characteristics of the wind turbine, the influence to the connection point, and voltage quality of the near nodes were simulated and analyzed. Finally, in view of the adverse effects of connection of wind turbine on the distribution network, measures to improve the voltage quality of the distribution network are given, including reasonable selection of the capacity of the wind power plant, suitable location and connection mode, load switching, and using of proper reactive compensation equipment.

2 Analysis and Evaluation of the Impact on Supply Voltage Quality

2.1 Voltage Quality Indicator

Distribution network voltage quality problems caused by distributed power supply mainly include voltage deviation and voltage fluctuation, which are described as follows [11].

The voltage deviation refers to the difference between the actual voltage value and the rated voltage at a certain point in the distribution network. Generally, the difference is defined by the ratio of this difference to the rated voltage:

$$\Delta U(\%) = \frac{U - U_N}{U_N} \times 100\% \tag{1}$$

where U is the effective value of the actual voltage; U_N is the rated voltage value of the point. *GB/T_12325-2008* stipulates that the voltage deviation of the power supply and distribution network of 10 kV and below is less than $\pm 7\%$.

Voltage fluctuations refer to a series of variations or continuous changes in the voltage value, usually expressed as voltage variations. Voltage variation is the difference between two adjacent extremes on the voltage curve, expressed as a percentage of the system's nominal voltage:

$$d = \frac{U_{\max} - U_{\min}}{U_N} \times 100\% \tag{2}$$

where $U_{\max}$ and $U_{\min}$ are respectively the extreme values adjacent to the effective value of the measuring point voltage. *GB/T 12326-2008* specifies that the distribution network of 10 kV and below allows the supply voltage fluctuation to be 1.25–4%.

2.2 Analysis of Factors Affecting Power Supply Quality

The voltage distribution of the power grid is determined by the power flow of the power grid. Once the injection power and load consumption in the power grid change, the voltage deviation and voltage fluctuation of the nodes of the power grid will be generated. The root cause of distributed wind power affecting the quality of the supply voltage is the fluctuation of the distributed power.

Since the distributed power supply generally has the greatest impact on the grid-connected point (PCC), the PCC is selected as the research object, and the principle of voltage fluctuation caused by grid-connected DG power fluctuation is analyzed [10].

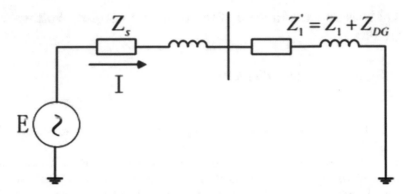

Fig. 1 Equivalent circuit transformed through Thevenin theorem

The system with distributed power is transformed into its equivalent circuit through Thevenin theorem at the grid point, as shown in Fig. 1.

When the distributed power source is incorporated into the distribution network, the original system injection power is changed, and the current variation ΔI on the line is used to describe $\Delta I = \Delta I_p + \Delta I_q$. According to the equivalent circuit in the above figure, the amount of voltage change at the point where the DG merges is derived as follows:

$$\Delta U_{DGF} = (R_k + jX_k) \cdot \left(\Delta I_p + j\Delta I_q\right) = |Z_k|(\cos\varphi + j\sin\varphi) \cdot$$
$$|\Delta I|(\cos\theta + j\sin\theta) = \frac{U^2}{S_k}\frac{\Delta S}{U}[\cos(\varphi + \theta) + j\sin(\varphi + \theta)] \qquad (3)$$

where U is the rated voltage of the DG integration point; S_k is the short-circuit capacity at the DG access point; S_n is the variation of the distributed injection power; $Z_K = R_K + jX_K$ is the equivalent impedance of the Thevenin grid from the DG integration point. φ is the grid impedance angle; θ is the DG power factor angle.

Usually, the phase shift at both ends of the line is not large, so the vertical component of the voltage change is ignored, and the horizontal component is used to approximate ΔU_{DGF}, thereby calculating the relative voltage change rate.

$$d = \frac{\Delta U_{DGF}}{U} \approx \frac{\Delta S\cos(\varphi + \theta)}{S_k} \times 100\% \qquad (4)$$

From the above analysis, the power supply voltage impact of a single distributed power source on the system is related to three factors: the change in injection power, the short-circuit capacity of the incorporated system, and the power factor of the distributed power source.

2.3 Calculation of Short-Circuit Capacity

The short-circuit capacity refers to the apparent power of the power system in the specified operating mode, when the three-phase short circuit is concerned. The system short-circuit capacity is a sign of the system voltage strength, and if the short-circuit capacity is large, the network is strong, then the starts and stops of distributed power supply, the fluctuations of power generation fluctuations and other issues will not cause too much voltage change. Three-phase short-circuit capacity is calculated as follows [12]:

$$S_k = \sqrt{3} U_N I_f \tag{5}$$

where S_k is the three-phase short-circuit capacity; U_N is the effective value of the rated line voltage before the fault point fault; I_f is the effective value of the fault point line current. For a certain system, U_N is a constant, and the current value is usually used directly to characterize the breaking capacity of the circuit breaker.

3 Case Study Analysis

3.1 Simulation Test System and Parameters

The simulation tool used in this paper is the transient analysis software PSCAD. The example system used is the IEEE distribution network 13-node test system [13]. The example has two voltage levels of 4.16 and 0.48 kV, with an asymmetric load, and the nodes 645 and 646 are both two phases. 652, 692, 611 is a single-phase system with obvious asymmetry, which is appropriate enough to represent the typicality of distribution network (Fig. 2).

The research object considered is the asynchronous wind turbine. The output is 1 MW, the gear ratio is 68, the fixed pitch control, no reactive compensation reimbursement device. The wind speed model uses a four-component model [14], which uses the average wind, array wind, slope wind and random wind to simulate natural changes in wind speed, average wind under the action of 13 m/s, the fan reaches the rated power.

3.2 Influence of Wind Turbine Access on Power Quality During Steady-State Operation

In order to study the operating characteristics of the fan, the average wind speed is from 9 to the 13 m/s, records the fan output characteristics as shown in Table 1.

Fig. 2 IEEE power
distribution 13-node test
system

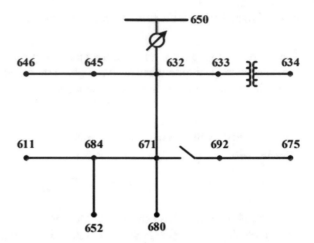

Simultaneously, take the rated power operation of the wind turbine and the half-rated power operation as examples. Figure 3 shows the effect of wind turbines in the system on the amplitude of the node voltage in the distribution network when the power system is in stable operation. During the simulation process, the wind turbine accessing points are all 671 nodes.

It can be seen from Table 1 that it sends out the active power to the distribution network after the fan starts, and a small amount of reactive power is also absorbed; meanwhile, the power factor also varies. From Fig. 3, in the absence of reactive compensation, due to reactive absorption, the voltage at the distribution network node drops and with the wind turbine sends out an increasing number of active power. As the power increases, the voltage at the distribution network node drops more. Therefore, when the active output of wind power increases continuously, it will cause the voltage at the grid connection of the wind turbine and other distribution network nodes to exceed the allowable lower limit, which will cause voltage deviation and affect the power quality of the distribution network.

Table 1 Fan output
characteristics under different
wind speeds

Wind speed m/s	Fan output active power (MW)	Fan absorbs reactive power (Mvar)	Power factor
9	0.30	0.20	0.82
10	0.50	0.27	0.87
11	0.64	0.36	0.88
12	0.82	0.47	0.86
13	1.00	0.62	0.85

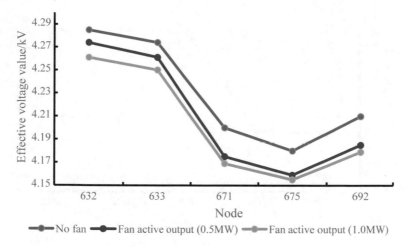

Fig. 3 Influence of steady-state operation of fan on distribution network voltage

3.3 Influence of Wind Turbine Power Fluctuation on Voltage Quality

The parameters of the wind turbine and access location are the same as in Sect. 3.2, considering the impact of the wind turbine on the distribution network [14–18] node under gust factors.

The output power characteristics of the wind turbine are shown in Fig. 4a. The voltage fluctuation of the node 671 is shown in Fig. 4b. Change the number of the wind turbines' simulation curve and other node fluctuations as shown in Fig. 5.

It can be seen from Fig. 4 that under the action of gusts, the active output of the asynchronous fan increases, and the reactive power of the absorption increases slowly. The effective value of the voltage at the grid connection point of the 671 node decreases and fluctuates. It can be seen from Fig. 5 that the change of the fan output has the most obvious influence on the load node (675 nodes) at the end of the feeder, and the upstream node of the feeder (the 632 node) has the smallest fluctuation. At

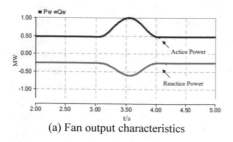

(a) Fan output characteristics

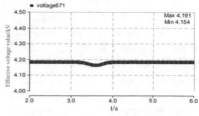

(b) No. 671 node voltage fluctuation

Fig. 4 Gust effect simulation curve

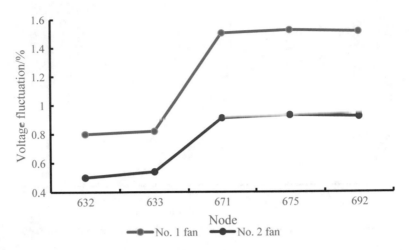

Fig. 5 Effect of total fan capacity on voltage fluctuations

the same time, with the increase of the number of fans, that is, the increase of the total installed capacity of the fan, the voltage fluctuations of the nodes in the distribution network become more and more obvious, which may exceed the upper limit of the fluctuation, which affects the quality of the distribution network.

4 Research on Countermeasures to Improve Voltage Quality

4.1 Impact of Fan Access Location on Voltage Quality

According to the theoretical analysis in Sect. 2.2, the short-circuit capacity of the access point will affect the voltage impact of the DG on the access point. The short-circuit capacity of different nodes of the distribution network is different. Therefore, it is theoretically possible to improve the voltage fluctuation by changing the access position. In order to make the distributed power access location more representative, the 632, 671, and 675 nodes are selected to study the influence of different access locations of the wind turbine on voltage fluctuations. The three nodes are the upstream nodes of the same feeder in the distribution network. The intermediate node and the end load node can represent the typicality of the DG access location. According to the short-circuit capacity calculation formula (5), the short-circuit capacity values of the three nodes are obtained, as shown in Table 2. In the simulation process, the fan power fluctuation is the same as that in Fig. 4b; that is, the wind turbine generates 0.5 MW power fluctuation, and the voltage fluctuation of the three nodes is recorded. The simulation data is shown in Fig. 6.

Table 2 Node short-circuit capacity

Node	Short-circuit capacity(MVA)
632	48.5
671	24.7
675	22.1

As can be seen from Fig. 5, when the fan access position is different and when the same power fluctuation occurs, the DG node at the end of the distribution network 675 has the greatest influence on the voltage quality of each node, and the upstream node of the access 632 has little influence on the node voltage. It can be seen from Table 2 that the short-circuit capacity of the 675 node is the smallest, and the short-circuit capacity of the 632 is the largest, which is in accordance with the analysis results in Sect. 2.2. Therefore, when selecting a distributed power access node, it is recommended to select the fan access point as a node with a short-circuit capacity that is close to the upstream of the feeder of the distribution network.

4.2   Influence of Fan Access Mode on Voltage Quality

There are many ways to access distributed power. This paper focuses on two methods: centralized access and distributed access. In the simulation process, firstly, two fans are connected to the 671 node to record the voltage fluctuations, and then simulate the voltage fluctuation caused by the single fan connected to the 671 node and the 632

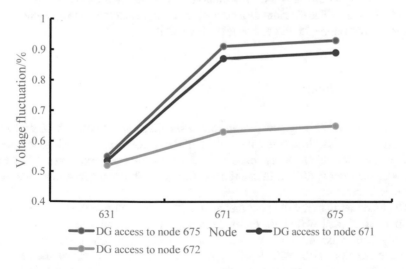

Fig. 6 Voltage fluctuations caused by different location access

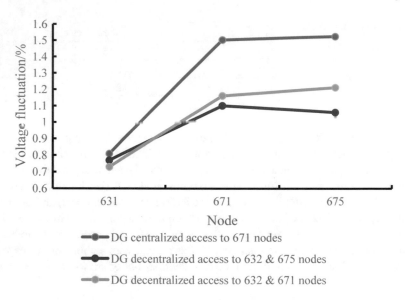

Fig. 7 Impact of different access methods on voltage fluctuations

or 675 nodes, respectively. The power fluctuation of the single fan is 2.3 (Fig. 4a). The simulation results are shown in Fig. 6.

As can be seen from Fig. 7, the distributed access of the fan can effectively reduce the voltage fluctuation amplitude compared to the centralized access mode. At the same time, it can be seen that the determination of the distributed access location also has an impact on the voltage fluctuation. The simulation results are the same as those in Sect. 3.1. The distributed access should also select the node upstream of the feeder and the node with large short-circuit capacity.

4.3 Load Switching

Load switching can change the power flow of the distribution network. Therefore, it is theoretically possible to adjust the voltage of the distribution network node through load control. Set the fan access node and the load switching node to be 671 nodes. Consider the impact of load fluctuation on the grid. The simulation parameters are shown in Table 3.

The load fluctuation will affect the voltage effective value of the fan access node, and the cutoff load will cause the node voltage to rise, while increasing the load capacity will cause the node voltage to drop. Therefore, when the wind farm is connected to change the voltage of the grid point, the voltage can be increased (or lowered) by switching the load near the fan.

4.4 Reasonable Choice of Installed Capacity

It can be seen from Sects. 3.2 and 3.3 that the voltage fluctuation of the grid connection point of the distribution network is related to the total installed capacity of the wind turbine. The greater the total active output of the wind turbine, the more obvious the impact on the voltage of the grid connection point. Therefore, if the wind farm is to be reduced, for the influence of voltage fluctuations at the grid connection point, it is necessary to limit the installed capacity of the fan. For this reason, in the planning stage of the wind farm, the maximum installed capacity of the wind turbine can be determined according to the capacity of the distribution network, so that the influence of the wind turbine access on the power quality of the distribution network is in line with national standards. According to the provisions of GB 12325-2008, the allowable limit of the voltage deviation of the distribution network of 10 kV and below is ±7%; that is, if the voltage of the distribution network node exceeds ±7%, the voltage of the node is considered to be exceeded. This can be used to plan the maximum grid-connected capacity of the wind turbine. Of course, other factors must be considered, such as the surge of the wind turbine and the voltage sag caused by the fan start.

4.5 Input Reactive Power Compensation

When the asynchronous fan is running, it can both output reactive power and absorb reactive power. The power grid requires the fan to operate under a higher power factor. Therefore, it is usually necessary to install a reactive power compensation device for the fan. The simulation installs a capacitor bank at the fan port for group switching. The fan access point is node 671, and the reactive power compensation value is changed. The simulation data is shown in Table 4.

It can be seen from Table 4 that after the compensation capacitor is input, the reactive power required by the fan can be effectively compensated, the power factor of the fan is increased, and the voltage of the grid point is increased. At the same time, it can be obtained from the fan power factor and the grid-connected point voltage data. The influence of the asynchronous fan on the access voltage deviation during

Table 3 Load switching simulation parameters	Object	Original load capacity/MVA	Load capacity after switching/MVA	Load power factor
	Cutting load	1.33	0	0.87
	Increase load	1.33	1.5	0.87

Table 4 Simulation data of different capacitance compensation capacities

Reactive power compensation capacity (Mvar)	Uncompensated voltage (kV)	Active/reactive power from the fan before compensation (MW/Mvar)	Compensated voltage (kV)	Precompensation power factor	Compensated power factor
0.5	4.17	1.0/−0.6	4.22	0.85	0.99
0.3	4.17	1.0/−0.6	4.21	0.85	0.97
0.1	4.17	1.0/−0.6	4.19	0.85	0.91

steady-state operation is related to the wind farm power factor, which is the same as the theoretical analysis conclusion in Sect. 2.2.

5 Conclusions

This paper first analyzes the influencing factors of distributed power supply on the voltage impulse; then, based on the 13-node example of distribution network, using PSCAD software, the operation characteristics of wind turbine grid connection and the voltage quality of distribution network node were analyzed. Finally, in view of the adverse effects of fan access on the distribution network, measures to improve the voltage quality of the distribution network are given. The main conclusions are as follows:

(1) The power supply voltage impact of the distributed power supply on the system is related to three factors: the change of the injected power, the short-circuit capacity of the incorporated system, and the power factor of the distributed power supply, which can be estimated by Eq. (4).

(2) In the absence of a reactive power compensation device, the asynchronous fan access during steady-state operation will reduce the voltage of the distribution network node, and the magnitude of the voltage drop is related to the operating state of the fan.

(3) The change of the output power of the wind turbine will cause the voltage fluctuation of the grid connection point and the adjacent nodes, and it has less influence on the upstream node of the feeder, and the impact on the end load node is the most serious.

(4) To suppress the influence of fluctuations in power generation on the supply voltage of the distribution network, it is necessary to select a reasonable access mode, access location, and access capacity. The access location should select the upstream of the power flow direction of the distribution network feeder, and the node with large short-circuit capacity, and distributed power is distributed

across different nodes. At the same time, load switching and dynamic reactive power compensation equipment input can also effectively improve the voltage quality.

References

1. Li, F., Wang, B., Tu, S., et al.: Peaking characteristics of wind power generation grid connection in Beijing-Tianjin-Tangshan Power Grid Analysis. Power Grid Technol. **18**(3), 128–132 (2009)
2. Lei, A.: Research topics related to wind power integration. Power Syst. Autom. **08**(12), 84–89 (2003)
3. Shi, K., Huang, W., Hu, Y., et al.: Stator double winding induction motor wind power generation system Analysis of low voltage ride through characteristics. Power Syst. Autom. **17**(4), 28–33 (2012)
4. Dong, W., Bai, X., Zhu, N., et al.: Intermittent power supply in grid-connected environment quality problem research. Power Grid Technol. **37**(5), 1265–1271 (2013)
5. Yu, J., Chi, F., Xu, K., et al.: The impact of distributed power access on the grid Analysis. J. Power Syst. Autom. **24**(1), 138–141 (2012)
6. Sun, T., Wang, W., Dai, H., et al.: Voltage fluctuations and flicker caused by wind power generation. Power Grid Technol. **27**(12), 62–66 (2003)
7. Hu, W., Wei, W., Xia, X., et al.: Multiple distributed power sources considering voltage adjustment constraints calculation of power input. Proc. CSEE **19**(25), 13–17 (2006)
8. Wang, J., Yan, N., Song, K., et al.: Study on the admission capacity of distributed power supply considering relay protection action in distribution network. Proc. CSEE **22**(6), 37–43 (2010)
9. Fei, P., Kun, S., Li, K., et al.: Influence and improvement of distributed power supply on distribution network voltage quality. Proc. CSEE **13**(2), 152–157 (2008)
10. Xiao, X.: Analysis and Control of Power Quality. China Electric Power Press, Beijing (2004)
11. Deng, G., Sun, Y., Xu, J.: A voltage stability analysis method considering bus short circuit capacity. Autom. Electric Power Syst. **8**(3), 15–19 (2009)
12. IEEE Distribution Planning Working Group: Radial distribution test feeders **6**(3), 975–985 (1991)
13. Liu, S.: Study on the Operating Characteristics and Influence of Asynchronous Fans. School of Electrical and Electronic Engineering, North China Electric Power University, Beijing (2011)
14. Dong, G., Ma, H.: Voltage quality evaluation of distribution network based on probabilistic load flow. Int. J. Mod. Educ. Comput. Sci. (IJMECS) **10**(8), 55–62 (2018). https://doi.org/10.5815/ijmecs.2018.08.06
15. Afzalan, M., Taghikhani, M.A.: Placement and sizing of DG using PSO&HBMO Algorithms in radial distribution networks. Int. J. Intell. Syst. Appl. (IJISA) **4**(10), 43–49 (2012)
16. Manju, M., Leena, G., Saxena, N.S.: Distribution network reconfiguration for power loss minimization using bacterial foraging optimization algorithm. Int. J. Eng. Manuf. (IJEM) **6**(2), 18–32 (2016). https://doi.org/10.5815/ijem.2016.02.03
17. Elmamlouk, W.M., Mostafa, H.E., El-Sharkawy, M.A.: PSO-based PI controller for shunt APF in distribution network of multi harmonic sources. Int. J. Intell. Syst. Appl. (IJISA) **5**(8), 54–66 (2013). https://doi.org/10.5815/ijisa.2013.08.07
18. Baghipour, R., Hosseini, S.M.: Placement of DG and capacitor for loss reduction, reliability and voltage improvement in distribution networks using BPSO. Int. J. Intell. Syst. Appl. (IJISA) **4**(12), 57–64 (2012)

Author Index

A

Albota, Solomiia, 485
Aleksander, Marek, 561
Aleshin, A. K., 683

B

Baldin, A. V., 197
Balonin, Nikolay A., 151
Bardenhagen, A., 495
Bazhenov, Ruslan I., 37
Bilushchak, Tetiana, 661
Bin, Liu, 197
Bobrovskaia, Anna S., 261
Bogatikov, V. N., 549
Bo, Wang, 713
Bowen, Zheng, 713
Brezhnev, Alexey, 293
Brezhneva, Alexandra, 293
Bryzgalov, E. A., 415

C

Chen, Daojun, 425
Chen, Xiaolu, 729, 759
Chen, Zhongming, 449
Cho, K. M., 15
Cui, Ting, 425

D

Darvas, György, 47
Dashevskiy, I. N., 305
Demishkevich, Eduard B., 315
Deng, Wei, 771
Devyatov, Andrey V., 221
Dli, M., 379

Dneprovskaya, Natalia V., 641
Dontsov, V. I., 211
Dosko, S. I., 197
Dovbnenko, M. S., 415
Dychka, Andrii, 271
Dychka, Ivan, 283

E

Ehrlich, Lev I., 173
Eremeev, Sergey, 81
Eremeykin, P. A., 539
Eventov, Victor L., 345

F

Fan, Jiaqi, 729
Fedoseev, Victor B., 405
Fedushko, Solomia, 333, 527
Filist, Sergey, 293
Fimmel, Elena, 117
Finn, Victor K., 173
Firsov, G. I., 69, 241

G

Ganiev, R. F., 415
Gao, He, 729, 749
Gavriushin, Sergey S., 261, 315, 539, 705
Gizun, Andrii, 571
Glazunov, V. A., 683
Gnatyuk, Sergiy, 561, 571, 581
Gorbatyuk, Yana, 571
Gorshenin, Andrey, 629
Gourary, M. M., 161
Grachev, Vladimir I., 57
Gribov, D. A., 305

785

Gribova, Valeriya, 3
Guo, Hu, 425
Guoli, Zhang, 713

H
Han, Yingkun, 739
He, Gao, 713
He, Matthew, 25
Hrytsak, Anatoliy, 581
Huang, Yuan Ji, 771
Hui, Qiao, 713
Hu, Z. B., 25

I
Igamberdiev, Abir U., 173

J
Jianqiang, Lu, 713
Ji, Sungchul, 185

K
Kaminskyj, Roman, 617
Kemaykin, V. K., 673
Kinzeryavyy, Vasyl, 561, 571, 581
Kireyenkova, M., 379
Kirichenko, A. V., 231
Kirillov, I. E., 549
Kiseleva, Anna A., 251
Kleschev, Alexander, 3
Koganov, A. V., 139
Kokoreva, Elena V., 693
Komleva, Nina V., 641
Korolev, Victor, 629
Kotelianets, Vitalii, 571
Kovalev, Nikita S., 355
Kozhukhin, I. V., 673
Kozlova, Yu. G., 673
Kozyrev, Sergei V., 93
Krut'ko, V. N., 211
Kucherov, K. V., 197
Kuznetsov, V. N., 651
Kyrychenko, Karina, 561

L
Laktyunkin, Alexander, 459
Laryushkin, P. A., 231
Lee, Moon Ho, 15
Lee, Sung Kook, 15
Liao, Rongtao, 771
Li, Chenkun, 425
Liuchuang, Wu, 713

Liu, Jierong, 449
Liu, Weimng, 739
Li, Xiangxin, 759
Lu, Jianqiang, 729, 749
Luzhnov, Petr V., 251

M
Markova, A. M., 211
Markovets, Oleksandr, 485
Mastykash, Oleg, 439
Mazurov, M., 393
Meshchikhin, I. A., 705
Mitronin, Alexander V., 261
Moskalenko, Philip, 3
Murashov, Mikhail V., 601
Mutovkina, N. Yu., 591, 651
Myna, Zhanna, 661

N
Nazyrov, Firuz G., 221

O
Odarchenko, Roman, 561

P
Pakhomov, Andrey A., 57
Paliukh, B. V., 673
Panchelyuga, Maria S., 345
Panchelyuga, Victor A., 345
Pang, Jingzhi, 449
Panin, S. S., 415
Park, Beum Jun, 185
Pecheykina, M., 495
Peleshchyshyn, Andriy, 439, 473, 485, 661
Petoukhov, Sergey V., 25, 47, 103, 117
Petukhova, Elena S., 103
Potapov, Alexander A., 57, 405, 459
Prorokov, A. E., 549
Puchkov, A., 379

Q
Qing, Chuan, 449
Qingwu, Gong, 713

R
Radchenko, Yevgen, 283
Rakcheeva, T., 127, 323
Rakcheeva, T. A., 139
Rakov, D., 495
Rashoyan, G. V., 683
Reid, John Stuart, 185

Ren, Zheng, 729, 759
Romanchuk, Vitaliy A., 37
Rusakov, S. G., 161

S
Sanayeva, G. N., 549
Sapozhnik, Tetyana, 581
Savkiv, Lidia, 617
Sayapin, Sergey N., 221
Seilova, Nurgul, 581
Seltsova, Ekaterina, 81
Sergeev, Alexander M., 355
Sergeev, Mikhail B., 151
Shakhovska, Khrystyna, 527
Shakhovska, Nataliya, 333, 527, 617
Shalfeeva, Elena, 3
Shalyukhin, K. A., 683
Shamaev, Dmitry M., 251
Shishulin, Alexander V., 405
Shkapov, Pavel M., 221
Shklovskiy-Kordi, Nikita E., 173
Shkurat, Oksana, 271, 283
Shurygina, Ksenia I., 693
Skvorchevsky, Anatoly K., 355
Skvortsov, S. A., 683
Solovyev, Anatoliy A., 517
Sopilko, Iryna, 581
Spasenov, A. Yu., 197
Statnikov, I. N., 69, 241
Stepanenko, Iryna, 571
Sulema, Yevgeniya, 271, 283
Suschuk-Sliusarenko, Viktoriya, 271, 283
Syerov, Yuriy, 333

T
Tang, Jiaqi, 449
Timchenko, Vadim, 3
Tingting, Zheng, 713
Tomakova, Rimma, 293
Trach, Olha, 473

U
Urintsov, Arkadiy I., 641
Utenkov, V. M., 197

V
Valuev, Andrey M., 507, 517
Vasilieva, Maria, 629
Vent, D. P., 549
Veynberg, Roman, 293
Volkova, Ludmila P., 507
Vus, Volodymyr, 485

W
Wang, Rui, 739
Wang, Ziye, 369
Wenhui, Chen, 713
Wenxiao, Qian, 713

X
Xu, Cai, 739
Xu, Liu, 713

Y
Yang, Pengwe, 739
Yarka, Ulyana, 661
Yubuzova, Khalicha, 561
Yu, Miaomiao, 771

Z
Zhang, Mengya, 369
Zhang, Qingying, 369
Zhang, Yaoyu, 449
Zhargalova, A. D., 539
Zheng, Bowen, 729, 749
Zheng, Tingting, 759
Zhou, Nianguang, 425
Zhuk, D. M., 197
Zhu, Liang, 771